A Specialist Periodical Report

Mass Spectrometry
Volume 3

A Review of the Literature Published
between July 1972 and June 1974

Senior Reporter
R. A. W. Johnstone, *Robert Robinson Laboratories,
University of Liverpool*

Reporters

T. W. Bentley, *University College of Swansea*

J. H. Bowie, *University of Adelaide, Australia*

C. J. W. Brooks, *University of Glasgow*

A. Dell, *University Chemical Laboratory, Cambridge*

D. E. Games, *University College of Cardiff*

B. N. McMaster, *Robert Robinson Laboratories, University of Liverpool*

F. A. Mellon, *Robert Robinson Laboratories, University of Liverpool*

B. S. Middleditch, *Baylor College of Medicine, Texas*

B. J. Millard, *University of London*

H. R. Morris, *University Chemical Laboratory, Cambridge*

T. R. Spalding, *Ahmado Bello University, Zaria, Nigeria*

J. M. Wilson, *University of Manchester*

ISBN 0 85186 2780
© Copyright 1975

The Chemical Society
Burlington House, London, W1V 0BN

ISBN: 0 85186 2780

Set in Times on Monophoto Filmsetter and printed offset by
J. W. Arrowsmith Ltd., Bristol, England

Made in Great Britain

Foreword

The old team of reporters responsible for Volumes 1 and 2 of this Specialist Periodical Report has been largely replaced for Volume 3. We take this opportunity of thanking our predecessors for the effort put into those volumes and hope we can maintain, if not surpass their standards. In the main, the lay-out of Volume 3 follows the pattern of the previous volumes, and the chapters run through theory to methods to results. Because of the tremendous enthusiasm with which mass spectrometry is being applied to a wide variety of problems, division into chapters may appear somewhat artificial in places, depending on one's viewpoint, but we have tried to bear in mind the needs of broad groups of research workers. Thus, inorganic and organic sections are included separate from a gas chromatographic–mass spectrometric section. For the first time, we have treated the use of mass spectrometry in studies of drug metabolism as a separate unit for the benefit of the large and fast-increasing number of people working in this area. Also, as a departure from previous practice, a short critical appraisal of a small but important area of use of mass spectrometry has been included, namely the sequencing of peptides and carbohydrates. The aim of this last section is not a full literature coverage, which can be obtained from various parts of Volumes 1—3, but an assessment of the best methods currently in use, and an indication of future trends; it is hoped that future volumes will cover other selected areas. The mention of literature leads naturally to the thought that Volume 3 alone contains thousands, rather than hundreds of references. Since this situation applies also to Volumes 1 and 2, there is considerable merit in the suggestion voiced by several colleagues and acquaintances, that a cumulative index would be extremely useful for literature searching. Accordingly, in Volume 4 we hope to have such an index. If readers feel there are other changes they would like to see in the style or treatment of mass spectrometry in this series, the Senior Reporter would be pleased to have discussions with them.

Finally, I would like to thank most sincerely all the contributors to the present volume for their diligent coverage of the literature, and for forwarding their manuscripts before the editorial deadline. It has been a real pleasure working with them.

R. A. W. Johnstone

Contents

Chapter 1 Theory and Energetics of Mass Spectra   1
           By B. N. McMaster

    1 Introduction   1

    2 Calculations of Ion Structures and Energies   1
        Ab initio Self-consistent Field Calculations including
            Electron Correlation   2
        Ab initio Self-consistent Field Calculations neglecting
            Electron Correlation   5
        Semi-empirical Calculations   6

    3 Ionization Processes and Energy Deposition Functions   7
        Direct Ionization   7
        Indirect Ionization   8
        Energy Deposition Functions   10
        Electroionization   11
        Charge-transfer Ionization   12
        Penning Ionization   13

    4 Unimolecular Decay Processes   14
        Radiationless Transitions   14
            Vibrational Relaxation   14
            Vibronic Relaxation   15
            Unimolecular Reactions   16
        Dynamical Models   16
            Potential Energy Surfaces   17
            Statistical Approximation   18
            Tests of Dynamical Models   21
        Unimolecular Ion Decompositions   22
            Recent Theoretical Developments   22
            Application of RRKM Models   23

5 Experimental Methods 26
Photoelectron–Photoion Coincidence Spectroscopy 26
Charge-transfer Mass Spectra 29
Lifetime Measurements of Ion Decompositions 33
   Field Ionization Kinetics 33
   Other Studies 41
Ion Kinetic Energy Spectroscopy 42

6 Appearance Potential Measurements 46
Deconvolution Methods 47
Empirical Methods 48
Results 51
   Accurate Measurements 51
   Approximate Measurements 53
   Negative Ion Studies 57

Chapter 2 Structure and Mechanism in Mass Spectrometry 59
By T. W. Bentley

1 Introduction 59

2 Metastable Ions 60
Consecutive Processes 60
Intensities 61
Collisional Activation 62
Kinetic Energies 65

3 Reactions of Isotopically Labelled Species 67
Hydrogen Rearrangements 67
Carbon Scrambling 70

4 Substituent Effects 72
Linear Free Energy Relationships 72
Retro-Diels–Alder Reactions 73

5 Mechanistic Interpretations 74
Comparison with Carbonium Ion Chemistry 74
Charge Localization 77

6 Molecular Orbital Calculations 79
Ion Structures 79
Reactions 80

7 Energetics of Fragmentation 82

8 Ion Cyclotron Resonance 82

9 Comparison with Other Chemical Processes 83

10 Other Approaches 84

Chapter 3 Alternative Methods of Ionization and Analysis 86
By J. M. Wilson

1 Introduction 86
2 Photoionization 86
3 Negative Ion Production 88
4 Field Ionization 90
5 Chemical Ionization 96
6 Ion Cyclotron Resonance 102
7 Ion–Molecule Reactions 105
   Acid–Base Equilibria 105
   Ion–Solvent Interactions 109
   Structures of Ions in the Gas Phase 110
8 Plasma Chromatography and Atmospheric Pressure Ionization 113
9 Other Methods of Ionization 116

Chapter 4 Computerized Data Acquisition and Interpretation 117
By F. A. Mellon

1 Introduction 117
2 On-line Data Acquisition and Control 118
   System Interface and Configuration 118
   On-line Computer Control 120
   Accuracy of Mass and Intensity Measurement 123
3 Interpretation of Mass Spectra by Heuristic, Pattern Recognition, and Library Search Techniques 126
   Pattern Recognition 127
   Heuristic Methods 132
      Low Resolution 132
      High Resolution 133
   Comparison of Spectra and Library Searching 134
4 Miscellaneous Applications 139
   Gas Chromatography and Mass Spectrometry 139
   Multiple Ion Detection without Feedback Control 141
   Ionization Efficiency and Metastable Ion Data 141
   Analysis of Spectra of Mixtures 142
   General 142

Chapter 5 Organometallic, Co-ordination, and Inorganic Compounds 143
By T. R. Spalding

1 Introduction 143
   Books and Reviews 143
   New Techniques of General Interest 144

2 Main-group Organometallics 144
   Group II 144
   Group III 146
      Compounds containing B—O, B—S, or B—N Bonds, including Boron Heterocycles 146
      Carbaboranes and Derivatives 149
      Compounds of Other Elements 150
   Group IV 151
      Silicon 151
      Silicon Heterocycles 153
      Silicon–Silicon Bonded Compounds 154
      Compounds containing Si—H Bonds 155
      Compounds containing Si—N or Si—P Bonds 155
      Compounds containing Si—O or Si—Main Group VI Bonds 156
      Compounds containing Si—Halogen Bonds 158
      Germanium 158
      Germanium Heterocycles 158
      Other Germanium Compounds 159
      Tin 160
      Tin Heterocycles 160
      Compounds containing Tin–Tin Bonds 162
      Compounds containing Sn—N Bonds 162
      Compounds containing Sn—O or Sn—S Bonds 163
      Lead 163
   Group V
      Phosphorus 165
      Phosphorus Heterocycles 166
      Compounds with P—P or P—Main Group V Bonds 167
      Compounds with P—O or P—S Bonds 167
      Compounds with P—Halogen Bonds 171
      Arsenic 171
      Antimony and Bismuth 174
   Group VI 174

3 Transition-metal Organometallics 176
   Results of General Interest 176
   Ion–Molecule Reactions 176

Chemical Ionization Spectra	177
Distinction of Geometrical Isomers	178
Metal Carbonyls and Carbonyl Hydrides	178
Metal Nitrosyls	179
Complexes containing Metal–Carbon σ-Bonds	180
Hydrocarbon–Metal π-Complexes and Related Complexes	182
Acetylenes and Olefins	182
Dienes and Polyenes	184
Allyls	186
Cyclobutadienes	186
Cyclopentadienyls	187
Ferrocenes and Related Compounds	188
Arenes and Related Ligands	189
Complexes containing C_7 and C_8 Rings	191
Complexes with Donor Ligands	192
Carbyne and Carbene Complexes	192
Nitrogen Ligands	193
Phosphorus Ligands	194
Arsenic and Antimony Ligands	198
Main Group VI Ligands	198
Transition-metal Cluster Compounds	199
Compounds with Bonds to Main-group Elements	201
Boron-containing Ligands, including Carbaboranes	203
Miscellaneous Ligands	203
4 Co-ordination Compounds	**204**
Compounds with Metal–Oxygen Bonds	204
β-Diketonato-complexes	204
Carboxylates	206
Alkoxides	207
Compounds with Metal–Nitrogen Bonds	208
Porphyrins and Related Compounds	209
Compounds with Metal–Oxygen and Metal–Nitrogen Bonds	210
Schiff-base and Related Compounds	210
Compounds with Metal–Sulphur Bonds	211
Compounds with Metal–Oxygen and Metal–Sulphur Bonds	212
5 Inorganic Compounds	**213**
Group I	213
Group II	213
Group III	214
Boron Hydrides and Related Compounds	214
Boron–Nitrogen Compounds, including Borazines	215

Group IV	216
Group V	217
Phosphorus–Nitrogen Compounds, including Phosphonitrile Derivatives	219
Groups VI, VII, and Noble Gas Compounds	220
Transition-metal Groups	221

Chapter 6 Natural Products 224
By D. E. Games

1 Introduction	224
2 Alkaloids	225
3 Oxygen Heterocycles and Phenols	226
4 Isoprenoids	229
5 Steroids	231
6 Antibiotics	235
7 Nucleic Acid Components	240
8 Pyrrole Pigments	243
9 Carbohydrates	246
10 Amino-acids, Biogenic Amines, and Peptides	248
Amino-acids and Biogenic Amines	249
Peptides	250
11 Lipids	256
Simple Lipids	256
Complex Lipids	258
Prostaglandins	260

Chapter 7 Reactions of Organic Functional Groups: Positive and Negative Ions 262
By J. H. Bowie

1 General Introduction	262
2 Reactions of Positive Ions	263
Hydrocarbons (including Hydrocarbon Cations)	263
Alkanes and Alkenes	263
Aromatic	265
Halides	267
Alcohols and Phenols	268
Aldehydes, Ketones, and Quinones	270
Aliphatic	270
Aromatic	272

Acids, Esters, Anhydrides, and Carbonates	273
Ethers	275
Amines, Amides, and Related Systems	276
The $-C=N$, $-CN$, $>N-N<$ and $-N=N-$ Groups	278
The $N-O$ Group	279
Heterocyclic Systems (excluding Sulphur Compounds)	280
Three, Four- and Five-membered Rings	280
Six- and Seven-membered Rings	282
Sulphur Compounds	284

3 Doubly-charged Positive Ions   287

4 Reactions of Negative Ions   288
General Introduction   288
Modes of Formation of Negative Ions   289
Instrumentation   291
Metastable Ions and Collision-induced Dissociations   292
Formation and Fragmentation of Negative Ions   293

Chapter 8 Gas Chromatography—Mass Spectrometry   296
By C. J. W. Brooks and B. S. Middleditch

1 General Considerations   296
Introduction   296
Practical Aspects   298
Stable Isotopes   301
Functional Derivatives   302
Current Trends   303

2 Applications   304
Hydrocarbons   304
Long-chain compounds   305
Prostaglandins   307
Sphingosine Derivatives   309
Carbohydrates   310
Oxygenated Terpenoids   313
Steroids: (A) Reference Compounds   315
   Alcohols   315
   Ketones   316
   Corticosteroids   317
   Other Steriods   317
Steroids: (B) in Biological Material   318
   Sterols   318
   Steroidal Acids   320
   Hormonal Steroids and Metabolites in the Human   320
   Hormonal Steroids of Animals or Plants   321
   Metabolism of Exogenous Steroids   321

Amines: Reference Compounds and Reaction Products	322
Amino-acids and Peptides	324
Natural Metabolites and Drugs: (A) Reference Compounds	326
Natural Metabolites and Drugs: (B) in Biological Material	327
Non-nitrogenous Compounds	327
Nitrogenous Compounds	328
Insect Pheromones and Other Secretions	330
Food Flavours and Aromas	331
Pesticides and Pollutants	334
Organic Geochemistry	336
Miscellaneous	338

Chapter 9 Drug Metabolism   339
By B. J. Millard

1 Introduction	339
2 Low-resolution Mass Spectrometry	340
Reference Compounds	340
Metabolites Separated by Thin-layer Chromatography	342
Metabolites Separated by Solvent Extraction	352
Metabolites Trapped from a Gas Chromatograph	353
3 Quantitative Gas Chromatography–Mass Spectrometry	353
Single Ion Monitoring	353
Multiple Ion Monitoring	354
Non-labelled Standards	354
Labelled Standards	356
4 High-resolution Mass Spectrometry	358
Qualitative Applications	358
Quantitative Applications	360
5 Chemical Ionization	360

Chapter 10 Protein and Carbohydrate Sequence Analysis   362
By H. R. Morris and A. Dell

1 Introduction	362
2 Protein Structure	362
Derivatization	362

Sequence Analysis 364
Small Peptides 364
Polypeptide and Protein Structures 368
Miscellaneous 370

3 Carbohydrate Sequence Analysis 371

Author Index 377

1
Theory and Energetics of Mass Spectra

BY B. N. McMASTER

1 Introduction

The quasi-equilibrium theory remains the cornerstone of discussions of mass spectral fragmentations in the literature, and has been well reviewed by Wahraftig.[1] However, very significant advances have been made in the theoretical treatments of chemical dynamics in closely related fields. More sophisticated experimental methods are now being applied to studies of ion decompositions and promise to fill some of the wide gaps in our knowledge. Most emphasis has been therefore placed on these aspects and a review style has been adopted, since this is felt to be more useful in the midst of such developments. Where clarification of concepts or methods seems necessary, criticism has been made in a spirit of stimulating further work and discussion.

The topics reviewed follow a logical development. Theoretical calculations of ion structures and energies are briefly discussed, followed by ionization processes and their important relationship to energy deposition functions. Current developments in theories of unimolecular rate processes are reviewed and their relevance to ion decompositions is stressed. Results from photoelectron–photoion coincidence, charge-transfer mass spectra, field ionization kinetics, and metastable ion kinetic energy measurements are discussed with particular emphasis on fundamental aspects. Finally, methods of determining appearance potentials are critically evaluated, and their uses in thermochemical calculations are reported.

Although the literature coverage is as comprehensive as possible, some degree of selection has been necessary. Numerous reviews on specific topics are cited in the appropriate sections, but more general coverage of related work in mass spectrometry can be found in Vol. 2 of this series[2] and in a very comprehensive literature survey covering 1972—73.[3]

---

[1] A. L. Wahraftig, in 'Mass Spectrometry', ed. A. Maccoll (M.T.P. International Review of Science), Physical Chemistry, Series 1, Butterworths, London, 1972, Vol. 5, p. 1.
[2] J. M. Wilson, in 'Mass Spectrometry', ed. D. H. Williams (Specialist Periodical Reports), The Chemical Society, London, 1973, Vol. 2, p. 1; I. Howe, ibid., p. 33.
[3] A. L. Burlingame, R. E. Cox, and P. J. Derrick, Analyt. Chem., 1974, 46, 248R.

2 Calculations of Ion Structures and Energies

Ab initio **Self-consistent Field Calculations including Electron Correlation.**—An important distinction must be drawn between *ab initio* SCF calculations using extended basis sets and those using minimal basis sets.[4,5] Only the former are expected to give results of Hartree–Fock accuracy in the single-particle approximation (which ignores electron correlation). These calculations are very expensive for polyatomic molecules, and minimal basis set calculations are therefore more generally performed. With judicious choice of basis functions for particular systems, results within a few kcal mol^{-1} of the Hartree–Fock limit can be obtained using minimal basis sets.[5] But even with results of this accuracy erroneous conclusions may be drawn because electron correlation is ignored in the Hartree–Fock treatment.[6,7] Theoretical calculations of electronic energies are generally used to determine the energy difference between species. If the correlation energy of each species is the same, then comparisons at the Hartree–Fock level will be quite accurate. However if it differs, even slightly, the relative energies may be drastically changed because the correlation energy is of a similar order of magnitude to the total chemical binding energy.

A number of techniques for estimating the correlation energy, or at least that part which differs between related species, have recently been developed to correct the Hartree–Fock energies. Some of these methods and their application to diatomic molecules have been reviewed by Wahl,[7] including the molecular orbital configuration interaction (MO–CI),[8] multiconfiguration self-consistent field (MCSCF),[9] and independent electron-pair approximation (IEPA)[10] treatments. The calculations of electron correlation effects reported below have used extended basis sets, unless stated otherwise.

Kutzelnigg *et al.* have calculated correlation energy corrections for vinyl and ethyl ions with classical and non-classical (H-bridged) structures (1), (2) and (3), (4) respectively using the IEPA method.[10] They used the geometries corresponding to local energy minima obtained by slightly-extended basis set calculations,[11] and found that the non-classical structures were more stable by *ca.* 8 kcal mol^{-1}. Although the quantitative magnitude of the stability is uncertain because of the remaining small errors, its qualitative validity is assured and contrasts with calculations ignoring electron correlation which predicted the classical

---

[4] J. L. Whitten, *Accounts Chem. Res.*, 1973, **6**, 238.
[5] L. Radom and J. A. Pople, in 'Theoretical Chemistry' ed. W. Byers Brown (M.T.P. International Review of Science), Physical Chemistry, Series 1, Butterworths, London, 1972, Vol. 1, p. 71.
[6] W. A. Goddard, T. H. Dunning, W. J. Hunt, and P. J. Hay, *Accounts Chem. Res.*, 1973, **6**, 368.
[7] A. C. Wahl, in 'Theoretical Chemistry', ed. W. Byers Brown (M.T.P. International Review of Science), Physical Chemistry, Series 1, Butterworths, London, 1972, Vol. 1, p. 41.
[8] G. C. Lie and E. Clementi, *J. Chem. Phys.*, 1974, **60**, 1275, 1288.
[9] G. Das, *J. Chem. Phys.*, 1973, **58**, 5104.
[10] B. Zurawski, R. Ahlrichs, and W. Kutzelnigg, *Chem. Phys. Letters*, 1973, **21**, 309.
[11] P. C. Hariharan, W. A. Lathan, and J. A. Pople, *Chem. Phys. Letters*, 1972, **14**, 385.

structure to be more stable in the case of the $C_2H_3^+$ ions,[11,12] and about equally stable in the case of the $C_2H_5^+$ ion.[11] These important differences in predictions have been attributed to the overestimation of electron repulsion in the Hartree–Fock approximation.[10]

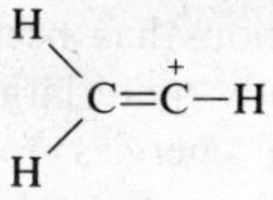

(1) 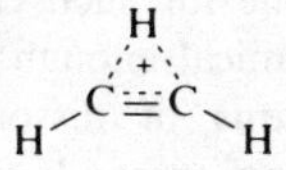   (2) 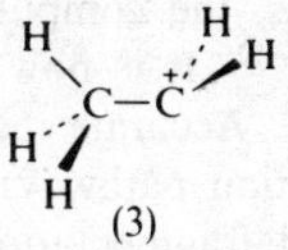   (3) 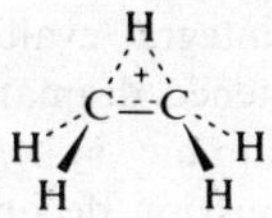   (4)

Similar calculations for the CH_5^+ ion indicated that the C_s (5) and C_{2v} (6) structures have almost identical energies, while the C_{4v} (7) and D_{3h} (8) structures were less stable by about 6 and 20 kcal mol^{-1} respectively.[13] Minimal basis set calculations which ignored the correlation energy predicted all these structures to have very similar energies,[5] although later calculations with extended basis sets predicted significant differences.[11] A dissociation energy of 40 kcal mol^{-1} was obtained for the reaction $CH_5^+ \rightarrow CH_3^+ + H_2$ when the correlation

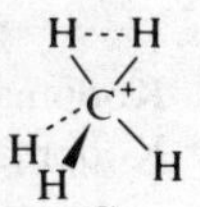

(5) 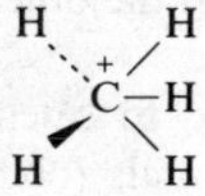   (6) 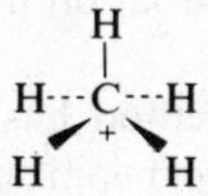   (7) 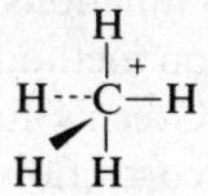   (8)

energy was treated.[13] The importance of correlation energy has also been noted in the derivation of accurate potential energy curves for proton transfer in $H_5O_2^+$,[14] and electrocyclic transformations of cyclopropyl and allyl cations, radicals, and anions.[15]

The π-electron states of benzene have been extensively studied by Hay and Shavitt using an MO–CI method with a frozen σ-core.[16] Excitation energies were obtained for many singlet, triplet, quintet, and Rydberg states of the molecule and several doublet states of the molecular ion. Good agreement with experiment was observed over all and the molecular ion states were well discussed. Very accurate calculations of correlation effects in the ground and ionized states of methane have been reported by Meyer using a pseudo-natural orbital (PNO–CI) treatment, and an interesting discussion of the energy surface of CH_4^+ has been given.[17] Other CI calculations have been reported for neutral and ionic states of the silyl radical using a minimal basis set.[18]

[12] A. C. Hopkinson, K. Yates, and I. G. Csizmadia, *J. Chem. Phys.*, 1971, **55**, 3835.
[13] V. Dyczmons and W. Kutzelnigg, *Theor. Chim. Acta*, 1974, **33**, 239.
[14] W. Meyer, W. Jakubetz, and P. Schuster, *Chem. Phys. Letters*, 1973, **21**, 97.
[15] P. Merlet, S. D. Peyerimhoff, R. J. Buenker, and S. Shih, *J. Amer. Chem. Soc.*, 1974, **96**, 959.
[16] P. J. Hay and I. Shavitt, *J. Chem. Phys.*, 1974, **60**, 2865.
[17] W. Meyer, *J. Chem. Phys.*, 1973, **58**, 1017.
[18] B. Wirsam, *Chem. Phys. Letters*, 1973, **18**, 578.

For some particular cases, which include the dissociation of a doublet molecular ion to a doublet and a singlet state product, the correlation energy does not change significantly with change in geometry, and calculations of Hartree–Fock accuracy then provide quite reliable dissociation energies and potential energy surfaces.[7,19] By use of various approximations[20] and symmetries[21] in the integral evaluations, the computing time for such calculations has been shortened dramatically.[22] It is now economical enough to study quite large polyatomic systems.[23] Accurate calculations of dissociation energies for various ion decomposition pathways by these methods would be very useful in evaluating proposed fragmentation mechanisms, and would also provide important data for dynamical calculations using potential energy surfaces (see Section 4).

Electron correlation effects on energies may also be calculated by a conceptually different approach based on many-body theories of electronic structure.[24] Rather than perform complete calculations of the different states to determine their energy differences, these methods evaluate excitation energies, ionization potentials, or electron affinities directly from the Hartree–Fock wavefunctions of the initial state. Two approaches are currently being developed, one using Green's functions and perturbation theory[25] and the other using the equations-of-motion method.[26,27]

The Green's function method has been used to calculate the 'Koopman's defect' corrections to vertical ionization potentials (obtained by Koopman's theorem), resulting from electronic reorganization and electron correlation differences in the ion compared with the molecule. The agreement between calculated values and photoelectron spectroscopy measurements for F_2 and N_2,[28] HOF,[29] H_2O,[30] and H_2CO[31] is impressive. For the last two examples the method was extended to calculate vibrational structure as well, and again gave excellent agreement with experimental vibrational energies and oscillator strengths. In the initial applications of the equations-of-motion method the excitation energies of formaldehyde[32] and benzene[33] were calculated. It has also been used to calculate vertical ionization potentials of N_2 in good agreement with experiment.[34]

[19] R. C. Raffenetti and K. Ruedenburg, *J. Chem. Phys.*, 1973, **59**, 5978.
[20] D. L. Wilhite and R. N. Euwema, *Chem. Phys. Letters*, 1973, **20**, 610.
[21] R. M. Pitzer, *J. Chem. Phys.*, 1973, **58**, 3111.
[22] E. Clementi and H. Popkie, *J. Chem. Phys.*, 1973, **58**, 4699.
[23] G. C. Lie and E. Clementi, *J. Chem. Phys.*, 1974, **60**, 3005.
[24] K. F. Freed, *Ann. Rev. Phys. Chem.*, 1971, **22**, 313.
[25] L. S. Cederbaum, G. Hohlneicher, and W. von Niessen, *Mol. Phys.*, 1973, **26**, 1405.
[26] J. Simons and W. D. Smith, *J. Chem. Phys.*, 1973, **58**, 4899.
[27] J. Simons, *Chem. Phys. Letters*, 1974, **25**, 122.
[28] L. S. Cederbaum, *Chem. Phys. Letters*, 1974, **25**, 562.
[29] D. P. Chong, F. G. Herring, and D. McWilliams, *Chem. Phys. Letters*, 1974, **25**, 568.
[30] L. S. Cederbaum and W. Domcke, *Chem. Phys. Letters*, 1974, **25**, 357.
[31] L. S. Cederbaum and W. Domcke, *J. Chem. Phys.*, 1974, **60**, 2878.
[32] D. L. Yeager and V. McKoy, *J. Chem. Phys.*, 1974, **60**, 2714.
[33] J. B. Rose, T.-I. Shibuya, and V. McKoy, *J. Chem. Phys.*, 1974, **60**, 2700.
[34] T.-T. Chen, W. D. Smith, and J. Simons, *Chem. Phys. Letters*, 1974, **26**, 296.

The great advantage of these direct methods is that only the wavefunctions of the initial state are required to calculate the energies and oscillator strengths of any other states, thus leading to huge computational savings. For the best results the wavefunctions should be of Hartree–Fock quality, but reliable estimates of the ordering of excited states (at least) could be obtained with less exact wavefunctions. These new methods will be extremely useful for calculating the energies and oscillator strengths of excited states of ions.

Methods for the calculation of Rydberg levels using model potentials[35] and the multiple-scattering-$X\alpha$ technique[36] have been applied to several polyatomic molecules. Rydberg states lying above the first ionization potential are likely to autoionize, and it is important to have some knowledge of their positions and oscillator strengths. It is becoming increasingly apparent that autoionization contributes substantially to the total ionization cross-section of polyatomic molecules, especially near the lower ionization potentials, and can have a marked influence on decomposition processes (see Sections 3 and 5).

Ab initio Self-consistent Field Calculations neglecting Electron Correlation.— Many of these calculations on polyatomic ions have used the minimal basis set method (STO–3G) developed by Pople and co-workers, who have made a full review of this field;[5] more recent results using extended basis sets have been collected in a useful report.[37] Studies have concentrated on the important problem of establishing the relative stabilities of structural isomers of several carbonium ions; e.g. $C_3H_5^+$,[38,39] $C_3H_7^+$,[40] $C_5H_5^+$,[41] and cyclopropylcarbinyl cations.[42] Open (classical) structures have been found in most cases to be more stable than the corresponding bridged (non-classical) structures. However, relative stabilities favouring open structures which amount to only several kcal mol^{-1} must be considered uncertain, since electron correlation effects have been shown to favour bridged over open structures by this amount (see above).

It is also important to note that the accuracy of these calculations, with respect to the Hartree–Fock limit, depends critically on the basis set employed. Extended basis sets, including polarization functions, are absolutely necessary for best accuracy. Calculations on $C_2H_2^{+}$ near the Hartree–Fock limit indicate that minimal basis set functions do not provide an adequate description of the severe electron reorganization between neutral and ionic states.[43] The transferability

---

[35] T. C. Betts and V. McKoy, *J. Chem. Phys.*, 1974, **60**, 2947.
[36] F. W. Averill, T. E. H. Walker, and J. T. Waber, *J. Chem. Phys.*, 1974, **60**, 2907.
[37] W. A. Lathan, L. A. Curtiss, W. J. Hehre, J. B. Lisle, and J. A. Pople, *Progr. Phys. Org. Chem.*, 1974, **11**, 175.
[38] L. Radom, P. C. Hariharan, J. A. Pople, and P. von R. Schleyer, *J. Amer. Chem. Soc.*, 1973, **95**, 6531.
[39] L. Radom, J. A. Pople, and P. von R. Schleyer, *J. Amer. Chem. Soc.*, 1973, **95**, 8193.
[40] P. C. Hariharan, L. Radom, J. A. Pople, and P. von R. Schleyer, *J. Amer. Chem. Soc.*, 1974, **96**, 599.
[41] W. J. Hehre and P. von R. Schleyer, *J. Amer. Chem. Soc.*, 1973, **95**, 5837.
[42] W. J. Hehre and P. C. Hiberty, *J. Amer. Chem. Soc.*, 1974, **96**, 302.
[43] S. Y. Chu, I. Ozkan, and L. Goodman, *J. Chem. Phys.*, 1974, **60**, 1268.

of σ-orbitals to the ionic states is particularly poor. Other calculations (STO–3G) have been performed for haloethyl cations[44] and acylium ions.[45]

Semi-empirical Calculations.—Because of their great computational economy, semi-empirical molecular orbital treatments[46] (*e.g.* MINDO, INDO, CNDO, EHT) are widely used but their accuracy is difficult, if not impossible, to determine *a priori*.[47] These methods supposedly take some account of electron correlation effects, but their success depends entirely on the parametrization used. They do not generally reproduce observed ionization potentials very well, because of the large changes which occur in electron correlation and reorganization upon ionization, and the reliability of conclusions based on the simpler approximations is very suspect.

The heats of formation of the C_1—C_4 alkyl carbonium ion structures have been calculated by Dewar using his MINDO/2 method, and good agreement with experimental values was obtained for the higher homologues.[48] The most stable forms of the $C_3H_7^+$, $C_4H_9^+$, and $C_5H_{11}^+$ ions are predicted to have a bridged protonated cyclopropane structure. The $C_5H_5^+$ ions have been studied using the more recent MINDO/3 version,[49] and also by the CNDO[50] and EHT[51] treatments, providing interesting comparisons with the *ab initio* calculations.[41] Semi-empirical treatments predict structures (9) and (10) to be local energy minima, but differ in the ordering of relative stabilities; MINDO/3 ordering

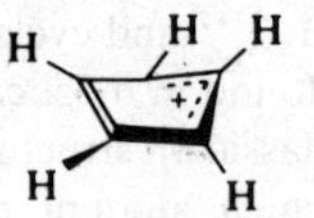

(9) 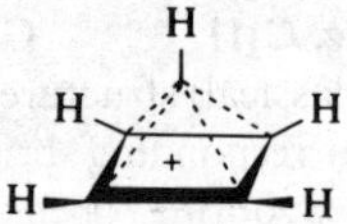   (10)

agrees with the *ab initio* results. However, these calculations were only performed for a singlet state, whereas the *ab initio* calculations suggest that a triplet state cyclopentadienyl structure (D_{5h}) is more stable than any singlet state structures.[41] CNDO studies of several oxocarbonium ions,[52] and INDO studies of fluorovinyl cations[53] and 2-substituted allyl cations[54] have also been reported.

Several workers have attempted to explain various features of mass spectral fragmentations from semi-empirical calculations of charge densities or bond densities at the (assumed) equilibrium geometry of the molecular ion, either in

[44] W. J. Hehre and P. C. Hiberty, *J. Amer. Chem. Soc.*, 1974, **96**, 2665.
[45] L. Radom, *Austral. J. Chem.*, 1974, **27**, 231.
[46] B. J. Nicholson, *Adv. Chem. Phys.*, 1970, **18**, 249.
[47] K. F. Freed, *J. Chem. Phys.*, 1974, **60**, 1765.
[48] N. Bodor, M. J. S. Dewar, and D. H. Lo, *J. Amer. Chem. Soc.*, 1972, **94**, 5303.
[49] M. J. S. Dewar and R. C. Haddon, *J. Amer. Chem. Soc.*, 1973, **95**, 5836.
[50] H. Kollmar, H. O. Smith, and P. von R. Schleyer, *J. Amer. Chem. Soc.*, 1973, **95**, 5834.
[51] W.-D. Stohrer and R. Hoffman, *J. Amer. Chem. Soc.*, 1972, **94**, 1661.
[52] J. D. Pulfer and M. A. Whitehead, *Canad. J. Chem.*, 1973, **51**, 2220.
[53] C. U. Pittman, L. D. Kispert, and T. B. Patterson, *J. Phys. Chem.*, 1973, **77**, 494.
[54] B. K. Carpenter, *J.C.S. Perkin II*, 1973, 1.

its ground state[55-59] or 'pseudo-excited' states.[59,60] These approaches take no explicit account of the dynamics of ion decompositions, and assume they are controlled by the properties of the ion at its equilibrium geometry. A clear presentation of the pitfalls awaiting such approaches has been given by Krier *et al.*, with specific reference to the decomposition of the ethylamine molecular ion.[61] A brief reiteration of their main points is worthwhile. Because the electron distribution changes markedly as the ion dissociates, both the charge densities and bond densities will also change, often in an unpredictable matter. Therefore the relation of the initial (equilibrium) charge or bond densities to the eventual fragmentation is by no means straightforward. On the other hand the potential energy surface of the molecular ion does control the decomposition process. However, the accuracy of current semi-empirical methods is quite inadequate to determine even the dissociation energy (let alone the surface) with any reliability. This is illustrated by some EHT and Iterative EHT calculations, where even the ordering of the dissociation energies changes with the method.[62-64]

3 Ionization Processes and Energy Deposition Functions

It is vital that the various ionization processes and the behaviour of their cross-sections with energy be properly understood.[65] Electron spectroscopy has contributed most towards our understanding of ionization phenomena, and Brion has comprehensively reviewed recent developments.[66] Major points illustrated here involve photoionization (much more detailed information is available on this process), and particular reference is made to a brief, but excellent, discussion of photoionization processes by Chupka.[67]

Direct Ionization.—The magnitude of the transition moment for photoionization depends on the spatial spread and nodal character of the vibronic overlap integral between the initial and final electronic states compared with the phase and wavelength of the photoelectron wavefunction.[68] As the photoelectron energy from

[55] C. E. Parker, M. M. Bursey, and L. G. Pedersen, *Org. Mass Spectrometry*, 1973, 7, 1077.
[56] C. E. Twine, C. E. Parker, and M. M. Bursey, *Org. Mass Spectrometry*, 1973, 7, 1179.
[57] C. E. Parker, J. R. Hass, M. M. Bursey, and L. G. Pedersen, *Org. Mass Spectrometry*, 1973, 7, 1189.
[58] C. E. Parker, M. M. Bursey, and L. G. Pedersen, *Org. Mass Spectrometry*, 1974, 9, 204.
[59] G. Loew, M. Chadwick, and D. Smith, *Org. Mass Spectrometry*, 1973, 7, 1241.
[60] M. Yamomoto, I. Fujita, M. Itoh, and K. Hirota, *Bull. Chem. Soc. Japan*, 1972, 45, 3520.
[61] C. Krier, J. C. Lorquet, and A. Berlingin, *Org. Mass Spectrometry*, 1974, 8, 387.
[62] M. Ogata and H. Ichikawa, *Bull. Chem. Soc. Japan.*, 1972, 45, 3231.
[63] H. Ichikawa and M. Ogata, *J. Amer. Chem. Soc.*, 1973, 95, 806.
[64] H. Ichikawa and M. Ogata, *Bull. Chem. Soc. Japan.*, 1973, 46, 1873.
[65] L. G. Christophorou, 'Atomic and Molecular Radiation Physics', Wiley-Interscience, New York, 1971, pp. 55, 126, 163, 374.
[66] C. E. Brion, in ref. 1, p. 55.
[67] W. A. Chupka, in 'Ion–Molecule Reactions', ed. J. L. Franklin, Butterworths, London, 1972, vol. 1, p. 33.
[68] W. C. Price, A. W. Potts, and D. G. Streets, in 'Electron Spectroscopy', ed. D. A. Shirley, North-Holland, Amsterdam, 1972, p. 187.

a particular state rises (with increasing photon energy), the transition moment for s-electrons becomes progressively larger compared with p-electrons. At the ionization threshold, however, the transition moment is generally much larger for p-electrons than for s-electrons. Similar conclusions apply to the ionization of the corresponding σ- and π-type molecular orbitals.

The behaviour of the cross-section for direct photoionization with increasing photon energy (*i.e.* the 'threshold law') is observed to have the following general characteristics.[69] There is an abrupt step at the ionization threshold and it then rises gradually to a maximum at photon energies around 20—40 eV, thereafter falling to zero asymptotically. Although the cross-sections often have very similar magnitudes at their respective maxima, their values at the threshold differ widely according to the type of orbital (*e.g.* σ, π) being ionized. The relative heights of the bands in photoelectron spectra can therefore change markedly with photon energy, and these observations are now being used to assign their orbital character.[68,69]

It is clear that photoelectron spectra (PES) will give a reliable estimate of the photoionization energy deposition function only for the particular photon energy which is used (*e.g.* 21.21 eV for He^I PES). The energy deposition function will change with photon energy following the change in the corresponding photoelectron spectrum. Therefore when relative ion intensities are obtained from a calculated (or measured) breakdown graph and the PES, a meaningful comparison with the experimental photoionization mass spectrum can only be made at the same photon energy. Meisels and Emmel obtained best agreement between calculated and experimental 14.6 eV photoionization mass spectra of the lower alkanes when threshold photoelectron spectra were used rather than 21.21 eV PES.[70] The former are expected to give a much better estimate of the photoionization energy deposition function at 14.6 eV.

When the photoelectron spectra are used to estimate energy deposition functions, it is necessary to make corrections for the energy-dependent transmission function of the electron energy analyser.[71] However, the exact form of the correction is not accurately established and may differ between instruments, leading to uncertainties in the derived energy distribution function.

Indirect Ionization.—So far we have ignored the complications caused by indirect ionization processes. These occur by initial excitation of the molecule to a superexcited state, defined simply as an excited state lying above the lowest ionic state (and possibly above several excited ionic states). This superexcited molecule may radiate, predissociate, or autoionize.[65,67] The radiative transition is much slower (lifetime $\approx 10^{-8}$ s) than either of the alternative processes for polyatomic molecules and may be ignored.

[69] A. Katrib, T. P. Debies, R. J. Cotton, T. H. Lee, and J. W. Rabalais, *Chem. Phys. Letters*, 1973, **22**, 196.
[70] G. G. Meisels and R. H. Emmel, *Internat. J. Mass Spectrometry Ion Phys.*, 1973, **11**, 455.
[71] J. Berkowitz and P. M. Guyon, *Internat. J. Mass Spectrometry Ion Phys.*, 1971, **6**, 302.

Autoionization generally occurs from Rydberg levels of the molecule as a result of energy exchange between the excited state of the 'ion-core' and the Rydberg electron, which departs with excess kinetic energy leaving the molecular ion in a lower (excited) state (Figure 1). A molecular ion formed by autoionization will generally possess a range of internal energies lying well below the energy

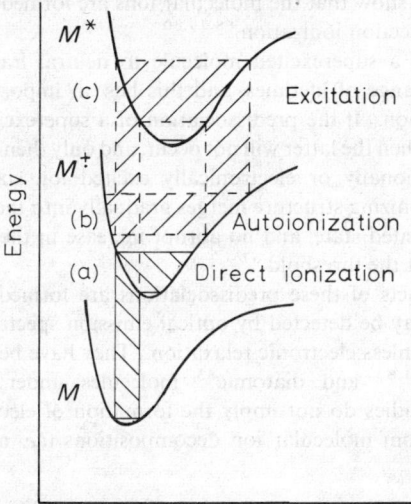

Figure 1 *Illustration of directly ionized* (a) *and autoionized* (b) *molecular ion states, and the autoionizing molecular states* (c).

of the particular autoionizing level. For the most important type of autoionization process, the transition probabilities depend on the relative strengths of the configuration interaction between the Rydberg level and the various lower lying electronic states of the molecular ion. The vibrational energy distribution in a particular electronic state will be determined by the Franck–Condon factors between that state and the Rydberg level. Since the geometry of the Rydberg level will be similar to that of an excited ionic state, these Franck–Condon factors may differ substantially from those for direct ionization to the same molecular ion state. The energy deposition function resulting from autoionization can therefore be expected to differ substantially, and unpredictably, from that due to direct ionization.

Autoionization is a resonance process and usually gives an asymmetric peak in the photoionization cross-section. The width of the peak is inversely proportional to the lifetime of the autoionizing state, and merges into the direct ionization continuum at the lower limit of the lifetime.[65,72] Both photoionization and

---

[72] P. M. Dehmer, J. Berkowitz, and W. A. Chupka, *J. Chem. Phys.*, 1974, **60**, 2676; 1973, **59**, 5777.

electroionization[73—78] have been used to determine the positions of autoionizing levels, especially in diatomic and triatomic molecules where the autoionization structure is readily apparent. But knowledge of the autoionizing levels does not directly provide information about the energy deposition function of the resulting molecular ion. This information can be obtained by measuring the photoelectron spectra at photon energies equal to the autoionizing levels, and such studies of oxygen, for example, show that the molecular ions are formed in vibronic levels not populated by direction ionization.[68,79,80]

Predissociation of a superexcited molecule to neutral fragments may also occur with a wide range of lifetimes, and this has an important effect on the ionization cross-section. If the predissociation of a superexcited state is faster than autoionization then the latter will not occur, and only then can the ionization thresholds of vibrationally or electronically excited ion states be observed. Otherwise the autoionizing structure merges gradually into the direct ionization continuum of the excited state, and no abrupt increase in the ionization cross-section is observed at the threshold.[67]

The neutral products of these predissociations are formed in electronically excited states and may be detected by optical emission spectroscopy if they do not undergo radiationless electronic relaxation. They have been observed from several polyatomic[81,82] and diatomic[83] molecules under electron impact excitation. These studies do not imply the formation of electronically excited neutral fragments from molecular ion decompositions (*i.e.* not as stated in a recent review[3]).

Energy Deposition Functions.—Experimental energy deposition functions (EDF) applicable to photoionization mass spectra could be obtained by measuring photoelectron spectra at all photon energies. Such studies would clearly be limited to interesting model systems, to formulate general principles regarding the variation of the EDF with photon energy. Furthermore they would require the use of intense broad-band photon sources (*e.g.* synchrotron radiation), which are not widely accessible, to obtain good data using monochromators. The work discussed in the preceding sections indicates that the EDF can change quite

---

[73] P. Marmet, E. Bolduc, and J. J. Quéméner, *J. Chem. Phys.*, 1972, **56**, 3463.
[74] P. Marmet, E. Bolduc, and J. J. Quéméner, *J. Chem. Phys.*, 1972, **57**, 1957.
[75] R. Carbonneau and P. Marmet, *Internat. J. Mass Spectrometry Ion Phys.*, 1972, **10**, 143.
[76] R. Carbonneau and P. Marmet, *Phys. Rev.* (*A*), 1974, **9**, 1898.
[77] J. D. Morrison and J. C. Traegar, *Internat. J. Mass Spectrometry Ion Phys.*, 1973, **7**, 391.
[78] R. Cabaud, A. Hoareau, P. Nounou, and R. Uzan, *Internat. J. Mass Spectrometry Ion Phys.*, 1972, **8**, 181.
[79] J. A. Kinsinger and J. W. Taylor, *Internat. J. Mass Spectrometry Ion Phys.*, 1973, **11**, 461.
[80] K. Tanaka and I. Tanaka, *J. Chem. Phys.*, 1973, **59**, 5042.
[81] K. Fukui, I. Fujita, and K. Kuwata, *Bull. Chem. Soc. Japan*, 1972, **45**, 2278.
[82] T. Ogawa, M. Tsuji, M. Toyoda, and N. Ishibashi, *Bull. Chem. Soc. Japan*, 1973, **46**, 2637, 3380.
[83] K. C. Smyth, J. A. Schiavone, and R. S. Freund, *J. Chem. Phys.*, 1973, **59**, 5225; 1974, **60**, 1358.

substantially with photon energies up to 24—40 eV, but varies only slightly at higher energies.

The technique of threshold photoelectron spectroscopy provides important information about the EDF for direct ionization.[84] The yield of photoelectrons ejected with zero kinetic energy is measured as a function of the photon energy, thus discriminating against autoionization and giving the relative threshold cross-sections for all the direct ionization processes. Comparison of these threshold PES with PES obtained at several photon energies affords a direct determination of the behaviour of the threshold law(s) for photoionization, which may vary for different states. Unfortunately PES are not yet available at enough photon energies to allow detailed comparisons, but it is already clear that the commonly assumed step function behaviour is only a crude approximation. The extent of autoionization can be determined by comparing threshold PES with photoionization yield measurements if the threshold laws are known. Qualitative comparisons, assuming a step-function threshold law, strongly suggest that substantial autoionization occurs in the lower alkanes.[84]

The usefulness of the 'optical approximation' for determining the energy deposition functions applicable to 70 eV mass spectra[85] is severely limited by the present lack of accurate ionization cross-sections for individual states as a function of ionizing energy; a more basic limitation is the neglect of multiple ionization and autoionization processes, which may introduce a significant error for 70 eV electroionization.

Electroionization.—The same types of ionization processes occur in electron and photon impact but in electroionization optically forbidden transitions occur and additional ionic states are populated. Experimental investigation of electroionization processes is much more difficult than that for photoionization, but coincidence techniques promise to provide important information.[86] Measurements of electroionization efficiency curves can in principle reveal the various transitions occurring, but the different nature of the threshold laws makes analysis difficult.[87] A good example of the difficulties involved is provided by a careful measurement of the electroionization cross-section for argon.[88] This work indicates that the structure in the cross-section is due to autoionization rather than 'competition' between single and double ionization as had been suggested earlier.[89]

The electroionization cross-section behaviour might be regarded as the integral of the corresponding photoionization cross-section to a first approximation. The behaviour of the latter implies that the common assumption of a linear threshold law for direct electroionization processes is not a good approximation. The

---

[84] R. Stockbauer and M. G. Inghram, *J. Chem. Phys.*, 1971, **54**, 2242.
[85] G. G. Meisels, C. T. Chen, B. G. Giessner, and R. H. Emmel, *J. Chem. Phys.*, 1972, **56**, 793.
[86] H. Ehrhadt, M. Schulz, T. Tekaat, and K. Willman, *Phys. Rev. Letters*, 1969, **22**, 89.
[87] J. D. Morrison, in ref. 1, p. 25.
[88] A. Crowe, J. A. Preston, and J. W. McConkey, *J. Chem. Phys.*, 1972, **57**, 1620.
[89] C. Lifschitz, *J. Chem. Phys.*, 1971, **55**, 4155.

simple power function threshold laws break down completely for electroionization at 10—20 eV above threshold, but it is often claimed that they hold sufficiently well near threshold to be quantitatively useful.[87] However, this is just the region where photoionization threshold laws deviate most strongly from simple step function behaviour, and experimental determinations of the threshold laws for electroionization of polyatomic species are urgently needed. Metastable ions should provide a good criterion since they are only formed by a narrow band of molecular ion states, and their electroionization efficiency curves would not suffer from interference by other states. The behaviour of the threshold law could therefore be readily determined from sufficiently accurate data.

The derivative method of analysing ionization efficiency curves[87] can provide very useful information about the presence of excited or autoionizing states, provided its assumptions are remembered. Scheppele *et al.* have measured the second derivative ionization efficiency (SDIE) curves of the major fragment and parent ions of *p*-methylbenzil.[90] The presence of non-trivial structure in these curves certainly indicates the presence of excited or autoionizing states, but their further conclusions based on a breakdown graph constructed from normalized SDIE curves are less secure. This approach assumes that the threshold law is linear and that autoionization is negligible. They claim that the former assumption is reasonable because the integrated derivative curves give the correct experimental relative ion intensities at energies below 20 eV. This finding only shows that the derivative curves were measured with fair accuracy over the appropriate energy range; it proves nothing about the nature of the threshold law.

PI and EI mass spectra of toluene[91] and *cis*- and *trans*-decene[92] have been compared at similar ionizing energies. Such direct comparisons of relative ion yields from photoionization and electroionization must take account of the different threshold laws in each case. Electroionization gives a higher weighting to the relative ion yields of decompositions with lower threshold energies, because of the approximately linear threshold law, whereas photoionization gives an equal weighting to the relative ion yields of all decompositions, whatever their threshold energies, because of the approximately constant threshold law. Hence PI mass spectra contain higher relative fragment ion yields than EI mass spectra at the same ionizing energy; this was observed in these studies. However, the conclusion[91] that molecular ions formed by photoionization must therefore possess higher internal energies than those formed by electroionization does not follow from this observation.

Charge-transfer Ionization.—Charge-transfer provides a powerful method of investigating the ion decompositions following direct ionization.[93] In principle

---

[90] S. E. Scheppele, R. K. Mitchum, K. F. Kinneberg, G. G. Meisels, and R. H. Emmel, *J. Amer. Chem. Soc.*, 1973, **95**, 5105.
[91] W. L. Stebbings and J. W. Taylor, *Internat. J. Mass Spectrometry Ion Phys.*, 1972, **9**, 471.
[92] B. M. Johnson and J. W. Taylor, *Internat. J. Mass Spectrometry Ion Phys.*, 1972, **10**, 1.
[93] E. Lindholm, in 'Ion–Molecule Reactions', ed. J. L. Franklin, Butterworths, London, 1972, vol. 2, p. 457.

the molecular ion is prepared at a single, well-defined recombination energy (RE) and the normalized ion yields in the resulting mass spectrum give the branching ratios for ion decompositions at that particular energy. The entire breakdown graph for the molecular ion can be constructed by measuring the branching ratios over a wide range of recombination energies. However, this ideal situation is seldom realized in practice for two major reasons; virtually all ions have several recombination energies of differing cross-sections, and charge-transfer ionization of polyatomic molecules is a non-specific resonance process.

Although recombination energies are readily derived from spectroscopic data, their relative cross-sections can only be determined by indirect methods which make unreasonable assumptions.[93] Furthermore, the relative cross-sections for the recombination energies of a particular ion will also depend on the molecule being ionized. The charge-transfer ionization observed in a perpendicular instrument occurs by a vertical transition at large separations of the two species, since close-collision processes which would result in momentum transfer are strongly discriminated.[93] The charge-transfer cross-section will therefore be significant only when the recombination energy falls within a Franck–Condon region for direct ionization of the molecule. The several recombination energies of the primary ion will, therefore, make different contributions to the total charge-transfer ionization for different molecules, and one can seldom be certain which RE is more nearly correct.

The resonance condition assumes that all the exothermicity (RE − IP) of the charge-transfer process is converted into internal energy of the molecular ion. If non-specific resonance occurs, some exothermicity may appear as translational energy of the products leaving the molecular ion with correspondingly less internal energy. Recent experiments by Rayborn and Bailey clearly show that such non-specific resonance makes a major contribution to the total charge-transfer cross-section.[94] They measured the perpendicular translational energy distribution of CO_2^{+} ions formed by charge-transfer with Ar^{+}, and found that it covered all values up to the maximum exothermicity (2.16 eV) with a cross-section around 70% of that for the resonance process.

A careful experimental investigation of charge-transfer ionization of polyatomic molecules by Lifshitz and Tiernan strongly supported the conclusion that molecular ions are often formed with a wide range of internal energies extending below the recombination energy.[95] Many uncertainties are therefore involved in constructing the breakdown graphs from charge-transfer mass spectra, but future work can be expected to clarify the situation. Measurements of the translational energies of the product ions will be particularly important.

Penning Ionization.—Penning ionization possesses a particular advantage over photoionization and electroionization. The ionization products can be studied by both optical emission and electron spectroscopy, and important information about the formation of electronically excited states can sometimes be obtained.

[94] G. H. Rayborn and T. L. Bailey, *J. Chem. Phys.*, 1974, **60**, 813.
[95] C. Lifshitz and T. O. Tiernan, *J. Chem. Phys.*, 1972, **57**, 1515.

The technique has been reviewed extensively by Brion[66] and more recently by Niehaus.[96] Cermák has reported high-resolution electron spectroscopic studies which confirm the vertical nature of Penning ionization of some polyatomic molecules.[97] The emission spectra of several diatomics[98] and carbonyl compounds have been observed by Setser, who showed that electronically excited $CO^{+\cdot}$ $(A^2\Pi)$ ions are formed in the decompositions of $COS^{+\cdot}$, $H_2CO^{+\cdot}$, $(HCO)_2^{+\cdot}$, and $HCO_2H^{+\cdot}$ molecular ions, but not from acrolein, acetaldehyde, or benzaldehyde.[99] Similar optical emission studies of diatomic ions and neutrals, excited by ion-neutral collisions, have been reviewed.[100]

4 Unimolecular Decay Processes

The molecular ion may decay from its initial excited state(s) in several ways. Decomposition reactions are of course our main concern, but it would be naïve to ignore other decay processes since they have important consequences. Only isolated species undergoing unimolecular processes are considered.

Radiationless Transitions.—Radiative transitions are generally much slower then radiationless transitions in polyatomic molecular ions and may be ignored. In some diatomic and triatomic species the radiationless transitions are slower, and then luminescence studies can provide very detailed information about excited state structures and lifetimes of ions; the radiative lifetimes are usually longer than 10^{-8} s.[101] Three types of radiationless transitions may be distinguished for our purposes: vibrational relaxation within an electronic state, electronic relaxation together with vibrational relaxation (vibronic relaxation), and unimolecular reactions including dissociation and isomerization. The theoretical[102—105] and experimental[106,107] studies of these processes have developed enormously in the last few years. The main purpose of this section is to report leading references and reviews, and to indicate the significance of these studies to mass spectral processes.

Vibrational Relaxation. The rate at which vibrational modes exchange energy is determined by their anharmonicity.[108,109] Thus C–H stretching modes are expected to exchange energy faster (*ca.* 10^{14} s^{-1}) than skeletal stretching or

---

[96] A. Niehaus, *Ber. Bunsengesellschaft Phys. Chem.*, 1973, **77**, 632.
[97] V. Cermák and J. B. Ozenne, *Internat. J. Mass Spectrometry Ion Phys.*, 1971, **7**, 399.
[98] W. C. Richardson and D. W. Setser, *J. Chem. Phys.*, 1973, **58**, 1809.
[99] D. W. Setser, *Internat. J. Mass Spectrometry Ion Phys.*, 1973, **11**, 301.
[100] D. Brandt, C. Ottinger, and J. Simonis, *Ber. Bunsengesellschaft Phys. Chem.*, 1973, **77**, 648.
[101] P. Erman, J. Brzozowski, and B. Sigfridsson, *Nucl. Instr. Methods*, 1973, **110**, 471.
[102] K. F. Freed, *Topics Current Chem.*, 1972, **31**, 105.
[103] E. W. Schlag, S. Schneider, and S. F. Fischer, *Ann. Rev. Phys. Chem.*, 1971, **22**, 465.
[104] R. Lefebvre and J. Savolainen, *J. Chem. Phys.*, 1974, **60**, 2509.
[105] R. Lefebvre and J. A. Beswick, *Mol. Phys.*, 1972, **23**, 1223.
[106] C. S. Parmenter, in 'Spectroscopy', ed. D. A. Ramsay (M.T.P. International Review of Science), Physical Chemistry, Series 1, Butterworths, London, 1972, Vol. 3, p. 297.
[107] T. L. Netzel, W. S. Struve, and P. M. Rentzepis, *Ann. Rev. Phys. Chem.*, 1973, **24**, 473.
[108] G. R. Fleming, O. L. J. Gijzeman, and S. H. Lin, *J.C.S. Faraday II*, 1974, **70**, 37.
[109] K. G. Kay and S. A. Rice, *J. Chem. Phys.*, 1973, **58**, 4852.

deformation modes. This is of obvious importance to the quasi-equilibrium hypothesis. At high internal energies the decomposition rates may be so fast that the vibrational energy does not have time to become completely randomized prior to decomposition; the rate constants calculated from quasi-equilibrium models will then be too high. Chemical activation experiments indicate that about 10^{-12} s is required for vibrational energy to become completely randomized.[110–112]

Vibronic Relaxation. Spurred by the recent development of spectroscopic techniques which allow vibronic relaxation processes to be studied in great detail,[103,106,107] theoreticians are making prodigious efforts to specify and calculate the controlling factors.[102,103,113–117] The relaxation rate depends on the vibronic coupling between the quasi-degenerate states,[118–120] which to a first approximation is given by the product of the electronic transition moment, the vibrational overlap integral (Franck–Condon factor), and the final state degeneracy. If the electronic energy gap between the states is so large that the vibrational levels of the lower (final) state form a quasi-continuum, then the degeneracy factor is given by the vibrational state density.

In molecules of high symmetry the electronic transition may be symmetry-forbidden, causing the vibronic relaxation rate to be slow enough that some excited electronic states are isolated from lower ones over a certain time scale. The relaxation is not strictly forbidden owing to the so-called 'breakdown' of the Born–Oppenheimer approximation. The vibrational overlap integral depends on the relative separation and displacement of the electronic states, and is strongly dependent on the anharmonicities of the communicating vibrational modes. A large proportion of the transferred energy is therefore accepted initially by the most anharmonic vibrational modes, particularly the C–H stretching mode.[103] Interference effects between vibrational and vibronic relaxation may occur when both processes have similar relaxation rates, and lead to a decay function which does not follow first-order kinetics.[113,121]

The lifetimes of the lowest excited states undergoing vibronic relaxation have been measured for some molecules and range from picoseconds to microseconds, covering the whole range of observable values.[103,106,107,122] Although the highly excited vibronic states of molecular ions will usually relax very rapidly,

---

[110] B. S. Rabinovitch, J. F. Meager, K.-J. Chao, and J. R. Barker, *J. Chem. Phys.*, 1974, **60**, 2932.
[111] K. C. Kim and D. W. Setser, *J. Phys. Chem.*, 1973, **77**, 2021.
[112] J. G. Moehlmann and J. D. McDonald, *J. Chem. Phys.*, 1973, **59**, 6683.
[113] S. H. Lin, *J. Chem. Phys.*, 1973, **58**, 5760.
[114] D. F. Heller, K. F. Freed, and W. M. Gelbart, *J. Chem. Phys.*, 1972, **56**, 2309.
[115] A. Nitzan and J. Jortner, *J. Chem. Phys.*, 1972, **56**, 2079.
[116] J. R. Christie and D. P. Craig, *Mol. Phys.*, 1972, **23**, 345.
[117] K. Jug and H. von Weyssenhoff, *J. Chem. Phys.*, 1972, **56**, 517.
[118] O. L. J. Gijzeman, *Chem. Phys. Letters*, 1974, **26**, 152.
[119] O. Atabek, A. Hardisson, and R. Lefebvre, *Chem. Phys. Letters*, 1973, **20**, 40.
[120] G. Orlandi and W. Siebrand, *Chem. Phys. Letters*, 1973, **21**, 217.
[121] S. F. Fischer, *Chem. Phys. Letters*, 1972, **17**, 25.
[122] C. S. Parmenter, *Adv. Chem. Phys.*, 1972, **22**, 365.

within say 10^{-13}—10^{-14} s, some lower excited electronic states are likely to be isolated, to the extent that the molecular ions dissociate rather than vibronically relax to the ground state. This competition between vibronic relaxation and decomposition will be strongly dependent on the internal energy available within the electronic state.

Unimolecular Reactions. Unimolecular isomerizations represent a particular type of vibronic relaxation, whereas dissociations are radiationless transitions whose final states are the continua associated with the two fragments. Several important quantum mechanical treatments of these processes have been given in various formulations, often using the concepts and methods of resonance scattering theory.[123—132] The results of these treatments can provide a very comprehensive description of the photochemical and photophysical processes occurring in simple polyatomic molecules, whose excited-state properties can be studied in minute detail with modern spectroscopic techniques.[133,134] Much will be learnt from these exacting studies, but such detailed treatments cannot be extended to larger molecules, and in more highly excited states much of the detail is removed by relaxation processes. Gross mathematical simplifications can therefore be made with considerable justification, and the importance of these theoretical approaches lies in showing when quantum mechanical effects might become important. For example, if the decay rates for vibronic relaxation and decomposition are comparable, then interference effects can give rise to a decay function which does not follow first-order kinetics.[128]

Dynamical Models.—More advanced and detailed discussions of the concepts outlined here may be found in Levine's lucid monograph on the underlying quantum mechanics,[135] and in some excellent reviews.[136—139] It is impossible at present to perform complete quantum mechanical calculations of the dynamical evolution of a molecular species which is initially prepared in a state prone to decomposition. Various approximate models must therefore be devised; their dynamical attributes can then be calculated—principally the rate of formation

[123] K. G. Kay, *J. Chem. Phys.*, 1974, **60**, 2370.
[124] K. G. Kay and S. A. Rice, *J. Chem. Phys.*, 1972, **57**, 3041.
[125] W. M. Gelbart, S. A. Rice, and K. F. Freed, *J. Chem. Phys.*, 1972, **57**, 4699.
[126] C. E. Caplan and M. S. Child, *Mol. Phys.*, 1972, **23**, 249.
[127] O. K. Rice, *J. Chem. Phys.*, 1971, **55**, 439.
[128] M. Shapiro, *J. Chem. Phys.*, 1972, **56**, 2582.
[129] R. A. van Santen, *J. Chem. Phys.*, 1972, **57**, 5418.
[130] Gr. Alexandru, *Internat. J. Mass Spectrometry Ion Phys.*, 1973, **10**, 39; **11**, 17; 1974, **13**, 313.
[131] S. J. Formosinho, *J.C.S. Faraday II*, 1974, **70**, 605.
[132] J. Brickman, *Z. Naturforsch*, 1973, **28a**, 1759.
[133] E. S. Yeung and C. B. Moore, *J. Chem. Phys.*, 1974, **60**, 2139.
[134] K. Evans, R. Scheps, S. A. Rice, and D. F. Heller, *J.C.S. Faraday II*, 1973, **69**, 856.
[135] (a) R. D. Levine, 'Quantum Mechanics of Molecular Rate Processes', Oxford University Press, London, 1969, pp. 215—219; (b) *ibid.*, pp. 239—245; (c) *ibid.*, pp. 246—267.
[136] R. D. Levine, *Ber. Bunsengesellschaft Phys. Chem.*, 1974, **78**, 111.
[137] R. D. Levine, in ref. 7, p. 229.
[138] T. F. George and J. Ross, *Ann. Rev. Phys. Chem.*, 1973, **24**, 263.
[139] J. C. Light, *Adv. Chem. Phys.*, 1970, **19**, 1.

of the products and their state distributions. The validity of the assumptions may then be assessed by comparison with experimental results, and some insight into the 'mechanism' of the reaction be gained. The validity of simpler approximations also may be assessed, by comparisons with more accurate models.

Potential Energy Surfaces. An important model assumes the reaction occurs on a potential energy hypersurface,[140—142] which may belong to a single electronic state (an adiabatic surface) or include other electronic states (a diabatic surface). The latter case is adopted when surface crossings occur.[143] Until recently potential energy surfaces were constructed by semi-empirical methods, but *ab initio* surfaces are now becoming available and will do so increasingly in the future.[144] *Ab initio* calculations of near-Hartree–Fock quality should give quite accurate potential energy surfaces for molecular ion decompositions, and these calculations are within current capabilities even for sizeable polyatomic species (see Section 2).

The dynamical behaviour of the species has been calculated by treating the motions of the atoms (or groups) on the potential energy surface classically,[145] since quantum mechanical treatments are not yet feasible. Only qualitative behaviour is obtained for single trajectories, but if the results of a sufficiently large number of trajectories are averaged these classical trajectory methods give accurate quantitative values for the *average* dynamical attributes of the system. These computations may be compared with experimental results, or used as a classical 'standard' for the evaluation of models based on simplified forces or dynamics. Thermal reactions,[146] hot atom reactions,[147,148] and ion-molecule reactions[149—151] have been studied, to mention only a few examples closely related to ion decompositions. Tully has recently extended the method to include non-adiabatic processes, such as charge-transfer reactions, by a 'trajectory-surface-hopping' technique which has given excellent agreement with experiment for ion–molecule reactions.[151]

The gross features of potential energy surfaces along a reaction path have often been predicted on the basis of orbital symmetry,[152] or orbital phase continuity[153]

[140] O. Sinanoglu, *Comments Atom. Mol. Phys.*, 1972, **3**, 53.
[141] J. C. Polanyi, *Accounts Chem. Res.*, 1972, **5**, 161.
[142] T. Carrington and J. C. Polanyi, in 'Chemical Kinetics', ed. J. C. Polanyi (M.T.P. International Review of Science), Physical Chemistry, Series 1, Butterworths, London, 1972, Vol. 9, p. 135.
[143] T. Carrington, *Accounts Chem. Res.*, 1974, 7, 20.
[144] S. D. Peyerimhoff and R. J. Buenker, *Ber. Bunsengesellschaft Phys. Chem.*, 1974, **78**, 119.
[145] D. L. Bunker, *Methods Comput. Phys.*, 1971, **10**, 287.
[146] D. L. Bunker and W. L. Hase, *J. Chem. Phys.*, 1973, **59**, 4621.
[147] L. M. Raff, *J. Chem. Phys.*, 1974, **60**, 2220.
[148] S. Chapman and R. J. Suplinskas, *J. Chem. Phys.*, 1974, **60**, 248.
[149] J. R. Krenos, R. K. Preston, R. Wolfgang, and J. C. Tully, *J. Chem. Phys.*, 1974, **60**, 1634.
[150] D. R. McLaughlin and D. L. Thompson, *J. Chem. Phys.*, 1973, **59**, 4393.
[151] J. C. Tully, *Ber. Bunsengesellschaft Phys. Chem.*, 1973, **77**, 557.
[152] R. B. Woodward and R. Hoffmann, 'The Conservation of Orbital Symmetry', Verlag Chemie, Berlin, 1971.
[153] W. A. Goddard, *J. Amer. Chem. Soc.*, 1972, **94**, 793.

principles. Mahan has applied these methods to ion–molecule reactions with success, particularly for small species where the molecular orbital correlation diagrams are straightforward.[154] However, George and Ross have critically discussed the theoretical basis for these rules, and indicated the situations where they may be expected to fail.[155] Ion decompositions are likely to come under the latter category because of the changes which occur in the nuclear orbital and and rotational angular momenta; the electronic orbital and spin angular momenta may not then be conserved.

Statistical Approximation. Levine has discussed in detail the theoretical background to the statistical approximation.[135c] It assumes that at a given energy E, the rate of a process can be expressed as the probability of forming a particular intermediate state from the initial state times the rate of its breakdown into the final state, summed over all possible intermediate states. This approximation can only provide an upper limit for the rate of a process. The quasi-equilibrium hypothesis introduces the further approximation that the aforementioned probability is independent of the initial state. The relationships between the various models using these, and further, approximations (*e.g.* the RRKM[156,157] and QET[158] transition-state theories) have been clearly exposed using the principle of microscopic reversibility and the concept of a yield function.[135a,c,159] QET assumes that the intermediate (transition) state is formed from a microcanonical ensemble,[158] but it is stressed this should be 'a microcanonical ensemble, *with a given value of all conserved quantum numbers*' (ref. 135, p. 251). Thus the total angular momentum J, and its projection M on a fixed axis, should be treated explicitly: the original QET formulation[158] fails to do this.

The quasi-equilibrium transition state theory does not imply (nor require) there to be an equilibrium between the initial (or final) states and the transition state.[160] It simply assumes that a 'configuration of no return' can be defined such that species which reach it will react, at a rate which can be calculated by statistical mechanical methods.[137] This critical configuration is almost universally taken to be the potential energy saddlepoint along a reaction co-ordinate, but this is not necessarily the best choice since the accompanying assumption that the transmission coefficient is unity may be grossly inaccurate. This is very likely if the density $\rho^{\pm}$ of exit states associated with the transition state ($\pm$) is not a minimum for the configurations along the reaction co-ordinate. In this case the system might be 'reflected' before or after passing through the transition state (depending on where the minimum lies) and return to an initial state, thereby causing the transmission coefficient to be much less than unity.

[154] B. H. Mahan, *J. Chem. Phys.*, 1971, **55**, 1436.
[155] T. F. George and J. Ross, *J. Chem. Phys.*, 1971, **55**, 3851.
[156] R. A. Marcus and O. K. Rice, *J. Phys. and Colloid Chem.*, 1951, **55**, 894.
[157] R. A. Marcus, *J. Chem. Phys.*, 1952, **20**, 359.
[158] H. M. Rosenstock, M. B. Wallenstein, A. L. Wahraftig, and H. Eyring, *Proc. Nat. Acad. Sci., U.S.A.*, 1952, **38**, 667.
[159] R. D. Levine and R. B. Bernstein, *J. Chem. Phys.*, 1972, **56**, 2281.
[160] J. B. Anderson, *J. Chem. Phys.*, 1973, **58**, 4684.

It would seem more reasonable to choose the transition state as the configuration q^+ having the minimum density of exit states along the reaction co-ordinate q.[161,162]

$$\frac{\partial \rho^+(E, J, q)}{\partial q} = 0, \qquad q = q^+ \tag{1}$$

This 'minimum local entropy' formulation gives a minimum value for the rate constant of the RRKM (QET) model,

$$k(E, J) = \int_0^{E-E_0} \rho^+[E - E_0 - \varepsilon^+, J, q^+(\varepsilon^+)] d\varepsilon^+ / h\rho(E, J) \tag{2}$$

and thus a *least upper limit* to the true rate constant (within the quasi-equilibrium approximation). Further support for this formulation is provided by the 'adiabatic open channel' model of Quack and Troe[163] (see below), who also find that the criterion

$$\frac{\partial W^+(E, J, q)}{\partial q} = 0, \qquad q = q^+ \tag{3}$$

gives almost identical results to (1), and is easier to apply because of the more continuous nature of W^+ compared with ρ^+. [But note that

$$W^+(E - E_0, J, q^+) \neq \int_0^{E-E_0} \rho^+[E - E_0 - \varepsilon^+, J, q^+(\varepsilon^+)] d\varepsilon^+$$

because of the dependence (1) of q^+ on ε^+ in the integrand, though the inequality of these functions is apparently small.] The configuration q^+ varies with the internal and rotational energy, and the minimum local entropy model corresponds to an RRKM model whose transition state becomes 'tighter' with increasing internal energy. It corrects a subtle deficiency of the latter models, employing a fixed transition state configuration, in which the transmission coefficient may change markedly with internal energy. Bunker,[161] Hase,[164] and others[163] have used this model with considerable success for thermal unimolecular reactions.

The 'adiabatic open channel' model assumes the initial and final states are correlated adiabatically, but in a manner that conserves good quantum numbers (E, J, M).[163] This is achieved through coupling of the particular vibrational and rotational states corresponding to the degrees of freedom of the reaction co-ordinate. A simple interpolation procedure is used to calculate the energies $V_a(q)$ of these adiabatic channels between the initial and final states as a function of the distance between the fragments (q). The rate constant is calculated from the total number of open adiabatic channels $W(E, J)$, defined as those channels whose maximum energies $V_a(q)_{max}$ are less than the total internal energy E,

[161] D. L. Bunker and M. D. Pattengill, *J. Chem. Phys.*, 1968, **48**, 772.
[162] W. H. Wong, *Canad. J. Chem.*, 1972, **50**, 3386.
[163] M. Quack and J. Troe, *Ber. Bunsengesellschaft Phys. Chem.*, 1974, **78**, 240.
[164] W. L. Hase, *J. Chem. Phys.*, 1972, **57**, 730.

assuming the transmission coefficient is unity for open channels and zero for closed channels $[V_a(q)_{max} > E]$.[163]

$$k(E, J) = W(E, J)/h\rho(E, J) \qquad (4)$$

The approach used in this model may appear directly opposed to the quasi-equilibrium models, which assume the transition is non-adiabatic with respect to the internal states. However, some of the assumptions made in this treatment lessen this contrast, and it in fact gives results very similar to the minimum local entropy model, illustrating the essential unity underlying these various statistical models. An important result is that both these models predict much stronger curvature of the $k(E, J)$ function than the RRKM (QET) model, with the rate constant increasing more slowly at higher energies.[163] Quack and Troe's paper also includes a good discussion of the relationship between the quantum scattering and transition state theories.[163]

The validity of the quasi-equilibrium hypothesis rests on the existence of a statistical equilibrium distribution of energy between all the internal degrees of freedom. This occurs only if there is strong coupling between them. For certain reaction channels the coupling between some internal degrees of freedom may be weak, and the energy distribution will then be non-equilibrium. In an important series of papers, Levine and Bernstein have proposed the concept of 'entropy deficiency' to characterize the deviation from statistical equilibrium of the properties of a dynamical system.[136,165-167] This approach uses the methods of non-equilibrium statistical mechanics to assign a thermodynamic weight W to a particular distribution. Equilibrium distributions in which all internal states are equally probable are assigned weights W_0, but non-equilibrium distributions where some internal states are not equally probable have smaller weights,

$$W = W_0 \exp(-N \Delta S/R) \qquad (5)$$

where N is Avogadro's number, R is the gas constant, and ΔS is the entropy deficiency (≥ 0).

In the transition state theory the unimolecular reaction rate constant is, then,

$$\begin{aligned} k(E) &= W^+(E^+)/h\rho(E), \qquad E^+ = E - E_0 \\ &= W_0^+(E^+) \exp[-N \Delta S^+(E^+)/R]/h\rho(E) \\ &= k_0(E) \exp[-N \Delta S^+(E^+)/R] \end{aligned} \qquad (6)$$

where $k_0(E)$ is the rate constant calculated from the quasi-equilibrium approximation. The entropy deficiency of the transition state $\Delta S^+(E^+)$ will generally be energy dependent, since the degree of coupling between internal degrees of freedom may change with the available internal energy. The specific reasons for any such entropy deficiency are not defined, but particularly important factors are expected to be the time-dependent relaxation phenomena associated with the coupling of internal states (see above). Resolution of these factors would require

---

[165] R. D. Levine and R. B. Bernstein, *Chem. Phys. Letters*, 1973, **22**, 217.
[166] R. B. Bernstein and R. D. Levine, *J. Chem. Phys.*, 1972, **57**, 434.
[167] A. Ben-Shaul, R. D. Levine, and R. B. Bernstein, *J. Chem. Phys.*, 1972, **57**, 5427.

the application of more complete dynamical theories. In determining $\Delta S^{\ddagger}(E^+)$ from experimental measurements it would seem best to use the minimum local entropy model to calculate the quasi-equilibrium value [e.g. $k_0(E)$], thus making some allowance for the inherent deficiencies of transition state models using a fixed critical configuration.

Tests of Dynamical Models. Perhaps the most obvious dynamical attribute to use in comparing theory and experiment is the microscopic rate constant. Very few accurate values have been measured for reactions of isolated species, although some new techniques for studying ion decomposition rates may provide important data (see Section 5). Henchman has given an excellent review of the problems involved, with particular reference to ion-molecule reactions.[168a] On the other hand, product energy distributions can be measured quite accurately and these provide a severe test of the dynamical model, especially when a complete analysis of the energy partitioning between the translational, vibrational, and rotational degrees of freedom of the products can be performed. Molecular beam experiments are necessary to achieve this and good reviews of ion-neutral[169,170] and neutral-neutral[171,172] beam reactions are available. The product energy distributions predicted by a statistical model,[173] and the translational energy distribution predicted by particular quasi-equilibrium transition state models[174,175] have been derived.

The product translational energy distributions measured for several ion-molecule reactions, occurring by unimolecular breakdown of long-lived collision complexes, are in good agreement with quasi-equilibrium predictions.[174] But the reaction

$$C_2H_4^+ + C_2H_4 \rightarrow [C_4H_8^+]^* \rightarrow CH_3 + C_3H_5^+ \rightarrow C_3H_3^+ + H_2$$

has a distribution which deviates markedly from the quasi-equilibrium predictions, and does so increasingly at higher total energies.[175] Other types of unimolecular reactions which often exhibit non-equilibrium translational energy distributions include the photodissociation reactions studied by Wilson *et al.*,[176,177] and the chemical activation reactions studied by Lee and co-workers.[178—180] Product rotational and vibrational energy distributions have

---

[168] (a) M. J. Henchman, in 'Ion–Molecule Reactions', ed. J. L. Franklin, Butterworths, London, 1972, Vol. 1, p. 101; (b) *ibid.*, p. 184.
[169] J. Dubrin and M. J. Henchman, in ref. 142, p. 213.
[170] Z. Herman and K. Birkinshaw, *Ber. Bunsengesellschaft Phys. Chem.*, 1973, **77**, 566.
[171] J. L. Kinsey, in ref. 142, p. 173.
[172] Y. T. Lee, *Ber. Bunsengesellschaft Phys. Chem.*, 1974, **78**, 135.
[173] W. H. Wong, *Canad. J. Chem.*, 1972, **50**, 633.
[174] S. A. Safron, N. D. Weinstein, D. R. Herschbach, and J. C. Tully, *Chem. Phys. Letters*, 1972, **12**, 564.
[175] A. Lee, R. L. Leroy, Z. Herman, R. Wolfgang, and J. C. Tully, *Chem. Phys. Letters*, 1972, **12**, 569.
[176] S. J. Riley and K. R. Wilson, *Discuss. Faraday Soc.*, 1972, **53**, 132.
[177] G. E. Busch and K. R. Wilson, *J. Chem. Phys.*, 1972, **56**, 3626, 3638, 3655.
[178] J. M. Parson and Y. T. Lee, *J. Chem. Phys.*, 1972, **56**, 4658.
[179] J. M. Parson, K. Shobatake, Y. T. Lee, and S. A. Rice, *J. Chem. Phys.*, 1973, **59**, 1402, 1416, 1427.
[180] K. Shobatake, Y. T. Lee, and S. A. Rice, *J. Chem. Phys.*, 1973, **59**, 1435, 2483, 6104.

been measured by i.r. chemiluminescence techniques[142] for some unimolecular chemical activation reactions.[181,182] Non-equilibrium distributions are often observed, even to the extent of population inversion, and these results have recently been characterized very successfully using the entropy deficiency model.[183]

Unimolecular Ion Decompositions.—*Recent Theoretical Developments.* Both Klotz[184-186] and Knewstubb[187-189] have proposed models which seek to correct the failure of the original QET formulation[158] to conserve the total angular momentum. The close connection between ion decomposition rate constants and ion-molecule reaction cross-sections through the microscopic reversibility principle has been exposed in these papers, but requires some comment. As shown by Klotz[184] the problem of defining an unknown transition state configuration can be avoided, by instead calculating the cross-section for the formation of a collision-complex from the separated fragments whose properties can be measured. However, this collision-complex must be equivalent to the decomposing ion, and the calculated cross-section must therefore refer to a collision involving strong coupling between the degrees of freedom of the fragments which correlate with the dissociating bond. The simple Langevin model used by Klotz[184,185] calculates a close-collision cross-section from a consideration of only the long-range isotropic ion-induced dipole potential; other long- and medium-range forces which may be important are ignored.[190] Henchman has critically discussed these aspects in a very penetrating review, and concluded that the Langevin model provided no better than a qualitative upper bound to ion-molecule close-collision cross-sections at thermal relative translational energies.[168b] Furthermore a close-collision does not necessarily imply that a strong coupling collision occurs, since the behaviour of the system after crossing the centrifugal barrier is determined by the short-range bonding forces which are also ignored in the Langevin model. Knewstubb's 'ballistic' model attempts to make some allowance for this aspect in a straightforward empirical manner which retains much of the mathematical simplicity of the Langevin model.[187]

The main value of these approximate treatments is that they enable the major qualitative effects arising from the conservation of total angular momentum to be studied systematically. This is primarily a matter of determining the rotational energies for species of appropriate symmetries, and taking proper account of the correlation between the orbital and rotational momenta. Formulae have

---

[181] K. C. Kim and D. W. Setser, *J. Phys. Chem.*, 1973, **77**, 2493.
[182] H. W. Chang and D. W. Setser, *J. Chem. Phys.*, 1973, **58**, 2298.
[183] R. D. Levine, B. R. Johnson, and R. B. Bernstein, *Chem. Phys. Letters*, 1973, **19**, 1.
[184] C. E. Klotz, *J. Phys. Chem.*, 1971, **75**, 1526.
[185] C. E. Klotz, *Z. Naturforsch.*, 1972, **27a**, 553.
[186] C. E. Klotz, *J. Chem. Phys.*, 1973, **58**, 5364.
[187] P. F. Knewstubb, *Internat. J. Mass Spectrometry Ion Phys.*, 1971, **6**, 217.
[188] P. F. Knewstubb, *J.C.S. Faraday II*, 1972, **68**, 1196.
[189] P. F. Knewstubb, *Internat. J. Mass Spectrometry Ion Phys.*, 1973, **10**, 371.
[190] D. Hyatt and L. Stanton, *J.C.S. Faraday II*, 1973, **69**, 340.

been derived for various rotational effects on the rate constants, and for the average translational and rotational energies of the products under particular assumptions which should be carefully noted.[185,188] The rotational effects are largest when the difference between the precursor moments of inertia and the sum of the product moments of inertia is greatest (e.g. when an ion fragments in half), and are more pronounced near threshold.[187] They are minor for the loss of the relatively light fragments (e.g. hydrogen) unless the precursor is itself small.[184,185,187,188]

A more subtle point concerns the range of validity of these models. It was stressed above that the Langevin model is only applicable to thermal relative translational energies (e.g. $\leqslant 0.1$ eV).[168a] Qualitative arguments suggest that decompositions occurring with the release of higher kinetic energies will have lower rate constants than predicted by this model. Knewstubb has focused attention on this point in his consideration of a 'hypopectic' transition state.[188] Decompositions occurring with high separation velocities are likely to have only a few strongly coupled internal degrees of freedom, because of the shorter times involved. A high entropy deficiency therefore ensues, and the rate constant is lower than predicted by a quasi-equilibrium model. Such effects are expected to be most apparent in decompositions where light fragments are lost, especially with a high release of kinetic energy.[188] These considerations may explain the anomalously slow rise of the rate constant at higher internal energies, which has been observed for fragmentations ejecting H or H_2[256] (see Section 5).

Another general point arising from these theoretical studies concerns the observation of metastable ions. The original QET formulation predicted that the rate constant at threshold was often large for small ions (e.g. $> 10^6$ s^{-1}), but this is simply an artefact of that model. The true rate constant at threshold is close to zero but may rise very rapidly. Hence, in principle, metastable ions may be detected for every fragmentation. If they are not detected it implies that the probability of observing them is very low because:
(i) $k(E)$ rises extremely rapidly, or
(ii) a faster fragmentation is competing over the range of internal energies corresponding to the interval of rate constants for metastable ions.

Applications of RRKM Models. The classification, RRKM model, is used here simply to avoid any confusion of terminology between QET calculations and the more specific quasi-equilibrium hypothesis. It seems appropriate since the RRKM formulation[156,157] is invariably used in QET calculations. Good reviews of these models have been presented by Wahraftig[1] and Setser,[191] to which little need be added. The development of a very efficient computer algorithm to calculate energy level sums and densities by direct counting is most timely.[192,193] Any type of oscillator (e.g. anharmonic, hindered internal rotor) can be treated exactly, and it will be most useful for small molecules whose energy levels are

---

[191] D. W. Setser, in ref. 142, p. 1.
[192] S. E. Stein and B. S. Rabinovitch, *J. Chem. Phys.*, 1973, **58**, 2438.
[193] S. E. Stein and B. S. Rabinovitch, *J. Chem. Phys.*, 1974, **60**, 908.

known accurately from spectroscopic measurements. Its main application for large polyatomic molecules will be to test and improve the accuracy of approximate analytical methods[194–196] for real systems which deviate from harmonic oscillator behaviour. Useful guidelines for assigning frequency factors may also be found in the extensive series of studies by Rabinovitch and co-workers, who have determined an internally consistent set of transition state parameters for unimolecular decompositions of chemically activated alkyl radicals.[191,197,198]

The application of theoretical calculations to experimental problems has always been hindered by a severe lack of accurate information about the energetics of fragmentations, their breakdown graphs, and the energy deposition functions of the molecular ions. Hopefully the techniques discussed in Sections 5 and 6 will overcome some of these deficiencies. There are two major applications for calculated rate constants; one is to confirm proposed fragmentation patterns, and the other to obtain information about the detailed mechanism of ion decompositions. In the former case, the problem resides in determining the precise electronic state(s) involved in each fragmentation. Experimental breakdown graphs and accurate energetic information often enable this to be decided, but not always. Simulation of ion yield data can then provide useful confirmatory evidence, particularly about whether fragmentations are competing or not. The simplifications that need to be made in treating other factors affecting ion yields (e.g. energy deposition functions, threshold laws) undermine the decisiveness of such calculations, although direct comparison with breakdown graphs (if available) avoids this difficulty. Except in the threshold region, breakdown graphs are determined only by relative rate constants and agreement with experiment may therefore result from a cancellation of systematic errors. This might be considered useful in the former application, since fairly simple model calculations (e.g. RRKM) may be used with hopes of not going too far wrong, at least for similar competing fragmentations, but it is fatal in the 'mechanistic' application.

If the $k(E)$ functions are systematically in error then comparisons with breakdown graphs or relative ion yields may easily lead to erroneous conclusions about the mechanism being drawn from the simulated frequency factor and activation energy. This type of application should therefore be restricted to situations where the absolute magnitude of $k(E)$ is measured as a function of ionizing energy (see Section 5), or has a critical effect, as in metastable ion decompositions. Furthermore, only the most accurate theoretical models should be used in such studies, involving as few *ad hoc* assumptions as possible, to obtain unambiguous results.

A detailed reinvestigation of the benzene ion exemplifies the problems involved in determining precise fragmentation paths, even for one of the very few cases

[194] W. Forst, *Chem. Rev.*, 1971, **71**, 339.
[195] M. R. Hoare and P. Pal, *Mol. Phys.*, 1971, **20**, 695.
[196] S. H. Lin, *Mol. Phys.*, 1971, **20**, 953.
[197] C. W. Larson, P. T. Chua, and B. S. Rabinovitch, *J. Phys. Chem.*, 1972, **76**, 2507.
[198] E. A. Hardwidge, B. S. Rabinovitch, and R. C. Ireton, *J. Chem. Phys.*, 1973, **58**, 340.

where much accurate information is available.[199] This particular ion is intriguing, but its behaviour is perhaps not typical. Rosenstock *et al.* concluded that at energies up to 14.5 eV, $C_6H_5^+$ and $C_6H_4^+$ were formed competitively from the ground state, whereas $C_4H_4^{+\cdot}$ and $C_3H_3^{+\cdot}$ were formed competitively from an isolated state lying 2.25 eV higher. This state is not necessarily the first excited state, but perhaps a structural isomer having the same heat of formation. Briefly, their approach was to calculate the dissociation energies from known (or estimated) heats of formation of the fragment ions, and then vary the respective frequency factors to obtain best fits with the measured photoion yield curves, but some important discrepancies remained. There was poor agreement between the calculated[199] and experimental[256] $k(E)$ curves for $C_6H_5^+$ formation (a point discussed further in Section 5), and this immediately throws into question the validity of the QET model for this fragmentation and has serious implications for some of their arguments. Additionally, the abrupt decrease in the slope of the photoion yield curves for $C_6H_5^+$ and $C_6H_4^+$ could not be explained by the proposed fragmentation processes; it suggests that different channels for the formation of these ions open up at 14.6 eV, the adiabatic IP of the $1b_{2u}$ state. (The observation that the molecular ion of hexa-1,5-diyne loses H· from an isolated state[200] may be not entirely unrelated to the problem of the ionic decomposition processes of benzene). Perhaps the most important general conclusion arising from this work,[199] which is the best of its type reported, concerns the information which can be gleaned from detailed studies of the shapes of accurately measured IE curves near threshold. In this particular case the kinetic shift appears to be very large, being accentuated by the short residence time of the ion source (*ca.* 10^{-7} s).

Calculated branching ratios were used in one estimation of dissociation energies for the formation of (11) relative to CH_3^+ from the collision-complexes (12), produced by ion-molecule reactions of H_3^+ with CH_3NH_2, CH_3OH, and CH_3SH.[201]

$[CH_3XH_n]^{+\cdot}$ $[CH_3XH_{n+2}]^{+\cdot}$ $X = N, O, S$
    (11)   (12) $(n = 1, 0, 0)$

Fair to good agreement with experiment was obtained over *ca.* 1 eV range of internal energy, but some assumptions were required to determine the internal energy of (12) as a function of H_2 pressure. The diagnostic usefulness of such approximate calculations was emphasized.

Similar approaches have been used to estimate activation energies for the loss of Br· from the *cis*- and *trans*-isomers of 2-substituted cyclopentyl (13) and cyclohexyl bromides (14),[202] and for competing rearrangement and simple

---

[199] H. M. Rosenstock, J. T. Larkins, and J. A. Walker, *Internat. J. Mass Spectrometry Ion Phys.*, 1973, **11**, 309.
[200] M. L. Gross and R. J. Aerni, *J. Amer. Chem. Soc.*, 1973, **95**, 7875.
[201] M. T. Bowers, W. J. Chesnavich, and W. T. Huntress, *Internat. J. Mass Spectrometry Ion Phys.*, 1973, **12**, 357.
[202] Y. Grounelle, J.-M. Péchiné, and D. Solgadi, *Org. Mass Spectrometry*, 1973, **7**, 1287.

cleavage fragmentations of *threo*- and *erythro*-2,3-diphenylbutane.[203] Calculated relative ion intensities were compared with experimental measurements over several eV and the activation energies were adjusted to obtain best agreement, using estimated frequency factors. A number of further simplifying approximations

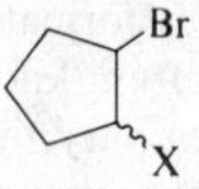

(13) 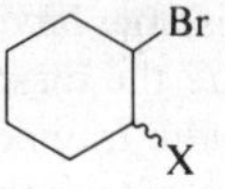   (14)   X = F, Cl, Br, OH

were made in the ion yield formulae, and for the EDF and threshold law (assumed linear[202]) which are unlikely to remain invariant over such a large energy range (9 eV). The high degree of approximation involved in these studies emphasizes the difficulties involved, but also diminishes confidence in the quantitative validity of the results.

Other calculations of isotope effects[204,205] and ortho-effects[206,207] using the simple, but very inaccurate, classical approximation for $k(E)$ cannot be considered any better than qualitative. The use of this approximation in any computations is strongly discouraged; certain other approximations are just as easy to program and of much better accuracy.[194] Furthermore, it is not advisable to use measured AP's to estimate activation energies,[206—208] unless there is good evidence that the kinetic shift is negligible.[209] This applies especially to semi-log determinations which usually contain a significant (and variable) systematic error (see Section 6).

In summary, only when the necessary accurate data is available are theoretical calculations expected to provide any reliable evidence about the detailed mechanisms of ion decompositions (*e.g.* frequency factor, rotational effects, entropy deficiency); and then only with the more advanced models discussed earlier. Nevertheless, there would appear to be a very useful place for the simpler models in defining the overall fragmentation patterns of molecules, and especially to help determine whether excited (isolated) states are involved.

5 Experimental Methods

Photoelectron–Photoion Coincidence Spectroscopy.—Photoelectron–photoion coincidence spectroscopy was developed by Brehm and von Puttkamer,[210] and ranks as the best technique for studying the energy-dependence of ion decompositions as described by their breakdown graphs. Eland has discussed the

---

[203] J.-M. Péchiné, *Org. Mass Spectrometry*, 1972, **6**, 805.
[204] M. Corval and P. Masclet, *Org. Mass Spectrometry*, 1972, **6**, 511.
[205] I. Howe, N. A. Uccella, and D. H. Williams, *J.C.S. Perkin II*, 1973, 76.
[206] S. A. Benezra and M. M. Bursey, *J.C.S. Perkin II*, 1972, 1537.
[207] M. M. Bursey and C. E. Parker, *Tetrahedron Letters*, 1972, 2211.
[208] H. Hoshino, S. Tajima, and T. Tsuchiya, *Bull. Chem. Soc. Japan*, 1973, **46**, 3043.
[209] G. Bouchoux, M. Fétizon, and H.-E. Audier, *Org. Mass Spectrometry*, 1974, **9**, 12.
[210] B. Brehm and E. von Puttkamer, *Z. Naturforsch.*, 1967, **22a**, 8.

basic principles of the method[211] and although extensive development and improvement can be expected in the future, important results are already being obtained with the 'first-generation' instruments.[212–217] By this method the decomposition processes of excited states of ions may be investigated directly, and different aspects may be studied depending on the particular experimental arrangement. Most work has been performed using a single photon energy, usually the He^1 584 Å line, so that autoionization is generally negligible and only processes resulting from direct ionization are observed.

The usual coincidence technique measures the photoelectron spectrum coincident with a particular ion. The differential PES directly measures the probability of formation of the ion as a function of the ionization energy, the 'breakdown curve'.[211–213] A complete breakdown graph for the molecular ion can be obtained by measuring the breakdown curves for each ion and normalizing with respect to their total. An improvement of about one order of magnitude in the signal-to-noise ratios of the coincident PES, which is necessary to determine accurate breakdown graphs,[213] now appears feasible[214] and decisive results can soon be expected. The early results[212] have already enabled quite detailed comparisons to be made with photoionization and charge-transfer measurements for methanol and ethanol.[213] These comparisons show that autoionization makes important contributions to the photoionization mass spectra, and the consequences have been well discussed by Brehm et al.[213]

Other interesting features concern the fragmentation reactions. In methanol the threshold energies for CH_2OH^+ formation differ in the photoionization (11.69 eV)[218] and coincidence (12.0 eV) studies. Autoionization processes probably cause the fragmentation below 12.0 eV because structure is observed between 11.2—11.8 eV, whereas the PES shows this region is not populated by direct ionization.[213] The PES band for the first excited state of CH_3OH^+ extends from 12.0—13.6 eV but the coincidence spectra indicate that this state decomposes entirely to CH_2OH^+ (Figure 2), thus explaining the appearance potential of 12.0 eV in both the coincidence and charge-transfer mass spectra (Figure 3). This dissociation presumably occurs by a very fast vibronic relaxation to the ground state (which then contains enough internal energy to dissociate), explaining the lack of vibrational structure in the first excited state PES band. In contrast with methanol, autoionization makes little observable contribution to the fragmentation of ethanol, but both major fragment ions, masses 45 and 31, seem to be formed by two non-competing processes.[213]

---

[211] J. H. D. Eland, *Internat. J. Mass Spectrometry Ion Phys.*, 1972, **8**, 143.
[212] E. von Puttkamer, *Z. Naturforsch.*, 1970, **25a**, 1062.
[213] B. Brehm, V. Fuchs, and P. Kebarle, *Internat. J. Mass Spectrometry Ion Phys.*, 1971, **6**, 279.
[214] B. Brehm, J. H. D. Eland, R. Frey, and A. Küstler, *Internat. J. Mass Spectrometry Ion Phys.*, 1973, **12**, 197, 213.
[215] B. Brehm, R. Frey, A. Küstler, and J. H. D. Eland, *Internat. J. Mass Spectrometry Ion Phys.*, 1974, **13**, 251.
[216] J. H. D. Eland, *Internat. J. Mass Spectrometry Ion Phys.*, 1973, **12**, 389.
[217] C. J. Danby and J. H. D. Eland, *Internat. J. Mass Spectrometry Ion Phys.*, 1972, **8**, 153.
[218] K. M. A. Refaey and W. A. Chupka, *J. Chem. Phys.*, 1968, **48**, 5205.

A novel variation of the coincidence technique measures the kinetic energy distributions of the fragment ions formed at a particular ionizing energy.[211] This can be done using a time-of-flight mass selector, but ions of similar masses interfere and the resolution is low.[217] Better results have been obtained with a

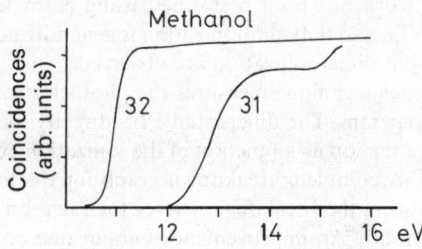

Figure 2 *Integral photoelectron-photoion coincidence spectra of methanol* (21.21 eV photons)
(Reproduced by permission from *Internat. J. Mass Spectrometry Ion Phys.*, 1971, **6**, 279)

magnetic sector mass selector combined with time-of-flight detection methods.[214] These experiments on the fragmentations of SO_2^+, CS_2^+,[214] N_2O^+, COS^+, and CF_4^+ [215] constitute a crucial test of theoretical models of ion decompositions. Neither quasi-equilibrium nor direct dissociation models correctly predict the observed kinetic energy distributions. It is not surprising that quasi-equilibrium models fail to explain the results, since the coupling between different internal

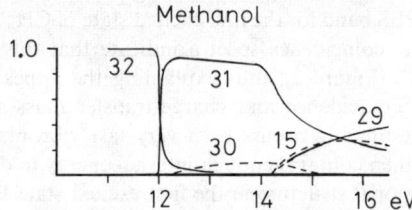

Figure 3 *Charge-transfer breakdown graph of methanol*
(Reproduced by permission from *Arkiv Fyzik*, 1962, **21**, 97)

degrees of freedom is expected to differ quite markedly in these small molecules. Coincident PES experiments also show that the rates of the vibronic relaxation processes from the excited states vary dramatically with the ionizing energy in these molecular ions,[214—216] thus explaining why the direct dissociation model also fails. Photoion kinetic energy analysis (without coincident photoelectron detection) has also been used, in conjunction with other coincidence measurements, to study the decomposition of CO_2^+ ions.[219]

[219] J. H. D. Eland, *Internat. J. Mass Spectrometry Ion Phys.*, 1972, **9**, 397.

The photoelectron–photoion coincidence spectra of perfluoroethane indicated that the first excited state of the molecular ion was isolated from the ground state.[220] The excited state was observed to decompose to $C_2F_5^+$, with CF_3^+ formation becoming dominant at higher energies (probably *via* secondary fragmentation of $C_2F_5^+$). On the other hand, the ground state was observed to form only CF_3^+. This contradicts an earlier investigation of the charge-transfer mass spectra of C_2F_6, where both $C_2F_5^+$ and CF_3^+ were formed when $Xe^{+\cdot}$ [Recombination Energy (RE) = 12.13, 13.44 eV] was used, but only CF_3^+ when $Kr^{+\cdot}$ (RE = 14.00, 14.67 eV) was used.[221] It seems likely that the formation of $C_2F_5^+$ by charge-transfer with $Xe^{+\cdot}$ is anomalous since, from thermochemical arguments, only ion-pair formation could give this ion below *ca.* 15 eV; a reinvestigation would be desirable to settle this discrepancy.

For some time the existence of isolated states in the benzene molecular ion has been discussed in the literature (see *e.g.* ref. 199). Eland has recently published preliminary results of a photoelectron–photoion coincidence study, and found no clear evidence for the formation of fragment ions (of differing carbon number) from isolated states.[222] As he pointed out, these results do not permit an unambiguous interpretation, and the question of isolated states in $C_6H_6^{+\cdot}$ remains open.

Stockbauer has developed a threshold photoelectron–photoion coincidence technique which permits breakdown graphs to be measured for direct photoionization, eliminating any autoionization contributions.[223] The energy resolution and signal-to-noise ratio of the reported breakdown graphs for CH_4, CD_4, C_2H_6, and C_2D_6 were very good, and enabled several discrepancies between QET predictions and the results of some other investigations to be resolved. QET was found to give a good qualitative description of the fragmentations of these molecular ions, and it was considered that more accurate calculations would give quantitative agreement.[223] The discrepancies of other photoionization studies were shown to be due to autoionization or threshold law effects, and possibly due to experimental artifacts in the case of earlier photoelectron–photoion coincidence[212] and charge-transfer measurements.

The sensitivity problems arising from low photoionization yields may be alleviated by the use of intense monochromatized synchrotron radiation sources, and coincidence experiments currently being set up should provide improved results in the near future.[224]

Charge-transfer Mass Spectra.—Lindholm and co-workers have published a series of extensive studies of the electronic structures of several aromatic molecules. Photoelectron, ultraviolet, and electron spectroscopy, and charge-transfer

---

[220] I. G. Simm, C. J. Danby, and J. H. D. Eland, *J.C.S. Chem. Comm.*, 1973, 832.
[221] R. E. Marcotte and T. O. Tiernan, *J. Chem. Phys.*, 1971, **54**, 3385.
[222] J. H. D. Eland, *Internat. J. Mass Spectrometry Ion Phys.*, 1974, **13**, 457.
[223] R. Stockbauer, *J. Chem. Phys.*, 1973, **58**, 3800.
[224] P. M. Guyon, *Adv. Mass Spectrometry*, 1974, **6**, in the press.

mass spectra were used as aids to interpretation.[225–234] Of particular interest here is the interpretation of the charge-transfer mass spectra and the fragment ion appearance potentials, which often matched the adiabatic ionization potentials of electronically excited states determined from PES. These interesting observations were interpreted as implying that fragmentation occurred from the excited states of the molecular ions, and also that the basic postulates of the QET did not apply to these decompositions.[93] If this interpretation is correct it has important consequences, but objections have been raised.[1] An alternative interpretation, which is consistent with all the data and with current theories of unimolecular decay processes, is outlined below.

It is important that the points concerning charge-transfer ionization discussed in Section 3 be remembered. Methanol is a well-studied example[213,219,234] and is considered here (Figures 2 and 3); for recombination energies (RE) below 12.0 eV only the molecular ion is formed. The ground state band in the PES extends over 10.9—11.5 eV so vertical charge-transfer ionization is only expected to occur within this range, which lies below the first dissociation limit of 11.69 eV for H˙ loss.[213] When the first excited state (12.0—13.5 eV) is ionized with RE > 12.0 eV, it undergoes rapid vibronic relaxation (internal conversion) to the ground state which now has sufficient internal energy to dissociate completely. Only the CH_2OH^+ ion is formed from the vibrationally excited ground state because no other dissociation process can compete at this energy, in accord with QET. Ionization to the second excited state (14.1—16.8 eV) is followed by vibronic relaxation to the ground state and the formation of CH_2OH^+ again, now however possessing sufficient internal energy to decompose further to CHO^+. The cross-over in the breakdown curves for these ions over a wide energy range supports this interpretation of a secondary fragmentation process (Figure 3). The formation of CH_3^+ at these energies probably occurs by competitive decomposition of the highly vibrationally excited molecular ion. Although the dissociation energy for this process is higher than that for CH_2OH^+

---

[225] P. J. Derrick, L. Asbrink, O. Edqvist, B.-O. Jonsson, and E. Lindholm, *Internat. J. Mass Spectrometry Ion Phys.*, 1971, **6**, 161.
[226] P. J. Derrick, L. Asbrink, O. Edqvist, B.-O. Jonsson, and E. Lindholm, *Internat. J. Mass Spectrometry Ion Phys.*, 1971, **6**, 177.
[227] P. J. Derrick, L. Asbrink, O. Edqvist, B.-O. Jonsson, and E. Lindholm, *Internat. J. Mass Spectrometry Ion Phys.*, 1971, **6**, 191.
[228] P. J. Derrick, L. Asbrink, O. Edqvist, B.-O. Jonsson, and E. Lindholm, *Internat. J. Mass Spectrometry Ion Phys.*, 1971, **6**, 203.
[229] C. Fridh, L. Asbrink, B.-O. Jonsson, and E. Lindholm, *Internat. J. Mass Spectrometry Ion Phys.*, 1972, **8**, 85.
[230] C. Fridh, L. Asbrink, B.-O. Jonsson, and E. Lindholm, *Internat. J. Mass Spectrometry Ion Phys.*, 1972, **8**, 101.
[231] L. Asbrink, C. Fridh, B.-O. Jonsson, and E. Lindholm, *Internat. J. Mass Spectrometry Ion Phys.*, 1972, **8**, 215.
[232] L. Asbrink, C. Fridh, B.-O. Jonsson, and E. Lindholm, *Internat. J. Mass Spectrometry Ion Phys.*, 1972, **8**, 229.
[233] C. Fridh, L. Asbrink, B.-O. Jonsson, and E. Lindholm, *Internat. J. Mass Spectrometry Ion Phys.*, 1972, **9**, 485.
[234] P. Wilmenius and E. Lindholm, *Arkiv Fyzik*, 1962, **21**, 97.

formation, its frequency factor is likely to be higher enabling it to compete effectively at high internal energies.

These charge-transfer mass spectra can thus be interpreted in terms of the three principal factors controlling other ion decompositions. Firstly, the vertical nature of the charge-transfer ionization process, which determines the energy deposition function of the molecular ion. Secondly, the rapid vibronic relaxation to lower ionic states, not necessarily the ground state, which are isolated on the mass spectrometer time scale, or in the sense that further vibronic relaxation is slow compared with decomposition. And finally, the competitive decompositions from the isolated state(s) are governed by the usual factors controlling these processes (see Section 4). The striking correspondence between charge-transfer appearance potentials and higher ionization potentials merely reflects the nature of the energy deposition function in direct ionization. The same effects are observed in direct photoionization using the coincidence technique.[213] Remembering the variable nature of the recombination energy for charge-transfer mass spectra, all the data[225-234] can be interpreted in this fashion, and some further interesting points arise.

It is clearly of importance to be able to predict whether an excited ionic state undergoes rapid vibronic relaxation, or whether it is isolated and leads to decomposition. Molecular ions formed in excited states whose high resolution PES bands show no distinct vibrational structure, undergo very rapid vibronic relaxation to lower states at rates greater than $ca.\ 10^{14}\ s^{-1}$. A close inspection of the spectra obtained by Lindholm and co-workers[225-234] suggests that if an excited state does exhibit vibrational structure in the high resolution photoelectron spectrum then it is likely to be isolated. Strictly speaking, this criterion only requires that the rates of the relaxation (or dissociation) processes be slow compared with vibrational periods (10^{-13}—10^{-14} s), but it may provide a useful 'rule of thumb' for identifying possible isolated states.

sym-Triazine provides an interesting example (Figure 4).[229] The first excited state ($1e''$, $IP_{ad} = 11.69$ eV) has strong vibrational structure up to about 12.0 eV where it becomes weaker and broadens, merging into a smooth continuum above 12.2 eV. The breakdown curve for the molecular ion drops to zero above $RE \approx 12$ eV, whereas that for $C_2H_2N_2^{+\cdot}$ ($M^{+\cdot}$ – HCN) rises to unity. This striking behaviour suggests that the molecular ion in the $1e''$ state undergoes slow vibronic relaxation below ionization energies of 12 eV, but does so rapidly above this energy and completely dissociates from the vibrationally excited ground state. This explanation is supported by the broadening observed in the vibrational structure, which implies that the vibronic relaxation rate rises through a value of 10^{13}—$10^{14}\ s^{-1}$ at 12 eV. An alternative interpretation, that decomposition occurs from the isolated $1e''$ state with a dissociation energy of 12 eV,[229] implies an unusually low activation energy (< 0.3 eV) for this ring-opening fragmentation.

The above discussion of charge-transfer appearance potentials, and the underlying processes which lead to them, casts doubt on the previous interpretations.[93,225-234] In particular, the use of charge-transfer mass spectra to aid the assignment of electronic structures of excited states of molecular ions is misleading. The observed ion decompositions are determined by the electronic structure of a lower (isolated) state, which is often the ground state, not by the excited state. The latter only determines which (isolated) state it relaxes to, and how much vibrational energy that state will then contain.

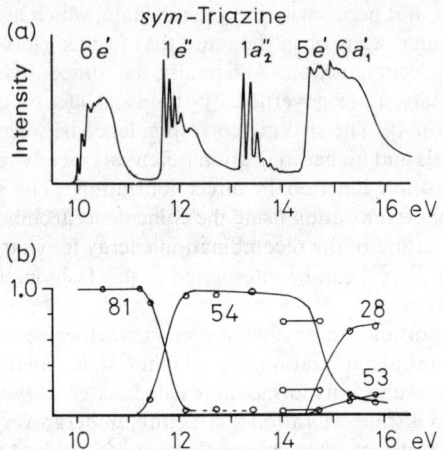

Figure 4 *Photoelectron spectrum* (21.21 eV *photons*) (a) *and charge-transfer breakdown graph* (b) *of* sym-*triazine*
(Reproduced by permission from *Internat. J. Mass Spectrometry Ion Phys.*, 1972, **8**, 85)

Photoionization and electroionization occur by indirect, as well as direct ionization processes, and will often result in ionization outside the direct Franck–Condon regions. Therefore, the appearance potentials determined accurately by these methods are more likely to yield significant dissociation energies, as Lindholm has noted.[93] Furthermore, the breakdown curves determined from charge-transfer mass spectra will not be appropriate for photoionization and electroionization mass spectra if the appearance potentials for charge-transfer ionization are higher than for the latter. In these cases, the charge-transfer breakdown curves will underestimate the extent of decomposition of the molecular ion, because some regions where decomposition is energetically possible are not accessible by charge-transfer ionization. This may explain the observations of Meisels and Emmel that charge-transfer breakdown graphs overestimate the molecular ion intensity in photoionization mass spectra.[70] However, charge-transfer breakdown graphs will be appropriate for mass spectra obtained by direct (vertical) ionization processes, provided the correct energy deposition function is also known. The EDF is zero outside the vertical Franck–Condon ionization regions and errors in the breakdown graphs at such energies become irrelevant. In this context the correct assignment of recombination energies is particularly important (see Section 3).

Investigations of the temperature dependence of charge-transfer mass spectra have shown that internal thermal energy is effective in promoting ion decompositions.[95] Using a primary ion with a well-defined recombination energy, the interesting threshold regions of the breakdown curves for several alkanes were studied by varying the total internal energy with temperature, and the results were compatible with QET predictions.[235]

Other measurements of the charge-transfer mass spectra of methylamine,[236] methanol,[237] and the lower alkanes[238] used projectile ions of high kinetic energy (1.8 keV), and the recombination energies are consequently poorly defined.

Lifetime Measurements of Ion Decompositions.—*Field Ionization Kinetics.* Field ionization kinetics (FIK) is the most important new technique developed for studying the kinetics of gas-phase ion decompositions. It is the only method which enables these processes to be investigated in the picosecond–nanosecond time region, but it is of course limited to decompositions which occur at the low ionizing energies attainable with FI (generally less than *ca.* 12 eV). The technique was originally devised by Beckey and co-workers using a single-focusing mass spectrometer,[239—242] but more recently the decisive advantages of using a double-focusing mass spectrometer have been discussed and exploited by Falick, Derrick, and Burlingame.[243] Unfortunately, it appears that earlier derivations of the relationship between ion decomposition rate constants and FIK curves are inadequate, and in some important aspects are incorrect; confusion has therefore arisen in the interpretation of these curves. Criticisms discussed below should in no way detract from the achievements of these workers in establishing this technique, but instead reflect on the great importance attached to it here.

In the double-focusing method the fragment ion current, $I'_f(V)$, is measured with increasing emitter potential V at a constant electric sector voltage, and can be transformed to a function of molecular ion decomposition time, t.[243] Accurate numerical calculations of the electrostatic potential distribution in the FI source, and the resulting average molecular ion trajectory, are used to determine decomposition times as a function of V.[244] More approximate analytical calculations of decomposition times[245] have recently been criticized and may be an order of

---

[235] C. Lifshitz and T. O. Tiernan, *J. Chem. Phys.*, 1973, **59**, 6143.
[236] T. Nagatani, K. Yoshihara, and T. Shiokawa, *Bull. Chem. Soc. Japan.*, 1973, **46**, 1306, 1628.
[237] T. Nagatani, K. Yoshihara, and T. Shiokawa, *Bull. Chem. Soc. Japan*, 1973, **46**, 1450.
[238] S. Ikuta, K. Yoshihara, and T. Shiokawa, *Bull. Chem. Soc. Japan*, 1973, **46**, 3648.
[239] H. D. Beckey, H. Key, K. Levsen, and G. Tenschert, *Internat. J. Mass Spectrometry Ion Phys.*, 1969, **2**, 101.
[240] K. Levsen and H. D. Beckey, *Internat. J. Mass Spectrometry Ion Phys.*, 1971, **7**, 341.
[241] K. Levsen and H. D. Beckey, *Internat. J. Mass Spectrometry Ion Phys.*, 1972, **9**, 51.
[242] K. Levsen and H. D. Beckey, *Internat. J. Mass Spectrometry Ion Phys.*, 1972, **9**, 63.
[243] A. M. Falick, P. J. Derrick, and A. L. Burlingame, *Internat. J. Mass Spectrometry Ion Phys.*, 1973, **12**, 101.
[244] J.-P. Pfeiffer, A. M. Falick, and A. L. Burlingame, *Internat. J. Mass Spectrometry Ion Phys.*, 1973, **11**, 345.
[245] B. W. Viney, *Internat. J. Mass Spectrometry Ion Phys.*, 1972, **8**, 417.

magnitude wrong.³ These $I'_f(t)$ curves are often presented in the literature[247–251] as 'rates of formation', which is strictly incorrect since they are merely differential ion currents as a function of time [cf. (7)] and also need to be corrected for the increase in total ionization with increasing emitter potential.[243,247]

Presuming this is done, to obtain $I_f(t)$, these curves can then be related to the rate of ion formation as a function of t.

$$\frac{dN_f(t)}{dt} = \frac{I_f(t)}{K \Delta t}, \quad 10^{-11} < t < 10^{-9} \text{ s} \qquad (7)$$

K represents the ion transmission and detection efficiency which remains constant with time in this technique, and Δt is the time resolution which varies continuously between about 10^{-10} and 10^{-11} s, becoming smaller more rapidly at shorter times.[243] In a single-focusing instrument, the 'fast metastable' fragment ion intensity as a function of apparent mass position can similarly be transformed to obtain $I_f(t)/\Delta t$,[241,245,246] but the corresponding K is not constant with time and is difficult to estimate accurately.

An average rate constant may be determined from these ion formation rates (7) if the number of precursor ions decomposing to a particular fragment ion is known as a function of time (assuming the kinetic processes are first-order). Since secondary fragmentations are seldom observed under FI the only precursor is the molecular ion so that,

$$\frac{dN_f(t)}{dt} = \tilde{k}_f(t) N'(t) \qquad (8)$$

where $N'(t)$ is the number of molecular ions of the *particular electronic state and structure* which decompose to form the fragment ion, and $\tilde{k}_f(t)$ represents an average rate constant (see below). Following Beckey et al.[239] it has invariably been assumed that $N'(t)$ may be replaced by the (constant) molecular ion current $I_p = N_p/K$ (in the normal FI mass spectrum) to obtain a 'phenomenological rate constant'.[243,247]

$$\bar{k}_f(t) = \frac{I_f(t)}{I_p \Delta t} = \frac{dN_f(t)/dt}{N_p} \qquad (9)$$

But, N_p represents the total number of molecular ions which have *not* decomposed within 10^{-5} s, and is not related to the measured ion formation rate $dN_f(t)/dt$ through $\tilde{k}_f(t)$ (8). It is simply an arbitrary normalizing factor which puts all the $\bar{k}(t)$ curves on the same relative (arbitrary) scale. The $\bar{k}(t)$ curves (9) thus represent *relative rates* of formation $dN_f(t)/dt$, not rate constants—'phenomenological' or otherwise. The arbitrary nature of the $\bar{k}(t)$ scale is obvious in some examples[250]

---

[246] P. J. Derrick and A. J. B. Robertson, *Proc. Roy. Soc.*, 1971, **A234**, 491.
[247] P. J. Derrick, A. M. Falick, and A. L. Burlingame, *J. Amer. Chem. Soc.*, 1972, **94**, 6794.
[248] P. J. Derrick, A. M. Falick, and A. L. Burlingame, *J. Amer. Chem. Soc.*, 1973, **95**, 437.
[249] K. Levsen and H. D. Beckey, *Internat. J. Mass Spectrometry Ion Phys.*, 1974, **14**, 45.
[250] P. J. Derrick, A. M. Falick, S. Lewis, and A. L. Burlingame, *Org. Mass Spectrometry*, 1973, **7**, 887.
[251] P. J. Derrick, A. M. Falick, A. L. Burlingame, and C. Djerassi, *J. Amer. Chem. Soc.*, 1974, **96**, 1054.

where values of ca. 10^5 s^{-1} are obtained for decompositions occurring within 10^{-10} s!

The time-dependence or 'shape' of FIK curves $\bar{k}_f(t) \propto dN_f(t)/dt$ is directly dependent on that of their precursor $N'(t)$ through the rate constant $\tilde{k}_f(t)$ (8), and immediately distinguishes direct (or concerted) decompositions from consecutive decompositions following prior rearrangement(s) (or perhaps slow vibronic relaxation) of the initially formed molecular ion state(s) (Figure 5). If the FIK curve has its maximum at 'zero' time (10^{-11} s) then the decomposition is direct, at least on the picosecond time scale (processes occurring within 10^{-12} s will not be resolved), and $N'(t)$ follows an inverse exponential function [Figure 5(a)]. Otherwise the FIK curve rises from zero to a maximum before falling off,

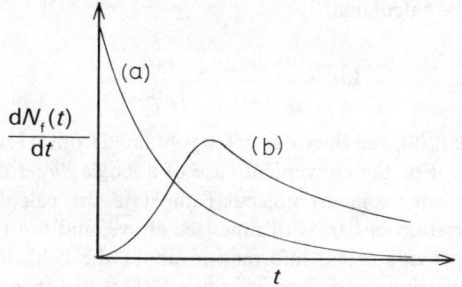

Figure 5 *Rates of ion formation for direct (or concerted)* (a) *and consecutive decompositions* (b) *illustrating their differing time-dependence*

corresponding to a consecutive decomposition where $N'(t)$ is given by the difference between inverse exponential functions [Figure 5(b)]. The time dependence of the exponential fall-offs is complicated still further by the nature of the rate constant, $\tilde{k}_f(t)$, which represents an average of the microscopic rate constants, $k_f(E)$, over the internal energy distribution $P'(E, t)$ of the ensemble of *decomposing* molecular ions, $N'(t)$. Since the more energetic molecular ions decompose faster, $P'(E, t)$ is continually shifting to lower energies and $\tilde{k}_f(t)$ therefore continually decreases with time—but to different extents for different $k(E)$ curves. The above considerations give some idea of the complexities involved in a complete analysis of FIK curves in terms of microscopic rate constants, $k(E)$, quite apart from the erroneous interpretations of the nature of the $\bar{k}(t)$ curves.

In general, a number of precursors may contribute to the formation of a particular fragment ion. In the present context, for example, D-labelled molecular ions may scramble their labels by various competing, consecutive H–D-transfer processes. Fragment ions, distinguished only by a particular number of labels, may be formed from several molecular ion species having different labelling patterns. These will all have different decay functions, $N'(t)$, owing to the different pathways and kinetic isotope effects involved in their formation from the initial molecular ion structure. In this case, it is impossible to determine the decay function, $N'(t)$, for the specific molecular ion state (and structure) decomposing to a fragment ion of specific state and structure, which is the only situation where (8) applies. Hence, $\tilde{k}_f(t)$ cannot generally be determined through (8).

In the particular case that all fragment ions are formed by competing direct decompositions of the same initially formed molecular ion state, $N_p(t)$ may be obtained by integration of the measured FIK curves.

$$N_p(t) = \frac{I_p(t)}{K} = \sum_i \int_t^{10^{-9}\,s} \frac{I_i(t')}{K\,\Delta t'} dt' \qquad (10)$$

Assuming no mass discrimination occurs (*i.e.* K is the same for all ions) and that decomposition is negligible at times $t > 10^{-9}$ s (which will not always be true), then $\tilde{k}_f(t)$ can be calculated.

$$\tilde{k}_f(t) = \frac{dN_f(t)/dt}{N_p(t)} = \frac{I_f(t)}{I_p(t)\Delta t} \qquad (11)$$

$I_p(t)$ does *not* include I_p (9), and the value of $\tilde{k}_f(t)$ obtained from (11) has the correct absolute magnitude. For the convenient case of a single *direct* decomposition occurring from only one (isolated) molecular ion state, this calculation is easily performed. The variation of $\tilde{k}_f(t)$ with time [see above, and not to be confused with $\bar{k}_t(t)$], should provide useful information about the behaviour of $k_f(E)$ if the initial $P'(E, O)$ distribution is known, since $P'(E, t)$ can then be calculated. Studies of the temperature dependence of FIK curves for such decompositions would be particularly valuable in this respect.[241,242]

It has become common practice to plot $\bar{k}(t)$ curves as $\log \bar{k}(t)$ *vs.* $\log t$ simply to enable the inclusion of odd points for metastable ion decompositions occurring at much longer times. These points only represent the tail end of the decay function $N'(t)$ and are of minor interest compared to the FIK curves at short times. The slope of $\log \bar{k}(t)$ *vs.* $\log t$ plots is not directly related to $\tilde{k}(t)$. These log-log plots serve no useful purpose and in fact disguise the differences in the shapes of the FIK curves, which can provide important qualitative information. The interpretation of FIK curves is quite straightforward if they are simply plotted as $\bar{k}_f(t)$ *vs.* t which is equivalent to the *relative rates* $dN_f(t)/dt$ *vs.* t (9). They may then be directly interpreted in terms of the precursor decay functions, $N'(t)$ (8), whose detailed differences in 'shape' are determined both by the prior rearrangement processes, whose rate constants mainly control the rising portion, and the subsequent fragmentation step whose rate constant $\tilde{k}_f(t)$ controls the fall-off behaviour. The small additional effect on the *relative* FIK curves due to differing time-dependences of the $\tilde{k}_f(t)$ junctions, which will alter the relative magnitudes of $dN_f(t)/dt$, *cf.* $N'(t)$ (8), only becomes apparent with increasing time.

In some instances, the use of $\log \bar{k}(t)$ *vs.* $\log t$ plots has led to incorrect calculations of $I_p(t)$. Derrick *et al.* determined the molecular ion intensity at 'zero' time $I_M = I_p(10^{-11} \text{ s})$ as the 'area under a log-log plot'.[247,248,250]

$$I_M = \int_{10^{-11}}^{10^{-5}} \log \bar{k}(t) \, d\log t = \int_{10^{-11}}^{10^{-5}} \log \left[\frac{I_f(t)}{N_p \, \Delta t}\right] \frac{dt}{2.3t}$$

$$\propto \int_{10^{-11}}^{10^{-5}} \frac{\log [I_f(t)/\Delta t]}{t} \, dt$$

This procedure is wrong [cf. integrand in (10)].

Suggestions for the presentation of FIK data arising from the preceding discussion are summarized below.

(i) FIK data should be presented as relative $I_f(t)/\Delta t$ vs. t plots (on an arbitrary ordinate scale), rather than $I'_f(t)$ vs. $\log t$ or $\log \bar{k}_f(t)$ vs. $\log t$ plots; these curves then represent the relative rates of ion formation, $I_f(t)/\Delta t \propto dN_f(t)/dt$, as a function of decomposition time (7). They might conveniently be termed 'decay curves', $\bar{d}_f(t)$, to distinguish them from any rate constants they are related to (8). In particular, this recommendation implies that *any* published FIK plots should be corrected for the increase in total ionization with increasing emitter potential, and for the variation in Δt with time (which differs for ions of different mass). Measured $I'_f(t)$ curves are of little value since neither correction has been applied; [in some circumstances plots of $\bar{d}_f(t) \equiv I_f(t)/\Delta t$ vs. $\log t$ may usefully accentuate differences in FIK curves at short times].

(ii) The exact position of the maximum in an FIK curve has a crucial bearing on the overall interpretation of the ion decomposition kinetics, especially if it occurs at 'zero' time (10^{-11} s) in which case it is decisive evidence for a direct (or concerted) fragmentation process. In such cases, it is extremely important to correct for the resolution (natural peak width) of the instrument, preferably by deconvolution[246] if the signal-to-noise ratio is sufficiently good.

Virtually all the published FIK measurements have used plots of $I'_f(t)$ vs. $\log t$ [or $I'_f(t)$ vs. t] and $\log \bar{k}(t)$ vs. $\log t$, and much of the information necessary for a detailed analysis is unavailable or hidden in the log-log transformation. For example, in the interesting study of $[3,3,6,6\text{-}^2H_4]$cyclohexene[247] it is vital to know exactly where the FIK curve maxima occur. But it is not easy to decide from the points given on the $\log \bar{k}(t)$ vs. $\log t$ plot, and impossible from the $I'_f(t)$ vs. $\log t$ curves which have not been corrected for Δt nor the total ionization. Nevertheless, a qualitative discussion of some results will indicate aspects where the points discussed above are important, and pinpoint areas where the interpretations seem inadequate.

To continue with the example cited above, Derrick et al. discussed the elimination of ethylene from $[3,3,6,6\text{-}^2H_4]$cyclohexene in terms of only four isotopically distinguished molecular ion structures (Scheme I, ref. 247). But even assuming that only 1,3-allylic H–D-transfers occur and ignoring axial–equatorial differences,[247] at least eleven distinguishable molecular ion structures may contribute in different degrees to the formation of the fragment ions, $[M - C_3H_5]^+$, $[M - C_2H_4]^{+\cdot}$, $[M - CH_3]^+$ and their isotopic variants. The authors have taken no account of kinetic isotope effects on the H–D-scrambling processes which will affect the shapes of the FIK curves, and their claim that 'there is apparently no significant isotope effect on these rearrangements' seems quite unjustified since the shapes do clearly differ. Their reasons for advancing certain specific mechanisms (*e.g.* for methyl loss; Scheme II, ref. 247) could equally well be explained by kinetic isotope effects on prior H–D-rearrangements, and their main conclusions seem to stem from an incomplete analysis of the kinetics. Perhaps unwittingly they chose an extremely complicated system, and a complete interpretation would require a much more extensive investigation including other deuteriated derivatives. This is not the first time that one of the early examples investigated with a new technique paradoxically transpired to be very complicated. More importantly, it illustrates beautifully the wealth of detail about H–D-randomization processes that may eventually be exposed by the FIK technique.

Similar remarks apply to their investigation of 3,3-[$^2H_2$]- and 4,4-[$^2H_2$]-hexanol where they assumed the loss of H_2O/HDO was fast compared with prior H–D rearrangement in the molecular ion.[248] The shapes of the measured $I'_f(t)$ *vs.* log t curves do not support this assumption, though they do clearly implicate a consecutive fragmentation process. A more detailed kinetic analysis would be required before any conclusions could be drawn on the relative average rate constants (and isotope effects) for H–D-transfer from C-3 *cf.* C-4, and for H_2O–HDO loss from the respective molecular ion structures so formed.

Very interesting information about H–D rearrangement processes has also been obtained from FIK studies of the McLafferty rearrangement in carbonyl derivatives. Beckey and Levsen reported very detailed and stimulating observations on 'fast metastable' peaks for such fragment ions and their variation with temperature,[240–242] but their theoretical analysis of the derived FIK curves is unable to account for all the observations. Their analysis of the kinetics assumes that all the fragmentations are direct (concerted) processes, which immediately implies that the FIK curve has a maximum at zero time. To explain the observed peak shifts they were forced to postulate that the distribution of rate constants had a maximum value, k_{max}.[240] A peak shift is only obtained for $k_{max} \approx 10^{11}$ s^{-1}; for higher values no peak shift occurs, while for lower values the FIK curves have no maximum and suffer a sharp cut-off at decomposition times, $t = k_{max}^{-1}$ [Figure 5(a), ref. 240]. Such behaviour is not observed experimentally, and moreover they experienced difficulty in accounting for the relatively short decomposition times (*ca.* 10^{-11} s) at which FIK maxima often occur.[240]

The reason for this small peak shift (10^{-12}–10^{-11} s)[242] is rather more subtle than this, and requires some elaboration. It is expected that the initially formed molecular ion states which undergo genuine gas-phase decompositions will also decompose by various field-induced dissociations occurring near the emitter. These 'direct' decompositions are very fast because of the large distortion of the molecular ion potential energy surfaces by the high electrostatic fields within *ca.* 100 nm of the emitter surface.[243,244] Therefore, the gas-phase ion decompositions will not be observable until the rates of the 'direct' decompositions become slower than the former. This will occur when the molecular ions have left the high-field region, which takes 10^{-12}–10^{-11} s,[243] thus explaining why the FIK maxima occur at *minimum* times of 10^{-11}—10^{-12} s.[242] (If the field-induced decomposition of a particular molecular ion state occurs by the same process as the gas-phase decomposition of that state, *e.g.* especially *via* simple bond cleavages, then tailing of the 'normal' fragment ion peak will be observed,[242] rather than shifted peak maxima).

A 'zero' time of *ca.* 10^{-11} s can thus be assigned to genuine gas-phase ion decompositions, and the explanation of FIK maxima which occur at longer times is quite obvious in terms of consecutive (first-order) kinetic processes, as discussed above. There it was stated that a maximum in the FIK curve at 10^{-11} s implied a direct or concerted fragmentation process, but with the important proviso that a consecutive decomposition, in which prior rearrangements occurred within 10^{-12} s, would be kinetically indistinguishable from this. Therefore, strictly speaking, FIK curves can only provide conclusive evidence *against* a direct fragmentation process when the maximum occurs at times $t > 10^{-11}$ s.

Beckey and Levsen's study of the temperature-dependence of the FIK curve for the McLafferty rearrangement in menthone[241] constitutes decisive evidence for the consecutive (step-wise) nature of this fragmentation. With increasing temperature the maximum shifts to shorter times owing to the increase in the average rate constants for both the rearrangement (γ-H transfer) $\tilde{k}_r$ and the subsequent fragmentation $\tilde{k}_f$. Judging from the variations in the behaviour of the leading and trailing sections with temperature (Figure 5, ref. 241), it appears that $\tilde{k}_r$ increases more rapidly with temperature than $\tilde{k}_f$, implying that $k_r(E)$ rises more steeply with internal energy E than $k_f(E)$. The FIK curves for the double hydrogen rearrangement in butyl benzoate also provide excellent evidence for consecutive decomposition, but here the temperature dependence of $\tilde{k}_f$ seems to be much smaller than that of $\tilde{k}_r$. The loss of OD˙ from $C_6D_5CO_2H^{+}_{\cdot}$ is also found to be a consecutive process, whereas the loss of OH˙ appears to be direct.[249]

In the examples of fragmentations with hydrogen rearrangement given by Beckey and Levsen in their review paper,[242] it is interesting that the positions of the FIK maxima depend on both the type of compound and the type of hydrogen transferred. In their terminology, 'slow' implies FIK maxima at long times ($> 10^{-8}$ s) and 'delayed' at 10^{-11} s. Considering only γ-H (McLafferty) rearrangements, the fragmentations in aliphatic esters involving primary γ-H transfer are 'slow', whereas secondary γ-H transfer is 'delayed'. In alkyl phenyl ketones the fragmentations are 'delayed' for both primary and secondary hydrogens.[242] Derrick *et al.* observed that the FIK curves for McLafferty rearrangements in aliphatic ketones have maxima near 10^{-11} s for both tertiary and secondary γ-H's,[250—252] whereas menthone which fragments by primary γ-H transfer is an intermediate case.[241] All these observations reflect the dependence of the average rate constant for γ-H transfer ($\tilde{k}_H$) on the precise nature of the transition state, since it is $\tilde{k}_H$ which controls the 'rise-time' of the FIK curve. It appears that γ-H transfer occurs within *ca.* 10^{-12} s for most carbonyl derivatives even at the low internal energies attained by FI, except for some esters and

---

[252] P. J. Derrick, A. M. Falick, and A. L. Burlingame, *J. Amer. Chem. Soc.*, 1974, **96**, 615.

aliphatic ketones containing primary γ-H's. For a particular class of compound, one might tentatively conclude that $\tilde{k}_\mathrm{H}$ for γ-H transfer follows the (expected) order: primary < secondary < tertiary, although no direct evidence is yet available for the latter comparison.

Recently reported FIK studies of the loss of acetic acid from β-substituted ethyl acetates (12), presumably *via* a secondary γ-H transfer, also indicate that

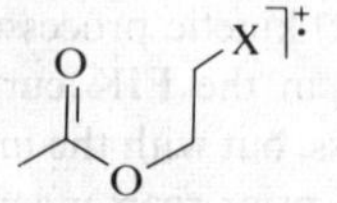

$$\text{[structure]} \rightarrow \text{CH}_2\text{CHX}]^{\ddagger} + \text{MeCO}_2\text{H} \tag{12}$$

$$X = \text{Ph, Me, CH}_2\text{Ph, NMe}_2, \text{CH}_2\text{NMe}_2$$

the substituents have a marked effect on $\tilde{k}_\mathrm{H}$.[253] The authors interpreted their results in terms of charge-localization effects on the fragmentation step, but close inspection of their log $\bar{k}(t)$ vs. log t plots reveals that the maxima occur at different times, reflecting differences in $\tilde{k}_\mathrm{H}$. The observed peak shifts vary from 10^{-11}—10^{-10} s and indicate that $\tilde{k}_\mathrm{H}$ follows the order:

$$X = \text{Ph} > \text{Me} > \text{CH}_2\text{Ph} > \text{NMe}_2 > \text{CH}_2\text{NMe}_2$$

(no rearrangement was observed in the last case).[253] It is also likely that the substituents have a considerable effect on the consecutive fragmentation rate constant $\tilde{k}_\mathrm{f}$, but it is difficult to analyse this from the published log $\bar{k}(t)$ vs. log t plots.

The results of Derrick *et al.* have also provided useful evidence for the factors controlling the competition between H–D-randomization and the McLafferty rearrangement in D-labelled alkyl ketones.[250,251] Tertiary hydrogens randomize much more rapidly than secondary (or primary) hydrogens, but there would not appear to be conclusive evidence favouring any particular mechanism of H–D-randomization. Their dismissal[251] of the simultaneous 1,2-H–D exchange mechanism does not allow for the possibility that α-methyl substitution increases the fragmentation rate sufficiently to preclude any competitive H–D-randomization, at least at shorter times (higher internal energies).

Intriguing results were obtained in an FIK study of the McLafferty rearrangement in [4,4-^{2}H$_2$]hexanal (Figure 6).[252] The time-dependence of the $I'_\mathrm{f}(t)$ curves seems to differ too much for $[\text{C}_2\text{H}_3\text{DO}]^{\ddagger}$ and $[\text{C}_4\text{H}_7\text{D}]^{\ddagger}$ to be formed from the same precursor (even allowing for differences in the effects of total ionization and Δt corrections). Therefore it may be concluded that different precursors are involved, which might be either:
(i) different isolated electronic states, in which case both decompositions may occur by initial γ-D transfer with (most probably) consecutive fragmentation from the same molecular ion structures; or
(ii) different molecular ion structures formed from the same (initial) electronic state.

[253] G. Wood, A. M. Falick, and A. L. Burlingame, *Org. Mass Spectrometry*, 1974, **8**, 279.

The former possibility seems more plausible, and accurate appearance potential measurements would provide very useful further information. The authors assumed that both fragmentations occurred from the ground state, suggesting that only a few tenths of an electron volt of internal energy was available by FI

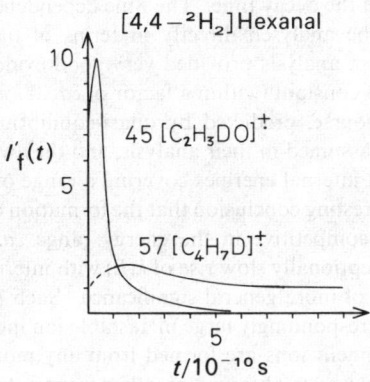

Figure 6 *Measured FIK curves $I'_f(t)$ for γ-H rearrangements of $[4,4-{}^2H_2]$hexanal* (Reproduced by permission from *J. Amer. Chem. Soc.*, 1974, **96**, 615)

and therefore the first excited state lying 1.5 eV higher was unlikely to be populated. Their reasons are not compelling, since FI is commonly believed to reach ionization energies of *ca.* 12 eV and the first IP of hexanal lies at 9.8 eV. Other comments are made in Chapter 2, Section 5.

Other Studies. Investigations at longer decomposition times ($>10^{-9}$ s) generally measure just the fall-off region in the fragment ion formation rate, and seldom reveal the consecutive nature of numerous ion decompositions.[254,255] Tatarczyk and von Zahn were therefore able to analyse their results (from electroionization), covering the period 5—500 µs, in terms of direct fragmentation processes determined only by the fragmentation rate constant $\bar{k}_f(t)$. The time dependence of the decay curves constituted direct evidence for a continuous distribution of microscopic rate constants $k_f(E)$. Since the decay curves represent a (time-dependent) average over this distribution, it was difficult to extract information about the individual $k_f(E)$ functions.[254] Furthermore, at such long times the decay curves only contain the contributions of the lower $k(E)$ values which generally correspond to low internal energies, and hence give limited information about the variation in $k(E)$ with internal energy. On the other hand, these studies do pertain to the threshold energy region for many fragmentations, and can therefore provide very interesting information about the distribution of $k(E)$ values near threshold.

[254] H. Tatarczyk and U. von Zahn, *Z. Naturforsch.*, 1972, **27a**, 1646.
[255] P. W. Ryan, J. H. Futrell, and M. L. Vestal, *Chem. Phys. Letters*, 1973, **18**, 329.

These experiments have used electroionization,[254,255] and also field ionization,[239] but Andlauer and Ottinger obtained very significant results using charge-transfer ionization.[256] Their apparatus measured decay curves in the nanosecond to microsecond range, but unfortunately some of the primary ions could not be prevented from entering the drift-region and causing charge-transfer ionization throughout the decay time. The time dependence of the decay curves could not therefore be analysed directly in terms of the ion decomposition kinetics. Their indirect analysis provided very good evidence for the existence of only a discrete rate constant (within a factor of ca. 3) for a particular internal energy. This is, of course, predicted by quasi-equilibrium theories, but it is stressed this was not assumed in their analysis, and they were able to determine $k(E)$ at a few different internal energies covering a range of 1—2 eV.

Apart from the interesting conclusion that the formation of $C_6H_5^+$ and $C_4H_4^{+\cdot}$ from $C_6H_6^{+\cdot}$ is not competitive in the energy range ca. 14.0—16.0 eV, their observation of an exceptionally slow rise of $k(E)$ with internal energy for $C_6H_5^+$ formation is perhaps of more general significance. Such behaviour of the $k(E)$ function implies a correspondingly large metastable ion intensity, and in general when $[M - H]^+$ fragment ions are formed from any molecular ion, relatively intense metastable peaks are observed cf. other fragmentations. The failure of QET calculations of $k(E)$ for $C_6H_5^+$ formation from benzene to agree with the experimental curve, in contrast to that for $C_6H_4^{+\cdot}$ formation[199] (see Section 4), further suggests that ion decomposition by the loss of H· is atypical. It appears that the rate constant is much less than predicted by quasi-equilibrium models, particularly at higher internal energies, and thus involves an increasingly severe entropy deficiency with increasing energy. Possible reasons for this effect in fragmentations occurring by the loss of light fragments are mentioned in Section 4.

Ion Kinetic Energy Spectroscopy.—Peak shapes can be used to determine the kinetic energy released in metastable ion decompositions,[257—259] and very accurate measurements are possible using ion kinetic energy spectroscopy (IKES).[260] Terwilliger et al. have recently described a signal averaging system, using a reversed-geometry mass spectrometer,[261] for ensemble averaging mass-analysed ion kinetic energy spectra (MIKES) by scanning the electric sector voltage under computer control.[262] The signal-to-noise ratio is much improved over manual techniques, with some gain in energy resolution, and it would now appear feasible to determine accurate kinetic energy *distributions* rather than

---

[256] B. Andlauer and C. Ottinger, Z. Naturforsch., 1972, **27a**, 293.
[257] R. G. Cooks, J. H. Beynon, R. M. Caprioli, and G. R. Lester, 'Metastable Ions', Elsevier, Amsterdam, 1973.
[258] V. W. Maslen, Internat. J. Mass Spectrometry Ion Phys., 1974, **13**, 207.
[259] A. L. Harkness, Internat. J. Mass Spectrometry Ion Phys., 1973, **10**, 267.
[260] J. H. Beynon, A. E. Fontaine, and G. R. Lester, Internat. J. Mass Spectrometry Ion Phys., 1972, **8**, 341.
[261] J. H. Beynon, R. G. Cooks, J. W. Amy, W. E. Baitinger, and T. Y. Ridley, Analyt. Chem., 1973, **45**, 1023A.
[262] D. T. Terwilliger, J. H. Beynon, and R. G. Cooks, Internat. J. Mass Spectrometry Ion Phys., 1974, **14**, 15.

just the average kinetic energy (T). The important relationship of such measurements to theories of ion decompositions is discussed in Section 4. Details of the procedures used to determine T from metastable peak widths have been described,[263] and values ranging from a few tenths of a millielectron volt to a few tens of electron volts have been observed. This large variation (over five orders of magnitude) has led, understandably, to the diagnostic use of these measurements in distinguishing metastable ion structures,[263–265] and some examples are mentioned in Chapter 2, Section 2. The assumptions involved in drawing conclusions about ion structures from observed kinetic energy releases require close scrutiny.

Jones et al. have stressed that T is generally composed of contributions $T^{\ddagger}$ and T^r, representing the average kinetic energies released from the excess internal energy $\varepsilon^{\ddagger}$ and the reverse activation energy ε_0^r respectively.[266] Assuming for the moment that T^r is zero (e.g. in simple bond cleavages where ε_0^r is expected to be negligible), the observed value of $T^{\ddagger}$ depends on the magnitude of $\varepsilon^{\ddagger}$ in the metastable ions which is determined by the behaviour of $k(E)$ near threshold. If $k(E)$ rises rapidly $\varepsilon^{\ddagger}$ will be small and $T^{\ddagger}$ is expected to be correspondingly small, but if $k(E)$ rises slowly $\varepsilon^{\ddagger}$ may be quite large (Figure 7b) resulting in a correspondingly high value of $T^{\ddagger}$. Klotz discussed this point in an interesting paper and showed how information about the behaviour of $k(E)$ might

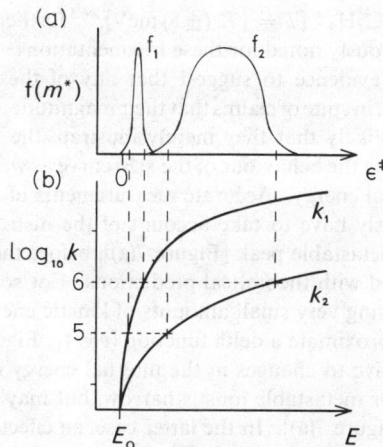

Figure 7 *Probability of metastable ion formation as a function of the excess of internal energy* $\varepsilon^{\ddagger} = E - E_0$ *(a), for different $k(E)$ functions (b)*

[263] E. G. Jones, L. E. Bauman, J. H. Beynon, and R. G. Cooks, *Org. Mass Spectrometry*, 1973, **7**, 185.
[264] J. H. Beynon, M. Bertrand, and R. G. Cooks, *Org. Mass Spectrometry*, 1973, **7**, 785.
[265] R. G. Cooks, J. H. Beynon, M. Bertrand, and M. K. Hoffman, *Org. Mass Spectrometry*, 1973, **7**, 1303.
[266] E. G. Jones, J. H. Beynon, and R. G. Cooks, *J. Chem. Phys.*, 1972, **57**, 2652.

be obtained experimentally.[186] This approach can usefully test the predictions of quasiequilibrium models and might reveal any major deviations due to significant entropy deficiencies. But it should be remembered that if T^r makes a significant contribution to T, then the formulae relating T to ε^+ are not applicable.

H–D-isotope effects do not cause a very large shift of the $k(E)$ curves in an absolute sense, and their influence on T is usually observed to be small with $T_H/T_D = 1.0\ (\pm 0.05)$.[267,268] When T is itself small, they have a relatively large effect; e.g., in the loss of HCN from $C_3H_3N^{+\cdot}$ in [$^2H_4$]pyrimidine and [$^2H_4$]-pyrazine, $T_H/T_D = 2.2$ because T_H is only 0.9 meV.[269] The decompositions (13), (14) gave quite different values of T; around 0.2—0.5 meV for (13) compared with

$$PhCOCH_2X]^{+\cdot} \rightarrow PhCO]^+ + CH_2X^{\cdot}; \quad X = F, Cl, Br \quad (13)$$

$$PhCOCX_3]^{+\cdot} \rightarrow PhCO]^+ + CX_3^{\cdot}; \quad X = H, F. \quad (14)$$

8 and 130 meV for X = H, F respectively (14).[270] The authors invoked a slow predissociation to explain the former results, but on flimsy arguments. It is quite possible that differences in activation energies and rotational effects, which exert the major influence on the shapes of $k(E)$ curves near threshold, would explain these observations. The difference in the rotational symmetries of the substituted methyl fragments from (13) compared with (14) is noteworthy. Large values of T are observed for the loss of H$^{\cdot}$ from the molecular ions, $C_6H_6^{+\cdot}$ ($T = 63$ meV)[263] and $C_7H_8^{+\cdot}$ [$T = 172\ (\pm 5)$ meV],[265] further illustrating the very slow rise of $k(E)$ previously noted for these fragmentations.

There is no good evidence to suggest that any of these examples involve contributions from T^r, in spite of claims that their magnitude alone implies this.[265] It would seem more likely that they merely illustrate the range of values T^+ may take, depending on the behaviour of the $k(E)$ curve as well as the partitioning of ε^+ into translational energy. Accurate measurements of translational energy *distributions* will clearly have to take account of the distribution of ε^+ values contributing to the metastable peak [Figure 7(a)], before the observed distributions can be compared with theoretical predictions. For some decompositions, especially those releasing very small amounts of kinetic energy, the distribution of ε^+ may closely approximate a delta function [e.g. f_1, Figure 7(a)].

T^+ will be insensitive to changes in the internal energy distribution, $P(E)$, if the ε^+ distribution for metastable ions is narrow, but may be strongly affected if it is wide [e.g. f_2, Figure 7(a)]. In the latter case, an effect on T^+ will only be observed if $P(E)$ is altered over the range of internal energies contributing to the metastable ions. Thus variations in ion source temperature are expected to have a greater effect on $P(E)$ at low internal energies than changes in the ionizing voltage, and T was found to vary significantly with the former but not the latter.[263]

[267] M. Bertrand, J. H. Beynon, and R. G. Cooks, *Org. Mass Spectrometry*, 1973, **7**, 193.
[268] M. Bertrand, J. H. Beynon, and R. G. Cooks, *Internat. J. Mass Spectrometry Ion Phys.*, 1972, **9**, 346.
[269] J. H. Beynon, R. M. Caprioli, and T. Ast, *Org. Mass Spectrometry*, 1971, **5**, 229.
[270] R. G. Cooks, K. C. Kim, and J. H. Beynon, *Chem. Phys. Letters*, 1974, **26**, 131.

On the other hand, the internal energy distribution in fragment ions is also determined by the energy fluctuation effect, and a fragment ion of a particular structure may be formed with markedly different internal energy distributions from different precursors. This is particularly likely to happen when the precursor fragmentations have reverse activation energies, since then some portion of ε_0^r will be partitioned into internal energy of the fragments and their internal energy distribution will vary, perhaps markedly with different precursors. Corresponding variations in $T^{\ne}$ would then be expected for metastable decompositions of these fragment ions.

The considerable differences in T observed for a particular metastable fragment ion formed from different precursors have been attributed to a 'kinetic delay' due to the prior fragmentation.[263] However, for fragment ions to be formed with sufficient internal energy to decompose (as metastable ions), the prior fragmentation must be occurring at energies well above its metastable region (*i.e.* at rates much faster than 10^6 s^{-1}). Therefore, any 'delay' will be entirely negligible on the time scale of the subsequent (metastable) fragmentation (10^{-5}—10^{-6} s).

Decompositions involving a reverse activation energy are likely to convert a more or less discreet proportion of ε_0^r into translational energy T^r. In this case the metastable peak has 'wings', with the contribution from $T^{\ne}$ superimposed.[257] Many metastable peaks for rearrangement processes exhibit this general behaviour, and the observed values of T are often large—up to an electron volt.[263—266,271—273] But $T^{\ne}$ will be negligible compared with T^r only if the latter is very large, because rearrangement decompositions usually have $k(E)$ curves which rise quite slowly in which case $T^{\ne}$ can be considerable. Furthermore, for metastable fragment ions the contribution of $T^{\ne}$ may vary significantly (see above). Therefore, in comparing T for an ion from various precursors, only large differences can be attributed to different decomposition processes with any certainty. Even then it is always possible that the differences are due to formation of the ion in different (isolated) electronic states rather than different structures, although the latter is perhaps more likely.

Interesting IKES studies of rearrangements in substituted anisoles (15)[271] and nitrobenzenes (16)[272] have been reported. For both fragmentations, metastable peaks with two components were observed and the relative amounts of

$$XC_6H_4OCH_3]^{\ddag} \rightarrow XC_6H_5]^{\ddag} + CH_2O$$
$$X = m\text{- or } p\text{- H, Me, Cl, Br, CN}$$
(15)

$$XC_6H_4NO_2]^{\ddag} \rightarrow XC_6H_4O]^+ + NO^{\cdot}$$
$$X = p\text{- NH}_2, \text{OMe, OH, Me, H, F, Cl, CN, HCO}$$
(16)

[271] R. G. Cooks, M. Bertrand, J. H. Beynon, M. E. Rennekamp, and D. W. Setser, *J. Amer. Chem. Soc.*, 1973, **95**, 1732.
[272] J. H. Beynon, M. Bertrand, and R. G. Cooks, *J. Amer. Chem. Soc.*, 1973, **95**, 1739.
[273] T. Keough, J. H. Beynon, R. G. Cooks, C. Chang, and R. H. Shapiro, *Z. Naturforsch.*, 1974, **29a**, 507.

each component varied with the substituent. The component with large T was more favoured by electron-donating substituents, and the two decomposition processes appeared to be competitive. Further analysis of the energy partitioning seems somewhat illusory. The necessary heats of formation were determined from semi-log AP's,[271] which are not accurate enough for such purposes (see Section 6). Where these were unavailable (or gave impossible results),[272] some astounding assumptions were made to obtain the 'total excess energy' ε_{xs} ($= \varepsilon^{\ddagger} + \varepsilon_0^r$).

Accurate analysis of the partitioning of ε_0^r into T^r will be difficult. Firstly, very accurate heats of formation must be available for all the species; this is seldom the case at present. Secondly, accurate appearance potentials are required for the metastable ions; some of the techniques discussed in Section 6 promise to provide such data. At this point ε_{xs} can be determined, but the true dissociation energy is needed for the determination of ε_0^r, since ε_{xs} includes $\varepsilon^{\ddagger}$. Furthermore the contribution of $T^{\ddagger}$ should be separated from T to obtain T^r. If this could be done, perhaps by analysing the kinetic energy distribution, then $\varepsilon^{\ddagger}$ might be roughly estimated from $T^{\ddagger}$ (using QET formulae)[186] to obtain ε_0^r. Obviously, severe problems have yet to be overcome before a meaningful analysis of the energy partitioning is possible for ion decompositions.

Translational energies of fragment ions have been measured by several techniques[274–277] to obtain average kinetic energies $\bar{\varepsilon}_t$.[278,279] The proposal that $\bar{\varepsilon}_t$ measured near threshold could be used to determine corrections to AP's for the excess internal energy $\varepsilon^{\ddagger}$,[280] has been severely criticized.[266,186] It was found that the extrapolated values of $\bar{\varepsilon}_t$ greatly overestimated the magnitude of $T^{\ddagger}$ at the threshold,[266] and moreover, the simple empirical relation (17)[280] is not applicable if $\bar{\varepsilon}_t$ (or T) includes contributions from any reverse activation energy.

$$\varepsilon^{\ddagger} = \alpha N \bar{\varepsilon}_t \qquad (17)$$

Corrections to accurate AP's using (17) are not advisable. The improved heats of formation obtained in certain cases[280] would seem to reflect the inaccuracy of AP's determined by empirical methods, rather than the validity of such corrections.

6 Appearance Potential Measurements

Appearance potentials can be determined most accurately from ionization efficiency curves obtained with monoenergetic photon[281] or electron[66,87] beams,

[274] D. K. Sen Sharma and J. L. Franklin, *Internat. J. Mass Spectrometry Ion Phys.*, 1974, **13**, 139.
[275] P. W. Harland, J. L. Franklin, and D. E. Carter, *J. Chem. Phys.*, 1973, **58**, 1430.
[276] G. Levin and I. Platzner, *J. Chem. Phys.*, 1974, **60**, 2007.
[277] R. Fuchs, *Internat. J. Mass Spectrometry Ion Phys.*, 1972, **8**, 193.
[278] N. D. Weinstein, *J. Chem. Phys.*, 1973, **58**, 408.
[279] J. L. Franklin and D. K. Sen Sharma, *J. Chem. Phys.*, 1973, **58**, 409.
[280] M. A. Haney and J. L. Franklin, *J. Chem. Phys.*, 1968, **48**, 4093.
[281] N. W. Reid, *Internat. J. Mass Spectrometry Ion Phys.*, 1971, **6**, 1.

and values determined by these methods should always be preferred. Accurate dissociation energies can be obtained if corrections are made for the effects of thermal energy, which are usually significant in polyatomic molecules.[282] Appearance potentials are more commonly determined from ionization efficiency curves measured with unfiltered electron beams, and these less accurate methods are discussed in some detail below.

Deconvolution Methods.—The most direct method of removing the effect of the wide electron energy distribution from ionization efficiency (IE) curves is by Fourier transform deconvolution, but the data must have a very high signal-to-noise ratio (S/N) for it to be successful. Morrison and co-workers have achieved this by ensemble averaging using computerized data acquisition,[283] and such systems are expected to be more widely used in the future. Recent work in the Reporter's laboratory has demonstrated the feasibility of this approach using a commercial mass spectrometer-computer system.[284] Alternative deconvolution techniques using iterative methods are also available, and the difference method has been successfully used by Thynne et al. in their studies of negative ions (see below). More recently an iterative ratio method, which is claimed to have advantages over the difference method, has been described[285] and would appear worthy of investigation for IE curve analysis.

Approximate deconvolution methods have also been devised for some particular, assumed electron energy distributions using 'inverse convolvers'.[286] The energy-distribution-difference (EDD) methods use this principle and perform exact deconvolutions if the electron energy distribution has an inverse exponential form $[m(U) = \exp(-cU)]$ for the single-EDD method,[287] or a Maxwellian form $[m(U) = U \exp(-cU)]$ for the double-EDD method.[288] The former method has been shown to be simply a variation of the critical slope method and is not reliable for appearance potential determinations,[289] although it has given ionization potentials in quite good agreement with PES values for particular cases.[290] The double-EDD method is much more reliable but its success depends on how closely the actual electron energy distribution approximates the Maxwellian function, and IE data of quite high S/N are required.[289]

The RPD-technique is a well-known experimental method which achieves some deconvolution of IE curves, and modulation,[291] double-modulation,[292]

---

[282] A. S. Werner, B. P. Tsai, and T. Baer, *J. Chem. Phys.*, 1974, **60**, 3650.
[283] R. G. Dromey, J. D. Morrison, and J. C. Traegar, *Internat. J. Mass Spectrometry Ion Phys.*, 1971, **6**, 57.
[284] R. A. W. Johnstone and B. N. McMaster, to be published.
[285] P. E. Siska, *J. Chem. Phys.*, 1973, **59**, 6052.
[286] R. G. Dromey and J. D. Morrison, *Internat. J. Mass Spectrometry Ion Phys.*, 1971, **6**, 253.
[287] R. E. Winters, J. H. Collins, and W. L. Courchene, *J. Chem. Phys.*, 1966, **45**, 1931.
[288] J. Vogt and C. Pascual, *Internat. J. Mass Spectrometry Ion Phys.*, 1972, **9**, 441.
[289] R. A. W. Johnstone and B. N. McMaster, *Adv. Mass Spectrometry*, 1974, **6**, 451.
[290] R. A. W. Johnstone and F. A. Mellon, *J.C.S. Faraday II*, 1972, **68**, 1209.
[291] D. E. Golden, N. G. Koepnick, and L. Fornari, *Rev. Sci. Instr.*, 1972, **43**, 1249.
[292] F. D. Schowengerdt and D. E. Golden, *Rev. Sci. Instr.*, 1974, **45**, 391.

and automation[293,294] improvements have been reported. Very careful design and tuning of the RPD-ion source is necessary to obtain accurate results.[295]

Empirical Methods.—By far the most widely used empirical method for determining appearance potentials is the semi-logarithmic (SL),[296] or energy compensation (EC)[297] method, sometimes supplemented by the extrapolated voltage difference (EVD)[298] method when the semi-logarithmic IE curves are not parallel. It is commonly assumed that if the semi-logarithmic IE curves of the sample and reference ion are parallel near threshold (*e.g.* ca. 0.1—1 % I_{50}), then the derived appearance potential is reliable within the reproducibility (*e.g.* ±0.1—0.2 eV). However, the necessary condition is that the curves should not only be parallel in this region, but also have exactly the same shape from onset up to the normalization energy, usually 50 eV. It is generally impossible to ensure this condition holds for fragment ions, and is not even an easy task for molecular ions.

Figure 8 shows two IE curves normalized at 50 eV and having the same appearance potential V_0 but different shapes, typical behaviour for molecular

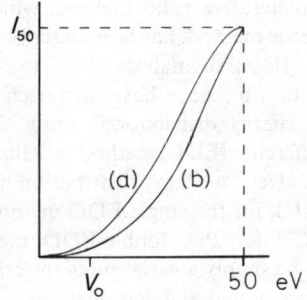

Figure 8 *IE curves with the same true AP (V_0), but different shapes normalized at 50 eV*

or metastable ions (a) and fragment ions (b). The latter generally exhibit a longer foot which is lost earlier in the noise level (at about 0.01 % I_{50} in the best cases for manually collected data but often much higher than this). Below the true threshold (V_0), the curves have similar shapes since they are primarily determined by the electron energy distribution, and the semi-logarithmic curves have similar slopes if measured to low enough values (Figure 9). However, because of the different shapes of the IE curves above V_0, the semi-log curve (b) is displaced below (a) and a difference ΔV is obtained between the appearance potentials

---

[293] C. Lifshitz, J. Agam, A. Weinberg, D. Kantor, U. Shainok, and M. Peres, *Internat. J. Mass Spectrometry Ion Phys.*, 1973, **11**, 243.
[294] M.-J. Hubin-Franskin, J. Heinesch, and J. E. Collin, *Internat. J. Mass Spectrometry Ion Phys.*, 1974, **13**, 131.
[295] S. M. Gordon, P. C. Haarhoff, and G. J. Krige, *Internat. J. Mass Spectrometry Ion Phys.*, 1969, **3**, 13.
[296] F. P. Lossing, A. W. Tickner, and W. A. Bryce, *J. Chem. Phys.*, 1951, **19**, 1254.
[297] E. W. Kiser and E. J. Gallegos, *J. Phys. Chem.*, 1962, **66**, 947.
[298] J. W. Warren, *Nature*, 1950, **165**, 811.

using this method. The problem clearly stems from the arbitrary, if convenient, nature of the normalization procedure, and it appears that many questionable conclusions have been drawn from the results of this method.

Occolowitz *et al.* have presented a very useful comparison of the appearance potentials (AP) obtained from computed IE curves using the SL, EC, EVD,

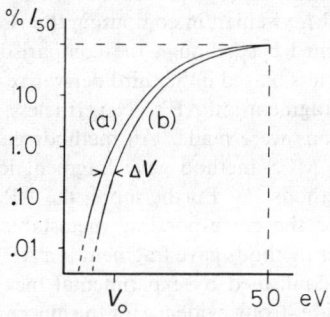

Figure 9 *Semi-logarithmic IE curves of Figure 8 (see text)*

second derivative (SD), critical slope (CS), and modified critical slope (MCS) methods.[299] In considering their results, it is convenient to use the concept of an 'effective threshold law'[300,301]

$$I(E) = (E - E_0)^s \qquad (18)$$

which is simply the power function giving the closest fit to the true ionization cross-section near threshold measured with an exactly monoenergetic electron beam. The IE curve $i(V)$ is then the convolution of $I(E)$ with the electron energy distribution function $m(U)$.

$$i(V) = \int_0^\infty I(V + U)m(U) \, dU \qquad (19)$$

The basis of the derivative method[302,303] is that if s (18) has an integral value n, then $i^{(n+1)}(V)$ gives the reversed electron energy distribution and its maximum provides a convenient point from which the AP may be determined. It is exceptional for s to be integral, and the derivative method gives systematic errors when s is not integral.[300] The critical slope method[304,305] assumes that $m(U)$ is

---

[299] J. L. Occolowitz, B. J. Cerimele, and P. Brown, *Org. Mass Spectrometry*, 1974, **8**, 61.
[300] G. G. Meisels and B. G. Giessner, *Internat. J. Mass Spectrometry Ion Phys.*, 1971, **7**, 489.
[301] C. D. Finney, J. P. Sung, and K. A. Finney, *Internat. J. Mass Spectrometry Ion Phys.*, 1974, **13**, 459.
[302] J. D. Morrison, *J. Chem. Phys.*, 1954, **21**, 1767.
[303] C. D. Finney and A. G. Harrison, *Internat. J. Mass Spectrometry Ion Phys.*, 1972, **9**, 221.
[304] R. E. Honig, *J. Chem. Phys.*, 1948, **16**, 105.
[305] A. L. Barfield and A. L. Wahraftig, *J. Chem. Phys.*, 1964, **41**, 2947.

Maxwellian and that $s = 2$. A modified CS method has been proposed that assumes $s = 1$ for molecular and metastable ions, and $s = 2$ for fragment ions.[299] This modification is rather arbitrary, but it does make some allowance for the differences observed between the IE curves for these respective types of ions.

It was found that none of these methods was generally capable of determining accurate AP's and in the worst cases gave values over 1 eV too high, even though $m(U)$ was assumed to be Maxwellian in computing the IE curves.[299] The MCS method generally performed best, though the comparison with the derivative method would have been less biased if the third derivative had been correspondingly used to determine fragment ion AP's. Nevertheless, some very interesting and important observations were made. All methods gave the same AP's for metastable ions, but the MCS method gave fragment ion AP's about 0.5 eV lower than the other methods.[299] Furthermore, the AP for the fragment ion was slightly less than for the corresponding metastable ion using the MCS method, whereas the other methods gave fragment ion AP's about 0.4 eV higher. These conclusions were confirmed by experimental measurements of over 30 fragmentations, and provide strong evidence for the inaccuracy of most empirical methods for determining appearance potentials.

A new empirical method has been described which avoids the assumptions made in the methods discussed above.[289,306] It assumes that the effective threshold law can be described by a simple power function (18) near the onset, but determines the value of s from the 'critical slope function' $i(V)/i'(V)$, and does not require it to be an integer. The electron energy distribution is assumed to be given by the function,

$$m(U) = U^p \exp(-qU^r) \qquad (20)$$

which is sufficiently flexible to describe the distributions likely to be encountered in most standard ion sources. The parameters p, q, r are also determined from $i(V)/i'(V)$, or from direct measurements of the electron energy distribution. Appearance potentials measured with both this method[306] and the double-EDD technique[288] are generally found to be within 0.05 eV of accurate photoionization measurements.[284,289] The former method only requires a good enough S/N to measure accurate first-derivative IE curves to obtain $i(V)/i'(V)$ $\{\equiv [d \ln i(V)/dV]^{-1}\}$, whereas the double-EDD method is dependent on the S/N of the second-derivative IE curve.

It is clear that appearance potentials of sufficient *accuracy* (rather than mere reproducibility) for reliable thermochemical calculations can be determined from unfiltered electron impact IE curves, but only if the data has a high S/N. It is impossible to achieve the necessary precision by slow (and tedious) manual measurements of IE curves, and full advantage should be taken of the electronically-controlled data acquisition capabilities which are often available in mass spectrometer installations. A very necessary order of magnitude improvement in the accuracy of electron impact appearance potentials can be expected using these improved methods.[283,288,289,306]

[306] R. A. W. Johnstone and B. N. McMaster, *J.C.S. Chem. Comm.*, 1973, 730.

Results.—This section is primarily intended as a collation, for reference purposes, of the numerous appearance potential measurements reported in the last two years. Although heats of formation of fragment ions have been calculated in most papers, it is recommended that only those values determined by accurate AP measurements be used with any confidence in thermochemical calculations. The coverage is selective and mainly concentrates on organic ions; inorganic ions are covered in Chapter 5. Electron spectroscopy measurements and ionization studies without mass analysis are not reported.

Accurate Measurements. Very detailed photoionization studies have been reported for hydrogen,[307] water,[308] carbon dioxide,[309] nitrous oxide,[310,311] nitric oxide,[312] hydrogen oxyfluoride,[313] oxygen difluoride,[314] and xenon fluorides.[315] Photoionization techniques have also been used to study ion decompositions in acetylene,[316,317] cyanoacetylene,[318] formic and acetic acid monomers,[319] methanol and formaldehyde,[320] and vinyl chloride, vinyl fluoride, and 1,1-difluoroethylene.[321] These studies have provided accurate ion dissociation energies but the amount of structure apparent in the photoionization efficiency curves also attests to the complexity of dissociative ionization processes. Little understanding of these details has yet been reached. The importance of measuring photoionization yield data at low enough pressures to preclude contributions from ion-molecule reactions has been stressed.[319] A series of important determinations of the heats of formation of various hydrocarbon ions has been published by Lossing using his energy-selected electron beam instrument. Accuracies of ± 2 kcal mol^{-1} were generally achieved and the following ions have been studied: $CH_3{}^+$, $C_2H_5{}^+$, $C_3H_7{}^+$, $C_4H_9{}^+$;[322] $C_2H_3{}^+$, $C_3H_5{}^+$, $C_7H_7{}^+$;[323] $C_3H_3{}^+$, $C_3H_5{}^+$, $C_4H_7{}^+$;[324] $C_4H_5{}^+$, $C_5H_7{}^+$, $C_5H_8{}^+$.[325] The importance of such measurements in determining ion structures is discussed in Chapter 2, Section 6, and they have greatly assisted the interpretation of more conventional mass spectral studies based on relative ion abundance

---

[307] K. E. McCulloh and J. A. Walker, *Chem. Phys. Letters*, 1974, **25**, 439.
[308] K. E. McCulloh, *J. Chem. Phys.*, 1973, **59**, 4250.
[309] N. Bose and W. Sroka, *Z. Naturforsch*, 1973, **28a**, 22.
[310] P. Coppens, J. Smets, M. G. Fishel, and J. Drowart, *Internat. J. Mass Spectrometry Ion Phys.*, 1974, **14**, 57.
[311] W. Sroka and R. Zietz, *Z. Naturforsch.*, 1973, **28a**, 794.
[312] P. C. Killgoar, G. E. Leroi, J. Berkowitz, and W. A. Chupka, *J. Chem. Phys.*, 1973, **58**, 803.
[313] J. Berkowitz, E. H. Appelman, and W. A. Chupka, *J. Chem. Phys.*, 1973, **58**, 1950.
[314] J. Berkowitz, P. M. Dehmer, and W. A. Chupka, *J. Chem. Phys.*, 1973, **59**, 925.
[315] J. Berkowitz, W. A. Chupka, P. M. Guyon, J. H. Holloway, and R. Spohr, *J. Phys. Chem.*, 1971, **75**, 1461.
[316] V. H. Dibeler, J. A. Walker, and K. E. McCulloh, *J. Chem. Phys.*, 1973, **59**, 2264.
[317] V. H. Dibeler and J. A. Walker, *Internat. J. Mass Spectrometry Ion Phys.*, 1973, **11**, 49.
[318] H. Okabe and V. H. Dibeler, *J. Chem. Phys.*, 1973, **59**, 2430.
[319] D. J. Knowles and A. J. C. Nicholson, *J. Chem. Phys.*, 1974, **60**, 1180.
[320] P. Warneck, *Z. Naturforsch.*, 1971, **26a**, 2047.
[321] D. Reinke, R. Kraessig, and H. Baumgartel, *Z. Naturforsch.*, 1973, **28a**, 1021.
[322] F. P. Lossing and G. P. Semeluk, *Canad. J. Chem.*, 1970, **48**, 955.
[323] F. P. Lossing, *Canad. J. Chem.*, 1971, **49**, 357.
[324] F. P. Lossing, *Canad. J. Chem.*, 1972, **50**, 3973.
[325] J. L. Holmes, *Org. Mass Spectrometry*, 1974, **8**, 247.

measurements.³²⁵ A brief report on the fragmentation of ethylene using a similar instrument has also appeared.³²⁶

Morrison and Traegar have reported electroionization studies of the ion yields from H_2O, H_2S,³²⁷ NH_3, PH_3,³²⁸ CH_4 and $SiH_4$³²⁹ up to ionization energies of 35—45 eV using their Fourier transform deconvolution technique. The first-derivative IE curves reported in this work dramatically illustrate the change in the ionization threshold law with increasing energy, which renders interpretation of the results very difficult. It is important to realize that the first-derivative IE curves are simply used as a convenient means of displaying the data, and it should not be imagined that these curves, normalized to their total, bear any relationship to breakdown graphs. IP's and AP's of $[M - H]^+$ ions of indene, fluorene, acenaphthalene, and fluoranthene have been measured by an automated RPD technique.³³⁰

Dunbar has developed a valuable new technique using a monochromated visible light source to induce photodissociation of vibrationally relaxed ions trapped in an ICR cell.³³¹⁻³³⁶ Ion dissociation rates are measured as a function of photon energy and the dissociation onsets may be readily determined, but the measured values are often well above thermochemical dissociation energies. For example, the onsets for photodissociation of isomeric $C_4H_8^{+\cdot}$ ions by loss of H· and $CH_3^·$ are up to 0.5 eV above the thermochemical thresholds;³³³ similar cases are found in the formation of $C_7H_7^+$, $C_8H_9^+$, or $C_9H_{11}^+$ from numerous alkylbenzene molecular ions.³³⁴ On the other hand, the photodissociation and thermochemical thresholds for the loss of H· from $C_7H_8^{+\cdot}$ (toluene) are quite close, but in this case an excited electronic state also lies at the same energy.³³⁵ These results arise because the photon energy can only be absorbed in an electronic transition to a higher state of the ion, which will often lie above the lowest dissociation energy of the ground state. If this excited state can vibronically relax to the ground state, the ion will then dissociate and the shape of the ion yield curve will be determined by the shape of the Franck–Condon envelope for the initial electronic transition. Other situations can also be envisaged; for example, the excited state may be isolated and the ion will only dissociate if the dissociation energy of the excited state lies below the Franck–Condon region. The great potential of this technique stems from the low dissociation energies

---

³²⁶ I. H. Suzuki and K. Maedi, *Internat. J. Mass Spectrometry Ion Phys.*, 1974, **13**, 293.
³²⁷ J. D. Morrison and J. C. Traegar, *Internat. J. Mass Spectrometry Ion Phys.*, 1973, **11**, 77.
³²⁸ J. D. Morrison and J. C. Traegar, *Internat. J. Mass Spectrometry Ion Phys.*, 1973, **11**, 277.
³²⁹ J. D. Morrison and J. C. Traegar, *Internat. J. Mass Spectrometry Ion Phys.*, 1973, **11**, 289.
³³⁰ W. F. Frey, R. N. Compton, W. T. Naff, and H. C. Schweinler, *Internat. J. Mass Spectrometry Ion Phys.*, 1973, **12**, 19.
³³¹ R. C. Dunbar, *J. Amer. Chem. Soc.*, 1971, **93**, 4354.
³³² R. C. Dunbar and J. M. Kramer, *J. Chem. Phys.*, 1973, **58**, 1266.
³³³ J. M. Kramer and R. C. Dunbar, *J. Chem. Phys.*, 1973, **59**, 3092.
³³⁴ R. C. Dunbar, *J. Amer. Chem. Soc.*, 1973, **95**, 6191.
³³⁵ R. C. Dunbar, *J. Amer. Chem. Soc.*, 1973, **95**, 472.
³³⁶ R. C. Dunbar and E. C. Fu, *J. Amer. Chem. Soc.*, 1973, **95**, 2716.

of (long-lived) ions, which lie well within the photon energies of visible light sources. The high energy resolution which can be achieved in this optical region should enable very accurate energetic studies of ion decompositions[335] and structure[336] to be carried out.

Measurements of the equilibrium constants for ion-molecule reactions can provide accurate ionic heats of formation of vibrationally relaxed ground state ions. These studies are performed in ion sources operating at fairly high pressures, often using ICR techniques, and are discussed in Chapter 3, Section 6. Results have been reported for carbonium ions,[337–341] ammonium ions,[342–345] aliphatic carboxylate anions,[346] and phenolate anions.[347]

Approximate Measurements. Several workers have been concerned with the determination of the magnitude of the kinetic shift predicted by statistical models of ion decompositions. Differences of up to 1 eV or more in the AP's of metastable and fragment ions obtained by the semi-log method have been attributed to the 'measurable' part of the kinetic shift (see *e.g.* ref. 2). A redetermination of the largest of these differences using the single-EDD method showed that in no case was it greater than the experimental error, which was fairly large for the metastable ions (± 0.2 eV) because of the low S/N then attainable.[348] These results confirm the conclusions of Occolowitz *et al.*[299] that most empirical methods severely overestimate the AP's of fragment ions as against metastable ions (see above), and strongly suggest that this 'measurable' part of the kinetic shift is negligible. Accurate photoionization measurements of the AP's of metastable and fragment ions also showed no differences in the examples studied so far.[218,349,350]

In many cases (including the latter examples) the kinetic shift is expected to be small, but even if it were large the AP's of metastable and fragment ions should not be markedly different when proper account is taken of the lifetime distributions and relative intensities at their respective onsets.[348] It will be necessary to measure the AP's of decompositions occurring at much longer lifetimes than the

---

[337] R. J. Blint, T. B. McMahon, and J. L. Beauchamp, *J. Amer. Chem. Soc.*, 1974, **96**, 1269.
[338] T. B. McMahon, R. J. Blint, D. P. Ridge, and J. L. Beauchamp, *J. Amer. Chem. Soc.*, 1972, **94**, 8934.
[339] J. H. J. Dawson, W. G. Henderson, R. M. O'Malley, and K. R. Jennings, *Internat. J. Mass Spectrometry Ion Phys.*, 1973, **11**, 61.
[340] S.-L. Chong and J. L. Franklin, *J. Amer. Chem. Soc.*, 1972, **94**, 6347.
[341] S.-L. Chong and J. L. Franklin, *J. Amer. Chem. Soc.*, 1972, **94**, 6632.
[342] D. H. Aue, H. M. Webb, and M. T. Bowers, *J. Amer. Chem. Soc.*, 1974, **94**, 4726.
[343] M. T. Bowers, D. H. Aue, H. M. Webb, and R. T. McIver, *J. Amer. Chem. Soc.*, 1971, **93**, 4314.
[344] W. G. Henderson, M. Taagepera, D. Holtz, R. T. McIver, J. L. Beauchamp, and R. W. Taft, *J. Amer. Chem. Soc.*, 1972, **94**, 4728.
[345] M. Taagepera, W. G. Henderson, R. T. C. Brownlee, J. L. Beauchamp, R. W. Taft, and D. Holtz, *J. Amer. Chem. Soc.*, 1972, **94**, 1369.
[346] R. Yamdagni and P. Kebarle, *Canad. J. Chem.*, 1974, **52**, 861.
[347] R. T. McIver and J. H. Silvers, *J. Amer. Chem. Soc.*, 1973, **95**, 8462.
[348] T. W. Bentley, R. A. W. Johnstone, and B. N. McMaster, *J.C.S. Chem. Comm.*, 1973, 510.
[349] W. A. Chupka and J. Berkowitz, *J. Chem. Phys.*, 1967, **47**, 2921.
[350] K. M. A. Refaey and W. A. Chupka, *J. Chem. Phys.*, 1968, **48**, 5205.

typical 10^{-5}—10^{-6} s of metastable ions in conventional mass spectrometric systems to determine reliable kinetic shifts. In this respect, Gross's suggestion that the ICR technique should be used is a worthy one.[351] However the *ad hoc* corrections which were necessary to obtain agreement between the measured IP's and known values,[351] indicate that AP measurements using this technique are not yet reliable nor accurate enough to be used with confidence. A more direct technique reported by Lifshitz and co-workers also suffers from experimental uncertainties at the present time.[352]

The AP's of the $C_6H_5CO^+$ and $C_6H_5^+$ ions from a series of alkyl phenyl ketones have been determined by the single-EDD method, and the increase in the derived heats of formation of $C_6H_5^+$ with larger alkyl groups was interpreted as a degrees-of-freedom effect on the internal energy partitioning in the primary fragmentation.[353] The heat of formation of $C_5H_5^+$ from a wide range of substituted benzenes was also determined, and suggested that the open chain 3-penten-1-yne isomer ($\Delta H_f \approx 277$ kcal mol^{-1}) was formed rather than the cyclopentadienyl ion ($\Delta H_f \approx 240$ kcal mol^{-1}), since all the measured values were greater than 280 kcal mol^{-1}.[354] The EDD method was also used to determine AP's of fragment ions from cyclobutanone.[355]

Investigations have continued into substituent effects on the IP's of aromatic compounds,[356-358] but it is disturbing that detailed conclusions have been drawn from these inaccurate (though reproducible) measurements rather than from the much superior PES technique. An interpretation of substituent effects on the IP's of disubstituted benzenes has been presented using PMO theory in the simple Hückel approximation.[359] The use of semi-log AP's in substituent effect correlations of the energetics of aromatic molecular ion decompositions[360,361] engenders little confidence in the conclusions drawn; other criticisms are given in Chapter 2, Section 4. Elaborate conclusions have been reached by some workers following uncritical estimates of the inaccuracies of semi-log AP's and an illustration of this is provided by a study of the formation of the benzoyl ion from several benzoyl derivatives.[362] On the basis of differences in the apparent $\Delta H_f[C_7H_5O^+]$ amounting to 1 eV at most, Benoit concluded that these ions were formed in excited electronic states from some precursors. However, only cursory account was taken of the errors of these AP's which may amount to several tenths of an electron volt in unfavourable cases (see ref. 299) and no

---

[351] M. L. Gross, *Org. Mass Spectrometry*, 1972, **6**, 827.
[352] C. Lifshitz, A. Mackenzie Peers, M. Weiss, and M. J. Weiss, *Adv. Mass Spectrometry*, 1974, **6**, in the press.
[353] S. Tajima, N. Wasada and T. Tsuchiya, *Bull. Chem. Soc. Japan*, 1973, **46**, 3687.
[354] S. Tajima and T. Tsuchiya, *Bull. Chem. Soc. Japan*, 1973, **46**, 3291.
[355] G. R. Branton and C. N. K. Pua, *Canad. J. Chem.*, 1973, **51**, 624.
[356] S. Pignataro, V. Mancini, G. Innorta, and G. Distefano, *Z. Naturforsch.*, 1972, **27a**, 534.
[357] G. Distefano, S. Pignataro, G. Innorta, and F. Fringuelli, *Chem. Phys. Letters*, 1973, **22**, 132.
[358] F. Benoit, *Org. Mass Spectrometry*, 1972, **6**, 1289.
[359] R. A. W. Johnstone and F. A. Mellon, *J.C.S. Faraday II*, 1973, **69**, 36.
[360] F. Benoit, *Org. Mass Spectrometry*, 1972, **6**, 1377.
[361] F. Benoit, *Org. Mass Spectrometry*, 1973, **7**, 295.
[362] F. Benoit, *Org. Mass Spectrometry*, 1973, **7**, 1407.

explicit allowance was made for possible variations in the excess of energy terms, or other sources of error, in deriving the heats of formation. The use, in this paper,[362] of canonical valence bond resonance forms to describe electronically excited states is incredible and incorrect. The dominating influence of the dissociation energy on the fragmentation routes in substituted diphenylmethanes has been discussed.[363]

Jahlonen and Pihlaja have reviewed[364] the use of IP's and AP's for analysing molecular structures.[365–367] While being aware of the inaccuracies of empirical determinations on an absolute scale, they claim that IP and AP differences can be measured accurately for similar compounds. The assumptions used in current empirical methods, as discussed above, make this claim questionable. It should not be confused with the precision of these methods, which is seldom better than ± 0.05 eV at the best of times. The low error limits quoted by these workers [e.g. $\pm(0.002—0.02)$ eV] stem from their assumption that the semi-log IE curves follow straight lines near the threshold $(0.35—0.7\% \; I_{50})$.[365] More precise data show that the slopes of semi-log IE curves change continuously, though slowly, near threshold;[284,306] their error estimates are therefore biased and underestimate the true error.

Leaving this point aside, the further assumptions which are necessary to equate IP or AP differences with differences in molecular heats of formation should be clearly recognized. Firstly, if the heats of formation of isomeric molecular ions, M_1^{+} and M_2^{+}, are identical, then

$$-[\Delta H_f(M_1) - \Delta H_f(M_2)] = \text{IP}(M_1) - \text{IP}(M_2) \tag{21}$$

where the ionization potentials refer to *adiabatic* values. There is no way of proving this assumption (unless the ΔH_f values (21) are already known!). Comparison of accurate IP's obtained by PES with known ΔH_f's suggests it is seldom, if ever, true (see e.g. ref. 368). Secondly, for isomeric molecular ions undergoing the 'same' fragmentation, losing a neutral species N to form fragment ions A_1^{+} and A_2^{+}, then

$$-[\Delta H_f(M_1) - \Delta H_f(M_2)] = \text{AP}(A_1^{+}) - \text{AP}(A_2^{+}) + [\Delta H_f(A_1^{+}) - \Delta H_f(A_2^{+})]$$
$$+ [E_1 - E_2] \tag{22}$$

where E includes all the excess of energy terms. To equate the LHS with the difference in AP's, the sum of the last two terms must be zero. The ions, A_1^{+} and A_2^{+}, must not only have identical heats of formation, but the energy partitioning between A^{+} and N as well as other factors (e.g. kinetic or competitive shifts) must be the same for each fragmentation. It would be extraordinarily fortuitous if all these assumptions held within the small error limits necessary

---

[363] G. Innorta, S. Torroni, S. Pignataro, and V. Mancini, *Org. Mass Spectrometry*, 1973, **7**, 1399.
[364] J. Jalonen and K. Pihlaja, *Org. Mass Spectrometry*, 1973, **7**, 1203.
[365] J. Jalonen and K. Pihlaja, *Org. Mass Spectrometry*, 1972, **6**, 1293.
[366] J. Jalonen, P. Pasanen, and K. Pihlaja, *Org. Mass Spectrometry*, 1973, **7**, 949.
[367] A. G. Loudon and R. Z. Mazengo, *Org. Mass Spectrometry*, 1974, **8**, 179.
[368] R. Botter, F. Menes, Y. Gounelle, J.-M. Pechine, and D. Solgadi, *Internat. J. Mass Spectrometry Ion Phys.*, 1973, **12**, 188.

to derive useful information about molecular structures, particularly for the small effects of non-bonded interactions.[364] Problems have already been noted for this method,[367] and it can scarcely hope to contend with direct thermochemical methods.

Undheim and co-workers have observed that thermal evaporation of several types of compounds from the direct insertion probe gives mass spectra in which particular ions have unusually low appearance potentials.[369–377] These ions often have the highest m/e value in the mass spectrum and have consequently been attributed to ionization of free radicals generated by pyrolysis, although the evidence is not always unambiguous. Where comparison of the measured AP's may be made with accurate free radical IP's the agreement is not very good, but this might be due to the use of the semi-log method. For example Hvistendahl et al. obtain $IP(C_7H_7^{\cdot}) = 6.74\,(\pm 0.05)\,eV^{369}$ compared with the photoionization value of $6.23\,(\pm 0.07)\,eV$.[378] The following compounds have been studied: tropylium halides,[369] anilinium oxides,[370,371] pyrilium salts,[372] cinnolinium iodides,[373] hydroxy- and mercapto-pyridines,[374] pyridinium-3-oxides,[375] N-arylpyridinium-3-oxides,[376] and 8-hydroxy-thiazolo[3,2-a]pyridinium-3-oxides.[377]

A number of other measurements of ionization and appearance potentials have been reported using the semi-log method, and in several cases approximate ionic heats of formation were derived. Errors involved in these determinations render any detailed discussion inconclusive, so it is therefore not pursued here. Organic compounds which have been studied include: isomeric hydrocarbons[379] (forming $C_9H_{11}^+$,[380] $C_9H_9^+$,[381] $C_9H_7^+$ and $C_{11}H_9^+$ ions[382]), methyl phenanthrenes,[365] n,n'-dimethyl-1,1-binaphthalenes and n,n'-dimethylbiphenyls,[367] aliphatic ozonides,[383] 1,3-dioxolan, 1,3-dithiolan, and 1,3-oxathiolan,[384] methyl-substituted 1,3-oxathians,[366] 1,3,6-dioxathiocan,[385] and piperidine amides and piperideine amides.[386] Some studies of organometallic and inorganic compounds which were reported include: ferrocene, cobaltocene, nickelocene,

[369] G. Hvistendahl, K. Undheim, and P. Györösi, *Org. Mass Spectrometry*, 1973, **7**, 903.
[370] G. Hvistendahl and K. Undheim, *Org. Mass Spectrometry*, 1972, **6**, 217.
[371] H. M. R. El-Monafi, G. Hvistendahl, and K. Undheim, *Org. Mass Spectrometry*, 1974, **9**, 350.
[372] G. Hvistendahl, P. Györösi, and K. Undheim, *Org. Mass Spectrometry*, 1974, **9**, 80.
[373] G. Hvistendahl and K. Undheim, *Tetrahedron*, 1972, **28**, 1737.
[374] T. Gronneberg and K. Undheim, *Org. Mass Spectrometry*, 1972, **6**, 823.
[375] T. Gronneberg and K. Undheim, *Org. Mass Spectrometry*, 1972, **6**, 225.
[376] K. Undheim and P. E. Hansen, *Org. Mass Spectrometry*, 1973, **7**, 635.
[377] P. E. Fjeldstad and K. Undheim, *Org. Mass Spectrometry*, 1973, **7**, 639.
[378] F. A. Elder and A. C. Parr, *J. Chem. Phys.*, 1969, **50**, 1027.
[379] C. Köppel, H. Schwarz, and F. Bohlmann, *Org. Mass Spectrometry*, 1974, **9**, 324.
[380] C. Köppel, H. Schwarz, and F. Bohlmann, *Org. Mass Spectrometry*, 1974, **9**, 321.
[381] C. Köppel, H. Schwarz, and F. Bohlmann, *Org. Mass Spectrometry*, 1974, **8**, 25.
[382] H. Schwarz and F. Bohlmann, *Org. Mass Spectrometry*, 1973, **7**, 395.
[383] M. Bertrand, J. Carles, S. Fliszar, and Y. Rousseau, *Org. Mass Spectrometry*, 1974, **9**, 297.
[384] G. Conde-Caprace and J. E. Collin, *Org. Mass Spectrometry*, 1972, **6**, 415.
[385] G. Conde-Caprace and J. E. Collin, *Org. Mass Spectrometry*, 1972, **6**, 341.
[386] H. Schwarz and F. Bohlmann, *Org. Mass Spectrometry*, 1973, **7**, 1197.

and magnesocene,[387] rhodium and iridium dicarbonyl ketonates,[388] manganese, chromium, and tungsten derivatives,[389] transition metal and metal-free phthalocyanines,[390] alkylthiophosphate esters,[391] and tetrafluorohydrazine.[392]

Heats of formation of free radicals have been estimated from the AP's of CH_3^+ ions formed from several hydrocarbons.[393] Ionization potentials of some dipositive ions from aromatic compounds were measured by a modified critical slope method,[394] and a new method of determining double and triple IP's by a collisional charge-stripping technique[395] has been described.[396]

Negative Ion Studies. Measurements of the lifetimes of molecular negative ions have been reported for some benzene derivatives, non-benzenoid aromatics, and non-aromatic organic molecules,[397] and the variation of autodetachment lifetimes with electron beam energy has also been investigated.[398] The RPD technique was used to measure cross-sections for O_2^- and C^- production in CO_2,[399] and also to obtain the negative IE curves for dissociative electron capture processes in the perfluorinated derivatives of cyclobutane, 2-butene, propylene, ethylene, and methane.[400] Thynne and co-workers have also investigated the same processes for perfluorocyclobutane,[401] perfluoropropylene, and perfluoropropane[402] using an iterative deconvolution method. These studies were aimed at determining accurate energetic data but the resonant peak maxima differ by 0.3—0.5 eV between each set of results, with those of Lifshitz and Grajower always being lower. This significant discrepancy between two supposedly accurate methods is very disconcerting, and seems to stem from differences in the energy calibration points used by both groups. Thynne *et al.* use their measured onset of the O^-/SO_2 peak at 4.2 eV, corresponding to a maximum at 4.9 eV,[403] whereas Lifshitz *et al.* claim that the maximum occurs at 4.6 eV[293]

---

[387] G. M. Begun and R. N. Compton, *J. Chem. Phys.*, 1973, **58**, 2271.
[388] F. Bonati, G. Distefano, S. Pignataro, and S. Torroni, *Org. Mass Spectrometry*, 1972, **6**, 971.
[389] G. Distefano, A. Foffani, G. Innorta, and S. Pignataro, *Internat. J. Mass Spectrometry Ion Phys.*, 1971, **7**, 383.
[390] D. D. Eley, D. J. Hazeldine, and T. F. Palmer, *J.C.S. Faraday II*, 1973, **69**, 1808.
[391] E. Santoro, *Org. Mass Spectrometry*, 1973, **7**, 589.
[392] S. N. Foner and R. L. Hudson, *J. Chem. Phys.*, 1973, **58**, 581.
[393] D. K. Sen Sharma and J. L. Franklin, *J. Amer. Chem. Soc.*, 1973, **95**, 6562.
[394] R. Engel, D. Halpern, and B.-A. Funk, *Org. Mass Spectrometry*, 1973, **7**, 177.
[395] J. H. Beynon, R. G. Cooks, and T. Keough, *Internat. J. Mass Spectrometry Ion Phys.*, 1974, **13**, 437.
[396] R. G. Cooks, T. Ast, and J. H. Beynon, *Internat. J. Mass Spectrometry Ion Phys.*, 1973, **11**, 490.
[397] A. Hadjiantoniou, L. G. Christophorou, and J. G. Carter, *J.C.S. Faraday II*, 1973, **69**, 1691, 1704.
[398] L. G. Christophorou, A. Hadjiantoniou, and J. G. Carter, *J.C.S. Faraday II*, 1973, **69**, 1713.
[399] D. Spence and G. J. Schulze, *J. Chem. Phys.*, 1974, **60**, 216.
[400] C. Lifshitz and R. Grajower, *Internat. J. Mass Spectrometry Ion Phys.*, 1972, **10**, 25.
[401] P. W. Harland and J. C. J. Thynne, *Internat. J. Mass Spectrometry Ion Phys.*, 1972, **10**, 11.
[402] P. W. Harland and J. C. J. Thynne, *Internat. J. Mass Spectrometry Ion Phys.*, 1972, **9**, 253.
[403] P. W. Harland and J. C. J. Thynne, *J. Phys. Chem.*, 1970, **74**, 52.

from a comparison with the established value of 9.8 eV for O^-/CO.[404] It is time a consensus was reached, and the best data would appear to favour the lower value.[293]

Measurements of dissociative electron attachment IE curves have also been reported by Thynne and co-workers for: perfluoro-n-butane,[405] perfluoropropionitrile, and perfluorobutyronitrile,[406] tungsten hexafluoride,[407] germanium tetrafluoride,[408] hexafluorodimethylperoxide,[409] and methanol and its deuteriated derivatives.[410] Several of these studies include measurements of autodetachment lifetimes of metastable molecular ions.[401,405–407] Negative IE curves have also been measured for acrylonitrile using the RPD technique,[411] and for tetracyanoethylene using an energy-selected electron beam (half-width ⩽ 0.1 eV).[412] The average kinetic energy of negative fragment ions from CF_4 and SiF_4 was determined and used to interpret the dissociative electron attachment processes,[413] and the effect of thermal energy on the IE curves of negative ions from SF_6 has been investigated.[414]

Collisional ionization with caesium atoms has recently been developed into a new method for studying dissociative electron attachment processes, and results have been reported for tetrafluorosuccinic anhydride, hexafluoroglutaric anhydride,[415] maleic anhydride, and succinic anhydride.[416]

---

[404] P. J. Chantry, *Phys. Rev.*, 1968, **172**, 125.
[405] P. W. Harland and J. C. J. Thynne, *Internat. J. Mass Spectrometry Ion Phys.*, 1973, **11**, 445.
[406] J. C. J. Thynne and P. W. Harland, *Internat. J. Mass Spectrometry Ion Phys.*, 1973, **11**, 399.
[407] J. C. J. Thynne and P. W. Harland, *Internat. J. Mass Spectrometry Ion Phys.*, 1973, **11**, 137.
[408] P. W. Harland, S. Cradock, and J. C. J. Thynne, *Internat. J. Mass Spectrometry Ion Phys.*, 1972, **10**, 169.
[409] K. A. G. MacNeill and J. C. J. Thynne, *Internat. J. Mass Spectrometry Ion Phys.*, 1972, **9**, 135.
[410] J. C. J. Thynne and P. W. Harland, *Internat. J. Mass Spectrometry Ion Phys.*, 1973, **11**, 127.
[411] S. Tsuda, A. Yokohata, and T. Umaba, *Bull. Chem. Soc. Japan*, 1973, **46**, 2273.
[412] C. E. Brion and L. A. R. Olsen, *Internat. J. Mass Spectrometry Ion Phys.*, 1972, **9**, 413.
[413] J. L.-F. Wang, J. L. Margrave, and J. L. Franklin, *J. Chem. Phys.*, 1973, **58**, 5417.
[414] C. Lifshitz and M. Weiss, *Chem. Phys. Letters*, 1972, **15**, 266.
[415] C. D. Cooper and R. N. Compton, *J. Chem. Phys.*, 1974, **60**, 2424.
[416] R. N. Compton, P. W. Reinhardt, and C. D. Cooper, *J. Chem. Phys.*, 1974, **60**, 2953.

2
Structure and Mechanism in Mass Spectrometry

BY T. W. BENTLEY

1 Introduction

'Proposed ion structures are substantiated by precise mass measurement.'*

Although, unfortunately, the attitude expressed in the above quotation still exists, the ideas of many authors about structure and mechanism in organic mass spectrometry may be described as 'cautiously optimistic' or 'healthily sceptical'. The former view has been adopted consistently by McLafferty and co-workers for over ten years.[1,2] Around 1968, it appeared to us that pointless speculation was reaching epidemic proportions, and our rather sceptical review was published in 1970.[3] The present Chapter is an updating of that review, with emphasis on recent advance, since much of the relevant background material has been well covered elsewhere.[4]

Over the past five years, increasingly significant experimental and theoretical approaches have been applied to the problem of carbonium ion structures in the gas phase. (A review on structure and reactivity of hydrocarbon ions covers the literature up to 1970.[5]) While it is still possible to criticize all current approaches, one hopes that several independent approaches will point to the same (correct) structure.

It is necessary to distinguish clearly between odd and even electron systems. Many even-electron ions can be generated by ionization of free radicals and accurate heats of formation can be calculated.[6] By the Franck–Condon principle, these ions initially will have the same structure as the radicals from which they were formed. In the case of simple hydrocarbon ions the experimental heats of formation are in good agreement with several independent MO calculations (see

---

[1] F. W. McLafferty, 'Mass Spectrometry of Organic Ions', Academic Press, 1963, Chap. 7.
[2] F. W. McLafferty, in 'Topics in Organic Mass Spectrometry', ed. A. L. Burlingame, Wiley-Interscience, New York, 1970.
[3] T. W. Bentley and R. A. W. Johnstone, *Adv. Phys. Org. Chem.*, 1970, **8**, 151.
[4] I. Howe, in 'Mass Spectrometry', ed. D. H. Williams (Specialist Periodical Reports), The Chemical Society, London, (*a*) 1971, Vol. 1, Chap. 2; (*b*) 1973, Vol. 2, Chap. 2.
[5] P. Ausloos and S. G. Lias, in 'Ion-Molecule Reactions', ed. J. L. Franklin, Butterworths, London, 1972, p. 707.
[6] Review: J. Jalonen and K. Pihlaja, *Org. Mass Spectrometry*, 1973, **7**, 1203.

* Variations of this statement appear in the literature not infrequently and it serves no useful purpose to quote the source of this particular example, published in 1972.

Section 6). However, in the case of odd-electron ions there is very little direct corroborating evidence, *e.g.* from MO calculations. Also, the well-established existence of even-electron species such as CH_5^+ suggests that one should very cautiously extrapolate conventional views of bonding in neutral organic molecules to ions in the gas phase.

2 Metastable Ions

Consecutive Processes.—In the early sixties, it was widely assumed that the observation of a metastable ion indicated a one-step fragmentation process with the ejection of one neutral fragment. Somewhat reluctantly, this idea began to be discarded in 1965,[7] but has not yet entirely disappeared. Perhaps the problem may be clarified by considering Figure 1: an ion $[ABC]^+$ fragments by loss of C as a neutral species, followed by loss of A as a neutral species to give an ion $[B]^+$.

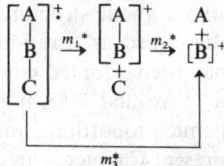

Figure 1 *General scheme for metastable ions for consecutive processes*

If C is ejected in the second field-free region, the normal metastable ion m_1^* may be observed. If A is ejected in the second field-free region, then the normal metastable ion m_2^* may be observed. If both processes occur in rapid succession in the second field-free region, then the normal metastable ion m_3^* may be observed. Considering that fragmentations may require $< 10^{-11}$ s, and that the ions may spend considerably longer than this in the second field-free region, there is ample time for two fragmentations. Thus, on a molecular time scale, there is no need to consider that the metastable ion, m_3^*, corresponds to a concerted one-step process. Two fast consecutive processes explain the results satisfactorily.

When all three metastable ions are observed, as in the fragmentation of genipin,[8] (see Figure 2) and the neutral fragments are both H_2O, there can be no reasonable doubt that the above interpretation is correct. It is inconceivable that H_4O_2 could be lost as one neutral fragment, or that two molecules of H_2O could be released from different parts of the molecule at *exactly* the same time.

Similarly, if two metastable ions, m_1^* and m_3^* in Figure 1, are observed, the most likely explanation is that formation of B^+ from $[A - B]^+$ is favourable and occurs rapidly. This conclusion greatly reduces the novelty of a recent publication

---

[7] A. H. Jackson, G. W. Kenner, K. M. Smith, R. T. Aplin, H. Budzikiewicz, and C. Djerassi, *Tetrahedron*, 1965, **21**, 2913; see also K. R. Jennings, *Chem. Comm.*, 1966, 283; R. A. W. Johnstone, B. J. Millard, F. M. Dean, and A. W. Hill, *J. Chem. Soc. (C)*, 1966, 1712; J. Seibl, *Helv. Chim. Acta*, 1967, **50**, 263.

[8] T. W. Bentley, Ph.D. Thesis, Liverpool University, 1969.

discussing 'concerted' loss of H_2CN.[9] The case where two metastable ions, m_1^* and m_2^*, are observed is trivial, as are cases where only one metastable ion, m_1^* or m_2^*, is observed.

However, there may be rare examples where only m_3^* is observed and the most reasonable explanation is that formation of $[B]^+$ from $[A - B]^+$ is *very* favourable and occurs *very* rapidly. In such cases it would be worthwhile to 'defocus' metastable ions from the first field-free region, or otherwise to increase the instrument's

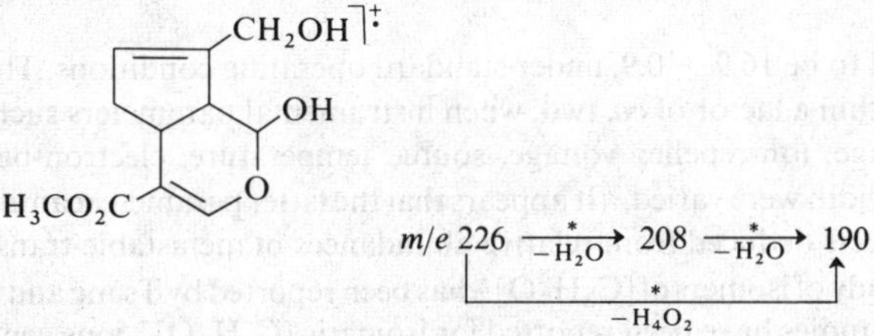

$$m/e\ 226 \xrightarrow{*}_{-H_2O} 208 \xrightarrow{*}_{-H_2O} 190$$
$$\xrightarrow{*}_{-H_4O_2}$$

Figure 2 *Consecutive fragmentation of genipin*

sensitivity to metastable ions. Situations corresponding to that represented in Figure 1, where all three metastable ions are observed, are very common,[7,8] particularly when metastable ions are detected by defocusing.[10]

Meyerson and co-workers[11] have recently pointed out that the interpretation of rapid consecutive metastable ions is consistent with work by Hipple in 1945.[12] Although these ideas have been rediscovered and republished several times since then, they do not yet appear to be generally known. In an entirely different context, misplaced idealism may also account for the still prevailing view that S_N1 reactions *necessarily* proceed *via* dissociated carbonium ions (as opposed to ion pairs). Similar difficulties in criticisms of the charge-localization ideology in mass spectrometry are discussed in Section 5.

Intensities.—The ratios of metastable ion intensities for competing fragmentation reactions have been used extensively to characterize gaseous ion structures.[3,4] To interpret the results, it is necessary to assume that the intensity ratio is more or less independent of the internal energy distributions of the fragmenting ions. Unfortunately, this ideal situation is not usually realized in practice, but this difficulty has now been largely solved by collision-induced decomposition of metastable ions (see subsection below on *Collisional Activation*). Alternatively, one can consider only large changes ($> \times 5$) in ratios of metastable ion intensities as reliable evidence for different structures.

---

[9] K. M. Baker and A. Frigerio, *J.C.S. Perkin II*, 1973, 648.
[10] M. Barber and R. M. Elliott, ASTM E.14, Conference on Mass Spectrometry, Montreal, Canada, June 1964; K. R. Jennings, *J. Chem. Phys.*, 1965, **43**, 4176; P. Schulze and A. L. Burlingame, *ibid.*, 1968, **49**, 4870, and refs. therein.
[11] S. Meyerson, R. W. Vander Haar, and E. K. Fields, *J. Org. Chem.*, 1972, **37**, 4114.
[12] J. A. Hipple and E. U. Condon, *Phys. Rev.*, 1945, **68**, 54.

The effect of internal energy distributions on ratios of metastable ion intensities, $I(m_1^*)/I(m_2^*)$, can be studied by varying instrumental parameters or by generating the ions from different precursors. Fragmentation of the ion $[C_7H_6F]^+$, generated from eleven different precursors, has been studied.[13] $I(m_1^*)/I(m_2^*)$ for equations (1) and (2):

$$[C_7H_6F]^+ \xrightarrow{m_1^*} [C_5H_4F]^+ + C_2H_2 \tag{1}$$

$$\xrightarrow{m_2^*} [C_7H_5]^+ + HF \tag{2}$$

were found to be 16.9 ± 0.9, under standard operating conditions. This ratio was altered, within a factor of ca. two, when instrumental parameters such as accelerating voltage, ion-repeller voltage, source temperature, electron-beam energy, and β-slit width were varied. (It appears that the latter parameter can also influence isotope effects deduced from relative abundances of metastable transitions.[14]) A detailed study of isomers of $[C_3H_7O]^+$ has been reported by Tsang and Harrison.[15]

Similar studies have been reported for isomeric $[C_4H_9O]^+$ ions generated from twelve different alcohols, ethers, acetals, or ketals,[16] and for $[C_3H_6]^{+\cdot}$ radical cations generated from fifteen precursors.[17]

Collisional Activation.—Collision with neutral molecules provides a convenient method of adding internal energy to ions in the field-free region of the mass spectrometer, and causes fragmentation of the ions with activation energies covering a broad range; for a particular ion such fragmentation can be viewed as its 'collisional activation (CA) spectrum'.[18] CA spectra may be obtained for each ion in the normal mass spectrum. Although the name and symbol are newly-coined, there have been several earlier reports of this phenomenon.[19] Recent detailed studies have now established the usefulness of CA spectra in studies of ion structures and ion fragmentation mechanisms, and in determination of molecular structures.

It appears that the internal energy of the precursor ion has a negligible effect on the ion's CA spectrum, except for product ions formed *via* the processes of lowest activation energy. Therefore, in studies of ion structures, these processes should be ignored. Thus, the major difficulty in comparing $I(m_1^*)/I(m_2^*)$ for different ions (see above) is largely overcome. CA spectra appear to follow the predictions of the quasi-equilibrium theory.

---

[13] K. R. Jennings and A. Whiting, *Org. Mass Spectrometry*, 1972, **6**, 917; see also D. J. McAdoo, P. F. Bente, M. L. Gross, and F. W. McLafferty, *Org. Mass Spectrometry*, 1974, **9**, 525.

[14] M. A. Baldwin and F. W. McLafferty, *Internat. J. Mass Spectrometry Ion Phys.*, 1973, **12**, 86; see also R. Neeter and N. M. M. Nibbering, *Org. Mass Spectrometry*, 1973, **7**, 1091.

[15] C. W. Tsang and A. G. Harrison, *Org. Mass Spectrometry*, 1973, **7**, 1377.

[16] T. J. Mead and D. H. Williams, *J.C.S. Perkin II*, 1972, 876.

[17] M. L. Gross and P.-H. Lin, *Org. Mass Spectrometry*, 1973, **7**, 795.

[18] F. W. McLafferty, P. F. Bente, R. Kornfeld, S.-C. Tsai, and I. Howe, *J. Amer. Chem. Soc.*, 1973, **95**, 2120.

[19] Literature cited in Ref. 18.

CA spectra can be measured easily and accurately with a double-focusing mass spectrometer, in which a mass-analysed beam enters the collision region and the products are energy analysed. Unfortunately, this is the reverse of the usual order of analysers and therefore requires a special instrument, having a reversed geometry.[20] Such instruments may form a widely-used 'new generation', because they appear to have no serious disadvantages over conventional double-focusing instruments in routine applications, and a possible advantage of CA mass spectra in molecular structure determination is that more simple cleavage reactions are observed.[21]

To obtain a pure CA spectrum, it is necessary to subtract the contributions of normal metastable ions (MI spectrum). The magnetic field is adjusted to select only the desired precursor ion mass and its unimolecular metastable decomposition products are scanned by reducing the electrostatic analyser (ESA) potential from its normal value to zero, while recording the resulting ion current at the collector. The pressure in the field-free region between the magnet and the ESA is then increased by adding helium until the intensity of the precursor ion is reduced to 10% of its original value. A second ESA scan produces the CA spectrum from which the MI spectrum is subtracted to give the pure CA spectrum.[21]

The CA spectra of $[C_2H_5O]^+$ ions from twenty-five precursors suggest that only two structures, presumably (1) and (2), are stable,[21] in agreement with earlier ICR measurements.[22] However, as both techniques[21,22] measure collision-induced decomposition of relatively long-lived ions, one cannot rule out the possibility that: (i) the ions do not have the supposed structures; (ii) the collision causes rearrangement of isomeric structures, e.g. (3); (iii) isomeric ions decompose similarly, and thus cannot be distinguished.

$$CH_3\overset{+}{O}=CH_2 \qquad CH_3CH=\overset{+}{O}H \qquad \underset{CH_2-CH_2}{\overset{\overset{H}{|}}{\overset{O^+}{\diagup\diagdown}}}$$

(1) \qquad\qquad (2) \qquad\qquad (3)

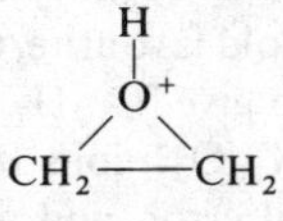

As internal energy effects do not appear to be significant, such techniques can be used to show clearly that two structures are different. However, the argument that two structures are the same is much less strong [see (ii) and (iii) above]. Also, having shown that two structures are different, one cannot be sure what the structures are, although the characteristic unimolecular decompositions,[21] or ion–molecule reactions,[22] may be helpful indications. A detailed study of the fragmentation of $[C_2H_5O]^+$ ions has also been reported.[23]

---

[20] T. Wachs, P. F. Bente, and F. W. McLafferty, *Internat. J. Mass Spectrometry Ion Phys.*, 1972, **9**, 333.
[21] F. W. McLafferty, R. Kornfeld, W. F. Haddon, K. Levsen, I. Sakai, P. F. Bente, S.-C. Tsai, and H. D. R. Schuddemage, *J. Amer. Chem. Soc.*, 1973, **95**, 3886.
[22] J. L. Beauchamp and R. C. Dunbar, *J. Amer. Chem. Soc.*, 1970, **92**, 1477; however, see also Section 8 of this chapter.
[23] B. G. Keyes and A. G. Harrison, *Org. Mass Spectrometry*, 1974, **9**, 221.

In an extension of the research discussed above, the CA spectra of $[C_3H_7O]^+$ ions from various precursors were measured. Extensive rearrangements were observed and it appears that structures (4)—(7) are the stable forms.[24]

$(CH_3)_2 \overset{+}{C}=OH$ $\qquad$ $CH_3CH_2CH=\overset{+}{O}H$ $\qquad$ $CH_2=\overset{+}{O}CH_2CH_3$

(4) $\qquad\qquad\qquad\qquad$ (5) $\qquad\qquad\qquad\qquad$ (6)

$CH_3CH=\overset{+}{O}-CH_3$

(7)

As might be expected, parallel work on the CA spectra of $[C_2H_6N]^+$ and $[C_3H_8N]^+$ ions suggest that the immonium structures (8), (9), and (10)—(14), respectively, are the stable structures. (Stabilization of carbonium ions by hetero-atoms is discussed in Section 6.)

$CH_3CH=\overset{+}{N}H_2$ $\qquad\qquad\qquad$ $CH_3\overset{+}{N}H=CH_2$

(8) $\qquad\qquad\qquad\qquad\qquad\qquad$ (9)

$CH_3CH_2\overset{+}{N}H=CH_2$ $\qquad$ $(CH_3)_2C=\overset{+}{N}H_2$ $\qquad$ $CH_3CH_2CH=\overset{+}{N}H_2$

(10) $\qquad\qquad\qquad\qquad$ (11) $\qquad\qquad\qquad\qquad$ (12)

$CH_3\overset{+}{N}H-CHCH_3$ $\qquad\qquad$ $(CH_3)_2\overset{+}{N}=CH_2$

(13) $\qquad\qquad\qquad\qquad\qquad$ (14)

By deuterium labelling, it was shown that a substantial proportion of the collison-induced decompositions of ions (8) and (9) did not involve loss of positional identity of the hydrogen atoms.[25]

An old favourite, the loss of CO_2 from the molecular ion of diphenyl carbonate (15) to give $[C_{12}H_{10}O]^{+\cdot}$ has also been studied by CA.[26] The CA spectra of the $[C_{12}H_{10}O]^{+\cdot}$ ions from (15) and (16) were found to be identical within experimental error, and were clearly different from the spectra of the three isomeric hydroxybiphenyls (17). The diphenylether structure corresponding to (16) may tentatively be proposed for this ion. The CA spectra of $[C_{12}H_9O]^+$ ions from all five compounds (15)—(17) are very similar.[26]

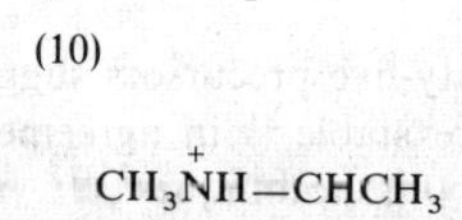

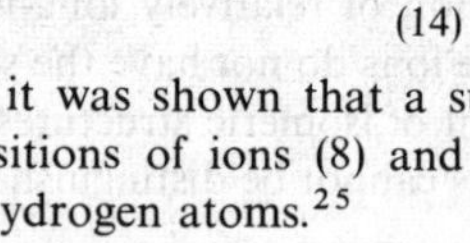

[24] F. W. McLafferty and I. Sakai, *Org. Mass Spectrometry*, 1973, **7**, 971.
[25] K. Levsen and F. W. McLafferty, *J. Amer. Chem. Soc.*, 1974, **96**, 139.
[26] K. Levsen and F. W. McLafferty, *Org. Mass Spectrometry*, 1974, **8**, 353.

Partial support for the above interpretation comes from the kinetic energy released in reaction (3), which was found to be the same for $[C_{12}H_{10}O]^{+\cdot}$ ions

$$[C_{12}H_{10}O]^{+\cdot} \rightarrow [C_{11}H_{10}]^{+\cdot} + CO \qquad (3)$$

from both (15) and (16).[27] However, as Levsen and McLafferty point out,[26] kinetic energy release and MI spectra are only characteristic of the structure of the ions decomposing in the metastable region and may be influenced by internal energy effects. Ions which decompose in the metastable region (not collision-induced) necessarily are of higher internal energy than normal ions collected without fragmentation and so could have isomerized before metastable decomposition. In contrast, the CA spectrum is characteristic of all the ions passing through the mass spectrometer, independent of their internal energies.

Perhaps the most famous ion in organic mass spectrometry, $[C_7H_7]^+$, has also been induced to decompose by collision. Three long-lived isomers, presumably (18)—(20) have been studied by CA.[28] Ions formed from $C_6H_5CH_2Y$

(18) 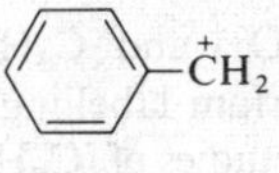   (19) 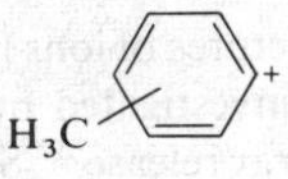   (20)

appear to give a mixture of (18) and (19), depending on the nature of Y: e.g. when Y = H, (18) appears to be the major product, but when Y = CH_2Ph, (19) appears to be the major product. Some ring-substituted toluenes appear to give rise to structure (20).

The CA spectra of other ions have also been reported: $[C_3H_6O]^{+\cdot}$,[21] a detailed study of $[C_7H_8]^{+\cdot}$,[21,29] and $[C_{13}H_9]^{+\cdot}$.[21]

Kinetic Energies.—Much attention has been paid to the kinetic energy released during fragmentation of metastable ions. As well as being an important thermochemical quantity, the kinetic energy released must also be one characteristic of ion structure—at least of the narrow band of metastable ions observed in such experiments; however, as discussed above, metastable ions are *not* typical ions.

It has been shown that the kinetic energy released in unimolecular reactions, as measured from the width of the corresponding metastable peak, shows only a small dependence on such instrumental parameters as source temperature, ion residence time in the source, and ion accelerating voltage. The widths of metastable peaks also depend on the width of the energy-resolving (β) slit, but satisfactory results can be obtained if the metastable peak width is extrapolated to zero slit width.[27]

Fragmenting ions generated from different members of an homologous series of molecular ions appear to release almost the same kinetic energy, and do not

---

[27] E. G. Jones, L. E. Bauman, J. H. Beynon, and R. G. Cooks, *Org. Mass Spectrometry*, 1973, **7**, 185.
[28] J. Winkler and F. W. McLafferty, *J. Amer. Chem. Soc.*, 1973, **95**, 7533; see also M. K. Hoffman and J. C. Wallace, *J. Amer. Chem. Soc.*, 1973, **95**, 5064.
[29] K. Levsen, F. W. McLafferty, and D. M. Jerina, *J. Amer. Chem. Soc.*, 1973, **95**, 6332.

exhibit a significant degrees-of-freedom effect analogous to that for metastable ion abundances. While increases in the average internal energies of the ions leads to larger releases of kinetic energy, the effect appears to be small.[27,30]

The shapes of the metastable peaks for loss of H_2CO from p- and m-chloroanisole are different; as the peak shape from the latter is the same as the peak shape from the ion formed by loss of HCN from benzaldehyde O-methyl ether (21), it was concluded that a five-membered cyclic intermediate was involved, followed by rearrangement of (22) to (23).[31] However, insufficient evidence was presented to justify the authors'[31] remarkable confidence in the above interpretation.

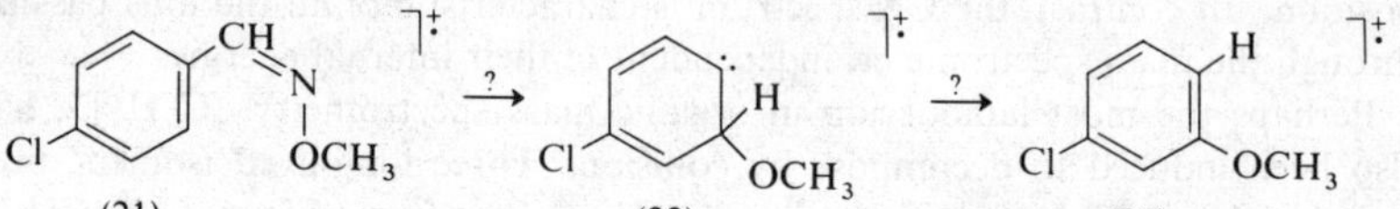

The structures of ions $[C_5H_{10}O]^{+\cdot}$ and $[C_3H_6O]^{+\cdot}$ from aliphatic ketones have also been investigated by deuterium labelling, metastable ion abundances and kinetic energy release.[32] Similar studies of $[C_7H_7]^+$, $[C_7H_8]^{+\cdot}$, and $[C_7H_7OCH_3]^{+\cdot}$ have been carried out.[33]

The 'significantly different' ion kinetic energy (IKE) spectra of isomeric fluorobenzenes and benzyl fluoride indicate incomplete substituent scrambling in the first field-free region of the mass spectrometer,[34] in contrast to results obtained in normal MI spectra.[13] As ions decomposing in the first field-free region presumably have higher internal energies than those decomposing in the second field-free region, it is perhaps surprising that *less* scrambling is observed in the IKE spectra. On the other hand, there is more time for scrambling in the longer-lived ions observed in MI spectra. Loss of H_2O and CO_2H from the molecular ion of benzyl benzoate (24) has been discussed and a mechanism consistent with deuterium labelling and IKE spectra has been proposed.[35]

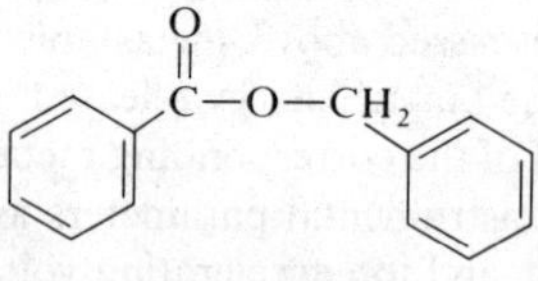

(24)

[30] E. G. Jones, J. H. Beynon, and R. G. Cooks, *J. Chem. Phys.*, 1972, **57**, 2652.
[31] J. H. Beynon, M. Bertrand, and R. G. Cooks, *Org. Mass Spectrometry*, 1973, **7**, 785.
[32] J. H. Beynon, R. M. Caprioli, and R. G. Cooks, *Org. Mass Spectrometry*, 1974, **9**, 1.
[33] R. G. Cooks, J. H. Beynon, M. Bertrand, and M. K. Hoffman, *Org. Mass Spectrometry*, 1973, **7**, 1303; see also T. Keough, J. H. Beynon, R. G. Cooks, C. W. J. Chang, and R. H. Shapiro, *Z. Naturforsch.*, 1974, **29a**, 507.
[34] S. Safe, W. D. Jamieson, and D. J. Embree, *Canad. J. Chem.*, 1974, **52**, 867; see also S. Safe, O. Hutzinger, and W. D. Jamieson, *Org. Mass Spectrometry*, 1973, **9**, 169.
[35] J. H. Beynon, R. M. Caprioli, R. H. Shapiro, K. B. Tomer, and C. W. J. Chang, *Org. Mass Spectrometry*, 1972, **6**, 863.

Deuterium isotope effects on kinetic energy release have also been discussed,[36,37] and there has been a warning that isotope effects may depend on slit width.[14]

Defocusing techniques for detecting metastable ions of complex molecules have been investigated.[38]

More detailed aspects of the energetics of fragmentation of metastable ions ('energy-partitioning') have been studied[39—41] (see also Chapter 1).

3 Reactions of Isotopically Labelled Species

Hydrogen Rearrangements.—As hydrogen rearrangements have recently been reviewed in great detail,[42,43] they will only be discussed briefly here.

Further evidence that loss of C_2H_2O from acetanilide (25) proceeds by hydrogen transfer to the nitrogen atom has been presented.[44,45] Super-high resolution techniques were required to separate the doublet at m/e 66 due to C_5H_4D and C_5H_6, but it appears that the fragment ions (26) decompose further by approximately equal losses of HNC or DNC.[46] This strongly suggests that the fragment ions with sufficient energy to decompose further contain the partial structure H—N—D, which is consistent with previous evidence. Long-lived fragment ions from similar reactions, studied by ICR spectroscopy, are discussed in Section 8.

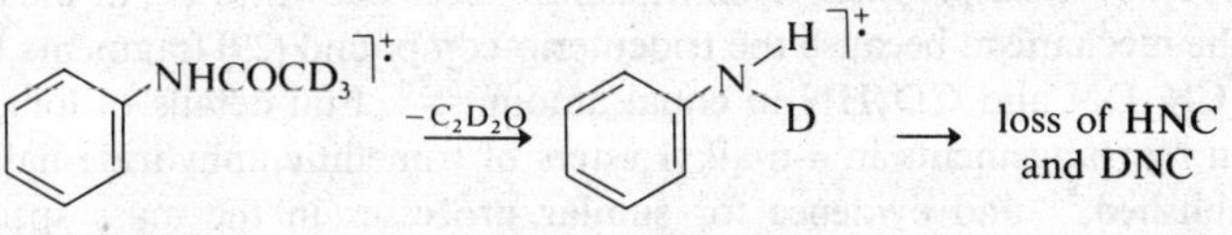

PhNHCOCD$_3$]$^{+\cdot}$ $\xrightarrow{-C_2D_2O}$ PhN(H)(D)]$^{+\cdot}$ $\longrightarrow$ loss of HNC and DNC

(25)   (26)

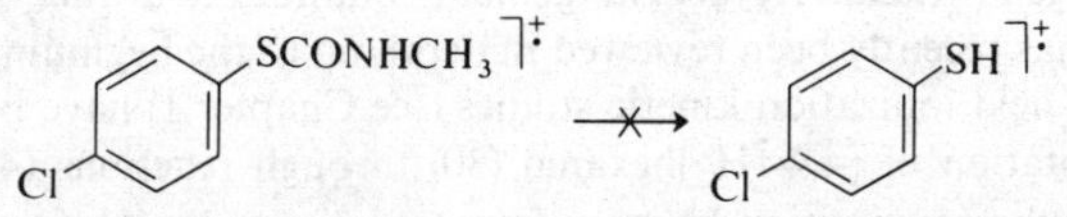

4-Cl-C$_6$H$_4$-SCONHCH$_3$]$^{+\cdot}$ $\xrightarrow{\times}$ 4-Cl-C$_6$H$_4$-SH]$^{+\cdot}$

(27)   (28)

Although metastable ion abundances show an isotope effect on the loss of CS from the molecular ion of *p*-chlorothiophenol (28), no such effect was noted for

---

[36] M. Bertrand, J. H. Beynon, and R. G. Cooks, *Internat. J. Mass Spectrometry Ion Phys.*, 1972, **9**, 346.
[37] M. Bertrand, J. H. Beynon, and R. G. Cooks, *Org. Mass Spectrometry*, 1973, **7**, 193.
[38] K. Aizawa, S. Yoshida, and N. Takahashi, *Org. Mass Spectrometry*, 1974, **9**, 470.
[39] R. G. Cooks, M. Bertrand, J. H. Beynon, M. E. Rennekamp, and D. W. Setser, *J. Amer. Chem. Soc.*, 1973, **95**, 1732.
[40] J. H. Beynon, M. Bertrand, and R. G. Cooks, *J. Amer. Chem. Soc.*, 1973, **95**, 1739.
[41] J. L. Holmes, D. McGillivray, and N. S. Isaacs, *Org. Mass Spectrometry*, 1974, **9**, 510.
[42] J. T. Bursey, M. M. Bursey, and D. G. I. Kingston, *Chem. Rev.*, 1973, **73**, 191.
[43] D. G. I. Kingston, J. T. Bursey, and M. M. Bursey, *Chem. Rev.*, 1974, **74**, 215.
[44] S. Hammerum and K. B. Tomer, *Org. Mass Spectrometry*, 1972, **6**, 1369.
[45] M. Ohashi, T. Kizaki, K. Tsujimoto, Y. Shida, and Y. Yamada, *Shitsuryo Bunseki*, 1973, **21**, 85 (*Chem. Abs.*, 1973, **79**, 3166).
[46] K. B. Tomer, S. Hammerum, and C. Djerassi, *Tetrahedron Letters*, 1973, 915.

the $[M - C_2H_3NO]^{\ddagger}$ fragment ion from S-p-chlorophenyl methylthiocarbamate (27). Thus it appears that the hydrogen atom is not transferred to sulphur in these metastable ions,[47] in contrast with similar processes, [e.g. (25) → (26)], in which the hydrogen atom appears to be transferred to the heteroatom.

Loss of C_2H_4 with hydrogen transfer from the molecular ion of S-ethylthiobenzoate has been assumed to generate the molecular ion of thiobenzoic acid, although that fragmentation is very weak (ca. 0.2% of base peak!)[48] In comparing the mass spectrum[48] with a previously published one,[49] it should be noted that the former paper adds seven small crosses to the bar graph of the mass spectrum to indicate that all seven major peaks are drawn at one-tenth of their true relative abundances. Even then the small peaks discussed in this paper[48] barely emerge above the base line.

$$\underset{(29)}{\text{pyridyl-N(CH}_3\text{)(CD}_3\text{)}}]^{\ddagger} \longrightarrow \begin{array}{l}[M - CH_2DN]^{\ddagger} \\ [M - CD_2HN]^{\ddagger}\end{array}$$

Many systems containing the part structure $-N=C-N(CH_3)_2$ fragment by loss of CH_3N. A simple methyl shift has now been shown to be, at most, only part of the mechanism because the trideuterio-compound (29) fragments by loss of both CH_2DN and CD_2HN in equal amounts.[50] Full details of long-range hydrogen rearrangements in 4-n-alkyl esters of trimellitic-anhydride have now been published,[51] and evidence for similar processes in the mass spectra of m- and p-(n-alkoxy)benzoic acids has also been found.[52]

The γ-hydrogen (McLafferty) rearrangement continues to attract considerable attention and has recently been reviewed in depth up to the beginning of 1973.[43] More recently, field ionization kinetic studies (see Chapter 1) have been applied to the fragmentation of $[4,4-^2H_2]$hexanal (30) through reactions (4) and (5).[53] It appears that the competition between fragmentations (4) and (5) is a sensitive function of time. At times of ca. 10^{-11} s, the 'rate' of formation of $[C_2H_3DO]^{\ddagger}$ is almost ten times greater than that of $[C_4H_7D]^{\ddagger}$. However, at times of the order 10^{-10} s, formation of $[C_4H_7D]^{\ddagger}$ exceeds that of $[C_2H_3DO]^{\ddagger}$ by a factor of about ten. The tendency for formation of $[C_4H_7D]^{\ddagger}$ to be favoured then continues, and at times of ca. 10^{-6} s it is at least one hundred times more intense than $[C_2H_3DO]^{\ddagger}$. In interpreting these results, it would be helpful to know whether the ions $[C_2H_3DO]^{\ddagger}$, formed after very short times $(10^{-11}$ s), decompose further.

[47] Lj. Stambolija and D. Stefanović, *Org. Mass Spectrometry*, 1973, 7, 1415.
[48] K. B. Tomer and C. Djerassi, *Org. Mass Spectrometry*, 1973, 7, 771.
[49] T. W. Bentley and R. A. W. Johnstone, *J. Chem. Soc.* (*B*), 1971, 1804.
[50] D. L. von Minden, J. G. Liehr, M. H. Wilson, and J. A. McCloskey, *J. Org. Chem.*, 1974, 39, 285.
[51] S. Meyerson, I. Puskas, and E. K. Fields, *J. Amer. Chem. Soc.*, 1973, 95, 6056.
[52] M. A. Winnik, C. K. Lee, and P. T. Y. Kwong, *J. Amer. Chem. Soc.*, 1974, 96, 2901.
[53] P. J. Derrick, A. M. Falick, and A. L. Burlingame, *J. Amer. Chem. Soc.*, 1974, 96, 615.

Another difficulty with the interpretation of these results is that the IP of vinyl alcohol was taken to be 9.5 eV.[53] This value is certainly not reliable and we prefer a lower one of about 9.25 eV.[54] Hence, the reaction (4) which appears to be favoured at very short reaction times may involve formation of the neutral but-1-ene (IP = 9.6 eV), and the more favoured reaction (5) at longer times may

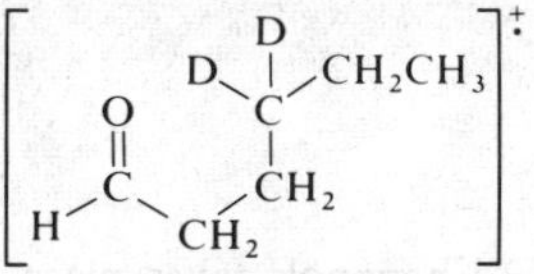

$$\begin{bmatrix} & D & \\ & D\diagdown \overset{|}{C}\diagup CH_2CH_3 \\ O & C & \\ \parallel & \diagup CH_2 & \\ H\diagup C\diagdown CH_2 & \end{bmatrix}^{\dot{+}} \rightarrow [C_2H_3DO]^{\dot{+}} + C_4H_7D \quad (4)$$
$$\rightarrow [C_4H_7D]^{\dot{+}} + C_2H_3DO \quad (5)$$

(30)

involve further hydrogen rearrangement and formation of ionized but-2-ene (IP = 9.1 eV).

The authors[53] discussed the above interpretation but preferred an explanation involving competing concerted and stepwise γ-deuterium transfer processes. The concerted process involved γ-deuterium transfer with simultaneous loss of the alkene, whereas the stepwise process involved initial γ-deuterium transfer to the oxygen atom. Although the stepwise process is the generally accepted mechanism of the McLafferty rearrangement,[43] the authors[53] assumed on the basis of 'chemical intuition' that the concerted reaction represented the faster process leading to $[C_2H_3DO]^{\dot{+}}$. The reverse argument is often applied to explain why rearrangements compete more successfully with simple cleavage reactions at low electron-beam energies, and it is difficult to see why the transition state for the completely concerted reaction should be 'looser' (*i.e.* of higher entropy) than that for the γ-hydrogen transfer.

Other recent aspects of the McLafferty rearrangement include a study of bicyclic ketones[55] and an investigation of the mechanism of tautomerization of enolic ions.[56]

Field ionization studies of deuteriated n-hexanols show that loss of H_2O (or HDO) occurs both from C-3, apparently *via* a five-membered transition state, and from C-4, apparently *via* a six-membered transition state. At short times (*ca.* 10^{-10} s) the five- and the six-membered transition states are involved to about equal extents, and at the very highest energies available by FI the five-membered transition state appears to be favoured over the six.[57] Again, this very interesting technique has produced results which challenge existing views.

If a stable carbonium ion is formed, it has been suggested that loss of H_2O may proceed by consecutive loss of OH and an adjacent H, *e.g.* 1,2-elimination from the molecular ion of the bicyclic alcohol (31).[58] However, in a paper where other

[54] T. W. Bentley and R. A. W. Johnstone, *Adv. Phys. Org. Chem.*, 1970, **8**, 242.
[55] J. D. Henion and D. G. I. Kingston, *J. Amer. Chem. Soc.*, 1974, **96**, 2532.
[56] G. Eadon, *Org. Mass Spectrometry*, 1973, **7**, 1345.
[57] P. J. Derrick, A. M. Falick, and A. L. Burlingame, *J. Amer. Chem. Soc.*, 1973, **95**, 437.
[58] D. R. Dimmel and J. M. Seipenbusch, *J. Amer. Chem. Soc.*, 1972, **94**, 6211.

1,2-eliminations were discussed, it was mentioned that this interpretation may be incorrect.[59] 1,1-Eliminations have also been discussed.[60—62]

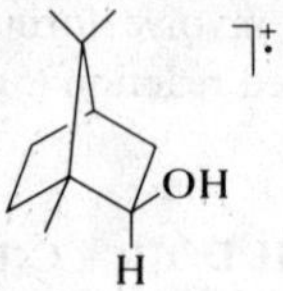

$$[M - OH]^+ \xrightarrow{?} [M - OH - H]^{+\cdot}$$

(31)

Loss of H_2O from a series of bicyclo[3,3,1]nonanols takes place in a substantially stereospecific manner; the position of abstraction of the hydrogen atom by the hydroxy-group was related to the proximity effects in the ground-state conformations.[63]

Carbon Scrambling.—As carbon scrambling in molecular ions of aromatic compounds has been the subject of many detailed studies, it is important to note that up to 32% scrambling may occur when the labelled 1-methylcyclohexene (32) is dehydrogenated to toluene (33).[64] Clearly, 'unambiguous' syntheses will have to be carefully checked by ^{13}C n.m.r. spectroscopy to determine the position(s) of the label.

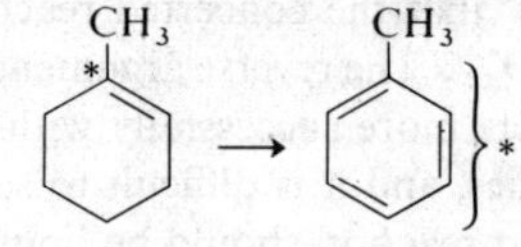

(32)   (33)

The mechanism by which the positional identity of the carbon atoms is lost in $[C_7H_7]^+$ ions has had to be modified on several occasions as a result of ^{13}C-labelling studies. Isomerization of benzyl cations to form tropylium cations (or the corresponding set of equilibrating cations[65]) does not appear to occur exclusively by random or specific insertion of the benzylic carbon into the benzene ring, because [2,6-$^{13}C_2$]toluene (34) fragments to yield some unlabelled $[C_5H_5]^+$ and some $[^{13}C_2H_2]^{+\cdot}$. Similar results have recently been obtained for the tropylium salt (35),[66] and for doubly-labelled cycloheptatriene,[67] which suggest that

---

[59] R. Robbiani and J. Seibl, *Org. Mass Spectrometry*, 1973, **7**, 1153.
[60] A. Kalir, E. Sali, and A. Mandelbaum, *J.C.S. Perkin II*, 1972, 2262.
[61] S. Meyerson and G. J. Karabatsos, *Org. Mass Spectrometry*, 1974, **8**, 289.
[62] T. A. Molenaar-Langeveld and N. M. M. Nibbering, *Org. Mass Spectrometry*, 1974, **9**, 257.
[63] J. Cable, J. K. Macleod, M. R. Vegar, and R. J. Wells, *Org. Mass Spectrometry*, 1973, **7**, 1137.
[64] J. L. Marshall, D. E. Miiller, and A. M. Ihrig, *Tetrahedron Letters*, 1973, 3491.
[65] T. W. Bentley and R. A. W. Johnstone, *Adv. Phys. Org. Chem.*, 1970, **8**, 197.
[66] A. Siegel, *J. Amer. Chem. Soc.*, 1974, **96**, 1251.
[67] R. A. Davidson and P. S. Skell, *J. Amer. Chem. Soc.*, 1973, **95**, 6843.

the near total loss of positional identity of the carbon atoms in toluene may not occur solely, if at all, in the process leading to formation of $[C_7H_7]^+$ ions but rather, as we pointed out earlier,[65] after their formation. Parallel studies of the loss of positional identity in phenylcarbene, (36) ⇌ (37) have also been discussed.[68] Carbenes, like molecular ions in mass spectrometry, are electron deficient and may behave similarly.

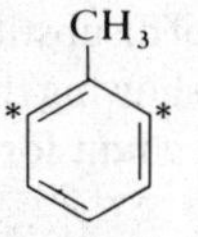

(34) 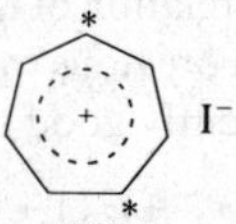   (35) 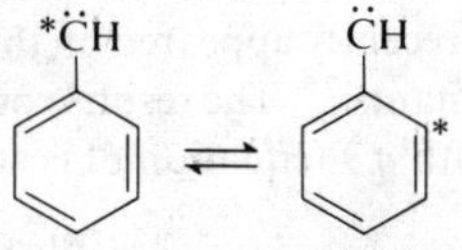   (36)   (37)

Related studies of the labelled cycloheptatriene (38) suggest that both the molecular ion and the $[M - H]^+$ ion fragment in the ion source by loss of $H^{13}CN$, whereas $H^{12}CN$ is eliminated from longer-lived ions.[69] The perfluorinated toluene (39) fragments to $[C_6F_6]^{+\cdot}$ (62% of label retained), $[C_6F_5]^+$ (76% of label retained), and $[CF_3]^+$ (82% of label retained).[70]

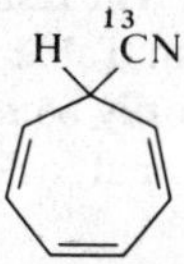

(38) 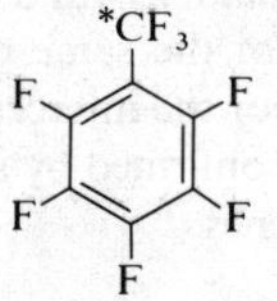   (39)

The molecular ions of diphenyl ether (16) and diphenyl sulphide, specifically labelled with ^{13}C, undergo substituent isomerization prior to the loss of CO and CS, respectively. Although the process is not specific, loss of CH_3 from diphenyl sulphide involves preferential loss of C-1.[71] Other investigations of scrambling include an examination of the mass spectrum of isobutene using both D- and ^{13}C-labelling,[72] and a similar study of $[C_3H_8N]^+$ ions.[73]

---

[68] W. D. Crow and M. N. Paddon-Row, *J. Amer. Chem. Soc.*, 1972, **94**, 4746; see also W. E. Billups, L. P. Lin, and W. Y. Chow, *ibid.*, 1974, **96**, 4026, and refs. therein.
[69] A. Venema, N. M. M. Nibbering, and Th. J. de Boer, *Org. Mass Spectrometry*, 1972, **6**, 675; see also C. Köppel, H. Schwarz, and F. Bohlmann, *Org. Mass Spectrometry*, 1974, **9**, 332, and F. Bohlmann, C. Köppel, B. Müller, H. Schwarz, and P. Weyerstahl, *Tetrahedron*, 1974, **30**, 1011, and previous papers in this series.
[70] M. I. Gorfinkel, T. D. Petrova, V. E. Platonov, and I. S. Isaev, *Zhur. Org. Khim.*, 1973, **9**, 867 (*Chem. Abs.*, 1973, **79**, 41534).
[71] J. B. Henion and D. G. I. Kingston, *J. Amer. Chem. Soc.*, 1973, **95**, 8358.
[72] M. S.-H. Lin and A. G. Harrison, *Canad. J. Chem.*, 1974, **52**, 1813; see also J. L. Holmes, *Org. Mass Spectrometry*, 1974, **8**, 247.
[73] G. Cum, G. Romeo, and N. Uccella, *Org. Mass Spectrometry*, 1974, **9**, 365; see also G. S. d'Alcontres, G. Cum, and N. Uccella, *Org. Mass Spectrometry*, 1973, **7**, 1173.

4 Substituent Effects

Linear Free Energy Relationships.—Since the number of papers on this topic continues to decline, recent reviews well summarize the current state of the art.[4,74] There is now general agreement that the ionization potential is a dominant factor in correlations of ionic abundances with Hammett σ-constants, because it influences the number of ions with insufficient energy to decompose and is a component of the activation energy (AP − IP) for decomposition. Further studies have recently appeared on the correlation of IP's of disubstituted benzenes with σ^+-constants.[75] The results confirm previous work showing that the correlation of IP with σ^+ (equation 6) is sufficiently good to account for the Hammett

$$(IP)_{1,2} = k(\sigma_1^+ + \sigma_2^+) + C \tag{6}$$

correlations observed during mass spectrometric fragmentation. In equation (6) the subscripts 1 and 2 refer to the substituents in the benzene ring, and k and C are constants.

However, this recent work[75] makes additional assertions which are, at best, difficult to justify. It was suggested that the values of k and C in equation (6) were constants for various series of disubstituted benzenes. This conclusion was assured by drawing the best straight line through all the points for series of disubstituted nitro-, methoxy-, and chloro-benzenes. The results of a more detailed statistical analysis of the same data are shown in Table 1, which shows that neither the slopes nor the intercepts may satisfactorily be regarded as constants. This conclusion is confirmed by a more extensive analysis of the data, including trisubstituted benzenes.[76]

Table 1 *Correlation of ionization potentials of substituted benzenes with $\Sigma \sigma_p^{+a}$*

Series of substituted benzenes	Slope	Intercept /eV	Correlation coefficient	Reference to source of data
p-Disubstituted nitro-	0.89 ± 0.07	9.08 ± 0.06	0.971	b
p-Disubstituted methoxy-	0.64 ± 0.06	8.97 ± 0.07	0.961	c
p-Disubstituted chloro-	0.88 ± 0.05	9.14 ± 0.03	0.988	d
All three above	0.80 ± 0.03	9.14 ± 0.03	0.984	b–d
All three above[e]	0.77	9.1		
Monosubstituted	0.97 ± 0.11	9.33 ± 0.08	0.929	f
Monosubstituted	0.83 ± 0.08	9.50 ± 0.06	0.954	g

(a) $\Sigma \sigma_p^+$ is the sum of the σ_p^+ constants for each substituent: *cf.* equation (6); (b) P. Brown, *Org. Mass Spectrometry*, 1970, **4**, 533; (c) P. Brown, *ibid.*, p. 519; (d) P. Brown, *ibid.*, 1970, **3**, 639; (e) average value given in ref. 75; (f) vertical IP, determined by photoelectron spectroscopy, taken from A. D. Baker, D. P. May, and D. W. Turner, *J. Chem. Soc. (B)*, 1968, 22; (g) electron impact values, excluding substituents COMe, CMe$_3$, CO$_2$Me, Et, and CHO, taken from I. Howe and D. H. Williams, *J. Amer. Chem. Soc.*, 1969, **91**, 7137.

[74] M. M. Bursey, 'Advances in Liner Free Energy Relationships', ed. J. Shorter, Plenum, 1972.
[75] F. Benoit, *Org. Mass Spectrometry*, 1972, **6**, 1289, 1377.
[76] H. W. Gibson, *Canad. J. Chem.*, 1973, **51**, 3065.

It should also be noted that the correlation of the IP's of monosubstituted benzenes with σ^+, while sufficient to explain the rather scattered results obtained for Hammett correlations in mass spectrometry, is not particularly good as shown by the correlation coefficients in Table 1; some substituents, e.g. I and OH, deviate markedly.[76,77] A PMO treatment has also been applied to the calculation of the IP's of substituted benzenes.[78]

Parallel studies of the correlation of IP's of aliphatic compounds with σ^* or σ_I have been carried out. The latest paper, in a series containing fifteen very short papers, shows that the first IP's of n-alkanes are a linear function of $\Sigma\sigma_I$.[79]

Related aspects are discussed in refs. 80 and 81, Section 6 of this Chapter, and also Chapters 1, 3, and 7. Neighbouring group effects in mass spectrometry have also been discussed.[82—84]

Retro-Diels–Alder Reactions.—The *cis*-isomer of the rather complex derivative of cyclohexene (40) fragments by the retro-Diels–Alder reaction (rDA), whereas similar fragmentation of the *trans*-isomer appears to be negligible.[85] A possible explanation is that the rDA is stereospecific, concerted, and follows thermal selection rules. According to this hypothesis,[85] there is no rearrangement to a common structure before fragmentation.

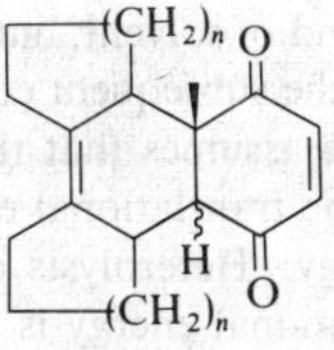

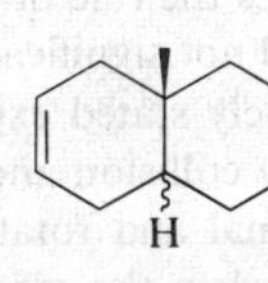

However, there is now evidence[86] for a previous suggestion that hydrogen rearrangement may occur prior to fragmentation,[87] and this may depend on the stereochemistry. Also in less complex systems e.g. (41), there appears to be no stereospecificity in the rDA.[88] Thus, at least in molecules such as (41), the rDA appears to be stepwise, possibly accompanied by hydrogen rearrangement.[86,88] In these circumstances, one should be cautious about making mechanistic

[77] A. Streitwieser, jun., *Progr. Phys. Org. Chem.*, 1963, **1**, 1.
[78] R. A. W. Johnstone and F. A. Mellon, *J.C.S. Faraday II*, 1973, 36.
[79] H. F. Widing and L. S. Levitt, *Tetrahedron*, 1974, **30**, 611.
[80] H. Kuschel and H.-F. Grützmacher, *Org. Mass Spectrometry*, 1974, **9**, 395, 403.
[81] T. W. Bentley, R. A. W. Johnstone, and A. F. Neville, *J.C.S. Perkin I*, 1973, 449.
[82] R. G. Cooks, R. N. McDonald, P. T. Cranor, H. E. Petty, and N. E. Wolfe, *J. Org. Chem.*, 1973, **38**, 1114.
[83] J. K. Kim, M. C. Findlay, W. G. Henderson, and M. C. Caserio, *J. Amer. Chem. Soc.*, 1973, **95**, 2184.
[84] H. Schwarz and F. Bohlmann, *Org. Mass Spectrometry*, 1974, **9**, 283.
[85] A. Karpati, A. Rave, J. Deutsch, and A. Mandelbaum, *J. Amer. Chem. Soc.*, 1973, **95**, 4244.
[86] P. J. Derrick, A. M. Falick, and A. L. Burlingame, *J. Amer. Chem. Soc.*, 1972, **94**, 6794.
[87] T. W. Bentley and R. A. W. Johnstone, *Adv. Phys. Org. Chem.*, 1970, **8**, 257.
[88] S. Hammerum and C. Djerassi, *J. Amer. Chem. Soc.*, 1973, **95**, 5806.

generalizations from a few observations because in neutral molecules, for example in the Cope rearrangement of biallyl derivatives, the reaction mechanism appears to be markedly dependent on the substitution pattern.[89]

From a study of the rDA of variously substituted steroids it was concluded that it was unnecessary to invoke quasi-thermal decomposition to explain the fragmentation behaviour.[90]

5 Mechanistic Interpretations

Comparison with Carbonium Ion Chemistry.—The physical organic chemistry of carbonium ion reactions in solution is one of the best-established areas of the subject, whereas mechanistic interpretations of organic mass spectrometry have only developed over the past ten years. The following discussion provides a comparison of these two areas in the hope that mechanistic interpretations in mass spectrometry may be placed in clearer perspective.

To simplify the comparison it will be assumed for the purposes of this discussion that S_N1 reactions of alkyl halides (RX) in solution occur by ionization to a free carbonium ion with ejection of an anion, as shown in equation (7):

$$R-X \rightarrow R^+ + X^- \qquad (7)$$

This equation ignores the role of ion pairs and of solvent, but inclusion of these complications would not significantly affect the subsequent conclusions.

Although it is rarely stated explicitly, one assumes that the substrate RX is initially activated by collision and part of the translational energy is then converted into vibrational and rotational energy. Heterolysis of the R—X bond presumably occurs when the excess of vibrational energy is in the appropriate degree(s) of freedom. Such reactions are regarded as endothermic and the energy surface is often represented as shown in Figure 3. Clearly this diagram is a considerable simplification. In particular, assuming that it takes a very small but finite amount of time to convert translational energy into vibrational energy and that only vibrational energy can cause heterolysis of the R—X bond,[91] then the vibrationally-activated molecules [R—X]* may undergo a thermoneutral heterolysis to the transition state [R⋯X]‡, followed by collapse to the products R^+ and X^- as shown in equation (8):

$$[R-X] \xrightarrow{(1)} [R-X]^* \xrightarrow{(2)} [R\cdots X]^{\ddagger} \xrightarrow{(3)} R^+ + X^- \qquad (8)$$

$$[R-X]^{+\cdot} \xrightarrow{(1)} [R-X]^{*+\cdot} \xrightarrow{(2)} [R\cdots X]^{\ddagger +\cdot} \xrightarrow{(3)} R^+ + X\cdot \qquad (9)$$

A comparable mass spectrometric fragmentation is shown in equation (9). The most obvious difference between equations (8) and (9) is that mass spectrometric processes begin with ions or ion–radicals and in the above example the product X· is formed instead of X^-. Also the mechanism of formation of vibrationally activated species is different; in mass spectrometry, collisions between

---

[89] M. J. S. Dewar and L. E. Wade, *J. Amer. Chem. Soc.*, 1973, **95**, 290.
[90] H. Budzikiewicz and M. Linscheid, *Org. Mass Spectrometry*, 1974, **9**, 88.
[91] R. A. Firestone and B. G. Christensen, *Tetrahedron Letters*, 1973, 389.

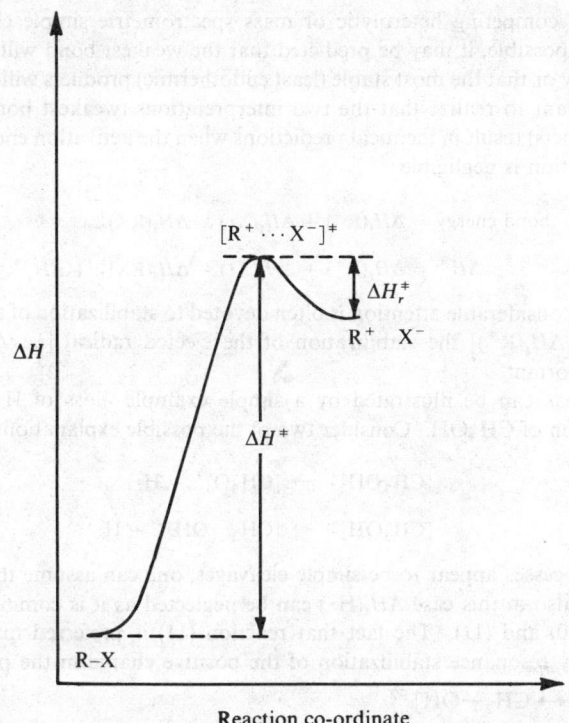

Figure 3 *Section through an energy surface for heterolytic bond cleavage* (*equation* 7)

ions and molecules are rare, and vibrational activation is thought to be caused by radiationless decay from higher electronically excited states of the ions and by collision of the neutral molecules with the walls of the ion source before ionization occurs.

However, although the detailed mechanisms of formation of vibrationally activated species [R—X]* and [R—X]*⁺· are different, there are close similarities between the two subsequent processes—equations (8) and (9). By distributing their excess vibrational energies in the appropriate degrees of freedom, both [R—X]* and [R—X]*⁺· may form the appropriate activated complex for decomposition, [R···X]⁺, *without external influences* in a thermoneutral reaction. Thus, according to this simple analysis, the key steps (2) in both equations (8) and (9) are essentially identical. Also step (3), collapse of the activated complex to products, probably occurs in a similar manner for both reactions, with the activation energy for the reverse reaction ($\Delta H_r^{\neq}$, Figure 3) being very small. Both reactions may be regarded as endothermic, with the transition state resembling the final product (*cf.* Hammond's postulate).

If several competing heterolytic or mass spectrometric simple cleavage reactions are possible, it may be predicted that the weakest bond will be broken more readily or that the most stable (least endothermic) products will be formed. It is important to realize that the two interpretations (weakest bond or most stable products) result in identical predictions when the activation energy for the reverse reaction is negligible:

$$\text{bond energy} = \Delta H_f(R^+) + \Delta H_f(X\cdot) - \Delta H_f(RX)^{+\cdot}$$

$$\Delta H^{\neq} = \Delta H_f(R^+) + \Delta H_f(X\cdot) - \Delta H_f(RX)^{+\cdot} + \Delta H_r^{\neq}$$

Also, while considerable attention is often devoted to stabilization of the positive charge [*i.e.* $\Delta H_f(R^+)$] the stabilization of the ejected radical [*i.e.* $\Delta H_f(R\cdot)$] is equally important.

These ideas can be illustrated by a simple example—loss of H$\cdot$ from the molecular ion of CH_3OH. Consider two of the possible explanations:

$$[CH_3OH]^{+\cdot} \rightarrow [CH_3O]^+ + H\cdot \qquad (10)$$

$$[CH_3OH]^{+\cdot} \rightarrow [CH_2-OH]^+ + H\cdot \qquad (11)$$

As both processes appear to be simple cleavages, one can assume that $\Delta H_r^{\neq}$ is negligible; also in this case $\Delta H_f(H\cdot)$ can be neglected as it is common to both reactions (10) and (11). The fact that reaction (11) is preferred may then be explained by resonance stabilization of the positive charge in the product ion $[CH_2\overset{+}{=}OH \leftrightarrow \overset{+}{C}H_2-OH]$.[92]

A more frequent interpretation in organic mass spectrometry involves formation of the charge-localized canonical form of the molecular ion (*e.g.* $CH_3\overset{+\cdot}{O}H$), which 'triggers' fragmentation. This is discussed in the next section, but two other comparisons with carbonium ion chemistry are worthwhile. Firstly, in the thousands of papers published on S_N1 reactions over the past forty years, the reviewer is not aware of any author who found it necessary to postulate that heterolysis of the R—X bond was triggered from a charge-localized canonical form of RX (*i.e.* R^+X^-). Secondly, the difficulties in questioning an accepted theory (the context of the following case is that of non-classical ions) have been expressed:[93]

'Many theoretical proposals are advanced and subjected to experimental test. The uncovering of unfavourable experimental data results in the ready revision of the proposal or its withdrawal.

'The situation is quite different for a theory that has reached the stage of wide acceptance. Such a theory appears in the textbooks, is taught to a new generation of students, and becomes accepted as a fixed part of the fabric of established

---

[92] See also J. M. Wilson, MTP International Review of Science, Organic Chemistry, Series 1, Vol. 1, p. 42.
[93] H. C. Brown, 'Boranes in Organic Chemistry', Cornell University Press, London, 1972, p. 141.

chemical theory. It becomes exceedingly difficult to question such an accepted concept.'

Charge Localization.—Undoubtedly the most firmly entrenched dogma in mechanistic interpretations of organic mass spectrometry is the necessity of charge or radical sites to cause ('trigger') fragmentation. This view continues to be held, even though it appears to be inconsistent with the enormous numbers of neutral molecules which undergo bond cleavages. Furthermore, it is occasionally stated that charge localization is 'not inconsistent' with the quasi-equilibrium theory of mass spectrometry. One cannot avoid the impression that the aim of the exercise is to create a selective kind of self-consistency within mass spectrometry, with the inference that other areas of chemistry are different and therefore not relevant to theories of mass spectrometry; this attitude was thought to have disappeared with the vital force theory of organic chemistry. Alternative explanations to charge localization have been given in the previous section and elsewhere.[3,92,94]

One of the most serious criticisms of charge localization is that the conventional lines, representing two-electron bonds and used to show the structures of organic molecules, should only be used for all-electron (total energy) properties such as stereochemistry, heat of formation, and dipole moment. Such pictures of electron distributions are not suitable for explaining one-electron properties such as ionization potentials. A simple example, methane, illustrates this point because, using the localized bond model (42), one would predict that all eight outer-shell valence electrons in methane were identical and thus only one energy level should be observed by photoelectron spectroscopy. Experimentally, at least two ionization levels may be detected, as expected from MO calculations.[95]

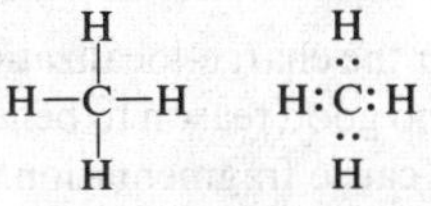

$$\begin{array}{cc} H & H \\ | & \cdot\cdot \\ H-C-H & H:C:H \\ | & \cdot\cdot \\ H & H \end{array}$$

(42)

Perhaps the inadequacies of localized bond pictures of organic structures are under-emphasized in undergraduate chemistry courses. It is particularly unfortunate to make statements such as 'ionization of the ethane molecule may result in loss of an electron from the C—C bond'.[96] These authors[96] go so far as to suggest that this interpretation may be more precise than expressing the ion as $[C_2H_6]^{+\cdot}$; actually there are at least five energy levels for ionization of the outer-shell electrons of ethane.[95]

As molecules containing heteroatoms with non-bonded electrons often show specific fragmentations, the concept of charge localization is most frequently invoked in such cases. However, as Krier, Lorquet, and Berlingin have

[94] T. W. Bentley, R. A. W. Johnstone, and F. A. Mellon, *J. Chem. Soc. (B)*, 1971, 1800.
[95] J. N. Murrell and W. Schmidt, *J.C.S. Faraday II*, 1972, **68**, 1709.
[96] M. M. Campbell and O. Runquist, *J. Chem. Educ.*, 1972, **49**, 104.

emphasized,[97] so-called non-bonding electrons are not localized in the neutral molecule and *electronic reorganization takes place after ionization*. For ionization of ammonia, they calculated that *ca.* 30% of the charge came from the nitrogen atom and 23% from each hydrogen.[97] Similar calculations have been carried out for steroidal hormones,[98] and similar conclusions were reached.

Another serious difficulty for the charge-localization-*cum*-triggering hypothesis is the experimental evidence that a molecule may fragment with an appearance potential less than the 'isolated' ionization potential of the functional group involved.[94,99,100] Also, consecutive McLafferty rearrangements in 'isolated' dibutyrophenones[101,102] can be explained readily if the idea that a charge or radical site is a prerequisite for fragmentation is discarded, and other evidence can be explained similarly.[103,104] (Studies of apparent charge localizations are reported in refs. 105—108.)

A fundamental question, raised by a referee and discussed by Krier, Lorquet, and Berlingin[97] and also raised during correspondence about our paper,[94] is whether structures such as (43) and (44) represent, to a first approximation, isolated electronic states or canonical forms of the same molecular ion. Thus, the comparison with carbonium ion chemistry, discussed above,[93] becomes more acute; the subtlety of the distinction may be comparable with that in the classical–non-classical ion controversy. If after ionization and *subsequent electron reorganization*, there are several low-lying electronic states of the molecular ion, they may be readily accessible and much less widely separated in energy than is suggested by the 'isolated' ionization potentials of the separate functional groups, *i.e.* X and Y in (43) and (44).

$$\overset{+\cdot}{X}-(CH_2)_n-\overset{\cdot\cdot}{Y} \qquad \overset{\cdot\cdot}{X}-(CH_2)_n-\overset{+\cdot}{Y}$$

(43) (44)

In summary, the evidence for the charge-localization-*cum*-triggering hypothesis has been demolished; there is no good reason to believe that proximity of a charge or radical site is necessary to cause fragmentation. However, one cannot deny the usefulness of the idea as a simple, infinitely flexible, mechanistic picture, which should be used cautiously. 'By keeping in mind its limitations, they (organic chemists) might be less inclined to derive a hasty conclusion in a tricky case.'[97]

[97] C. Krier, J. C. Lorquet, and A. Berlingin, *Org. Mass Spectrometry*, 1974, **8**, 387.
[98] G. Loew, M. Chadwick, and D. Smith, *Org. Mass Spectrometry*, 1973, **7**, 1241.
[99] J. L. Occolowitz, *Austral. J. Chem.*, 1967, **20**, 2387.
[100] See also M. Meot-Ner and F. H. Field, *J. Amer. Chem. Soc.*, 1973, **95**, 7207.
[101] A. Mandelbaum and K. Biemann, *J. Amer. Chem. Soc.*, 1968, **90**, 2975.
[102] A. Tatematsu, S. Naga, H. Sakurai, T. Goto, and H. Nakata, *Bull. Chem. Soc. Japan*, 1971, **44**, 3450.
[103] I. Lengyel, D. B. Uliss, and R. V. Mark, *J. Org. Chem.*, 1970, **35**, 4077.
[104] K. G. Das and S. K. Saudi, *Indian J. Chem.*, 1973, **11**, 443, 552.
[105] C. J. Fortin, M. Forest, C. Vaziri, D. Gravel, and Y. Rousseau, *Canad. J. Chem.*, 1973, **51**, 3445.
[106] C. J. Fortin and Y. Rousseau, *Canad. J. Chem.*, 1973, **51**, 3457.
[107] G. Wood, A. M. Falick, and A. L. Burlingame, *Org. Mass Spectrometry*, 1974, **8**, 279.
[108] E. Rouvier and A. Cambon, *Org. Mass Spectrometry*, 1974, **9**, 453.

6 Molecular Orbital Calculations

Ion Structures.—Because of the high cost of accurate MO calculations, the problem of calculating structures and energies of organic molecules and ions involves simplifying assumptions. It is unfortunate that many so-called *ab initio* calculations of the energies of carbonium ions have been carried out by *assuming* key aspects of the structures, such as symmetry, bond lengths, and bond angles. As no experimental data are available for carbonium ion structures, the input data into these sophisticated computer programmes were often no more than a guess; the colourful American expression, 'garbage in, garbage out' emphasizes the importance of presenting computers with soundly-based input data.

A geometry search, in which the geometric co-ordinates of the molecule are systematically varied to obtain the structure of minimum energy, is now recognized to be an important adjunct to the MO calculation itself. The results of two independent calculations are shown in Table 2. Although detailed discussion of these results is beyond the scope of this review, the convergence of experimental and theoretical data towards substantial agreement on the relative energies of simple carbonium ions is highly encouraging.

The likelihood of rapid developments in this field is highlighted by the fact that the computer programmes used for results given in Table 2, published in

Table 2 *Heats of formation of carbonium ions*

		$\Delta H_f(R^+)$/kcal mol^{-1}		
R	Expt.[a]	Expt.[b]	MINDO/2[d]	ab initio[f]
CH_3	261	260	276	—
C_2H_5	219	219	225	219[g]
n-C_3H_7	208	209	216[e]	209
s-C_3H_7	192	190	191	190
n-C_4H_9	201	218	209[e]	202
i-C_4H_9	199	—	—	199
s-C_4H_9	183	183	183	181
t-C_4H_9	167	170	171	163
t-C_5H_{11}	159	163[c]	—	152

(a) F. P. Lossing and G. P. Semeluk, *Canad. J. Chem.*, 1970, **48**, 955; (b) 'best values' from J. L. Franklin, J. G. Dillard, H. M. Rosenstock, J. T. Herron, K. Draxl, and F. H. Field, 'Ionization Potentials, Appearance Potentials, and Heats of Formation of Gaseous Positive Ions, National Bureau of Standards, Washington D. C., 1969; (c) from J. J. Solomon and F. H. Field, *J. Amer. Chem. Soc.*, 1973, **95**, 4483, assuming that $\Delta H_f(t-C_4H_9^+) = 167$ kcal mol^{-1}; (d) N. Bodor, M. J. S. Dewar, and D. H. Lo, *J. Amer. Chem. Soc.*, 1972, **94**, 5303; (e) geometry search yielded edge-protonated cyclopropanes—energy for classical ion calculated by fixing the terminal $\angle$CCC to 110.5°; (f) L. Radom, J. A. Pople, and P. von R. Schleyer, *J. Amer. Chem. Soc.*, 1972, **94**, 5935; standard bond lengths assumed from optimized C—C and C—H bond lengths of ethyl cation; (g) assumed to calibrate the relative energy scale.

1972, have now been superceded. Dewar's group has recently published calculations using MINDO/3,[109] and Pople's group uses a more extended basis set (6 − 31G*) to calculate energies, employing the optimized geometries calculated using the minimal STO − 3G basis set.[110] Even more sophisticated calculations including electron correlation (the tendency of electrons to correlate their motions in such a way as to reduce their mutual repulsions) have been carried out on $[CH_3]^+$,[111] and $[CH_5]^+$ [112] (see also Chapter 1).

There is much less agreement between independent methods when the structures and energies of cyclic hydrocarbons and ions are calculated. Many *ab initio* calculations appear to underestimate the stabilities of small rings,[110] whereas semi-empirical calculations such as MINDO overestimate their stabilities.[109] This is well illustrated by the various calculations of the cyclic isomers of $[C_5H_5]^+$;[109,113—115] mass spectrometric evidence for a cyclic structure for $[C_5H_5]^+$ has been deduced from charge-exchange experiments.[116]

Reactions.—In agreement with *ab initio* MO calculations,[117] hydride transfer occurs very readily in the ethyl cation.[118,119] For scrambling in the molecular ion of ethane, INDO calculations suggest that the activated complex equivalent to diborane is more stable than one in which hydrogens on the same side of the molecule exchange positions.[120]

The ability of a nitrogen atom to direct mass spectrometric fragmentation processes is well known. It has been calculated that the stabilization of a carbonium ion by an adjacent NH_2 group is 89 kcal mol^{-1}, of which 69 kcal mol^{-1} was due to resonance stabilization with 23 kcal mol^{-1} due to inductive stabilization. In contrast, an OH group was calculated to provide resonance stabilization of 48 kcal mol^{-1}, but slight destabilization by the inductive effect. Apparently the ion $[CH_2F]^+$ experiences counterbalancing resonance and inductive effects, each of *ca.* 30 kcal mol^{-1}.[121]

Primary mass spectral cracking patterns of saturated hydrocarbons and their derivatives have been treated extensively by MO methods. The basic assumption is that the distribution of charge in the highest occupied MO of the molecular

[109] M. J. S. Dewar and R. C. Haddon, *J. Amer. Chem. Soc.*, 1973, **95**, 5836.
[110] P. C. Hariharan, L. Radom, J. A. Pople, and P. von R. Schleyer, *J. Amer. Chem. Soc.*, 1974, **96**, 599.
[111] F. Driessler, R. Ahlrichs, V. Staemmler, and W. Kutzelnigg, *Theor. Chim. Acta*, 1973, **30**, 315.
[112] V. Dyczmons and W. Kutzelnigg, *Theor. Chim. Acta*, 1974, **33**, 239.
[113] H. Kollmar, H. O. Smith, and P. von R. Schleyer, *J. Amer. Chem. Soc.*, 1973, **95**, 5834.
[114] W. J. Hehre and P. von R. Schleyer, *J. Amer. Chem. Soc.*, 1973, **95**, 5837.
[115] M. J. S. Dewar and R. C. Haddon, *J. Amer. Chem. Soc.*, 1974, **96**, 255.
[116] T. Keough, J. H. Beynon, and R. G. Cooks, *J. Amer. Chem. Soc.*, 1973, **95**, 1695.
[117] P. C. Hariharan, W. A. Lathan, and J. A. Pople, *Chem. Phys. Letters*, 1972, **14**, 385.
[118] P. Ausloos, R. E. Rebbert, L. W. Sieck, and T. O. Tiernan, *J. Amer. Chem. Soc.*, 1972, **94**, 8939.
[119] J. H. Vorachek, G. G. Meisels, R. A. Geanangel, and R. H. Emmel, *J. Amer. Chem. Soc.*, 1973, **95**, 4078.
[120] C. E. Parker, M. M. Bursey, and L. G. Pedersen, *Org. Mass Spectrometry*, 1973, **7**, 1077.
[121] P. A. Kollman, W. F. Trager, S. Rothenberg, and J. E. Williams, *J. Amer. Chem. Soc.*, 1973, **95**, 458.

ion determines the fragmentation pattern.[122] Additional research on this approach has been carried out,[123,124] and further criticisms have been voiced.[97,125]

More sophisticated interpretations, based on INDO semi-empirical MO theory, have been applied to substituent effects in molecular ions of aromatic compounds. Perhaps the most celebrated fragmentation in this field is decomposition of molecular ions of substituted acetophenones, and the effect of *p*-phenyl substitution has now been re-examined.[126] The benzene rings of both (45) and (46) were calculated to be more nearly coplanar in the +1 molecular ion than in the neutral molecule, which accounts for the apparent increase in the resonance stabilization during mass spectrometric fragmentation.

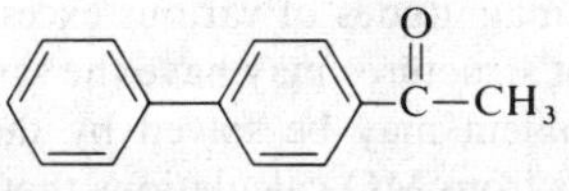

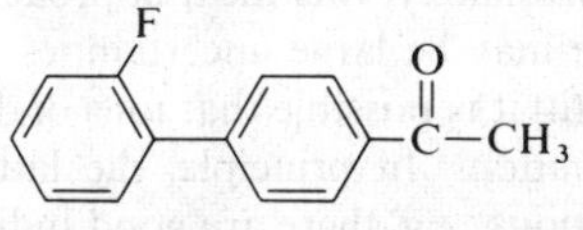

Unusual results have been obtained in the mass spectra of *o*- and *p*-halogeno-substituted phenylacetates (47) and phenetoles (48).[127] INDO calculations are consistent with bond formation between the carbonyl oxygen and *ortho*-halogen in the molecular ions of *o*-fluoro- and *o*-chloro-phenylacetates, whereas phenetoles do not undergo such cyclizations; this accounts for the observed variations in frequency factors for decomposition.[128] INDO calculations[129] are also consistent with previous interpretations of hydrogen scrambling in molecular ions of substituted benzoic acids.[130]

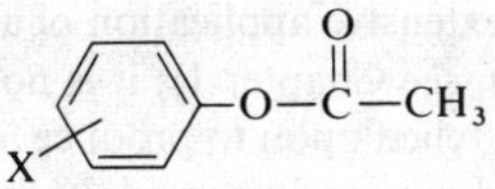

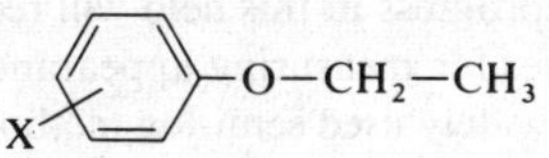

---

[122] W. C. Herndon, *Progr. Phys. Org. Chem.*, 1972, **9**, p. 136.
[123] M. Yamamoto, I. Fujita, M. Itoh, and K. Hirota, *Bull. Chem. Soc. Japan*, 1972, **45**, 3520.
[124] S. Tajima, Y. Niwa, N. Wasada, and T. Tsuchiya, *Bull. Chem. Soc. Japan*, 1972, **45**, 1250.
[125] S. Ikuta, K. Yoshihara, and T. Shiokawa, *Bull. Chem. Soc. Japan*, 1973, **46**, 3648.
[126] C. E. Twine, jun., C. E. Parker, and M. M. Bursey, *Org. Mass Spectrometery*, 1973, **7**, 1179; see also C. E. Twine, jun. and M. M. Bursey, *J. Org. Chem.*, 1974, **39**, 1290.
[127] M. M. Bursey and C. E. Parker, *Tetrahedron Letters*, 1972, 2211.
[128] C. E. Parker, J. R. Hass, M. M. Bursey, and L. G. Pedersen, *Org. Mass Spectrometry*, 1973, **7**, 1189.
[129] C. E. Parker, M. M. Bursey, and L. G. Pedersen, *Org. Mass Spectrometry*, 1974, **9**, 204.
[130] F. Benoit, *Org. Mass Spectrometry*, 1973, **7**, 295.

7 Energetics of Fragmentation

Although detailed discussion of this topic is given in Chapter 1, a brief discussion is presented here with emphasis on the relevance of energetics of fragmentation to studies of ion structures. Ideally, one would like to calculate the heat of formation of an ion M^+, formed by ionization of a neutral molecule M (or radical) of known structure. In this way one measures a property of an ion of the same known structure, as the ionization process of the neutral molecule M occurs faster than movement of nuclei. This heat of formation can then be compared with the heat of formation of fragment ions calculated from measurements of appearance potentials. If the two heats of formation agree, one might deduce that the two ions have the same structure.

Unfortunately, this ideal approach is difficult to realize in practice because: (i) there may be large uncertainties in the magnitudes of various excess energy terms; (ii) it is possible that ions of different structures may have the same heats of formation. In principle, the latter problem may be solved by theoretical calculations; *e.g.* there are good indications from MO calculations that isomers of $[s\text{-}C_3H_7]^+$ are of different and significantly higher energies.[110] However, the reliability of current MO technology for larger odd-electron systems is doubtful.

There has been some progress in accounting for excess energies. It now appears that the so-called 'kinetic shift', the excess energy required to drive a fragmentation fast enough for the ion to decompose in the mass spectrometer ($k \geqslant 10^{+5}\,\text{s}^{-1}$) is at least partly an artefact of the method of measuring appearance potentials.[131] Further progress in this field will require more extensive application of accurate techniques for measuring appearance potentials (see Chapter 1); it is now clear that the widely used semi-log method cannot be relied upon to produce *accurate* appearance potential data, although the method may yield *reproducible* results.

Another major difficulty is estimating the activation energy for the reverse reaction, particularly for rearrangement processes; it seems unlikely that it will be negligible as in simple cleavage processes or that it will all appear as excess of kinetic energy in the products of fragmentation.[39,40]

8 Ion Cyclotron Resonance (ICR)

Detailed discussion of ICR is presented in Chapter 3, and the following remarks emphasize only aspects relevant to studies of ion structures.

The logic and the drawbacks of ICR studies of ion structures are the same as those for collisional activation spectra (Section 2). It can be logically deduced that two ions have different structures if they undergo different ion–molecule reactions; unfortunately, as the ions are long-lived and therefore have had ample time for rearrangements, one cannot be sure what these different structures are.

If two ions undergo the same ion–molecule reactions, nothing can be reliably deduced, as there are three possible explanations: (i) the two ions have the same structure; (ii) the two ions may have different structures but the ion–molecule

---

[131] T. W. Bentley, R. A. W. Johnstone, and B. N. McMaster, *J.C.S. Chem. Comm.*, 1973, 510.

reactions of the two isomeric structures are identical (*e.g.* differently substituted benzenes); (iii) the two ions may have different structures but the ion–molecule collision causes one structure to isomerize to the other. It is difficult to assess the likelihood of the latter possibility; however, the ion–molecule reaction:

$$[C_3H_6]^+ + ND_3 \rightarrow C_2H_4D + [CH_2D_2]^+ \qquad (12)$$

appears to be selective, indicating that hydrogen scrambling did not occur in the collision complex.[132]

Two examples illustrate the technique's utility in studies of ion structures; in in the loss of C_2H_3O from the molecular ion of phenyl actate (47; X = H).[133] and loss of C_2H_4 from the molecular ion of phenetole (48; X = H),[134] both reactions appear to yield fragment ions having the structure of phenol.

As the residence time of ions studied by ICR is much longer (*ca.* 10^{-3} s) than in conventional mass spectrometry ($\leqslant 10^{-6}$ s), additional information about kinetics and energetics can be obtained, *e.g.* the 'kinetic shift' of appearance potentials should be reduced.[135] Also it appears that the molecular ion of hexa-1,5-diyne, which is isomeric with benzene, undergoes an exceptionally slow loss of H having a remarkably low activation energy; fragmentation appears to be preceded by hydrogen scrambling.[136]

Other mechanistically useful information can be obtained using trapped-ion ICR techniques to determine gas-phase equlibria, and hence the relative stabilities of carbonium ions.[137] From proton affinities, heats of formation can be obtained, *e.g.* using $\Delta H_f(CH_3CH\overset{+}{=}OH)$ = 143 kcal mol^{-1}, ΔH_f for protonated ethylene oxide (3) is calculated to be 169 kcal mol^{-1}.[138] In comparing the isomers of $[C_2H_5O]^+$, (1)—(3), (3) appears to be the least stable, and has recently been identified from its characteristic ion–molecule reaction with PH_3.[139]

9 Comparison with Other Chemical Processes

Since there are many mass spectrometric fragmentation processes and many other chemical processes, it is statistically likely that there will be some apparent similarities between mass spectrometry and other chemical processes. However, in the case of flash vacuum thermolysis[140] (very low pressure pyrolysis[141]) of bicyclic aromatic molecules, apparent similarities occur quite frequently. Both mass spectrometry and thermolytic reactions of cyclic aromatic systems involve

---

[132] M. L. Gross, *J. Amer. Chem. Soc.*, 1972, **94**, 3744.
[133] K. B. Tomer and C. Djerassi, *Tetrahedron*, 1973, **29**, 3491.
[134] N. M. M. Nibbering, *Tetrahedron*, 1973, **29**, 385.
[135] M. L. Gross, *Org. Mass Spectrometry*, 1972, **6**, 827.
[136] M. L. Gross and R. J. Aerni, *J. Amer. Chem. Soc.*, 1973, **95**, 7875.
[137] T. B. McMahon, R. J. Blint, D. P. Ridge, and J. L. Beauchamp, *J. Amer. Chem. Soc.*, 1972, **94**, 8934.
[138] T. B. McMahon and J. L. Beauchamp, *Rev. Sci. Instr.*, 1972, **43**, 509.
[139] R. H. Staley, R. R. Corderman, M. S. Foster, and J. L. Beauchamp, *J. Amer. Chem. Soc.*, 1974, **96**, 1260.
[140] H. J. Hageman and U. E. Wiersum, *Chem. in Britain*, 1973, **9**, 206.
[141] D. M. Golden, G. N. Spokes, and S. W. Benson, *Angew. Chem. Internat. Edn.*, 1973, **12**, 534.

endothermic fragmentation with ejection of stable neutral fragments (*e.g.* CO, CO_2, SO, SO_2, N_2) and the heats of formation of the neutral aromatic fragments may closely parallel those for the corresponding molecular ions. Hence, as discussed in Section 5, the bond strengths in the non-aromatic ring of the bicyclic aromatic may closely parallel those in the corresponding molecular ions.

This topic has recently been reviewed[140,141,141a] and later work is reported in refs. 11 and 142–148. In many published thermolyses and mass spectrometric fragmentations, the neutral species SO appears to be lost. The ease of loss of SO and CO_2 is emphasized by the hypothetical equilibrium:

$$SO + CO_2 \rightleftharpoons SO_2 + CO \qquad (13)$$
$$\Delta H_f \quad -15 \quad -94 \quad -71 \quad -27 \quad \text{kcal mol}^{-1}$$

in which, according to standard heats of formation, the left-hand side would be favoured by 11 kcal mol^{-1}. Thus it is reasonable that loss of SO may occur rather than SO_2.

As discussed above (Section 4), the preferred stereochemistry of the retro-Diels–Alder reaction appears to be in agreement with the selection rules for a thermal reaction.[85] Also the electron-deficient behaviour of carbenes may be similar to that of carbonium ions (see Section 3).

10 Other Approaches

As well as the previously reported $E/2$ and $2E$ mass spectra, preliminary results have now been published on $-E$ mass spectra, in which positively charged ions are converted into negatively charged ions in the analyser of the mass spectrometer.[116] Relative abundances of ions in $-E$ mass spectra appear to reflect expected stabilities of negative ions, *e.g.* $[C_5H_5]^-$ and $[C_6H_5O]^-$ ions are abundant in the $-E$ mass spectra of anisole, but are not abundant in the published negative ion mass spectrum! It was assumed that $[C_5H_5]^-$ probably has a cyclic structure, that the positive ion $[C_5H_5]^+$ from which it was derived would not have had time to rearrange, and therefore that the ion $[C_5H_5]^+$ in the normal mass spectrum of anisole has a cyclic structure.[116] However, before the above conclusion can be accepted, the assumptions used to derive it must be tested.

A similar argument has been applied to the negative form of the molecular ion of benzene. Assuming that $[C_6H_6]^{2+}$ has an acyclic structure and/or that charge

---

[141a] R. C. Dougherty, *Fortschr. Chem. Forsch.*, 1974, **45**, 93.
[142] C. Wentrup and P. Müller, *Tetrahedron Letters*, 1973, 2915; C. Wentrup, *ibid.*, p. 2919.
[143] D. C. DeJongh and M. L. Thomson, *J. Org. Chem.*, 1972, **37**, 1135.
[144] M. L. Thomson and D. C. DeJongh, *Can. J. Chem.*, 1973, **51**, 3313.
[145] D. C. DeJongh and M. L. Thomson, *J. Org. Chem.*, 1973, **38**, 1356.
[146] H. Heaney and A. P. Price, *J.C.S. Perkin I*, 1972, 2911.
[147] U. E. Wiersum and T. Nieuwenhuis, *Tetrahedron Letters*, 1973, 2581.
[148] J. C. Tou and C.-S. Wang, *J. Organometallic Chem.*, 1972, **34**, 141.

exchange yields acyclic $[C_6H_6]^{+\cdot}$, it has been argued that the reactive form of the ion $[C_6H_6]^{+\cdot}$ of benzene formed by direct ionization must be acyclic.[149] Such arguments could be simply syllogisms.

Molecular ions containing two α-naphthyl groups, $[N-(CH_2)_n-N]^{+\cdot}$ where N = α-naphthyl, appear to have unusual stability and it has been proposed that the ions cyclize.[150]

[149] T. Keough, T. Ast, J. H. Beynon, and R. G. Cooks, *Org. Mass Spectrometry*, 1973, 7, 245.
[150] P. Caluwe, K. Shimada, and M. Szwarc, *J. Amer. Chem. Soc.*, 1973, 95, 1433.

3
Alternative Methods of Ionization and Analysis

BY J. M. WILSON

1 Introduction

Since the chapter under this title in Volume 2 was written there has been a considerable increase in the use of the techniques described therein. Much of this chapter is devoted to further developments in the applications of these methods. This has resulted in considerable overlap of techniques, particularly in the field of ion–molecule reactions where the same process may be observed using Ion Cyclotron Resonance (ICR) spectrometers or using mass spectrometers of more conventional design operating at higher pressures than the ICR cell. Both Electron Impact (EI) and Photoionization (PI) are used in high-pressure mass spectrometry. The results of studies by these methods are in many cases complementary and therefore a section has been included on ion–molecule reactions, devoted to three topics on which results have been obtained using various instrumental techniques: acid–base equilibria in the gas phase, ion–solvent equilibria, and the structure and reactivity of gaseous ions.

An important recent development in the field of ion–molecule reactions and their applications to analytical chemistry has been the use of Atmospheric Pressure Ionization and Plasma Chromatography. These use the same ionization process but different methods of mass analysis. Although this development is relatively recent, it has attracted considerable attention particularly because of the inherent sensitivity of the method.

Both CI and FI now appear to be established as accepted methods for deriving otherwise unobtainable molecular weight information and reports of such determinations are not covered.

2 Photoionization

Photoion–photoelectron coincidence measurements are beginning to throw some light on the problems of unimolecular decomposition of ions. In the mass spectrum of C_2F_6 the two most abundant fragment ions are $C_2F_5^+$ (onset 15.3 eV) and CF_3^+ (onset 13.7 eV).[1] Ions formed in the first excited state at *ca.* 16 eV decompose mainly by the high-energy route to $C_2F_5^+$. This result suggests that $C_2F_6^{+\cdot}$ in its first excited state decomposes directly without any

[1] I. G. Simm, C. J. Danby, and J. H. D. Eland, *J.C.S. Chem. Comm.*, 1973, 832.

equilibration of electronic energy in contradiction to the present usage of quasi-equilibrium theory in which electronic states other than the ground state are normally ignored. By contrast $CS_2^{+\cdot}$ decomposes from both accessible states by all possible routes with a very fast energy interconversion immediately after ionization.[2]

Much of the difficulty in testing any quantitative theory of mass spectra lies in the absence of knowledge of the energy-deposition function for EI or for PI. With such a function it is possible to use a fragmentation breakdown graph to reproduce a mass spectrum. In an assessment of two methods, it was found that the first derivative of the total ionization efficiency curve gave a better match with the experimental mass spectrum than did the photoelectron spectrum.[3] In a similar examination of methods for obtaining breakdown curves for fragment ions, those from ionization efficiency curves of fragments gave a better fit than those obtained from charge-exchange experiments in tandem mass spectrometers. Both for toluene[4] and for dec-2-ene[5] it has been shown that there is more fragmentation of the molecular ion by PI than by EI. There are, however, advantages in temperature control and energy control and a number of laboratories prefer to use PI for the study of both unimolecular decompositions and ion–molecule reactions.

In a further study of the t-butylcyclohexanols both helium (21.22 eV) and hydrogen (Lyman α–10.19 eV) lamps have been used.[6] In addition to the large difference between the intensity ratios $[M^+]/[M^+ - H_2O]$ for the two isomers there is an ion of m/e 110 present in the *trans*- and absent in the *cis*-isomer. This difference can be explained if water elimination from the *trans*-isomer is mainly 1,4 and from the *cis*-isomer specifically 1,3. The 1,4 elimination product (1) could lose ethylene to form (2) but this is impossible with the 1,3-product.

(1)   (2)

Free radicals in the gas phase can be identified using Ar, Kr, and Xe resonance lamps. These allow selective ionization of species with different ionization potentials, and the ionization of radicals in the absence of fragment ions of the same mass but higher appearance potential. In the reaction of oxygen atoms with acetylene both HCCO· and CH_2 species have been identified, and steady-state

---

[2] B. Brehm, J. H. D. Eland, R. Frey, and A. Kustler, *Internat. J. Mass Spectrometry Ion Phys.*, 1973, **12**, 213.
[3] G. G. Meisels and R. H. Emmel, *Internat. J. Mass Spectrometry Ion Phys.*, 1973, **11**, 455.
[4] W. L. Stebbings and J. W. Taylor, *Internat. J. Mass Spectrometry Ion Phys.*, 1972, **9**, 471.
[5] B. M. Johnson and J. W. Taylor, *Internat. J. Mass Spectrometry Ion Phys.*, 1972/3, **10**, 1.
[6] Z. M. Akhtar, C. E. Brion, and L. D. Hall, *Org. Mass Spectrometry*, 1973, **7**, 647.

concentrations of 10^{-10} mol l^{-1} are detectable.[7] The same method has been used to detect $C_4H_8O_3$ as an intermediate in the reaction of ozone with but-2-ene.[8]

3 Negative Ion Production

Reports of negative ion mass spectra indicate that three methods are generally used: low energy electron beams, higher energy electron beams in conventional ion sources using secondary electrons for attachment, and plasma sources of slow electrons. Much of the work done with plasma sources has been described in a book recently published.[9]

Beynon has described a further method for producing negative ions, using a collision method.[10] If a collision gas (N) is introduced into the analyser of a double-focusing mass spectrometer and the electrostatic analyser field is reversed the following process may be observed:

$$M^+ + N \rightarrow M^- + N^{2+}$$

Such spectra ($-E$ spectra) can be observed if the collision gas has a relatively low double ionization potential: benzene gave good results. Ions observed are stable negative ions of the same mass as positive ions in the normal spectrum, e.g. for anisole the most abundant ions in the $-E$ spectrum are $C_6H_5O^-$ and $C_5H_5^-$. Some aromatic molecules produce very stable molecular negative ions which do not fragment appreciably. In such cases, introduction of a collision gas into the field-free region of a negative ion mass spectrometer allows the observation of collision-induced metastable ions and thus the detection of fragments. For phthalic anhydride, $[M - CO]^-$ and $[M - C_2O_3]^-$ fragments were observed by such a method.[11]

In the negative ion mass spectra of carbaboranes $[M - H]^-$ ions were observed with greater abundance when bridging hydrogen atoms were present.[12] Labelling studies in bridged compounds suggest that the bridging hydrogen is specifically involved.[13] The spectra of phosphorus ylides can be rationalized in terms of the formation of either neutral or ionic tervalent phosphorus species,[14] as shown in Scheme 1.

Nitroaromatic compounds usually exhibit negative molecular ions and fragmentation can be related to molecular structure. In the nitroacenaphthenes[15] the loss of ·OH from the molecular ion is observed when there is a saturated

---

[7] I. T. N. Jones and K. D. Bayes, *J. Amer. Chem. Soc.*, 1972, **94**, 6869.
[8] R. Atkinson, B. J. Finlayson, and J. N. Pitts, jun., *J. Amer. Chem. Soc.*, 1973, **95**, 7592.
[9] M. von Ardenne, K. Steinfelder, and R. Tümmler, 'Elektronanlagerungs-Massenspektrographie organischer Substanzen', Springer-Verlag, Berlin, 1971.
[10] T. Keough, J. H. Beynon, and R. G. Cooks, *J. Amer. Chem. Soc.*, 1973, **95**, 1695.
[11] J. H. Bowie, *J. Amer. Chem. Soc.*, 1973, **95**, 5795.
[12] T. Onak, J. Howard, and C. L. Brown, *J.C.S. Dalton*, 1973, 76.
[13] C. L. Brown, K. P. Gross, and T. Onak, *J. Amer. Chem. Soc.*, 1972, **94**, 8055.
[14] R. G. Alexander, D. B. Bigley, and J. F. J. Todd, *Org. Mass Spectrometry*, 1973, **7**, 963; *J.C.S. Chem. Comm.*, 1972, 553.
[15] J. F. J. Todd, R. B. Turner, B. C. Webb, and C. H. J. Wells, *J.C.S. Perkin II*, 1973, 1167.

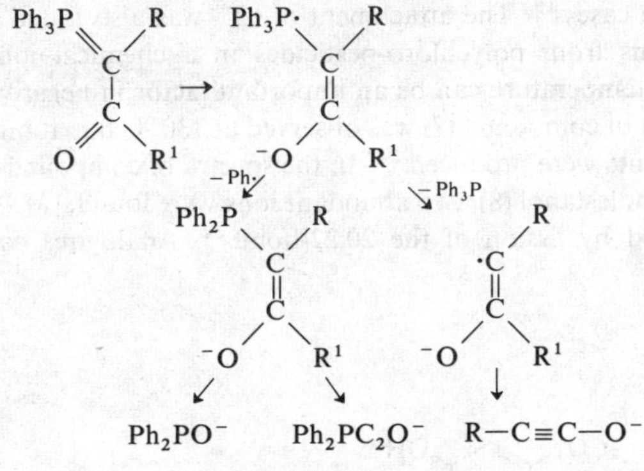

Scheme 1

C–H bond *ortho* to a nitro-group. Of the nine isomeric nitrophenyl-1,2,4-triazoles, only (3) and (4) do not have distinctive negative ion spectra.[16] The spectrum of trinitrotoluene varies considerably with electron energy. At 2 eV there are abundant ions at high mass but at 6 eV the majority of ions are NO_2^-. It is

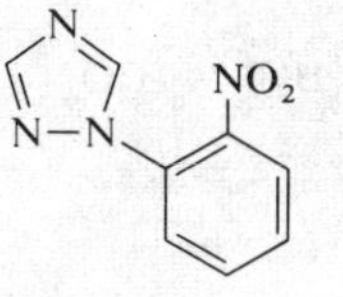

(3) 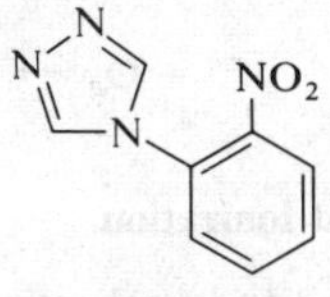   (4)

suggested that sensitivity at this electron energy is good enough to allow detection of TNT at 20 °C (equilibrium vapour pressure at this temperature is 1 p.p.m.), and that monitoring of m/e 46 by this method would be an efficient method of detection of nitro-organic explosives in airline baggage.[17]

The negative ion spectra of some trienes and tetraenes appear to be much more sensitive to double-bond position than are positive ion spectra. Ions of m/e 67 and 81 are present in the spectrum of compound (5) but absent in that of (6); it appears that fragmentation of vinylic and allylic bonds is allowed but not fission of the central bond of a diene.[18] Electron-attachment mass spectra of 20-chloropregnane derivatives show, in addition to fragment ions, an $[M + Cl]^-$

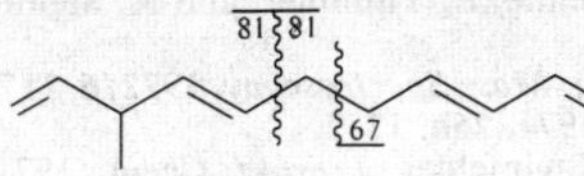

(5) 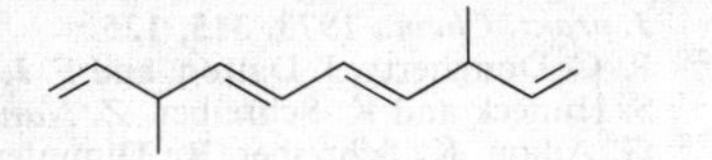   (6)

[16] J. H. Bowie and A. J. Blackman, *Austral. J. Chem.*, 1972, **25**, 1335.
[17] J. Yinon, H. G. Boettger, and W. P. Weber, *Analyt. Chem.*, 1972, **44**, 2235.
[18] V. I. Khvostenko, V. P. Yur'ev, I. Kh. Aminev, G. A. Tolstikov, and S. R. Rafikov, *Proc. Acad. Sci. (USSR)*, 1972, **202**, 128.

ion in some cases.[19] The attachment of Cl^- was also found in the analysis of negative ions from polychloro-pesticides in a chemical ionization source.[20] The source temperature can be an important factor in negative ion production. The $M^{+\cdot}$ ion of compound (7) was observed at 130 °C but at higher temperatures only fragments were produced.[21] In the spectra of compounds related to 22,26-epi-imino-cholestanol (8), two abundant ions were found, $[M - H]^-$ and a fragment formed by fission of the 20,22-bond.[22] Analogous compounds with a

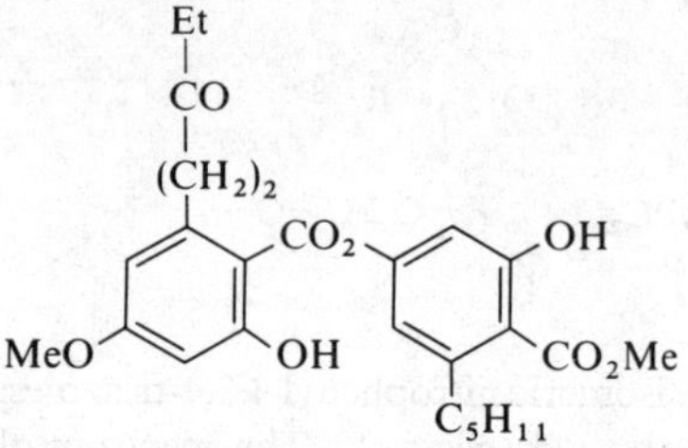

(7)

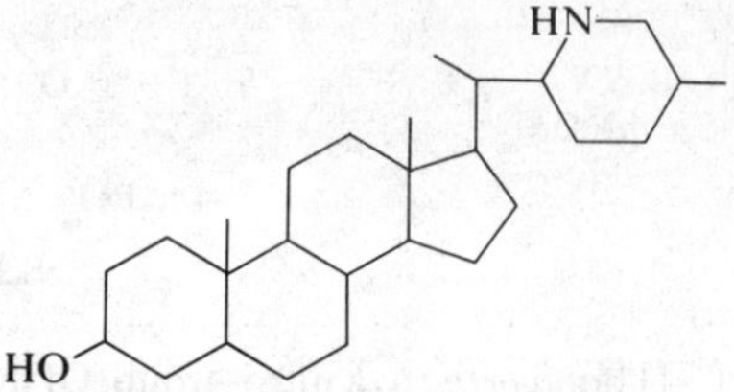

(8)

$C{=}N$ bond exhibit an $[M + 14]^-$ ion which is claimed to be formed by the ion–molecule reaction

$$M^- + O_2 \rightarrow [M + O - H_2]^- + H_2O$$

4 Field Ionization

The general method and applications have been reviewed in a book published in 1971.[23] Since there is currently much interest in the metastable ions in field ion mass spectra, an attempt has been made to increase the sensitivity of detection of small ion currents by using a scintillation detector;[24] this detector also makes possible the discrimination of ions of differing kinetic energy. FI sources have been coupled with quadrupole mass analysers. This can be done using an ion-focusing lens to decelerate the ions and refocus them.[25] Deceleration of ions can however lead to secondary electron production by collisions at lens electrodes but this source of noise can be avoided by using a suitable deflection angle between source and analyser with electrostatic deflection of ions.[26] Pulse techniques can be used for surface studies. If a voltage pulse is applied to the cathode, this

---

[19] G. Adam, D. Voigt, K. Schreiber, M. von Ardenne, R. Tümmler, and K. Steinfelder, *J. prakt. Chem.*, 1973, **315**, 125.
[20] R. C. Dougherty, J. Dalton, and F. J. Biros, *Org. Mass Spectroscopy*, 1972, **6**, 1171.
[21] S. Huneck and K. Schreiber, *Z. Naturforsch.*, 1971, **26b**, 1357.
[22] G. Adam, K. Schreiber, R. Tümmler, and K. Steinfelder, *J. prakt. Chem.*, 1971, **313**, 1051.
[23] H. D. Beckey, 'Field Ionization Mass Spectrometry', Pergamon Press, Oxford, 1971.
[24] G. G. Wanless, *Internat. J. Mass Spectrometry Ion Phys.*, 1972, **10**, 85.
[25] A. Martin and J. Block, *Messtechnik*, 1973, **81**, 149.
[26] H. J. Heinen, Ch. Hötzel, and H. D. Beckey, *Internat. J. Mass Spectrometry Ion Phys.*, 1974, **13**, 55.

increased field causes a desorption pulse in the spectrum, and an increase in the current of stongly adsorbed ions can be observed.[27]

The study of surfaces of field ion emitters has been reviewed.[28] Benzene ions are usually strongly adsorbed and the presence of such adsorbed ions increases the field strength at the emitter. This can be shown by the enhancement of field-induced dissociation of n-heptane in the presence of benzene, and the increase in the relative and absolute abundance of $C_3H_6^{2+}$ from acetone.[29] It is considered that doubly-charged ion production at the emitter is due to the presence of unsaturated linkages at the surface of organic micro-needles. This hypothesis is shown in Scheme 2 which also gives a mechanism for the suppression of double ionization by water molecules.[30]

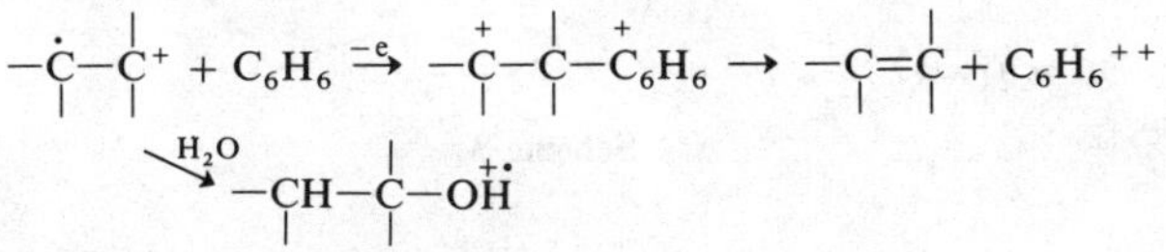

$$-\overset{\cdot}{\underset{|}{C}}-\underset{|}{C^+} + C_6H_6 \xrightarrow{-e} -\underset{|}{\overset{+}{C}}-\underset{|}{\overset{+}{C}}-C_6H_6 \rightarrow -\underset{|}{C}=\underset{|}{C} + C_6H_6^{++}$$

$$\xrightarrow{H_2O} -CH-\underset{|}{\overset{+\cdot}{C}}-OH$$

Scheme 2

When an emitter is used which incorporates particles of zeolite, the spectrum of t-butyl alcohol shows that dehydration has taken place to a considerable extent. This is a catalytic process and the abundance of unsaturated ions produced by it is less sensitive to field changes than the $[M - CH_3]^+$ ions produced by field-induced dissociation.[31] Zeolite surfaces promote the formation of the Ph_3C^+ ion from Ph_3C-X molecules (where $X = Br, Cl,$ or OH) whereas at platinum surfaces the molecular ions are formed. Desorption of the carbonium ions requires high temperatures but low fields and the ionization process is considered to be reaction with Lewis acid sites in the zeolite surface. It can also be enhanced by adsorbed tetracyanoethylene.[32]

Field ion spectra of mixtures of acetone with ether or with hydrocarbons may contain ions which are not present in spectra of the pure components. These are considered to be products of reaction of adsorbed acetone ions and Scheme 3 is a possible explanation of their formation.[33]

Graphite emitters have been used for field ionization and molecular ions of various compounds could be observed using fairly low fields.[34]

The study of metastable ions formed in the FI source continues to produce interesting results. Such unimolecular processes take place after desorption and should be relevant to the study of ions produced by electron or photon impact.

[27] F. W. Rollgen and H. D. Beckey, *Messtechnik*, 1972, **80**, 115.
[28] H. D. Beckey and F. W. Rollgen, *J. Vacuum Sci. Technol.*, 1972, **9**, 471.
[29] F. W. Rollgen and H. D. Beckey, *Ber. Bunsengesellschaft phys. Chem.*, 1972, **76**, 661.
[30] F. W. Rollgen and H. D. Beckey, *Z. phys. Chem. (Frankfurt)*, 1972, **82**, 161.
[31] M. S. Zei, *Ber. Bunsengesellschaft phys. Chem.*, 1974, **78**, 443.
[32] J. H. Block and M. S. Zei, *Surface Sci.*, 1971, **27**, 419.
[33] F. W. Rollgen and H. D. Beckey, *Ber. Bunsengesellschaft phys. Chem.*, 1971, **75**, 988.
[34] P. L. Gutshall and P. J. Bryant, *J. Vacuum Sci. Technol.*, 1972, **9**, 498.

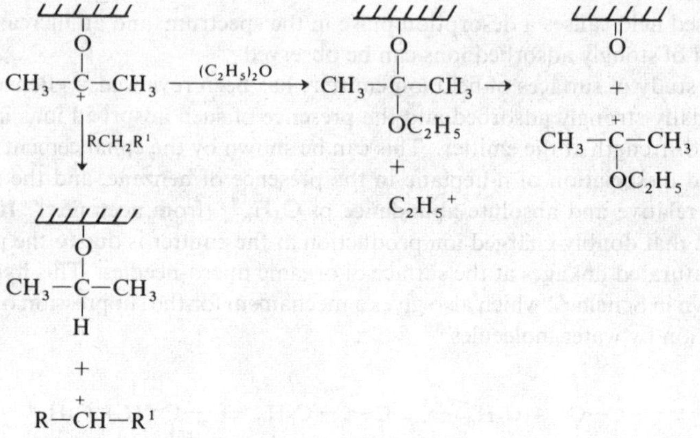

Scheme 3

Both 'fast' and 'normal' metastable ions are observed for charge-separation fragmentations of doubly charged ions. For the process

$$C_6H_6^{2+} \rightarrow C_5H_3^+ + CH_3^+$$

the kinetic energy release measured is the same as that found for the same process after electron impact.[35] The observation of 'fast' metastable ions makes it possible to study the kinetics of fragmentation processes if the appropriate source parameters are known. A theoretical expression for flight times in a source with a blade emitter has been calculated.[36] Similar calculations on ion trajectories show that at times greater than 10^{-11} s the field is independent of emitter surface irregularities.[37]

Such calculations make possible the analysis of some very fast processes. In the loss of water from n-hexanol, ions of low internal energy (slow decomposition, $k \sim 10^9 \, s^{-1}$) decompose by a 1,4 process. At higher energies ($k \sim 10^{11} \, s^{-1}$), 1,3- and 1,4-elimination are of equal importance. This is consistent with the 70 eV electron impact results. These show that the process is predominantly 1,4 for the ions which survive as $C_6H_{12}^{+\cdot}$. Ions of higher energy, decomposing by 1,3 and 1,4 processes will have an excess of energy to decompose further to C_4 and C_3 ions.[38] In the mass spectrum of cyclohexene the process

$$C_6H_{10}^{+\cdot} \rightarrow C_4H_6^{+\cdot} + C_2H_4$$

is considered to be a retro-Diels–Alder fragmentation but in the electron impact source the randomization of all hydrogen and deuterium in labelled species is

---

[35] H. D. Beckey, M. D. Migahed, and F. W. Rollgen, *Internat. J. Mass Spectrometry Ion Phys.*, 1972/3, **10**, 471.
[36] B. W. Viney, *Internat. J. Mass Spectrometry Ion Phys.*, 1972, **8**, 417.
[37] J.-P. Pfeiffer, A. M. Falick, and A. L. Burlingame, *Internat. J. Mass Spectrometry Ion Phys.*, 1973, **11**, 345.
[38] P. J. Derrick, A. M. Falick, and A. L. Burlingame, *J. Amer. Chem. Soc.*, 1973, **95**, 437.

essentially complete before fragmentation. In the FI mass spectrum of [3,3,6,6-$^2H_4$]cyclohexene the ion $C_4H_2D_4^+$ predominates at times of ca. 10^{-11} s but randomization is complete by 10^{-9} s.[39]

In the spectrum of [4,4-$^2H_2$]hexanal both products, $C_2H_3OD^{+\cdot}$ and $C_4H_7D^{+\cdot}$, of the McLafferty rearrangement are observed as ions. The enol ion predominates at times less than 10^{-10} s but at longer times the hydrocarbon ion is more abundant. Either two distinct mechanisms are involved or the intermediate (9)

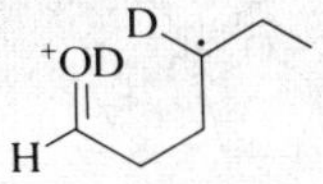

(9)

decomposes in two ways, (i) by a fast cleavage of the 2,3-bond to form $C_2H_3OD^{+\cdot}$ or (ii) with a hydrogen rearrangement to form ionized but-2-ene, which should have a lower ionization potential than the enol of acetaldehyde.[40] The latter explanation is consistent with the result of studies on labelled analogues of n-heptanal from which $C_2H_4O^{+\cdot}$ is formed without randomization by EI and by FI whereas $C_5H_{10}^{+\cdot}$ is formed with considerable randomization of hydrogens by EI but with a minor degree of randomization by FI.[41] In the FIMS of $CD_3COCD_2C_5H_{11}$, both $CD_3C(OH)CD_2^{+\cdot}$ and $[M - CD_3]^+$ are formed with no hydrogen randomization at times shorter than 7×10^{-10} s.[42]

The technique of Field Desorption (FD) continues to be extended. Anodes for this work can be activated at temperatures as high as 1200 °C and this produces an emitter surface of high chemical and mechanical stability.[43] They are stable to attack by dilute mineral acids and alkali and by fluorocarbons under FI conditions. Deposition of purine bases on such emitters from dilute HCl solution gave more intense spectra than were obtained following deposition from neutral aqueous solution. Guanine deposited from dilute NaOH solution gave a spectrum consisting of M^+, $[M + H]^+$ and $[M + Na]^+$. An emitter dipped in sodium acetate solution gave a spectrum in which the most abundant ion was Na_2OAc^+.[44] FD spectra can be used in conjunction with liquid chromatography. As yet it is an off-line combination using collected fractions; 0.001 mol l^{-1} solutions can be used.[45]

Nucleosides and nucleotides give characteristic FD spectra. From adenosine (10) the ions M^+, $[M + H]^+$, $[B + H]^+$, $[B + 2H]^+$ and S^+ are formed. Guanosine, which decomposes close to evaporation temperature, does not give an abundant FI spectrum but does give an abundant M^+ and $[M + H]^+$ by FD.

[39] P. J. Derrick, A. M. Falick, and A. L. Burlingame, *J. Amer. Chem. Soc.*, 1972, **94**, 6794.
[40] P. J. Derrick, A. M. Falick, and A. L. Burlingame, *J. Amer. Chem. Soc.*, 1974, **96**, 616.
[41] P. Brown and C. Fenselau, *Org. Mass Spectrometry*, 1973, **7**, 305.
[42] P. J. Derrick, A. M. Falick, S. Lewis, and A. L. Burlingame, *Org. Mass Spectrometry*, 1973, **7**, 890.
[43] H.-R. Schulten and H. D. Beckey, *Messtechnik*, 1973, **81**, 121.
[44] H.-R. Schulten and H. D. Beckey, *Org. Mass Spectrometry*, 1972, **6**, 885.
[45] H.-R. Schulten and H. D. Beckey, *J. Chromatog.*, 1973, **83**, 315.

In the FD spectrum of 5'-adenosine monophosphate (11) the ions $[M + H]^+$, $[M - PO_3H_2]^+$, $[B + H]^+$, and $[B + H_2]^+$ can be recognized thus allowing the three constituent units of the molecule to be identified by mass.[46] The disodium

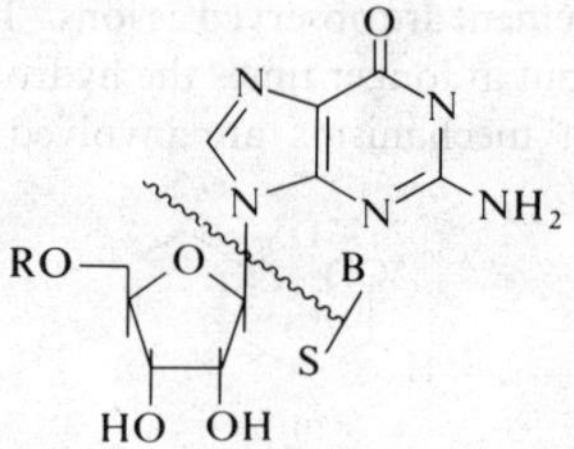

(10) R = H
(11) R = PO_3H_2

salts of glucose-6-phosphate and the deoxyfluoroglucose-6-phosphates all have abundant $[M + H]^+$ ions in their FD spectra. Ions found at lower mass correspond to Na–H exchange followed by protonation, some inorganic salt ions and a low abundance of organic fragments such as $C_2H_5O_2^+$ and $C_2H_4FO^+$.[47]

An analysis of some antibiotics by FD and other methods demonstrates the value of this technique for compounds which are both involatile and thermally unstable.[48] In the mass spectra of novobiocin (12) a molecular ion is the base peak

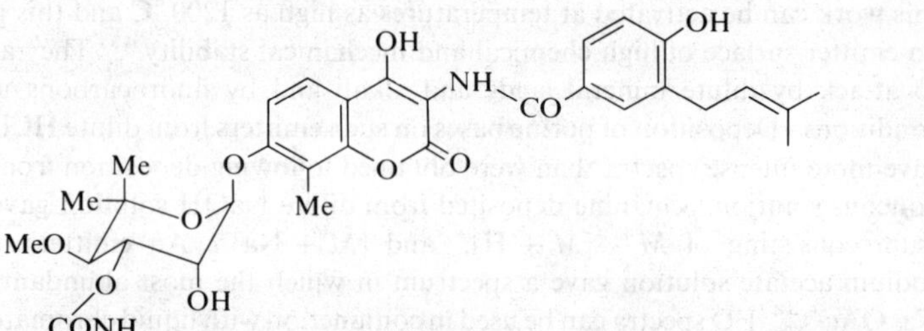

in FD and is completely missing in the EI and CI (methane and isobutane) spectra. The next most abundant ion is $[M + H]^+$ which is also absent from the CI spectra. When the emitter is heated by passing 15 mA current through it, the molecular ion is still the base peak but fragments corresponding to fission of the glycosidic bond can be observed. One of the most spectacular applications was to the analysis of filipin which is a mixture, the principal component being filipin III (13). The FD spectrum contains peaks corresponding to the molecular

[46] H.-R. Schulten and H. D. Beckey, *Org. Mass Spectrometry*, 1973, **7**, 861.
[47] H.-R. Schulten, H. D. Beckey, E. M. Bessell, H. B. Foster, M. Jarman, and J. H. Westwood, *J.C.S. Chem. Comm.*, 1973, 416.
[48] K. L. Rinehart, jun., J. C. Cook, jun., K. H. Maurer, and U. Rapp, *J. Antibiotics*, 1974, **27**, 1.

ions of (13) and its contaminants, filipins I, II, and IV which are deoxy-analogues. The most abundant ions of this mixture are the $[M - H_2O]^{+\cdot}$ ions. By combining the intensities of $M^{+\cdot}$ and $[M - H_2O]^{+\cdot}$ peaks for each constituent the composition of the mixture can be calculated; results are in close agreement with those obtained by chromatographic separation.

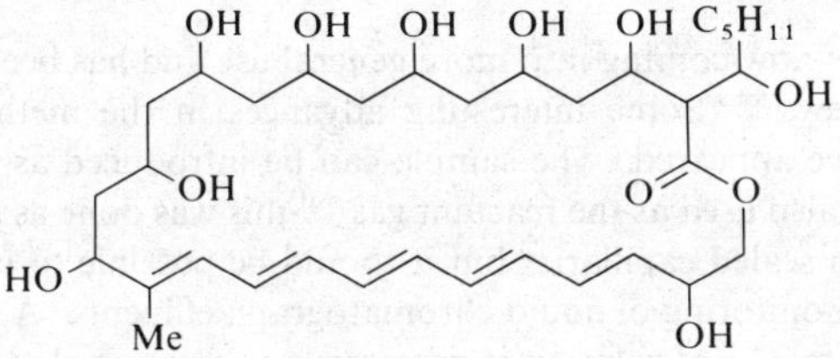

(13)

A further refinement of the field desorption technique has been achieved by indirect heating of the sample by near i.r. irradiation instead of passing current through the emitter wire. The irradiation method produces a more uniform temperature distribution over the surfaces of the micro-needles and results in more abundant molecular ions and less abundant fragment ions in the spectra of unstable compounds.[49] Molecular ions or protonated molecular ions are observed for all the naturally occurring amino-acids by FD, including cystine and arginine, which give difficulty with EI or with CI.[50] The FD spectra of aryl glucosides show abundant molecular ions and the intensity ratios of fragment ions allow a much better distinction to be made between anomers than is possible with FI or with EI.[51]

FI is generally useful in the analysis of complex mixtures. Over 200 lines were observed in the FIMS of a bacterial pyrolysate and had elemental formulae assigned.[52] The authors hope to use the pyrolysis–m.s. method for taxonomic studies. The g.c.–FIMS combination has been used to check the homogeneity of peaks in g.c.[53] Samples of amber from different sources give characteristically different FI spectra. The EI spectra[54] are dominated by common fragment ions and do not allow this differentiation. Bisarene chromium complexes give FI spectra consisting of essentially the molecular ions only and this method can be used for the quantitative analysis of mixtures of such compounds.[55] The FI spectra of volatile inorganic hydrides at a platinum knife-edge emitter have been reported. Protonated molecular ions were observed for hydrides of Group IV,

[49] H. U. Winkler and H. D. Beckey, *Org. Mass Spectrometry*, 1973, **7**, 1007.
[50] H. U. Winkler and H. D. Beckey, *Org. Mass Spectrometry*, 1972, **6**, 655.
[51] W. D. Lehmann, H.-R. Schulten, and H. D. Beckey, *Org. Mass Spectrometry*, 1973, **7**, 1103.
[52] H.-R. Schulten, H. D. Beckey, H. L. C. Meuzelaar, and A. J. H. Boerboom, *Analyt. Chem.*, 1973, **45**, 191.
[53] D. E. Games, A. H. Jackson, and D. S. Millington, *Tetrahedron Letters*, 1973, 3063.
[54] H. J. Eichhoff and G. Mischer, *Z. Naturforsch.*, 1972, **27b**, 380.
[55] I. L. Agafonov and V. I. Faerman, *Zhur. analit. Khim.*, 1972, **27**, 196.

V, and VI elements.[56] Earlier FI studies of aniline–nitrobenzene mixtures suggested that the mixed dimer ion observed might be a charge-transfer complex[57] but later studies with methylated anilines suggest that a hydrogen bond must be involved.[58]

5 Chemical Ionization

This technique is now coming into more general use and has been the subject of two recent reviews.[59] Some interesting advances in the methods of sample introduction have appeared. The sample can be introduced as a solution in a solvent which is also used as the reactant gas;[60] this was done as a batch process with solutions in sealed capillaries but it should be possible to use this method for continuous monitoring of liquid chromatograph effluents. A flow rate of the order of 0.01 ml min^{-1} of solution is necessary to maintain chemical ionization conditions. Ammonia, methanol, water, and hexane are all suitable solvents. As reactant gases, water and methanol have the slight disadvantage that they form oligomeric ions which may interfere with the ions that are characteristic of the sample. Direct exposure of the sample to the ion plasma in the source allows analysis of samples at much lower temperatures than normally used, and spectra may be obtained from unprotected peptides.[61] The limits of this system appear to have been met with the peptides (Ala)$_6$ and Pro-Phe-His-Leu-Leu, for which it was possible to record a spectrum in a few seconds, before appreciable decomposition had taken place.

A three-dimensional quadrupole ion storage trap has been described which allows ions to be stored in the source region for up to 4 ms.[62] Using this source and a quadrupole mass analyser, CI spectra can be observed at pressures as low as 10^{-4} Torr.[63] Negative-ion spectra are obtained in good yield from organochlorine insecticides in a CI source using methane. The reactant ions are Cl$^-$, H$_2$OCl$^-$, and HCl$_2^-$ from the sample. The ions produced are $[M + Cl]^-$, $[M + ClO]^-$ and $[M + HCl_2]^-$. The fragment ions observed are due to elimination of peripheral functional groups rather than fragmentation of the carbon skeleton.[20]

An obvious development has been the use of different gases for selective ionization or to control the amount of fragmentation in the spectrum. Isobutane CI is used for analysis of nitrogen isotopes in ammonia. The t-C$_4$H$_9^+$ ion will protonate ammonia but not water so H$_3$O$^+$ cannot interfere in the analysis of

[56] G. G. Devyatykh, I. L. Agafonov, and V. I. Faerman, *Russ. J. Inorg. Chem.*, 1971, **16**, 1689.
[57] B. E. Job and W. R. Patterson, in 'Some Newer Methods in Structural Chemistry', ed. R. Bonnet and J. G. Davis, United Trades Press, London, 1967, p. 129.
[58] H. D. Beckey and M. D. Migahead, *Z. Naturforsch.*, 1971, **26a**, 2063.
[59] F. H. Field, in 'Mass Spectrometry', ed. A. Maccoll, M.T.P. International Review of Science, Series 1, Vol. 5, Butterworths, London, 1972.
[60] M. A. Baldwin and F. W. McLafferty, *Org. Mass Spectrometry*, 1973, **7**, 1111.
[61] M. A. Baldwin and F. W. McLafferty, *Org. Mass Spectrometry*, 1973, **7**, 1353.
[62] G. Lawson, R. F. Bonner, and J. F. J. Todd, *J. Sci. Instr.*, 1973, **6**, 357.
[63] R. F. Bonner, G. Lawson, and J. F. J. Todd, *J.C.S. Chem. Comm.*, 1972, 1179.

$^{15}NH_4^+$.[64] Ammonia is a useful reactant gas, giving the ions NH_4^+ and $[NH_3]_nH^+$. It will protonate $\alpha\beta$-unsaturated ketones but not saturated ketones, which appear in the spectrum as $[M + NH_4]^+$.[65] In an analysis of mono- and di-saccharides by CI, ammonia and methane were used separately. The ammonia CI spectrum gives molecular weight information only while with methane typical fragments can be observed.[66] Tetramethylsilane forms the ions $SiMe_3^+$ and Me_3Si–$SiMe_4^+$; these react with a variety of compounds with n- or π-electrons to give $[M + SiMe_3]^+$ ions. It, like ammonia, is a useful gas for the analysis of compounds which do not give a stable protonated molecular ion with methane or isobutane,[67] e.g. n-tetradecanol. A more detailed study of the reactions of the ions of $SiMe_4$ with the trimethylsilyl ether of n-tetradecanol shows that the process is not simple. When $C_{14}H_{29}OSi(CD_3)_3$ is used, the $[M + SiMe_3]^+$ ion is formed mainly without exchange but there is a significant proportion of ions in which there has been exchange of a methyl or of a trimethylsilyl group during the reaction.[68] It can be explained in terms of the decomposition of the collision complex (14). Trimethylsilyl ethers of secondary alcohols react

$$\left[\begin{array}{c} R \\ Me_3Si \cdots \overset{|}{\underset{Si\cdots CH_3}{O}} \cdots Si(CD_3)_3 \\ Me_3 \end{array} \right]^+ \longrightarrow \begin{array}{c} R \\ \underset{SiMe_3}{\overset{+}{O}} \diagdown SiMe_3 \end{array} + MeSi(CD_3)_3$$

(14)

differently and the most abundant ion observed is $[M + H]^+$. The determination of active hydrogen is readily performed in a CI source using D_2O as reactant gas.[69] Complete exchange of –OH, –SH and –NH takes place during or preceding ionization by $(D_2O)_nD^+$; only slight exchange is observed in the α-positions of aldehydes and ketones.

Nitric oxide has interesting properties as a reactant gas. In addition to charge exchange,[70] it condenses with ketones, esters, and carboxylic acids to give $[M + NO]^+$ and abstracts hydride from aldehydes and ethers.[71] The use of more than one reactant gas can provide confirmatory evidence of structure. The identification of 1-methylcyclohex-1-en-3-ol produced by Douglas Fir beetles prior to mating was achieved using both helium and methane CI spectra.[72]

Mixed gases provide an advantage for work with gas chromatography. If the major component can be an inert gas or nitrogen then the second component can

---

[64] C. V. Lundeen, A. S. Viscomi, and F. H. Field, *Analyt. Chem.*, 1973, **45**, 1288.
[65] I. Dzidic and J. A. McCloskey, *Org. Mass Spectroscopy*, 1972, **6**, 939.
[66] A. M. Hogg and T. L. Nagabhushan, *Tetrahedron Letters*, 1972, 4827.
[67] T. J. Odiorne, P. Vouros, and D. J. Harvey, *J. Phys. Chem.*, 1972, **76**, 3217.
[68] T. J. Odiorne, D. J. Harvey, and P. Vouros, *J. Org. Chem.*, 1973, **38**, 4274.
[69] D. F. Hunt, C. N. McEwen, and R. A. Upham, *Analyt. Chem.*, 1972, **44**, 1292.
[70] N. Einolf and B. Munson, *Internat. J. Mass Spectrometry Ion Phys.*, 1972, **9**, 141.
[71] D. F. Hunt and J. F. Ryan, *J.C.S. Chem. Comm.*, 1972, 620.
[72] J. P. Vite, G. B. Pitman, A. F. Fentiman, and G. W. Kinzer, *Naturwiss.*, 1972, **59**, 469.

be added after passing through the column thereby avoiding any difficulties produced by the use of a non-standard carrier gas. Helium–water mixtures have been used in such a system, the He^+ ions causing dissociative charge exchange and the H_3O^+ or $H(H_2O)_n^+$ ions protonating the molecule.[73] Argon–water mixtures have similar properties. This system, used in the analysis of barbiturates, gives abundant $[M + H]^+$ ions, whereas the EI spectra have a very low abundance of $M^{+\cdot}$; charge exchange with the argon ions produces fragments which allow a distinction to be made between (15) and (16), which is not possible using

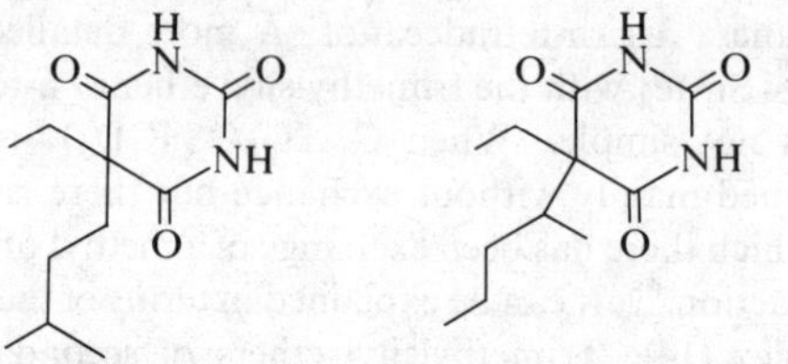

(15)   (16)

methane or isobutane.[74] Another potentially useful mixed gas system is N_2–NO. The fragment ions produced by dissociative charge exchange from N_2^+ are of the same mass as those produced by electron impact but the molecular ion produced by charge exchange from NO^+ is usually abundant. The resultant spectrum is rather like an EI spectrum with an enhanced molecular ion. Using this system the $M^{+\cdot}$ has been observed for trimethylsilyl ethers of bile acid methyl esters which exhibit no molecular ion by EI.[75]

Investigation of the mechanism of chemical ionization processes continues. Ion-ejection techniques in ICR show that both CH_5^+ and $C_2H_5^+$ contribute to all ions in the methane–hexane system, with the exception of $C_5H_{11}^+$ for which CH_5^+ is the only precursor.[76] This is consistent with the suggestion that only the methanonium ion is a strong enough acid to protonate a methyl group in hexane. In the methane CI spectra of specifically deuterated n-decane analogues, there is a considerable amount of isotope scrambling in the formation of alkyl fragment ions.[77] Although dissociative protonation is responsible for some of the fragments the principal process is hydride abstraction followed by unimolecular decomposition as shown. The latter process is accompanied or preceded by hydrogen scrambling.

$$C_{10}H_{22} \xrightarrow{CH_5^+} C_{10}H_{21}^+ \rightarrow C_nH_{2n+1}^+ + C_{10-n}H_{20-2n}$$

The CI spectra of simple amino-acids have been obtained using both methane and isobutane. Using the latter gas, the protonated molecular ion predominates at 440 K. At 660 K it dissociates, mostly by elimination of HCO_2H. The

[73] G. P. Arsenault, J. Amer. Chem. Soc., 1972, **94**, 8241.
[74] D. F. Hunt and J. F. Ryan, Analyt. Chem., 1972, **44**, 1306.
[75] B. Jelus, B. Munson, and C. Fenselau, Analyt. Chem., 1974, **46**, 729.
[76] R. P. Clow and J. H. Futrell, J. Amer. Chem. Soc., 1972, **94**, 3748.
[77] D. F. Hunt and C. N. McEwen, Org. Mass Spectrometry, 1973, **7**, 441.

Arrhenius plot for this fragmentation process gives a value for log A consistent with a tight transition state.[78] Comparison of the spectra of $C_4H_9COLeuOH$ with the two gases at various temperatures suggests that ions formed by protonation by CH_5^+ and $C_2H_5^+$ have 'temperatures' ca. 160 °C in excess of source temperature. The methane CI spectra of glycosides of general formula R—O—S (where S = monosaccharide residue) often exhibit an ion ROH_2^+. The origin of such ions has been investigated using as a model compound the tetrahydropyranyl ether (17).[79] The deuterium content of the fragment ion (18) is consistent with its formation by a unimolecular decomposition of the protonated molecular ion. The isomeric nucleosides (19) and (20) can be readily distinguished by intensity ratios as shown in Table 1.[80]

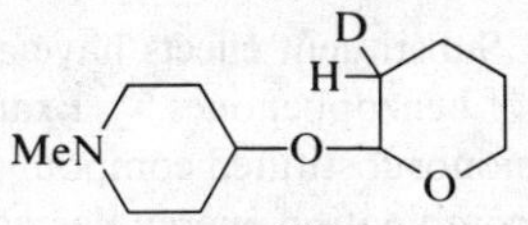

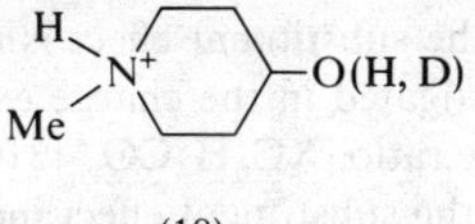

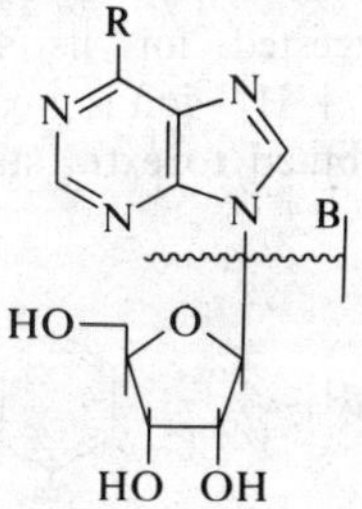

Table 1

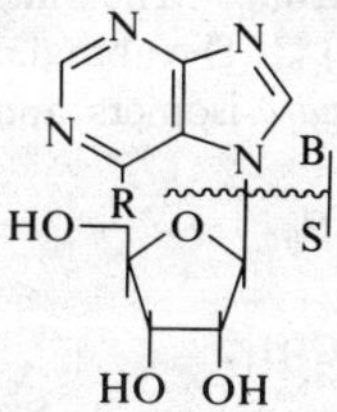

R	$[M + H]^+/[BH_2]^+$ (19)	$[M + H]^+/[BH_2]^+$ (20)
H	2.9	0.5
NH_2	10	0.1
$NHCH_3$	27	0.6
$N(CH_3)_2$	3.7	0.4
SCH_3	1.2	0.3

In the isobutane CI spectra of esters the process

$$R-CO-\overset{+}{O}HR^1 \rightarrow R^{1+} + RCO_2H$$

has been investigated in considerable detail. For t-amyl acetate its rate is independent of pressure between 0.5 and 4 Torr.[81] For a series of t-amyl esters the rate is

---

[78] M. Meot-Ner and F. H. Field, *J. Amer. Chem. Soc.*, 1973, **95**, 7207.
[79] R. L. Foltz, A. F. Fentiman, jun., L. A. Mitscher, and H. D. H. Showalter, *J.C.S. Chem. Comm.*, 1973, 872.
[80] J. A. McCloskey, J. H. Futrell, T. A. Elwood, K. H. Schramm, R. P. Banzica, and L. B. Townsend, *J. Amer. Chem. Soc.*, 1973, **95**, 5762.
[81] W. A. Laurie and F. H. Field, *J. Phys. Chem.*, 1972, **76**, 3917.

independent of the length of the acid chain.[82] The rate of formation of (22) from the protonated molecular ion of (21) by methane CI is in the order R = H > Me > Et > Pr^n. This is the same order as for aqueous acid-catalysed solvolysis

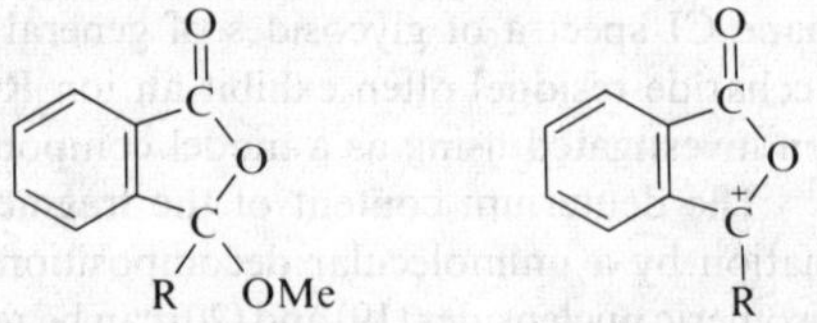

(21)   (22)

but the substituent effect is much smaller.[83] Substituent effects have also been investigated in the charge exchange spectra of benzophenones.[84] Examination of the ratio $[XC_6H_4CO^+] : [C_6H_5CO^+]$ for monosubstituted compounds shows that the substituent effect increases as the recombination energy decreases.

In the methane CI spectrum of phthalate esters the m/e 149 ion is found, as it is in the EI spectrum. The mechanism suggested for its formation is (23) → (24) → (25).[85] The abundance of the $[M + H]^+$ ion is greater than that of its *meta*- and *para*-isomers and this is attributed to extra stability of the

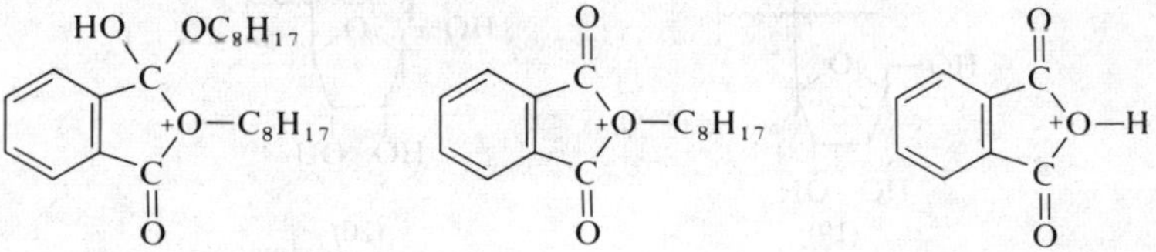

(23)   (24)   (25)

orthoester ion (23). It could however be due to the formation of a stable hydrogen bond between two oxygen atoms. Similar arguments are applied to the spectra of esters of maleic and fumaric acids.

Analytical applications of chemical ionization other than molecular weight determination are becoming more common. A chloroform extract of human gastric contents gives directly the CI spectrum of basic drugs in the mixture. Other components, *e.g.* lipids, do not interfere as they are not in the same mass range as the more common narcotics.[86] CI has also been used to analyse the phenylthiohydantoins from the Edman degradation of peptides.[87] Using the $[M + H]^+$ ion the method can be 10—100 times as sensitive as EI using the $M^{+\cdot}$

[82] W. A. Laurie and F. H. Field, *J. Amer. Chem. Soc.*, 1972, **94**, 3359.
[83] D. P. Weeks and F. H. Field, *J. Org. Chem.*, 1973, **38**, 3380.
[84] N. Einolf and B. Munson, *Org. Mass Spectrometry*, 1973, **7**, 155.
[85] H. M. Fales, G. W. A. Milne, and R. S. Nicholson, *Analyt. Chem.*, 1971, **43**, 1785.
[86] G. W. A. Milne, H. M. Fales, and T. Axenrod, *Analyt. Chem.*, 1971, **43**, 1815.
[87] H. M. Fales, Y. Nagai, G. W. A. Milne, H. B. Brewer, T. J. Bronzert, and J. J. Pisano, *Analyt. Biochem.*, 1971, **43**, 288.

ion. Quantification of the method is more difficult by mass spectrometry than by gas chromatography, but it can be done, in this case using [^{2}H$_5$]phenylisothiocyanate to produce labelled standards rather than using ^{15}N labels. This method has an advantage over gas chromatography in that Glu, Ser, Thr, Lys, Asp, Gln, and Asn can all be analysed as phenythiohydantoins without further derivatization.

One of the better mass spectrometric methods of determining double bond position in alkenes and unsaturated esters involves hydroxylation with OsO$_4$ followed by trimethyl silylation of the diol. The EI mass spectrum of the TMS ether shows the diagnostic fragments in abundance, but the molecular ion is often absent. Both position and molecular weight information can be obtained from the methane CI spectrum.[88] The polychloroalicyclic pesticides related to aldrin have been analysed both by positive[89] and negative ion[20] CI. Both give molecular weight information, but fragmentation in the positive ion spectra is more sensitive to structural and stereochemical differences. Particularly interesting are the spectra of isodrin (26) and aldrin (27). The $[M - Cl]^+$ ion from (26) undergoes a retro-Diels–Alder fragmentation to produce C$_7$H$_2$Cl$_5{}^+$ much more readily than does the isomeric ion from (27).

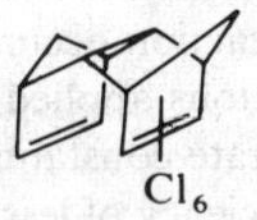

Cl$_6$
(26)

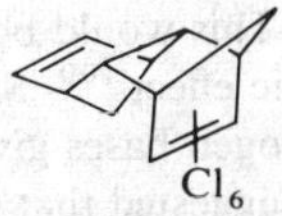

Cl$_6$
(27)

Chemical ionization is also being used for the analysis of inorganic and organometallic compounds. Diene complexes of the type LFe(CO)$_3$ have characteristic methane CI spectra with abundant protonated molecular ions. At higher sample pressures [L$_2$Fe$_2$(CO)$_3$]H$^+$ may also be formed.[90] Lanthanide complexes with β-diketones are difficult to analyse by electron impact because fragments of the heavier metal complexes will interfere with the molecular ions containing lighter metals. There is in addition considerable overlap in isotope patterns between adjacent members of the series. The isobutane CI spectra have effectively no fragment ions for the series M(thd)$_3$ where thd = 2,2,6,6-tetramethylheptane-3,5-dionato and M varies from La to Lu.[91] In a similar study, lanthanide complexes with 1,1,1,2,2,3,3-heptafluoro-7,7-dimethylocta-4,6-dione gave less simple spectra and in all cases there were ions appearing 18 mass units below the protonated molecular ion. These are believed to be formed by H–F exchange in the collision complex.[92] Further discussion of the mass spectra of inorganic compounds appears in Chapter 6.

[88] W. Blum and W. J. Richter, *Tetrahedron Letters*, 1973, 835.
[89] F. J. Biros, R. C. Dougherty, and J. Dalton, *Org. Mass Spectrometry*, 1972, **6**, 1161.
[90] D. F. Hunt, J. W. Russell, and R. L. Torian, *J. Organometallic Chem.*, 1972, **43**, 175.
[91] T. H. Risby, P. C. Jurs, F. W. Lampe, and A. L. Yergey, *Analyt. Chem.*, 1974, **46**, 161.
[92] T. H. Risby, P. C. Jurs, F. W. Lampe, and A. L. Yergey, *Analyt. Chem.*, 1974, **46**, 726.

6 Ion Cyclotron Resonance

This technique[93] has been extremely productive of results in the past two years. There has been considerable discussion of the expressions used for the calculation of rate constants from power absorptions.[94,95] Equations of motion have been derived for ions in a trapping field in pulsed ICR experiments.[96] Collision frequencies can be measured in a pulsed system by producing coherent motion of ions by an r.f. pulse and measuring the rate of decay of this motion caused by ion–molecule collisions.[97] Collision rates increase with decreasing temperature;[98] this is considered to be an effect of the rotational energy of the molecule. A pulsed technique has also been used to measure ion transit times.[99] The electron voltage is pulsed (short pulse) above the ionization potential, followed by a trapping pulse of variable length. Ion current will be registered only if the trapping pulse is long enough for ions to reach the monitor. The plot of ion current against trapping pulse time is a curve with an onset (first ions reach monitor) and a maximum (last ions have reached monitor). The mean arrival time is at half maximum current.

An average dipole orientation (ADO) theory has been applied to ion–polar molecule reactions. For proton transfer from CH_5^+ to alkyl halides there is an exact correspondence between experimental rate constants and those calculated from ADO theory. This would be expected if reaction occurs at every collision and there is no steric effect.[100] Similar calculations applied to proton transfer from $C_4H_9^+$ to nitrogen bases give theoretical rate constants larger than those measured.[101] It is suggested that there is an efficiency of less than unity for this process because C—N bond formation takes place first and the transfer of a proton from C to N competes with the back reaction.

At low pressures ICR spectra do not normally tend to differ from electron impact spectra run using conventional mass spectrometers, since there are few fragmentation processes observed with rate constants less than $10^5 \, s^{-1}$. An exception is hexa-1,5-diyne where the loss of a hydrogen atom takes place at a very slow rate although the measured activation energy is only 0.34 eV.[102]

Hydrogen bonds are important in consideration of the stability of products of ion–molecule reactions involving compounds with O–H and N–H bonds. The exothermicity of processes of the type

$$ROH_2^+ + ROH \rightarrow (ROH)_2H$$

[93] C. J. Drewery, G. C. Goode and K. R. Jennings, in 'Mass Spectrometry', ed. A. Maccoll, M.T.P. International Review of Science, Series 1, Vol. 5, Butterworths, London, 1972.
[94] T. McAllister, *Internat. J. Mass Spectrometry Ion Phys.*, 1972, **8**, 162; 1972, **9**, 127.
[95] V. Anicich and M. T. Bowers, *Internat. J. Mass Spectrometry Ion Phys.*, 1973, **11**, 329.
[96] T. E. Sharp, J. R. Eyler, and E. Li, *Internat. J. Mass Spectrometry Ion Phys.*, 1972, **9**, 421.
[97] C. A. Lieder, R. W. Wien, and R. T. McIver, *J. Chem. Phys.*, 1972, **56**, 5184.
[98] S. E. Buttrill, *J. Chem. Phys.*, 1973, **58**, 656.
[99] T. B. McMahon and J. L. Beauchamp, *Rev. Sci. Instr.*, 1971, **42**, 1632.
[100] T. Su and M. T. Bowers, *J. Amer. Chem. Soc.*, 1973, **95**, 7609.
[101] T. Su and M. T. Bowers, *J. Amer. Chem. Soc.*, 1973, **95**, 7611.
[102] M. L. Gross and R. J. Aerni, *J. Amer. Chem. Soc.*, 1973, **95**, 7875.

is usually of the order of 30 kcal mol^{-1},[103] and at the pressures common in ICR experiments the energy gained in this process may be lost by bond fission, as in the processes

$$C_2H_5OH_2^{+\cdot} + H_2O \rightarrow CH_3\cdot + CH_2=O\cdots\overset{+}{H}\cdots OH_2$$

and

$$C_2H_5OH_2^{+\cdot} + CH_3F \rightarrow CH_3\cdot + CH_2=O\cdots\overset{+}{H}\cdots FCH_3$$

The energy released from the hydrogen bond formation may be used in a dehydration process. Such processes have been well documented for cations and are now reported for anions, as shown in Scheme 4.[104] This could be considered to be a gas-phase analogy of a base-catalysed elimination reaction.

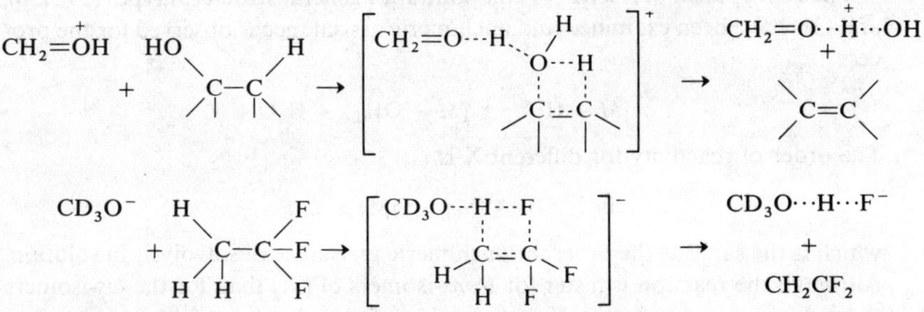

Scheme 4

Other gas-phase analogies for ionic reactions in solution have been observed. The reaction of protonated acetic acid with methanol has been observed using double-resonance techniques. The product (28) is considered to be a protonated

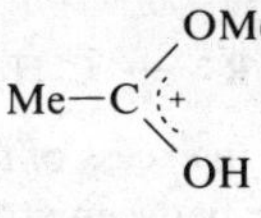

$$Me-C\begin{smallmatrix}OMe\\+\\OH\end{smallmatrix}$$

(28)

ester.[105] No such esterification is observed with mixtures of methanol and formic acid. In this case the nucleophile, methanol, has a higher proton affinity than the acid and only proton transfer is observed.

Gas-phase analogies of aromatic substitution have been investigated using double-resonance techniques. Acetylation of monosubstituted benzenes with $(CH_3CO)_2^{+\cdot}$ shows substituent effects qualitatively similar to those observed in

[103] D. P. Ridge and J. L. Beauchamp, *J. Amer. Chem. Soc.*, 1971, **93**, 5925.
[104] D. P. Ridge and J. L. Beauchamp, *J. Amer. Chem. Soc.*, 1974, **96**, 637.
[105] P. W. Tiedemann and J. M. Riveros, *J. Amer. Chem. Soc.*, 1974, **96**, 185.

solution, but the reaction

$$CH_2ONO_2^+ + C_6H_5R \rightarrow RC_6H_5NO_2^+ + CH_2O$$

has an inverted substituent effect.[106] These reactions should correspond to the electrophilic addition step in aromatic substitution. A methylenation reaction has been observed in which both addition to the aromatic ring and removal of a ring hydrogen atom are involved.[107]

$$CH_3OCH_2^+ + C_6H_5R \rightarrow R \cdot C_7H_6^+ + CH_3OH$$

This reaction shows the expected order of substituent effects on rate constants but no positional selectivity in the removal of CH_3OH or CH_3OD when *ortho-*, *meta-*, or *para-*deuteriotoluene is used as substrate.

The ICR spectra of a series of compounds of general structural types (29), (30), and (31) have been examined and anchimeric assisatance is observed for the process

$$[M + H]^+ \rightarrow [M - OH]^+ + H_2O$$

The order of reactivity for different X is

Br > SH > SMe > OMe ≫ Cl,F

which is the same as the order for anchimeric assistance of solvolysis in solution. Although the reaction is faster for *trans-*isomers of (31) than for the *cis-*isomers there does appear to be some participation of *cis-*substituents.[108]

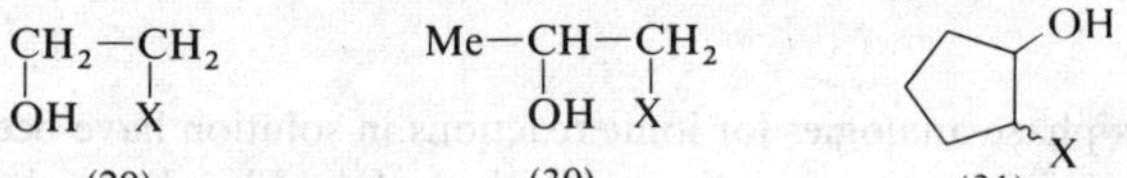

$$\begin{array}{cc} CH_2-CH_2 \\ | \quad\quad | \\ OH \quad X \end{array} \quad\quad \begin{array}{cc} Me-CH-CH_2 \\ | \quad\quad | \\ OH \quad X \end{array}$$

(29)   (30)   (31)

Photochemical reactions of ions have been observed by ICR. The molecular ions of alkylbenzenes can photodissociate by the process

$$C_6H_5CH_2R^{+\cdot} \rightarrow C_7H_7^+ + R\cdot$$

The onset of this process is greatly in excess of the thermochemical requirements but is close to the energy difference between ground state and an excited state of the molecular ion as derived from photoelectron spectra.[109] The plot of fragment-ion current against wavelength (photodissociation curve) is thus an approximation to an electronic spectrum of the ion and may be characteristic of ion structure. The photodissociation curves of $C_7H_8^{+\cdot}$ from toluene, cycloheptatriene, and norbornadiene are quite different.[110] Experiments with labelled

---

[106] R. C. Dunbar, J. Shen, and G. A. Olah, *J. Amer. Chem. Soc.*, 1972, **94**, 6862.
[107] R. C. Dunbar, J. Shen, E. Melby, and G. A. Olah, *J. Amer. Chem. Soc.*, 1973, **95**, 7200.
[108] J. K. Kim, M. C. Findlay, W. G. Henderson, and M. C. Caserio, *J. Amer. Chem. Soc.*, 1973, **95**, 2184.
[109] R. C. Dunbar, *J. Amer. Chem. Soc.*, 1973, **95**, 6191.
[110] R. C. Dunbar and E. W. Fu, *J. Amer. Chem. Soc.*, 1973, **95**, 2718.

toluene, however, show that complete scrambling takes place during photo-dissociation.[111] These results taken together suggest that the toluene ion in its ground state retains the gross structure of the parent molecule, but ions in the first excited state, which have enough energy to decompose, undergo isomerization and scrambling. A bimolecular photochemical reaction has been observed with $C_3H_5^+$ ions in the ICR spectrum of ethylene.[112]

$$C_3H_5^+ + C_2H_4 \xrightarrow{h\nu} C_3H_7^+ + C_2H_2$$

The onset of this process (520 nm) suggests that there is a state of $C_3H_5^+$ which is 2.5 eV above the ground state, although calculations of the π-electron energies (ignoring σ-electrons) gave a value of 5 eV for the first singlet state.

7 Ion–Molecule Reactions

This section covers a number of processes which have been studied using various techniques, including ICR and conventional mass analysis, EI and PI sources, and shows how the various techniques can produce similar or complementary results.

Acid–Base Equilibria.—The most accurate set of proton affinities (PA) of amines has been produced by Yamdagni and Kebarle[113] using a high-pressure ion source and direct measurement of equilibrium constants. These are shown in Table 2.

Table 2 Proton affinities/kcal mol^{-1}

NH_3	207	Cyclohexylamine	227
CH_3CONH_2	210	$C_6H_5N(CH_3)_2$	229
Pyrrole	214	$(CH_3)_3N$	230
$C_6H_5NH_2$	216	$C_6H_5N(CH_3)C_2H_5$	231
CH_3NH_2	218	Pyrrolidine	231
$C_6H_5NHCH_3$	222	$C_6H_5N(C_2H_5)_2$	234
$o\text{-}CH_3C_6H_4NH_2$	222	$H_2NCH_2CH_2NH_2$	235
$(CH_3)_2NH$	225	$H_2N(CH_2)_3NH_2$	243
$C_6H_5NHC_2H_5$	225	$H_2N(CH_2)_5NH_2$	243
Pyridine	226	$H_2N(CH_2)_7NH_2$	243

The last values quoted in the table, for the aliphatic diamines, are anomalously high and are considered to be due to the formation of internal hydrogen bonds. The energy gain from this hydrogen-bond formation is expressed as ΔH_{cycl} where

$$\Delta H_{cycl} = PA[H_2N(CH_2)_nNH_2] - PA[CH_3(CH_2)_nNH_2]$$

The strain energy of the cyclic hydrogen-bonded monomer is taken to be the difference between ΔH_{cycl} and the enthalpy of formation of an acyclic proton-bound dimer. The values of these parameters for diamines are shown in Table 3.

[111] R. C. Dunbar, *J. Amer. Chem. Soc.*, 1973, **95**, 472.
[112] J. M. Kramer and R. C. Dunbar, *J. Amer. Chem. Soc.*, 1972, **94**, 4347.
[113] R. Yamdagni and P. Kebarle, *J. Amer. Chem. Soc.*, 1973, **95**, 3504.

A study of an extended series of compounds of the same type by ICR shows that ΔH_{cycl} has a maximum (and hence strain energy a minimum) when $n = 4$.[114]

Table 3

n	ΔH_{cycl}	Strain Energy/kcal mol^{-1}
2	12.6	10.4
3	20.5	2.5
5	20.1	2.9
7	20.0	3.0

An alternative method which has been used to derive orders of reactivity and a measure of thermodynamic properties is the bracketing technique. This can be done in ICR by double resonance where proton transfer might be observed from the conjugate acid of the weaker base to the stronger base and not *vice versa*. Observation of the presence or absence of proton transfer in binary mixtures of bases allows unknown compounds to be bracketed between stronger and weaker bases of known proton affinity. The use of different bases as reactant gases in CI allows bracketing of samples. Using dimethylamine, methylamine, and ammonia as reactant gases, the following order of basicity was found:

$$C_6H_5N(CH_3)_2 > (CH_3)_2NH > C_6H_5NHCH_3 > CH_3NH_2 > C_6H_5NH_2 > NH_3$$

This order is in agreement with the values of proton affinities in Table 2.[115] The effect of alkyl substitution upon gas-phase basicity shows a pronounced increase in basicity on going from ammonia to trimethylamine and only a slight one on going from methylamine to t-butylamine.[116] There is no simple explanation of the difference between the solution order of basicity of amines and the gas-phase order. All the contributing thermodynamic properties change in an orderly way with substitution, but the inverted order of aqueous base strength arises from slight differences in the rate of change of these properties in response to progressive alkyl substitution.[117] A plot of gaseous PA against heat of protonation in solution gives three separate straight lines for primary, secondary, and tertiary amines. The difference between the groups is principally in the number of $\overset{+}{N}-H\cdots OH_2$ bonds which can be formed in the solvated ion.[118]

The proton affinities of a large number of organic oxygen compounds have been measured using the bracketing technique with a pulsed-mode time-of-flight mass spectrometer with a delay time of up to 7 μs. The effect of increasing the length of the alkyl chain diminishes to a constant value for any functional group when the chain is longer than propyl, and limits for the proton affinities of some

---

[114] D. H. Aue, H. M. Webb, and M. T. Bowers, *J. Amer. Chem. Soc.*, 1973, **95**, 3504.
[115] I. Dzidic, *J. Amer. Chem. Soc.*, 1972, **94**, 8333.
[116] W. G. Henderson, M. Taagepera, D. Holtz, R. T. McIver, J. L. Beauchamp, and R. W. Taft, *J. Amer. Chem. Soc.*, 1972, **94**, 4728.
[117] E. M. Arnett, F. M. Jones, tert., M. Taagepera, W. G. Henderson, J. L. Beauchamp, D. Holtz, and R. W. Taft, *J. Amer. Chem. Soc.*, 1972, **94**, 4724.
[118] D. H. Aue, H. M. Webb, and M. T. Bowers, *J. Amer. Chem. Soc.*, 1972, **94**, 4726.

simple oxygen functional groups are shown in Table 4.[119] The order of basicities found by this method has been confirmed by an ICR study using double resonance and pressure variation.[120] In the ICR spectra of the ethers $CH_3O(CH_2)_n$-OCH_3 the formation of a proton-bound dimer is inhibited in comparison with

Table 4 Proton affinities /kcal mol^{-1} for $R > C_3H_7$

RCH_2OH	189	CH_3CO_2R	207
R_2CHOH	197	RCO_2CH_3	205
R_3COH	204	$RCO_2C_2H_5$	208
HCO_2R	198	RCO_2R	210

monoethers when $n > 2$. This is due to the formation of a cyclic hydrogen-bonded ion,[121] similar to the species formed by diamines. Proton affinities of aromatic hydrocarbons[122] and N-nitrosamines[123] have also been measured.

Accurate measurements of equilibria can be made using pulsed ICR. If a pulse of ions is trapped in the cell for up to 1 s it is possible to observe the formation and decay of F^-, CN^-, and SH^- in a mixture of NF_3, HCN, and H_2S. The reactions observed are

$$e + NF_3 \rightarrow F^- + NF_2$$
$$F^- + HCN \rightarrow HF + CN^-$$
$$F^- + H_2S \rightarrow HF + SH^-$$
$$CN^- + H_2S \rightleftharpoons HCN + SH^-$$

Using long trapping times it is possible to measure the equilibrium directly. It is also possible to measure the rates of forward and reverse reactions separately by ejecting SH^- and CN^-.[124] Yamdagni and Kebarle have also measured relative acidities of some carboxylic and halogen acids in the gas phase.[125] The thermodynamic measurement used by these workers, $D(AH) - EA(A)$ is the endothermicity of the process

$$AH + e \rightarrow A^- + H^{\cdot}$$

where $D(AH)$ is the A–H bond dissociation energy and $EA(A)$, the electron affinity of A; values are shown in Table 5. Acid–anion equilibria can also be used to calculate electron affinities, providing the A–H bond energy is known.[126] The gas-phase acidity of monosubstituted phenols has been measured, and the substituent effects found parallel those observed in solution.[127] Acidities of

[119] J. Long and B. Munson, *J. Amer. Chem. Soc.*, 1973, **75**, 2427.
[120] P. C. Isolani, J. M. Riveros, and P. W. Tiedemann, *J.C.S. Faraday II*, 1973, **69**, 1023.
[121] T. H. Morton and J. L. Beauchamp, *J. Amer. Chem. Soc.*, 1972, **94**, 3671.
[122] S.-L. Chong and J. L. Franklin, *J. Amer. Chem. Soc.*, 1972, **94**, 6630.
[123] S. Billets, H. H. Jaffe, and F. Kaplan, *Org. Mass Spectrometry*, 1973, 7, 431.
[124] R. T. McIver jun. and J. R. Eyler, *J. Amer. Chem. Soc.*, 1971, **93**, 6334.
[125] R. Yamdagni and P. Kebarle, *J. Amer. Chem. Soc.*, 1973, **95**, 4050.
[126] R. Yamdagni and P. Kebarle, *Ber. Bunsengesellschaft phys. Chem.*, 1974, **78**, 181.
[127] R. T. McIver jun. and J. H. Silvers, *J. Amer. Chem. Soc.*, 1973, **95**, 8462.

hydrocarbons are difficult to measure because of low kinetic acidities. Of the following three processes:

$$CH_3OH + CH_2=CH-CH_2^- \xrightarrow{k_1} CH_3O^- + C_3H_6 \quad (1)$$
$$C_6H_5CH_3 + CH_3O^- \xrightarrow{k_2} CH_3OH + C_6H_5CH_2^- \quad (2)$$
$$C_6H_5CH_3 + CH_2=CH-CH_2^- \xrightarrow{k_3} C_6H_5CH_2^- + C_3H_6 \quad (3)$$

only the first two are fast enough to be observed on the ICR time scale: k_3 is too small.[128] In mixtures of all three components, since both reactions (1) and (2) can occur, the formation of $C_6H_5CH_2^-$ is essentially the methanol-catalysed reaction (3).

Table 5

	$D(AH) - EA(A)$		$D(AH) - EA(A)$
HF	56.3	HCl	20.0
CH_3CO_2H	31.8	$ClCH_2CO_2H$	19.0
$C_2H_5CO_2H$	30.6	Cl_2CHCO_2H	12.0
n-$C_3H_7CO_2H$	29.7	HBr	10.0
HCO_2H	28.6	HI	0.7

The reaction of a basic molecule with Lewis acid provides a different measure of basicity. From analysis of reactions of the general type

$$RX + CH_3X^1H^+ \rightarrow RXCH_3^+ + X^1H$$

it is possible to determine the methyl cation affinity (MCA) of RX.[129] This is effectively the basicity of RX towards a soft acid whereas the proton affinity is basicity towards a hard acid. The measured order of basicities is as follows:

PA $H_2O > CH_3F > HI > CO > HCl > HF > N_2$
MCA $CO > HI > H_2O > HCl > CH_3F > N_2 > HF$

In dialkyl ketones the general reaction,

$$RCOR^{1\ddagger} + RCOR^1 \rightarrow RCO(RCOR^1)^+ + R^{1\cdot}$$

is observed. From the relative rates of such reactions in a number of ketones the acyl affinities of the parent ketones can be measured.[130]

It is not always possible to decide the relative stabilities of carbonium ions by the presence or absence of the hydride transfer reaction, $R^+ + R^1H \rightarrow R^{1+} + RH$, as observed by double resonance techniques, since the rates of such processes are often too slow on the ICR time scale. It is however possible to obtain such information by the investigation of halide transfer reactions.[131] In the ICR

---

[128] J. I. Brauman, C. A. Lieder, and M. J. White, *J. Amer. Chem. Soc.*, 1973, **95**, 927.
[129] J. L. Beauchamp, D. Holtz, S. D. Woodgate, and S. L. Patt, *J. Amer. Chem. Soc.*, 1972, **94**, 2800.
[130] P. W. Tiedemann and J. M. Riveros, *J. Amer. Chem. Soc.*, 1973, **95**, 3140.
[131] T. B. McMahon, R. J. Blint, D. P. Ridge, and J. L. Beauchamp, *J. Amer. Chem. Soc.*, 1972, **94**, 8935.

spectrum of difluoromethane the following processes are observed:

$$CH_2F_2^{+\cdot} \rightarrow CH_2F^+ + F\cdot$$
$$CH_2F_2^{+\cdot} \rightarrow CHF_2^+ + H\cdot$$
$$CHF_2^+ + CH_2F_2 \rightleftharpoons CH_2F^+ + CHF_3$$

From the observation of equilibria in mixtures of fluorinated methanes the following order of fluoride affinities is observed:[131] $CF_3^+ > CH_3^+ > CHF_2^+ > CH_2F^+$. This is not the real order of carbonium ion stability because C–F bond energies vary considerably with the number of fluorine atoms present in the parent molecule. Combining the known C–F bond energies with the measured fluoride affinities leads to an order of carbonium ion stabilities as follows: $CHF_2^+ > CH_2F^+ > CF_3^+ > CH_3^+$. The effect on carbonium ion stability of successive replacement of hydrogen by fluorine is not straightforward, but the replacement of fluorine by chlorine increases the carbonium ion stability as would be predicted from the change in polarizability.[133]

Ion–Solvent Interactions.—It is expected that the study of the stability of ion–solvent clusters will throw some light on the general topic of ion solvation.[134] Investigations using ICR techniques have been limited to the interaction of an ion with a single solvent molecule. Studies of larger clusters rely on measurements at higher pressures than can be achieved in ICR operation.

The association of hydrocarbon ions with parent molecules can in general be observed only below room temperature.[135] In mixtures of gas, equilibrium constants for reactions of the type

$$X^+ + CH_4 \rightleftharpoons [XCH_4]^+$$

can be measured at variable temperatures, and the values of ΔH obtained from these measurements are comparable to those calculated using an ion-induced-dipole model.[136] In hydrogen, equilibrium constants for

$$H_3^+ + H_2 \rightleftharpoons H_5^+$$

and
$$H_5^+ + H_2 \rightleftharpoons H_7^+$$

have also been measured.[137]

Equilibria involving polar solvents are of more general interest but as of yet there is still disagreement between the results of Field and Kebarle[138] on the

---

[132] R. J. Blint, T. B. McMahon, and J. L. Beauchamp, *J. Amer. Chem. Soc.*, 1974, **96**, 1269.
[133] J. H. J. Dawson, W. G. Henderson, R. M. O'Malley, and K. R. Jennings, *Internat. J. Mass Spectrometry Ion Phys.*, 1973, **11**, 61.
[134] P. Kebarle, in 'Ions and Ion Pairs in Organic Reactions', ed. M. Szwarc, Wiley-Interscience, New York, 1972.
[135] D. P. Beggs and F. H. Field, *J. Amer. Chem. Soc.*, 1971, **93**, 1576, 1585; L. W. Seick, S. K. Searles, and P. Ausloos, *J. Chem. Phys.*, 1970, **53**, 849, *ibid.*, 1971, **54**, 91; *J. Res. Nat. Bur. Stand., Sect. A*, 1971, **75**, 147; S. L. Bennett, S. G. Lias, and F. H. Field, *J. Phys. Chem.*, 1972, **76**, 3919.
[136] S. L. Bennett and F. H. Field, *J. Amer. Chem. Soc.*, 1972, **94**, 5188, 6305.
[137] S. L. Bennett and F. H. Field, *J. Amer. Chem. Soc.*, 1972, **94**, 8669.
[138] S. L. Bennett and F. H. Field, *J. Amer. Chem. Soc.*, 1972, **94**, 5186; A. J. Cunningham, J. D. Payzant, and P. Kebarle, *ibid.*, p. 7627.

value of $K_{1,2}$ for the $H(H_2O)_n{}^+$ system, with the two values differing by a factor of 10^{10}. High-pressure equilibrium studies show that the strength of the bond between an anion and a single acidic solvent molecule increases both with the gas-phase basicity of the anion and with the gas-phase acidity of the solvent.[139] The association of alkoxide ions with alcohol molecules show that the order of solvating ability

$$t\text{-}C_4H_9OH > i\text{-}C_3H_7OH > C_2H_5OH > CH_3OH$$

is the same as the order of gas-phase acidity.[140] These measurements were made by ICR using an indirect method of forming the cluster ion.

$$RO^- + HCO_2R^1 \rightarrow [RO\cdots H\cdots OR^1]^- + CO$$

A similar method is used to form clustered chloride ions, and compare the relative strengths of solvent–chloride bonds.[141]

$$COCl_2 + e \rightarrow COCl^- + Cl^{\cdot}$$
$$COCl^- + M \rightarrow Cl(M)^- + CO$$
$$Cl(M)^- + M^1 \rightleftharpoons Cl(M^1)^- + M$$

It is of some interest that CO can be displaced from the $COCl^-$ ion by xenon.[142] In another ICR study of nucleophilic reactions of xenon, the dissociation energy of the Xe–C bond in $XeCH_3{}^+$ was found to be 43 ± 8 kcal mol^{-1}.[143]

When larger clusters are investigated, using higher pressures, the acid–base strength relationship no longer continues to be valid. The equilibrium

$$(R_2O)_2H^+ + R_2O \rightleftharpoons (R_2O)_3H^+$$

has been studied for water, methanol and dimethyl ether and when a comparison is made with the 2,1-equilibrium it is found that the free energy of association decreases as the cluster size increases. This decrease is greater for methanol than for water and much greater for dimethyl ether.[144] It is therefore necessary to examine tendencies as cluster size increases in order to make extrapolations from gas-phase solvating ability to liquid-phase properties. In acetone the clusters $(CH_3COCH_3)_nH^+$, $(CH_3COCH_3)_nCH_3{}^+$, and $(CH_3COCH_3)_nCOCH_3{}^+$ were observed at pressures of up to 0.6 Torr.[145] Attempts to measure the stabilities of these clusters were unsuccessful because equilibrium was not attained at this pressure.

Structures of Ions in the Gas Phase.—Both ICR and high-pressure techniques have been used to investigate reactions which provide indirect evidence about the structures of the reacting ions. In the ICR spectrum of $(CD_3CH_2)_2N_2O$ the

[139] R. Yamdagni and P. Kebarle, *J. Amer. Chem. Soc.*, 1971, **93**, 7139.
[140] R. T. McIver, jun., J. A. Scott, and J. M. Riveros, *J. Amer. Chem. Soc.*, 1973, **95**, 2706; L. K. Blair, P. C. Isolani, and J. M. Riveros, *J. Amer. Chem. Soc.*, 1973, **95**, 1057.
[141] J. M. Riveros, A. C. Breda, and L. K. Blair, *J. Amer. Chem. Soc.*, 1973, **95**, 4066.
[142] J. M. Riveros, P. W. Tiedemann, and A. C. Breda, *Chem. Phys. Letters*, 1973, **20**, 345.
[143] D. Holtz and J. L. Beauchamp, *Science*, 1971, **173**, 1237.
[144] E. Grimsrud and P. Kebarle, *J. Amer. Chem. Soc.*, 1973, **95**, 7939.
[145] K. A. G. MacNeil and J. H. Futrell, *J. Phys. Chem.*, 1972, **76**, 409.

$C_2H_2D_3^+$ ion transfers deuterium only to the parent molecule.[146] This can be most easily interpreted as a reaction of a classical ethyl cation. The reactivity of $C_3H_6^{+\cdot}$ ion with ammonia has been investigated in a photoionization source. The ion from propylene gives product ratios different from those obtained using cyclopropane. The molecular ion of the latter compound transfers a proton to ammonia with increasing efficiency as photon energy is increased, suggesting that NH_4^+ is the product from an open-chain $C_3H_6^{+\cdot}$ ion whereas $CH_2NH_2^{+\cdot}$ and $CH_2NH_3^{+\cdot}$ are formed from the cyclic ion.[147] In an examination of the same process by ICR, it is found that for the reactions

$$C_3H_6^{+\cdot} + ND_3 \rightarrow CH_2ND_3^{+\cdot} + C_2H_4$$

and

$$C_3H_6^{+\cdot} + ND_3 \rightarrow CH_2ND_2^+ + C_2H_4D\cdot$$

there is very little randomization of H and D in the collision complex. These reactions are also observed for the $C_3H_4D_2^{+\cdot}$ ions produced by the following reaction

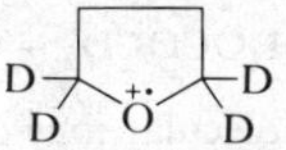

$$\text{D}\underset{\underset{D}{|}}{\overset{+\cdot}{\underset{O}{\square}}}\text{D} \rightarrow C_3H_4D_2^{+\cdot} + CD_2O$$

In the reaction of this ion with ammonia the three CX_2 groups are equivalent but there appears to be no randomization of H and D atoms between the carbon atoms and a cyclopropane structure has been tentatively assigned.[148]

A common feature of the mass spectra of olefins at higher pressures is the formation of a dimeric ion, which may be stabilized by a third body collision or may decompose. From an analysis of the ICR spectra of $CH_3CH=CH_2$, $CD_3CH=CH_2$, and $CD_3CD=CD_2$ it was suggested that the intermediates (32), (33), and (34) may be involved in the formation of the $C_4H_7^+$ and $C_5H_9^+$ fragments.[149] In a high pressure PI experiment using propylene, the collisionally

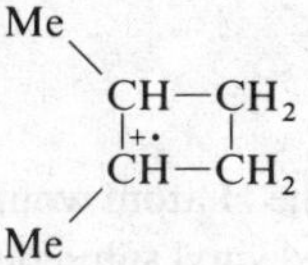

$$\begin{matrix} Me & & Me & & Me & Me \\ \diagdown & & \diagdown & & \diagdown & \diagup \\ CH-CH_2 & & CH-CH_2{}^+ & & C \\ |+\cdot \; \; | & & | & & | \\ CH-CH_2 & & CH-CH_2{}^{\cdot} & & C \\ \diagup & & \diagup & & \diagup \diagdown \\ Me & & Me & & Me \; \; \; \; Me \\ (32) & & (33) & & (34) \end{matrix}$$

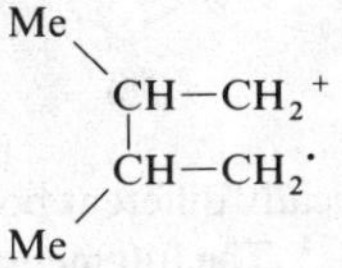

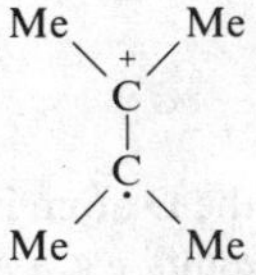

stabilized $C_6H_{12}^{+\cdot}$ ion was observed and this undergoes H_2 transfer to propylene.[150]

$$C_6H_{12}^{+\cdot} + C_3H_6 \rightarrow C_6H_{10}^{+\cdot} + C_3H_8$$

---

[146] H. H. Jaffe and S. Billets, *J. Amer. Chem. Soc.*, 1972, **94**, 674.
[147] L. W. Sieck, R. Gorden jun., and P. Ausloos, *J. Amer. Chem. Soc.*, 1972, **94**, 7157.
[148] M. L. Gross, *J. Amer. Chem. Soc.*, 1972, **94**, 3744.
[149] M. T. Bowers, D. H. Aue, and D. D. Elleman, *J. Amer. Chem. Soc.*, 1972, **94**, 4255.
[150] L. W. Sieck and P. Ausloos, *J. Res. Nat. Bur. Stand., Sect. A.*, 1972, **76**, 253; *J. Chem. Phys.*, 1972, **56**, 1010.

The reactions of molecular ions of fluorinated ethylenes follow the same general pathway as found in simple olefins, in that the secondary products can be explained in terms of a dimer which may or may not be stable.[151] The dimeric ion from 1,1-difluoroethylene is stable but the dimers from tri- and tetra-fluoro-ethylene decompose by elimination of a CF_3 radical. The reactions of fluorinated ethylene ions with ethylene all give unstable dimers and the products are those that would be expected from a cyclobutane ion in which there has been no randomization of H and F.[152]

Reactivity under ICR conditions provides useful evidence about the possible structures of fragment ions. The ion $C_5H_8O^{+\cdot}$ formed by McLafferty rearrangement of 2-propylcyclopentanone gives the following reaction at short residence times

$$C_5H_8O^{+\cdot} + CD_3COCD_3 \rightarrow CD_3C(OH)CD_3^+ + C_5H_7O\cdot$$

This is a reaction typical of enols. At longer residence times the predominant process is

$$C_5H_8O^{+\cdot} + CD_3COCD_3 \rightarrow C_5H_8\overset{+}{O}COCD_3 + CD_3\cdot$$

Such a process is usually observed with the molecular ions of ketones. The ion initially formed must be cyclopentenol which ketonizes within a time scale of 10^{-3} s. Such ketonization was not observed in ions from acyclic ketones, even when ions were trapped for up to 10^{-1} s.[153] The $C_6H_5DO^{+\cdot}$ fragment ion from (35) transfers D^+ but not H^+ to bases. This result was taken as evidence that the structure of the ion is (36) rather than (37) but the only real conclusion from this type of experiment is that the structure cannot be (37). Any other structure in

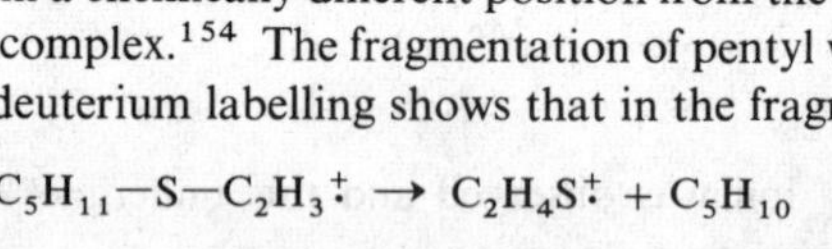

      OCH_2CD_3 OD

      (35)   (36)   (37)

which the D atom is in a chemically different position from the H atom would fit these results, e.g. a π-complex.[154] The fragmentation of pentyl vinyl sulphide has been examined and deuterium labelling shows that in the fragmentation,

$$C_5H_{11}-S-C_2H_3^{+\cdot} \rightarrow C_2H_4S^{+\cdot} + C_5H_{10}$$

hydrogen is transferred to the C_2H_3S group from positions 2, 3, and 4 of the pentyl group. The reactivity of $C_2H_3DS^{+\cdot}$ ions from 2,2-dideuteriopentyl vinyl

---

[151] R. M. O'Malley, K. R. Jennings, M. T. Bowers, and V. G. Anicich, *Internat. J. Mass Spectrometry Ion Phys.*, 1973, **11**, 89, 99.

[152] A. J. Ferrer-Correia and K. R. Jennings, *Internat. J. Mass Spectrometry Ion Phys.*, 1973, **11**, 111.

[153] J. R. Hass, M. M. Bursey, D. G. I. Kingston, and H. P. Tannenbaum, *J. Amer. Chem. Soc.*, 1972, **94**, 5095.

[154] N. M. M. Nibbering, *Tetrahedron*, 1973, **29**, 385.

sulphide is different from that of the ion of the same composition formed from 3- or 4-labelled compounds. Comparison of these reactivities with the reactivities of $C_2H_3DS^{+\cdot}$ ions from other molecules suggested that two different structures, (38) and (39) are involved.[155]

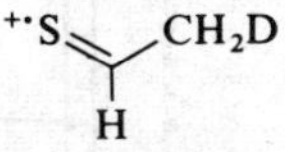

$$\underset{(38)}{\overset{+\cdot}{S}=\underset{H}{C}-CH_2D} \qquad \underset{(39)}{D-\overset{S^{+\cdot}}{\underset{H}{C}}=CH_2}$$

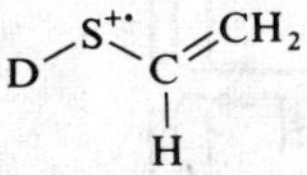

The toluene molecular ion formed by electron impact between 20 and 30 eV reacts with alkyl nitrates:

$$C_6H_5CH_3^{+\cdot} + RONO_2 \rightarrow CH_3C_6H_5\overset{+}{N}O_2 + RO\cdot$$

This reaction could not be detected when cycloheptatriene or norbornadiene were used in place of toluene.[156] This confirms the conclusion drawn from photodissociation studies (see p. 87) that toluene ions in their ground state, which do not have enough energy to decompose, do not rearrange to cycloheptatriene.

8 Plasma Chromatography and Atmospheric Pressure Ionization

Although the mass spectrometric sampling of ions produced at atmospheric pressure by radioactive sources has been used in the past to study ionic processes in radiation chemistry, it is only recently that such an ionization method has been applied to analytical problems. There have been two approaches to the problem of mass analysis of the ions produced; extraction of ions into the vacuum system of a mass spectrometer (API m.s.), and the measurement of ion-drift velocities in gas at atmospheric pressure (Plasma Chromatography).

A schematic diagram of a Plasma Chromatograph cell is shown in Figure 1. The gas molecules are ionized by radiation from a ^{63}Ni source. In the plasma of negatively and positively charged species in 'dry' air the most abundant ions are $O_2^-(H_2O)_n$ and $H^+(H_2O)_n$. A pulse of ions is allowed to pass through the shutter grid. After a time lapse a pulse is applied to the gate grid which allows those ions of a given drift time to be observed. A complete plasmagram is generated by plotting ion current against delay time. The delay time can be converted to a more generally applicable parameter, the reduced mobility K_0, which is the speed of an ion in an electric field of 1 V cm^{-1} in a gas at 273 K, 760 Torr. Using a simple apparatus as described, observable currents have been detected for 10^{-12} molar parts of dimethyl sulphoxide in air.[157]

[155] K. B. Tomer and C. Djerassi, *J. Amer. Chem. Soc.*, 1973, **95**, 5335.
[156] M. K. Hofmann and M. M. Bursey, *Tetrahedron Letters*, 1971, 2539.
[157] M. J. Cohen and F. W. Karasek, *J. Chromatog. Sci.*, 1970, **8**, 330.

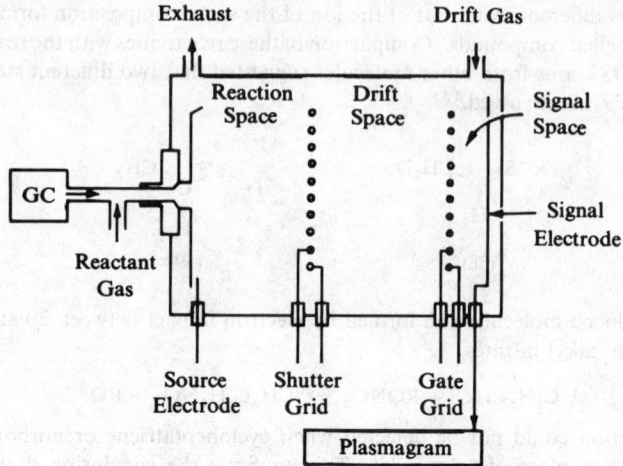

Figure 1 *Schematic diagram of Plasma Chromatograph cell*
(Reproduced by permission from *J. Chromatog. Sci.*, 1970, **8**, 330)

In this and a number of other publications on PC the identification of mass has been attempted by interpolation in a mass–mobility curve in which ion–molecule systems which have been studied by API m.s. are used for calibration. A detailed examination of such relationships has been carried out which throws some doubt on interpolations of this nature.[158] The relationship between mobility and mass can be described by the equation

$$K = \frac{3e}{16N}\left[\frac{1}{m} + \frac{1}{M}\right]^{1/2}\left[\frac{2\pi}{kT}\right]^{1/2}\frac{1}{\Omega_D}$$

where N = neutral density, e = charge, M = neutral mass, m = ion mass and Ω_D = collision cross section. When m becomes much greater than M the $[1/M + 1/m]^{1/2}$ term becomes insignificant and the cross-section term becomes more important. In general, mass-mobility correlations can only be expected to be accurate within $\pm 20\%$ but for a similar group of ions, e.g. a homologous series the accuracy may be as good as 2%. It was also suggested that it may be more useful to consider ion mobilities in the same manner as retention times are treated in other forms of chromatography.

A number of applications have been suggested. It is possible to identify n-octanol and n-nonanol in air by the presence of $(C_nH_{2n+2}O)_2H^+$ or $(C_nH_{2n+2}O)_2O_2^-$ ions.[159] The appearance of the chromatograph changes with sample pressure; the principal ions observed from phenylethyl alcohol were

---

[158] G. W. Griffin, I. Dzidic, D. I. Carroll, R. N. Stillwell, and E. C. Horning, *Analyt. Chem.*, 1973, **45**, 1204.
[159] F. W. Karasek and M. J. Cohen, *J. Chromatog. Sci.*, 1971, **9**, 390.

$C_8H_{10}O(H_2O)_3H^+$ and $(C_8H_{10}O)_2H^+$ at low sample pressure and $(C_8H_{10}O)_3$-H_3O^+ at higher pressures. The plasmagram of benzoic acid was more intense in the negative ion mode than in the positive ion mode, whereas the reverse was true of naphthalene.[160] The mono- and di-chlorobiphenyls gave intense positive responses but no negative response, but the tetra-, hexa-, octa-, and deca-substituted compounds gave increasingly intense negative responses, corresponding to non-dissociative electron capture,[161] although halide ions were observed for all the monohalogenobenzenes except fluorobenzene.[162] PC has been used to investigate the effect of air on the response of electron capture detectors in g.c. Such detectors measure the decrease in electron current when electron acceptors are present. The presence of air reduces the electron current and produces competitive ionizers, $(H_2O)_nO_2^-$.[163] In the positive plasmagram of air both NO^+ and H_2ONO^+ are formed in measurable amounts.[164] The addition of nitric oxide to the reactant gas increases the abundance of these ions and also increases the sensitivity of the system to aliphatic and aromatic hydrocarbons owing to the reactions

$$R-CH_2-R^1 + NO^+ \rightarrow R-\overset{+}{CH}-R^1 + NOH$$
$$ArH + NO(H_2O)^+ \rightarrow ArHNO^+ + H_2O$$

An atmospheric-pressure ionization mass spectrometer has been described which uses the same ionization process, but extracts the ions into a vacuum system and uses a quadrupole mass analyser.[165] Using this system it was possible to detect nicotine in the unconcentrated benzene extracts of the urine of smokers, and of non-smokers who had previously shared the same room. Nicotine was also detected in the air of the room. Some confirmation of the identity of m/e 163 as the protonated molecular ion of nicotine was obtained by adding pyridine to the air. This molecule is more basic than most of the other molecules present in small amounts in air and suppresses the ionization of everything except nicotine.[166] When benzene solutions are used in the API source the ions formed from the carrier gas (usually N_2) ionize benzene molecules by charge exchange. The reactant ions which ionize molecules of the solute are $C_6H_6^{+\cdot}$ and $C_6H_7^+$ which react by charge transfer and proton transfer respectively. One of the most interesting features of API mass spectrometry is its absolute sensitivity. Using single-ion monitoring, benzene solutions containing down to 150 fg of 2,6-dimethyl-γ-pyrone could be analysed.[167]

---

[160] F. W. Karasek, W. D. Kilpatrick, and M. J. Cohen, *Analyt. Chem.*, 1971, **43**, 1441.
[161] F. W. Karasek, *Analyt. Chem.*, 1971, **43**, 1982.
[162] F. W. Karasek and O. S. Tatone, *Analyt. Chem.*, 1972, **44**, 1758.
[163] F. W. Karasek and D. M. Kane, *Analyt. Chem.*, 1973, **45**, 576.
[164] F. W. Karasek and D. W. Denney, *Analyt. Chem.*, 1974, **46**, 633.
[165] E. C. Horning, M. G. Horning, D. I. Carroll, I. Dzidic, and R. N. Stillwell, *Analyt. Chem.*, 1973, **45**, 936.
[165] E. C. Horning, M. G. Horning, D. I. Carroll, R. N. Stillwell, and I. Dzidic, *Life Science*, 1973, **13**, 1331.
[167] D. I. Carroll, I. Dzidic, R. N. Stillwell, M. G. Horning, and E. C. Horning, *Analyt. Chem.*, 1974, **46**, 707.

9 Other Methods of Ionization

Although a number of methods involving bombardment of solid surfaces have been used in mass spectrometry, they have usually been restricted to the production of atomic ions and the applications have been confined to elemental analysis of bulk or surface materials. An attempt has been made to apply laser ionization techniques to the analysis of involatile organic solids. In such spectra of sodium salts of alkylsulphonic acids the most abundant ion observed is $[M + Na]^+$. Similar molecular weight information can be obtained from the laser ionization spectra of sodium alkylthiosulphates. In addition to $[M + Na]^+$ ions, some inorganic fragments were observed, such as $Na_3SO_3^+$.[168] The ion source uses a 0.1 J, 694.3 nm ruby laser, but the authors made no suggestions as to the ionization mechanism. Although this technique may give equally useful results to those obtained by FD for salts, the laser method has not been applied to compounds of low thermal stability. It may be that local variations in temperature at the bombarded surface might lead to decomposition. Attempts to obtain mass spectra of high molecular weight materials by the electrospray technique have continued; in this technique solutions volatilized in a jet to form a supersonic beam of gas phase macroions are then analysed by a repeller grid system. Polystyrene of average molecular weight 5.1×10^4 gives ions of unit charge often containing more than one macromolecule. Polystyrene of higher molecular weight gives spectra which can be interpreted in terms either of formation of ions with between 2 and 5 charges per polymer molecule or of degradation of the polymer to smaller, unit charge fragments. It would therefore appear that this method is very much in the experimental stage.[169]

---

[168] R. O. Mumma and F. J. Vastola, *Org. Mass Spectrometry*, 1972, **6**, 1373.
[169] D. Teer and M. Dole, 165th National Meeting of the American Chemical Society, Dallas, April 1973, Physical Chemistry Abstracts, No. 14.

4
Computerized Data Acquisition and Interpretation

BY F. A. MELLON

1 Introduction

The linking of computers to mass spectrometers has been one of the most fruitful developments in analytical organic chemistry and biochemistry in the past ten years. The use of mass spectrometer-computer systems, a relative rarity in Britain even as late as 1969, is now an accepted feature of many research and analytical laboratories. A variety of systems is currently being marketed, with many different types of mass spectrometer (magnetic sector, Nier–Johnson, Mattauch–Herzog, quadrupole, time-of-flight, and some specialized geometries) interfaced to various data systems. The data systems vary considerably in degree of sophistication, and range in price from as low as £8000 to an average upper limit in the range £40 000—50 000, so that most laboratories are now able to obtain a computer tailored to their particular problems and budgets. The programs (or 'software') provided with commercial mass spectrometer–computer systems are now reasonably standard, allowing for one or two special features found in some configurations, and are generally written in a form which requires minimum operator intervention in processing mass spectroscopic data. This object of minimum intervention can be taken to extremes with a resulting loss of flexibility which can lead to disadvantages when users wish to incorporate their own particular modifications into a system.

It was noted in previous Reports[1] that the routine use of computers in conjunction with mass spectrometers and the concomitant increase in mass spectroscopic information had resulted in an expanding interest in methods whereby large amounts of data might also be interpreted by computer. This trend continues, as shown by the fact that the largest number of papers published in the mass spectrometer–data processing field in the past two years has been concerned with heuristic, pattern recognition, and library matching techniques, all of which help to relieve the chemist of the psychological disorientation and acute myopia induced by long hours of examining too many mass spectra.

---

[1] S. D. Ward, in 'Mass Spectrometry', ed. D. H. Williams (Specialist Periodical Reports), The Chemical Society, London, 1971, Vol. 1, 1973, Vol. 2.

Reviews concerned with aspects of mass spectrometer–computer techniques have appeared, including some in recent books.[2,3] Review coverage has ranged from interpretive use of computers in mass spectrometry,[4—7] to more general reviews of data acquisition and processing in this field.[8—11] Several important conferences on mass spectrometry have included a large number of contributions on computer applications.[12,13]

2 On-line Data Acquisition and Control

System Interface and Configuration.—The interface between a mass spectrometer and computer, necessary to convert the continuous (analogue) signal emerging from the mass spectrometer ion detector into a computer-compatible (digital) form, is known as an analogue-to-digital converter (ADC). The design of ADC's for use in mass spectrometry no longer presents problems and there has been little interest in this area during the period covered by this Report. One notable exception has been the successful development of an ADC which enhances the mass spectrometer signal by an integration technique.[14] When a high-speed ADC is used with a sample-and-hold amplifier a single analogue value on a peak envelope is converted into a single digital value. The time required for the completion of this process is typically *ca.* 1 μs. In the case of small peaks, unfavourable noise and ion statistical effects may result in the conversion of an unrepresentative measurement and as a consequence the peak may either not be recognized at all or be poorly resolved. A solution to this problem was accomplished by integrating between analogue-to-digital conversions and using this integrated value as a digital sample. The circuit designed to perform this function (called an 'incremental integration system') gave improved reliability for small-peak detection provided the noise period was short relative to integration time, *i.e.* the method works with an electron multiplier detector, but not a Faraday cup collector.[14] The design and construction of ADC's has been greatly simplified by

---

[2] 'The Applications of Computer Techniques in Chemical Research', ed. P. Hepple, Institute of Petroleum, London, 1972.
[3] 'Computers in Chemistry', Vol. 39 of Topics in Current Chemistry, Springer-Verlag, Berlin, 1973.
[4] F. W. Karasek, *Analyt. Chem.*, 1972, **44**, 32A.
[5] T. L. Isenhour and J. B. Justice, Sixth International Mass Spectrometry Conference, Edinburgh, Sept. 1973 (to be published in *Adv. Mass Spectrometry*).
[6] T. Clerk and F. Erni, Ref. 3, pp. 91–107; *Fortschr. Chem. Forsch.*, 1973, **39**, 91.
[7] A. B. Delfino and A. Buchs, *Fortschr. Chem. Forsch.*, 1973, **39**, 109.
[8] K. Biemann, ref. 2, p. 5.
[9] D. Henneberg, K. Casper, E. Ziegler, and B. Weimann, *Angew. Chem., Internat. Edn.*, 1922, **11**, 357.
[10] A. B. Delfino and A. Buchs, ref. 3, pp. 109–137.
[11] S. Sasaki and Y. Ishida, *Kagaku No Ryoiki Zokau*, 1972, **98**, 43.
[12] Twentieth and Twenty-first Annual Conferences on Mass Spectrometry and Allied Topics, 1972 and 1973, organized by the American Society for Testing and Materials, Committee E-14.
[13] Sixth Triennial International Mass Spectrometry Conference, Edinburgh, 1973. Papers contributed to be published in *Adv. Mass Spectrometry*, Vol. 6.
[14] R. P. Page and A. V. Nowak, Proceedings of the Twenty-first Annual Conference on Mass Spectrometry and Allied Topics, San Francisco, 1973, p. 49; R. P. Page, A. V. Nowak, and R. Wertzler, *Analyt. Chem.*, 1973, **45**, 994.

modern solid-state technology, particularly by the development of medium- and large-scale integrated circuits and other devices. Design considerations allowing the construction of ADC's easily and inexpensively using such devices have been described.[15] A device which digitizes mass spectra recorded on oscillograph charts has been reported,[16] and the electronic set-up for a real-time mass spectrometer–data system discussed.[17]

Five basic classes of computer–mass spectrometer system configuration were described in previous Reports.[1] This classification is still fundamentally valid, but the configurations have been upgraded considerably in some cases. For example, the simple mass spectrometer–ADC–computer design has been expanded by the attachment of a second computer.[18] The mass spectrometer was interfaced to a PDP-8 computer and the latter was then coupled to an IBM 1800 situated 500 ft away. The remote computer was used to process high- and low-resolution mass spectra and was able to accept the full mass spectrometer output with no simultaneous interference to four other computer users. A system (LOGOS-II), based on a large XDS Sigma-7 computer which supports a number of mass spectrometers on-line, has been developed.[19] The computer and mass spectrometers use a specialized data acquisition interface and have supporting cathode ray display terminals with graphic capabilities. Configurations have been described in which the mass spectrometer is only one of several instruments coupled to a computer.[20–24]

Turning to less complex (and cheaper!) spheres, it has been shown that a small, versatile process computer can be used successfully for recording and evaluating low-resolution mass spectra.[25] If the computer was operated as an intermediary between the mass spectrometer and a larger computer, paper tape was used for intermediate storage. The system was designed specifically for processing hydrocarbon mixtures. Configurations in which the output of the mass spectrometer is recorded on magnetic tape[26,27] or disc[28] prior to computer processing have

[15] R. E. Dessy and J. Titus, *Analyt. Chem.*, 1974, **46**, 249A.
[16] G. J. Leferink and P. A. Leclercq, *Analyt. Chem.*, 1973, **45**, 625.
[17] G. Goby and L.-D. Nguyen, *Internat. J. Mass Spectrometry Ion Phys.*, 1973, **11**, 215.
[18] W. K. Rohwedder, R. O. Butterfield, D. J. Wolf, and H. J. Dutton, Proceedings of the Twentieth Annual Conference on Mass Spectrometry and Allied Topics, Dallas, 1972, p. 382.
[19] A. L. Burlingame, R. W. Olsen, and R. V. McPherron, Proceedings of the Twenty-first Annual Conference on Mass Spectrometry and Allied Topics, San Francisco, 1973; Sixth International Mass Spectrometry Conference, Edinburgh, 1973, paper 119.
[20] J. T. Clerc, C. Jost, T. Meier, and R. Schwarzenbach, *Chimia*, 1973, **27**, 665.
[21] R. M. Haynes, *Canad. Res. Develop.*, 1972, **5**, 30, 55.
[22] P. Briggs, D. Dix, D. Glover, and R. Kleinman, *Amer. Lab.*, 1972, **4**, 57.
[23] M. R. Fitzgerald and L. E. Bruggeman, *Amer. Chem. Soc. Div. Petrol. Chem. Prepr.*, 1971, **16**, 15.
[24] D. Dell Glover, Proceedings of the International Symposium on Gas Chromatography and Mass Spectrometry, ed. A. Frigerio, 1972, p. 245.
[25] H. D. Engelman and D. Severin, *Chem.-Ing.-Tech.*, 1973, **45**, 1166.
[26] P. J. Black and A. I. Cohen, Proceedings of the Twentieth Annual Conference on Mass Spectrometry and Allied Topics, Dallas, 1972, p. 380.
[27] M. W. Johnson, B. J. Gordon, and R. Self, *Lab. Practice*, 1973, **22**, 267.
[28] W. H. Christie, D. H. Smith, and H. S. McKown, *Chem. Instr.*, 1973, **5**, 43.

been described. One of these tape systems[26] records mass spectra in three intensity channels on Frequency Modulated (FM) analogue tape at a speed of 60 i.p.s. The FM tapes are played back into the computer (*via* an appropriate ADC) at much slower speeds, 3.75 i.p.s. for high resolution, and 7.5 i.p.s. for low-resolution spectra and rapid sampling rates with very fast computer cycle times are unnecessary. This data acquisition system was coupled to a mass spectrometer of Nier–Johnson geometry operating at a scan speed of 16 s per decade. The scope of a gas chromatograph–mass spectrometer–computer system has been detailed,[29] and the problems encountered in coupling a mass spectrometer on-line to a computer have been discussed with reference to a practical example.[30] A mass spectrometer–computer specifically designed for use in air pollution monitoring studies,[31] and a user-designed g.c.m.s.–computer system[32] have been reported. A useful guide to the rational use of a dynamic on-line data processing system for high- and low-resolution mass spectrometry has appeared.[33] The latter report concludes that the best configuration for straightforward g.c.m.s. data handling is a high- and low-resolution instrument attached to a 4 or 8 K store computer equipped with a visual display unit, fast paper-tape punch, and teleprinter. A more flexible system required to perform several analytical functions would require a 16 K core store, disc backing store, visual display, and line printer as an optimum configuration. The other main requirement was found to be fast communication in the integrated on-line mode, *i.e.*, a rapid and flexible means of allowing operator intervention.

In a somewhat unique mass spectrometer–data system, an off-line computer was used to obtain accurate mass data by detecting the matching condition when mass measurement was performed under static conditions.[34] Improved precision and more objective results were obtained in less operating time than required by manual methods, and it was possible to compare the experimental and theoretical limits of accuracy of the measurements from the resolution of the mass spectrometer and the number of ions detected.

On-line Computer Control.—Most of the examples of computer-controlled mass spectrometers quoted in previous Reports[1] concerned descriptions of quadrupole mass filters in which the control voltages of the mass analyser were generated by the data system, *via* a suitable voltage amplifier. Further examples of such systems have appeared,[35,36] including one combination which operates by push-

---

[29] R. Venkataraghavan and F. W. McLafferty, *Chem. Tech.*, 1972, **2**, 364.
[30] J. F. Guilbot, P. Pahaut, L. Jacob, N. Sellier, and G. Guichon, *Chim. Ind. Genie Chim.*, 1972, **105**, 955.
[31] D. Schnetzle, A. L. Crittenden, and R. J. Charlson, *J. Air Pollut. Contr. Ass.*, 1973, **23**, 704.
[32] R. Ryhage, Sixth International Mass Spectrometry Conference, Edinburgh, 1973, paper 118.
[33] H. A. Van't Klooster, J. S. Vaarkamp-Lijnse, and G. Dijkstra, *Org. Mass Spectrometry*, 1974, **8**, 303.
[34] J. O. Meredith, F. C. G. Southon, R. C. Barber, P. Williams, and H. E. Duckworth, *Internat. J. Mass Spectrometry Ion Phys.*, 1973, **10**, 359.
[35] H. P. Hotz and V. L. Dagragnano, Proceedings of the Twenty-first Annual Conference on Mass Spectrometry and Allied Topics, San Francisco, 1973, p. 79.
[36] J. Bargery, J. Farren, R. Hallett, J. Hawke, D. A. Smith, and J. W. Youren, ref. 2, p.

button function selection and includes an oscilloscope display.[35] The computer controls the quadrupole instrument by feeding in mass set voltages through a digital-to-analogue converter (DAC). The resulting ion-intensity output of the mass spectrometer is then digitized and fed into the computer. Mass spectra formed in this way consist of a series of counts per channel, each channel being equivalent to 1 a.m.u. The integration time is determined by the computer and mass spectra are stored on a disc. Half the 2048 data channels in the computer are used to acquire mass spectral line intensities, and the remainder for chromatographic information stored during a g.c.m.s. run. A similar computer-controlled quadrupole instrument has been used to monitor polychlorinated biphenyls.[37] An increasing amount of activity in the area of on-line control in mass spectrometry is now being concentrated on magnetic sector rather than quadrupole mass spectrometers. This has come about because of the widespread use of 'multiple ion detection' (MID) monitoring, mostly applied to g.c.m.s. experiments. MID is a technique for monitoring a limited number of ions by changing the accelerating voltage in the ion source of a magnetic sector mass spectrometer. MID methods are applied to the qualitative and quantitative detection of known compounds, e.g. the detection of drugs and their metabolites, and yield greatly enhanced sensitivity (generally 100–1000 times higher) compared with conventional magnetic scanning. Of course, the information contained in a total mass spectrum is lost, but this is of little consequence when only a few characteristic ions of a known compound are needed for its successful identification. Until fairly recently, MID was achieved by analogue recording of the ions (i.e. a chart recorder) with generation and switching of the ion-accelerating voltage by conventional electronic circuitry, but recently a number of computer controlled systems have been developed. The arrangement of a computer-controlled MID system is usually similar to that shown in the Figure, condensed from the appropriate references.[38—42]

50; H. Egli, W. K. Huber, H. Selhofer, and R. Vogelburg, Sixth International Mass Spectrometry Conference, Edinburgh, 1973, paper 38; W. F. Fies and M. Story, ibid., paper 22.

[37] J. W. Eichelberger, L. E. Harris, and W. L. Budde, Analyt. Chem., 1974, **46**, 227; Proceedings of the Twenty-first Annual Conference on Mass Spectrometry and Allied Topics, San Francisco 1973, p. 334.

[38] D. R. Pelster and J. T. Watson, Proceedings of the Twenty-first Annual Conference on Mass Spectrometry and Allied Topics, San Francisco, 1973, p. 61; J. T. Watson, D. R. Pelster, B. J. Sweetman, J. C. Frolich, and J. A. Oates, Analyt. Chem., 1973, **45**, 2071; J. H. Hengstmann, F. C. Falkner, J. T. Watson, and J. A. Oates, ibid., 1974, **46**, 34.

[39] W. H. Holland, B. Shore, and W. F. Holmes, Proceedings of the Twenty-first Annual Conference on Mass Spectrometry and Allied Topics, San Francisco, 1973, p. 444; W. F. Holmes, W. H. Holland, B. L. Shore, D. M. Bier, and W. R. Sherman, Analyt. Chem., 1973, **45**, 2063.

[40] C. C. Sweeley, N. D. Young, R. E. Teets, M. A. Bieber, and J. F. Holland, Proceedings of the Twenty-first Annual Conference on Mass Spectrometry and Allied Topics, San Francisco, 1973, p. 303; J. F. Holland, C. C. Sweeley, R. E. Thrush, R. E. Teets, and M. A. Bieber, Analyt. Chem., 1973, **45**, 308.

[41] K. Elkin, L. Pierrou, U. G. Ahlbor, B. Holmstedt, and J. E. Lindgren, J. Chromatography, 1973, **81**, 47.

[42] W. G. Stillwell, M. G. Horning, and R. N. Stillwell, Proceedings of the Twenty-first Annual Conference on Mass Spectrometry and Allied Topics, San Francisco, 1973, p. 269.

A representative example of such an arrangement is provided by the work of Elkin et al.[41] The hardware comprises voltage, baseline, and gain controls which are applied sequentially to each mass channel selected. The ion-accelerating voltage of the mass spectrometer is set by the computer through a 14-bit DAC and a voltage booster situated in the voltage control unit of the mass spectrometer. The computer selects the required masses by adding a potential in the range 0—1000 V to the accelerating voltage. This allows a mass range of 30% from the lowest mass to be covered. An eight-bit DAC permits the addition of

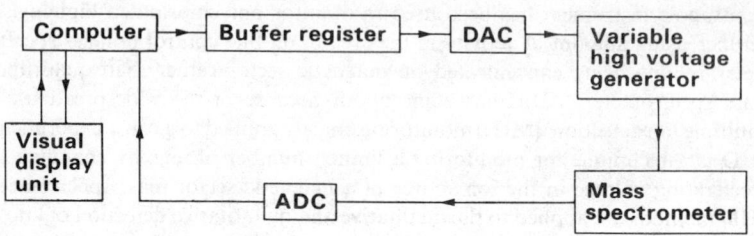

Figure *Typical configuration of a computer-controlled multiple ion detector system (see refs. 38—42)*

d.c. voltage to the analogue mass spectrometer output to cancel out unwanted signal due to g.c. column bleed in the g.c.m.s. combination. The analogue signal from the electron multiplier of the mass spectrometer is fed to a 10-bit ADC *via* a six-bit programmable gain-ranging amplifier, and is then stored by the 8 K computer. The system software permits choice of certain options by the operator *via* an oscilloscope display and keyboard input. The user specifies masses to be monitored, baseline level, sampling rate, and gain per channel. Spectra may be stored on magnetic tape for later examination. Programs calculate peak areas, peak heights, and retention times on completion of scanning, reject spikes, and smooth data by the Savitzky–Golay[43] technique. Mass fragmentograms (graphs of individual ion intensity against time) are displayed on an oscilloscope screen, the number of channels monitored being dependent only on mass storage limitations and data collection rate. Real-time monitoring of peaks emerging from the gas chromatograph is possible, with a teletype providing hard copy output. The usefulness of the system was exemplified by work on the metabolism of pentachlorophenol. In another computer-controlled MID system, required masses were selected after a rapid 1000 V electrical scan.[38] Up to 12 masses could be selected and fine tuning of each mass was obtained by displaying the peak on the oscilloscope output of the data system and adjusting a mass–charge cursor on the screen to a position where it coincided with the centre of a peak of interest. These exact values of mass-to-charge ratio were stored by the computer and

---

[43] A. Savitzky and M. J. E. Golay, *Analyt. Chem.*, 1964, **36**, 1627.

used when cycling through the selected ions. A stable carbon isotope ratio measuring system, operating by computerized peak switching and used in conjunction with an ion-counting detector system, has been reported.[44]

Short, repetitive sweeps of limited mass ranges by computer control of magnet scanning have been successfully used to obtain g.c.m.s.[45] and ^{13}C incorporation data.[46] In the g.c.m.s. application,[45] the operator set the selected masses and the number of scans by using toggle switches on the computer interface. The computer caused the magnet to cycle through a mass range of 20—30 a.m.u. and analysed the incoming ion intensity data. In this way, quantitative estimation of drugs down to 200–300 pg was possible but still at least an order of magnitude away from the higher sensitivity achievable by MID. This particular adaptation was used, instead of controlling the accelerating voltage as in MID, because of voltage drift problems often encountered with this technique. Measurement of stable carbon isotope ratio was accomplished by magnetic scanning of up to 5% of the selected mass range; the computer triggered each scan and stored the accumulated data in two 16-bit words.[46] In this way an intensity ratio of up to 1800 : 1 was measurable.

Some time ago, it was proposed that increased accuracy of mass measurement with a high-resolution double-focusing mass spectrometer could be attained by using the large spaces between peaks. In these spaces, no data are being digitized and it is possible to envisage the rescanning of the peaks themselves during this time by variation of the electrostatic analyser (ESA) voltage.[47] This proposal has now been realized by scanning the ESA voltage in real-time under direct computer feedback control.[48] The process is described as Signal Enhancement in Real-Time (SERT) and, as this title suggests, it gives improved signal-to-noise ratio as well as greater accuracy of mass measurement.

Accuracy of Mass and Intensity Measurements.—Contradictory conclusions concerning the sources of error in high-resolution mass measurement from data obtained by dynamic scanning were reported in Volume 2.[1] One group of workers had claimed that increased instrumental resolving power resulted in increased accuracy of mass measurement.[49] This was contradicted by another group which concluded that accuracy of mass measurement was a function of the product of resolution and sensitivity and was virtually constant over a wide

---

[44] J. M. Hayes and D. A. Schoeller, Proceedings of the Twnety-first Annual Conference on Mass Spectrometry and Allied Topics, San Francisco, 1973, p. 53.
[45] L. Baczynski, D. J. Duchamp, J. F. Zieserl, jun., and U. Axen, *Analyt. Chem.*, 1973, **45**, 479.
[46] R. J. Sykes, W. J. McMurray, J. Gage, P. Arneson, H. H. Wasserman, and S. R. Lipsky, Proceedings of the Twentieth Annual Conference on Mass Spectrometry and Allied Topics, Dallas, 1972, p. 173.
[47] F. W. McLafferty, J. A. G. Michnowicz, R. Venkataraghavan, P. Rogerson, and B. Giessner, *Analyt. Chem.*, 1972, **44**, 2282.
[48] F. W. McLafferty, J. A. G. Michnowicz, and R. Venkataraghavan, Proceedings of the Twentieth Annual Conference on Mass Spectrometry and Allied Topics, Dallas, 1972, p. 379.
[49] R. J. Klimowski, R. Venkataraghavan, and F. W. McLafferty, *Org. Mass Spectrometry*, 1970, **4**, 17.

range of resolving powers.[50] Certainly, theoretical considerations based on the number of ions per peak support the latter conclusion. The high resolving powers (5000—10 000) required in conventional mass spectrometers are necessary in order to resolve the peaks of the unknown and reference compound (usually perfluorokerosene). The advent of the double-beam mass spectrometer has made possible measurements[51] which confirm the deductions made by Smith et al.[50] The double-beam instrument allows unequivocal separation of compound and reference peaks without high-resolution conditions by electronically offsetting the separate reference and sample ion intensity output. The operation of such a double-beam mass spectrometer in conjunction with an on-line data system has been investigated with particular reference to mass measurement accuracy.[51] Samples were introduced into the mass spectrometer through a gas chromatograph, with the resolution of the mass spectrometer in the double beam mode set to 1600. Magnetic scans were done at 10 s per decade. For all peaks of magnitude down to 2% of the base peak with a 1 μg sample, and down to 20% of the base peak with 100 ng, 70% of the peaks were measured with an accuracy of better than 10 p.p.m. and 96% better than 15 p.p.m. Of particular interest was the comparison made between measurements in the single- and double-beam modes (Table 1).

The increased sensitivity available for high-resolution mass measurements in the double-beam mode is clearly visible, and the accuracy of the measurements entirely vindicates earlier conclusions on the relation between accuracy of mass measurement and resolution.[50]

Determinations of the accuracy of mass measurement on single-beam mass spectrometer–computer systems have been reported.[52,53] In one of these studies,

Table 1 *Comparison of accuracy of mass measurement between double- and single-beam modes of a double-beam mass spectrometer—computer system (data from ref. 51)*

Atomic composition (dichlorobenzene molecular ion)	Relative intensity	Calculated mass	Error^a m.m.u.	Error^b m.m.u.
$^{12}C_6H_4^{37}Cl_2$	8.4	149.9630	−0.5	Not observed
$^{12}C_6H_4^{35}Cl^{37}Cl$	57.2	147.9660	0.8	1.4
$^{13}C^{12}C_5H_4^{35}Cl_2$	8.0	146.9723	−1.2	3.1
$^{12}C_6H_4^{35}Cl_2$	100.0	145.9689	−1.2	2.4

(a) 50 ng on the g.c. column, 1500 resolution, double-beam mode
(b) 250 ng on the g.c. column, 10 000 resolution, single-beam mode

[50] D. H. Smith, R. W. Olsen, E. C. Walls, and A. L. Burlingame, *Analyt. Chem.*, 1971, **43**, 1796.
[51] M. L. Aspinal, J. R. Chapman, K. R. Compson, D. Hazelby, and A. Riddoch, Proceedings of the Twenty-first Annual Conference on Mass Spectrometry and Allied Topics, San Francisco, 1973, p. 471.
[52] R. S. Gohlke, G. P. Happ, D. P. Maier, and D. W. Stewart, *Analyt. Chem.*, 1972, **44**, 1484.
[53] R. M. Hilmer and J. W. Taylor, *Analyt. Chem.*, 1973, **45**, 1031.

the average errors in mass for major ions in each scan were found to lie between 3.1 and 5.7 p.p.m. when care was taken over adjustment of the instrument.[52] The other report discusses important instrumental characteristics of a double-focusing mass spectrometer which influence the choice of method for on-line data acquisition.[53] An average error better than 10 p.p.m. was found, and peaks as small as 0.05% of the base peak were detected. Accuracy was improved greatly (to better than 3 p.p.m.) when the mass spectrometer was isolated from vibration effects.

An on-line data system has been used to compare high-resolution photographic recording with electrical recording for the acquisition of g.c.m.s. data.[54] The instrument used had Mattauch–Herzog geometry and was equipped for both photographic and electrical recording. In the electrical recording mode, scan speed and resolution were varied in the range 8—24 s per decade and 5000—10 000, respectively. The resolution in the photographic mode was 40 000. The comparison showed that, for recording times < 10 s, photographic recording allowed more accurate determination of mass in a larger mass range with four times the resolution and 20 times less sample than did electrical recording. For injections of less than 20 μg, reasonable accuracy in mass measurement could only be achieved by photographic recording. In spite of this, most g.c.m.s. users will find the convenience of electrical recording an overwhelming advantage over photographic recording. Furthermore, photographic recording of g.c.m.s. runs containing a large number of closely separated g.c. peaks is presently impossible because of the time required to change photographic plates.

Apart from instrumental considerations, accuracy of mass and intensity measurement may also be improved by using carefully chosen computer algorithms. The problem of detecting mass peaks in digital ion current samples is usually overcome by specifying an intensity threshold and minimum number of digital samples; when these are exceeded a peak is defined and the computer calculates centroid times and peak areas. An alternative scheme, which has been applied to the output of a quadrupole mass spectrometer scanning at a rate of 325 a.m.u. s^{-1}, is to use decision processes (already established in the communications field) to recognize peaks.[55] The decision function was based on a modified cross-correlation technique[56] which is used to recognize peak shapes similar to a reference shape. The reference shape is defined by a sequence of numbers, f_t, and the shape adopted for skewed quadrupole mass peaks was the integer sequence 1, 3, 4, 3, 2, 1. This peak shape was cross-correlated with stored digital ion current values represented by $g_{t+\tau}$. When the reference peak shape chosen is as described, the discrete cross-correlation function C_τ, given by the equation:

$$C_\tau = \sum_{t=0}^{m-1} f_t g_{t+\tau}$$

[54] K. Habfast, K. H. Maurer, and G. Hoffman, Proceedings of the Twentieth Annual Conference on Mass Spectrometry and Allied Topics, Dallas, 1972, p. 414.
[55] N. W. Bell, Proceedings of the Twenty-first Annual Conference on Mass Spectrometry and Allied Topics, San Francisco, 1973, p. 193.
[56] J. S. Bendat and A. G. Piersol, 'Measurement and Analysis of Random Data', Wiley, 1966.

is the function $g_0 + 3g_1 + 4g_2 + 3g_3 + 2g_4 + g_5$. When the signal and reference sequences match, the cross-correlation function rises sharply. In order to eliminate noise the cross-correlation function was modified by subtracting the average of the stored digital samples from each digital value, thus giving the modified detection function, $-g_0 + g_1 + 2g_2 + g_3 - g_5$.[57] In practice, sampling at $\frac{1}{10}$ a.m.u. intervals followed by computer processing resulted in better noise immunity and more accurate mass and intensity values in a wider dynamic range. Other workers have noted, however, that problems can arise when peak shapes are assumed because of distortion of the peaks from the assumed shape.[53] This comment was confined to calculation of peak areas and centroid times by conventional methods and may not apply strictly to cross-correlation functions.

Following an earlier idea of Dromey and Morrison,[58] a computer technique for removing instrumental 'smearing' has been described.[59] The distortion of an observed mass spectrometer ion peak from an 'ideal' shape can be represented by the Fourier convolution integral:

$$F(x') = \int_{-\infty}^{\infty} T(x)A(x' - x)\,dx$$

or $F = AT$ in matrix form. It is possible to enhance the effective resolution of a mass spectrometer by solving this equation for T. A number of synthetic problems were generated by computer methods and then solved using either modified Fourier or iterative techniques. In a practical test, spectra from a defocused time-of-flight mass spectrometer in the limited mass range of 36—46 a.m.u. were processed. This mass region was chosen because the instrument function could be obtained using the spectra of pure argon and CO_2. The iterative technique successfully resolved all the peaks in this region with adequate quantitative accuracy. Fourier methods, however, would not yield quantitative results for unresolved peaks varying by more than a factor of two or three. The results are promising and should be improved upon once noise degradation and other problems are solved.

3 Interpretation of Mass Spectra by Heuristic, Pattern Recognition, and Library Search Techniques

The increased use of on-line mass spectrometer–computer systems has resulted in an enormous expansion in the output of mass spectroscopic data. The consequence of this has been a sort of 'data bottleneck' where the flow of information greatly exceeds the rate at which the chemist is capable of interpreting it. As a result, there has been a great deal of interest in the development of computerized methods to identify a compound, or at least its main structural features, from its mass spectrum. Three conceptually different approaches have been used. The

---

[57] W. W. Black, *Nuclear Instr. and Methods*, 1969, **71**, 317.
[58] R. G. Dromey and J. D. Morrison, *Internat. J. Mass Spectrometry Ion Phys.* 1971, **6**, 253.
[59] M. F. Zabielski and T. M. McHugh, Proceedings of the Twenty-first Annual Conference on Mass Spectrometry and Allied Topics, San Francisco, 1973, p. 198.

simplest of these attempts to match an unknown mass spectrum against a reference file of mass spectra of known compounds, *i.e.* the library search method. Parameters are calculated which indicate the degree to which the unknown spectrum resembles those in the library. Another approach uses *a priori* knowledge about the way mass spectra are formed, *i.e.* the computer is programmed to follow the same steps a human interpreter might take. This is the technique of heuristic analysis and may be exemplified, at a very simple level, by the programs which calculate empirical formulae, or produce element maps, from high-resolution mass spectra. At a more complex level, programs which simulate the fragmentation of molecules are devised. Probably the first mass spectroscopic heuristic program to follow this approach was a peptide-sequencing algorithm.[60] Finally, completely empirical methods for recognizing patterns in mass spectral data have been formulated using general approaches devised by information scientists. Each of these techniques is now discussed further.

Pattern Recognition.—Several excellent texts which explain pattern recognition methods in detail are available,[61] but a brief description of these techniques is appropriate. The pattern recognition techniques which have been applied to the processing of mass spectroscopic data represent low-resolution mass spectra of known compounds (known as the training set) as points or vectors in multi-dimensional space. The co-ordinates of each of these points are the ionic intensities in the corresponding mass spectrum. Mass spectra with certain structural features then (hopefully) cluster in this hyperspace in such a way that boundaries between different classes of compound may be defined. For example, all the compounds containing a phenyl group should fall on one side of the boundary, all those which do not should fall on the other. In order to define these boundary conditions (known as pattern classifiers, or decision surfaces), a number of different approaches have been devised. Having derived a pattern classifier from the set of known mass spectra, the classifier is then applied to the mass spectrum of an unknown; the latter is classified according to whichever side of the decision surface it appears in the hyperspace. The use of a number of different (but more or less related) pattern classification schemes faces the mass spectroscopist with the problem of selecting the most reliable one. The appearance of a report which analyses the ability of six different pattern recognition schemes to extract structurally significant information from a set of mass spectra is therefore especially welcome.[62] The six methods compared for their ability correctly to identify selected structural features from mass spectra were: sum spectra, normalized sum spectra, binary spectra, non-linear transformation, nearest neighbour, and learning machine. The 'sum spectra' approach calculates the 'average' spectrum of each of two

---

[60] K. Biemann, C. Cone, B. R. Webster, and C. P. Arsenault, *J. Amer. Chem. Soc.*, 1966, **88**, 5598; M. Senn and F. W. McLafferty, *Biochem. Biophys. Res. Comm.*, 1966, **23**, 4.
[61] N. J. Nilsson, 'Learning Machines', McGraw-Hill, New York, 1965; K. S. Fu, 'Sequential Methods in Pattern Recognition and Machine Learning', Academic Press, New York, 1968; W. S. Meisel, 'Computer Oriented Approaches to Pattern Recognition', Academic Press, New York and London, 1972.
[62] J. B. Justice and T. L. Isenhour, *Analyt. Chem.*, 1974, **46**, 223.

different classes of compound. An unknown is then classified by measuring its relative distance (in the hyperspace) from the means of the two classes considered and this is equivalent to using a decision function which passes between the two average spectra in the hyperspace. The technique can be improved by ensuring that each spectrum in the training set contributes equally to the average spectrum — the 'normalized sum spectra' approach. 'Binary spectra' are mass spectra reduced to a 'peak or no peak' form, with no intensity data.[63] 'Mean spectra' were calculated from sets of these binary spectra (appropriately categorized), these means having the significance that each mass position value corresponds to the probability of a peak being present in a spectrum of that particular class. The 'non-linear transformation' method is also based on the average spectrum approach, except that a generalized Walsh transform:

$$T(I) = (e^{2\pi i/m})^I$$

(where I = intensity of peak and m = maximum intensity + 1) is used to modify the data prior to summation.[64] This transformation is especially appropriate to the separation of data sets which are not *linearly* related to the desired classifications. Furthermore, it does not require the excessive computer time and storage demands of previously suggested non-linear transforms. The 'nearest neighbour' scheme compares unknown spectra with *each* spectrum in the set of known compounds. The Euclidean distance in n-dimensional space (where n is the number of mass positions) is measured between the unknown and each of the known spectra in the data set. The unknown is then assigned to the class in which its nearest neighbour appears. A modification of this technique is the 'k-nearest neighbour' method in which the unknown is classified according to whichever compound class is most strongly represented among its k-nearest neighbours (this method has already been applied to the interpretation of n.m.r. spectra).[65] The last classification scheme to be compared, the 'learning machine' approach, uses a vector **W** specially developed by an iterative 'training' program to classify mass spectra. The training program produces a function which defines a hyperplane in n-dimensional space separating the two categories of data. The product of the vector **W** (known as the Binary Pattern Classifier) and the unknown mass spectrum is positive if the unknown has the characteristic described by **W**. All these pattern recognition schemes were applied to 630 low-resolution mass spectra in order to predict 16 structural features. The order of predictive ability was found to be: sum spectra < binary spectra < normalized sum spectra < non-linear transform = learning machine < nearest neighbour.[62] The nearest neighbour method worked best with only one nearest neighbour; inclusion of more than one degraded the selectivity of the technique because of the increased probability of including compounds belonging to other classes. The better

[63] S. L. Grotch, *Analyt. Chem.*, 1971, **43**, 1362.
[64] J. B. Justice, D. N. Anderson, T. L. Isenhour, and J. C. Marshall, *Analyt. Chem.*, 1972, **44**, 2087.
[65] B. R. Kowalski and C. F. Bender, *Analyt. Chem.*, 1972, **44**, 1405.

predictive ability of 'binary spectra' compared with 'sum spectra' was thought to suggest the presence of considerable 'noise' in the original spectra (as far as structural identification is concerned). The easiest structural features to recognize were phenyl groups (97% correct predictions). This is hardly surprising as these are probably the groups most rapidly identified by human interpreters. Compounds containing phenyl groups are generally characterized by a few intense ions, some at low mass being especially diagnostic. The relative lack of fragmentation in the mass spectra of compounds containing phenyl groups compared with aliphatic systems means that structurally significant ions are less likely to be obscured by 'noise', i.e. overlapping fragmentation processes. In terms of pattern classification, phenyl groups are said to be linearly related to structural information in the mass spectra. Other easily identified types of functionality were the presence of 3 or more double bonds, of nitrogen and of amine, and the correct ratio of carbon to hydrogen. A general conclusion of the report was that mass spectra contain information which is stored non-linearly but which can be approximated to a linear function with an accuracy dependent on the type of information sought.[62]

Learning machine methods require a preliminary decision as to the number and types of categories to be looked for. Once the types of structural information sought have been decided upon, a technique for making multi-category decisions must be devised. As examples, a tree-branching arrangement of pattern classifiers has been used to determine empirical formulae,[66] and parallel arrangements of classifiers with different cut-off points have also been employed.[67] Finally, the use of 'Hamming-type' binary codes with possible self-correction capabilities has been suggested.[68] The efficacies of these three multi-category prediction techniques have been compared for their utility in processing mass spectra, using a set of C_3—C_{10} and a subset of C_4—C_{10} hydrocarbons.[69] Special attention was focused on the question of whether the weight vectors for the pattern classifiers could be successfully trained. Branching-tree methods of multi-category classification are virtually self-explanatory, successive classifiers narrowing the choice of different structural features to increasingly better defined categories *via* a branching network in which each branch point separates two categories of data. The parallel classification scheme categorizes the pattern vectors (mass spectra) into one of two categories separated by a cut-off point. A number of pattern classifiers are trained, each with a different cut-off point, in order to classify correctly the data into its most well-defined category. The binary code scheme involves the use of binary numbers to designate the number of carbon atoms in a molecule. A molecule with nine carbons would be represented by the binary number 1001. If four-bit representation is used, up to 15 carbon atoms can be trained, each bit being paralleled by a training set corresponding

---

[66] P. C. Jurs, B. R. Kowalski, and T. L. Isenhour, *Analyt. Chem.*, 1969, **41**, 21.
[67] P. C. Jurs, B. R. Kowalski, T. L. Isenhour, and C. N. Reilley, *Analyt. Chem.*, 1970, **42**, 1387.
[68] F. E. Lytle, *Analyt. Chem.*, 1972, **44**, 1367.
[69] W. L. Felty and P. C. Jurs, *Analyt. Chem.*, 1973, **45**, 885.

to a classifier. Thus only four classifiers are required for 15 carbons (binary 15 = 1111). The accuracy of this binary representation can be increased by incorporating extra digits to form an error-correcting Hamming code. With a data set of 600 C_3—C_{10} hydrocarbon mass spectra the highest predictive ability, 95.4%, was found for both the branching-tree and Hamming seven-bit binary coding schemes. With the entire set of spectra (including compounds with up to four oxygens and/or two nitrogens) in the training set, the branching-tree method gave the most reliable prediction (76.3%). The results therefore favour the use of binary coding with self-correction facilities, provided the training set is initially sorted into subsets (a hydrocarbon subset in this example) because the binary method requires a smaller number of pattern classifiers than the branching-tree and parallel schemes and saves computer time.

One of the main problems associated with the pattern recognition technique is that the standard representation of spectral information in hyperspace does not always yield linearly separable subspace. This problem can be overcome by improving the classification rule or by changing the representation of the information. An approach which follows the former course by defining optimum variables for spectral interpretation has been reported.[70] The approach works by defining new variables which are approximate class separators and by constructing sub-pattern spaces containing two classes only. Instead of the intensities of each m/e value in the mass spectra, moments were used giving ten variables per spectrum. The average degradation in classification efficiency of 4% was more than offset by the considerable saving in computer time and storage.[71] Using the k-nearest neighbour rule (with $k = 1$ and 3) a minimum of 9% increase in classifier efficiency despite a great reduction in the number of variables was obtained using the optimization method described.[70] An extension of the aforementioned techniques has resulted in the development of an efficient multi-class linear classifier by compromising linear separability for interclass separability.[72] The accuracy of this process was found to be comparable with the computationally more expensive k-nearest neighbour method. The alternative approach to improving pattern recognition, changing the representation of spectra in hyperspace, has also yielded encouraging results. Such transform techniques have already been mentioned,[64] but other transformation processes have been suggested, such as Hadamard transforms based on Walsh functions.[73] These Walsh functions are analogous to Fourier series (they have the appearance of infinitely clipped Fourier functions) and Hadamard transforms are simply discretely sampled Walsh functions. It was possible to generate Hadamard sequence energy spectra and use these in the k-nearest neighbour pattern classification scheme. An improvement in classification performance of *ca.* 6% over untransformed mass spectral data was found. Furthermore, the Hadamard transformed mass spectra

---

[70] C. F. Bender and B. R. Kowalski, *Analyt. Chem.*, 1973, **45**, 590.
[71] C. F. Bender, H. D. Shepherd, and B. R. Kowalski, *Analyt. Chem.*, 1973, **45**, 617.
[72] C. F. Bender and B. R. Kowalski, *Analyt. Chem.*, 1944, **46**, 294.
[73] B. R. Kowalski and C. F. Bender, *Analyt. Chem.*, 1973, **45**, 590.

could still operate usefully in pattern recognition processes even with a considerably reduced number of variables.

In an interesting application of pattern recognition approaches, programs were devised which generated simulated mass spectra of small organic molecules.[74] Molecular structures were represented in computer-compatible form by using fragmentation codes which designate specific groups of atoms and/or bonds within molecules. These *descriptive* pattern classifiers were then used to predict the presence or absence of mass spectral peaks at 60 nominal m/e values, and also to predict the intensities of eleven of these peaks. To test the simulation, a number of complete mass spectra were developed and these were found to give 93% of classifications correctly. The simulation technique was subsequently improved by training the pattern classifiers using an interactive least-squares procedure and an 'attribute inclusion algorithm' *i.e.* a program which extracted features from the molecular descriptions and investigated the importance of multiple features.[75]

Another novel use of pattern recognition methods has been an attempt to correlate the pharmacological activity of certain drugs with their mass spectra.[76] A number of different pattern recognition approaches were applied to the mass spectra of 30 sedatives and 36 tranquillizers. The methods used were the k-nearest neighbour technique (see above), a variant of the latter (non-linear mapping[77]), weighting of peaks,[76] and 'Fisher direction' approaches.[78] Non-linear mapping involves reduction of the n-dimensional points which represent the mass spectra to points in two- or three-dimensional space in which all the interpoint distances are preserved as well as possible. Judgements are then made by a human investigator who views this plot and attempts to classify unknowns by observing which cluster they appear closest to. Weighting of peaks, as its name suggests, is a method whereby structurally significant ions are given added prominence,[76] and the 'Fisher direction' techniques are similar to non-linear mapping except that the n-dimensional points are *projected* onto a two-dimensional plane. The k-nearest neighbour approach yielded an 83% successful classification, but non-linear mapping was not particularly successful as only a moderate degree of clustering was observed. In the other experiments correct classifications were obtained, although ambiguity was sometimes possible. A notable success was the correct classification of structurally diverse types. The results can only be regarded as preliminary at present, having already provoked a vigorous correspondence[79] on the merits and demerits of this approach to drug recognition, and final resolution of the problem awaits comprehensive testing on considerably larger data sets, with more diverse pharmacological activities. If the

[74] J. Schechter and P. C. Jurs, *Appl. Spectroscopy*, 1973, **27**, 30.
[75] J. Schechter and P. C. Jurs, *Appl. Spectroscopy*, 1973, **27**, 225.
[76] K.-L. H. Ting, R. C. T. Lee, G. W. A. Milne, M. Shapiro, and A. M. Guarino, *Science*, 1973, **180**, 417.
[77] J. W. Sammon, *IEEE Trans. Comput.*, 1969, **C18**, 401; B. R. Kowalski and C. F. Bender, *J. Amer. Chem. Soc.*, 1972, **94**, 5632.
[78] J. W. Sammon, *IEEE Trans. Comput.*, 1970, **C19**, 826.
[79] C. L. Perrin, *Science*, 1973, **183**, 551; K.-L. H. Ting, *Science*, 1973, **183**, 552.

technique proves to be valid, its potential is certainly very exciting, and any proven relation between mass spectra and pharmacological activity should have interesting consequences for current theories relating both properties to energetic and stereochemical data.

The molecular structures of low molecular weight organic compounds have been determined by applying linear binary pattern classifiers to low-resolution mass spectra.[80] The best preprocessing technique to apply to the mass spectra was adjudged to be a logarithmic transform. Because of limitations in the size of some of the training data sets, only ten molecular classes and elements out of 31 were found to be suitable for automatic interpretation. A simplified learning machine allowing data compression before training and yielding a reliable classifier even with linearly inseparable data in the training set has been used to form an eight-feature pattern classifier able to identify compounds of the structure $(RO)_2P(=X)Y$, where $R = H$, Me, or Et, $X = O$ or S, and Y is any functional group.[81] Alkanes and alkenes have been structurally identified with moderate success by classification into seven patterns by clustering and by using only the ten most intense peaks in their mass spectra.[82] Similar techniques were employed in an attempt to apply pattern recognition criteria to the mass spectra of C_{6-15}-O_{2-6} compounds with hydrogen deficiencies of four to six by predicting partial structures.[83] A success rate of 85% on a collection of 238 mass spectra was found. Learning machines have been applied to interpretation of the mass spectra of aliphatic alcohols[84] and to the qualitative analysis of low-resolution mass spectra by successive pattern identification.[85] A particularly novel learning method for classifying mass spectra into broad chemical groups by using digital learning nets consisting of stored-logic adaptive microcircuit elements has appeared.[86] Finally, learning machines have been shown to predict successfully structural features of molecules, with abilities in the range 83—95%, even when using modulo-14 or periodicity spectra.[87]

Heuristic Methods.—*Low Resolution.* Examples of the application of heuristic methods to the interpretation of low-resolution mass spectra are becoming more frequent. A four-part mass spectrometric data-processing program consisting of a preliminary filter, structure generator, structure fragmenter, and evaluation segment has been described.[88] The filter part examines the mass spectrum for certain functional groups and limits structural details, thus yielding parameters which are fed into a structure generator. Structures so generated are then fragmented according to appropriate rules of mass spectral decomposition, and the

[80] K. Varmuza and P. Krenmayr, *Z. analyt. Chem.*, 1973, **266**, 274.
[81] R. J. Mathews, *Austral. J. Chem.*, 1973, **26**, 1955.
[82] S. Sasaki and Y. Ishida, *Bunseki Kagaku*, 1972, **21**, 1029.
[83] S. Sasaki and A. Hidetsugu, *Shitsuryo Bunseki*, 1972, **20**, 131.
[84] P. Krenmayr and K. Varmuza, *Allg. Prakt. Chem.*, 1972, **23**, 289.
[85] M. S. Khots, *Zhur. analit. Khim.*, 1973, **28**, 797.
[86] T. J. Stonham, L. Aleksander, M. Camp, M. A. Shaw, and W. T. Pike, *Electron Letters*, 1973, **9**, 391.
[87] J. Franzen and H. Hillig, Sixth International Mass Spectrometry Conference, Edinburgh, 1973, paper 109.
[88] D. Koo and R. Sedgewick, Sixth International Mass Spectrometry Conference, Edinburgh, 1973, paper 115.

'generated' fragments are compared with those of the observed mass spectrum. The program then prints a number of candidate structures and gives them a 'score' depending on the similarity between observed and candidate structures. This approach, which appears to be similar in some of its features to that adopted by the group developing the DENDRAL programs,[89] has been applied to aliphatic alcohols, ethers, ketones, and amines. The success rate was 50% on the highest scoring structure and 80% in the highest three scoring cases in a sample size of 130 compounds.

Heuristic programming techniques have been used to devise possible fragmentation mechanisms from an examination of low resolution mass spectra.[90] The program, ION GENERATOR, is based on the charge-localization and electron 'bookkeeping' technique of rationalizing mass spectra. The program simulates the following sequence of operations: (i) ionization with charge localization; (ii) bond homolysis β to a radical site; (iii) bond formation between two adjacent radical sites; and (iv) hydrogen transfer to a radical site *via* appropriately sized transition states. Structures and low-resolution mass spectra are submitted to the program which then follows the sequence described and prints out possible fragmentation mechanisms. The authors of the paper hope that new mechanisms may be discovered by future applications of the program. The usefulness of heuristic techniques for making mechanistic deductions is of course directly dependent on the arguable validity of the general mechanistic ground rules employed in the program.

The analysis of mass spectra is greatly facilitated by examining isotope patterns in the spectra of compounds which contain polyisotopic species. Such an approach has been computerized in order to aid mass spectral interpretation where precise mass measurements are not available.[91] The program uses the uniqueness of isotopic abundance patterns by comparing observed isotopic clusters with all the permutations possible. Valency restrictions and central atom concepts were also imposed on the calculations. The inclusion of metastable peaks for assigning daughter ion compositions was another feature of the program. In the event of the failure of these methods to assign fragment ion compositions the 'small neutral' concept was invoked as a last resort. The scheme successfully identified major peaks, as long as significant impurities were absent. Less sophisticated computer-aided schemes have also been applied to the analysis of polyisotopic mass spectra.[92]

High Resolution. The extension of the DENDRAL programs[74,93] for heuristic interpretation of high-resolution mass spectra continues. This artificial intelligence

---

[89] A. Buchs, A. M. Duffield, G. Schroll, C. Djerassi, A. B. Delfino, B. G. Buchanan, G. L. Sutherland, E. A. Feigenbaum, and J. Lederberg, *J. Amer. Chem. Soc.*, 1970, **92**, 6831.
[90] A. B. Delfino and A. Buchs, *Helv. Chim. Acta*, 1972, **55**, 2017.
[91] R. W. Kiser and B. L. Bruner, Proceedings of the Twenty-first Annual Conference on Mass Spectroscopy and Allied Topics, San Francisco, 1973, p. 200.
[92] L. R. Crawford, *Internat. J. Mass Spectrometry Ion Phys.*, 1973, **10**, 279; J. D. Lee, *Talanta*, 1973, **20**, 1029; O. Borgen and B. Larsen, *Acta Chem. Scand.*, 1972, **26**, 2291.
[93] B. G. Buchanan and J. Lederberg, 'Information Processing', ed. C. V. Freiman, North-Holland, Amsterdam, 1972, Vol. 6, pp. 179–188.

technique has now been applied to the interpretation of the high-resolution mass spectra of 43 oestrogenic steroids.[94] Results were comparable to the performance of a trained mass spectroscopist, except that the computer was considerably faster. High-resolution mass spectra of oestrogenic steroids have also been analysed by a program (INTSUM) which systematically interprets and summarizes evidence found for mass spectral fragmentation processes in these molecules.[95] The program follows a route consisting of three basic tasks. Firstly, having been provided with the basic skeleton common to the set of related compounds, a list (ALLBREAKS) of all possible fragmentations resulting in smaller unique fragments is generated. In the next stage, each structure–spectrum pair is correlated with fragmentations in the ALLBREAKS list, allowing for hydrogen transfer. Finally, evidence for all the structure–spectrum pairs is correlated, and common fragmentation modes are grouped. The computer suggested some new fragmentation processes and the program was found to be a powerful aid to high-resolution spectral interpretation.

An interactive data-processing system based on keyboard communication with the computer and using a visual display unit has been successfully applied to the analysis of peptides and peptide mixtures.[96] The system functions by processing the high-resolution mass spectra of these compounds. Peptide mixtures have been similarly analysed by a program which also searches for metastable transitions from one sequence ion to any other sequence ion.[97]

Comparison of Spectra and Library Searching.—Automatic matching of the mass spectra of unknown compounds against a reference library continues to be one of the most popular computer methods for the identification of organic molecules probably because of the mathematical simplicity of such techniques, together with the possibility of unequivocal spectral assignment. These factors tend to out-outweigh the disadvantages of the large backing store and lengthy computation time required by library methods, relative to pattern recognition algorithms. In any case, these drawbacks are rapidly being overcome by a number of spectrum-coding techniques designed to reduce the interdependent computer time and storage demands of library matching procedures. The potential of spectral comparison techniques may also be increased by developing more efficient matching algorithms; several workers have claimed success in this area.

A theory which is capable of comparing the efficiencies of different library search techniques in a semi-quantitative manner has been described.[98] The

---

[94] D. H. Smith, A. M. Duffield, and C. Djerassi, Proceedings of the Twentieth Annual Conference on Mass Spectrometry and Allied Topics, Dallas, 1972, p. 371; D. H. Smith, B. G. Buchanan, R. S. Engelmore, A. M. Duffield, A. Yeo, E. A. Feigenbaum, J. Lederberg, and C. Djerassi, *J. Amer. Chem. Soc.*, 1973, **94**, 5963.

[95] D. H. Smith, B. G. Buchanan, W. C. White, E. A. Feigenbaum, J. Lederberg, and C. Djerassi, *Tetrahedron*, 1973, **29**, 3117.

[96] H. A. Van't Klooster, J. S. Vaarkamp-Lijnse, and G. Dijkstra, Sixth International Mass Spectrometry Conference, Edinburgh, 1973, paper 116.

[97] H.-K. Wipf, P. Irving, M. McCamish, R. Venkataraghavan, and F. W. McLafferty, *J. Amer. Chem. Soc.*, 1973, **95**, 3369.

[98] S. L. Grotch, Proceedings of the Twenty-first Annual Conference on Mass Spectrometry and Allied Topics, San Francisco, 1973, p. 69.

effect of coding errors and the use of procedures which improved matching algorithms were assessed using this mathematical theory for file searching. It was possible to deduce the threshold required to identify correctly unknown compounds with 'little' ambiguity. The application of principles derived from information theory has resulted in the development of a statistical library matching method.[99] Even when the search procedure was restricted to eight peaks in the mass spectrum, a large number of successful identifications was possible.

Two diagnostic functions and a binary coding technique have been used to provide a single-valued representation of a mass spectrum which can be stored easily in a reference file.[100] The approaches described were derived from set theory, which has been applied to mass spectral identification by other workers,[99,101] and were based on a Khinchine entropy function and divergence values.[100] The Khinchine function η is calculated according to the equation

$$-\eta = \sum_{i=1}^{n} p_i \log p_i$$

where p_i is the intensity of the ith peak in the mass spectrum. The η values derived from the mass spectra of known compounds were stored in a reference file, and the η of an unknown was then matched against all other η's in the file. The Khinchine function was found to be uniquely diagnostic in 90% of cases. In the remaining 10%, where the entropy functions did not differ by a value sufficient for unequivocal matching, divergence values were used to differentiate the mass spectra. The divergence values of a given compound in an aliphatic hydrocarbon series were referred to a normal alkane with the same carbon number and were given by

$$J(1, 2) = N_1 \sum_{i}^{n} (P_{1i} - P_i) \ln \frac{P_{1i}}{P_i} + N_2 \sum_{i}^{n} (P_{2i} - P_i) \ln \frac{P_{2i}}{P_i}$$

for compounds 1 and 2, where $P_i = (P_{1i} + P_{2i})/2$. A group of subsets which corresponded to compounds with the same carbon number was established, thus greatly reducing the amount of data to be searched through by the computer, and an octal coding scheme was used to store the spectral representations.[102] Data compression by judicious selection of spectral coding techniques has been one of

[99] S. Farbman, R. I. Reed, D. H. Robertson, and M. E. F. Silva, *Internat. J. Mass Spectrometry Ion Phys.*, 1973, **12**, 123.
[100] C. Merritt, jun., D. H. Robertson, R. A. Graham, and T. L. Nichols, Proceedings of the Twentieth Annual Conference on Mass Spectrometry and Allied Topics, Dallas, 1972, p. 355.
[101] D. H. Robertson and R. I. Reed, Proceedings of the Nineteenth Annual Conference on Mass Spectrometry and Allied Topics, Atlanta, 1971, p. 68.
[102] D. H. Robertson, J. Cavagnaro, J. B. Holz, and C. Merritt, jun., Proceedings of the Twentieth Annual Conference on Mass Spectrometry and Allied Topics, Dallas, 1972, p. 359.

the most successful ways in which library matching methods have been improved. In the octal coding scheme, mass ranges were divided into multiple groups of seven.[102,103] The number corresponding to the peak with the highest intensity in each range was coded as a three-bit binary number, a zero denoting absence of peaks, thus giving eight possible values (octal coding). It was possible to encode information covering 35 a.m.u. in a single 16-bit word (the most common word length in small computers). As many words as were needed were used to encode the spectrum, starting at m/e 23. Compounds which required the same number of words for coding could be organized into subsets or sublibraries, thus reducing the number of spectra which need be searched. It should be noted that this subset search, if rigorously applied, might not result in successful matching if for some reason a weak molecular ion was not detected in the mass spectrum of an unknown, but was present in one of the library spectra. If the ion of highest mass in the unknown to be coded corresponds to a fragmentation in which more than 35 a.m.u. were lost, the spectrum would be encoded in too small a number of words and the wrong subset would be searched even though the spectrum contained sufficient information for a correct match to be made in the absence of a molecular ion. This possibility can be guarded against by searching the next highest subset (or next lowest, in the event of contamination by high mass background peaks), or even the entire file of spectra in the event of no match being found on the first search. Coding schemes using one peak in 14, one in seven, and one in 14 plus two-bit intensity coding have been compared.[103] The first of these encoders required an average of 64-bits per spectrum, and the second two required 96-bits per spectrum. The effect of starting masses on the success of the coding schemes was also considered. A combination of statistical analysis and mass spectral heuristics was proposed as a general solution to some of the problems encountered in identifying compounds from their mass spectra. Another study has reported the use of two peaks out of every 14.[104] Four-bits were used to encode masses and two-bits intensities.[105] The average number of bits required to encode mass spectra in the Aldermaston collection was 48.[105] This coding scheme proved to be remarkably effective for identification purposes and only 10 s were required to search a library of 7000 spectra using an IBM 360/44 computer. Some particularly valuable warnings concerning the use of file searching methods were included in this work.[105b]

The probability of correct matching of unknown with reference mass spectra may be greatly increased by careful selection of data known to have high structural significance.[106] A program, called a 'Self-Training Interpretive and Retrieval

---

[103] D. H. Robertson and C. Merritt, jun., Proceedings of the Twenty-first Annual Conference on Mass Spectrometry and Allied Topics, San Francisco, 1973, p. 65.
[104] H. S. Hertz, R. A. Hites, and K. Biemann, *Analyt. Chem.*, 1971, **43**, 681.
[105] (a) S. L. Grotch, Proceedings of the Twentieth Annual Conference on Mass Spectrometry and Allied Topics, Dallas, 1972, p. 362; (b) S. L. Grotch, *Analyt. Chem.*, 1973, **45**, 2.
[106] R. Venkataraghavan, K.-S. Kwok, G. Pesyna, and F. W. McLafferty, Proceedings of the Twenty-first Annual Conference on Mass Spectrometry and Allied Topics, San Francisco, 1973, p. 197.

System' (STIRS), based on such an approach seeks to combine the advantageous features of library matching, learning machine, and heuristic techniques.[107] The system works by matching specific regions in reference spectra. In addition, current knowledge of mass spectral fragmentation processes is used to instruct the computer to recognize the classes of mass spectral data which should contain the most structural information. Further programs which attempt to identify the molecular ion, determine elemental compositions from isotope patterns, and match a spectrum against a reference file are incorporated in STIRS. The classes of spectral data compared in unknown and reference mass spectra are: (i) ion series spectra (see below); (ii) characteristic ions of low mass; (iii) characteristic ions of medium mass; (iv) characteristic ions of high mass; (v) losses of small primary neutral fragments; (vi) losses of large primary neutral fragments; (vii) secondary losses of neutral fragments from the most abundant ions of odd mass; (viii) secondary losses of neutral fragments from the most abundant ions of even mass. Data from the unknown spectrum in this class is matched against data from class 5 of the reference spectrum; (ix) fingerprint ions.

Classes (i)—(vi) are used to obtain an overall match factor (MF), given by [MF(i) + MF(ii) + 2MF(iii) + 2MF(iv) + 4MF(v) + 2MF(vi)]/12, by using the individual match factors of these spectral data classes [MF(i) to MF(vi) respectively]. A reference file of 13 000 mass spectra formed the basis of STIRS, and the system was tested on the mass spectra of 110 structurally diverse organic compounds. A number of structural features were correctly identified with a high degree of success, by comparison of the various data classes *e.g.* class (i) yielded 94% correct predictions of selected molecular sub-units. The overall match factor was found to be surprisingly independent of errors in the spectroscopic data in a number of cases, and it was felt that STIRS yielded some information which even a trained mass spectroscopist could not have deduced.

The concept of library search techniques based on ion series spectra (also used in STIRS) was first devised in order to provide a preliminary classification of unknown mass spectra.[108–110] Ion series spectra are formed by arranging m/e values into a rectangular array of 14 columns, numerically equivalent to the values for the hydrocarbon ions C_nH_{2n-11} to C_nH_{2n+2}.[108,111] The columns are then summed to give the ion series spectrum. The ion series spectra of different classes of compound were reduced to a 'mean' (the ion series spectra of members of a class were found to vary little within that class) and this was used as a correlation spectrum for that class. The ion series spectrum of an unknown was then computed and this was compared with the correlation set using a criterion given

---

[107] K.-S. Kwok, R. Venkataraghavan, and F. W. McLafferty, *J. Amer. Chem. Soc.*, 1973, **95**, 4185; F. W. McLafferty, R. Venkataraghavan, K.-S. Kwok, and G. Pesyna, Sixth International Mass Spectrometry Conference, Edinburgh, 1973, paper 110.
[108] D. H. Smith, *Analyt. Chem.*, 1972, **44**, 536.
[109] D. H. Smith and G. Eglinton, *Nature*, 1972, **235**, 325.
[110] D. H. Smith, N. A. B. Gray, C. T. Pillinger, B. J. Kimble, and G. Eglinton, 'Advances in Organic Geochemistry', ed. H. R. Von Gaertner, Pergamon, Oxford, 1972, p. 249.
[111] L. R. Crawford and J. D. Morrison, *Analyt. Chem.*, 1968, **40**, 1469.

by

$$\text{MISMATCH} = \sum_{n=1}^{14} |\text{Correlation set series}_n - \text{unknown series}_n|$$

A perfect match yields a MISMATCH value of 0, a perfect mismatch gives a value of 200.[108] This technique proved to be a useful method for the preliminary characterization of the components of complex mixtures, analysed by g.c.m.s. Using a program called COMSOC (Classification Of Mass Spectra On Computers) it was possible to classify correctly g.c.m.s. spectra run on samples from geochemical and environmental sources.[109] Following this preliminary sorting, further analysis by routines appropriate to the particular class of compound was possible.[110]

A reliable matching technique for mass spectra, designated Data Limiting Identification, has been reported to search a limited library of mass spectra successfully with no operator judgement.[112] The system is novel in that it performs the search in a reverse sense by extracting ion intensities from the unknown which correspond to masses in each individually selected mass spectrum in the library and then compares these. It can therefore operate under conditions of substantial interference from overlapping peaks in, for example, a g.c.m.s. run. The library consists of reduced spectra, stored in files of 100 mass spectra, and relies on reference spectra of assured quality—generated by the user in this case. This is a wise precaution as many of the commercially available libraries of mass spectra contain substantial errors due to contaminants or to human error during compilation.

An international mass spectral data retrieval system which operates over telephone lines has been described.[113] The system is an interactive one, *viz.* the chemist communicates directly with the computer *via* a teletype terminal, and is based on a number of search options.[114,115] Seven individual options were programmed into the retrieval system:[115] (i) a peak and intensity search; (ii) a molecular weight search; (iii) a molecular formula search—(a) complete (b) embedded; (iv) a molecular weight and peak search; (v) a molecular formula and peak search; (vi) a molecular weight and molecular formula search; (vii) spectrum printout.

The reference spectra were stored in a file of spectra abbreviated by the 'two out of 14' method, starting at $m/e = 6$. Using these options very few data points (two to four usually) were needed to reduce the number of possible answers to a tract-

---

[112] F. P. Abramson, Proceedings of the Twenty-first Annual Conference on Mass Spectrometry and Allied Topics, San Francisco, 1973, p. 76.
[113] S. R. Heller, H. M. Fales, G. W. A. Milne, R. J. Feldman, N. R. Daly, D. C. Maxwell, and A. McCormick, Proceedings of the Twenty-first Annual Conference on Mass Spectrometry and Allied Topics, San Francisco, 1973, p. 192; Sixth International Mass Spectrometry Conference, Edinburgh, 1973, paper 117; S. R. Heller, H. M. Fales, and G. W. A. Milne, *J. Chem. Documentation*, 1973, **13**, 130.
[114] S. R. Heller, H. M. Fales, and G. W. A. Milne, Proceedings of the Twentieth Annual Conference on Mass Spectrometry and Allied Topics, Dallas, 1972, p. 353.
[115] S. R. Heller, H. M. Fales, and G. W. A. Milne, *Org. Mass Spectrometry*, 1973, **7**, 107.

able size, indicating that mass spectra are generally grossly overdetermined. Normal searches required 2—6 s of central processor unit time on a PDP 10 computer, and the average time spent by the user at the computer terminal (teletype) was 5—15 min.[115]

A library search technique based on 'three out of 14' coding of reference spectra has been reported.[116] Difficulties encountered in matching mass spectra obtained on quadrupole mass spectrometers with spectra formed by magnetic deflection instruments have been overcome by using the 'two out of 14' coding method.[117] A generalized routine for library searching incorporating feature selection has been described,[118] and it has been shown that the probability of correct matching can be increased by incorporating n.m.r. and i.r. spectral data as well as mass spectral data into binary coded reference files.[119] A system for identifying drug overdoses by matching mass spectroscopic data, obtained during a g.c.m.s. run, against a library of drug and drug metabolite mass spectra has been developed,[120] and similar techniques have been applied to the analysis of body fluids in order to detect metabolic disorders.[121,122] Library matching techniques have also been used to detect steroids in biological material,[123,124] polycyclic aromatics[125] and other air pollutants,[126] and the products of thermal analysis of polymeric and other materials.[127] Computer-aided identification of monosaccharide stereo-isomers has been achieved by matching eight intensity ratios in unknowns with the corresponding ratios in reference spectra.[128]

4 Miscellaneous Applications

Gas Chromatography and Mass Spectrometry.—A recent review of integrated g.c.m.s includes a discussion of the use of computers in this field.[129] The benefits

---

[116] N. W. Bell, Proceedings of the Twentieth Annual Conference on Mass Spectrometry and Allied Topics, Dallas, 1972, p. 353.
[117] J. M. McGuire, A. L. Alford, and M. H. Carter, Proceedings of the Twentieth Annual Conference on Mass Spectrometry and Allied Topics, Dallas, 1972, p. 366.
[118] J. T. Clerc, F. Erni, C. Jost, T. Meili, D. Nageli, and R. Schwarzenbach, *Z. analyt. Chem.*, 1973, **264**, 192.
[119] F. Erni and J. T. Clerc, *Helv. Chim. Acta*, 1972, **55**, 489.
[120] C. E. Costello, T. Sakai, and K. Biemann, Proceedings of the Twentieth Annual Conference on Mass Spectrometry and Allied Topics, Dallas, 1972, p. 107.
[121] S. P. Markey, W. G. Urban, and A. J. Keyser, Proceedings of the Twenty-first Annual Conference on Mass Spectrometry and Allied Topics, San Francisco, 1973, p. 71.
[122] E. Jellum, O. Stokke, and L. Eldjan, *Analyt. Chem.*, 1973, **45**, 1099.
[123] R. Reimendal and J. B. Sjovall, *Analyt. Chem.*, 1973, **45**, 1083.
[124] J. D. Baty and A. P. Wade, *Analyt. Biochem.*, 1974, **57**, 27.
[125] R. C. Lao, R. S. Thomas, H. Oja, J. L. Monkman, and R. F. Pottie, Proceedings of the Twentieth Annual Conference on Mass Spectrometry and Allied Topics, Dallas, 1972, p. 131.
[126] J. Hardy and I. Jardine, Sixth International Mass Spectrometry Conference, Edinburgh, 1973, paper 120.
[127] D. L. Geiger and G. A. Kleineberg, Proceedings of the Twenty-first Annual Conference on Mass Spectrometry and Allied Topics, San Francisco, 1973, p. 82.
[128] J. Vink, J. H. W. Bruins Slot, J. J. de Ridder, J. P. Kamerling, and J. F. G. Vliegenthart, *J. Amer. Chem. Soc.*, 1972, **94**, 2542.
[129] R. Ryhage, *Quart. Rev. Biophys.*, 1973, **6**, 311.

derived from the application of computers to the on-line acquisition of g.c.m.s. data are probably the strongest driving forces behind the phenomenal expansion in the use of data systems coupled to mass spectrometers. As a consequence of this, examples of more sophisticated g.c.m.s.–computer techniques have appeared, some having been mentioned already in previous sections. Further examples include the assignment of the retention indices of all the compounds eluted during a g.c.m.s.–computer run by a data system.[130] The computer identifies three standard compounds coinjected with the sample into the g.c.m.s. combination and uses their positions in the total ion-current chromatogram to calculate retention data for all the sample peaks. At the end of a g.c.m.s. run the user specifies the calibration compounds used (e.g. n-alkanes) and also instructs the computer to search a limited number of scans in the total ion-current plot where these standards are expected to appear. The mass spectrum most similar to the standard compound is selected by a matching technique and the centre of the corresponding chromatographic peak is determined by finding the maximum summed intensity of characteristic ions of the standard in a narrow window. When the standard compounds have been identified the computer calculates retention indices throughout the chromatogram with the aid of a calibration table. Combined use of retention indices and mass spectra has allowed particularly rapid and reliable automatic identification of compounds in such areas as drug identification in comatose patients, characterization of amino-acids, and analysis of oligo- and poly peptide mixtures, and the technique has been used as a protein sequenator.[131] A computer algorithm which selects important peaks within an automated g.c.m.s. analysis has been described.[132] The program prevents storage of spectra taken on the total ion-current chromatogram baseline and on the flat parts of peak-tailing. It also recognizes changes in the baseline and is able to identify background spectra. Computer techniques for the resolution of overlapping g.c.m.s. peaks by eigenvector analysis[133] and the use of matched filters[134] have been reported.

Sensitivity enhancement of mass spectral data by computer processing of a g.c.m.s. run using a vector analogy technique has been attained.[135] Low-resolution mass spectra can be considered as being equivalent to a multi-dimensional vector (a representation used in pattern recognition). The angle between an

---

[130] H. Nau and K. Biemann, *Analyt. Letters*, 1973, **6**, 1071; *Analyt. Chem.*, 1974, **46**, 426.

[131] H. Nau, J. A. Kelley, H.-J. Förster, J. E. Biller, T. R. Smith, and K. Biemann, Proceedings of the Twenty-first Annual Conference on Mass Spectrometry and Allied Topics, San Francisco, 1973, p. 293; H. Nau and K. Biemann, 'Chemistry and Biology of Peptides', ed. J. Meienhofer, Ann Arbor Science, Ann Arbor, Michigan, 1972, pp. 679–686.

[132] D. Henneberg, K. Casper, and E. Ziegler, *Chromatographia*, 1972, **5**, 83, 205.

[133] J. E. Davis and L. B. Rogers, Abstracts of the Pittsburgh Conference on Analytical Chemistry and Applied Spectroscopy, Cleveland, 1973, p. 163.

[134] J. N. Haber and D. J. Jenden, Proceedings of the Twenty-first Annual Conference on Mass Spectrometry and Allied Topics, San Francisco, 1973, p. 78.

[135] D. Rosenthal and J. T. Bursey, Proceedings of the Twentieth Annual Conference on Mass Spectrometry and Allied Topics, Dallas, 1972, p. 419.

unknown spectrum $\vec{U} = (U_1, U_2, \ldots, U_n)$ and a standard spectrum $\vec{S} = (S_1, S_2, \ldots, S_n)$ in this vector representation is given by

$$\theta = \cos^{-1}\left\{\frac{\sum_{i=1}^{n} U_i S_i}{\sqrt{\left(\sum_{i=1}^{n} U_i - \sum_{i=1}^{n} S_i\right)}}\right\}$$

and the projection of $\vec{U}$ on $\vec{S}$ by

$$U_{\text{proj}} = |\vec{U}| \cos \theta$$

Plots of θ and U_{proj} against spectrum number were inspected for maxima and minima (after mathematical smoothing). Increased sensitivity and better signal-to-noise ratio (enhanced by 25 : 1) were observed using this technique. Furthermore, the projection of the effluent spectrum on the standard yielded a quantitative estimate of that component.

On-line g.c.m.s. data acquisition and processing in an analytical laboratory have been described,[21] and the use of a quadrupole g.c.m.s.–computer system in the analysis of samples from environmental sources has been reported.[136]

Multiple Ion Detection without Feedback Control.—Computer-controlled MID systems are described in Section 2. Less sophisticated applications where the computer is used simply as an ion monitoring device, without a feedback loop for voltage control, have also been reported.[137–140] The systems were employed in g.c.m.s.[137,138] and isotope ratio measurements.[139,140]

Ionization Efficiency and Metastable Ion Data.—The greatly improved accuracy and reproducibility of measurement attainable when ionization and appearance potentials are determined using a conventional electron impact ion source with computer acquisition of ionization efficiency data has been conclusively demonstrated.[141] Applications of the technique have yielded more accurate appearance potentials for metastable peaks,[142] resulting in a revision of the conclusions of other workers as to the meaning of values previously obtained. Computerized

---

[136] R. C. Lao, H. Oja, R. S. Thomas, and J. L. Monkman, Proceedings of the Twenty-first Annual Conference on Mass Spectrometry and Allied Topics, San Francisco, 1973, p. 316.

[137] J. T. Watson, D. R. Pelster, B. J. Sweetman, J. T. Frolich, and J. A. Oates, *Analyt. Chem.*, 1973, **45**, 2071.

[138] P. D. Klein, J. R. Haumann, and W. J. Eisler, *Analyt. Chem.*, 1972, **44**, 490.

[139] W. F. Haddon, H. C. Lukens, and M. Crim, Proceedings of the Twentieth Annual Conference on Mass Spectrometry and Allied Topics, Dallas, 1972, p. 172.

[140] B. H. Albrecht, J. R. Plattner, D. Hagerman, S. Markey, and R. C. Murphy, Proceedings of the Twentieth Annual Conference on Mass Spectrometry and Allied Topics, Dallas, 1972, p. 174.

[141] R. A. W. Johnstone and F. A. Mellon, *J.C.S. Faraday II*, 1972, **68**, 1209.

[142] R. A. W. Johnstone and B. N. McMaster, *J.C.S. Chem. Comm.*, 1973, 730.

collection of ionization efficiency data from a quadrupole mass spectrometer has been reported.[143]

Detection of metastable peaks by computerized high- and low-resolution mass spectrometry has been shown to be feasible.[144] Under scan conditions of 10 s per decade (10 000 resolution) and 1 s per decade (1000 resolution) it was possible to detect metastable peaks in the intensity range 0.5—1 % of the base peak.

Analysis of Spectra of Mixtures.—A program for the analysis of mixtures, coded in FORTRAN IV, has been used as a first filter in the processing of the mass spectra of mixtures.[145] The algorithm was designed to find all the possible components of a mixture by determining whether a certain specified number of major peaks in standard spectra were also present in the unknown. Computer memory requirements were minimized by storing the most intense peaks of each standard in decreasing order in an array.

General.—A computer program which simulates the behaviour of a field ionization source has been reported[146] and flexible computer algorithms, written in FORTRAN IV, which are capable of representing mass spectra on a computer plotter, have been described.[147] To show that not all data processing systems require sophisticated and expensive hardware it should be noted that a programmable desk calculator has been used for processing high-resolution mass spectra, in order to obtain the exact masses, intensities, and empirical formulae of ions recorded on a spectrographic plate.[148]

---

[143] B. R. Conrad, P. M. Fast, D. L. Nordlund, and P. W. Gilles, *Nasa Star*, 1972, **10**, 176.
[144] A. Carrick, Proceedings of the Twentieth Annual Conference on Mass Spectrometry and Allied Topics, Dallas, 1972, p. 407; A. Carrick and H. M. Paisley, *Org. Mass Spectrometry*, 1974, **8**, 229.
[145] T. L. Isenhour, *Analyt. Chem.*, 1973, **45**, 2153.
[146] J. P. Pfeifer, A. M. Falick, and A. L. Burlingame, *Internat. J. Mass Spectrometry Ion Phys.*, 1973, **11**, 345.
[147] Z. Tashma, *Org. Mass Spectrometry*, 1973, **7**, 249.
[148] J. H. Weber, D. G. Earnshaw, D. W. K. Severin, and A. W. Decora, Proceedings of the Twenty-first Annual Conference on Mass Spectrometry and Allied Topics, San Francisco, 1973, p. 60.

5
Organometallic, Co-ordination, and Inorganic Compounds

BY T. R. SPALDING

1 Introduction

The publications covered in this report appeared between March 1972 and March 1974. The majority of the articles reviewed contained reference to the characterization of the compounds studied, *i.e.* only the parent molecular ion or a few fragment ions were reported. The choice of which of these references to include was often difficult and must to some extent reflect the bias of the reporter. Far fewer detailed discussions of mass spectra, fragmentation mechanisms, or the energetics of the processes involved were published.

Some changes in the format of this chapter from those in previous Reports[1,2] have been adopted. Apart from the sections of Main-group Organometallics, Transition-metal Organometallics, and Co-ordination Compounds, there is also a section on Inorganic Compounds. Ionization potential data are referred to in the text and not tabulated, and articles which do not discuss spectra in detail but which would previously be found in the Appendix are now mentioned in the text. The emphasis of the report remains on the discussion of fragmentation behaviour.*

Books and Reviews.—The first book devoted to the mass spectra of organometallic and inorganic compounds has been published;[3] it lists more than 1500 references from the literature up to the beginning of 1971. Reviews of the fragmentation behaviour of transition-metal organometallics[4] and the concept of valency-change in the spectra of metal complexes[5] have appeared. A general article on

---

[1] M. I. Bruce, in 'Mass Spectrometry', ed. D. H. Williams (Specialist Periodical Reports), The Chemical Society, London, 1971, Vol. 1, p. 182.
[2] M. I. Bruce, in 'Mass Spectrometry', ed. D. H. Williams (Specialist Periodical Reports), The Chemical Society, London, 1973, Vol. 2, p. 193.
[3] M. R. Litzow and T. R. Spalding, 'Mass Spectrometry of Inorganic and Organometallic Compounds', Elsevier, Amsterdam, 1973.
[4] J. Muller, *Angew. Chem. Internat. Edn.*, 1972, **11**, 653.
[5] M. J. Lacey and J. S. Shannon, *Org. Mass Spectrometry*, 1972, **6**, 931.

* Throughout this chapter the ion derived from the parent molecule (pm) of a compound will be designated 'pmi' (parent molecular ion). A positive ion is assumed unless otherwise stated. This avoids any confusion which may arise if other symbols, *e.g.* P^+ or M^+ are used. In this chapter the former symbol refers to ionized phosphorus, and the latter to a metal ion.
Ions from a compound of A, XYZA say, which do not contain the central atom A have been ignored when reckoning 'total ion current' and base peaks ions.

aspects of the spectra of organometallic and inorganic compounds,[6] and reviews of intramolecular H transfers in the spectra of Main Group IV and V compounds,[7] negative-ion spectra of organometallic and inorganic compounds,[8] and organometallic compounds of Main Group IV, particularly cyclic siloxans,[9] have been published. It is becoming increasingly common to find sections in books and review articles dealing with particular classes of compounds which specifically discuss the mass spectra of the compounds concerned (see e.g. refs. 10—12).

New Techniques of General Interest.—A field desorption study of disodium deoxyfluoro-D-glucose-hexaphosphates allowed the $[pm + H]^+$ ion to be observed.[13] Numerous organometallic and inorganic compounds which are difficult to handle with other methods of ionization await study with this technique.

A method has been developed for direct presentation of consecutive metastable ion transitions in a double focusing spectrometer.[14] The name MAMIE (Metastable ions Arising from Metastable ions in Ion Energy spectra) has been suggested for the technique.

$$[pm - H]^+ \rightarrow m/e\ 81 \begin{array}{l} \nearrow m/e\ 65 \\ \rightarrow m/e\ 54 \\ \searrow m/e\ 41 \end{array}$$

Scheme 1

In the spectrum of N-trimethylborazine several consecutive reactions were observed (Scheme 1), the separate steps of which were confirmed by MIKE spectra.

2 Main-group Organometallics

Group II.—Although studies concerning organometallic compounds of lithium and the Group I elements have not beeen forthcoming during the period covered by this report, a number have appeared dealing with compounds of beryllium and mercury.

The spectrum of diphenylberyllium at source temperatures of 200—240 °C shows evidence of parent trimer and monomer species.[15] Fragmentation of metal-containing ions was commonly by loss of hydrogen atoms or molecules,

---

[6] A. L. Burlingame and G. A. Johanson, *Analyt. Chem.*, 1972, **44**, 337R.
[7] J. T. Bursey, M. M. Bursey, and D. G. I. Kingston, *Chem. Rev.*, 1973, **73**, 191.
[8] J. G. Dillard, *Chem. Rev.*, 1973, **73**, 589.
[9] V. Yu. Orlov, *Russ. Chem. Rev.*, 1973, **42**, 529.
[10] See, for instance, *J. Organometallic Chem.*, Subject Reviews and Annual Surveys.
[11] D. C. Bradley, *Adv. Inorg. Chem. Radiochem.*, 1972, **15**, 259.
[12] 'Organic Selenium Compounds', ed. D. L. Klayman and W. H. H. Gunther, Wiley-Interscience, New York, 1973.
[13] H.-R. Schutten, H. D. Beckey, E. M. Besell, A. B. Foster, M. Jarman, and J. H. Westwood, *J.C.S. Chem. Comm.*, 1973, 416.
[14] J. M. Miller and G. L. Wilson, *Internat. J. Mass Spectrometry Ion Phys.*, 1973, **12**, 225.
[15] F. Glockling, R. J. Morrison, and J. W. Wilson, *J.C.S. Dalton*, 1973, 95.

acetylene, HBe, and Be atoms. The initial fragmentations of $[Ph_2Be]^{\ddot{+}}$ are shown in Scheme 2.

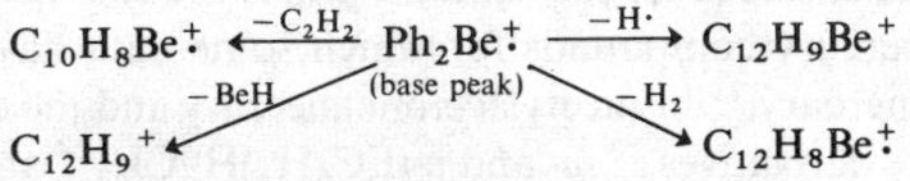

$$C_{10}H_8Be^{\ddot{+}} \xleftarrow{-C_2H_2} Ph_2Be^{\ddot{+}} \xrightarrow{-H\cdot} C_{12}H_9Be^+$$
$$\text{(base peak)}$$
$$C_{12}H_9^+ \xleftarrow{-BeH} \qquad \xrightarrow{-H_2} C_{12}H_8Be^{\ddot{+}}$$

Scheme 2

From the ionization potential of $[Ph_2Be]^{\ddot{+}}$ (9.2 ± 0.1 eV) and the appearance potential of $[PhBe]^+$ (13.4 ± 0.2 eV), the ionic bond dissociation energy $D(PhBe-Ph)^+$ was calculated as 412 kJ mol^{-1}. This is much higher than corresponding values for dialkyl compounds which lie in the range 180—192 kJ mol^{-1}.

An ICR spectrometer was used to observe the products of reactions between species derived from Me_2Hg and ethylene.[16] One, the ethylene mercurinium ion $[C_2H_4HgCH_3]^+$, was suggested as having structure (1).

The spectrum of $(Me_3SiCH_2)_2Hg$ showed a pmi and base peak due to loss of CH_3.[17] A large number of ionic species did not contain Hg but 47.5% of the total ion current was carried by ions containing both silicon and mercury. A doubly-charged ion (2) was found for which there was no corresponding singly-charged ion.

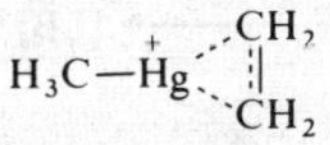

$$H_3C-\overset{+}{Hg}\cdots\overset{CH_2}{\underset{CH_2}{\|}} \qquad\qquad [\{(CH_3)_2SiCH_2\}_2Hg]^{2+}$$
$$(1) \qquad\qquad\qquad\qquad (2)$$

The compounds $MeHgC(N_2)C(O)R$ (R = Ph or Me) both show pmi's but fragment quite differently.[18] The phenyl derivative initially loses N_2 to give the base peak ion (3) which then loses CO, from a rearranged ion such as (4). The

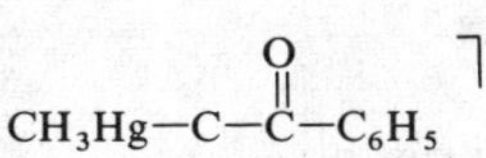

$$CH_3Hg-C-\overset{O}{\underset{\|}{C}}-C_6H_5 \Big]^{\ddot{+}} \qquad CH_3Hg-\overset{C_6H_5}{\underset{+\cdot}{C}}-C=O$$
$$(3) \qquad\qquad\qquad\qquad (4)$$

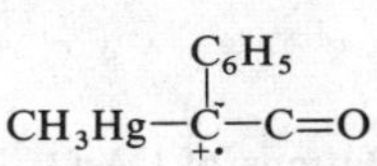

methyl compound fragments first by loss of either CH_3 or N_2 or $OCCH_3$. The ion formed by loss of N_2 then loses a CH_3 radical before CO is eliminated. The spectrum of $Hg[C(N_2)C(O)Me]_2$ shows a pmi and fragments due to loss of N, N_2, and two N_2. The corresponding phenyl compound does not exhibit a pmi but shows loss of N_2 and two N_2 and, unexpectedly, a doubly-charged molecular ion as the second most abundant ion in the spectrum. Other diazo-compounds reported include $MeHgC(N_2)CO_2R$ (R = Me, Et, Pr^i, Bu^t, PhCH$_2$, or Ph) and $MeHgC(N_2)R$ (R = Me or CN).[19] All show pmi's and ions due to loss of N_2.

[16] R. D. Back, J. Ganglhofer, and L. Kevan, *J. Amer. Chem. Soc.*, 1972, **94**, 6860.
[17] F. Glockling, S. R. Stobart, and J. J. Sweeney, *J.C.S. Dalton*, 1973, 2029.
[18] J. Lorberth, F. Schmock, and G. Lange, *J. Organometallic Chem.*, 1973, **54**, 23.
[19] S. J. Valenty and P. S. Skell, *J. Org. Chem.*, 1973, **38**, 3937.

Methylmercurynitronate gives a pmi and ions due to loss of CH_3 and O_2NCH_2 groups as well as a possible methylmercuryfulminate ion $[CH_3-Hg-CNO]^+$.[20] The ethyl derivative shows fewer ions and the pmi is the base peak.

Other organomercury compounds for which some data have been reported are di-(3-butenyl)mercury,[21] bis(acetylacetone)mercury and the chloromercury[22] and mercuriacetate derivatives,[23] m- and p-$(FC_6H_4)HgCF_3$[24] and C_5Cl_5HgCl.[25] Mercury derivatives of some steroidal ketones have been studied with the aid of high-resolution mass spectrometry.[26]

Group III.—The spectra and some appearance potential data of the triphenyl derivatives of boron and the Group III elements have been recorded.[27] With source temperatures of 105—360 °C, ions from dimeric species were found for Al and Ga only. Fragmentation was mainly by loss of Ph, C_6H_6, and $C_{12}H_{10}$ groups. Doubly-charged ions were observed in the spectrum of Ph_3B.

Compounds containing B—O, B—S, or B—N Bonds, including Boron Hetero-cycles. A study of the metastable peak widths relating to the elimination of >C=O species from alkoxy-compounds has led to the suggestion of two general mechanisms requiring similar activation energies, Schemes 3(a) and 3(b), 3(b) being favoured for bulky R groups.[28]

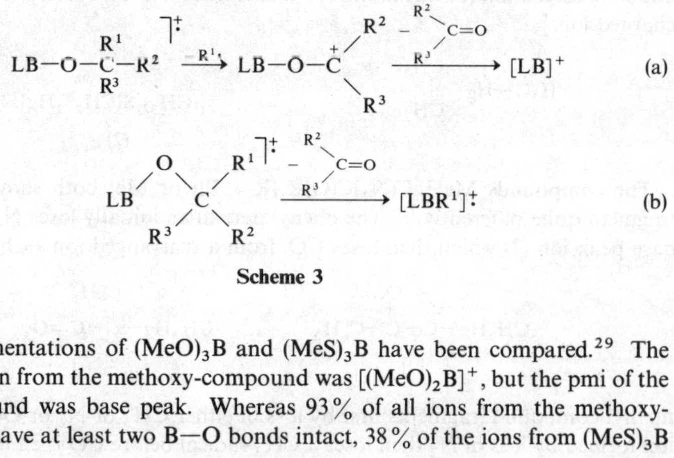

$$\left[LB-O-\underset{R^3}{\overset{R^1}{C}}-R^2 \right]^{+\cdot} \xrightarrow{R^1\cdot} LB-O-\overset{+}{C}\begin{matrix}R^2\\R^3\end{matrix} \xrightarrow[R^3]{R^2\cdots C=O} [LB]^+ \quad (a)$$

$$\left[\underset{R^3}{\overset{LB}{\diagdown}}\overset{O}{\underset{\diagup}{C}}\underset{R^2}{\overset{R^1}{\diagup}} \right]^{+\cdot} \xrightarrow[R^3]{R^2\cdots C=O} [LBR^1]^{+\cdot} \quad (b)$$

Scheme 3

The fragmentations of $(MeO)_3B$ and $(MeS)_3B$ have been compared.[29] The base peak ion from the methoxy-compound was $[(MeO)_2B]^+$, but the pmi of the thio-compound was base peak. Whereas 93% of all ions from the methoxy-compound have at least two B—O bonds intact, 38% of the ions from $(MeS)_3B$

---

[20] J. Lorberth and G. Lange, *J. Organometallic Chem.*, 1973, **54**, 165.
[21] J. B. Smart, R. Hogan, P. A. Scherr, M. T. Emerson, and J. P. Oliver, *J. Organometallic Chem.*, 1973, **64**, 1.
[22] R. Allmann, K. Flatau, and H. Musso, *Chem. Ber.*, 1972, **105**, 3067.
[23] R. H. Fish, R. E. Lundin, and W. F. Haddon, *Tetrahedron Letters*, 1972, 921.
[24] D. Seyferth, S. P. Hopper, and G. J. Murphy, *J. Organometallic Chem.*, 1972, **46**, 201.
[25] G. Wulfsberg, R. West, and V. N. M. Rao, *J. Amer. Chem. Soc.*, 1973, **95**, 8658.
[26] R. G. Smith, H. E. Emsley, and H. E. Smith, *J. Org. Chem.*, 1972, **37**, 4430; H. E. Smith and R. G. Smith, *Org. Mass Spectrometery*, 1973, **7**, 1019.
[27] F. Glockling and J. G. Irwin, *J.C.S. Dalton*, 1973, 1424.
[28] A. T. T. Hsieh, *Inorg. Nuclear Chem. Letters*, 1973, **9**, 801.
[29] R. H. Cragg, J. F. J. Todd, and A. F. Weston, *J.C.S. Dalton*, 1972, 1373.

have no B—S bonds. The thio-compound fragments by loss of Me, SMe, C_2H_4, and boron-containing groups such as MeSB and BS(SH). Other thioboranes have been characterized including $(RS)_3B$, $PhB(SR)_2$, and $Ph_2B(SR)$ (R = Me, Et, Pr, or Bu^t).[30]

A previous conclusion that cyclic borenium ions occurred in low abundance because they were thermodynamically unstable relative to their acyclic counterparts[31] has been re-examined. It now appears that if the exocyclic group attached to boron is either a dialkylamino- or alkylthio-group, then the cyclic borenium can have considerable stability.[32] Compounds (5)—(8) may produce cyclic

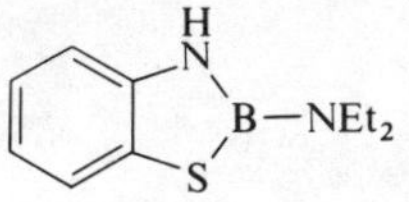

(5)

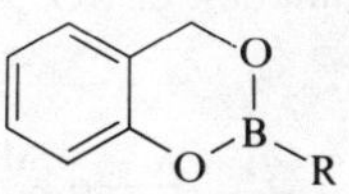

(6) R = NEt_2
(7) R = SEt

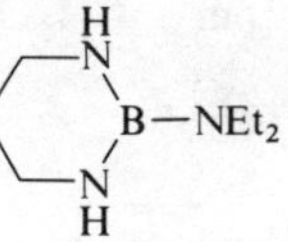

(8)

borenium ions of between 13 and 20% abundance of the base peak. It was pointed out that in all previous cases the exocyclic group was Ph or OR and was strongly bound to boron. The low abundances of the 'cyclic borenium' ions could largely be a function of this.

The spectra of triphenylboroxine and phenylboronic acid have been reported.[33] The former compound showed a marked preference to form ions by cleavage of B—O rather than B—C bonds. The acid showed both B—O and B—C cleavage but the major decompositions, including the elimination of BO_2H from the pmi to give $[C_6H_6]^{+\cdot}$, involved B—C cleavage. Triphenylborthin, $(PhBS)_3$, shows the pmi in relatively much higher abundance than the pmi's in the spectra of $(XBS)_3$ (X = Br, MeS, or MeO).[34] It has been suggested that charge delocalization is partly responsible. The effect also occurs in (9), formed from $[PhBS]_3^{+\cdot}$ by loss of first Ph and then H_2 groups.

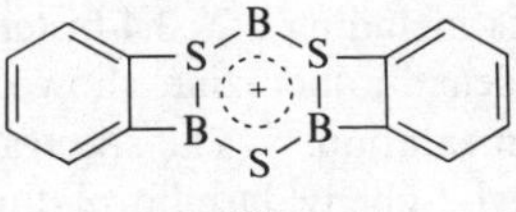

(9)

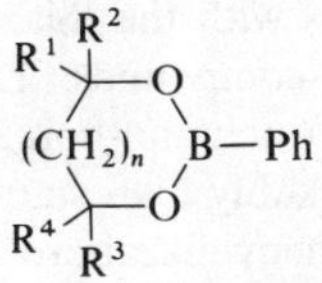

(10) R = e.g. H, Me, CH_2OH
n = 0—2

[30] R. H. Cragg, J. P. N. Husband, and A. F. Weston, *J. Inorg. Nuclear Chem.*, 1973, **35**, 3685.
[31] J. C. Kotz, R. J. Vanderzanden, and R. G. Cooks, *Chem. Comm.*, 1970, 923.
[32] R. H. Cragg, M. Nazery, J. F. J. Todd, and A. F. Weston, *J.C.S. Chem. Comm.*, 1973, 386.
[33] R. H. Cragg, J. F. J. Todd, and A. F. Weston, *Org. Mass Spectrometry*, 1972, **6**, 1077.
[34] R. H. Cragg and A. F. Weston, *J.C.S. Chem. Comm.*, 1974, 22.

The fragmentation of cyclic phenylboronate derivatives (10) of diols and polyols has been discussed.[35] The elimination of R^3R^4CO, $\cdot R^1$, or fragmentation to hydrocarbon ions were common decompositions from the pmi's of all compounds studied. When $n = 1$, specific fragmentations from pmi's leading to the ions $[PhBO]^{+\cdot}$ and $[R^1R^2CCH_2]^{+\cdot}$ or $[R^3R^4CCH_2]^{+\cdot}$ were observed and were suggested to be characteristic of the six-membered phenylboronate ring systems.

The ready formation of hydrocarbon ions including $[C_nH_n]^+$ (n = 8 or 7) from some phenylborolans and compounds (11)—(13) has been studied.[36] With phenyldioxaborolan (11), $[C_7H_7]^+$ is formed both directly from the pmi and from $[pm - CH_2O]^{+\cdot}$ by elimination of BO.

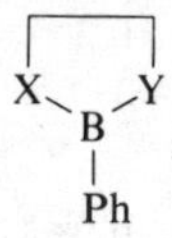

```
   ┌───┐         ┌───┐Me      ┌───┐
  X   Y        S   S         HN   S
   \B/          \B/            \B/
    |            |              |
    Ph           X              X

   (11)(12)(13)
X =  O   S   S    (14) X = Ph, Cl, SEt, NEt₂, or NHBu   (15) X = Ph, NEt₂, or
Y =  O   O   S                                                        NPr₂
```

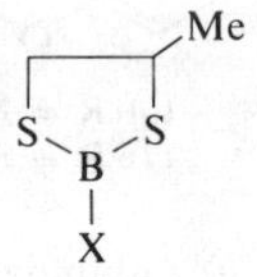

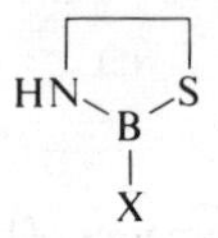

Spectra of some B-substituted 4-methyl-1,3,2-dithiaborolans (14) have been discussed.[37] The phenyl derivative has the pmi as base peak and initially loses C_3H_6, CH_3, and PhBS species. The fragmentation of the diethylamino-compound is more complex. Initial losses of C_2H_5, H, Et_2NBS, and CH_3 groups from the pmi are observed; the base peak is $[pm - CH_3]^+$. Thiazaborolidine derivatives (15) show abundant pmi's, and in the spectrum of the phenyl compound it is the base peak.[38] Initial loss of SH was found in all three spectra but only the phenyl derivative showed loss of H from the pmi. Comparatively more fragment ions with the thiazaborolidine system intact were observed with the dialkylamino-compounds. Larger ring systems including 1,2 : 3,4-bis(ethylenethio)-5,6-ethyleneimino-borazine and tris(ethylenethio)borazine showed pmi's with a remarkably high stability towards fragmentation.[38] The spectra of 1,3-dimethyl-2-phenyldiazaborolin and 1,3-dimethyl-2-phenyldiazaborolidine show interesting contrasts.[39] In the former the pmi is the base peak, with the ion $[pm - H]^+$ only 33% as abundant. In the latter the base peak is $[pm - H]^+$ and the pmi is 71% as abundant. The differences were attributed to the extra aromatic character of the former compound.

---

[35] I. R. McKinley and H. Weigel, *J.C.S. Chem. Comm.*, 1972, 1051.
[36] R. H. Cragg, G. Lawson, and J. F. J. Todd, *J.C.S. Dalton*, 1972, 878.
[37] R. H. Cragg, J. P. N. Husband, and A. F. Weston, *J.C.S. Dalton*, 1973, 568.
[38] R. H. Cragg and A. F. Weston, *J.C.S. Dalton*, 1973, 1054.
[39] J. S. Merriam and K. Niedenzu, *J. Organometallic Chem.*, 1973, **51**, C1.

Other heterocyclic compounds for which data have been reported include benzodioxaborin derivatives,[40] benzodioxaboroles,[41] diazadiborines,[42] benzodiazaborinone compounds,[43] N-methyltetrahydro-2,2-borazirene and N-methyl-2,1-borazarene,[44] diaza-diboranaphthalenes,[45] and diboradiazarobenzene derivatives.[46] Nearly all of the compounds mentioned above show pmi's. Aspects of the spectra of the following compounds have been reported; toluidinophenylchloroborane,[45] substituted anilinodimesitylboranes,[47] pyridine-(2-biphenylborane) and related compounds.[48] The (N-dimethylboryl)methylaminoboranes $\{(Me_2B)(Me)N\}_nB\{N(Me)SiMe_3\}_{3-n}$ ($n = 1$—3), show as base peaks the ion, $[Me_3Si]^+$, from the $n = 1$ compound but the rearranged ion, $[Me_3B]^{+\cdot}$, from the compounds for $n = 2$ or 3.[49]

Carbaboranes and Derivatives. Negative ions from some *closo*-[50] and *nido*-carbaboranes[51] have been recorded. The *closo*-carbaboranes 1,5-$C_2B_3H_5$, 1,2-$C_2B_4H_6$, 1,6-$C_2B_4H_6$, and 2,4-$C_2B_5H_7$ all showed [pm − H]$^-$ ions of much greater abundance than the negatively charged pmi, indicating the ease with which the conjugate base anions are formed. From both *closo*-$C_2B_4H_6$ isomers and *nido*-2,4-$C_2B_4H_8$, the loss of BH_3 units is strongly suggested. Although this may be expected for the *nido*-compounds where a B atom is attached to three H atoms, it is unexpected for the *closo*-derivatives and must involve rearrangement.[50] The *nido*-compounds 2,3-$C_2B_4H_8$, 2-Me-*nido*-2,3-$C_2B_4H_7$, and 2,3-$Me_2C_2B_4H_6$ were studied. All had [pm − H]$^-$ as base peak.

Other carbaboranes for which data have been reported include *closo*-1,5-dicarbapentaborane derivatives,[52] *nido*-1,2-dicarbapentaborane(7),[53] 2,3,4,5-tetracarbahexaborane(6),[54] *closo*-2,3-dimethylcarbaheptaborane(7),[55] *closo*-1,2-dicarbadecaborane(10),[56] *nido*-5,6-dicarbadecaborane(12) and 2,3-$C_2B_9H_{11}$ and derivatives of the 7,9-isomer[57].

A number of reports concern carbaborane derivatives containing Main Group atoms. They include 1-MeM$C_2B_4H_6$ (M = Ga or In),[58] compounds containing

[40] R. H. Cragg and M. Nazery, *J.C.S. Dalton*, 1974, 162.
[41] S. Jerumanis, P. A. Begin, and D. Yu Cong, *Canad. J. Chem.*, 1972, **50**, 1675.
[42] B. Asgarouladi, R. Full, K.-J. Schaper, and W. Siebert, *Chem. Ber.*, 1974, **107**, 34.
[43] W. L. Cook and K. Niedenzu, *Synth. Inorg. Metalorg. Chem.*, 1973, **3**, 229.
[44] H. Wille and J. Goubeau, *Chem. Ber.*, 1974, **107**, 110.
[45] J. R. Blackborow and J. C. Lockhart, *J.C.S. Dalton*, 1973, 1303.
[46] B. Frange, *Bull. Soc. chim. France*, 1973, 2165.
[47] M. C. Glogouski, P. J. Grisdale, J. R. L. Williams, and T. H. Regan, *J. Organometallic Chem.*, 1973, **54**, 51.
[48] R. Van Veen and F. Bickelhaupt, *J. Organometallic Chem.*, 1973, **43**, 33.
[49] H. Noth and W. Storch, *Chem. Ber.*, 1974, **107**, 1028.
[50] T. Onak, J. Howard, and C. Brown, *J.C.S. Dalton*, 1973, 76.
[51] C. L. Brown, K. P. Cross, and T. Onak, *J. Amer. Chem. Soc.*, 1972, **94**, 8055.
[52] A. B. Burg and T. J. Reilly, *Inorg. Chem.*, 1972, **11**, 1962.
[53] D. A. Franz, V. R. Miller, and R. N. Grimes, *J. Amer. Chem. Soc.*, 1972, **94**, 412.
[54] V. R. Miller and R. N. Grimes, *Inorg. Chem.*, 1972, **11**, 862.
[55] R. R. Rietz and R. Schaeffer, *J. Amer. Chem. Soc.*, 1973, **95**, 6254.
[56] R. R. Rietz, R. Schaeffer, and E. Walter, *J. Organometallic Chem.*, 1973, **63**, 1.
[57] V. Chowdhry, W. R. Pretzer, D. N. Rai, and R. W. Redolph, *J. Amer. Chem. Soc.*, 1973, **95**, 4560.
[58] R. N. Grimes, W. J. Rademaker, M. L. Denniston, R. F. Bryan, and P. T. Greene, *J. Amer. Chem. Soc.*, 1972, **94**, 1865.

Group IV elements,[59–62] phosphorus and arsenic derivatives of 2-carba-3-germa-*closo*-dodecaborane(10).[63]

Carbaboranes with iodo,[64] hydroxy,[65] hydroxymethyl, and acetate groups[66] attached have been characterized. The fluorinated carbaborane $C_2B_{10}F_{12}$ is reported to show a pmi and to fragment by loss of fluorine atoms, CF_4, and boron fluoride species.[67]

Compounds of Other Elements. Di-t-butylaluminium fluoride has been shown to exist in a trimeric form in the gas phase, although in benzene solution it is known to be tetrameric.[68] Gas-phase monomers were indicated for a series of quinolinato-complexes of dialkyl-aluminium, -gallium, and -indium.[69] However the spectra of piperidine complexes of aluminium, gallium, and indium dialkyls all showed ions derived from dimeric species.[70] Although $(Me_2AlNHMe)_3$ shows ions from the trimeric species, $(Me_2AlNHEt)_3$ and $(Et_2AlNHMe)_2$ show only dimeric ions.[71] Other compounds with Al—N bonds which have been characterized are $Et_2IAlNMe_2H$ and $(EtIAlNMe_2)_2$.[72]

A number of gallium compounds have been reported including $MeGa_2X_5$ (X = Cl or I) and $MeGa_2Y_4Z$ (Y = Cl or Br, Z = I; Y = Cl, Z = Br or I),[73] and $Me_2GaB_3H_8$.[74]

Cyclopentadienylindium compounds, C_5H_5In and $(MeC_5H_4)In$, both show pmi's as base peak,[75] but the compounds $(C_5H_5)_3In$ and $(MeC_5H_4)_3In$ do not show pmi's, having the highest m/e ion due to $[pm - C_5H_5]^+$ and $[pm - MeC_5H_4]^+$, respectively. Some adducts of cyclopentadienylindium, $C_5H_5InBX_3$ (X = F, Cl, Br, or Me),[76] and $C_5H_5InI_2$ and $(C_5H_5)_2InI$[77] have been characterized. A number of alkylindium dihalides including $(RInX_2)_n$ (R = Me, Pr, or Bu, X = Br;[78] R = Me, Et, or Bu, X = I[78,79]) have all been reported to give dimeric ions. In one case, $EtInBr_2$, ions derived from a tetramer were observed. Dimers

---

[59] A. Tabereaux and R. N. Grimes, *J. Amer. Chem. Soc.*, 1972, **94**, 4768.
[60] A. Tabereaux and R. N. Grimes, *Inorg. Chem.*, 1973, **12**, 792.
[61] M. L. Thompson and R. N. Grimes, *Inorg. Chem.*, 1972, **11**, 1925.
[62] C. G. Savory and M. G. H. Wallbridge, *J.C.S. Dalton*, 1972, 918.
[63] D. C. Beer and L. J. Todd, *J. Organometallic Chem.*, 1973, **50**, 93.
[64] T. J. Reilly and A. B. Burg, *Inorg. Chem.*, 1973, **12**, 1450.
[65] G. D. Mercer and F. R. Scholer, *Inorg. Chem.*, 1973, **12**, 2102.
[66] A. F. Zhigach, V. T. Laptev, V. N. Bochkarev, A. B. Petrunin, B. P. Parfenov, and A. N. Polivanov, *J. Gen. Chem. U.S.S.R.*, 1973, **43**, 866.
[67] R. J. Lagow and J. L. Margrave, *J. Inorg. Nuclear Chem.*, 1973, **35**, 2084.
[68] H. Lehrakuhl, O. Olbrysch, and H. Nehl, *Annalen*, 1973, **4**, 708.
[69] B. Sen and G. L. White, *J. Inorg. Nuclear Chem.*, 1973, **35**, 497.
[70] B. Sen and G. L. White, *J. Inorg. Nuclear Chem.*, 1973, **35**, 2207.
[71] K. J. Alford, K. Gosling, and J. D. Smith, *J.C.S. Dalton*, 1972, 2203.
[72] K. Gosling and A. L. Bhuiyan, *Inorg. Nuclear Chem. Letters*, 1972, **8**, 329.
[73] W. Lind and I. J. Worrall, *J. Organometallic Chem.*, 1972, **35**, 40.
[74] J.-J. Barlin and D. F. Gaines, *J. Amer. Chem. Soc.*, 1972, **94**, 1367.
[75] J. S. Poland and D. J. Tuck, *J. Organometallic Chem.*, 1972, **42**, 307.
[76] J. G. Contreras and D. J. Tuck, *Inorg. Chem.*, 1973, **12**, 2596.
[77] J. G. Contreras and D. J. Tuck, *J. Organometallic Chem.*, 1974, **66**, 405.
[78] M. J. S. Gynane, L. G. Waterworth, and I. J. Worrall, *J. Organometallic Chem.*, 1972, **43**, 257.
[79] J. S. Poland and D. J. Tuck, *J. Organometallic Chem.*, 1972, **42**, 315.

are present in the vapour of dimethylindium acetate but not indium triacetate or diacetate.[80] Methylindium toluene-3,4-dithiolate has been shown to be dimeric in the gas phase.[81]

A small pmi is observed from dimethylthallium succinimide.[82] It fragments by loss of the succinimide group followed by one and then both methyl groups to give [Tl]$^+$ as base peak. The phenylthallium dithiophosphates Ph$_n$Tl{SP(S)-(OMe)$_2$}$_{3-n}$ ($n = 2$ or 1) do not show pmi's but TlSP(S)(OMe)$_2$ does.[83]

Group IV.—Comparison of the ionization potentials of PhSMMe$_3$ (M = C, Si, Ge, Sn, or Pb) with those of the MMe$_4$ compounds, shows that the values are lower and in a narrower range in the former series.[84] In addition, PhSCMe$_3$ does not follow the trend found for the analogous Si to Pb compounds. It was argued that $(p-d)\pi$-bonding between S and M (M $\neq$ C) can be invoked to explain the results. This proposal was in agreement with other data from the series S(MMe$_3$)$_2$ (M $\neq$ Pb) and the compounds PhSSiMe$_2$Cl and PhSSiCl$_3$.

Fragmentation of C$_6$F$_5$OMR$_3$ and C$_6$F$_5$SMR$_3$ (R = Me or Ph and M = Si, Ge, or Sn) has been compared with the C$_6$F$_5$MR$_3$ derivatives.[85] The oxygen and sulphur compounds showed fewer rearranged M—F containing ions than in the spectra of the perfluorophenyl derivatives. Base peak ions were [R$_3$M]$^+$ from all the compounds except Me$_3$SiXC$_6$F$_5$ which had, for X = O, [Me$_2$SiF]$^+$, and for X = S, [Me$_3$Si]$^+$ and [H$_2$SiC$_5$F$_3$]$^+$, as base peaks.

Spectra of Me$_4$M (M = Si, Ge, Sn, or Pb) have again been reported,[86] this time at ionizing voltages between 15 and 100 eV. Trends in ion yields and stabilities of major ions were discussed.

Silicon. Tetramethylsilane has been used as a CI reagent gas. At the relatively high pressures used, there were reactions between [Me$_3$Si]$^+$ and the sample molecules which gave abundant quasimolecular ions [pm + Me$_3$Si]$^+$ in most cases.[87]

Fragmentation of vinyltrimethylsilane and vinyltri([^{2}H$_3$]methyl)silane over a range 9—70 eV has been observed with a quadrupole mass spectrometer.[88] Initial losses of CH$_3$, C$_2$H$_4$, and C$_2$H$_3$ groups from the pmi is followed by loss of C$_2$H$_2$, C$_2$H$_4$, and H$_2$ and, to a much smaller extent, CH$_2$.

The spectra of mono-, bis-, tris- and tetrakis-trimethylsilylmethanes and some chloromethanes have been compared.[89] Parent molecular ions were weak or absent from the silanes and base peaks were [pm − Me]$^+$ ions. The α-chloro-

---

[80] J. J. Habeeb and D. J. Tuck, *J.C.S. Dalton*, 1973, 243.
[81] A. F. Berniaz and D. J. Tuck, *J. Organometallic Chem.*, 1972, **46**, 243.
[82] B. Walther and C. Rockstroh, *J. Organometallic Chem.*, 1972, **42**, 41.
[83] B. Walther, *Z. anorg. Chem.*, 1973, **395**, 112.
[84] G. Distefano, A. Ricci, R. Danieli, A. Foffani, G. Innorta, and S. Torroni, *J. Organometallic Chem.*, 1974, **65**, 205.
[85] G. F. Lanthier, J. M. Miller, and A. J. Oliver, *Canad. J. Chem.*, 1973, **51**, 1945.
[85] K. G. Heumann, K. Buckmann, E. Kubassek, and K. H. Lieser, *Z. Naturforsch.*, 1973, **28b**, 107.
[87] T. J. Odiome, D. J. Harvey, and P. Vouros, *J. Phys. Chem.*, 1972, **76**, 3217.
[88] G. Smolinsky and M. J. Vasile, *Org. Mass Spectrometry*, 1973, **7**, 1069.
[89] D. R. Dimmel, C. A. Wilkie, and F. Ramon, *J. Org. Chem.*, 1972, **37**, 2665.

silanes had [Me$_3$Si]$^+$ as base peak and showed rearranged ions containing Si—Cl bonds such as [MeSiCl$_2$]$^+$. The reaction proposed for the formation of this ion from Me$_3$SiCCl$_3$ is given in Scheme 4. Allylic ions were also a feature in the spectra of the α-chlorosilanes.

$$\text{Me}_3\text{SiCCl}_3 \overset{\rceil^{\ddagger}}{\xrightarrow{-\text{Cl}\cdot}} \text{Me}_3\overset{+}{\text{SiCCl}}_2 \xrightarrow{-\text{C}_3\text{H}_6} \text{Me}\overset{+}{\text{SiCl}}_2$$

Scheme 4

Mass spectrometry has been used to characterize the products from reactions of carbon vapour and Me$_3$SiH including 1,3-bis(trimethylsilyl)allene and 1,2-bis(trimethylsilyl)ethylene.[90] Similar compounds which give pmi's are (Me$_3$Si)-PhC=C=C(SiMe$_3$)H,[91] and Et$_3$Si(C≡C)$_n$SiEt$_3$ (n = 3 or 4).[92] The products of photochemical reactions of silylbuta-1,3-dienes[93] and rearrangements of silylallylic and silylpropynylic compounds[94] have been analysed. The characterization of some trimethylsilyl derivatives of acids including (MeSi)$_3$CCO$_2$H,[95] and (Me$_3$Si)$_3$C(SiMe$_2$)CH$_2$CO$_2$H[96] has been recorded.

The spectra of some (*gem*-2,2-difluorocyclopropyl)silanes (16) have been compared with the (*gem*-1,1-difluoroallyl)silanes (17).[97] Generally, the pmi's of compounds (16) are weak or not observed whilst (17) give stronger pmi's. Otherwise the spectra are quite similar, suggesting interconversion of corresponding fragment ions from (16) and (17) derivatives. The major fragmentation to

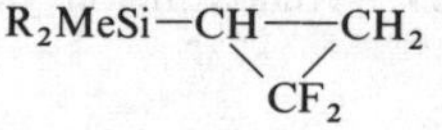

R$_2$MeSi—CH——CH$_2$ R$_2$MeSi—CH$_2$—CH=CF$_2$
          \\   /
           CF$_2$

(16)   R = Me, Cl, F, or H   (17)

Si-containing ions is loss of the C$_2$H$_3$CF$_2$ group. Abundant ions are also produced after initial loss of R$_2$MeSiF. The spectra of some 2,2,3,3-tetrafluorocyclobutylsilanes are similar to the cyclopropyl compounds (16) in that pmi's are weak or absent and loss of R$_2$MeSiF groups is characteristic.[98]

The ionization potentials and aspects of the spectra of some 1-R-substituted indanyls and indenyls [R = Me$_3$Si, Me$_2$Si(OEt), Me$_2$SiPh, or SnMe$_3$] have been recorded.[99]

---

[90] P. P. Skell and P. W. Owen, *J. Amer. Chem. Soc.*, 1972, **94**, 1578.
[91] G. Merault, J. P. Picard, P. Bourgeois, J. Dunogues, and N. Duffaut, *J. Organometallic Chem.*, 1972, **42**, C80.
[92] R. Eastmond, T. R. Johnson, and D. R. M. Walton, *Tetrahedron*, 1972, **28**, 4601.
[93] J. W. Connolly, *J. Organometallic Chem.*, 1974, **65**, 343.
[94] J. Slutsky and H. Kwart, *J. Amer. Chem. Soc.*, 1973, **95**, 8678.
[95] O. W. Steward, J. S. Johnson, and C. Eaborn, *J. Organometallic Chem.*, 1972, **46**, 96.
[96] O. W. Steward and J. S. Johnson, *J. Organometallic Chem.*, 1973, **55**, 209.
[97] V. N. Bochkarev, V. D. Sheludyakov, A. N. Polivanov, V. V. Shcherbinin, and V. F. Mironov, *J. Gen. Chem. U.S.S.R.*, 1973, **43**, 618.
[98] V. N. Bochkarev, A. N. Polivanov, V. D. Sheludyakev, and V. V. Shcherbinin, *J. Gen. Chem. U.S.S.R.*, 1973, **43**, 1639.
[99] P. E. Rakita, M. K. Hoffman, M. N. Andrew, and M. M. Bursey, *J. Organometallic Chem.*, 1973, **49**, 213.

Other compounds containing trimethylsilyl groups attached to cyclic systems reported to show pmi's include 2,5-di(trimethylsilyl)-3,4-dichlorothiophen[100] and 2-trimethylsilyl-4-NN-dimethylthienylamine.[101]

The spectra of doubly-charged ions from Ph_4Si and Ph_3SiCl have been recorded.[102] They did not contain the characteristic rearrangement peaks which are produced from the corresponding singly-charged pmi's, *e.g.* the ions $[pm - C_{12}H_{10}]^{2+}$ from Ph_4Si and $[pm - PhSiCl]^{2+}$ from Ph_3SiCl were absent. The base peak from Ph_4Si was $[pm - Ph - H_2]^{2+}$ whereas the singly-charged base peak was $[Ph_3Si]^+$.

Other compounds for which data have been reported include silyl-substituted $\alpha\beta$-unsaturated ketones,[103] dimethyl-4,4-diphenyl-4-sila-1,7-heptanedioate and related compounds,[104] $Me_3SiC_6Br_5$,[105] and $(Me_3Si)_nCCl_{4-n}$ ($n = 2$ or 4) and $(Me_3Si)_nCHCl_{3-n}$ ($n = 3$—1).[106]

Silicon Heterocycles. The first stable silacyclopropanes including dispiro-{bicyclo[4,1,0]heptane}-7,2'-silacyclopropane-3',7''-bicyclo[4,1,0]heptane[107] and some related silacyclopropanes[108] have been reported to show pmi's. A number of 7-silabicyclo[2,2,1]-hept-5-ene derivatives,[109] a hepta-2,5-diene,[110] and a 7-silabicyclo[2,2,2]octa-2,5-diene derivative[111] have been characterized. Silacyclobutane compounds $(CH_2)_3SiRX$ (R = X = Me or Cl; R = Me, X = Cl) and 4-silaspiro[3,3]heptane all show pmi's.[112]

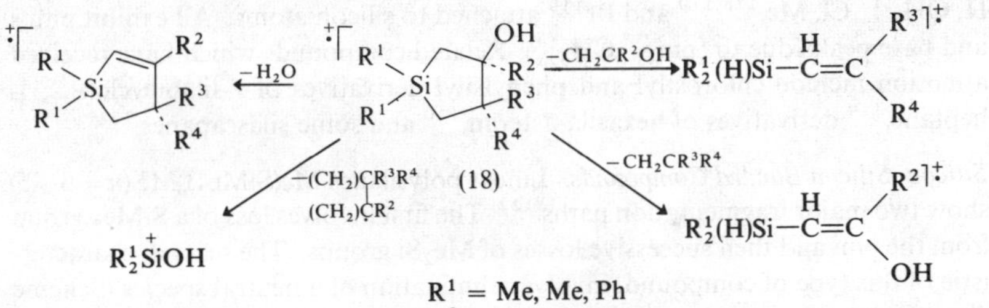

Scheme 5

R^1 = Me, Me, Ph
R^2 = H, Me, H
R^3 = H, H, H
R^4 = H, Me, H

[100] M. R. Smith and H. Gilman, *J. Organometallic Chem.*, 1972, **42**, 1.
[101] S. W. Slocum and P. L. Gierer, *J. Org. Chem.*, 1973, **38**, 4189.
[102] T. Blumenthal and J. H. Bowie, *Org. Mass Spectrometry*, 1972, **6**, 1083.
[103] A. G. Brook and J. M. Duff, *Canad. J. Chem.*, 1973, **51**, 2024.
[104] R. A. Felix and W. P. Weber, *J. Org. Chem.*, 1973, **37**, 2323.
[105] F. Smith, G. J. Moore, and C. Tamborski, *J. Organometallic Chem.*, 1972, **42**, 257.
[106] D. R. Dimmel, C. A. Wilkie, and F. Ramon, *J. Org. Chem.*, 1972, **37**, 2662.
[107] R. L. Lambert and D. Seyferth, *J. Amer. Chem. Soc.*, 1972, **94**, 9246.
[108] D. Seyferth, C. K. Haas, and D. Annarelli, *J. Organometallic Chem.*, 1973, **56**, C7.
[109] H. Balasubramanian and M. V. George, *Tetrahedron*, 1973, **29**, 2395.
[110] T. J. Barton, J. L. Witiak, and C. L. McIntosh, *J. Amer. Chem. Soc.*, 1972, **94**, 6229.
[111] T. J. Barton and E. Kline, *J. Organometallic Chem.*, 1972, **42**, C21.
[112] R. Damrauer, R. A. Davis, M. T. Burke, R. A. Karn, and G. T. Goodman, *J. Organometallic Chem.*, 1972, **43**, 121.

A study of a series of silacyclopentanols (18) has shown that the major fragmentations involve hydrogen transfer leading to elimination of H_2O, migration of OH to silicon, and elimination of unsaturated C-2—C-3 and C-4—C-5 species (Scheme 5).[113]

The initial fragmentation of 1,1-dimethyl-2-methyl-1-silacyclopentane is loss of C_2H_4 or CH_3.[114] The base peak ion is [pm − C_2H_4]$^{+\cdot}$. The spectrum of 1,1-dimethyl-2,5-dimethyl-1-silacyclopentane shows a base peak, [pm − $CH_2CHMe − CH_3]^+$.[115]

Fragmentation of silacyclopent-3-enes (19) depends mainly on the nature of X.[116] For X = Cl, F, or OMe, loss of SiX_2 is important. Whereas loss of X is

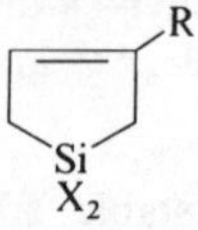

(19) R = H or Me
X = H, F, Cl, Me, Ph, OMe, or NMe_2

characteristic of most compounds, loss of HX is typical of X = H, Cl, Ph, OMe, or NMe_2. Elimination of 1,3-diene species is common for X = Ph, OMe, and NMe_2 derivatives.

A number of tetrasila-adamantane derivatives have been characterized with H, OH,[117] Cl, Me,[118,119] and Br[119] attached to silicon atoms. All exhibit pmi's and base peaks due to [pm − CH_3]$^+$. Related compounds which have received attention include chlorosilyl and phenylsilyl derivatives of 1-silabicyclo[2,2,1]-heptane,[119] derivatives of hexasila-asteran,[120] and some silascapanes.[121]

Silicon–Silicon Bonded Compounds. Linear polysilanes $Me(SiMe_2)_nMe$ (n = 3—5) show two major fragmentation paths.[122] The first involves loss of a $SiMe_3$ group from the pmi and then successive losses of Me_2Si groups. The second, characteristic of this type of compound, involves elimination of a neutral species (Scheme 6).

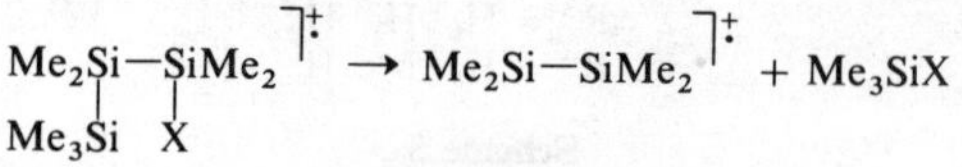

$$\begin{matrix} Me_2Si-SiMe_2 \\ | \\ Me_3Si \quad X \end{matrix}\Bigg]^{+\cdot} \rightarrow \Big[Me_2Si-SiMe_2\Big]^{+\cdot} + Me_3SiX$$

(X = Me, $SiMe_3$, Si_2Me_5)

Scheme 6

[113] C. Lageot, J. C. Maire, P. Mazerolles, and G. Manuel, *J. Organometallic Chem.*, 1973, **60**, 55.
[114] T. W. Dolzine, A. K. Holand, and J. P. Oliver, *J. Organometallic Chem.*, 1974, **65**, C1.
[115] R. T. Fessenden and W. D. Kray, *J. Org. Chem.*, 1973, **38**, 87.
[116] V. N. Bochkarev, A. N. Polivanov, N. G. Komalenkova, S. A. Bashkirova, and E. A. Chernyshev, *J. Gen. Chem. U.S.S.R.*, 1973, **43**, 784.
[117] C. L. Frye and J. M. Klosowski, *J. Amer. Chem. Soc.*, 1972, **94**, 7187.
[118] G. D. Homer and L. H. Sommer, *J. Amer. Chem. Soc.*, 1973, **95**, 7700.
[119] G. Fritz and M. Hahnke, *Z. anorg. Chem.*, 1972, **390**, 191.
[120] G. Fritz, H. J. Dannappel, and E. Matern, *Z. anorg. Chem.*, 1973, **399**, 263.
[121] G. Fritz, G. Marquardt, and H. Scheer, *Angew. Chem. Internat. Edn.*, 1973, **12**, 654.
[122] Y. Nakadaira, Y. Kobayashi, and H. Sakurai, *J. Organometallic Chem.*, 1973, **63**, 79.

Organometallic, Co-ordination, and Inorganic Compounds 155

The ratio of abundances of the ions $[XC_6H_4SiMe_2]^+/[SiMe_3]^+$ from $PhMe_5Si_2$ and some *p*-substituted phenyl compounds was found to increase with the electron donating ability of the *p* substituent.[123] In all cases the base peak ion was $[XC_6H_4SiMe_2]^+$.

Cyclic polysilanes $Si_nMe_{2n-1}R$ ($n = 5$, R = H, F, Cl, or Ph) are suggested to fragment mainly by loss of R and then an $SiMe_2$ group.[124] When $n = 6$ (and R = Cl, Ph, OEt, or $SiMe_2Cl$), initial fragmentation is suggested to involve ring contraction to a Si_5 species followed by loss of $SiMe_2R$ or formation of the $[RSiMe_2]^+$ ion. Other cyclic compounds studied include permethylcyclopolysilanes Si_nMe_{2n-2} ($n = 8$—13).[125]

Compounds containing Si—H Bonds. A number of compounds containing alkyl or aryl groups and Si—H bonds have been characterized, including $(Me_2SiH)CH_2SiMe_3$,[126] $(Me_2SiH)SiH_2(SiMe_2H)$,[127] $(PhSiH_2)CHPhOSiMe_3$,[128] and $(PhSiH_2)SiH(SiH_3)_2$.[129]

Compounds containing Si—N or Si—P Bonds. Parent molecular ions are absent from the spectra of dialkylmethyleneamidosilanes such as $Bu^t_2CNSiX_nY_{3-n}$ (X = Me, Y = Cl for $n = 0$—3).[130] Ions at highest *m/e* are due to $[pm - CH_3]^+$ or $[pm - Cl]^+$. Diphenylmethyleneamidosilanes do give pmi's and some rearranged ions containing Si—Ph species.[130] Bis(trifluoromethyl)methyleneamidosilane derivatives, $Me_{4-n}Si\{NC(CF_3)_2\}_n$ ($n = 4, 2$, or 1), and some germanium and tin compounds all show pmi's.[131]

Other compounds which have been characterized include $X^1Me_2SiCH_2X^2$ (X^1 and $X^2 = Me_2N$ or Me_2P),[132] $Me_3SiN(Me)PCl_2$,[133] $(Me_3Si)_2NP(NSiMe_3)$,[134] $Me_3Si(Ph)P$—$P(Ph)SiMe_3$,[135] $Ph(SiMe_3)_2P$,[136] $(Me_3SiN)_2S(NMe_2)_2$,[137] and $(Me_3SiN)_2SO$,[138] and related compounds. Silylated hydrazines of the type $R^1_3SiN(R^2)NR^2H$ (R^1 = Ph or Et, $R^2 = CO_2Me$ or CO_2Et) and $Ph_3SiN(Ph)NPhH$ are all reported to show pmi's.[139] Initial loss of R or ROH, or both groups is common to all the compounds.[139] The lithium derivative, LiNCPh-

---

[123] H. Sakurai, M. Kira, and T. Sato, *J. Organometallic Chem.*, 1972, **42**, C24.
[124] C. Lageot, J. C. Maire, M. Ishikawa, and M. Kumada, *J. Organometallic Chem.*, 1973, **57**, C39.
[125] R. West and A. Indriksons, *J. Amer. Chem. Soc.*, 1972, **94**, 6110.
[126] A. W. P. Jarvie and R. J. Rowley, *J. Organometallic Chem.*, 1973, **57**, 261.
[127] P. S. Skell and P. W. Owen, *J. Amer. Chem. Soc.*, 1972, **94**, 5434.
[128] H. Watanabe, J. Kogure, and Y. Nagai, *J. Organometallic Chem.*, 1972, **43**, 285.
[129] F. Feher and R. Freund, *Inorg. Nuclear Chem. Letters*, 1973, **9**, 937.
[130] J. B. Farmer, R. Snaith, and K. Wade, *J.C.S. Dalton*, 1972, 1501.
[131] M. F. Lappert and D. E. Palmer, *J.C.S. Dalton*, 1973, 157.
[132] J. Grobe and G. Heyer, *J. Organometallic Chem.*, 1973, **61**, 133.
[133] R. Jefferson, J. F. Nixon, T. M. Painter, R. Keat, and L. Stobbs, *J.C.S. Dalton*, 1973, 1414.
[134] E. Neike and W. Flick, *Angew. Chem. Internat. Edn.*, 1973, **12**, 585.
[135] M. Baudler, M. Hallab, A. Zarkadas, and E. Tolls, *Chem. Ber.*, 1973, **106**, 3962.
[136] M. Baudler and A. Zarkadas, *Chem. Ber.*, 1973, **106**, 3970.
[137] O. Glemser, J. Wegener, and R. Hofer, *Chem. Ber.*, 1972, **105**, 475.
[138] O. Glemser, M. F. Feser, S. P. van Halasz, and H. Saran, *Inorg. Nuclear Chem. Letters*, 1972, **8**, 321.
[139] K.-H. Linke and J.-J. Gohansen, *Chem. Ber.*, 1973, **106**, 3438.

{N(SiMe$_3$)$_2$}, shows the [pm − Li]$^+$ ion at highest m/e value.[140] Related compounds Me$_3$SiNCPh{NR$_2$} (R = Me or Me$_3$Si) both exhibit pmi's.

The cyclic compounds (20)—(23) show pmi's and initial decomposition by loss of two R groups, in some cases successively, (20) and (21), and in others, (22) and (23), simultaneously.[141] A number of rearranged ions were observed including [Me$_2$SiSiMePh$_2$]$^+$ from (21) and [(MeO)$_4$Si]$^{+\cdot}$ from (23). Other cyclic

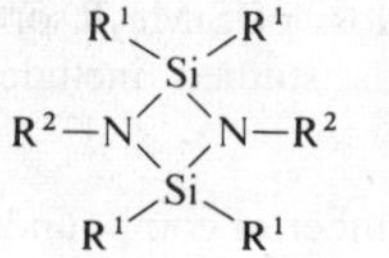

	R^1	R^2
(20)	Me	Ph
(21)	Me	SiPh$_2$Me
(22)	NHMe	Si(NHMe)$_3$
(23)	OMe	Si(OMe)$_3$

compounds reported are the eight- and twelve-membered cyclothiazasilanes (Me$_2$SNSiR$_2$N)$_{2\ or\ 3}$,[142] some 1-aza-2-silacyclopentanes,[143] and Me$_3$SiN-(MePS)$_2$S.[144] The last compound fragmented by loss of CH$_3$, SCH$_3$, S$_2$CH$_3$, and two S groups. Trimethylsilyl derivatives of purines,[145] pyrimidines,[145,146] pteridines and quinolines[146] are not discussed here (see Chapters 6 and 8).

Compounds containing Si—O *or* Si—*Main Group VI Bonds.* Trimethylsilylcyclopropyl ether derivatives (24) are reported to fragment initially by loss of H or

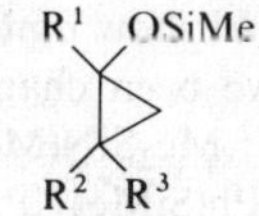

(24)

CH$_3$ groups.[147] The base peak was [Me$_3$Si]$^+$, and [Me$_2$SiOH]$^+$ was abundant in all the spectra. *N*-Trifluoroacetyl-α-amino-acid trimethylsilyl esters (25) show an interesting reaction leading to elimination of CO$_2$ from pmi (Scheme 7), which

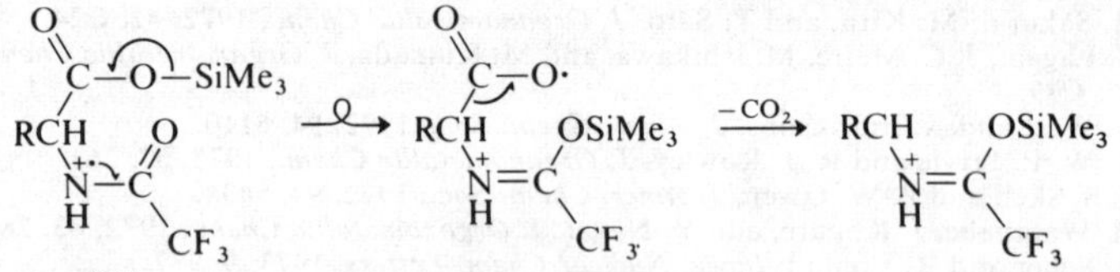

Scheme 7

[140] A. R. Sanger, *Inorg. Nuclear Chem. Letters*, 1973, **9**, 351.
[141] C. Lageot and J. C. Maire, *J. Organometallic Chem.*, 1972, **37**, 239.
[142] R. Appel and I. Ruppert, *Chem. Ber.*, 1973, **106**, 902.
[143] T. Tsai and C. J. Marshall, *J. Org. Chem.*, 1972, **37**, 596.
[144] H. W. Roesky and M. Dietl, *Angew. Chem. Internat. Edn.*, 1973, **12**, 425.
[145] E. White, V. P. M. Krueger, and J. A. McCloskley, *J. Org. Chem.*, 1972, **37**, 430.
[146] W. J. A. VandenHeuval, J. L. Smith, P. Haug, and J. L. Beck, *J. Heterocyclic Chem.*, 1972, **9**, 451.
[147] G. M. Rubottom and M. I. Lopez, *J. Org. Chem.*, 1973, **38**, 2097.

is uncommon in other derivatives of amino-acids.[148] The [pm − CO_2]$^{+\cdot}$ and subsequent ions are abundant when R is an aliphatic group.

Other trimethylsiloxy-compounds for which data have been reported include alkyltrimethylsilyl ketals, $R_2C=C(OMe)OSiMe_3$, and related compounds, $R^1R^2C=C(OSiMe_3)CR^1R^2CO_2Me$,[149] $R^1R^2C=C(OSiR_3^3)_2$ and $R^1R^2C=C(OSiMe_3)CR^1R^2CO_2SiMe_3$,[150] $(NC)_3COSiMe_3$, $\{(NC)_2COSiMe_3\}_2$ and a series of O-trimethylsilylcyanohydrins,[151] carbamate and thiocarbamate derivatives of hydrazines $Me_2NN(R)C(X)XSiMe_3$ (X = O or S)[152] and related compounds.

Cyclic 1,3,2-benzodioxasilols fragment by the sequential loss of R and CH_3 from the same silicon atom.[153] This is unusual and in the absence of any possible skeletal rearrangements, geminal cleavage was proposed (Scheme 8).

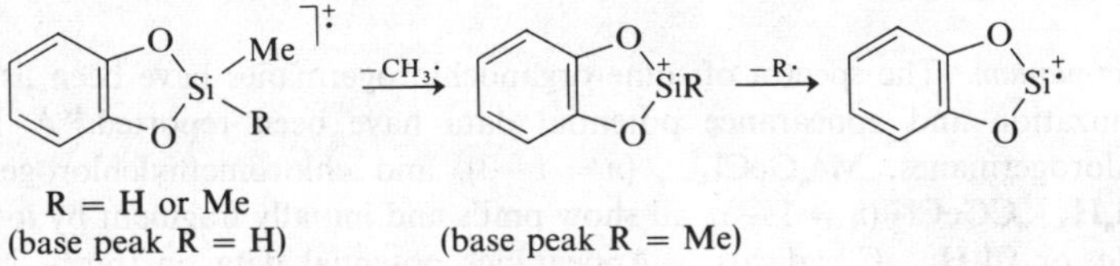

R = H or Me
(base peak R = H)   (base peak R = Me)

Scheme 8

Other compounds containing Si–Main Group VI bonds reported to show pmi's include $(CF_3)_2PSSiMe_3$,[154] $Me_3SiSeMe$ and $Me_2Si(SeMe)_2$,[155] and some cyclosilathianes $(R^1R^2SiS)_{2\ or\ 3}$ (R^1 = Me, R^2 = Et or vinyl).[156]

Fragmentation patterns of a series of phenoxy- and toloxy-phenylsilanes have been reported.[157] The pmi was the base peak for each compound studied and the loss of an OR group was an important decomposition process. Further fragmentation was by loss of C_6H_6 for $Ph_3(m-tolO)Si$ and $Ph_2(m-tolO)_2Si$, or ROH from $Ph(m-tolO)_3Si$, $Ph(o-tolO)_3Si$, and $Ph(PhO)_3Si$, giving abundant ions possibly with cyclic structures (Scheme 9). In some cases loss of H_2O competed with loss of ROH.

Compounds containing Si—Halogen Bonds. A number of compounds have been characterized, including CF_3SiF_3,[158] $R^1R^2C=C=C(R^3)(SiF_2)_2C(R^4):CR^5R^6$ (R = H or alkyl) and the cyclic compound $F_2\overline{SiR^1C:CR^2SiF_2}$,[159] and related

---

[148] L. Birkofer and G. Schmidtberg, *Org. Mass Spectrometry*, 1973, **7**, 1387.
[149] C. Ainsworth, F. Chen, and Y.-N. Kuo, *J. Organometallic Chem.*, 1972, **46**, 59.
[150] C. Ainsworth and Y.-N. Kuo, *J. Organometallic Chem.*, 1972, **46**, 73.
[151] W. Lidy and W. Sundermeyer, *Chem. Ber.*, 1973, **106**, 587.
[152] L. K. Peterson and K. I. The, *Canad. J. Chem.*, 1972, **50**, 562.
[153] J. G. Liehr and W. J. Richter, *Org. Mass Spectrometry*, 1973, **7**, 53.
[154] K. Gosling and J. L. Miller, *Inorg. Nuclear Chem. Letters*, 1973, **9**, 355.
[155] J. W. Anderson, G. K. Barker, J. E. Drake, and M. Rodger, *J.C.S. Dalton*, 1973, 1716.
[156] M. M. Millard, L. J. Pazdernik, W. F. Haddon, and R. E. Lundin, *J. Organometallic Chem.*, 1973, **52**, 283.
[157] V. N. Bochkarev, A. N. Polivanov, G. Ya. Zhigalin, M. V. Sobolevskii, and E. A. Chernyshev, *J. Gen. Chem. U.S.S.R.*, 1973, **43**, 1278.
[158] K. G. Sharp and T. D. Coyle, *Inorg. Chem.*, 1972, **11**, 1259.
[159] C. S. Liu, J. L. Margrave, and J. C. Thompson, *Canad. J. Chem.*, 1972, **50**, 465.

compounds. A series of eleven p- or m-fluorophenylsilanes $(FC_6H_4)SiXYZ$ (X, Y, Z = Me, F, Cl, Br or OMe) are all reported to show pmi's.[160]

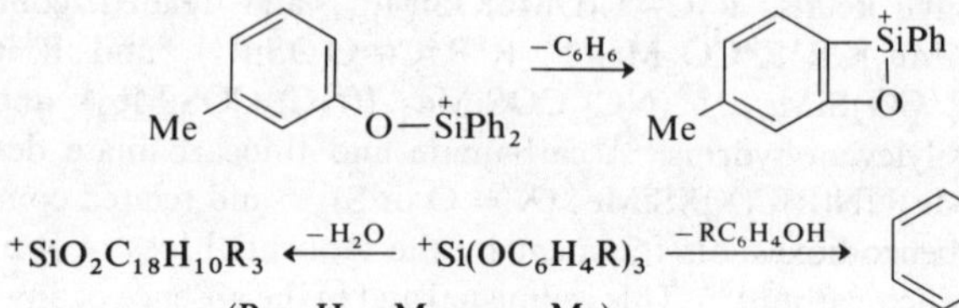

Scheme 9

$^+SiO_2C_{18}H_{10}R_3 \xleftarrow{-H_2O} {}^+Si(OC_6H_4R)_3 \xrightarrow{-RC_6H_4OH}$

(R = H, o-Me, or m-Me)

Germanium. The spectra of some organochlorogermanes have been listed and ionization and appearance potential data have been reported.[161] Methylchlorogermanes, Me_nGeCl_{4-n} (n = 1—3), and chloromethylchlorogermanes, $Cl_nH_{3-n}CGeCl_3$ (n = 1—3), all show pmi's and initially fragment by loss of Cl, Me, or $Cl_nH_{3-n}C$ radicals. Appearance potential data on $[pm - Me]^+$ or $[pm - Cl]^+$ from the series Me_nGeCl_{4-n} (n = 0—4), and $[Cl_3C]^+$ and $[GeCl_3]^+$ from Cl_3CGeCl_3, were used to calculate a number of bond dissociation energies in the molecules and ions. It was found that, although the bond dissociation energy, $D(Me—GeXYZ)^+$, in the pmi's (X, Y, Z = Me or Cl) increased with the number of Cl groups attached to Ge from 0.76 eV in Me_4Ge to 1.11 eV in $MeGeCl_3$, the $D(Cl—GeXYZ)^+$ values decreased sharply from 2.13 to 0.44 eV going from Me_3GeCl to $GeCl_4$.

The spectrum of $(C_5H_5)_2Ge$ shows a pmi. Initial loss of C_5H_5 is followed by loss of C_2H_2 molecules.[162]

Germanium Heterocycles. A series of germacyclopentanols (26) show some fragmentations similar to those of the silicon analogues (18).[163] Loss of H_2O, C-4—C-5 containing species, and OH transfer to the germanium atom are

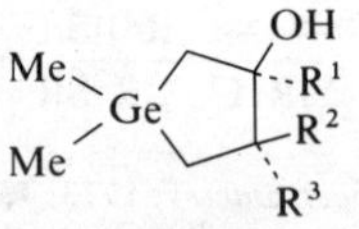

(26) R^1 = H, Me, H, Me, Me
R^2 = H, H, H, Me, H
R^3 = H, H, Me, H, Me

observed. Further fragmentation by loss of CH_3 from $[pm - H_2O]^{+\cdot}$ and $[pm - CH_2CR_2R_3]^{+\cdot}$ occurs. All derivatives show cleavage of Ge—C-2 and (C-3—C-4) bonds to give $[H_2C=CR^1OH]^{+\cdot}$.

[160] J. Lipowitz, *J. Amer. Chem. Soc.*, 1972, **94**, 1582.
[161] J. Tamas, G. Czira, A. K. Maltsev, and O. M. Nefedov, *J. Organometallic Chem.*, 1972, **40**, 311.
[162] J. V. Scribelli and D. M. Curtis, *J. Amer. Chem. Soc.*, 1973, **95**, 924.
[163] C. Lageot, J. C. Maire, G. Manuel, and P. Mazerolles, *J. Organometallic Chem.*, 1973, **54**, 131.

The fragmentation behaviour of some germacyclopentenes (27)—(29) has been studied.[164] Ions produced by the elimination of GeX and R groups and of C_6H_6 from phenyl derivatives were commonly observed. Decomposition of the pmi of the methoxy-derivative of (27) by expulsion of GeHPh (or Ge and C_6H_6?) was

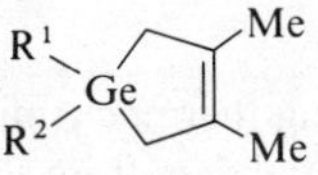

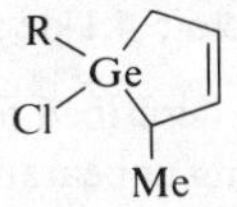

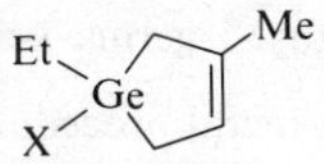

(27) R^1 = Ph
R^2 = H, F, Cl, Br, I, or OMe

(28) R = Ph or Et

(29) X = H or Cl

a novel feature of this spectrum. The three major fragmentation processes observed in the spectra of dibenzo[b,f]germepins (30) and dihydro-derivatives were loss of R groups, elimination of neutral molecules such as C_6H_6, and loss of Ge-containing species.[165]

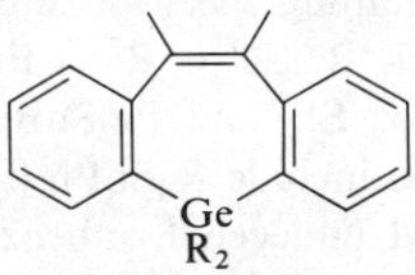

(30) R = Me or Ph

Other Germanium Compounds. 5-Germylcyclopentadiene and related compounds have been characterized.[166]

Compounds containing Ge—N bonds reported include linear and cyclic substituted germylhydrazines such as $Me_2NNHGeMe_3$ and $(Me_2NNGeMe_2)_3$,[167] and $Ph_3GeNRNRH$ (R = CO_2Me or CO_2Et),[168] and pseudohalogenomethyl-germanes $MeGeH_2X$ (X = N_3, NCO, or NCS).[169]

The compounds Ph_5Ge_2X (X = F, Cl, I, or CO_2CF_3), $Ph_4Ge_2X_2$ (X = F, Cl, Br, I, H, or CO_2CF_3), and $(Ph_2GeO)_3$ all show pmi's,[170] but $Ph_4Ge_2(CO_2CCl_3)_2$ does not. Typically, fragmentation of the tetraphenyl compounds produces $[Ph_3Ge]^+$ and $[Ph_4Ge]^{+\cdot}$ ions (X = F, Cl, I, or CO_2CF_3). The trimeric oxide showed the loss of H_2O from the pmi. More than 56% of the total ion current was carried by Ge_3O_3 ions.

---

[164] C. Lageot, J. C. Maire, P. Riviere, M. Massmol, and J. Barrau, *J. Organometallic Chem.*, 1974, **66**, 49.
[165] J. Y. Corey, E. R. Corey, M. D. Glick, and J. S. Dueber, *J. Heterocyclic Chem.*, 1972, **9**, 1379.
[166] P. C. Angus and S. R. Stobart, *J.C.S. Dalton*, 1973, 2374.
[167] L. K. Peterson and K. I. The, *Canad. J. Chem.*, 1972, **50**, 553.
[168] K.-H. Linke, H. J. Gohausen, and G. Wrobel, *Chem. Ber.*, 1972, **105**, 1780.
[169] J. E. Drake and R. T. Hemmings, *Canad. J. Chem.*, 1973, **51**, 302.
[170] F. Glockling and R. E. Houston, *J.C.S. Dalton*, 1973, 1357.

The $R_3^1GeO_2SR^2$ derivatives (R^1 = Et or Ph, R^2 = Ph or p-tol) and Me_3GeO_2SMe all show pmi's.[171] The initial fragmentation of the triethylgermyl compounds involves loss of SO_2R or Et radicals. Further loss of SO_2 and two C_2H_4 molecules sequentially from [pm − Et]$^+$ was observed. Other compounds studied include $-O+Ge(Bu_2)OCH(CH_2)_4CH-O+_n$ (n = 2, 4, or 6),[172] and methyl-1-germa-1-oxa-12-bicyclo[5,4,1]dodecane.[173]

Tin. A general procedure has been described for deriving thermochemical data for ions, radicals, and molecules using appearance potential data from a series of compounds.[174] Results for the series Me_nSnCl_{4-n} (n = 0—4) were given and compared with data obtained by other methods.

The spectra of three series of alkyltin compounds RMe_2BuSn (R = Et, Pr, Pr^i, Bu, Bu^i, Bu^t, 2-pentyl, 3-pentyl, cyclopentyl, cyclohexyl, benzyl, neophyl, or Ph), $RMe_2\{Ph(CH_3)CH\}Sn$ (R = Et, Pr, Pr^i, Bu^i, Bu^s, Bu^t, isopentyl, neophyl, or Ph), and RBu^i_3Sn (R = Me, Pr, or Pr^i) have been listed.[175] As had been observed previously, initial fragmentation was by loss of an organic radical. This was followed by elimination of one or more olefin groups for (R ≠ Me or Ph) giving rise to tin hydride-containing species. Similar results were reported for the butyltin compounds $Bu^tSnR_2^1R^2$ ($R^1 = R^2$ = Et or Pr; R^1 = Bu^t, R^2 = Me), $Bu_2^tSnR^1R^2$ ($R^1 = R^2$ = Me or Et), and $Pr_3^iSnBu^t$.[176] Other compounds for which data have been reported include R_3SnPh (R = Bu^i, Bu^t, or neophyl),[177] Ph_3SnR (R = butenyl, pentenyl, undecenyl, or benzyl), Me_3SnR (R = cyclohexyl or benzyl),[178] $CH_3HC=C=CH(SnMe_3)$,[179] $Sn(CF_3)_4$,[180] and the Sn^{II} derivatives $Sn(CH_2SiMe_3)_2$[181] and $(MeC_5H_4)_2Sn$.[182] The spectra of a number of trialkyltin halides have been listed including Me_3SnX,[175] Pr_3^iSnX (X = F, Cl, Br, or I) $Bu_2^tMeSnBr$, Bu^tMe_2SnBr,[176] $(C_5H_5)_2MeSnI$,[183] and $(C_5H_5)SnCl$.[184]

Tin Heterocycles. From a study of some trimethyltin norborn-2-ene derivatives several novel fragmentation routes associated with the presence of the bicyclic system and/or double bonds were found.[185] Both *endo-* and *exo-*5-trimethyltin, and *syn-* and *anti-*2-trimethyltin isomers and 3-trimethylstannylnortricyclene were found to fragment *via* retro-Diels–Alder type reactions to produce $[C_5H_6]^+$ and $[Me_2SnCH=CH_2]^+$ ions (*e.g.* Scheme 10). Other decompositions found

---

[171] E. Lindner and K. Schrardt, *J. Organometallic Chem.*, 1972, **44**, 111; 1973, **50**, C33.
[172] A. N. Sara, *J. Organometallic Chem.*, 1973, **47**, 331.
[173] P. Mazerolles and A. Faucher, *J. Organometallic Chem.*, 1973, **63**, 195.
[174] T. R. Spalding, *J. Organometallic Chem.*, 1973, **56**, C65.
[175] M. Gielen and G. Mayence, *J. Organometallic Chem.*, 1972, **46**, 281.
[176] M. Gielen and M. De Clercq, *J. Organometallic Chem.*, 1973, **47**, 351.
[177] H.-J. Gotze, *Chem. Ber.*, 1972, **105**, 1775.
[178] M. Gielen, B. de Poorter, R. Liberton, and M. T. Paelinck, *Bull. Soc. chim. belges*, 1973, **82**, 277.
[179] M. L. Poutsma and P. A. Ibarbia, *J. Amer. Chem. Soc.*, 1973, **95**, 6000.
[180] R. A. Jacob and R. L. Lagow, *J.C.S. Chem. Comm.*, 1973, 104.
[181] P. J. Davidson and M. F. Lappert, *J.C.S. Chem. Comm.*, 1973, 317.
[182] P. G. Harrison and M. A. Healy, *J. Organometallic Chem.*, 1973, **51**, 153.
[183] H.-J. Albert and U. Schroer, *J. Organometallic Chem.*, 1973, **60**, C6.
[184] K. D. Bos, E. J. Bulten, and J. G. Noltes, *J. Organometallic Chem.*, 1972, **39**, C53.
[185] J. D. Kennedy and H. G. Kuivila, *J.C.S. Perkin II*, 1972, 1812.

in the spectrum of the *anti*-7-compound were by loss of CH_3, Me_2Sn, C_2H_4, or both Me_2Sn and C_2H_4, and C_7H_9 (Scheme 11). In the cases of *syn*-2- and *anti*-7-derivatives, the retro-Diels–Alder reaction was so easy that it occurred

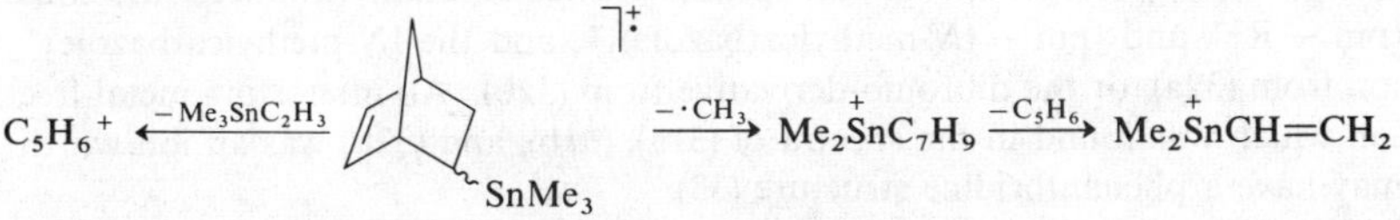

Scheme 10

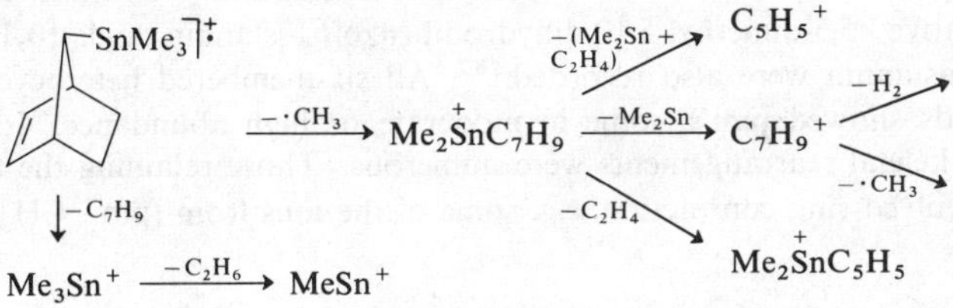

Scheme 11

from the pmi's but less readily for the *anti* compound, paralleling the difference in thermal behaviour. The *anti*-7-isomer shows a strong $[C_7H_9]^+$ ion, the ready formation of which may be due to anchimeric assistance by the double bond in the loss of Me_2Sn from $[Me_2SnC_7H_9]^+$. Aspects of the spectra of $Me_3Sn(CH_2)_2$-$CH=CH_2$, *cis*- and *trans*-$Me_3SnCH_2CH=CHMe$ and some trimethylsilyl norborn-2-ene derivatives were also discussed.

The fragmentation of four 5,10-dihydrophenazastannines (31) and (32) has been described.[186] Loss of R groups consecutively, R_2Sn, dibromo-*N*-methylcarbazole separately or with R_2 species are all observed from pmi's of (31). Subsequent

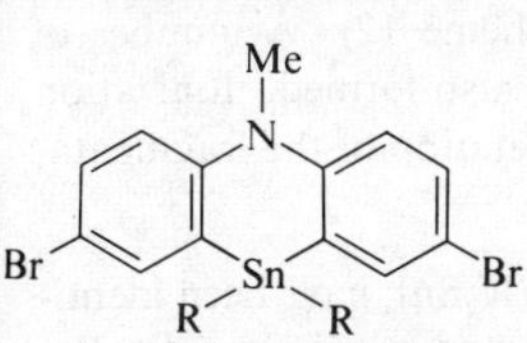

(31) a; R = CH_3 or CD_3
b; R = C_6H_5

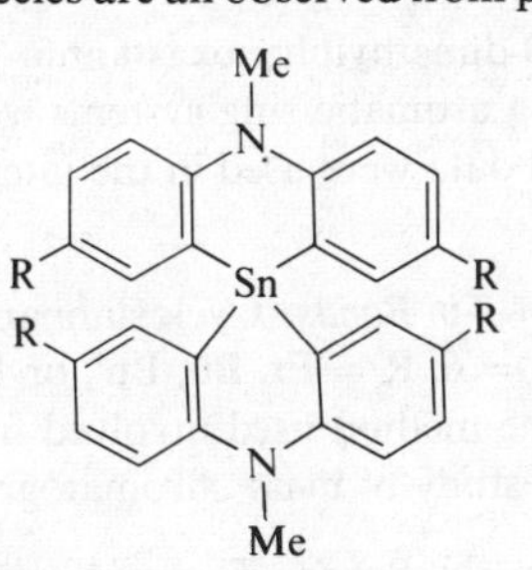

(32) a; R = H
b; R = Br

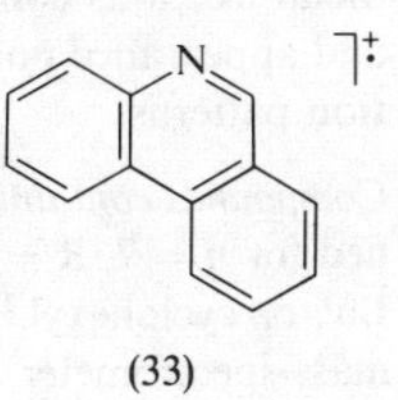

(33)

[186] I. Lengyel and M. J. Aaronson, *J. Organometallic Chem.*, 1972, **42**, 95; *Angew. Chem. Internat. Edn.*, 1972, **11**, 521.

fragmentations are accompanied by skeletal rearrangements and can involve hydrogen migrations. An unusual reaction was the formation of [SnBr]$^+$ from [pm − 2R]$^{+\cdot}$ which involves the migration of bromine to tin. In compounds (32) with no exocyclic groups on tin, primary fragmentation produced the ions [pm − R]$^+$ and [pm − (N-methylcarbazole)]$^{+\cdot}$, and the [N-methylcarbazole]$^+$ ion from (32a), or the dibromo-derivative from (32b). An interesting metal-free ion which was found in the spectra of (31a), (31b), and (32b) was an ion which may have a phenanthridine structure (33).

The spectra of 10,10-dimethyl- and 10,10-diethyl-phenoxastannin have been re-investigated[187] on the suggestion[187,188] that a previous report[189] actually concerned not six- but twelve-membered heterocyclic ring compounds. The spectra of related compounds 10,10-dimethylphenothiastannin and the 5,5-dioxide derivative, 5,5-dimethyl-5,10-dihydrodibenzo[b,e]stannin and 10,10'-spirobiphenoxastannin were also recorded.[187] All six-membered heterocyclic ring compounds showed pmi's, some in moderate or high abundance. Ions produced *via* skeletal rearrangements were numerous. Those retaining the tin atom often involved ring contraction, *e.g.* some of the ions from [pm−CH$_3$]$^+$

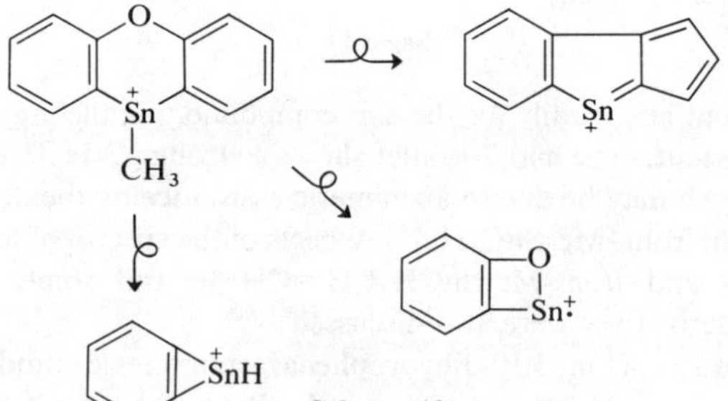

Scheme 12

in the spectrum of 10,10-dimethylphenoxastannin (Scheme 12). A number of metal-free ions containing aromatic ring systems were also formed. Ionization and appearance potential data were used in the interpretation of the fragmentation patterns.

Compounds containing Tin–Tin Bonds. Cyclostannanes (R$_2$Sn)$_n$ have been identified for $n = 7$, R = Et; $n = 6$, R = Et, Bu, Bu^i, or Ph; and $n = 5$, R = Et, Bu, Bu^i, or cyclohexyl.[190] The method used involved fractional evaporation in the mass spectrometer and a study of mass chromatograms.

Compounds containing Sn—N Bonds. The primary stannylamine, Bu$^t_3$SnNH$_2$, shows no pmi; [(C$_4$H$_9$)$_2$SnNH$_2$]$^+$ is formed as well as ions due to loss of C$_4$H$_8$,

---

[187] J. K. Terlouw, W. Heerma, P. C. M. Frintrop, G. Dijkstra, and H. A. Meinma, *J. Organometallic Chem.*, 1974, **64**, 205.
[188] H. A. Meinma and J. G. Noltes, *J. Organometallic Chem.*, 1973, **63**, 243.
[189] I. Lengyel, M. J. Aaronson, and J. P. Dillon, *J. Organometallic Chem.*, 1970, **25**, 403.
[190] H.-P. Ritter and W. P. Neumann, *J. Organometallic Chem.*, 1973, **56**, 199.

Bu, and NH_2 groups.[191] Some triethyltin triazenes $Et_3Sn(ArN_3)$[192] have been characterized.

Compounds containing Sn—O or Sn—S Bonds. The spectra of some O-tri-organotin hydroxylamine derivatives have been listed.[193] Dimeric ions were found from $Me_3SnON(R)COPh$ (R = H or Ph), but $Me_3SnONEt_2$ and $Ph_3SnON(Ph)COPh$ were monomeric in the gas phase. Loss of Me and $ONEt_2$ groups from $Me_3SnONEt_2$ was observed. The other derivatives showed loss of O and PhN groups from $[pm - R]^+$ ions (R = Me or Ph) with the concurrent formation of what may be four-membered heterocyclic ions. Organotin nitronates, $Me_3SnO_2NCR^1R^2$ (R^1 = H, R^2 = H or Me, $R^1 = R^2$ = Me), have been examined.[20] The spectrum of Me_3SnO_2NCHMe showed ions derived from a dimeric species and several rearranged ions including $[Me_3Sn_2O]^+$.

Studies of series of dimethylchlorotin carboxylates Me_2SnClO_2CR (R = CF_3, C_2F_5, C_3F_7, $CClF_2$, CH_3, CH_2Cl, $CHCl_2$, or CCl_3) and their hydrolysis products have been reported.[194] Evidence was found for gas-phase polymers; rearranged ions with Sn—X bonds were common from the halogenoalkyl derivatives. A number of dithiocarbamate compounds, $R^1_2SnCl(S_2CNR^2_2)$ (R^1 = alkyl or phenyl, R^2 = various groups), have been characterized.[195] Compounds with the dicyanoethylene-1,2-dithiolate ligand, $R_2Sn\{S_2(CCN)_2\}$, (R = Me or Ph) show pmi's and fragmentation by loss of one and two R groups;[196] some dimeric ions were reported. Both $Me_2Sn(O_2SMe)_2$,[197] and $Me_2Sn\overline{SNSN}$[198] show pmi's. Organotin sulphides $Me_2Sn(SMe)_2$, $(R_3Sn)_2S$, (R = Me or Ph), and $(R_2SnS)_3$, (R = Me, Bu, or Ph), have been characterized.[199] Rearranged ions such as $[R_4SnS]^+$ for (R = Me or Ph) were noteworthy features in the spectra of all the compounds.

Lead. An ICR spectrometer has been used to study ion–molecule reactions from $PbEt_4$.[200] A major ion produced from $PbEt_4$ and $[PbEt_3]^+$ was $[Pb_2Et_7]^+$.

Other compounds which have been studied include $Ph_3PbON(Ph)COPh$,[143] $Me_3PbO_2NCR^1R^2$,[20] and $Ph_2Pb\{S_2(CCN)_2\}$.[196]

Group V.—A comparative study of the spectra of $MeMH_2$, Me_2MH, Et_2MH, and some deuteriated analogues (M = N or P), Me_3M, Et_3M (M = N, P, As, Sb, or Bi), Pr_3M (M = Sb or Bi), Bu_3M (M = P or As), $(n-C_5H_{11})_3As$, $(n-C_6H_{13})_3As$, and Et_2AsBr has been reported.[201] In contrast to the amines and similar Main Group IV compounds, odd-electron fragments are of moderate to high abundance in the spectra of phosphorus, arsenic, antimony, and bismuth derivatives.

---

[191] H.-J. Gotze, *Angew. Chem. Internat. Edn.*, 1974, **13**, 88.
[192] J. Hollaeder, W. P. Neumann, and G. Alester, *Chem. Ber.*, 1972, **105**, 1540.
[193] P. G. Harrison, *Inorg. Chem.*, 1973, **12**, 1545.
[194] C. S.-C. Wang and J. M. Shreeve, *J. Organometallic Chem.*, 1973, **49**, 417.
[195] B. W. Fitzsimmons and A. C. Sawbridge, *J.C.S. Dalton*, 1972, 1678.
[196] E. S. Bretschneider and C. W. Allen, *Inorg. Chem.*, 1973, **12**, 623.
[197] E. Lindner and D. W. R. Frembs, *J. Organometallic Chem.*, 1973, **49**, 425.
[198] H. Roesky and H. Wieser, *Angew. Chem. Internat. Edn.*, 1973, **12**, 674.
[199] P. G. Harrison and S. R. Stobart, *J. Organometallic Chem.*, 1973, **47**, 89.
[200] R. C. Dunbar, J. F. Ennever, and J. P. Fackler, *Inorg. Chem.*, 1973, **12**, 2735.
[201] R. G. Kostyanovsky and V. G. Plekhanov, *Org. Mass Spectrometry*, 1972, **6**, 1183.

The major fragmentation path for amines is *via* α-cleavage giving rise to ions of the type [>N=CH$_2$]$^+$. This type of decomposition was insignificant for phosphorus derivatives and practically absent in the spectra of the arsenic, antimony, and bismuth compounds. The observed trend may be explained in terms of the decrease in (i) the ability of the central atom to stabilize α-cleavage ions and (ii) the M—C bond strengths. The elimination of M—H containing species was a feature of the spectra of phosphorus and amine compounds and gave rise to hydrocarbon ions, some of which were intense. Loss of olefin groups (R—H) for R > Me was a common fragmentation path in the spectra of the amines, and phosphorus, arsenic, and antimony derivatives. The bismuth compounds gave very simple spectra with [Bi]$^+$ as base peak, the only other important ions being the pmi and [R$_n$Bi]$^+$ (n = 1 or 2).

Some branched-chain aliphatic derivatives have been studied including Pr$^i_2$MH and Bu$^t_2$MH (M = N or P); R$_2$PCl (R = Pr^i or Bu^t), Bu$^t_2$P(O)X (X = Cl or H), and Bu$^t_3$Sb.[202] Compared with the straight-chain derivatives, the pmi's of the branched-chain compounds are less abundant. Common fragmentations from the phosphorus compounds were loss of CH$_3$ and olefin groups. Ions produced by α-cleavage were low in abundance from phosphorus derivatives and absent in the spectrum of Bu$^t_3$Sb. No simple loss of Cl was observed in the spectra of any of the chlorophosphines.

The compounds *cis*- and *trans*-(Ph$_2$M^1)CH=CH(M^{2}Ph$_2$) (M^1 = M^2 = P or As; M^1 = P, M^2 = As) and Ph$_2$P(CH$_2$)$_2$AsPh$_2$ have been studied and their spectra listed.[203] Always the pmi's were base peaks. The virtually identical spectra from *cis*- and *trans*-isomers suggest rotation about the C—C bond in the ions. During fragmentation, a phenyl group was transferred from one atom M^2 to the other M^1 producing ions derived from [Ph$_3$M^1]$^{+\cdot}$. The rearrangement only occurred from As to P in the compounds containing both elements. Apart from the ions derived from [Ph$_3$M]$^{+\cdot}$ the only other ions containing Main Group V elements were formed by loss of Ph, C$_6$H$_6$, and C$_2$H$_2$ (or C$_2$H$_4$ for the saturated compound) from the pmi's.

The spectra of a number of P, As, Sb, and Bi triphenyl-, phenyl-, and methyl-2,2'-biphenylene-(34), 2-biphenylyl-2,2'-biphenylene-, pentaphenyl-, and phenyl- and methyl-bis-2,2'-biphenylene-(35) derivatives have been listed and dis-

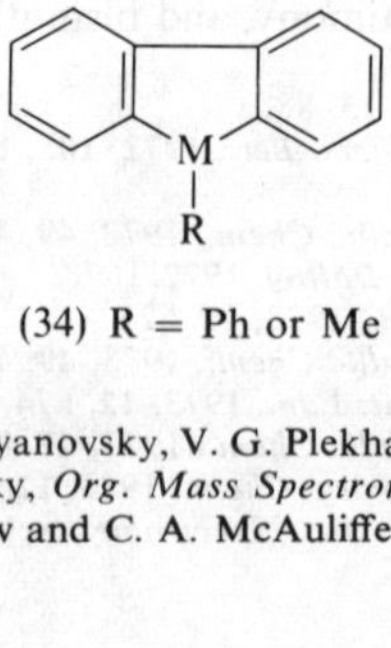

(34) R = Ph or Me 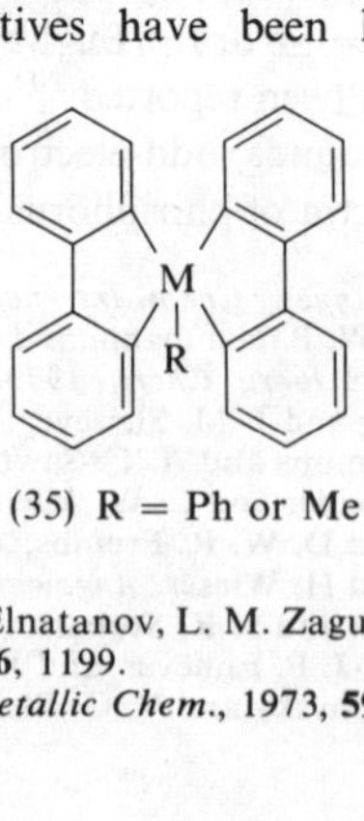   (35) R = Ph or Me

---

[202] R. G. Kostyanovsky, V. G. Plekhanov, Yu. I. Elnatanov, L. M. Zagurskaya, and V. N. Voznesensky, *Org. Mass Spectrometry*, 1972, **6**, 1199.
[203] K. K. Chow and C. A. McAuliffe, *J. Organometallic Chem.*, 1973, **59**, 247.

cussed.[204] All the compounds studied showed appreciable pmi's except Ph_3Bi and Ph_5M (M = P, As and Sb) for which these were weak, and Ph_5Bi which showed none. Most fragmentations involved loss of one or more H atoms and/or cleavage of P—C bonds with loss of R groups [R = Me, Ph, $C_{12}H_9$ (biphenyl), $C_{12}H_8$ (biphenylene)]. Secondary fragmentations may occur with expulsion of C_2H_2 or C_6H_4 species.

Phosphorus. The cleavage of the P—O bond is characteristic in the fragmentation of acyldialkylphosphines and related compounds such as $PhNHCOPPh_2$.[205] In this they are markedly different from the corresponding nitrogen compounds. In acyldimethylphosphines, this cleavage produces $[Me_2P]^+$ and the rearranged ion $[RPMe_2]^{+\cdot}$. Other decompositions involve loss of R and CO sequentially, (RCO − H), or Me_2P [Scheme 13, R = CH_3, CD_3,

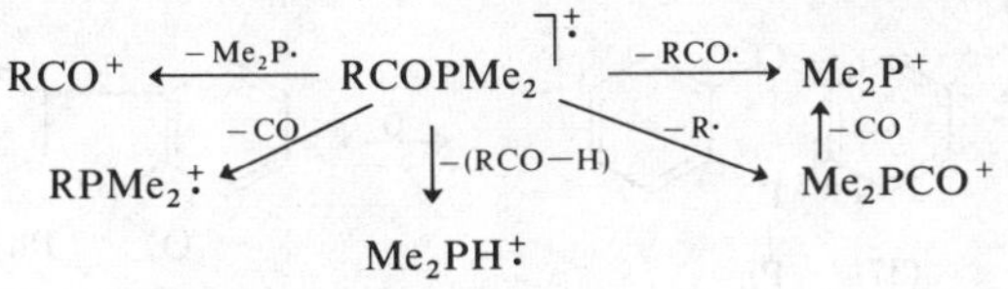

$$RCO^+ \xleftarrow{-Me_2P\cdot} \left[RCOPMe_2\right]^{+\cdot} \xrightarrow{-RCO\cdot} Me_2P^+$$

$$\Big\downarrow{-CO} \qquad \Big\downarrow{-(RCO-H)} \qquad \Big\downarrow{-R\cdot} \qquad \Big\uparrow{-CO}$$

$$RPMe_2^{+\cdot} \qquad Me_2PH^{+\cdot} \qquad Me_2PCO^+$$

Scheme 13

or $CH(CF_3)_2$]. With the bulkier alkyl groups, Pr^i or Bu^t, loss of olefin (R − H) becomes a major fragmentation mode. Loss of olefin is also significant for the pmi's of dialkylvinylphosphine derivatives, $R_2PCH=CHOCOMe$ (R = Pr^i or Bu^t), which also lose CH_3 and OCH_3 groups. Further fragmentation of [pm − CH_3]$^+$ by elimination of (R—H) molecules was thought to produce cyclized ions such as (36).

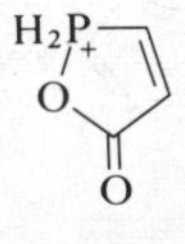

(36)

A number of phenylphosphine derivatives have been characterized including $Ph_nP\{(CH_2)_2PH_2\}_{3-n}$ (n = 0—2)[206] and related compounds.

Fragmentation of carbonyl-stabilized triphenyl phosphonium and arsonium compounds, $Ph_3MCRCOX$ (R = H, X = Me, Ph, or OMe; R = CN, X = OMe), have been compared,[207] and pmi's were reported for 1-phenyl-1-formyl-methylenetriphenylphosphine[208] and some phosphoniumpyrazolylylides.[209]

---

[204] D. Hellwinkel, C. Wunsche, and M. Bach, *Phosphorus*, 1973, **2**, 167.
[205] R. G. Kostyanovsky, V. G. Plekhanov, Kh. Khafizov, L. M. Zagurskaya, G. K. Kadorkina, and Yu. I. Elnatanov, *Org. Mass Spectrometry*, 1973, **7**, 1113.
[206] R. B. King, J. C. Lloyd, and P. N. Kapoor, *J.C.S. Perkin I*, 1973, 2226.
[207] A. J. Dale and P. Froyen, *Phosphorus*, 1973, **2**, 297.
[208] C. J. Devlin and B. J. Walker, *Tetrahedron*, 1972, **28**, 3501.
[209] E. Zbiral and E. Bauer, *Tetrahedron*, 1972, **28**, 4189.

Negative-ion mass spectra of some carbonyl-stabilized phosphoranes, Ph_3-PCR^1COR^2 ($R^1 = H$, $R^2 = Me$, Ph, or OEt; $R^1 = COMe$, $R^2 = Me$ or OEt and $R^1 = CO_2Et$, $R^2 = OEt$), have been described.[210] An interesting finding is that all compounds undergo fragmentation of pmi's by loss of a Ph radical to produce the base peak which loses $R^1C\equiv CR^2$ to give the $[Ph_2PO]^-$ ion. That this process has a direct counterpart in the positive-ion spectra of phosphoranes is very unusual.

Phosphorus Heterocycles. The spectra of 10-phenylphenoxaphosphine and some derivatives showed extensive formation of rearranged ions.[211] The pmi (37), or the rearranged pmi (37a) initially lost Ph or PhO radicals or a $C_{12}H_8O$ molecule. Further fragmentation of $[pm - Ph]^+$ involved the loss of the heteroatom (Scheme 14).

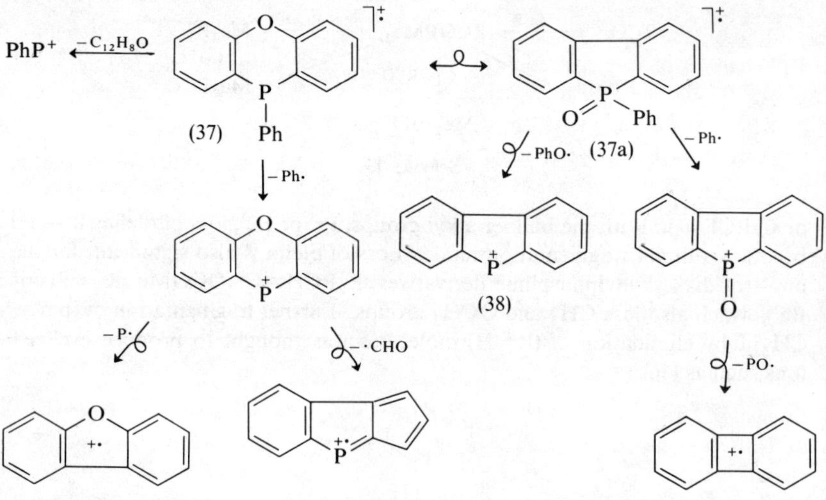

Scheme 14

Similar rearrangements were observed in the spectra of the phosphine oxide and sulphide derivatives and bis-(*o*-methoxyphenyl)phenylphosphine and its oxide. The phosphine showed a primary fragmentation by loss of OMe which probably involved rearrangement to a P—H containing ion since further fragmentation was by elimination either of C_6H_5OMe to give an ion of structure (38) or of C_6H_6 to give the methoxy-derivative of (38). Other primary fragmentations involved loss of $C_{12}H_{10}PO$ or C_7H_7 species. Differences in the decomposition modes of the pmi's from phenoxaphosphine (39a) and the phenthiaphosphine

---

[210] R. G. Alexander, D. B. Bigley, and J. F. J. Todd, *Org. Mass Spectrometry*, 1973, **7**, 963.
[211] I. Granoth, J. B. Levy, and C. Symmes, *J.C.S. Perkin II*, 1972, 697.

analogue (39b) have been observed.[212] Loss of Ph, PhP, and PhPH from the pmi of (39a) occurred both in the ion source and first field-free region of the mass spectrometer. For the pmi of (39b) the loss of Ph or PhS predominated in the ion source but loss of PhPH was the dominant reaction in the field-free region.

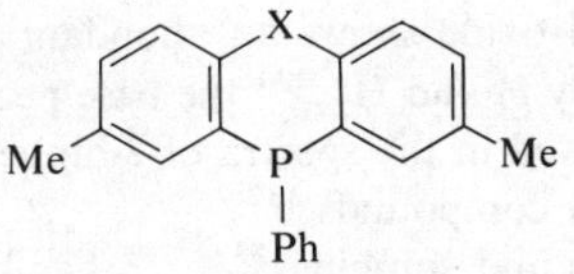

(39) a; X = O
b; X = S

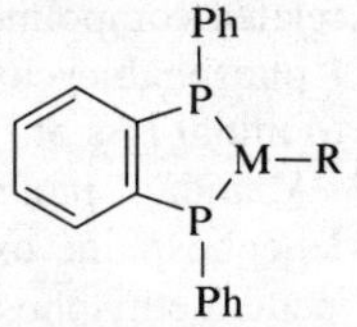

(40) a; M = P, R = Me, Et, or Ph
b; M = As, R = Me or Ph

Other heterocyclic derivatives for which data have been reported include 1-phenylphosphacyclohepta-2,6-diene and its As analogue,[213] C-alkylphospholes,[214] 1-alkylphosphacyclohexa-2,5-dienes[215] and related compounds,[216] phosphabenzoles and related compounds,[215–219] and some alkylphosphorane derivatives such as homocubyltriphenylphosphorane.[220]

Compounds with P—P or P—Main Group V Bonds. A number of cyclic compounds containing P—P or P—As bonds have been characterized. They include $(PR_F)_n$ ($n = 4$ or 5, $R_F = C_2F_5$ or C_3F_7),[221] and some benzotriphospholes (40a) and benzodiphospharsoles (40b) and related compounds.[222] The tri- and diphosphole derivatives fragmented by loss of R and then PhPMH to give ion (38). This ion was also seen in the spectra of some corresponding monosulphide compounds which fragmented by losing MR, Ph, SH, and P to give (38).

Compounds with P—O or P—S Bonds. (See also Chapters 6 and 7.) Numerous derivatives containing the $Ph_2P(O)$ group have been characterized using mass spectrometry. Among them are Pr^iPh_2PO; 3-diphenylphosphinyl derivatives of 3-methylbutan-2-ol and the corresponding butan-2-one;[223] 4-diphenylphosphinyl-2,3,4-trimethylpentan-2-ol and related compounds;[224] 2,2-dimethylvinyldiphenylphosphine oxide; 2-diphenylphosphinyl derivatives of propan-

[212] W. D. Weringa and I. Granoth, *Org. Mass Spectrometry*, 1973, **7**, 459.
[213] G. Markl and G. Dannhardt, *Tetrahedron Letters*, 1973, 1455.
[214] L. D. Quin, S. G. Borleshe, and J. E. Engel, *J. Org. Chem.*, 1973, **38**, 1858.
[215] G. Markl and D. E. Fischer, *Tetrahedron Letters*, 1973, 223; 1972, 4925.
[216] G. Markl and D. Matthes, *Angew. Chem. Internat. Edn.*, 1972, **11**, 1019.
[217] H. G. de Graaf, J. Dubbeldam, H. Vermeer, and F. Bickelhaupt, *Tetrahedron Letters*, 1973, 2397.
[218] G. Markl and K.-H. Heier, *Angew. Chem. Internat. Edn.*, 1972, **11**, 1017.
[219] G. Markl and F. Kneidl, *Angew. Chem. Internat. Edn.*, 1973, **12**, 931.
[220] E. W. Turnblom and T. J. Katz, *J.C.S. Chem. Comm.*, 1972, 1270; *J. Amer. Chem. Soc.*, 1973, **95**, 4292.
[221] H. G. Ang, M. E. Redwood, and B. O. West, *Austral. J. Chem.*, 1972, 25, 493.
[222] F. G. Mann and A. J. H. Mercer, *J.C.S. Perkin I*, 1972, 1631, 2548.
[223] D. Howells and S. Warren, *J.C.S. Perkin II*, 1973, 1492.
[224] D. Howells and S. Warren, *J.C.S. Perkin II*, 1973, 1645.

1-ol, 2-methylpropylamine, and related compounds,[225] and substituted alkyldiphenylphosphine oxides such as $Ph_2P(O)(CH_2)_2(MeO_2C)C=CHCO_2Me$.[226] Other phosphine oxides include $R_2P(O)C\equiv C(SMe)$ (R = Me or Et) and similar compounds,[227] 3,4-benzo-1-phospha-1-oxacyclohex-3-ene and related compounds,[228] phosphatricyclododecatrieneones,[229] and triphenylphosphiren oxide[230] and related compounds.

The novel 1-phosphabicyclo[2,2,1]heptane 1-oxide shows an abundant pmi and ions due to initial loss of C_2H_4 followed by H and H_2;[231] the base peak is due to $[PO]^+$. Abundant pmi's are also observed in the spectra of 2-biphenyl-2,2'-biphenylylenephosphine oxide and related compounds.[232]

A series of pentamethylphosphetan 1-oxides and sulphides[233—235] has been characterized. For the halogeno-compounds, pmi's were more abundant for Cl

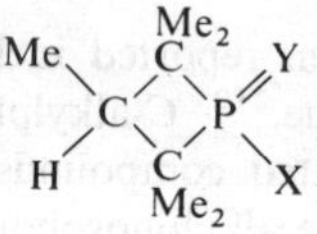

	X	Y
(41)	Cl or Br	O
	Cl or Br	S
	NHR or NR_2	O
	(R = Et, Pr, Ph, etc)	
	OR	
	(R = alkyl or aryl)	O

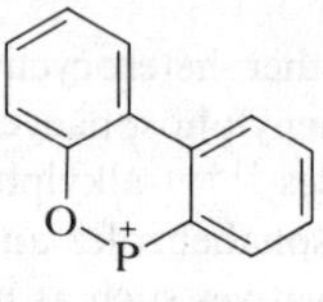

(42)

than Br and for S than O derivatives. The major fragmentation paths involved loss of hydrocarbon entities.

The spectrum of triphenylphosphite shows a pmi and base peak due to the ion $[(PhO)_2P]^+$, which decomposes by loss of H_2O to give (42).[236]

The absence of pmi's from a series of alkoxytetra-alkylphosphoranes has been reported[237] but the fragment ions were consistent with the suggested formulae. Pmi's were observed for a number of 1,3,2-dioxaphospholens.[238]

The spectra of some pyridyl-2- and pyridyl-4-phosphonates have been discussed.[239] Both (43) and (44) (X = H) show pmi's and fragment initially by losing C_2H_5, CH_3CHO, C_2H_4 (via a McLafferty rearrangement), and C_2H_3

[225] P. F. Cann, D. Howells, and S. Warren, *J.C.S. Perkin II*, 1972, 304.
[226] M. Davies, A. N. Hughes, and S. W. S. Jafry, *Canad. J. Chem.*, 1972, **50**, 3625.
[227] W. Hagens, H. J. T. Bos, and J. F. Arens, *Rec. Trav. Chim.*, 1973, **92**, 762.
[228] G. Markl and H. Baier, *Tetrahedron Letters*, 1972, 4439.
[229] Y. Kashman and O. Awerbouch, *Tetrahedron*, 1973, **29**, 191.
[230] E. W. Koos, J. P. Vander Kooi, E. E. Green, and J. K. Stille, *J.C.S. Chem. Comm.*, 1972, 1085.
[231] R. B. Wetzel and G. L. Kenyon, *J. Amer. Chem. Soc.*, 1972, **94**, 9230.
[232] D. Hellwinkel and H.-J. Wilfinger, *Chem. Ber.*, 1972, **105**, 3878.
[233] J. Emsley and J. K. Williams, *J.C.S. Dalton*, 1973, 1576.
[234] R. E. Ardrey, J. Emsley, A. J. B. Robertson, and J. K. Williams, *J.C.S. Dalton*, 1973, 2641.
[235] J. Emsley, T. B. Middleton, and J. K. Williams, *J.C.S. Dalton*, 1974, 633.
[236] V. N. Bochkarev, A. N. Polivanov, E. F. Bugerenko, V. I. Aksenov, and E. A. Chernyshev, *J. Gen. Chem. U.S.S.R.*, 1972, 42, 2345.
[237] H. Schmidbaur, H. Stuhler, and W. Buchner, *Chem. Ber.*, 1973, **106**, 1238.
[238] J. L. Dickstein and S. Trippett, *Tetrahedron Letters*, 1973, 2203.
[239] D. Redmore, *J. Org. Chem.*, 1973, **38**, 1306.

(involving a double hydrogen rearrangement), the P—C bond remaining intact in nearly all phosphorus-containing ions.

(43) (EtO)$_2$(O)P—pyridine with X at 4-position and X at 6-position

(44) (EtO)$_2$(O)P—pyridine with X at 3,5-positions

Two major fragmentation paths from pmi's, both involving H-transfer, have been observed in the spectra of O-ethyl-S-alkylmethanephosphonothioates (45).[240] One is a McLafferty-type rearrangement, Scheme 15(a), to give (46), the other involves a double-hydrogen rearrangement to give (47), Scheme 15(b). The ratio of abundances of (46) and (47) is dependent on the chain length and

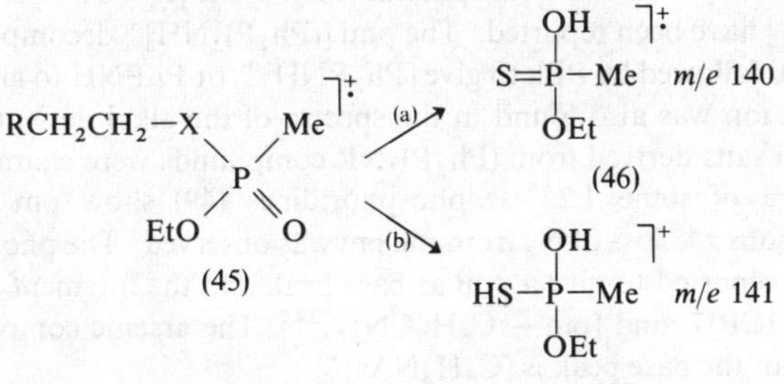

Scheme 15

(45) RCH$_2$CH$_2$—X, Me, P, EtO, O

(a) → (46) S=P(OH)(Me)(OEt), m/e 140

(b) → (47) HS—P(OH)(Me)(OEt), m/e 141

branching of the S-alkyl group. As R becomes larger, the McLafferty rearrangement becomes relatively less important and the ratio (46):(47) decreases. The derivatives of (45; X = O) were reported to show the analogous double-hydrogen rearrangement but not the McLafferty rearrangement.

Some 7-phosphono-7-arylnorcaradienes are reported to show pmi's which fragment by losing the P(OR)$_2$O group.[241] An analogous reaction was observed from 2-arylaziridin-2-ylphosphonates with a carbonyl group at C-3 (48; R = Me or Et, Ar = Ph or p-tol).[242]

Additional evidence showing that methyl, chloro, fluoro, and ethyl substituents exert a considerable influence on the fragmentation of phenoxaphosphinic acids has been discussed.[243] Usually, where there was no ethyl substituent, the base peak was the pmi; for ethyl derivatives, the ions [pm − CH$_3$]$^+$ formed the base

[240] Z. Tashma, J. Katzhendler, and J. Deutsch., *Org. Mass Spectrometry*, 1973, **7**, 955.
[241] H. Scherer, A. Hartmann, M. Regitz, B. D. Tunggal, and H. Gunther, *Chem. Ber.*, 1972, **105**, 3357.
[242] T. Nishiwaki, *Org. Mass Spectrometry*, 1972, **6**, 693.
[243] J. B. Levy, G. W. Whitehead, and I. Granoth, *Israel J. Chem.*, 1972, **10**, 27.

peak. All the compounds studied lost PO_2H from either the pmi or, in the case of the ethyl compounds, $[pm - CH_3]^+$. Chloro-substituted compounds lost chlorine, mainly as CClO, from $[pm - PO_2H]^{+\cdot}$. The expulsion of CH_3 from pm and fragment ions was important for compounds with at least two methyl

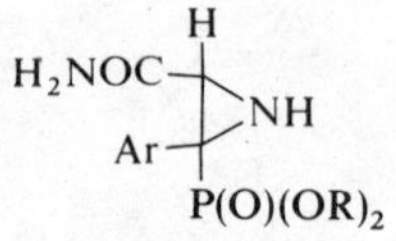

(48)

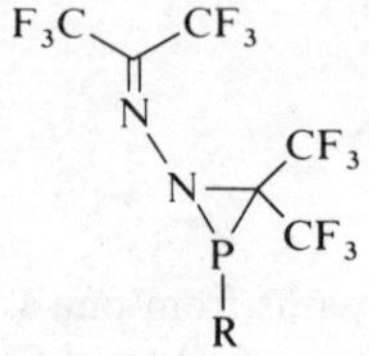

(49)

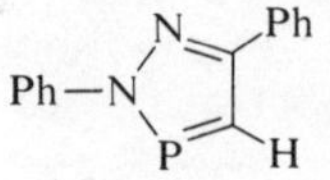

(50)

groups attached to one aromatic ring, and may involve ring expansion to give a substituted tropylium species.

Aspects of the spectra of the compounds $(Ph_2P)_2NR$ [R = H[244] or $(CH_2)_nCH_3$ (n = 1—4)[245]] have been reported. The pmi $[(Ph_2P)_2NH]^{+\cdot}$ decomposed by loss of a Ph radical followed by PPh to give $[Ph_2PNH]^+$, or PhPNH to give $[Ph_3P]^{+\cdot}$. The $[Ph_3P]^{+\cdot}$ ion was also found in the spectra of the alkyl derivatives. Some phosphonium salts derived from $(Ph_2P)_2NR$ compounds were characterized.[245]

The spectra of some $1,2\lambda^5$-azaphosphoridines (49) show pm and $[pm - NC(CF_3)_2]^+$ ions,[246] loss of N_2 from the pm was observed. The phosphadiazole (50) has been reported to give a pmi as base peak and the fragment ions $[pm - P]^+$, $[pm - HCP]^{+\cdot}$, and $[pm - C_6H_5CN]^{+\cdot}$.[247] The arsenic compound forms similar ions but the base peak is $[C_6H_3NAs]^+$.

The spectra of some 2-substituted-2-phospha-1,3-diazacyclohexanes (51) are typically marked by the loss of the 2-substituent group to give the base peak. However, the corresponding ion, $[pm - Ph]^+$, is not the base peak for 2-phenyl-2-phospha-1,3-dioxacyclohexane (51a), and is absent in the spectrum of the analogous thio-compound.[248]

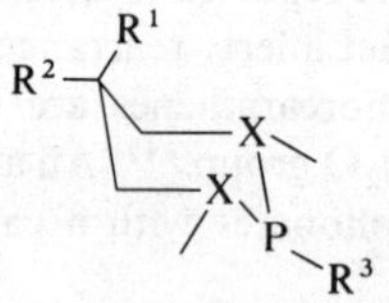

(51) X = N, R^1, R^2 = H or Me
R^3 = e.g. Ph, Et, Cl, OMe
(51a) X = O, R^1, R^2 = H, R^3 = Ph

[244] J. Ellerman and W. H. Gruber, Z. Naturforsch., 1973, 28b, 701.
[245] D. F. Clemens and W. E. Perkinson, Inorg. Chem., 1974, 13, 333.
[246] J. K. Burger, J. Fehn, and W. Thenn, Angew. Chem. Internat. Edn., 1973, 12, 502.
[247] G. Markl and C. Martin, Tetrahedron Letters, 1973, 4503.
[248] E. Maryanoff and R. O. Hutchins, J. Org. Chem., 1972, 37, 3475.

Other compounds for which data have been reported include diazadiphosphetidine di-imides,[249] some azaphospholodihydropyridines and related compounds,[250] and 2,2-dichloro-1,5-dimethyl-4,6-dioxo-1,4,5,6-tetrahydro-1,5,2λ^5-diazaphosphorine.[251]

The spectra of some phosphazenes (52) and tetrazatetraphosph(v)ocines showed an unusually high abundance of doubly-charged pmi's for certain alkyl groups. When R = Me or Et, these doubly-charged ions were respectively 17 and 20% of the abundance of the respective base peaks, [pm − H]$^+$ and [pm − Et]$^+$.[252] With R = cyclohexyl, the [pm]$^{2+}$ ion was absent.

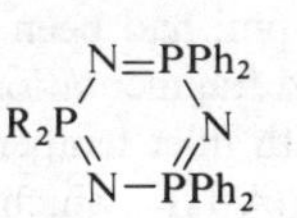

$$\begin{array}{c} N{=}PPh_2 \\ R_2PN \\ N{-}PPh_2 \end{array}$$

(52)

Compounds with P—Halogen Bonds. Neither Me$_4$PF nor Me$_4$SbF shows a pmi, but they do give expected fragment ions such as [Me$_3$MF]$^+$ and [Me$_4$M]$^+$.[253] Fragmentation by elimination of HF to give >M=CH$_2$ containing ions was more evident for the phosphorus compound; the antimony analogue lost R or F groups.

Other fluorophosphine compounds include difluoroprop-1-ynylphosphine, a borane adduct H$_3$BPF$_2$C$_3$H$_3$, F$_4$PC$_3$H$_3$,[254] and F$_2$P(CH$_2$)$_2$PF$_2$ and a diborane adduct.[255]

Arsenic. As well as a weak pmi, a base peak [Me$_3$As]$^{+}_{\cdot}$, and other ions of the type [Me$_n$As]$^+$ (n = 1—4), Me$_5$As has been reported to produce ions such as [Me$_3$As=CH$_2$]$^{+}_{\cdot}$ in low intensity.[256] The spectrum of C(CH$_2$AsMe$_2$)$_4$ shows a pmi which fragments initially by loss of H or CH$_3$.[257] The base peak, [pm − CH$_3$]$^+$, is suggested to contain an As—As bond in a cyclic structure (53). Further

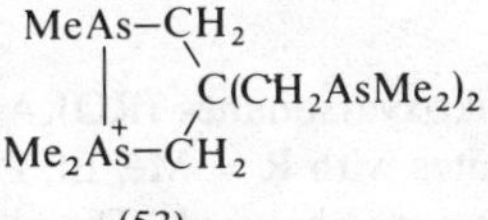

$$\begin{array}{c} MeAs{-}CH_2 \\ \backslash \\ C(CH_2AsMe_2)_2 \\ / \\ Me_2\overset{+}{As}{-}CH_2 \end{array}$$

(53)

[249] O. J. Scherer, P. Klusmann, and N. Kuhn, *Chem. Ber.*, 1974, **107**, 552.
[250] W. Zeiss and A. Schmidpeter, *Tetrahedron Letters*, 1972, 4229.
[251] N. Singh and H. P. Latscha, *Angew. Chem. Internat. Edn.*, 1972, **11**, 528.
[252] R. Appel and G. Saleh, *Chem. Ber.*, 1973, **106**, 3455.
[253] H. Schmidbaur, K.-H. Mitschke, W. Buchner, H. Stuhler, and J. Weidlein, *Chem. Ber.*, 1973, **106**, 1226.
[254] E. L. Lines and L. F. Centofanti, *Inorg. Chem.*, 1973, **12**, 598.
[255] K. Morse and J. G. Morse, *J. Amer. Chem. Soc.*, 1973, **95**, 8469.
[256] K.-H. Mitschke and H. Schmidbaur, *Chem. Ber.*, 1973, **106**, 3645.
[257] J. Ellermann, H. Schossner, A. Hagg, and H. Schodel, *J. Organometallic Chem.*, 1974, **65**, 33.

fragmentation of (53) by the loss of $AsMe_3$, As_2Me_4, $MeAsCH_2$, C_2H_6, and CH_4 groups produced ions, most of which retain the As—As bond. The effect of inlet temperature on the spectra of two alkyliodoarsines was also reported.[257] With inlet temperature of 10—15 °C and an ion source temperature of 150 °C, $MeAsI_2$ and $EtAsI_2$ showed pmi's and fragments due to loss of Me or I radicals, and H, C_2H_4, or C_2H_5 groups, respectively. With inlet and ion source temperatures both at 150 °C the spectra were remarkably different. Methyldi-iodoarsine showed $[(MeAs)_3]^{+\cdot}$ as the most abundant ion with $[MeAsI]^+$, the previous base peak, only 17% of the abundance of the trimeric species. The ethyl derivative gave ions derived from tetrameric $(EtAs)_4$ and a base peak, $[Et_2As_3]^+$, where previously the pmi had been the most abundant ion. Other compounds for which spectra and fragmentation schemes were discussed were: Me_2AsI, which showed a pmi with inlet temperature at 10 °C and $[pm - H]^+$ base peak; $CH_2(AsI_2)_2$ and $C_2H_4(AsI_2)_2$, which both showed pmi's and $[pm - I]^+$ base peaks; and $C(CH_2AsI_2)_4$, for which no pmi was observed, the ion of largest mass being $[C(CH_2AsI)_4]^{+\cdot}$.

Phenylarsine derivatives reported include some $Ph_3As=CXY$ compounds with $X,Y = CO_2Et$ and $X = H$, $Y = NO_2$.[258] The elimination of HNCO from $[Ph_3AsCHNO_2]^{+\cdot}$ produced $[Ph_3AsO]^{+\cdot}$ by rearrangement.

Heterocyclic compounds of arsenic for which data have been reported include 1-substituted-2,5-diphenylarsoles[259] and substituted arsabenzenes and related compounds.[219,260]

The spectra of some organo-arsenites, $(RO)_3As$, -thioarsenites $(RS)_3As$, $MeAs(OEt)_2$,[261] 2-substituted 1,3,2-dioxarsenanes (54),[262] trialkylarsenates

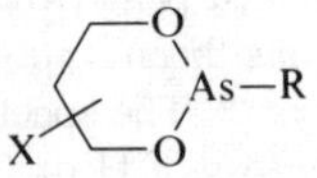

(54) X = H, R = Me, Et, Pr, Bu, or Ph
X = 5,5-dimethyl, R = Me or Ph
X = 4-methyl, R = Ph

$(RO)_3AsO$, and penta-alkoxyarsonanes $(RO)_5As$[263] have been illustrated and discussed. For the arsenites with R = Me, Et, Pr, or Ph and thioarsenites with R = Me, Et or Ph, pmi's were observed. The alkoxy-derivatives fragmented by losing one OR group [or Me from $MeAs(OEt)_2$] to give the base peak, $[(RX)_2As]^+$ (X = O or S). Further fragmentation for R > Me occurred by loss of (XR − H) and (R − H) groups. Triphenyl arsenite initially lost OPh then H_2O and C_6H_4, or PhOH, to give the base peak (Scheme 16).

[258] I. Gosney and D. Lloyd, *Tetrahedron*, 1973, **29**, 1697.
[259] G. Markl and H. Hauptmann, *Angew. Chem. Internat. Edn.*, 1972, **11**, 439.
[260] G. Markl, J. Advena, and H. Hauptmann, *Tetrahedron Letters*, 1974, 303.
[261] P. Froyen and J. Moller, *Org. Mass Spectrometry*, 1973, **7**, 73.
[262] P. Froyen and J. Moller, *Org. Mass Spectrometry*, 1973, **7**, 691.
[263] P. Froyen and J. Moller, *Org. Mass Spectrometry*, 1974, **9**, 132.

The base peak from $(PhS)_3As$ is $[(PhS)_2As]^+$ which fragments by loss of PhSH and C_6H_6 but not H_2S. Generally, comparing the spectra of the arsenic compounds with their phosphorus analogues, it was seen that a major difference is the almost complete absence of double-hydrogen rearrangements from the

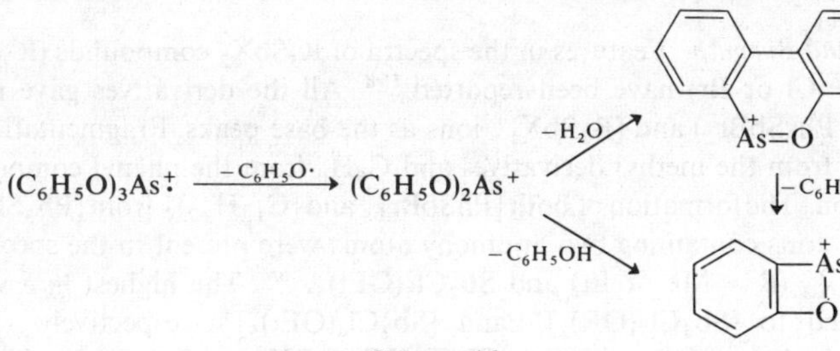

$$(C_6H_5O)_3As^{+\cdot} \xrightarrow{-C_6H_5O\cdot} (C_6H_5O)_2As^+ \begin{matrix} \xrightarrow{-H_2O} \\ \\ \xrightarrow{-C_6H_5OH} \end{matrix}$$

Scheme 16

arsenic compounds to produce ions of the type $[(RX)_2M(H)OH]^+$. This may in part be due to the relatively weaker As—OR compared with P—OR bonds. This aspect is also reflected in the very low abundance of ions of the type $[(RX)_2As=O]^+$ because, unlike the phosphorus analogues, the M—OR bond cleaves preferentially to the MO—R bond. Further, the thermodynamic stability of the As=O bond is much lower than P=O.

All compounds of type (54) showed pmi's.[262] The 2-alkyl derivative had $[pm - R]^+$ as base peak which fragmented analogously to the arsenites. The phenyl derivatives fragmented initially to give $[pm - Ph]^+$ or $[pm - R'CHO]^{+\cdot}$ ($R' = H$ or Me for 4-methyl compound) as well as producing $[PhAs]^+$ ions, the base peak for 4-methyl and 5,5-dimethyl compounds. Trialkylarsenates ($R = Me$, Et, Pr, Bu, or C_5H_{11}) did not show pmi's. The highest m/e values observed were due to $[pm - H]^+$ for ($R = Me$, Et, or Pr) $[pm - (R - CH_2)]^+$ for ($R = Bu$ or C_5H_{11}). Base peaks were $[(RO)AsOH]^+$ for ($R = Me$ or Et) and $[As(OH)_4]^+$ for the other compounds. Differences in fragmentation behaviour between trialkylarsenates and the corresponding phosphorus compounds include the sequence shown in Scheme 17, only from the arsenic derivatives. The $[pm - Me]^+$ ion from $(MeO)_2MeAsO$ also lost an oxygen atom to give $[(MeO)_2As]^+$.

$$[pm]^{+\cdot} \rightarrow [pm - OR]^+ \rightarrow [pm - OR - O]^+$$

Scheme 17

Penta-alkyloxyarsoranes ($R = Me$ or Et) show $[pm - OR]^+$ at highest mass. Generally, fragmentation appears to produce ions similar to those in the spectra of the corresponding trialkylarsenates.

A number of compounds with As—N bonds have been characterized, including some alkyl bis(dimethylarsino)amines,[264] bis(dialkylamino)methylarsines,[265]

[264] F. Kober, *Z. anorg. Chem.*, 1973, **401**, 243.
[265] F. Kober, *Z. anorg. Chem.*, 1973, **397**, 97.

MeAsSNSN,[266] $(Et_2As)_3N$, $R_4As_2N_2HCl$ (R = Me or Et), and $(Ph_2AsN)_4$.[267] The highest mass observed in the spectrum of $(Ph_2AsN)_4$ was due to $[pm + H]^+$ whilst the base peak was $[Ph_7As_4N_4]^+$. Fragmentation of the ion of highest m/e value initially occurred by loss of C_6H_5, $C_{12}H_{10}$, $C_{12}H_{10}N$, and C_6H_5 simultaneously with Ph_2AsN.

Antimony and Bismuth. Features in the spectra of R_3SbX_2 compounds (R = Me or Ph, X = Cl or Br) have been reported.[268] All the derivatives gave pmi's, (except for Ph_3SbBr_2) and $[R_3SbX]^+$ ions as the base peaks. Fragmentation by loss of HX from the methyl derivatives and C_6H_6 from the phenyl compounds was common. The formation of both $[PhSbBr]^+$ and $[C_{12}H_{10}]^{+\cdot}$ from $[Ph_3SbBr]^+$ was noted. Ions containing two antimony atoms were present in the spectra of $\{SbCl_4(OR)\}_2$ (R = Me or Et) and $Sb_2Cl_5(OEt)_5$.[269] The highest m/e values corresponded to $[Sb_2Cl_7(OR)_2]^+$ and $[Sb_2Cl_4(OEt)_5]^+$, respectively. Fragmentations by loss of OR radicals and RCl, HCl, ROH, and R_2O molecules were observed. Other compounds studied include compounds with $O(CH_2)_2XSb$-groups, (X = O or NH),[270] and $(C_6F_5)_3Bi$.[271]

Group VI.—Dialkyl selenides (R = Me, Pr, or Bu), selenacyclohexane, PhSeMe, and the diaryl selenides $(XC_6H_4)_2Se$ (X = *p*-Me, *p*-MeO, *p*-Cl, or *p*-Br) have been studied mass spectrometrically;[272] all show pmi's. Dimethyl selenide fragments initially by losing CH_3 or C_2H_4 to give $[SeH_2]^{+\cdot}$. The base peaks in the spectra of R = Pr or Bu are $[RSeH]^+$, formed from pmi's by loss of (R—H). Ions produced by loss of alkyl radicals on cleavage of C-1—C-2 and C-2—C-3 bonds are also present. Selenacyclohexane fragments by losing CH_3, C_2H_4, C_3H_6, and SeH. Methylphenyl selenide also loses SeH from its pmi to give $[C_7H_7]^+$, and CH_2Se to give $[C_6H_6]^{+\cdot}$. The loss of the Se atom is an important decomposition for diaryl selenides either directly from the pmi's or after the stepwise loss of both ring substituents.

The spectra of some Se^{IV} dibromides R_2SeBr_2 [R = Me, Pr, —$(CH_2)_4$—, or Ph], and $PhMeSeBr_2$ have been discussed, and features in the spectra of some corresponding dichlorides were mentioned.[273] Generally pmi's were absent and the highest m/e value corresponded to $[pm - X]^+$. Fragmentation of this ion by loss of X is the only one observed for diaryl compounds but for compounds containing alkyl groups loss of an R group or an olefin competes.

Perfluoroalkyl derivatives $C_4F_9SeC_2F_5$, $(R_F)_2Se_2$ ($R_F = C_2F_5$ or C_4F_9), $(C_2F_5)_2Se_3$ and $C_2F_5SeCF_2C(O)F$ have all been reported to give pmi's.[274]

---

[266] O. J. Scherer and R. Wies, *Angew. Chem. Internat. Edn.*, 1972, **11**, 529.
[267] L. K. Krannich, U. Thewalt, W. J. Cook, S. R. Jain, and H. H. Sisler, *Inorg. Chem.*, 1973, **12**, 2304.
[268] H. Preiss, *Z. anorg. Chem.*, 1972, **389**, 280.
[269] H. Preiss, *Z. anorg. Chem.*, 1972, **389**, 393.
[270] A. Kiennemann and R. Kieffer, *J. Organometallic Chem.*, 1973, **60**, 255.
[271] G. B. Deacon and I. K. Johnson, *Inorg. Nuclear Chem. Letters*, 1972, **8**, 271.
[272] E. Rebane, *Acta Chem. Scand.*, 1973, **27**, 2861.
[273] E. Rebane, *Acta Chem. Scand.*, 1973, **27**, 2870.
[274] C. D. Desjardins and J. Passmore, *J.C.S. Dalton*, 1973, 2314.

Fragmentation from $(C_2F_5)_2Se_2$ was by loss of F, C_2F_5, and C_2F_6. The tellurium compounds $(C_2F_5)_2Te$, $(C_2F_5)_2Te_2$, $(C_2F_5)TeCF_2C(O)F$, and $(C_2F_5Te)_2Hg$ also produce pmi's.[275]

The spectra of some selenoureas (55) and selenothiocarbamic esters (56) have been illustrated, and their fragmentation patterns compared with analogous S

$$R^1R^2N-\overset{\overset{Se}{\|}}{C}-NR^3R^4 \qquad R^1R^2N-\overset{\overset{Se}{\|}}{C}-SCH_3$$

(55) (56)

R^1, R^2, R^3, and R^4 are H, Me, or Ph groups

and O compounds.[276] In general, fragmentation of the selenium compounds more closely resembled their sulphur than their oxygen counterparts particularly with respect to the numerous migration reactions of H or CH_3 or Ph groups on to Se prior to decomposition.

The fragmentation paths in the spectra of some isosteres of phenanthrene (57) and (58) containing one S and one Se atom, or two Se atoms have been discussed.[277] A notable feature, which has also been found in the spectra of other

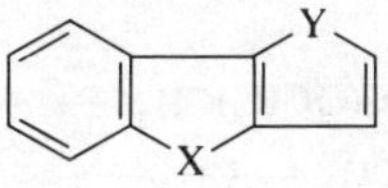

(57) 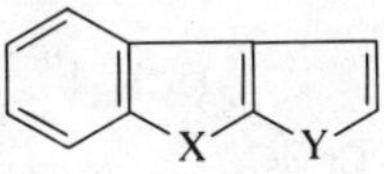 (58)

a; X = S, Y = Se
b; X = Se, Y = S
c; X = Y = Se

selenium heterocycles, is the loss of the selenium atom from the pmi's [or in the case of (57c) and (58c), sequential loss of two selenium atoms]. Other heterocyclic selenium compounds for which data have been reported include dicycloalkyl-1,4-diselenins and dicycloalkylselenophens ($n = 3$—6),[278] 3,3'-dioxo-4,4,4,'4'-tetramethyl-2,2-biselenolanylindene,[279] some benzoselenazoles and benzoisoselenazoles[280] and related compounds.

The spectra of a number of selenoxides have been discussed.[281] The compounds studied were R_2SeO (R = Me, Pr, Bu, or $PhCH_2$), the cyclic compounds $(CH_2)_nSe(O)$ ($n = 4$ or 5), PhSe(O)Me, and $(XC_6H_4)_2SeO$ (X = H, p-Me, p-Cl,

---

[275] H. L. Paige and J. Passmore, *Inorg. Nuclear Chem. Letters*, 1973, **9**, 277.
[276] A. M. Kirkien, R. J. Shine and J. R. Plimmer, *Org. Mass Spectrometry*, 1973, **7**, 233.
[277] P. Jacquignon, A. Croisy, M. Renson, E. Iteke, and L. Christiaens, *Org. Mass Spectrometry*, 1973, **7**, 1235.
[278] I. Lalezari, A. Shafiee, and M. Yalpani, *J. Heterocyclic Chem.*, 1972, **9**, 1411.
[279] L. Fitzer and W. Luthke, *Chem. Ber.*, 1972, **105**, 919.
[280] A. Croisy, P. Jacquignon, R. Weber, and M. Renson, *Org. Mass Spectrometry*, 1972, **6**, 1321.
[281] E. Rebane, *Chem. Scripta*, 1973, **4**, 219.

m-Cl or *p*-Br). A common feature of the spectra was the occurrence of a selenoxide to selenenate rearrangement of the pmi involving both alkyl and aryl group migration, followed either by Se—O bond fission or Se—O fission and a hydrogen migration. Aliphatic selenoxides exhibit H migration to oxygen followed by loss of OH or alkene. Diaryl selenoxides simply lose an oxygen atom. Elimination of SeOH is observed from the pmi of dibenzyl selenoxide and the cyclic selenoxides. Skeletal rearrangements in diaryl selenoxides can lead to the elimination of the Se atom or CO from the pmi's.

Other selenium compounds characterized include cycloalkeno-1,2,3-selenadiazoles.[278,282]

3 Transition-metal Organometallics

Results of General Interest.—In his excellent progress report of the study of the 'Decomposition of (Transition-metal) Organometallic Complexes in the Mass Spectrometer', Muller has discussed fragmentation reactions in terms of three basic processes.[4] First, decompositions involving simple metal–ligand bond cleavage; second, simple cleavage of bonds within the complex ligands; third, fragmentation reactions with rearrangements. Often competition between all three types is observed to occur from one ion (*e.g.* Scheme 18).

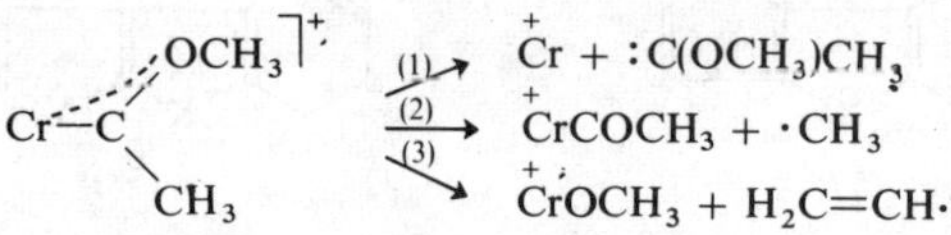

$$\text{Cr} \underset{\text{CH}_3}{\overset{\text{OCH}_3}{=}} \text{C} \Big]^{+\cdot} \xrightarrow[\text{(3)}]{\text{(1)}} \begin{array}{l} \overset{+}{\text{Cr}} + :\text{C(OCH}_3)\text{CH}_3 \\ \overset{+}{\text{CrCOCH}_3} + \cdot\text{CH}_3 \\ \overset{+}{\text{CrOCH}_3} + \text{H}_2\text{C}=\text{CH}\cdot \end{array}$$

Scheme 18

For complexes with several different ligands attached to the metal, a sequence of simple M—L cleavage can be found; ligands (*a*) CO, NO, N_2, H, PF_3, Me, CF_3, PH_3, C_2H_4, C_2H_2 are more readily removed than (*b*) F, Ph, alkyl, phosphanes, phosphites, amines, isocyanides, carbenes, R_2S, sulphoxides, cyclic oligo-olefins, and aromatic hydrocarbons. The positions of terminally bound Cl, Br, or I are difficult to assess. Of course, the gradations which occur within groups (*a*) and (*b*) are not independent of the central metal atom or other ligands and no rigorous series can be written. Simple cleavages of bonds within ligands are discussed with reference to removal of H, to C—O bond cleavage in carbonyls, and to removal of other entities X, or OR, R, NR_2 radicals. Rearrangements involving ring cleavage, migration of H and transfer of nucleophilic groups are discussed with examples.

Ion-Molecule Reactions.—The rates of the reactions between ferrocene molecules and the ions $[Fe]^+$ and $[C_5H_5Fe]^+$ have been studied.[283] The ionic products were $[(C_5H_5)_2Fe]^+$, formed from both reactions, and the adduct $[(C_5H_5)_3Fe_2]^+$ from

---

[282] H. Meier and E. Voight, *Tetrahedron*, 1972, **28**, 187.
[283] S. M. Schildcrout, *J. Amer. Chem. Soc.*, 1973, **95**, 3846.

the reaction of $[C_5H_5Fe]^+$ and ferrocene. A previous suggestion that $[(C_5H_5)_3Fe_2]^+$, which contributes ca. 3% of the total ion current, came from a dimeric $\{(\pi\text{-}C_5H_5)_2Fe\}_2$ species can how be discounted.

The existence of the ions $[Ar_2Cr_2(CO)_3]^+$ and $[Ar_2Cr_3(CO)_6]^+$ produced from arenechromium tricarbonyl derivatives by ion–molecule reactions has been confirmed with the help of appearance potential measurements and studies of variation of ion intensities with ion-repeller voltage and source pressure.[284] Their modes of formation are shown in Scheme 19.

$$[ArCr(CO)_3] + [ArCr(CO)_3]^+ \rightarrow [Ar_2Cr_2(CO)_3]^+ + 3CO$$
$$[ArCr(CO)_3] + [Ar_2Cr_2(CO)_3]^+ \rightarrow [Ar_2Cr_3(CO)_6]^+ + Ar$$

Scheme 19

Ion–molecule reactions in the spectrum of $(\pi\text{-}C_5H_5)NiNO$ produce a number of ions containing two nickel atoms including $[(C_5H_5)_2Ni_2NO]^+$, $[(C_5H_5)_2Ni_2]^+$, $[(C_5H_5)Ni_2(C_3H_3)]^+$, and $[(C_5H_5)Ni_2NO]^+$.[285] If simple organic molecules (L) are present, additional reactions give rise to ions of the type $[C_5H_5NiL]^+$ and to fragment ions derived therefrom (L may be e.g. an amine or a saturated or unsaturated hydrocarbon molecule such as cyclohexene; Scheme 20). The cross-sections for the ion–molecule reactions and the appearance

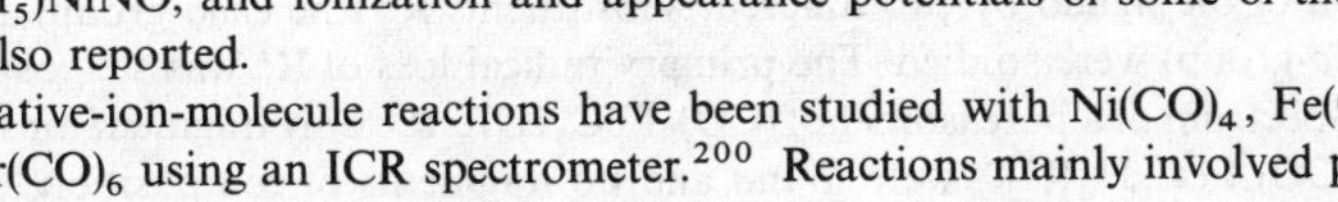

$C_5H_5\overset{+}{Ni}$—[cyclohexane] $\xrightarrow{-H_2}$ $C_5H_5\overset{+}{Ni}$—[cyclohexene] $\xrightarrow{-H_2}$ $C_5H_5\overset{+}{Ni}$—[benzene]

Scheme 20

potentials of some of the secondary ions were measured. The spectrum of $(\pi\text{-}C_5H_5)NiNO$, and ionization and appearance potentials of some of the ions were also reported.

Negative-ion-molecule reactions have been studied with $Ni(CO)_4$, $Fe(CO)_5$, and $Cr(CO)_6$ using an ICR spectrometer.[200] Reactions mainly involved parent molecules $M(CO)_n$ and $[M(CO)_{n-1}]^-$ ions.

Chemical Ionization Spectra.—Methane CI mass spectra using $[CH_5]^+$ and $[C_2H_5]^+$ reactant ions with different types of transition-metal compound have been reported.[286] Whereas the electron impact spectra of heptafulvene-, cycloheptatrienone-, and cycloheptatriene-carboxaldeiron tricarbonyl (59)—(61) do not show pmi's whilst those of cyclooctatetraene- and cyclobutadiene-iron tricarbonyl have weak pmi's, all these compounds readily accept a proton from $[CH_5]^+$ to give abundant, sometimes base peak, pmi's. Ions produced by stepwise loss of one, two, and three CO groups are observed from all these

---

[284] J. R. Gilbert, W. P. Leach, and J. R. Miller, *J. Organometallic Chem.*, 1973, **56**, 295; 1972, **42**, C51.
[285] J. Muller and W. Goll, *Chem. Ber.*, 1973, **106**, 1129.
[286] D. F. Hunt, J. W. Russell, and R. L. Torian, *J. Organometallic Chem.*, 1972, **43**, 175.

compounds as well as ions of greater mass than $[pm - H]^+$, corresponding to $[pm - H + FeL]^+$ [L = hydrocarbon ligand, except compound (59)].

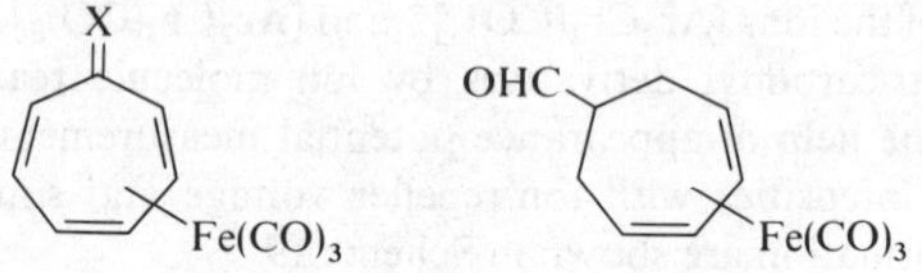

(59) X = CH$_2$
(60) X = O

(61)

The CI spectra of $(\pi\text{-}C_5H_5)_2M$, (M = Fe, Ru, Os, and Co) showed only three ions, namely $[pm]^+$, $[pm + H]^+$, and $[pm + C_2H_5]^+$. Nickelocene afforded the corresponding ions and also $[Ni]^+$, $[C_5H_5Ni]^+$, and $[(C_5H_5)_nNi_2]^+$ ($n = 2$ or 3). Although pmi's and $[pm + H]^+$ ions were observed in the spectra of $(\pi\text{-}C_5H_5)_2\text{-}MCl_2$ (M = Ti, Zr, or Hf), the elimination of HCl from $[pm + H]^+$ was facile and $[pm - Cl]^+$ ions were found as base peaks.

In contrast to the electron impact spectra of the carbonyls M(CO)$_6$ (M = Cr, Mo, or W) which show pmi's and stepwise loss of up to six carbonyls, the base peaks corresponded to $[pm + H]^+$ with the only other ions observed being $[pm]^+$, $[pm + H - CO]^+$, and in low abundance, $[2pm + H]^+$ and $[2pm + H - CO]^+$ from the Mo and W compounds.

Distinction of Geometrical Isomers.—In contrast to pairs of organic stereoisomers where often only minor differences in intensities of ions are detected, it seems it may be possible to assign an *exo*- or *endo*-isomer of a transition-metal π-hydrocarbon ligand complex solely on the basis of its mass spectrum.[287] A series of isomers of cyclopentadieneiron tricarbonyl complexes substituted at the 5-position of the ligand by two different substituents R^1 (the *endo*-group) and R^2 (the *exo*-group) were studied. The primary radical loss of R^2 was stereospecific. In the spectrum of a pure isomer of (CO)$_3$FeC$_5$H$_4$R^{1}R^2, only fragment ions of the type $[(CO)_n\text{FeC}_5H_4R^1]^+$ were found and no fragments of composition $[(CO)_n\text{-}FeC_5H_5R^2]^+$. However, for *endo*- and *exo*-isomers of a 2,5-diphenyl-1-silacyclopentadieneiron tricarbonyl complex no such clear-cut correlation was observed,[288] and the generality of the findings from the previous work[287] remains to be studied.

Metal Carbonyls and Carbonyl Hydrides.—Aspects of the spectra of the tetranuclear carbonyls $M_3^1M^2(CO)_{12}$ and $M_2^1M_2^2(CO)_{12}$ (M^1, M^2 = Co, Rh, or Ir) and Rh$_4$(CO)$_{12}$ have been discussed.[289] All the compounds studied showed pmi's and the ions $[M_nM_m(CO)_x]^+$, ($n + m = 4$ and $x = 0$—11). The mixed metal

---

[287] J. Muller, G. E. Herberich, and H. Muller, *J. Organometallic Chem.*, 1973, **55**, 165.
[288] W. Fink, *Helv. Chim. Acta*, 1974, **57**, 61.
[289] S. Martinengo, P. Chini, V. G. Albano, F. Cariati, and T. Salvatori, *J. Organometallic Chem.*, 1973, **59**, 379.

carbonyls, except for $Co_2Ir_2(CO)_{12}$, underwent redistribution reactions (Scheme 21), to produce new species with characteristic pmi's and ions due to loss of CO groups.

$$2M^1_3M^2(CO)_{12} \rightleftharpoons M^1_4(CO)_{12} + M^1_2M^2_2(CO)_{12}$$
$$2M^1_2M^2_2(CO)_{12} \rightleftharpoons M^1_3M^2(CO)_{12} + M^1M^2_3(CO)_{12}$$

Scheme 21

The dinuclear compounds, $MCo(CO)_9$ (M = Mn or Re) and $(CO)_5MnCo(CO)_3PPh_3$, have been reported to show pmi's, and ions due to loss of CO groups e.g. from $MnCo(CO)_9$, $[pm - n(CO)]^+$ ($n = 1$—9), $[Mn(CO)_n]^+$ ($n = 0$—5) and $[Co(CO)_n]^+$ ($n = 0$—3).[290]

A series of polynuclear carbonyl hydrides, including $H_2Os_5(CO)_{15}$, $H_2Os_5(CO)_{16}$, and $H_2Os_6(CO)_{18}$, show pmi's and the subsequent loss of several CO groups (5, 4, and 5, respectively) before losing 2H (or H_2).[291] The remaining CO groups are then lost to give the polynuclear ions $[Os_n]^+$. Other polynuclear carbonyl hydrides studied include $HMOs_2(CO)_{12}$, $HMOs_3(CO)_{16}$, and $H_3MOs_3(CO)_{13}$ (M = Mn or Re).[292] The loss of H from the monohydrides was found to be insignificant until there were relatively few CO groups remaining. The dinuclear complex $HReCr(CO)_{10}$ has been characterized.[293]

Metal Nitrosyls.—From chromium tetranitrosyl, the pmi and all $[Cr(NO)_n]^+$ ($n = 0$—3) ions were reported.[294] The base peak in the spectrum of $W(CO)_4(NO)Br$ is the pmi,[295] and fragmentation then occurs by loss of CO or NO groups to give $[W(CO)_nNOBr]^+$ and $[W(CO)_mBr]^+$ ($n, m = 0$—3). The compound $Re_2(CO)_4(NO)_2(NO_3)_4$ has been characterized.[296]

Compounds containing nitrosyl and π-cyclopentadienyl groups which have been examined include $(\pi$-$RC_5H_4)Mn(CO)_2NO$[297] and $(\pi$-$C_5H_5)Re(CO)NO$ and related compounds.[298]

The product from the reaction of WMe_6 with nitric oxide, $WMe_4\{ON(Me)NO\}_2$, showed no pmi, the ion at highest mass being $[pm - Me]^+$.[299] A number of ions of the type $[Me_xWO_y]^+$ were observed but none with O replaced by N, suggesting W—O rather than W—N bonding.

[290] G. Sbrignadello, G. Bor, and L. Maresca, *J. Organometallic Chem.*, 1972, **46**, 345.
[291] C. R. Eady, B. F. G. Johnson, and J. Lewis, *J. Organometallic Chem.*, 1973, **57**, C84.
[292] J. Knight and M. J. Mays, *J.C.S. Dalton*, 1972, 1022.
[293] A. S. Foust, W. A. G. Graham, and R. P. Stewart, *J. Organometallic Chem.*, 1973, **54**, C22.
[294] M. Heberhold and A. Razavi, *Angew. Chem. Internat. Edn.*, 1972, **11**, 1092.
[295] C. G. Barraclough, J. A. Bowden, R. Colton, and C. J. Commons, *Austral. J. Chem.*, 1973, **26**, 241.
[296] R. Davis, *J. Organometallic Chem.*, 1973, **60**, C22.
[297] H. Brunner and M. Langer, *J. Organometallic Chem.*, 1973, **54**, 221.
[298] R. P. Stewart, N. Okamoto, and W. A. G. Graham, *J. Organometallic Chem.*, 1972, **42**, C32.
[299] S. R. Fletcher, A. Shortland, A. C. Skapski, and G. Wilkinson, *J.C.S. Chem. Comm.*, 1972, 922.

The series of compounds $Fe(NO)_2\{P(OR)_3\}_2$ and $Co(NO)\{P(OR)_3\}_3$ (R = Me, Et, Pr, Pr^i, Bu, or Ph) and $Fe(NO)_2\{PF_2(OEt)\}_2$ all show pmi's.[300] The spectra of $Co(NO)\{P(OMe)_3\}_3$ and the iron fluorophosphine derivative were discussed. Both compounds initially fragment by losing NO, OR, or F (in the case of the iron compound), or the complete phosphite ligand. Further fragmentation involves further loss of such groups or of R radicals. The base peaks at 70 eV were $[Co\{P(OMe)_3\}_2]^+$ and $[Fe(NO)PF_2OEt]^+$. The spectrum of the iron compound contained some rearranged species including $[FeNOF]^+$, $[FeF]^+$, and $[FeO]^+$.

Other nitrosyl complexes containing phosphorus ligands for which data have been presented include $Cr(NO)(PF_3)_3$ and $Mn(NO)_3PF_3$,[301] and $LCo(NO)(CO)$, $L = P(CN)_n(CF_3)_{3-n}$ (n = 1—3).[302]

Complexes containing Metal–Carbon σ-Bonds.—Hexamethyltungsten shows no pmi but the series of ions $[(CH_3)_nW]^+$ (n = 0—5) was observed.[303] Although tetra-t-butyl–chromium showed no pmi with $[(C_4H_9)_3Cr]^+$ appearing as the ion of highest mass, the tetraneopentyl-[304] and tetrakis(1-norbornyl)- and tetrakis-(2,3,3-trimethylbicyclo[2,2,1]hept-1-yl)-chromium derivatives[305] did have pmi's. Tetrakis-(1-norbornyl)manganese had a pmi but the corresponding R_4M (M = Ti or Zr) compounds,[305] tetraneopentyltitanium[306] and tetrabenzylzirconium,[307] only gave $[pm - R]^+$ as highest mass.

The trimethylsilylmethyl derivative, $(Me_3SiCH_2Cu)_n$, was shown to be tetrameric in the gas phase;[308] fragmentation occurred initially by loss of Me, Me_4Si, CH_4, and Me_3SiCH_2 groups.

Both Me and PF_3 groups were lost from the pmi of $MeCo(PF_3)_4$, and series of ions, $[Co(PF_3)_n]^+$ and $[MeCo(PF_3)_n]^+$ (n = 0—4), were found as well as some rearranged ions containing Co—F bonds, such as $[FCoPF_3]^+$.[309] A feature of the spectrum of $EtCo(PF_3)_4$ was the formation of a series of $[HCo(PF_3)_n]^+$ (n = 1—4) ions produced by elimination of C_2H_4 from the corresponding ethylcobalt-containing ions.

Other transition-metal methyl compounds for which data have been reported include $(\pi\text{-}C_5H_5)_2MoMe_2$,[310] $(\pi\text{-}Me_5C_5)W(CO)_3Me$,[311] $Os(CO)_4Me_2$,[312] and $MeNbCl_4$.[313] Some thioether complexes of dimethylgold halides (Me_2AuX),

[300] T. Kruck, J. Waldmann, M. Hofler, G. Birkenhager, and C. Odenbrett, Z. anorg. Chem., 1973, **402**, 16.
[301] R. Middleton, J. R. Hull, S. R. Simpson, C. H. Tomlinson, and P. L. Timms, J.C.S. Dalton, 1973, 121.
[302] I. H. Sabherwal and A. B. Burg, Inorg. Chem., 1972, **11**, 3138.
[303] A. J. Shortland and G. Wilkinson, J.C.S. Dalton, 1973, 872.
[304] W. Kruse, J. Organometallic Chem., 1972, **42**, C39.
[305] B. K. Bower and H. G. Tennent, J. Amer. Chem. Soc., 1972, **94**, 2513.
[306] P. J. Davidson, M. F. Lappert, and R. Pearce, J. Organometallic Chem., 1973, **57**, 269.
[307] K.-H. Thiele, E. Kohler, and B. Adler, J. Organometallic Chem., 1973, **50**, 153.
[308] M. F. Lappert and R. Pearce, J.C.S. Chem. Comm., 1973, 24.
[309] T. Kruck, G. Sylvester, and I.-P. Kunau, Z. anorg. Chem., 1973, **396**, 165.
[310] J. L. Thomas, J. Amer. Chem. Soc., 1973, **95**, 1838.
[311] R. B. King and A. Efraty, J. Amer. Chem. Soc., 1972, **94**, 3773.
[312] R. D. George, S. A. R. Knox, and F. G. A. Stone, J.C.S. Dalton, 1973, 973.
[313] C. Santini-Scampucci and J. G. Reiss, J.C.S. Dalton, 1973, 2436.

MeS(CH$_2$)$_n$SMe(Me$_2$AuX) (n = 2 or 3, X = Cl, Br, or I), did not show pmi's. The highest m/e value was due to [Me$_4$Au$_2$X$_2$]$^+$.[314]

In the platinum azide complex, (Me$_3$PtN$_3$)$_4$, the base peak corresponded to the tetramer.[315]

The spectra of a number of compounds with groups more complicated than methyl have been reported, including allylic complexes of ruthenium,[316] some fluoroallylic cobalt complexes,[317] iron and manganese derivatives containing the $-\overline{\text{C}=\text{CRSO}_2\text{OCH}_2}$ (R = Me or Ph) group,[318] (CO)$_5$Mn$\overline{\text{C}=\text{CPhC(O)N(SO}_2\text{Cl)CH}_2}$,[319] and acetylenic derivatives of titanium and iron compounds.[320] The spectra of (CO)$_5$MnX (X = $-$CH$_2$C≡CPh; $-\overline{\text{C}=\text{CPhS(O)OCH}_2}$, a sultine; and $-\overline{\text{C}=\text{CPhS(O)}_2\text{OCH}_2}$, a sultone) have been compared.[318] The spectra of the acetylenic derivative and the sultine were similar but differed from that of the sultone. From the sultine, loss of SO$_2$ was comparatively easy, producing ions of the type [(CO)$_n$MnC$_9$H$_7$]$^+$ (found also in the acetylenic compound) whereas the sultone eliminated SO$_2$ only after losing at least two carbonyl groups; loss of SO$_3$ was even less favourable.

Heterocyclic complexes containing metal–alkyl bonds have been prepared using the ligand (62) and all gave pmi's. They include Me$_2$P{CH$_2$MCH$_2$}$_2$PMe$_2$ (M = Cu, Ag,[321] or Au[322]), {Me$_2$P(CH$_2$)$_2$}$_4$Ni$_2$, and {Me$_2$P(CH$_2$)$_2$}$_2$Ni,[323] and {Me$_2$P(CH$_2$)$_2$}$_3$Cr.[324] Other heterocyclic systems for which data have been reported include (CO)$_4\overline{\text{Fe(CH}_2\text{)}_3\text{SiMe}_2}$, which shows a pmi and ions corresponding to the loss of up to four CO groups and (4CO + C$_2$H$_4$),[325] (π-C$_5$H$_5$)$_2$-$\overline{\text{TiCH}_2\text{Si(Me)}_2\text{NSiMe}_3}$,[326] (CO)$_3$(PPh$_3$)$\overline{\text{Fe(CH}_2\text{)}_3\text{S}}$,[327] and octafluorodibenzole derivatives of zirconium, cobalt, rhodium, and iridium.[328]

Complexes containing fluorocarbon ligands have been the subjects of several investigations. Heptafluorobutene rhenium pentacarbonyl showed a pmi and three major series of ions, [C$_4$F$_7$Re(CO)$_n$]$^+$ (n = 1—4), [Re(CO)$_n$]$^+$ (n = 0—5)

---

[314] H. Schmidbaur and K. C. Dash, *Chem. Ber.*, 1972, **105**, 3662.
[315] K.-H. Von Dahlen and J. Lorberth, *J. Organometallic Chem.*, 1974, **65**, 267.
[316] T. Blackmore, M. I. Bruce, and F. G. A. Stone, *J.C.S. Dalton*, 1974, 106.
[317] W. R. Cullen and J. T. Jull, *Canad. J. Chem.*, 1973, **51**, 1521.
[318] D. W. Lichtenburg and A. Wojcicki, *Inorg. Chim. Acta*, 1973, **7**, 311.
[319] Y. Yamamoto and A. Wojcicki, *Inorg. Chem.*, 1973, **12**, 1779.
[320] K. Yasufuku and H. Yamazaki, *Bull. Chem. Soc. Japan*, 1972, **45**, 2664.
[321] H. Schmidbaur, J. Allkofe, and W. Buchner, *Angew. Chem. Internat. Edn.*, 1973, **12**, 415.
[322] H. Schmidbaur and R. Franke, *Angew. Chem. Internat. Edn.*, 1973, **12**, 416.
[323] H. H. Karsch and H. Schmidbaur, *Angew. Chem. Internat. Edn.*, 1973, **12**, 853.
[324] E. Kurras, U. Rosenthal, H. Mennenga, and G. Oehme, *Angew. Chem. Internat. Edn.*, 1973, **12**, 854.
[325] C. S. Cundy and M. F. Lappert, *J.C.S. Chem. Comm.*, 1972, 445.
[326] C. R. Bennett and D. C. Bradley, *J.C.S. Chem. Comm.*, 1974, 29.
[327] K. Takahashi, M. Iwanami, A. Tsai, P. L. Cheng, R. L. Harlow, L. E. Harris, J. E. McClaskie, C. E. Pfluger, and D. C. Dittener, *J. Amer. Chem. Soc.*, 1973, **95**, 6113.
[328] S. A. Gardner, H. B. Gordon, and M. D. Rausch, *J. Organometallic Chem.*, 1973, **60**, 179.

and the rearranged ions $[Re(CO)_nF]^+$ ($n = 0$—4) which included the base peak, $[Re(CO)_3F]^+$.[329] In the spectrum of $(\pi\text{-}C_5H_5)Fe(CO)_2C_4F_7$, ions formed by loss of CO and C_4F_7, and rearranged FeF-containing ions were found, as well as

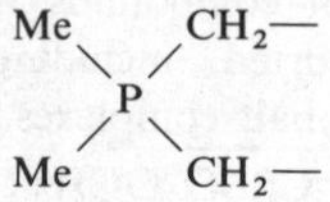

Me\\   /CH$_2$—
   P
Me/   \\CH$_2$—

(62)

some produced by elimination of HF, e.g. $[C_4F_6FeC_5H_4]^+$, and $[C_5H_4Fe]^+$ from $[C_5H_5FeF]^+$. Some chromium and cobalt complexes fragmented similarly. The spectra of $(\pi\text{-}C_5H_5)Co(CNR)(C_3F_7)I$ [R = C_6H_{11}, MeC_6H_4, $O_2NC_6H_4$, Pr^i, or $CH(Me)Ph$] have been listed.[330] For R = C_6H_{11} and MeC_6H_4, the C_3F_7 ligand is retained only in the pmi; for the other compounds, only two ions are found to retain it, the pmi and $[pm - I]^+$. The parent molecular ion from $(CF_3C{\equiv}CCF_3)Ir(CO)(PPh_3)_2C_6F_5$ fragments initially by losing CO, HF, and $CF_3C{\equiv}CCF_3$.[331]

Both derivatives of C_6F_6, 1,3- and 1,4-$\{(\pi\text{-}C_5H_5)Fe(CO)_2\}_2C_6F_4$, showed pmi's and base peaks due to $[pm - 4CO - FeF_2]^+$ ions.[332] The 1,4-derivative lost all four CO groups prior to other fragmentation whereas the 1,3-compound eliminated HF after loss of only two CO groups.

A number of complexes with metal–aryl σ-bonds and co-ordinating groups on the aryl ligand have been characterized. They include complexes (63),[333] (64),[334] some ruthenium and palladium complexes,[335] and the iridium complex (65) which gave $[pm - H]^+$ as base peak.[336]

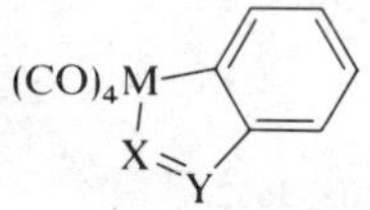

(63) M = Mn or Re; X=Y = O=CPh
(64) M = Mn or Re; X=Y = PhN=CH

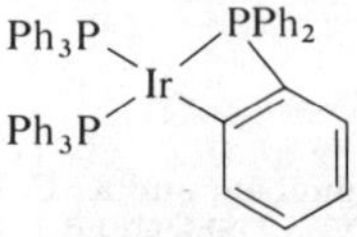

(65)

Hydrocarbon–Metal π-Complexes and Related Complexes.—*Acetylenes and Olefins.* Cobalt carbonyl complexes of acetylenes, $Co_2(CO)_6(C_2R_2)$, have been

---

[329] R. B. King and W. C. Zipperer, *Inorg. Chem.*, 1972, **11**, 2119.
[330] H. Brunner and W. Rambold, *J. Organometallic Chem.*, 1973, **60**, 351.
[331] R. L. Bennett, M. I. Bruce, and R. C. F. Gardner, *J.C.S. Dalton*, 1973, 2653.
[332] S. C. Cohen, *Org. Mass Spectrometry*, 1972, **6**, 1283.
[333] R. McKinney, G. Firestein, and H. D. Kaesz, *J. Amer. Chem. Soc.*, 1973, **95**, 7910.
[334] R. L. Bennett, M. I. Bruce, B. L. Goodall, M. Z. Iqbal, and F. G. A. Stone, *J.C.S. Dalton*, 1972, 1787.
[335] M. I. Bruce, B. L. Goodall, and F. G. A. Stone, *J. Organometallic Chem.*, 1973, **60**, 343.
[336] J. Schwarz and J. B. Cannon, *J. Amer. Chem. Soc.*, 1972, **94**, 6226.

studied with respect to the effect of R.[337] For R = Me or Ph, the major series of ions was $[Co_2(CO)_nC_2R_2]^+$ ($n = 0$—6). Few ions of significance involved fragmentation of the acetylenic group, and most of the ion current was carried by ions containing two cobalt atoms. An interesting successive loss of both cobalt atoms was observed from the phenyl compound (Scheme 22). For R = H, both series $[Co(CO)_nC_2H_2]^+$ ($n = 0$—6) and $[Co_2(CO)_n]^+$ ($n = 0$—4) were found and ions with fragmented acetylenic species became more important. Compounds in which R contained halogen (R = CH_2Cl or CF_3) showed a number of differences from the hydrocarbon derivatives. Both compounds gave the series $[Co(CO)_nC_2R_2]^+$ ($n = 0$—6) as in the previous compounds, but fragmentation of the C_2R_2 entity became much more significant. The complex with chloroacetylene showed $[Co_2(CO)_nC_4H_4Cl]^+$ ($n = 0$—5) and $[Co_2(CO)_nC_4H_4]^+$ ($n = 5, 4, 2,$ or 0) as well as rearranged ions e.g. $[Co_2Cl_2]^+$. The compound containing CF_3 showed $[Co_2(CO)_nC_4F_5]^+$ ($n = 0$—6), some abundant ions by loss of CF_2, rearranged ions such as $[Co_2F_2]^+$ and a fragmentation by loss of CoF_2 (Scheme 22). The spectrum of $Co_2(CO)_6C_2(CO_2Me)_2$ was compared with that of $Co_3(CO)_9CCO_2R$ (R = Me or Et). From the dicobalt compound, ions corresponding to loss of up to six carbonyl groups were found. Loss of two further CO groups suggests rearrangements involving methoxyl (Scheme 22). It appeared that these

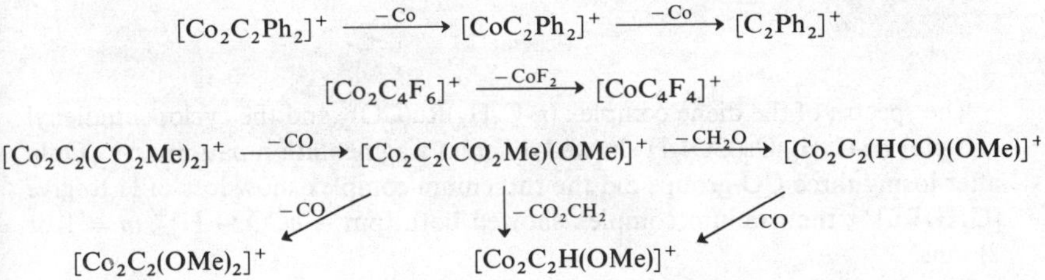

$$[Co_2C_2Ph_2]^+ \xrightarrow{-Co} [CoC_2Ph_2]^+ \xrightarrow{-Co} [C_2Ph_2]^+$$

$$[Co_2C_4F_6]^+ \xrightarrow{-CoF_2} [CoC_4F_4]^+$$

$$[Co_2C_2(CO_2Me)_2]^+ \xrightarrow{-CO} [Co_2C_2(CO_2Me)(OMe)]^+ \xrightarrow{-CH_2O} [Co_2C_2(HCO)(OMe)]^+$$

$$[Co_2C_2(OMe)_2]^+ \quad\quad [Co_2C_2H(OMe)]^+$$
(−CO, −CO₂CH₂, −CO)

Scheme 22

transfers, which also occurred in the Co_3 compounds, may involve both Co and C atoms as migration sites. Methyl transfer, which occurs in the free ligand with loss of CO_2, did not occur here. Further loss of CH_2O or CO_2CH_2 was observed with hydrogen rearrangements.

Other studied complexes containing acetylenic ligands include cyclopentadienyl derivatives of cobalt and rhodium,[338] $(Ar_2C_2)_2Os_2(CO)_6$ compounds,[339] and cobalt carbonyl derivatives of acetylenic phosphines, $(C_6F_5)_2PC_2R$ (R = Me or Ph).[340]

---

[337] O. Gambino, G. A. Vaglio, R. P. Ferrari, M. Valle and G. Cetini, *Org. Mass Spectrometry*, 1972, **6**, 723.
[338] R. S. Dickson and H. P. Kirsch, *Austral. J. Chem.*, 1973, **26**, 1911; 1974, **27**, 61.
[339] R. P. Ferrari, G. A. Vaglio, O. Gambino, M. Valle, and G. Cetini, *J.C.S. Dalton*, 1972, 1998.
[340] H. A. Patel, A. J. Carty, and N. K. Hota, *J. Organometallic Chem.*, 1973, **50**, 247.

Olefin complexes containing $Fe(CO)_4$ have been characterized, including $\{(Br)HC=CH(X)\}Fe(CO)_4$ (X = F, Br, or H),[341] thiete-1,1-dioxideiron tetracarbonyl,[342] and derivatives of bicyclic systems.[343] Some cyclopentadienylrhodium diolefin complexes are reported to show pmi's.[344]

Dienes and Polyenes. Trisbuta-1,3-diene derivatives of Mo and W show pmi's and base peaks due to $[pm - butadiene]^+$.[345] Features in the spectra of a series of $(\pi\text{-diene})Fe(PF_3)_3$ complexes have been discussed.[346] Two major series of ions were produced by loss of PF_3 or F groups, $[pm - n(PF_3)]^+$ ($n = 1\text{—}3$) and $[(\pi\text{-diene})Fe(PF_3)_{3-n} - F]^+$ ($n = 0\text{—}2$). Further fragmentation was affected by the nature of the diene ligand. 'Open' dienes lost olefin or acetylene fragments; 'closed' cyclic dienes fragmented first by losing one or more hydrogen molecules then hydrocarbon fragments C_nH_{2n} (Scheme 23).

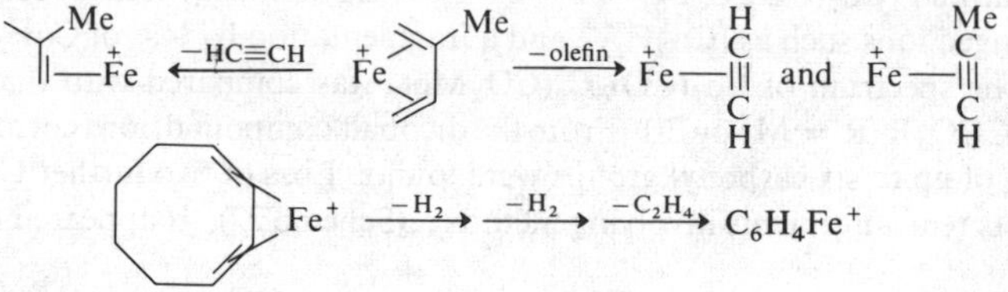

Scheme 23

The spectra of the diene complex $(\pi\text{-}C_5H_6)Ru(CO)_3$ and the cyclopentadienyl complex $(\pi\text{-}C_5H_5)Os(CO)_2H$ showed different fragmentation patterns.[347] Only after losing three CO groups did the ruthenium complex show loss of H to give $[C_5H_5Ru]^+$; the osmium complex showed both $[pm - nCO - H]^+$ ($n = 1$ or 2) ions.

Bicyclo[3,3,0]octa-2,4-dien-1-yl (cyclo-octa-1,5-diene)cobalt and related compounds show strong pmi's and fragment by losing H_2 and hydrocarbon species.[348] Abundant ions such as $[(C_8H_9)CoC_nH_n]^+$ ($n = 7$, 6, or 5), may contain aromatized C_nH_n ligands.

Other compounds for which data have been reported include (1,2-dichloro-1,3-diene)$Fe(CO)_3$,[349] $\{\pi\text{-}C_6H_5P(Ph)Et\}Fe(\pi\text{-diene})$ complexes,[350] cyclopentadienylduroquinone cobalt compounds,[351] some cyclopentadienylrhodium diene

---

[341] F. W. Grevels and E. K. von Gustorf, *Annalen*, 1973, 1821.
[342] J. E. McCaskie, P. L. Cheng, T. R. Nelson, and D. C. Dittmer, *J. Org. Chem.*, 1973, **38**, 3963.
[343] D. Ehntholt, A. Rosan, and M. Rosenblum, *J. Organometallic Chem.*, 1973, **56**, 315.
[344] Y. Wakatsuki and H. Yamazaki, *J. Organometallic Chem.*, 1974, **64**, 393.
[345] P. S. Skell, E. M. M. Van Dam, and M. P. Silvon, *J. Amer. Chem. Soc.*, 1974, **96**, 626.
[346] T. Kruck, L. Knoll, and J. Laufenberg, *Chem. Ber.*, 1973, **106**, 697.
[347] A. P. Humphries and S. A. R. Knox, *J.C.S. Chem. Comm.*, 1973, 326.
[348] S. Otsuka and T. Taketomi, *J.C.S. Dalton*, 1972, 1879.
[349] H. A. Brune, G. Horlbeck, and W. Schwab, *Tetrahedron*, 1972, **28**, 4455.
[350] E. K. von Gustorf, I. Frischler, J. Leitsch, and H. Dreeskamp, *Angew. Chem. Internat. Edn.*, 1972, **11**, 1088.
[351] D. W. Slocum and T. R. Engelmann, *J. Amer. Chem. Soc.*, 1972, **94**, 8597.

complexes,[352] other 1,3-diene compounds of rhodium,[353] and a cyclo-octa-1,5-dieneplatinum complex.[354]

The number of reports of diene complexes of $Fe(CO)_3$ is relatively very large. They include complexes with 'open'[355] and 'cyclic'[356] diene systems. Usually, only the appearance of a pmi, if found, and the ions $[pm - n(CO)]^+$ ($n = 1$—3) are noted. The elimination of oxygen-containing species from $[pm - 3(CO)]^+$ ions in the spectra of some cyclo-octatetraeneiron tricarbonyl derivatives, (66)—(69), has been studied.[357] Further loss of CO was observed from the $[pm - 3(CO)]^+$ ion of (66) to give $[C_8H_8Fe]^{+\cdot}$. Compounds (67) and (68) lost H_2O, with one H atom coming specifically from the ring system, to give $[C_8H_7RFe]^+$ (R = Me or H) which then lost C_2HR to give $[C_6H_6Fe]^+$. Loss of CH_2O from (69) occurred *via* a McLafferty rearrangement. Some pentalenediiron pentacarbonyl compounds have been characterized.[358,359]

The diene iron complexes $Fe_2(CO)_6C_4(CO_2R)_4$ (70) and $Fe_3(CO)_8C_4(CO_2R)_4$ (R = Me or Et)[360] undergo transfers of alkoxy-groups and hydrogen rearrangements similar to those found for the acetylene complexes of cobalt carbonyls discussed above[337] (Scheme 22). Compound (71) is reported to show a pmi.[361]

(66) R = CHO
(67) R = CH(Me)OH
(68) R = CH_2OH or CHDOH
(69) R = CH_2OMe or CHDOMe

(70) R^1 to R^4 = CO_2R(R = Me or Et)
(71) $(R^1 + R^4)$ and $(R^2 + R^3) = -(CH_2)_4-$

Complexes with 2,5-diphenyl-1-silacyclopentadiene ligands (72)—(74) have been reported to show pmi's. The iron tricarbonyl derivatives lost up to three CO molecules to produce the base peak.[288] Cobalt compounds showed base

[352] B. F. G. Johnson, J. Lewis, and D. J. Yarrow, *J.C.S. Dalton*, 1972, 2084.
[353] S. M. Nelson, M. Sloan, and M. G. B. Drew, *J.C.S. Dalton*, 1973, 2195.
[354] A. Vitagliano and C. Paiaro, *J. Organometallic Chem.*, 1973, 49, C49.
[355] A. Carbonaro and F. Cambisi, *J. Organometallic Chem.*, 1972, 44, 171; W. E. Billups, L. P. Lin, and B. A. Barker, *J. Organometallic Chem.*, 1973, 61, C55; R. Victor, R. Ben-Shoshan, and S. Sarel, *J. Org. Chem.*, 1972, 37, 1930.
[356] R. J. H. Cowles, B. F. G. Johnson, J. Lewis, and A. W. Parkins, *J.C.S. Dalton*, 1972, 1768; B. F. G. Johnson, J. Lewis, P. McArdle, and G. L. P. Randall, *J.C.S. Dalton*, 1972, 2076; R. C. Kerber and D. J. Ehntholt, *J. Amer. Chem. Soc.*, 1973, 95, 2927; G. Deganello, H. Maltz, and J. Kozarich, *J. Organometallic Chem.*, 1973, 60, 323; E. J. Reardon and M. Brookhart, *J. Amer. Chem. Soc.*, 1973; 95, 4311; W. Weidermuller and K. Hafner, *Angew. Chem. Internat. Edn.*, 1973, 12, 925; D. F. Hunt and J. W. Russell, *J. Amer. Chem. Soc.*, 1972, 94, 7198; A. J. Carty, R. F. Hobson, A. A. Patel, and V. Snieckus, *J. Amer. Chem. Soc.*, 1973, 95, 6835.
[357] J. E. Alsop and R. Davis, *J.C.S. Dalton*, 1973, 1687.
[358] D. F. Hunt and J. W. Russell, *J. Organometallic Chem.*, 1972, 46, C23; *J. Amer. Chem. Soc.*, 1972, 94, 7198.
[359] W. Weidemuller and K. Hafner, *Angew. Chem. Internat. Edn.*, 1973, 12, 925.
[360] O. Gambino, G. A. Vaglio, and G. Cetini, *Org. Mass Spectrometry*, 1972, 6, 1297.
[361] R. B. King and I. Haiduc, *J. Amer. Chem. Soc.*, 1972, 94, 4044.

peaks due to either pmi's or [pm − Me]⁺.²⁸⁸,³⁶² It was suggested that the stability of the latter ion was associated with aromatization of the heterocyclic ring system.

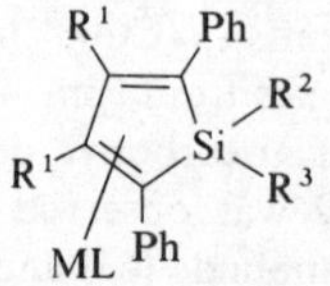

	ML	R¹	R²	R³
(72)	Fe(CO)₃	H	Me	Me or Ph
(73)	π-C₅H₅Co	H	Me	Me or Ph
		Ph	Me	
(74)	π-C₅H₅Co	Ph	Me	Me

A few iron tricarbonyl compounds containing substituted trimethylenemethane ligands have been characterized.³⁶³,³⁶⁴

Allyls. Complexes containing allylic ligands which have been examined include cobalt tricarbonyl derivatives,³⁶⁵ (π-2-methoxyallyl) iron tricarbonyl halides,³⁶⁶ bis(π-allyl)tropolonato-rhodium complexes,³⁶⁷ palladium hexafluoroacetylacetonato-derivatives,³⁶⁸ and compounds such as (75)³⁶⁹ and related complexes³⁷⁰,³⁷¹ containing iron tricarbonyl groups.

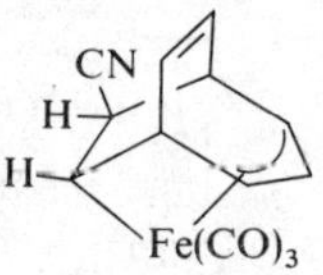

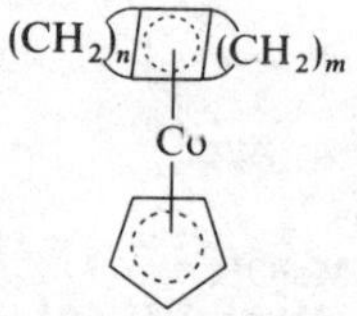

(76) $m = 4, 5,$ or 6
$n = 4$ or 5

Cyclobutadienes. Complexes with cyclobutadiene ligands continue to offer interesting variations in fragmentation behaviour. Whereas the cyclobutadiene ligand from the pmi of (π-C₅H₅)Co(π-C₄H₄) fragments by losing one and then two C₂H₂ units,³⁷² the sterically stabilized derivatives (76) lose neither C₂H₂ nor C₂H₂(CH₂)ₙ or ₘ units.³⁶¹,³⁷³ The trimethylsilyl-substituted cobalt complex (77)

---

³⁶² H. Sakurai and J. Hayashi, *J. Organometallic Chem.*, 1973, **63**, C10.
³⁶³ W. E. Billups, L.-P. Lin, and O. A. Gansow, *Angew. Chem. Internat. Edn.*, 1972, **11**, 637.
³⁶⁴ I. S. Krull, *J. Organometallic Chem.*, 1973, **57**, 363.
³⁶⁵ J. Clemens, M. Green, and F. G. A. Stone, *J.C.S. Dalton*, 1974, 93.
³⁶⁶ A. E. Hill and H. M. R. Hoffmann, *J.C.S. Chem. Comm.*, 1972, 574.
³⁶⁷ M. Green and G. J. Parker, *J.C.S. Dalton*, 1974, 333.
³⁶⁸ R. P. Hughes and J. Powell, *J. Organometallic Chem.*, 1972, **34**, C51; 1973, **60**, 387; 1973, **60**, 409.
³⁶⁹ A: Eisenstadt, *J. Organometallic Chem.*, 1973, **60**, 335.
³⁷⁰ V. Heil, B. F. G. Johnson, J. Lewis, and D. J. Thompson, *J.C.S. Chem. Comm.*, 1974, 270.
³⁷¹ R. Aumann and B. Lohmann, *J. Organometallic Chem.*, 1972, **44**, C51.
³⁷² M. Rosenblum, B. North, D. Wells, and W. P. Giering, *J. Amer. Chem. Soc.*, 1972, **94**, 1239.
³⁷³ R. B. King and A. Efraty, *J. Amer. Chem. Soc.*, 1972, **94**, 3021.

fragments by losing PhC≡CSiMe$_3$, but with the substituents in *cis*-positions, PhC≡CSiMe$_3$ as well as (Me$_3$SiC)$_2$ and (PhC)$_2$ are lost.[374]

Another sterically stabilized cyclobutadiene system (78) shows a pmi when MX$_2$ = NiBr$_2$ but not for PdCl$_2$.[375] Fragmentation occurs by initial loss of

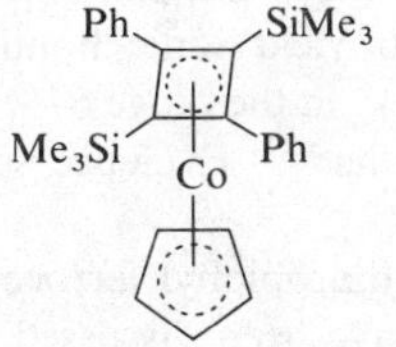

(77) 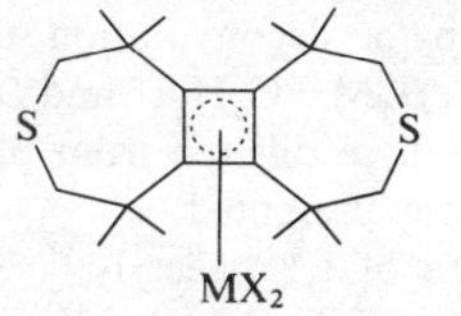   (78)

X or MX$_2$, and subsequently by elimination of C$_4$H$_8$ and Me groups. Other compounds reported include *p*-substituted phenylcyclobutadieneiron tricarbonyls,[376] a tricyclic cyclobutadiene (π-C$_5$Me$_5$)Co(C$_{12}$H$_{16}$),[377] (π-C$_5$H$_5$)Rh-(π-C$_4$Ph$_4$),[378] and (π-C$_5$H$_5$)Rh(π-C$_4$H$_3$COMe).[379]

Cyclopentadienyls. The spectra of the complexes (π-C$_5$H$_5$)$_2$TiX$_2$ (X = F, Cl, Br, or I), (π-C$_5$H$_5$)TiX$_3$, (π-C$_5$H$_4$Me)TiX$_3$, and (π-C$_5$Me$_5$)TiX$_3$ (X = Cl, Br, or OEt) have been listed.[380] Fragmentation of the pmi occurs predominantly by loss of cyclopentadienyl or X groups. Monocyclopentadienyl complexes often lose TiX$_3$ from pmi's and methylated ligands eliminate HX to give fragments which may contain a fulvene group.

Other compounds examined include bis(cyclopentadienyls) of molybdenum and tungsten,[310] cyclopentadienyl hydrides, halides, and carboxylates of tungsten,[381] (π-C$_5$Me$_5$) derivatives of chromium and manganese carbonyls,[311] dimethylsulphoniumcyclopentadienylmolybdenum tricarbonyl,[382] an iron carbonyl derivative of benzobenzvalene,[383] fulvalenedimanganese hexacarbonyl,[384] manganese, rhenium, and iron derivatives of the (π-C$_5$H$_4$CN) ligand,[385] and related compounds. The spectra of some di-π-cyclopentadienyl-μ-π-cyclopentadienyl dimetal complexes (M = Ni, Pd, or Pt) have been discussed.[386]

---

[374] H. Sakurai and J. Hayashi, *J. Organometallic Chem.*, 1972, **39**, 365.
[375] H. Kimling and A. Krebs, *Angew. Chem. Internat. Edn.*, 1972, **11**, 932.
[376] H. A. Brune, G. Horlbeck, H. Rottele, and U. Tanger, *Z. Naturforsch.*, 1973, **28b**, 68.
[377] R. B. King, A. Efraty, and W. M. Douglas, *J. Organometallic Chem.*, 1973, **56**, 345.
[378] G. C. Cash, J. F. Helling, M. Mathew, and G. J. Palenik, *J. Organometallic Chem.*, 1973, **50**, 277.
[379] S. R. Gardner and M. D. Rausch, *J. Organometallic Chem.*, 1973, **56**, 565.
[380] A. N. Nesmeyanov, Ya. S. Nekrasov, V. F. Sizoi, O. V. Nogina, V. A. Dubovitsky, and Ye. I. Sirotkina, *J. Organometallic Chem.*, 1973, **61**, 225.
[381] K. S. Chen, J. Kleinberg, and J. A. Landgrebe, *Inorg. Chem.*, 1973, **12**, 2826.
[382] V. I. Zolanovitch, A. Zh. Zhakaeva, V. N. Setkina, and D. N. Kursanov, *J. Organometallic Chem.*, 1974, **64**, C25.
[383] R. M. Moriarty, K. N. Chen, and J. L. Flippen, *J. Amer. Chem. Soc.*, 1973, **95**, 6489.
[384] R. F. Kovar and M. D. Rausch, *J. Org. Chem.*, 1973, **38**, 1918.
[385] R. E. Christopher and L. M. Venanzi, *Inorg. Chim. Acta*, 1973, **7**, 219.
[386] E. O. Fischer, P. Meyer, C. G. Kreiter, and J. Muller, *Chem. Ber.*, 1972, **105**, 3014.

Elimination of C_5H_6 from the pmi was followed by loss of M to give $[C_{10}H_{10}M]^+$ ions which then lost two C_5H_5 groups successively. Bis(pentalenyl) complexes of Fe, Co,[387] and Ni[388] are characterized by pmi's and such ions as $[C_8H_6M]^+$.

Ferrocenes and Related Compounds. Ionization efficiency curves for the pm, $[C_5H_5M]^+$, and $[M]^+$ ions from $(\pi$-$C_5H_5)_2M$ (M = Fe, Ru, or Ni) have been interpreted by a deconvolution technique to yield ionic bond dissociation energies $D(C_5H_5M$—$C_5H_5)^+$ and $D(C_5H_5$—$M)^+$ in the range 6—7 and 4—5 eV, respectively; these differed from previous results.[389] Evidence supporting the new results was presented.

The spectra of 1,1-diacetyl-[390,391] and 1,1-dipropionyl ferrocene and some deuteriated and ^{57}Fe labelled analogues[390] have been discussed. Initial fragmentation of the pmi was loss of R or COR. There followed from $[\text{pm} - R]^+$ the loss of two CO groups when R = Me or Et, or one CO and one C_2H_4 group R = Et (Scheme 24). This apparently involves interaction between the ring

$$[C_5H_4FeC_5H_4CHO]^+ \xleftarrow[R\,=\,Et]{-CO-C_2H_4} [\text{pm} - R]^+ \xrightarrow[R\,=\,Me\,or\,Et]{-2CO} [C_5H_4FeC_5H_4R]$$

Scheme 24

systems, an unusual fragmentation in the spectra of ferrocene compounds.[390] A more usual rearrangement involves the migration of a group from the side-chain to the iron atom. The $[FeCH_3]^+$ ion from the compound with R = Me was observed to be formed in this way, but no ions with a cyclopentadienyl group still attached to Fe were found. This contrasts with findings in the spectra of mono- and 1,1'-bis-(acetoacetyl)ferrocene and 1,1'-ferrocene dicarboxylic

$$\begin{array}{c}^+\lceil COCH_2COCH_3\\Fe\\(C_5H_4COCH_2COCH_3)\end{array} \xleftarrow{-C_5H_4} [(C_5H_4COCH_2COCH_3)_2Fe]^+ \xrightarrow{-C_5H_4CO} \begin{array}{c}^+\lceil CH_2COCH_3\\Fe\\(C_5H_4COCH_2CH_3)\end{array}$$

Scheme 25

acid,[391] where rearranged ions commonly retained the cyclopentadienyl group from which the transferred group had not come, *e.g.* Scheme 25.

Numerous ferrocene derivatives and related compounds have been characterized, including ferrocenylcyclopropanes,[392,393] $(\pi$-$C_5H_5)Fe(\pi$-$C_5H_4CR_2CN)$,[394]

---

[387] T. J. Katz, N. Acton, and J. McGinnis, *J. Amer. Chem. Soc.*, 1972, **94**, 6025.
[388] T. J. Katz and N. Acton, *J. Amer. Chem. Soc.*, 1972, **94**, 3281.
[389] G. D. Flesch, G. A. Junk, and H. J. Svec, *J.C.S. Dalton*, 1972, 1102.
[390] C. F. Sheley and D. L. Fischel, *Org. Mass Spectrometry*, 1972, **6**, 1131.
[391] H. Imai, *Bull. Chem. Soc. Japan*, 1972, **45**, 1264.
[392] G. W. Gokel, J. P. Shepherd, W. P. Weber, H. G. Boettger, J. L. Holwich, and D. J. McAdoo, *J. Org. Chem.*, 1973, **38**, 1913.
[393] J. A. Connor and J. P. Lloyd, *J.C.S. Perkin I*, 1973, 17.
[394] G. Marr and J. Ronayne, *J. Organometallic Chem.*, 1973, **47**, 417.

(π-C$_5$H$_5$)Fe{π-C$_5$H$_4$CH(OR)CN},[395] 1,2-bis(ferrocenyl)ethylene,[396] 2,3-ferrocenocyclopentanone,[397] and several alkoxy-substituted polychloro-ferrocenes and (π-C$_5$Cl$_4$I)$_2$Ru.[398] A number of ferrocenophanes and related compounds[399] and polyferrocenyl compounds have been reported.[400]

A study of cyclohexadienyl cyclopentadienyliron (79) partly deuteriated in the *exo*-position (X) indicates that H transfer to the iron atom is highly specific (*i.e.* from the *endo*-position, Scheme 26).[401] This lends support to a previous postulate

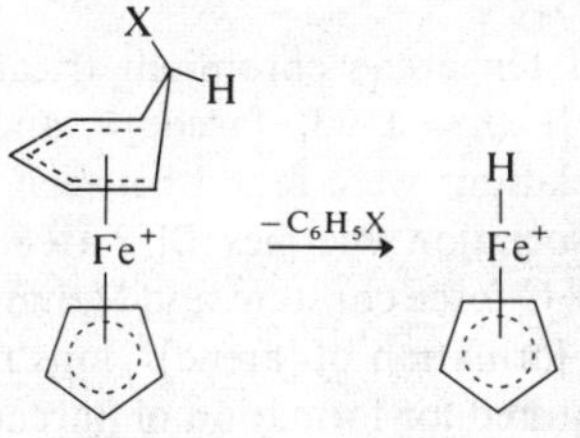

(79)

Scheme 26

that the loss of *endo*-H$_2$ from cyclohexadiene-Fe(CO)$_3$ occurs through initial transfer of hydrogen to the iron atom followed by cleavage of the iron hydride bonds.[402]

Arenes and Related Ligands. The spectrum of (π-C$_6$H$_6$)$_2$Mo differs from that of the chromium analogue in that in the former the pmi is base peak and no [C$_6$H$_6$M]$^+$ ion was observed.[403] Mean metal-(π-C$_6$H$_6$) bond dissociation energies, calculated using the appearance potential of [M]$^+$, were 196.5 and 246.6 kJ mol^{-1} for M = Cr and Mo, respectively. Apparently, metal–ligand bond strengths increase on substitution of the benzene ring by alkyl groups.[403,404] Bis(benzene)titanium shows pm, [pm − C$_6$H$_6$]$^+$, and [Ti]$^+$ ions.[405]

It has been suggested that the preference for loss of small stable molecules rather than radicals in the fragmentation of complexed π-organic ring systems can be

[395] T. Kondo, K. Yanomoto, and M. Kumada, *J. Organometallic Chem.*, 1973, 355.
[396] E. W. Neuse, *J. Organometallic Chem.*, 1973, **56**, 323.
[397] T. Shirafryi, A. Odaira, Y. Yamamoto, and H. Nozaki, *Bull. Chem. Soc. Japan*, 1972, **45**, 2884.
[398] F. L. Hedberg and H. Rosenberg, *J. Amer. Chem. Soc.*, 1973, **95**, 870.
[399] S. R. Carroll, J. L. Pflug, and J. A. Winstead, *Org. Mass Spectrometry*, 1972, **6**, 1279; M. Hisatome, S. Minagawa, and K. Yamakawa, *J. Organometallic Chem.*, 1973, **55**, C82; T. J. Katz, N. Acton and G. Martin, *J. Amer. Chem. Soc.*, 1973, **95**, 2935; S. Toma and M. Salisova, *J. Organometallic Chem.*, 1973, **57**, 191.
[400] E. W. Neuse, *J. Organometallic Chem.*, 1972, **40**, 387; E. W. Neuse and R. K. Crossland, *J. Organometallic Chem.*, 1972, **43**, 385.
[401] C. C. Lee, R. G. Sutherland, and B. J. Thompson, *Tetrahedron Letters*, 1972, 2625.
[402] T. H. Whitesides and R. W. Arhart, *Tetrahedron Letters*, 1972, 297.
[403] G. G. Devyatykh, P. E. Gaivoronskii, and N. V. Larin, *J. Gen. Chem. U.S.S.R.*, 1973, **43**, 1121.
[404] N. V. Larin, G. G. Devyatykh, and P. E. Gaivoronskii, *Russ. J. Inorg. Chem.*, 1972, **17**, 841.
[405] F. W. S. Benfield, M. L. H. Green, J. S. Ogden, and D. Young, *J.C.S. Chem. Comm.*, 1974, 866.

rationalized in terms of the activation energies of the various possible fragmentation paths from the pmi's.[406] With the example taken, (π-tetrahydronaphthalene)nonacarbonyltetracobalt, loss of one or more H_2 molecules but not H was found, whereas the 'free' hydrocarbon ligand shows abundant ions by loss of H. Appearance potential measurements supported the suggestion that H_2 loss was favoured, competing with loss of one or possibly two CO groups, and probably leading to the formation of new π-bonds in the hydrocarbon ligand. The base peak was [(naphthalene)Co_4]$^+$.

Ionization potentials of ten arene chromium tricarbonyls and appearance potentials of [pm – nCO]$^+$ (n = 1—3), [arene]$^+_\cdot$, and [Cr]$^+$ ions have been measured.[407] Good correlations were found between the first and second but not third ionic bond dissociation energies $D[(\pi\text{-arene})Cr(CO)_n - CO]^+$ (n = 0, 1, or 2), and both the C—O force constant and Hammett σ_p of the substituents on the benzene ring. The formation of [arene]$^+_\cdot$ ions required an energy intermediate between those required for formation of [(arene)Cr(CO)$_n$]$^+$ (n = 0 or 1) ions.

Some other (π-arene)M(CO)$_3$ (M = Cr, Mo, or W) derivatives for which data have been presented include σ-benzoanthracene chromium tricarbonyl compounds,[408] a triphenylboron derivative (π-Ph$_2$B—C$_6$H$_5$)Cr(CO)$_3$,[409] 1,3,5-triphenylbenzene derivatives (π-Ph$_3$C$_6$H$_3$){Cr(CO)$_3$}$_n$ (n = 1—3),[410] 2,6-diphenylpyridine, and 2,4,6-triphenylpyridine derivatives of chromium, molybdenum, and tungsten, such as (π-Ph$_3$C$_5$H$_2$N){Cr(CO)$_3$}$_n$ (n = 1—3),[411] and (π-2,4,6-triphenylphosphabenzene)M(CO)$_3$ (M = Cr or Mo).[412]

Other (π-arene) complexes which have been characterized include some bisallylic compounds containing the (π-C$_6$H$_6$)Mo group,[413] (π-C$_6$Me$_6$)Cr-(π-C$_5$Me$_5$),[414] (π-arene)Cr(PF$_3$)$_3$ compounds of C$_6$H$_6$, mesitylene and cumene,[415] (π-CH$_3$C$_6$H$_5$)Fe(PF$_3$)$_2$ and (π-CH$_3$C$_6$H$_5$)Fe(π-C$_4$H$_6$).[416] Among the complexes containing related ligands which have been examined are (7-exo-arylcycloheptatriene)chromium tricarbonyls,[417] bis(cycloheptatriene)vanadium which showed a pmi (ionization potential 6.79 ± 0.1 eV) as the base peak and fragmented by losing H, H_2, and C_2H_4,[418] heptafulvene complexes of chromium, molybdenum, and tungsten tricarbonyls,[419] and complexes with the (1-phenyl-

[406] G. Innorta, S. Pignataro, and G. Natile, *J. Organometallic Chem.*, 1974, **65**, 391.
[407] J. R. Gilbert, W. P. Leach, and J. R. Miller, *J. Organometallic Chem.*, 1973, **49**, 219.
[408] S. A. Gardner, R. J. Seyler, H. Veening, and B. R. Willeford, *J. Organometallic Chem.*, 1973, **60**, 271.
[409] J. Deberitz, K. Dirscherl, and H. Noth, *Chem. Ber.*, 1973, **106**, 2783.
[410] J. Deberitz and H. Noth, *J. Organometallic Chem.*, 1973, **55**, 153.
[411] J. Deberitz and H. Noth, *J. Organometallic Chem.*, 1973, **61**, 271.
[412] J. Deberitz and H. Noth, *Chem. Ber.*, 1973, **106**, 2222.
[413] M. L. H. Green, L. C. Mitchard, and W. E. Silverthorn, *J.C.S. Dalton*, 1973, 1952.
[414] H. Benn, G. Wilke, and D. Henneberg, *Angew. Chem. Internat. Edn.*, 1973, **12**, 1001.
[415] R. Middleton, J. R. Hull, S. R. Simpson, C. H. Tomlinson, and P. L. Timms, *J.C.S. Dalton*, 1973, 121.
[416] D. L. Williams-Smith, L. R. Wolf, and P. S. Skell, *J. Amer. Chem. Soc.*, 1972, **94**, 4042.
[417] M. I. Foreman, G. R. Knox, P. L. Pauson, K. H. Todd, and W. E. Watts, *J.C.S. Perkin II*, 1972, 1141.
[418] J. Muller and B. Mertschenk, *J. Organometallic Chem.*, 1972, **34**, C41.
[419] J. A. S. Howell, B. F. G. Johnson, and J. Lewis, *J.C.S. Dalton*, 1974, 293.

borinato) ligands (80)—(82).[420,421] Compounds (80) and (81) gave pmi's as base peak and fragmented initially by loss of H_2, then C_5H_6, C_6H_5, or C_6H_6 for R = Ph, but C_2H_2 and C_2H_4 for R = Me. The manganese compound showed

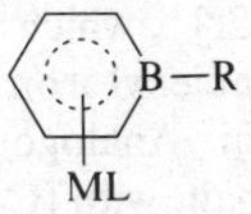

	ML	R
(80)	$(\pi-C_5H_5)$Co,	Me or Ph
(81)	$(\pi-C_5H_5BR)$Co,	Me or Ph
(82)	$(CO)_3$Mn,	Ph

a pmi, ions $[(C_5H_5BPh)Mn(CO)_n]^+$ ($n = 0$—2), and several rearranged ions such as $[C_6H_5Mn]^+$. A number of substituted borazine derivatives of chromium tricarbonyl have been characterized.[422]

Complexes containing C_7 and C_8 Rings. The spectra of the four derivatives $(\pi-C_5H_4R)(\pi-C_7H_6R)Ti$ (R = H or Me) have been listed.[423] Some fragmentation paths from the pmi's (base peaks) involving considerable interaction between the ring systems were observed, which could best be explained by assuming the transfer of a CH or CMe group from one ring to the other. The loss of C_6H_6 from $(\pi-C_5H_5)(\pi-C_7H_7)Ti$, both monomethyl derivatives, and the dimethyl compound (83) was observed (Scheme 27). A C_7H_8 (toluene?) group was lost from the pmi of the methyl derivatives, and C_8H_{10} (xylene?) from the dimethyl compound (Scheme 27). The elimination of C_2H_4 from the pmi's of both methylcycloheptatrienyl derivatives was also found.

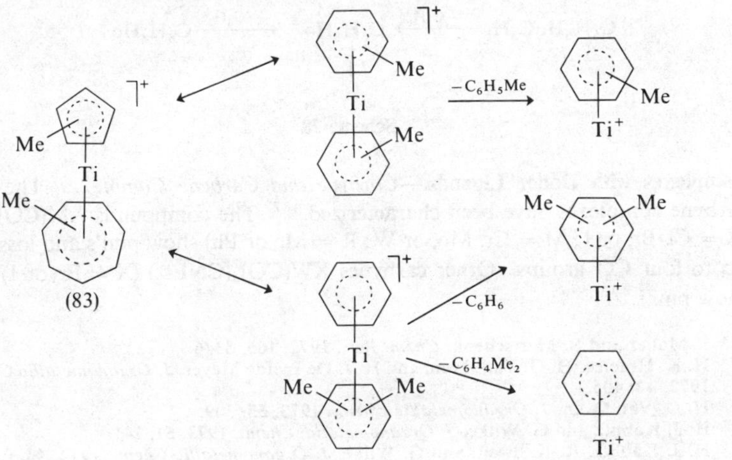

Scheme 27

---

[420] G. E. Herberich and G. Greiss, *Chem. Ber.*, 1972, **105**, 3413.
[421] G. E. Herberich and H.-J. Becker, *Angew. Chem. Internat. Edn.*, 1973, **12**, 764.
[422] J. L. Adcock and J. J. Lagowski, *Inorg. Chem.*, 1973, **12**, 2533.
[423] H. T. Verkouw and H. O. Van Oven, *J. Organometallic Chem.*, 1973, **59**, 259.

Cycloheptatrienylcycloheptadienyltitanium has a pmi as base peak and fragments initially by loss of H_2, C_7H_8, C_6H_6, and CH_3; dicycloheptatrienevanadium is similar and presumably ring interactions during fragmentation occur for both.[424] The pmi from cycloheptatrienylcyclohepta-1,3-dienechromium initially loses C_7H_{10} to give the base peak, $[C_7H_7Cr]^+$, which further decomposes by loss of C_7H_7, C_2H_2, or Cr. Dicycloheptadienyl iron produces both $[FeC_7H_8]^+$ (base peak) and $[C_7H_7Fe]^+$ from the pmi. Analogous ions are also found from cycloheptadienylcyclohepta-1,3-dienecobalt, with $[C_7H_7Co]^+$ being the base peak.

After losing ethylene from the pmi (base peak) of 1-methylallyl cyclo-octatetraene titanium, further fragmentation was by loss of C_2H_3 and C_2H_2 groups.[425] Cyclo-octatetraene tetraphenylcyclobutadiene titanium shows a pmi base peak and fragments by initially losing C_8H_{10}, C_2Ph_2, and C_2Ph_4 groups.[426] Other cyclo-octatetraene (cot) derivatives which have been characterized include $(cot)_2Zr$ and $(cot)ZrL$ (L = butadiene, $allyl_2$, or Cl_2),[427] $(cot)_2Hf$, and $(cot)HfCl_2$.[428]

Lanthanide derivatives, $(\pi\text{-}C_5H_5)M(\pi\text{-}C_8H_8)$ (M = Y, Nd, Sm, Ho, and Er), show pmi's and fragment by losing various hydrocarbon groups (Scheme 28).[429] The characterization of some derivatives $(\pi\text{-}C_8H_7R)_2U$ including (R = Et, Bu, Ph, or vinyl)[430] has been reported. Bis(cyclo-octatetraenyl)thorium gives a pmi.[431]

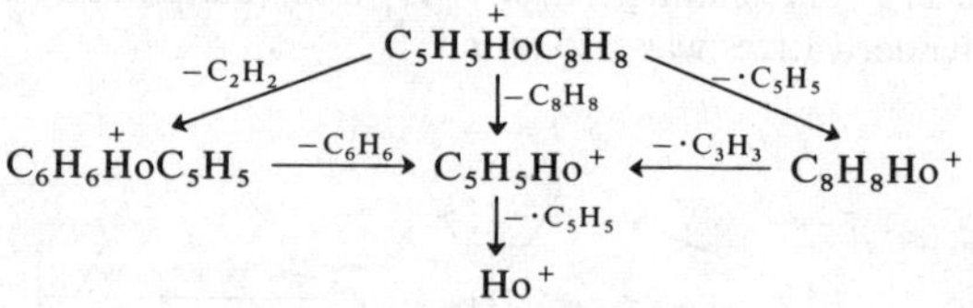

$$C_6H_6HoC_5H_5 \xrightarrow{-C_2H_2} C_5H_5HoC_8H_8 \xrightarrow{-C_5H_5} $$
$$\xrightarrow{-C_6H_6} C_5H_5Ho^+ \xleftarrow{-C_3H_3} C_8H_8Ho^+$$
$$\downarrow -\cdot C_5H_5$$
$$Ho^+$$

Scheme 28

Complexes with Donor Ligands.—*Carbyne and Carbene Complexes.* The first carbyne complexes have been characterized.[432] The compounds $XM(CO)_4CR$ (X = Cl, Br, or I; M = Cr, Mo, or W; R = Me or Ph) show pmi's and losses of up to four CO groups. Other carbynes $XW(CO)_4(CNEt_2)$ (X = Br or I) also show pmi's.[433]

---

[424] J. Muller and B. Mertschenk, *Chem. Ber.*, 1972, **105**, 3346.
[425] H. K. Hofetee, H. O. Van Oven, and H. J. De Liefde Meyer, *J. Organometallic Chem.*, 1972, **42**, 405.
[426] H. O. Van Oven, *J. Organometallic Chem.*, 1973, **55**, 309.
[427] H.-J. Kablitz and G. Wilke, *J. Organometallic Chem.*, 1973, **51**, 241.
[428] H.-J. Kablitz, R. Kallweit, and G. Wilke, *J. Organometallic Chem.*, 1972, **44**, C49.
[429] J. D. Jamerson, A. P. Masino, and J. Takats, *J. Organometallic Chem.*, 1974, **65**, C33.
[430] A. Streitwieser and C. A. Harmon, *Inorg. Chem.*, 1973, **12**, 1102.
[431] J. Goffart, J. Fuger, B. Gilbert, B. Kanellakopulos, and G. Duyckaerts, *Inorg. Nuclear Chem. Letters*, 1973, **8**, 403.
[432] E. O. Fischer, G. Kreis, C. G. Kreiter, J. Muller, G. Huttner, and H. Lorenz, *Angew. Chem. Internat. Edn.*, 1973, **12**, 564.
[433] E. O. Fischer, G. Kreis, F. R. Kreissl, W. Kalbfus, and E. Winkler, *J. Organometallic Chem.*, 1974, **65**, C53.

Carbene complexes to have been characterized include diphenylcarbene compounds of rhodium,[434] compounds containing a C(OX)R group (X = H or alkyl, R = alkyl or aryl group)[435] and phosphine adducts of such compounds,[436] and those with a C(NR^{1}R^2)R^3 group.[437] Other compounds have contained carbene ligands of the types C(NBu^{t}H)NR^{1}R^2,[438] C(OEt)NR^{1}R^2,[439] and C(ferrocenyl)X (X = OR or NR^{1}R^2).[440] Carbene derivatives of dimanganese-[441,442] and dirhenium-decacarbonyls,[442] (CO)$_9$MnRe{C(OMe)Ph},[442] and (CO)$_9$MnRe{C(OMe)Me}[443], have been reported. Evidence from i.r. and n.m.r. spectra showed the last-mentioned compound to have the carbene bonded to Mn, but no confirmative evidence could be found in the mass spectrum.

Ionization potentials have been reported for (CO)$_5$CrC(XMe)Me (X = O, S, Se, and NH),[444] and (CO)$_4$FeL [L = C(NHMe)$_2$, $\overline{\text{CN(Me)CH}}$=CHN(Me), and $\overline{\text{CN(Me)N}}$=NN(Me)].[445]

Nitrogen Ligands. Lack of pmi's and abundant ions derived from the amine ligand were observed with amine complexes of Pd, Pt,[446] and Ti.[447] Several unusual complexes with the bis(trimethylsilyl)amido ligand have been reported, including M{N(SiMe$_3$)$_2$}$_3$ (M = Sc, Ti, V, Cr, Fe,[448] and some lanthanides[449]), and the compounds M{N(SiMe$_3$)$_2$}$_2$(L) (M = Co or Zn, L = pyridine;[450] M = Co, L = Ph$_3$P[451]), and ThCl{N(SiMe$_3$)$_2$}$_3$.[452] Usually, strong pmi's were observed except for the zinc compound. The base peaks from the vanadium and titanium compounds were the pmi's; the other first-row transition metals showed [pm − HN(SiMe$_3$)$_2$]$^+$ as base peak. Fragmentation was initially by loss of Me or HN(SiMe$_3$)$_2$ groups, and then by loss of SiMe$_4$, NSiMe$_3$, and CH$_4$

---

[434] P. Hong, N. Mishii, K. Sonogashira, and N. Hagihara, *J.C.S. Chem. Comm.*, 1972, 993.
[435] J. A. Connor and E. M. Jones, *J.C.S. Dalton*, 1973, 2119; C. H. Game, M. Green, J. R. Moss, and F. G. A. Stone, *J.C.S. Dalton*, 1974, 351; J. R. Moss, M. Green, and F. G. A. Stone, *J.C.S. Dalton*, 1973, 975.
[436] F. R. Kreissl, E. O. Fischer, C. G. Kreiter, and H. Fischer, *Chem. Ber.*, 1973, **106**, 1262.
[437] E. O. Fischer and M. Leupold, *Chem. Ber.*, 1972, **105**, 599; J. A. Connor, P. D. Rose, and R. M. Turner, *J. Organometallic Chem.*, 1973, **55**, 111; K. Weiss and E. O. Fischer, *Chem. Ber.*, 1973, **106**, 1277.
[438] C. H. Davies, C. H. Game, M. Green, and F. G. A. Stone, *J.C.S. Dalton*, 1974, 357.
[439] E. O. Fischer, F. R. Kreissl, E. Winkler, and C. G. Kreiter, *Chem. Ber.*, 1972, **105**, 588.
[440] J. A. Connor and J. P. Lloyd, *J.C.S. Dalton*, 1972, 1470.
[441] C. P. Casey, R. A. Boggs, and R. L. Anderson, *J. Amer. Chem. Soc.*, 1972, **94**, 8947.
[442] E. O. Fischer, E. Offhaus, J. Muller, and D. Notke, *Chem. Ber.*, 1972, **105**, 3027.
[443] C. P. Casey and C. P. Cyr, *J. Organometallic Chem.*, 1973, **57**, C69.
[444] E. O. Fischer, G. Kreis, F. R. Kreissl, C. G. Kreiter, and J. Muller, *Chem. Ber.*, 1973, **106**, 3910.
[445] K. Ofele and C. G. Kreiter, *Chem. Ber.*, 1972, **105**, 329.
[446] I. Jardine and F. J. McQuillin, *Tetrahedron Letters*, 1972, 459.
[447] M. L. H. Green and C. R. Lucas, *J.C.S. Dalton*, 1972, 1000.
[448] E. C. Alyea, D. C. Bradley, and R. G. Copperthwaite, *J.C.S. Dalton*, 1972, 1580.
[449] D. C. Bradley, J. S. Ghotra, and F. A. Hart, *J.C.S. Dalton*, 1973, 1021.
[450] K. J. Fisher, *Inorg. Nuclear Chem. Letters*, 1973, **9**, 921.
[451] D. C. Bradley, M. B. Hursthouse, R. J. Smallwood, and A. J. Welch, *J.C.S. Chem. Comm.*, 1972, 872.
[452] D. C. Bradley, J. S. Ghotra, and F. A. Hart, *Inorg. Nuclear Chem. Letters*, 1974, **10**, 209.

species. A number of doubly-charged ions of relatively high abundance were noted.

Other compounds which have been investigated include $\{(CO)_5Cr\}_2N_2H_2$,[453] $\{(\pi\text{-}C_5H_5)Mn(CO)_2\}_2N_2H_2$,[454] $WCl_2\{(Ph_2PCH_2)_2\}_2N_2H(COR)$,[455] $(\pi\text{-}C_5H_5)\text{-}Re(CO)_2N_2$,[456] $Re(CO)_5NSOF_2$ and $Re(CO)_4(NSOF_2)_2$,[457] and $Re_2(CO)_8\text{-}(NCO)_2$.[458] Tetracarbonyl complexes of Cr, Mo, or W containing pyridine-2-carbaldehyde imine ligands (84) typically lost up to four CO groups before losing R and then HCN and pyridine to give $[M]^{+}$.[459] The spectra of some related cobalt complexes have been discussed.[460]

(84) (85)

Fragmentation of μ-(3,6-diphenylpyridazine)hexacarbonyldi-iron (85) from the pmi is by loss of up to six CO groups and then N_2 to give $[C_{16}H_{12}Fe_2]^{+}$.[461] Similar compounds have been characterized,[462] also some diazepine derivatives,[463] and some aza-allyl complexes of molybdenum and tungsten.[464]

Phosphorus Ligands. Two diphosphorane complexes, $\{(CO)_5Cr\}_2P_2H_4$ and $\{(\pi\text{-}C_5H_5)Mn(CO)_2\}_2P_2H_4$, have been reported.[465]

Characterization of the trialkylphosphine complexes (cyclohexyl)$_3$PAuX (X = Cl, Br, or I),[466] benzyldimethylphosphine complexes of Mn, Fe, Ru, and Rh,[467] and $(\pi\text{-}C_5H_5)Ni\text{-}(\mu\text{-}CO)_2Fe(\pi\text{-}C_5H_5)PR_3$ compounds has been reported. Evidence has been found for H/D exchange in the fragmentation of Os(D)Cl(CO)(cyclohexyl)$_3$P.[468,469] The spectra of 2,4,6-triphenylphosphorine complexes of Cr, Mo, and W pentacarbonyls show pmi's and loss of up to five

---

[453] D. Sellmann, A. Brandl, and R. Endall, *J. Organometallic Chem.*, 1973, **49**, C23.
[454] D. Sellmann, *J. Organometallic Chem.*, 1972, **44**, C46.
[455] J. Chatt, G. A. Heath, and G. J. Leigh, *J.C.S. Chem. Comm.*, 1972, 444.
[456] D. Sellmann, *J. Organometallic Chem.*, 1972, **36**, C27.
[457] R. Mews and O. Glemser, *J.C.S. Chem. Comm.*, 1973, 823.
[458] R. B. Saillant, *J. Organometallic Chem.*, 1972, **39**, C71.
[459] H. Brunner and W. A. Herrmann, *Chem. Ber.*, 1972, **105**, 770; *J. Organometallic Chem.*, 1973, **57**, 183.
[460] H. Brunner and W. Rambold, *J. Organometallic Chem.*, 1974, **64**, 373.
[461] H. A. Patel, A. J. Carty, M. Mathew, and G. J. Palenik, *J.C.S. Chem. Comm.*, 1972, 810.
[462] H. Alper, *Inorg. Chem.*, 1972, **11**, 976.
[463] D. P. Madden, A. J. Carty, and T. Birchall, *Inorg. Chem.*, 1972, **11**, 1453.
[464] H. R. Keable and M. Kilner, *J.C.S. Dalton*, 1972, **153**, 1535.
[465] D. Sellmann, *Angew. Chem. Internat. Edn.*, 1973, **12**, 1020.
[466] J. Bailey, *J. Inorg. Nuclear Chem.*, 1973, **35**, 1921.
[467] R. L. Bennett, M. I. Bruce, and F. G. A. Stone, *J. Organometallic Chem.*, 1972, **38**, 325.
[468] K. Yasufuku and H. Yamazaki, *J. Organometallic Chem.*, 1972, **38**, 367.
[469] F. G. Moers and J. P. Langhout, *Rec. Trav. chim.*, 1972, **91**, 591.

CO groups before fragmentation of the phosphorine ligand.[470] This sequence is also general for carbonyl phosphine complexes, e.g. in the spectra of π-C_7H_8Cr-$(CO)_2PR_3$,[471] $\{Cr(CO)_2PR_3\}_2$,[472] $(CO)_5M(R_2PCH_2)_2$ (M = Cr, Mo, or W; R = Me or Ph),[473] and some $PtCl_2(CO)PR_3$ compounds.[474]

The spectra of $M(CO)_5PR_3$ complexes (M = Cr or W; R = OMe, OEt, OBu, Bu, or Ph) have been listed and ionization and appearance potentials of the major ions measured.[475] The chromium complexes and the tungsten compounds with $P(OMe)_3$, $P(OEt)_3$, or PPh_3 ligands fragment mainly by losing CO groups. The spectra of the tungsten compounds with $P(OBu)_3$ or PBu_3 ligands, however, are dominated by ions produced by partial fragmentations of the phosphorus ligands which, apart from simple P—OR or P—R cleavage, are very different from the fragmentations of the free ligand. Behaviour of the first kind can be related to the higher activation energy required for the cleavage of P—X compared with M—CO bonds. The appearance of the second kind of fragmentation behaviour is due to a combination of energetic factors, the length of the R chain and, perhaps, the increased use of the heptaco-ordination capability of the tungsten atom.

Nineteen monosubstituted compounds $(CO)_5ML$ with M = Cr, Mo, or W and various ligands, L = PPh_3, $AsPh_3$, $SbPh_3$, $P(OPh)_3$, $Ph_2PC_6F_5$, $(p-FC_6H_4)_3P$, $(m-MeC_6H_4)_3P$, pyridine, $H_2NC_6H_{11}$, have been examined.[476] All complexes gave pmi's, and ions due to CO losses. Base peaks were generally $[ML]^+$ or $[M]^+$ ions. An unsuccessful attempt was made to relate major features in the spectra to Graham σ and π parameters. A feature of these complexes' spectra was several eliminations of two CO molecules, [Scheme 29(a)], supported by the presence of metastable ions.

$$[(CO)_5CrL]^+ \xrightarrow{-2CO} [(CO)_3CrL]^+ \quad L = AsPh_3 \text{ or } P(OPh)_3 \quad (a)$$

$$[\{(Me_2N)_3P\}_2Fe(CO)_3]^+ \xrightarrow{-2CO} [\{(Me_2N)_3P\}_2FeCO]^+ \quad (b)$$

Scheme 29

On the basis of metastable peaks and clastograms, similar reactions were postulated in the fragmentations of $\{(Me_2N)_3P\}_2Fe(CO)_3$ [Scheme 29(b)].[477] The spectra of this compound and several other tris(dimethylamino)phosphine derivatives, including $(Me_2N)_3PM(CO)_5$ and $\{(Me_2N)_3P\}_2M(CO)_4$ (M = Cr, Mo, or W), were studied and ionization and appearance potential data of major

---

[470] J. Deberitz and H. Noth, *J. Organometallic Chem.*, 1973, **49**, 453.
[471] W. P. Anderson, W. G. Blenderman, and K. A. Drews, *J. Organometallic Chem.*, 1972, **42**, 139.
[472] J. A. Bowden and R. Colton, *Austral. J. Chem.*, 1973, **26**, 43.
[473] J. A. Connor, J. P. Day, E. M. Jones, and G. K. McEwen, *J.C.S. Dalton*, 1973, 347.
[474] K. L. Klassen and N. V. Duffy, *J. Inorg. Nuclear Chem.*, 1973, **35**, 2602.
[475] S. Torroni, G. Innorta, A. Foffani, and G. Distefano, *J. Organometallic Chem.*, 1974, **65**, 209.
[476] S. T. Band and N. V. Duffy, *J. Inorg. Nuclear Chem.*, 1973, **35**, 3241.
[477] F. E. Saalfeld, J. J. DeCorpo, and M. V. McDowell, *J. Organometallic Chem.*, 1972, **44**, 333.

ions presented. From the latter information the average bond dissociation energies of the M—CO bonds were estimated.

A number of cis-disubstituted chromium, molybdenum, and tungsten carbonyls with ligands containing As, S, or N donor atoms have been studied.[478] They include cis-$L_2M(CO)_4$ [L = o-phenylenebis(dimethylarsine), 1,2-bis-(diphenylarsino)ethane, and 2,2,7,7-tetramethyl-3,6-dithiaoctane for M = Cr, Mo, or W] and several 2,5-dithiahexane, 2,2,8,8-tetramethyl-3,7-dithianonane, $NNN'N'$-tetramethylethylenediamine and $NNN'N'$-tetramethyl-1,3-diaminopropane complexes.

The spectra of the o-phenylenebis(dimethylarsine) complexes were examined in greater detail.[478] They showed competitive losses of CO and Me to give series of ions $[pm - n(CO)]^+$, $[pm - Me - n(CO)]^+$, and $[pm - 2Me - n(CO)]^+$. From complexes containing ethylene-bridged bidentate ligands loss of five groups of mass 28, corresponding to four CO molecules and one ethylene, were observed.

Derivatives with bridging phosphido groups which have been characterized include $(\pi-C_5H_5)(CO)_2Fe-\mu(PMe_2)Ni(CO)_3$, which lost five CO groups and then two Me radicals from the pmi to give $[C_5H_5FePNi]^+$ as base peak,[479] and $(\pi-C_5H_5)_2Ti(PPh)_3$, which initially fragmented by losing PPh from the pmi.[480]

The trifluorophosphine complexes of Cr, Mo, and W, $M(PF_3)_6$, fragment mainly by loss of PF_3 and F to give $[M(PF_3)_n]^+$ and $[M(PF_3)_nPF_2]^+$ ions ($n = 0$—5). The heavier metals also gave ions produced by loss of F_2 and PF_2 groups.[481]

Several $(\pi$-olefin$)Fe(PF_3)_4$ complexes have been examined.[482] As well as the series $[(\text{olefin})Fe(PF_3)_n]^+$ ($n = 0$—4) and $[(\text{olefin})Fe(PF_3)_nPF_2]^+$ ($n = 1$ or 2), the rearranged ions $[CNFe(PF_3)_n]^+$ ($n = 0$ or 1) were observed from the CN substituted olefins, $H_2C=CH(CN)$ and $H(Me)C=CH(CN)$. A series of $(\pi$-allyl$)Co(PF_3)_3$ complexes showed pm and fragment ions due to losses of up to three PF_3 groups and $[Co(PF_3)_n]^+$ ($n = 1$ or 2).[483] Apparently some similar rhodium complexes gave only $[pm - PF_3]^+$ as the ion of greatest mass.[484] Other π-allyl compounds reported include $(\pi-C_3H_5)Fe(PF_3)_2X$ (X = Br or I), which fragmented mainly by losing PF_3, F, and X and, to a lesser extent, H, H_2, and π-allyl groups.[485] The complex $(\pi-C_5H_6)Fe(PF_3)_3$ fragments by loss of PF_3 and F as well as H to form the ions $[(C_5H_5)Fe(PF_3)_n]^+$ ($n = 0$—2);[486] the base peak was $[C_5H_6Fe]^+$. The related compound, $(\pi-C_5H_5)Fe(PF_3)_2H$, also lost PF_3, H, and F but produced a strikingly different spectrum with a highly abundant pmi and $[C_5H_5Fe]^+$ as base peak.

---

[478] R. A. Brown, J. R. Parson, and G. R. Dobson, *Org. Mass Spectrometry*, 1973, **7**, 1059.
[479] W. Ehrl and H. Vahrenkamp, *J. Organometallic Chem.*, 1973, **63**, 389.
[480] K. Issleib, G. Wille, and F. Krech, *Angew. Chem. Internat. Edn.*, 1972, **11**, 527.
[481] K. M. Lee and R. E. Hester, *J. Organometallic Chem.*, 1973, **57**, 169.
[482] T. Kruck and L. Knoll, *Chem. Ber.*, 1973, **106**, 3578.
[483] T. Kruck, G. Sylvester, and I.-P. Kunau, *Z. Naturforsch.*, 1973, **28b**, 28.
[484] M. A. Cairns and J. F. Nixon, *J. Organometallic Chem.*, 1973, **51**, C27.
[485] T. Kruck and L. Knoll, *Z. Naturforsch*, 1973, **28b**, 34.
[486] T. Kruck and L. Knoll, *Chem. Ber.*, 1972, **105**, 3783.

Other PF_3 and related complexes for which data have been reported include $(\pi\text{-}C_5Me_5)M(PF_3)_2$ (M = Rh or Ir),[487] $\{RhCl(PF_3)L\}_2$ (L = CO or C_2H_4),[488] $(\pi\text{-}C_5H_8)Fe(PF_3)_3$, $(PF_3)_3Fe\text{-}\mu\text{-}PF_2\text{-}\mu\text{-}XFe(PF_3)_3$ (X = I or SEt),[489] $Fe(CO)_4\text{-}PF_2\text{-}\mu\text{-}O\text{-}PF_2Fe(CO)_4$,[490] $(\pi\text{-}C_5H_5)Mo(CO)(PF_2NMe_2)_2Cl$, $(\pi\text{-}C_5H_5)Fe(CO)\text{-}(COCH_3)(PF_2NC_5H_{10})$, and $(\pi\text{-}C_5H_5)Fe(PF_2NC_5H_{10})_2X$ (X = Br or I),[491] and $Rh_2X_2(PF_2NR_2)$ (X = Cl or I).[492] Both positive- and negative-ion spectra of $H_2Fe(PF_3)_4$ have been reported.[489] The pmi was observed only in the negative-ion spectrum; the series $[HFe(PF_3)_n]^+$ (n = 0—4) was found in the positive-ion spectrum.

The tetrafluorophosphorane complex, $Fe_2(CO)_6P_2F_4$, gave pm and fragment ions for loss of up to six CO molecules.[493] Further decomposition was by loss of F_2, F_2PFe, and PFe.

Initial fragmentation from pmi's of $(\pi\text{-}C_5H_5)NiP(OR)_3\{PO(OR)_2\}$, (R = Me or Et), was by elimination of $P(OR)_3$, $PO(OR)_2$ or OR.[494] Further decompositions occurred mainly from the remaining phosphorus-containing species. Other phosphite complexes and related compounds studied include $(OMe)_3\text{-}PAuNO_3$,[495] and $\{MeNi(PMe_3)(PMe_2O_2)\}_2$.[496] Complexes with $Me_2P(SH)$ as a ligand have included $(CO)_5MP(SH)Me_2$ (M = Cr or Mo)[497] and $(CO)_4\text{-}Mn(Br)P(SH)Me_2$ which lost HBr from the pmi and exhibited the series of ions $[pm - n(CO)]^+$ (n = 1—4) and $[Me_2PSMn(CO)_n]^+$ (n = 0—3).[498]

Some complexes containing trifluoromethylphosphine ligands have been studied. These include $\{(CF_3)_2PX\}_2Mo(CO)_4$ and $\{(CF_3)PX_2\}_2Mo(CO)_4$ (X = Cl, Br, or H),[499] $\{(CF_3)_2P_2C_2(CF_3)_2\}Fe(CO)_4$ and $(CF_3P)_4Fe_2(CO)_6$,[500] and $(\pi\text{-}C_5H_5)Fe(CO)_2(S)P(CF_3)_2$ and $(\pi\text{-}C_5H_5)_2Fe_3(CO)_2\{(O)P(CF_3)_2\}_4$.[501] The last compound initially fragmented by loss of CO followed by loss of $Fe\{(O)P(CF_3)_2\}_2$ and then the second CO group.

Pseudo-halide containing phosphine complexes $(\pi\text{-}C_5H_5)Mn(CO)_2(PhPX_2)$, (X = CN, NCO, NCS, and N_3), all show pmi's except for X = N_3, and $[pm - X]^+$ except for X = CN.[502] Further decomposition was usually by loss of CO or X. Rearranged ions from, for example, the compound with X = NCO, $[C_5H_5MnC_6H_5]^+$ and $[C_5H_5MnNCO]^+$, were common except from the dicyanophenylphosphine compound. Losses of CO, EtOH, and HNCO were

[487] R. B. King and A. Efraty, *J. Amer. Chem. Soc.*, 1972, **94**, 3768.
[488] M. A. Bennett and T. W. Turney, *Austral. J. Chem.*, 1973, **26**, 2321.
[489] T. Kruck and R. Kobelt, *Chem. Ber.*, 1972, **105**, 3765.
[490] W. M. Douglas, R. B. Johannesen, and J. K. Ruff, *Inorg. Chem.*, 1974, **13**, 371.
[491] R. B. King, W. C. Zipperer, and M. Ishaq, *Inorg. Chem.*, 1972, **11**, 1361.
[492] M. A. Bennett and T. W. Turney, *Austral. J. Chem.*, 1973, **26**, 2335.
[493] W. M. Douglas and J. K. Ruff, *Inorg. Chem.*, 1972, **11**, 901.
[494] V. Harder and H. Werner, *Helv. Chim. Acta*, 1973, **56**, 1621.
[495] H. Schmidbaur and R. Franke, *Chem. Ber.*, 1972, **105**, 2985.
[496] H. H. Karsch, H.-F. Klein, and H. Schmidbaur, *Chem. Ber.*, 1974, **107**, 93.
[497] E. Lindner and W.-P. Meier, *J. Organometallic Chem.*, 1973, **51**, C15.
[498] E. Lindner and H. Dreher, *J. Organometallic Chem.*, 1973, **55**, 347.
[499] J. F. Nixon and J. R. Swain, *J.C.S. Dalton*, 1972, 1038.
[500] A. H. Cowley and K. E. Hill, *Inorg. Chem.*, 1973, **12**, 1446.
[501] R. C. Dobbie and P. R. Mason, *J.C.S. Dalton*, 1973, 1125.
[502] M. Hofler and M. Schmitzler, *Chem. Ber.*, 1974, **107**, 194.

observed in the spectrum of $(\pi\text{-}C_5H_5)Mn(CO)_2\{PhP(NHCO_2Et)_2\}$.[502] Some cyclopentadienylmanganese dicarbonyl complexes of the ligands $PhPX_2$ (X = F, Cl, Br, I, or NEt_2) and PhPXY (X = F, Y = Cl or NEt_2) have been examined,[503] as has $(P_3N_3Cl_6)Cr(CO)_3$.[504]

Arsenic and Antimony Ligands. Iron and ruthenium complexes containing $(PhCH_2AsMe_2)$ ligands[467] and the tetracarbonyltungsten complex with o-styryl-dimethylarsine[505] have been reported. Tetraphenylarsole complexes of Mn and Re, $(Ph_4C_4As)M(CO)_5$, show pmi's and losses of up to five CO groups.[506] Complexes containing bidentate diarsine ligands such as $(CO)_4Cr\{Me_2AsCH_2\text{-}CHRAsMe_2\}$,[507] and $(CO)_{10}Fe_3\{Me_2AsCF_2CXYAsMe_2\}$[508] characteristically show pmi's and losses of CO groups.

Complexes with bridging $AsMe_2$ groups have been reported, including $(CO)_4Cr\{(AsMe_2)Mn(CO)_5\}_2$ and $\{(CO)_4MnAsMe_2\}_2$,[509] $(CO)_4Fe(AsMe_2)ML$ [ML = $Mn(CO)_5$,[510] $(\pi\text{-}C_5H_5)Mo(CO)_2$, $(\pi\text{-}C_5H_5)W(CO)_2$ and $Co(CO)_3$],[511] and $\{(CO)_5W(AsMe_2)\}X$ (X = O or NMe).[512] The carbonyl compounds generally fragment by losing from one to all CO groups and then Me to give abundant $[M - As - M']^+$ ions.

Main Group VI Ligands. Aspects of the spectra of $\{(\pi\text{-}C_5H_5)Mo(NO)Br(SR)\}_2$, $\{(\pi\text{-}C_5H_5)Mo(NO)SR_2\}_2$, and $\{(\pi\text{-}C_5H_5)Mo(NO)SR\}_2$ (R = various alkyl groups) have been discussed.[513] Compounds of the first type do not show pmi's and initially lose one and then later two R groups, followed by losses of one or both nitrosyls, and sulphur. Although $[C_5H_5MoNOBrSR]^+$ ions were found, no ions containing Br or SR alone were observed. Other compounds for which data have been presented include SCF_3 complexes of iron, molybdenum, and tungsten,[514] and related compounds,[515] some bis(trifluoromethyl)dithiolen-tricarbonyliron derivatives,[516] $\{Fe(CO)_3SeR\}_2$ (R = Me, Et, Pr^i, CF_3, or C_2F_5) and $\{(\pi\text{-}C_5H_5)Fe(CO)SeR\}_2$ (R = Et or Pr^n),[517] and $S_2\{Fe(CO)_3\}_3$.[342] The thiol-bridged complexes, $(CO)_4Mn(\mu\text{-}SH)_2\text{-}Mn(CO)_4$,[518] $(CO)_5Cr\text{-}\mu\text{-}SMe\text{-}M(CO)_n(\pi\text{-}C_5H_5)$, $[M(CO)_n = Fe(CO)_2$ and $W(CO)_3]$[519] afford pm and fragment

---

[503] M. Hofler and M. Schmitzler, *Chem. Ber.*, 1972, **105**, 1133.
[504] N. K. Hota and K. O. Harris, *J.C.S. Chem. Comm.*, 1972, 407.
[505] M. A. Bennett and I. B. Tomkins, *J. Organometallic Chem.*, 1973, **51**, 289.
[506] E. W. Abel, I. W. Nowell, A. G. J. Modinos, and C. Towers, *J.C.S. Chem. Comm.*, 1973, 258.
[507] W. R. Cullen, L. D. Hall, and J. E. H. Ward, *J. Amer. Chem. Soc.*, 1972, **94**, 5702.
[508] L. S. Chia, W. R. Cullen, J. R. Sams, and J. E. H. Ward, *Canad. J. Chem.*, 1973, **51**, 3223.
[509] W. Ehrl, R. Rinck, and H. Vahrenkamp, *J. Organometallic Chem.*, 1973, **56**, 285.
[510] W. Ehrl and H. Vahrenkamp, *Chem. Ber.*, 1973, **106**, 2556.
[511] W. Ehrl and H. Vahrenkamp, *Chem. Ber.*, 1973, **106**, 2563.
[512] H. Vahrenkamp, *Chem. Ber.*, 1972, **105**, 3574.
[513] J. A. McCleverty and D. Seddon, *J.C.S. Dalton*, 1972, 2588.
[514] J. L. Davidson and D. W. A. Sharp, *J.C.S. Dalton*, 1973, 1957.
[515] H. Alper and A. S. K. Chan, *J. Amer. Chem. Soc.*, 1973, **95**, 4905.
[516] C. J. Jones, J. A. McCleverty, and D. G. Orchard, *J.C.S. Dalton*, 1972, 1109.
[517] H. Rosenbuch and N. Welcman, *J.C.S. Dalton*, 1972, 1963.
[518] W. Beck, W. Danzer, and R. Hofer, *Angew. Chem. Internat. Edn.*, 1973, **12**, 77.
[519] W. Ehrl and H. Vahrenkamp, *Z. Naturforsch.*, 1973, **28b**, 365.

ions from loss of all CO groups. In the case of the tungsten derivative, loss of Me was observed as a process competitive with elimination of CO.

The spectra of nickel, palladium, and platinum complexes, $M(S_2PX_2)_2$ (X = Me, Ph, or F), and the related $\{S_2P(CF_3)_2\}_2M$ complexes of nickel and palladium have been examined.[520] For all compounds the pmi's were base peaks and initial fragmentation was by loss of X, followed by decomposition from the ligands to give ions of the type $[MS_nPX_m]^+$ ($n = 1$—3, $m < 4$). Some rearranged ions including $[MS_4P_2(CF_3)_2F]^+$ and $[MS_nPF]^+$ ($n = 0$—3) were observed from trifluoromethyl derivatives. A major fragmentation of the pmi of a platinum compound, $Pt_2S_6P_4(CF_3)_8$, gave the ion $[PtS_3P_2(CF_3)_4]^+$, corresponding to half the molecular weight. Aspects have been reported of the spectra of a number of derivatives, $M(S_2PX_2)_2$ (M = Mn, Fe, Co, Zn, Cd, or Hg; X = F or CF_3), and $M(S_2PX_2)_3$ (M = Fe or Co; X = F, CF_3, Me, or Ph),[521] $VO(S_2PX_2)_2$ (X = F, CF_3, Me, OEt, or Ph),[522] $Cl_3M(S_2PF_2)$ (M = Nb or Ta),[523] $Mo(S_2PF_2)_3$-$OMo(S_2PF_2)_2$ and related tungsten compounds,[524] $OV(S_2AsMe_2)_2$,[525] and $(Me_2AsS_2)M(CO)_4$ (M = Mn or Re).[526]

Other compounds containing Main Group VI ligands which have been studied include $(CO)_5MS(Me)CH(Me)Br$ (M = Cr or W),[527] $\{Ph_2P(CH_2)_3\}_2$-$PPhRhClX_2$ ($X_2 = O_2$ or S_2),[528] (π-$C_5H_5)_2Mo(O)_2SO_2$ and (π-$C_5H_5)_2MO$ (M = Mo or W),[529] (π-$C_5H_5)Mo(CO)_2(CSNMe_2)$, $(CO)_4Mn(CSNMe_2)$[530] and related compounds.

Transition-metal Cluster Compounds.—Tetrakis $\{\eta(\pi$-cyclopentadienyl)$\}$ tetranickel trihydride was reported to show pm and fragment ions including [pm − nH]$^+$ ($n = 1$—3), $[C_{15}H_{15}Ni_4]^+$, and $[C_{10}H_{10}Ni_2]^+$.[531] The base peak from the related cobalt compound (π-$C_5H_5)_4Co_4H_4$ was the pmi, and fragments due to loss of up to four H atoms were observed.[532]

A number of compounds of the type $HRu_3(CO)_x$(olefin) have been characterized, including $HRu_3(CO)_9(C_{12}H_{15})$ and $HRu_3(CO)_7(C_{24}H_{34})$ and also the related compounds $Ru_4(CO)_{10}(C_{12}H_{16})$,[533] and $Ru_3(CO)_9(C_6H_{10})$.[534] Both

---

[520] R. G. Cavell, W. Byers, E. D. Day, and P. M. Watkins, *Inorg. Chem.*, 1972, **11**, 1598.
[521] R. G. Cavell, E. D. Day, W. Byers, and P. M. Watkins, *Inorg. Chem.*, 1972, **11**, 1759.
[522] R. G. Cavell, E. D. Day, W. Byers, and P. M. Watkins, *Inorg. Chem.*, 1972, **11**, 1591.
[523] R. G. Cavell and A. R. Sanger, *Inorg. Chem.*, 1972, **11**, 2016.
[524] R. G. Cavell and A. R. Sanger, *Inorg. Chem.*, 1972, **11**, 2011.
[525] B. J. McCormick, J. L. Featherstone, H. J. Stoklosa, and J. R. Wasson, *Inorg. Chem.*, 1973, **12**, 692.
[526] E. Lindner and H.-M. Ebinger, *J. Organometallic Chem.*, 1974, **66**, 103.
[527] E. O. Fischer and G. Kreis, *Chem. Ber.*, 1972, **106**, 2310.
[528] T. E. Nappier, D. W. Meek, R. M. Kirchner, and J. A. Ibers, *J. Amer. Chem. Soc.*, 1973, **95**, 4194.
[529] M. H. L. Green, A. H. Lynch, and M. G. Swanwick, *J.C.S. Dalton*, 1972, 1445.
[530] W. K. Dean and P. M. Treichel, *J. Organometallic Chem.*, 1974, **66**, 87.
[531] J. Muller, H. Dorner, G. Huttner, and H. Lorenz, *Angew. Chem. Internat. Edn.*, 1973, **12**, 1005.
[532] J. Muller and H. Dorner, *Angew. Chem. Internat. Edn.*, 1973, **12**, 842.
[533] M. I. Bruce, M. A. Cairns, and M. Green, *J.C.S. Dalton*, 1972, 1293.
[534] M. Valle, O. Gambino, L. Milone, S. A. Vaglio, and G. Cetini, *J. Organometallic Chem.*, 1972, **38**, C47.

complexes, $H_2Ru_3(CO)_9C_2Ph_2$,[535] and $Ru_3(CO)_9X$ [X = $C_2Bu^t(H)$ or $C_2Ph(H)$][536] show pmi's and abundant ions containing Ru_3 (96, 98, and 95% of total ion current, respectively) including $[Ru_3(acetylene)]^+$ as base peak. The dihydro-complex fragments initially by loss of CO and (H + CO) groups. Loss of H or (H + CO) groups was not observed from Ru_3-containing ions from the other complexes.

Other complexes for which data have been presented include $RCCo_3(CO)_9$ (R = Ph or substituted phenyl groups),[537] $H_3Ru_3(CO)_9CCH_3$,[538] a series of $Os_3(CO)_8$(acetylene) compounds, $(C_7H_9)CCo_3(CO)_9$ and $(C_8H_9O)CCo_3$-$(CO)_9$,[539] $M_5C(CO)_{15}$ (M = Ru or Os),[540] some benzyne complexes containing Main Group V ligands such as $Os_3(CO)_7(PPh_2)_2C_6H_4$,[541] $HOs_3(CO)_9(MMe_2)$-C_6H_4 (M = P or As),[542] and the related complexes $Os_3(CO)_8(PPh_2)Ph$-$(PPhC_6H_4)$,[541] and $HOs_3(CO)_8(PMe_2)(Me_2PC_6H_4C_6H_3)$,[542] and $Fe_3(CO)_{9-n}$-$\{P(OMe)_3\}_n$ (n = 0—2).[543]

The spectra of μ_3-organoimido-complexes $(\pi\text{-}C_5H_5)_3Ni_3NR^1$ and $(\pi\text{-}C_5H_5)_2(\pi\text{-}C_5H_4R^1)Ni_3NR^2$ ($R^1 = R^2 = $ Ph or Bu^t) have abundant pmi's which are the base peaks in all except the phenylcyclopentadienyl complex.[544] Initial fragmentation of $(\pi\text{-}C_5H_5)_3Ni_3NPh$ was by loss of C_5H_5, C_5H_5Ni, $(C_5H_5)_2Ni$, C_5H_5NiNPh, and $PhNH_2$. Subsequently loss of C_5H_5 or Ni species occurred. The analogous butyl compound lost Me, C_4H_8, and C_5H_5NiH initially and then C_5H_6, $(C_5H_5)_2Ni$, or C_4H_8 groups. The ruthenium complexes, $Ru_3(CO)_{11-n}(NPh)_n$ (n = 1 or 2), $HRu_3(CO)_{10}NHPh$, and $H_2Ru_3(CO)_9NPh$ all gave pmi's and lost CO groups.[545] After losing either three or one CO, respectively, the hydrides also lost H or H_2, respectively, in competition with further loss of CO.

The sulphur-containing cluster compounds, $SCo_4(CO)_9\{PMe_2\}_2$ and $S_2Co_3(CO)_7(PMe_2)$, had pm and fragment ions formed by losses of up to nine or seven carbonyls, respectively.[546] The next major fragment ion in the spectrum of the tetracobalt compound was produced by loss of PMe_2, whereas the tricobalt complex lost both Me groups to give $[S_2Co_3P]^+$. Other sulphur-containing clusters include $(\pi\text{-}C_5H_5)_4Co_4S_4$,[547] $(\pi\text{-}C_5H_5)_3Mo_3S_4$, and $(\pi\text{-}C_5H_5)_2Mo_2S_4$.[548]

---

[535] O. Gambino, E. Sappa, and G. Cetini, *J. Organometallic Chem.*, 1972, **44**, 185.
[536] E. Sappa, O. Gambino, L. Milone, and G. Cetini, *J. Organometallic Chem.*, 1972, **39**, 169.
[537] R. Dolby and B. H. Robinson, *J.C.S. Dalton*, 1972, 2046; 1973, 1794.
[538] A. J. Canty, B. F. G. Johnson, J. Lewis, and J. R. Norton, *J.C.S. Dalton*, 1972, 1331.
[539] T. Kanigo, T. Kitzmura, N. Sakamoto, and T. Joh, *J. Organometallic Chem.*, 1973, **54**, 265.
[540] C. R. Eady, B. F. G. Johnson, J. Lewis, and T. Matheson, *J. Organometallic Chem.*, 1973, **57**, C82.
[541] C. W. Bradford and R. S. Nyholm, *J.C.S. Dalton*, 1973, 529.
[542] A. J. Deeming, R. E. Kimber, and M. Underhill, *J.C.S. Dalton*, 1973, 2589.
[543] P. M. Treichel, W. K. Dean, and W. M. Douglas, *Inorg. Chem.*, 1972, **11**, 1609.
[544] J. Muller, H. Dorner, and F. H. Kohler, *Chem. Ber.*, 1973, **106**, 1122.
[545] E. Sappa and L. Milone, *J. Organometallic Chem.*, 1973, **61**, 383.
[546] G. Natile, S. Pignataro, G. Innorta, and G. Bor, *J. Organometallic Chem.*, 1972, **40**, 215.
[547] G. L. Simon and L. F. Dahl, *J. Amer. Chem. Soc.*, 1973, **95**, 2165.
[548] W. Beck, W. Danzer, and G. Thiel, *Angew. Chem. Internat. Edn.*, 1973, **12**, 582.

The compound $Co_3(CO)_9(COBCl_2)NEt_3$ has been reported to give a pmi and the series of ions $[pm - n(CO)]^+$ and $[Co_3(CO)_nCOBCl_2]^+$ ($n = 1$—9).[549]

Compounds with Bonds to Main-group Elements.—The characterization of $\{MeOZnCo(CO)_4\}_4$,[550] $AlCo_3(CO)_9$,[551] $Ga_2\{Mn(CO)_5\}_4$,[552] and $ClIn\{Mn(CO)_5\}_{3-n}$ ($n = 0$—2)[553] has been reported.

Germyl–metal carbonyl derivatives to have been examined include compounds of manganese,[554] rhenium,[555] iron,[556] cobalt,[557] and osmium.[558] The fragmentation patterns of these compounds generally exhibited ions due to losses of CO, H, and GeH_x and to rearranged species, $[HM(CO)_n]^+$ (e.g. M = Mn, Re, and Co).

The spectra of $X_2Si\{Mn(CO)_5\}_2$ (X = H or Cl) showed that the majority of ions retained the $SiMn_2$ entity.[559] No ions of the type $[Mn_2(CO)_x]^+$ were observed, indicating that elimination of SiX_2 was not a favourable decomposition path. The spectra of several $(H_nCl_{3-n})Si$ derivatives were listed including $H_2ClSiMn(CO)_5$, $H_2ClSiCo(CO)_4$, $HCl_2SiCo(CO)_4$, and $HClSiMn(CO)_5Co(CO)_4$. In all cases loss of CO was observed, giving the complete series of ions $[pm - n(CO)]^+$. The compounds $(\pi\text{-}C_5H_5)(CO)_3MSiXYZ$ (M = Mo or W, X = Me, Y = Cl, Z = H; X = H, Y = Z = $Cl^{560,561}$; X = Me, Y = Z = Cl or X = Y = Z = Me)[561] have been examined.

The complexes $Cl_3SiCo(CO)_x(PF_3)_{4-x}$ (x = 2 or 3) show weak pmi's (0.3% total ion current) and fragment mainly by losing CO, PF_3, and Cl to give, for example from $Cl_3SiCo(CO)_3PF_3$, ions of the type $[Cl_3SiCo(CO)_nPF_3]^+$, $[Cl_3SiCo(CO)_n]^+$ (n = 0—3), and $[Cl_2SiCo(CO)_nPF_3]^+$ (n = 2 or 3).[562] Ionization and appearance potentials of the major ions were measured and the heats of formation of the compounds estimated. The bond dissociation energy $D(Cl_3Si\text{—}CoL_4)$ was estimated to be 401 ± 42 kJ mol^{-1} for both compounds.

Trifluorosilyl derivatives $F_3SiM(CO)_5$ (M = Mn or Re), $F_3SiMn(CO)_4PPh_3$, $F_3SiFe(\pi\text{-}C_5H_5)(CO)_2$, and $F_3SiM(\pi\text{-}C_5H_5)(CO)_3$ (M = Mo or W) fragmented by losses of CO and F and gave rearranged ions with M—F bonds.[563] Evidence for Re—H—Re bonding is suggested by the spectrum of $Cl_3Si(H)$-

---

[549] G. Schmid and V. Batzel, *J. Organometallic Chem.*, 1972, **46**, 149.
[550] J. M. Burlitch and S. E. Hayes, *J. Organometallic Chem.*, 1972, **42**, C13.
[551] K. E. Schwarzhaus and H. Steiger, *Angew. Chem. Internat. Edn.*, 1972, **11**, 535.
[552] H.-J. Haupt and F. Neumann, *Z. anorg. Chem.*, 1972, **394**, 67.
[553] A. T. T. Hsieh and M. J. Mays, *J.C.S. Dalton*, 1972, 516.
[554] R. D. George, K. M. Mackay, and S. R. Stobart, *J.C.S. Dalton*, 1972, 1505.
[555] K. M. Mackay and S. R. Stobart, *J.C.S. Dalton*, 1973, 214.
[556] S. R. Stobart, *J.C.S. Dalton*, 1972, 2442.
[557] R. D. George, K. M. Mackay, and S. R. Stobart, *J.C.S. Dalton*, 1972, 974.
[558] R. D. George, S. A. R. Knox, and F. G. A. Stone, *J.C.S. Dalton*, 1973, 973.
[559] K. M. Abraham and G. Urry, *Inorg. Chem.*, 1973, **12**, 2850.
[560] W. Malisch, H. Schmidbaur, and M. Kuhn, *Angew. Chem. Internat. Edn.*, 1972, **11**, 516.
[561] W. Malisch and M. Kuhn, *Chem. Ber.*, 1974, **107**, 979.
[562] F. E. Saalfeld, M. V. McDowell, A. G. MacDiarmid, and R. E. Highsmith, *Internat. J. Mass. Spectrometry, Ion Phys.*, 1972, **9**, 197.
[563] M. E. Redwood, B. E. Reichert, R. R. Schreike, and B. O. West, *Austral. J. Chem.*, 1973, **26**, 247.

$Re_2(CO)_9$.[564] Most ions retain the Re_2H group during fragmentation to give series of ions such as $[pm - n(CO)]^+$ and $[HRe_2(CO)_n]^+$ ($n = 1$—9), a type of behaviour also found in the compounds $HRe_3(CO)_{14}$ and $HMnRe_2(CO)_{14}$, in which H-bridging is known to occur. Dichloromethylsilyl and dichlorophenylsilyl derivatives give similar results.

A pmi was observed from $(\pi-C_5H_5)Co(NO)GeI_3$ but not from the analogous tin compounds.[565] The germanium derivative initially lost NO and I and then GeI_2 to give $[C_5H_5Co]^+$. The rearranged ion, $[C_5H_5CoNOI]^+$, was the base peak in the spectrum of the GeI_3 derivative. Other compounds reported include $(\pi-C_5H_5)_2W(Br)SnBr_3$,[529] $(\pi-C_5H_5)(PEt_3)NiMCl_3$ (M = Si, Ge, or Sn),[566] $(\pi-C_5H_5)Fe(CO)_2MX_2Ni(CO)(\pi-C_5H_5)$ (M = Ge or Sn, X = Cl or Br),[567] and $\{(CO)_5WSiI_2\}_2$.[568]

Ionization and appearance potential data from $Me_3MMn(CO)_5$ (M = Si, Ge, or Sn) have been used to calculate the ionic bond dissociation energies $D[Me_3M-Mn(CO)_5]^+$ and corresponding neutral metal–metal bond dissociation energies.[569] The trends observed were (strongest) Mn—Si > Mn—Sn > Mn—Ge; fragmentation patterns were discussed briefly. From some ions such as $[Me_3MMn(CO)_n]^+$, competitive loss of CO and Me was observed. Loss of Me was always found to be competitive with loss of a CO or PF_3 ligand in the spectra of the complexes, $Me_3SiMn(CO)_n(PF_3)_{3-n}$ ($n = 2$—5).[570] Ionization potentials of these complexes, as well as of $HMn(CO)_5$ and $Me_3SiMn(CO)_5$, were reported together with appearance potentials for several ions from $XMn(CO)_5$ (X = H or $SiMe_3$) and $Me_3SiMn(CO)_4PF_3$. Heats of formation for the compounds studied were calculated or estimated.

Other reported organo-Group IV metal derivatives include $Me_3MCo(CO)_4$,[571] $Mn(CO)_5SnMe_2X$ [X = I or $Mn(CO)_5$],[572] $\{Me_3SiFe(COR)(CO)_3\}_2$ (R = H or $SiMe_3$),[573] silacyclobutane derivatives of $(\pi-C_5H_5)Fe$ compounds,[574] $(Me_nCl_{3-n}Si)Co(CO)_4$ ($n = 1$ or 2),[575] (vinyl)ClGe-μ-Cl-Fe(CO)$_2$NO,[576] $(\pi-C_5H_5)_2M(X)SnMe_3$ (M = Mo or W, X = Cl, Br, or I),[577] $Os(CO)_4(MR_3)$ ($MR_3 = SnPh_3$ or $PbMe_3$),[558] Me_3Ge–ruthenium complexes containing the trihydropentalenyl ligand and related compounds,[578] a compound previously

[564] J. K. Hoyano and W. A. G. Graham, *Inorg. Chem.*, 1972, **11**, 1265.
[565] H. Brunner and S. Loskot, *Z. Naturforsch.*, 1973, **28b**, 314.
[566] F. Glockling and A. McGregor, *J. Inorg. Nuclear Chem.*, 1973, **35**, 1481.
[567] L. K. Thompson, E. Eisner, and M. J. Newlands, *J. Organometallic Chem.*, 1973, **56**, 327.
[568] G. Schmid and R. Boese, *Chem. Ber.*, 1972, **105**, 3307.
[569] R. A. Burnham and S. R. Stobart, *J.C.S. Dalton*, 1973, 1269.
[570] F. E. Saalfeld, M. V. McDowell, J. J. De Corpo, A. D. Berry, and A. G. MacDairmid, *Inorg. Chem.* 1973, **12**, 48.
[571] G. F. Bradley and S. R. Stobart, *J.C.S. Dalton*, 1974, 264.
[572] R. A. Burnham, F. Glockling, and S. R. Stobart, *J.C.S. Dalton*, 1972, 1991.
[573] M. A. Nasta, A. G. MacDairmid, and F. E. Saalfeld, *J. Amer. Chem. Soc.*, 1972, **94**, 2449.
[574] C. S. Cundy and M. F. Lappert, *J. Organometallic Chem.*, 1973, **57**, C72.
[575] A. P. Hagen, L. McAmis, and M. A. Stewart, *J. Organometallic Chem.*, 1974, **66**, 127.
[576] M. D. Curtis and R. C. Job, *J. Amer. Chem. Soc.*, 1972, **94**, 2153.
[577] D. H. Harris, S. A. Keppie, and M. F. Lappert, *J.C.S. Dalton*, 1973, 1653.
[578] S. A. R. Knox, R. P. Phillips, and F. G. A. Stone, *J.C.S. Chem. Comm.*, 1972, 1227; A. Brookes, J. Howard, S. A. R. Knox, F. G. A. Stone, and P. Woodward, *J.C.S. Chem. Comm.*, 1973, 587.

reported to be $(Me_3Si)_4Fe_2(CO)_8$ but shown by mass spectrometry to be $(Me_3SiOC)_4Fe_2(CO)_6$,[579] and the related compound $(Me_3SiOC)_4Co_2(CO)_4$.[580]
The spectra of $(\pi\text{-}C_5H_5)Fe(CO)_2Si(OR)_3$ (R = Me, Et, Pr, or Pr^i) show pmi's and $[C_5H_5FeSi(OR)_3]^+$ as base peaks but, for R = Bu^t, no pmi was observed, the ion of highest mass being $[C_5H_5Fe(CO)_2Si(OH)(OBu)_2]^+$ and the base peak ion $[C_5H_5FeSi(OH)_3]^+$.[581] The detailed spectrum of the compound for R = Pr^i was given; fragmentation occurred initially by losses of CO, then OR, and (R − H) species.

Boron-containing Ligands, including Carbaboranes. Borohydride complexes, $U(BH_4)_4$,[582] and $Cu_2\{HB(pyrazolyl)_3\}_2$,[583] afford pmi's as do some cobaltaborane complexes, $B_4H_8Co(\pi\text{-}C_5H_5)$, $\{1,2\text{-}B_4H_6\}Co_2(\pi\text{-}C_5H_5)_2$, and $5\text{-}B_9H_{13}Co\text{-}(\pi\text{-}C_5H_5)$,[584] and $Fe(CO)_4B_6H_{10}$.[585]

Numerous carbaborane complexes have been examined, usually to confirm their molecular weights, and generally no details of fragmentation are presented. Examples of the complexes reported include $(\pi\text{-}C_5H_5)Co(\pi\text{-}CB_7H_8)\text{-}(\pi\text{-}C_5H_5)$,[586] $(\pi\text{-}C_5H_5)Co(\pi\text{-}CB_7H_8)$,[587] $\{\pi\text{-}(1,7)\text{-}2,4\text{-}C_2B_3H_5\}Co_2\text{-}(\pi\text{-}C_5H_5)_2$,[588] derivatives containing $(\pi\text{-}2,3\text{-}C_2B_4H_6)$ and related carbaboranes with cyclopentadienyl iron and iron tricarbonyl groups,[589] $(\pi\text{-}C_5H_5)Co\text{-}(\pi\text{-}C_2B_7H_n)Fe(\pi\text{-}C_5H_5)$ (n = 9 or 11),[590] some *nido*-eleven atom metallocarbaboranes and their Lewis-base adducts such as $X\{9\text{-}(\eta\text{-}\pi\text{-}C_5H_5)\text{-}11\text{-}C_5H_5N\text{-}7,\text{-}8,9\text{-}C_2CoB_8H_{10}\}$,[591] $(\pi\text{-}C_5H_5)_2Co_2C_2B_{10}H_{12}$,[592] and the cobalt derivative of 1,2,4-tris(1′,1′,2′carbaboranyl)benzene, $Co_2(CO)_6B_{10}C_4H_{12}$.[593]

Miscellaneous Ligands. Complexes of chromium containing isocyanide ligands, $(RNC)Cr(CO)_5$ (R = Me, Et, Pr^i, Bu^t, *p*-tolyl, or *p*-ClC$_6$H$_4$), all gave reasonably abundant pmi's, losses of up to five CO groups, and base peaks $[CrCNR]^+$;[594] the ions $[CrCNH]^+$ and $[CrCN]^+$ were also observed. From alkyl isocyanides, (except R = Me), some evidence of fragmentation at the β-C atom was observed. Di-isocyanides $(RNC)_2M(CO)_4$ (M = Cr, R = Me or Et; M = Cr or Mo, R = Pr^i, Bu^t, *p*-tolyl, or *p*-ClC$_6$H$_4$), produced similar fragmentation patterns giving the series of ions, $[(CO)_nM(CNR)_2]^+$ (n = 0—4) and $[(CO)_nMCNR]^+$

---

[579] M. J. Bennett, W. A. G. Graham, R. A. Snaith, and R. P. Stewart, *J. Amer. Chem. Soc.*, 1973, **95**, 1684.
[580] W. M. Ingle, G. Preti, and A. G. MacDairmid, *J.C.S. Chem. Comm.*, 1973, 497.
[581] M. Hofler and J. Scheuren, *J. Organometallic Chem.*, 1973, **55**, 177.
[582] N. Davis, M. G. H. Wallbridge, B. E. Smith, and B. D. James, *J.C.S. Dalton*, 1973, 162.
[583] M. I. Bruce and A. P. P. Ostazewski, *J.C.S. Chem. Comm.*, 1972, 1124.
[584] V. R. Miller and R. N. Grimes, *J. Amer. Chem. Soc.*, 1973, **95**, 5079.
[585] A. Davidson, D. D. Traficante, and S. S. Wreford, *J.C.S. Chem. Comm.*, 1972, 1155.
[586] C. G. Salentine and M. F. Hawthorne, *J.C.S. Chem. Comm.*, 1973, 560.
[587] D. F. Dustin and M. F. Hawthorne, *Inorg. Chem.*, 1973, **12**, 1380.
[588] D. C. Beer, V. R. Miller, L. G. Sneddon, R. N. Grimes, M. Mathew, and G. J. Palenik, *J. Amer. Chem. Soc.*, 1973, **95**, 3048.
[589] L. G. Sneddon, D. C. Beer, and R. N. Grimes, *J. Amer. Chem. Soc.*, 1973, **95**, 6623.
[590] D. F. Dustin, W. J. Evans, and M. F. Hawthorne, *J.C.S. Chem. Comm.*, 1973, 805.
[591] C. J. Jones, J. N. Francis, and M. F. Hawthorne, *J. Amer. Chem. Soc.*, 1973, **95**, 7633.
[592] W. J. Evans and M. F. Hawthorne, *J.C.S. Chem. Comm.*, 1974, 38.
[593] K. P. Callahan and M. F. Hawthorne, *J. Amer. Chem. Soc.*, 1973, **95**, 4574.
[594] J. A. Connor, E. M. Jones, G. K. McEwen, M. K. Lloyd, and J. A. McCleverty, *J.C.S. Dalton*, 1972, 1246.

(n = 0—3), with break-up of the isocyanide group only after loss of all CO groups. Similar behaviour was found with $(RNC)_3M(CO)_3$ compounds.

Carbamoyl complexes, $(\pi\text{-}C_5H_4R)FeCO(L)CONH_2$ (R = CH_2Ph or H, L = CO; R = H, L = Ph_3P or Et_3P), yielded ions from loss of both carbonyl and carbamoyl CO groups. For (R = H, L = phosphine) ions, $[C_5H_5Fe(L)NH_2]^+$ were base peaks.[595] Nearly as abundant were $[C_5H_5Fe(L)]^+$ ions. The rhenium complex, $(CO)_5ReCONHMe$, fragments to give the ions $[(CO)_nReNHMe]^+$ (n = 1—6) and $[(CO)_nReNMe]^+$ (n = 1—4), with $[(CO)_5ReNHMe]^+$ and $[Re]^+$ as base peaks.[596] Carboxamido-complexes of manganese, $(\pi\text{-}C_5H_4X)Mn(CO)\text{-}(NO)CONHMe$ (X = H or Me), have also been reported.[597]

4 Co-ordination Compounds

A short review discussing the concept of valency change in the rationalization of spectra of metal complexes has appeared.[5] It is postulated that electron transfer may be possible between complexed atoms and the ligands in the ion so that the odd- or even-electron character of the ion can be interchanged. Most of the evidence discussed comes from studies of co-ordination compounds as for example, the fragmentation of aluminium and iron complexes of acetylacetone, (Scheme 30). The possibility of applying the valency change concept to inorganic and organometallic compounds was discussed.

$$[(acac)_3M^{III}]^{+\cdot} \xrightarrow[Al, Fe]{-acac\cdot} [(acac)_2M^{III}]^+ \leftrightarrow [(acac)_2M^{II}]^{+\cdot} \xrightarrow[Fe\ only]{-acac\cdot} [(acac)Fe^{II}]^+$$

Scheme 30

Compounds with Metal–Oxygen Bonds.—*β-Diketonato-complexes*. The relative ion intensities and the ionization and appearance potentials of the major ions from one iridium and several rhodium dicarbonyl acetylacetones (86a—g) have been measured.[598] As well as ions due to loss of one or both carbonyl groups, fragments characteristic of metal acetylacetonates were observed, *e.g.* [pm − 2CO − R]$^+$ from (86c) and (86e) and [pm − acac]$^+$ from all compounds except (86f).

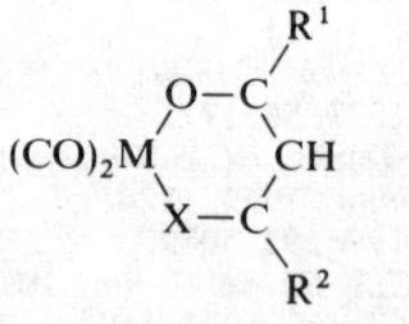

	a	b	c	d	e	f	g
R^1	Ph	Ph	Me	Me	CF_3	Me	CF_3
R^2	Ph	Me	Me	CF_3	CF_3	Me	CF_3
X	O	O	O	O	O	NPh	O

(86) $(CO)_2M$ with ring: O−C(R^1)=CH−C(R^2)=X

[595] J. Ellermann, H. Behrens, and H. Kohlberger, *J. Organometallic Chem.*, 1972, **46**, 119.
[596] R. W. Brink and R. J. Angelici, *Inorg. Chem.*, 1973, **12**, 1062.
[597] L. Busetto, A. Palazzi, D. Pietropaolo, and G. Dolcetti, *J. Organometallic Chem.*, 1974, **66**, 453.
[598] F. Bonati, G. Distefano, S. Pignataro, and S. Terroni, *Org. Mass Spectrometry*, 1972, **6**, 971.

Energetic data were used to interpret some aspects of the fragmentation patterns. For example, whilst the ionization potentials of the compounds varied between 7.5 and 9.2 eV, the energies required for the formation of $[pm - CO]^+$ and $[pm - 2(CO)]^+$ ions from the pmi's were the same within experimental error, indicating that loss of the carbonyl CO groups rather than loss of CO from the chelate ring was taking place. The energies for formation of $[pm - 3(CO)]^+$ and $[pm - 4(CO)]^+$ ions were, however, much more susceptible to the influence of the substituents R^1 and R^2.

The spectra of some 1,3-substituted acetylacetonato-complexes of rhodium 1,6-dichlorocyclo-octa-1,5-diene, $(acacR^1R^2)RhC_8H_{10}Cl_2$, have been described.[599] Substituents included Me, Ph, p-MeC$_6$H$_4$, p-ClC$_6$H$_4$, p-MeOC$_6$H$_4$, and p-NO$_2$C$_6$H$_4$. Fragmentation of all pmi's gave $[HacacR^1R^2]^+$ and $[RhC_8H_{10}Cl_2]^+$ but not $[acacR^1R^2Rh]^+$. Again, ionization and appearance potential measurements indicated that the energy required for the formation of $[RhC_8H_{10}Cl_2]^+$ from the pmi's was strongly influenced by the nature of R^1 and R^2. Alternative decompositions of the pmi's involved loss of Cl and HCl, or CO, Cl, and HCl simultaneously. The characterization of acetylacetonato-rhodium complexes of a number of cyclo-olefins has been reported.[600]

Hydrogen arrangements in the spectra of β-diketonates can occur specifically from one site but this is not always the case.[601] The metal atom and the ligand substituents can play important roles in such transfers. Copper, nickel, and oxovanadium compounds of pentane-2,4-dione, 3-methylpentane-2,4-dione, and 1,3-diphenyl-1,3-dione and deuteriated analogues were studied. For acetylacetonato-complexes of Cu and Ni, the H transferred in the formation of the ion $[Macac + H]^+$ from the pmi was almost exclusively from a methyl group (Scheme 31). However, for the OV(acac)$_2$ complex, the rearrangement occurred only ca. 90% specifically from the methyl groups and it seems possible that competitive H-exchange is occurring or specific site H-arrangement is followed by randomization. The 3-methyl derivative of Cu shows a site-specific formation of $[ML - H]^+$, L = ligand, (from the 3-methyl group) but the formation of $[ML + H]^+$ is not nearly so specific. Both ions are formed with less specificity in the Ni and OV compounds. The hydrogen atom lost in forming $[CuL - H]^+$ from the pmi of the diphenyl derivative comes specifically from the 2-position; the vanadium compound is closely similar. However, there appear to be at least

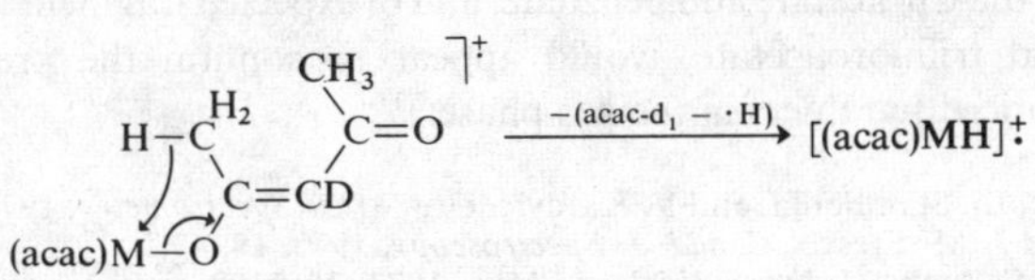

Scheme 31

[599] M. Ryska, K. Bourchal, and F. Hrabak, *J. Organometallic Chem.*, 1973, **51**, 353.
[600] R. Grigg and J. L. Jackson, *Tetrahedron*, 1973, **29**, 3903.
[601] M. J. Lacey, C. G. Macdonald, and J. S. Shannon, *Austral. J. Chem.*, 1972, **25**, 2559.

two possible modes of formation of the analogous ion from the nickel compound, one involving the phenyl ring.

An unusual rearrangement involving the transfer of three hydrogen atoms has been reported in the spectra of $MCl_2(OR)_2(acac)$ (M = Nb or Ta, R = Me or Et) (Scheme 32)[602] (see also section on alkoxides, below).

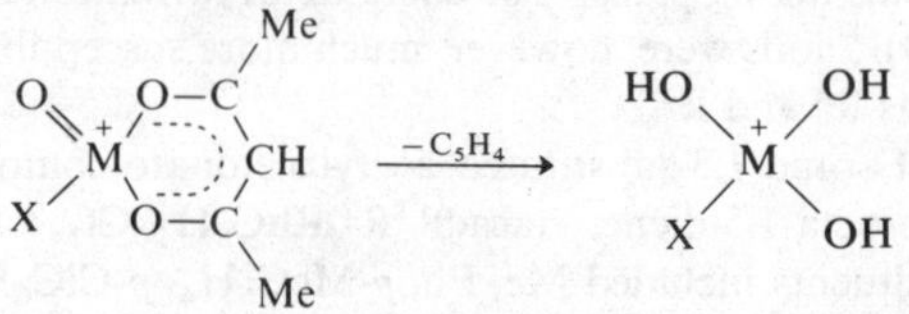

$$X = H \text{ or } Cl$$
Scheme 32

Other compounds for which data have been published include some dimeric copper and nickel chelates,[603] dipyridyl and pyrazine adducts of $M(acac)_2$ (M = Mn, Fe, Co, Ni, or Zn),[604] iridium complexes of trifluoromethyl-β-diketonates,[605] $X_2Re(acac)_2$ (X = Cl, Br, or I),[606] and Cl_2MoL_2 (L = various chelating groups including acetylacetonato).[607]

Carboxylates. Three reports have appeared concerning Cu^I carboxylates. The production of ions from dimeric species $\{(RCO_2)Cu\}_2$ (R = Me, F_3C,[608,609] Et, Pr, Pr^i, F_2HC,[608] Me_3C, $CH{=}CH_2$, $CH_2CH{=}CH_2$,[609] Ph,[608,609] p-FC_6H_4, p-ClC_6H_4, o-ClC_6H_4, and C_6F_5[608]) were reported in two papers, but the third[610] also reported finding ions from tetrameric species, $\{(RCO_2)Cu\}_4$ (R = H, Me, CF_3, and Ph), in low abundance. Carboxylates with saturated R groups showed pmi's and fragmented initially by loss of RCO_2, to produce the base peak, or of R. Further fragmentation tended to occur by elimination of even-electron species. From aryl derivatives the major fragmentation was the elimination of CO_2 and the formation of Cu—R bonds, with loss of RCO_2 occurring less importantly (Scheme 33). Compounds of acrylic and vinylacetic acids showed ions derived from pmi's by elimination of CO_2 and of RCO_2 as well as ions $[Cu_2(CH{=}CH_2)_2]^+$ from the acrylic compound. The ion $[Cu_2F]^+$ was abundant in the spectrum of the trifluoroacetate.

The reports of weak tetrameric pmi's and expected tetrameric fragment ions in the spectra of the Cu^I acetate and benzoate, and of expected fragment ions from the formate and trifluoroacetate, would appear to confirm the presence of previously unnoticed tetramers in the gas phase.[610]

[602] D. Stefanovic, Lj. Stambolija, and V. Katovic, *Org. Mass Spectrometry*, 1973, **7**, 1357.
[603] A. Malek and J. M. Fresco, *Canad. J. Spectroscopy*, 1973, **18**, 43.
[604] S. Ambe and F. Ambe, *J. Inorg. Nuclear Chem.*, 1973, **35**, 1109.
[605] G. M. Tanner, D. G. Tuck, and E. J. Wells, *Canad. J. Chem.*, 1972, **50**, 3950.
[606] W. D. Courrier, G. J. L. Lock, and G. Turner, *Canad. J. Chem.*, 1972, **50**, 1797.
[607] A. Van den Bergen, K. S. Murray, and B. O. West, *Austral. J. Chem.*, 1972, **25**, 705.
[608] D. C. K. Lin and J. B. Westmore, *Canad. J. Chem.*, 1973, **51**, 2999.
[609] T. Ogura and O. Fernando, *Inorg. Chem.*, 1973, **12**, 2611.
[610] D. A. Edwards and R. Richards, *Inorg. Nuclear Chem. Letters*, 1972, **8**, 783.

The spectra of some Cu^{II} carboxylates (R = Me, Et, Pr, or Bu) have been listed.[611] Tetrameric ions were reported from copper acetate and, since Cu^I carboxylates may be produced by heating the corresponding Cu^{II} salts, it seems possible that the source of these ions was the Cu^I compound. However, the

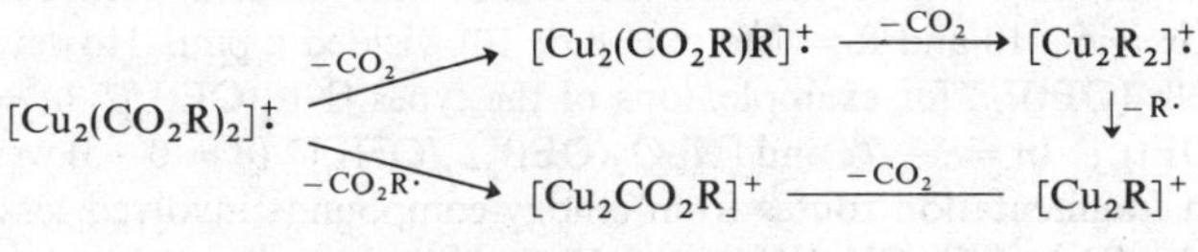

$$[Cu_2(CO_2R)_2]^{\ddagger} \xrightarrow{-CO_2} [Cu_2(CO_2R)R]^{\ddagger} \xrightarrow{-CO_2} [Cu_2R_2]^{\ddagger}$$
$$\downarrow -R\cdot$$
$$\xrightarrow{-CO_2R\cdot} [Cu_2CO_2R]^+ \xrightarrow{-CO_2} [Cu_2R]^+$$

Scheme 33

tetrameric pmi was not reported and about half the polymeric ions observed were not found in the spectrum of Cu^I acetate. A further complicating feature of the spectrum of Cu^{II} acetate was that the ion abundances varied according as to how the compound was prepared. In particular, the mononuclear ion abundances varied widely. It is clear that considerable care is required in interpreting the spectra of these compounds.

The general features found in the spectra of Cu^I carboxylates were also reflected in the spectra of $Tl(CO_2R)_3$ (R = Me, CF_3, or Ph).[612] Successive eliminations of two $MeCO_2$ groups from $[(MeCO_2)_2Tl]^+$ to give $[Tl]^+$, the base peak, were supported by the presence of metastables, as was the elimination of CO_2 from $[PhCO_2Tl]^+$. The formation of $[TlOF]^+$ in the spectrum of the trifluoroacetate and of $[Tl]^+$, the base peak, by the simultaneous ejection of two (CF_3CO_2) groups were novel features.

Other carboxylate compounds studied were $Cl_2Sn(O_2CH)_2$[613] and $Mo(O_2CCH_3)_2$,[614] both of which showed evidence of polymeric species in the gas phase.

Alkoxides. Two processes leading to the elimination of $>CO$ species from alkoxides, $LMOC(R^1R^2R^3)$, of boron and titanium have been suggested (Scheme 3, p. 146). The first involves removal of a radical R^1 followed immediately by elimination of R^2R^3CO to give $[LM]^+$ ions. The second involves the transfer of R^1 to the atom M with simultaneous elimination of R^2R^3CO to give $[LMR^1]^+$ ions. It was suggested that the first process was more favourable for the lower alkyl groups and the second for higher, including branched-chain groups. Examples were discussed from the spectra of $(RO)_4Ti$ (R = Me, Pr^i, and Bu^t).[28]

The spectra of the complexes $MCl_2(OR)_2L$ (M = Nb or Ta, R = Me or Et, L = acac or salicylaldehydrato), have been listed and features in the fragmentation patterns discussed.[602] Generally, these complexes initially lose Cl to give base peak ions, OR, and L groups. Further fragmentations involve losses of H_2O,

---

[611] Kh. Sh. Khariton, G. A. Popovich, and A. V. Ablov, *Doklady Akad. Nauk S.S.S.R.*, 1972, **207**, 1369.
[612] A. T. T. Hsieh, A. G. Lee, and P. L. Sears, *J. Org. Chem.*, 1972, **37**, 2637.
[613] A. N. Sara and K. Taugbol, *J. Inorg. Nuclear Chem.*, 1973, **35**, 1827.
[614] G. Holste, *Z. anorg. Chem.*, 1973, **398**, 249.

HCl, C_2H_6, C_2H_5OH, and C_2H_5Cl molecules. The acetylacetonates all show characteristic [pm – (OR – H)]$^+$ ions; the ethoxy-derivatives also show [pm – (R – H)]$^+$ and [pm – L – (R – H)]$^+$ ions. In contrast to the monomeric Nb compounds discussed above, the spectra of other alkoxyniobium complexes showed dimeric ions.[269] None of the compounds studied, $Nb_2(OR)_xCl_{10-x}$ (R = Et, x = 6—10 and R = Me, x = 8 or 10), yielded a pmi. However, in the case of $Nb_2(OEt)_{10}$, for example, ions of the types $[Nb_2(OEt)_n]^+$ (n = 8 or 9), $[Nb_2O(OEt)_n]^+$ (n = 5—7), and $[NbO_2(OEt)_{3-n}(OH)_n]^+$ (n = 0—3) were found. Common fragmentation routes from ethoxy-compounds involved losses of Cl, OR, C_2H_4, $(C_2H_5)_2O$, CH_3CHO, and H_2O.

Other metal–oxygen bonded compounds studied include $Re(OMe)_4$, which showed a pm and [pm – nMe]$^+$ (n = 1 – 3) ions,[615] and $\{(\pi\text{-}C_5Me_5)CrO\}_4$ and $\{(\pi\text{-}C_5Me_5)CrO_2\}_2$.[414]

Compounds with Metal–Nitrogen Bonds.—The spectrum of $\{(Me_3Si)_2N\}_2Mg$ shows only a monomeric pmi and no ions from polymeric species.[616] Several substituted pyrazole-gold complexes have been shown to be trimeric in the gas

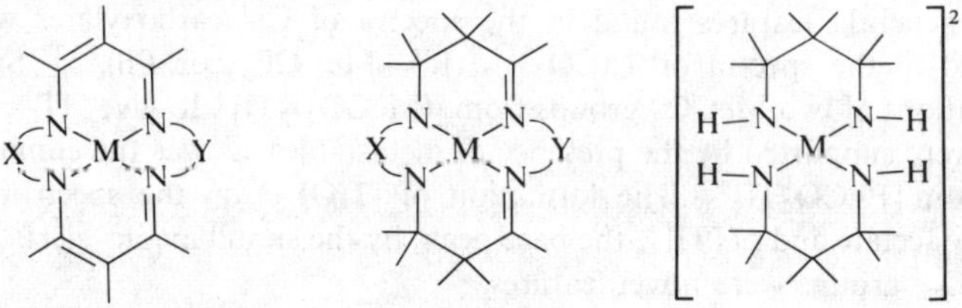

(87)   (88)   (89)

X, Y may be terminal, bridging —(CH$_2$)—, or multiply-bonded >C=C< or >C=N— groups

phase.[617] Dimethylglyoximato (dmg) complexes, $M(dmg)_3(BX)_2$ (M = Fe, X = F, OH, OMe, OEt, OPr^i, and OBu;[618] M = Co, X = F[619]) have abundant pmi's.

The characterizations of several compounds containing macrocyclic ligands with configurations about the metal atom of types (87),[620–622] (88),[623] (89),[624]

---

[615] K. Mertis, J. F. Gibson, and G. Wilkinson, *J.C.S. Chem. Comm.*, 1973, 93.
[616] U. Wannagat, H. Autzen, H. Kuckertz, and H.-J. Wismar, *Z. anorg. Chem.*, 1972, **394**, 254.
[617] F. Bonati, G. Minghetti, and G. Banditelli, *J.C.S. Chem. Comm.*, 1974, 88.
[618] S. C. Jackels, D. S. Dierdorf, N. J. Rose, and J. Zektzer, *J.C.S. Chem. Comm.*, 1972, 1291; S. C. Jackels and N. J. Rose, *Inorg. Chem.*, 1973, **12**, 1232.
[619] D. R. Boston and N. J. Rose, *J. Amer. Chem. Soc.*, 1973, **95**, 4163.
[620] B. M. Higson and E. D. McKenzie, *J.C.S. Dalton*, 1972, 269.
[621] D. St. C. Black and P. W. Kortt, *Austral. J. Chem.*, 1972, **25**, 281.
[622] T. J. Truex and R. H. Holm, *J. Amer. Chem. Soc.*, 1972, **94**, 4529.
[623] J. F. Myers and N. J. Rose, *Inorg. Chem.*, 1973, **12**, 1238.
[624] J. C. Dabrowiak, P. H. Merrell, and D. H. Busch, *Inorg. Chem.*, 1972, **11**, 1979.

(90),[625] and related ligands[626,627] have been reported. Complexes of tetra-azamacrocyclic ligands with both diene (91) and dieno (92) groups[628] and related compounds[629] have also been studied.

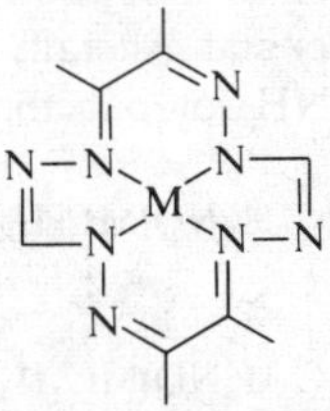

(90) 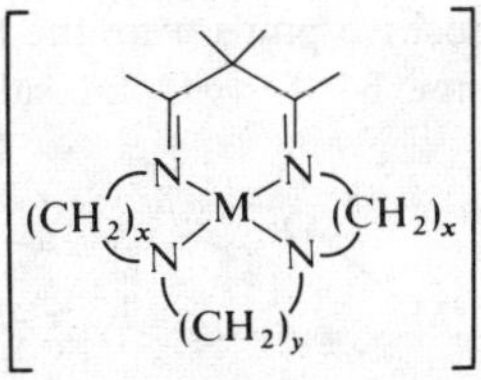   (91) 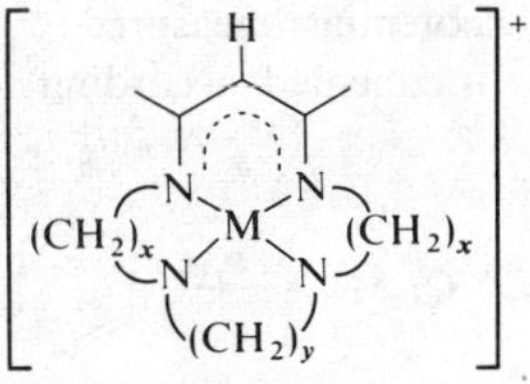   (92)

Porphyrins and Related Compounds. Octa-alkylporphyrins of Ru,[630] V,[631] Ni,[631,632] Cu,[632] Zn,[633] Tl,[634] and Hg[635] have been reported. The thallium compound, aquo-octaethylporphyrinatothallium hydroxide showed a pmi and $[pm - H_2O - OH]^+$ ion. Heterodinuclear metalloporphyrin derivatives of technetium and rhenium tricarbonyls have been characterized.[636] Octaethylporphin derivatives of Al, Ga, Si, Ge, Sn, Mn, Re, Mo, W, Fe[637] and Os[638] have been examined, as have a ruthenium derivative of tetraphenylporphin,[639] and some cobalt tetradehydrocorrins.[640]

Aspects of the spectra of several transition-metal phthalocyanines have been noted and some ionization potentials measured.[641] Commonly, pmi's were base peak with doubly- and triply-charged parent molecules in relatively high abundance, 17—58 and 0.8—1.2%, respectively, for Fe, Co, Ni, and Pt compounds.

Ion-molecule reactions between common perfluoro-compounds used as standard mass markers and metalloporphyrins have been reported to produce ions containing fluorocarbon residues.[642]

[625] V. L. Goedken and S.-M. Peng, *J.C.S. Chem. Comm.*, 1973, 63.
[626] S. C. Cummins and R. E. Sievers, *Inorg. Chem.*, 1972, **11**, 1483.
[627] S. Ogawa, T. Yamaguchi, and N. Gotoh, *J.C.S. Chem. Comm.*, 1972, 577.
[628] J. G. Martin and S. C. Cummins, *Inorg. Chem.*, 1973, **12**, 1477.
[629] C. M. Kerwin and G. A. Melson, *Inorg. Chem.*, 1973, **12**, 2410.
[630] G. W. Sovocool, F. R. Hopf, and D. G. Whitten, *J. Amer. Chem. Soc.*, 1972, **94**, 4350.
[631] R. Bonnett, P. Brewer, K. Noro, and T. Noro, *J.C.S. Chem. Comm.*, 1973, 562.
[632] C. O. Bender, R. Bonnett, and R. G. Smith, *J.C.S. Perkin I*, 1972, 771.
[633] G. H. Barnett, M. F. Hudson, S. W. McCombie, and K. M. Smith, *J.C.S. Perkin I*, 1973, 691.
[634] K. M. Smith, *Org. Mass Spectrometry*, 1972, **6**, 1401.
[635] M. F. Hudson and K. M. Smith, *J.C.S. Chem. Comm.*, 1973, 515.
[636] M. Tsutsui and C. P. Hrung, *J. Amer. Chem. Soc.*, 1973, **95**, 5777.
[637] J. W. Buchler, L. Puppe, K. Rohbock, and H. H. Schneehage, *Chem. Ber.*, 1973, **106**, 2710.
[638] J. W. Buchler and K. Rohbock, *J. Organometallic Chem.*, 1974, **65**, 223.
[639] J. J. Bonnet, S. S. Eaton, G. R. Eaton, R. H. Holm, and J. A. Ibers, *J. Amer. Chem. Soc.*, 1973, **95**, 2141.
[640] C. M. Elson, A. Hamilton, and A. W. Johnson, *J.C.S. Perkin I*, 1973, 775.
[641] D. D. Eley, D. J. Hazeldine, and T. F. Palmer, *J.C.S. Faraday II*, 1973, 1808.
[642] D. Rosenthal, F. R. Hopf, D. G. Whitten, and M. M. Bursey, *Org. Mass Spectrometry*, 1973, **7**, 497.

Compounds with Metal–Oxygen and Metal–Nitrogen Bonds.—*Schiff-base and Related Compounds.* Positive- and negative-ion spectra of Co^{II}, Ni^{II}, and Cu^{II} complexes of bis(salicylaldehyde)ethylenediamine, (salen), and bis(*o*-aminobenzylidene)ethylenediamine, (oaben), have been reported and ionization potentials measured.[643] The positive pmi's were the base peaks and generally fragmented according to Scheme 34 (Q = O for salen and NH for oaben).

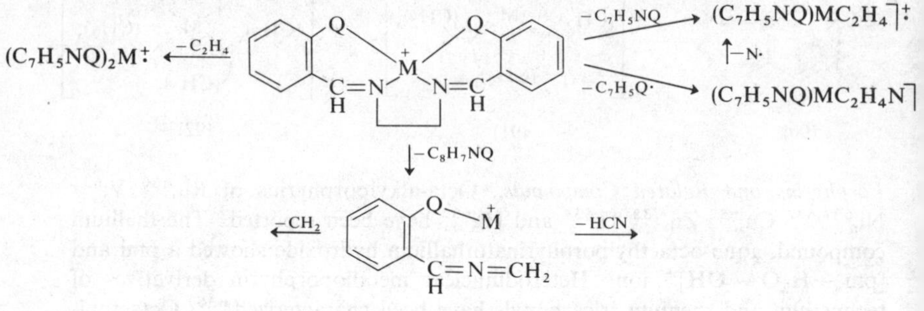
(C7H5NQ)2M+·
−C2H4
−C7H5NQ
(C7H5NQ)MC2H4]+·
−N·
−C7H5Q·
(C7H5NQ)MC2H4N]
−C8H7NQ
−CH2
−HCN
C=N=CH2

Scheme 34

Negative-ion spectra of Co and Cu complexes showed pmi's as base peaks but Ni compounds gave $[NiL_2 - H_2]^-$ as base peak ions.

The oxovanadium(IV) complex of bis(salicylaldehyde)isobutyl-1,2-diamine showed fragmentations of the pmi analogous to those in Scheme 34, and also loss of H_2O, O, and both Me groups successively.[644] Some Co and Cu derivatives of bis(salicylidene)-1,2-diaminoethylene have been characterized.[645]

The spectra of Ni and Cu compounds of salicylaldehyde-2-hydroxyanil and some related ligands show peaks due to dimeric species.[603] Very few ions were observed in the range between the dimeric and monomeric pmi's, indicating the facile cleavage of the bridging bonds.

Derivatives of salicylaldimines (93)[646] and (94)[647] (M = Pd or Pt) all show pmi's. The dimethylgold complexes exhibit loss of both Me groups simultaneously from the pmi, giving $[Au(salNR)]^+$. The palladium and

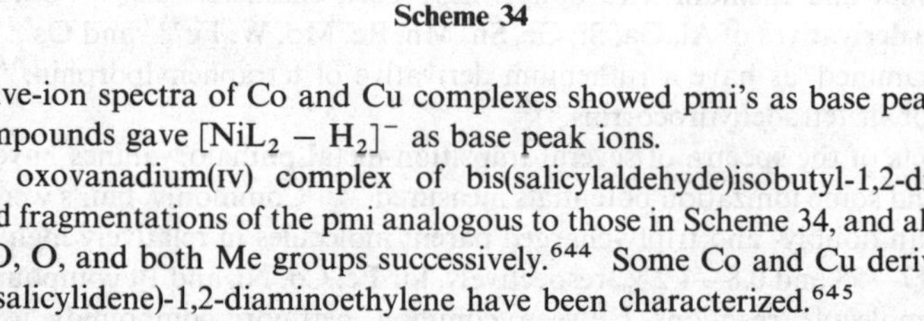
MR¹X
R²
(93) AuMe2
Me, C6H11 or Ph
(94) M−NMe2
Me, Et, C6H11, Ph or
p-ClC6H4
CH2

643 W. C. Gilbert, L. J. Taylor, and J. G. Dillard, *J. Amer. Chem. Soc.*, 1973, **95**, 2477.
644 K. S. Patel and J. C. Bailar, *J. Co-ordination Chem.*, 1973, **3**, 113.
645 H. Kanatomi and I. Murase, *Inorg. Chem.*, 1972, **11**, 1356.
646 K. S. Murray, B. E. Reichert, and B. O. West, *J. Organometallic Chem.*, 1973, **61**, 451.
647 B. E. Reichert and B. O. West, *J. Organometallic Chem.*, 1973, **54**, 391.

platinum complexes (94) give ions [pm − CH_3NCH_2]$^+$, [MeC_6H_4M]$^+$, and [M(salNR) − H]$^+$.[647]

Derivatives of bis(acetylacetonatoethylenedi-imine) with Cu,[648,649] Ni, Pd, and Pt,[648] and some fluorinated analogues and related compounds[650] have been examined. Some nickel ketazines (95) have been reported to show pmi's and

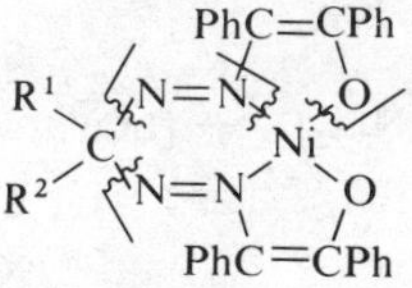

$$R^1 \underset{C}{\overset{\diagdown}{\diagup}} \underset{N=N}{\overset{N=N}{\diagdown}} \underset{Ni}{\overset{PhC=CPh}{\diagup}} \underset{O}{\overset{O}{\diagdown}}$$
$$R^2 \underset{}{\overset{}{\diagup}} \underset{N=N}{\overset{}{\diagdown}} \underset{PhC=CPh}{\overset{}{\diagup}} \underset{}{\overset{O}{\diagdown}}$$

(95) R^1 = Me, R^2 = Me, Et, Pr, Bu, or Ph
R^1 = Et, R^2 = Et

fragment by losing first N_2 then a CO group.[651] Other nickel-containing ions were produced by cleavage of Ni—O, Ni—N, and C—N bonds; the rearranged ions [NiXCPh$_2$]$^+$ (X = O or N) and [NiCPh$_2$]$^+$ were observed.

Chromium chelates of 2-formyl- and 2-acetyl-pyrrole (CrL$_3$) give pm and fragment ions by losses of H (or Me), L, and CO.[652] Metal oxine compounds, M(oxine)$_2$(M = Mn, Co, Ni, Cu, Zn, Pd, or Pb), all formed pmi's.[653] Both zinc and lead compounds fragmented by loss of one oxine group followed by elimination of the metal. The other derivatives lost an oxine group initially but then CO, followed by the metal (Scheme 35). Related compounds with (π-C_5H_5)-M(CO)$_2$ (M = Mo or W) groups have been investigated.[654]

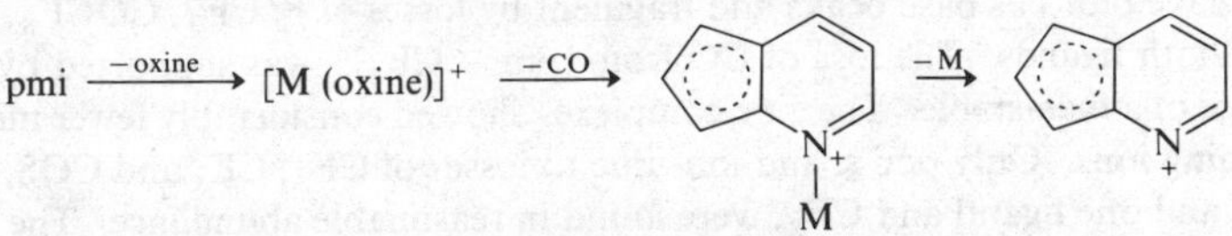

pmi $\xrightarrow{-\text{oxine}}$ [M (oxine)]$^+$ $\xrightarrow{-CO}$ [structure] $\xrightarrow{-M}$ [structure]

Scheme 35

Compounds with Metal–Sulphur Bonds.—The fragmentation pattern of Cu^{II} bis-dithiocarbamate, (S_2CNEt_2)$_2$Cu, was interpreted in the light of the valency-change concept (Scheme 36).[655] The pmi (96) fragmented mainly by losing even-electron species such as C_2H_4, C_2H_4S, and CS_2. The pmi from the corresponding

---

[648] B. Belcher, K. Blessel, T. Cardwell, M. Pravica, W. I. Stephen, and P. C. Uden, *J. Inorg. Nuclear Chem.*, 1973, **35**, 1127.
[649] P. C. Uden and K. Blessel, *Inorg. Chem.*, 1973, **12**, 352.
[650] E. F. Hasty, T. J. Colburn, and D. N. Hendrickson, *Inorg. Chem.*, 1973, **12**, 2414.
[651] C. M. Kerwin and G. A. Melson, *Inorg. Chem.*, 1972, **11**, 726.
[652] C. S. Davis and N. J. Gogan, *J. Inorg. Nuclear Chem.*, 1972, **34**, 2791.
[653] Y. Kidani, S. Naga, and H. Koike, *Bull. Chem. Soc. Japan*, 1973, **46**, 2105.
[654] H. Brunner and W. A. Herrmann, *Z. Naturforsch.*, 1973, **28b**, 606.
[655] J. F. Villa, D. A. Chatfield, M. M. Bursey, and W. E. Hatfield, *Inorg. Chim. Acta*, 1972, **6**, 332.

zinc compound, however, behaved as an odd-electron species, decomposing by loss of other odd-electron groups such as C_2H_5, C_2H_4S, and CHS_2 to give even-electron ions. Evidence of dimeric gas-phase species was found in the spectrum of the zinc derivative.

The spectra of trisdithiocarbamato-complexes of As, Sb, and Bi have been discussed.[656] Although pmi's were not always observed, [pm − S_2CNR_2]$^+$ ions

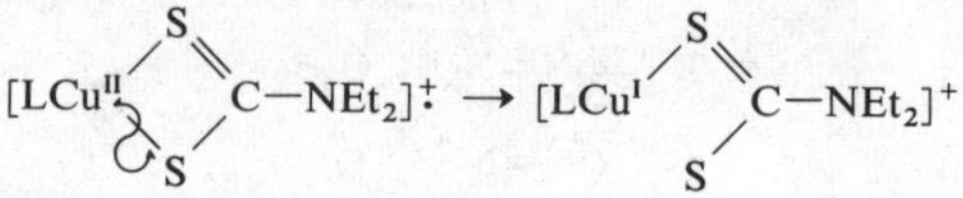

$$[LCu^{II} \underset{S}{\overset{S}{\diagdown}} C-NEt_2]^{\ddagger} \rightarrow [LCu^{I} \underset{S}{\overset{S}{\diagdown}} C-NEt_2]^+$$

(96)

Scheme 36

were; further fragmentation was by elimination of an MS group. The large number of ions derived from the ligands, and the observation of polynuclear sulphides up to M_4S_5, suggested that thermal decomposition of the complexes was occurring in the spectrometer.

Other thiocarbamato-complexes for which data have been reported include $M(S_2CNEt_2)_4$ (M = Ti, Zr, or V), $V(S_2CNEt_2)_3$ and $OV(S_2CNEt_2)_2$,[657] ONb, Mo_2O_3, and Re_2O_3 derivatives,[658] $(\pi\text{-}C_3H_5)Fe(CO)_2(S_2CNMe_2)$, $Rh(CO)_2\text{-}(S_2CNMe_2)$, and related compounds.[659]

Compounds with Metal–Oxygen and Metal–Sulphur Bonds.—The spectra of some Ni^{II} and Zn^{II} derivatives of monothio-β-diketones, $RC(SH)=CHCOCF_3$ (R = Ph, $p\text{-}MeC_6H_4$, 2-thienyl, or Me), have been presented.[660] The nickel complexes have pmi's as base peaks and fragment by losses of F, CF_3, $COCF_3$, and one or both ligands. The loss of CO from [pm − CF_3]$^+$ was supported by the presence of a metastable. The zinc complexes showed considerably fewer metal-containing ions. Only pmi's, and ions due to losses of CF_3, CF_3 and COS, one ligand, and one ligand and CF_2, were found in reasonable abundance. The loss of CF_2 from [pm − L]$^+$ was suggested to occur as shown in Scheme 37.

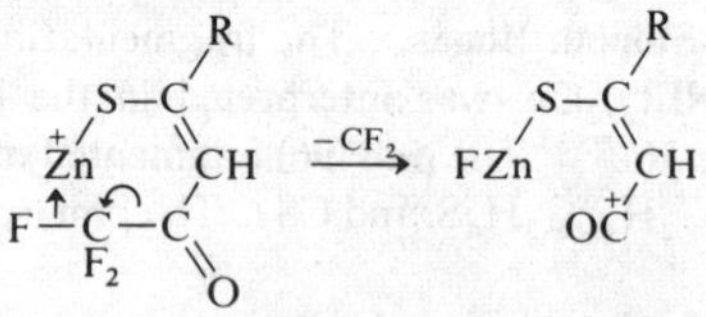

Scheme 37

[656] G. E. Manoussakis, E. D. Micromastoras, and C. A. Tsipis, *Z. anorg. Chem.*, 1974, **403**, 87.
[657] D. C. Bradley, I. F. Rendall, and K. D. Sales, *J.C.S. Dalton*, 1973, 2228.
[658] A. I. Casey, D. J. Mackey, R. L. Martin, and A. H. White, *Austral. J. Chem.*, 1972, **25**, 477.
[659] E. W. Abel and M. O. Dunster, *J.C.S. Dalton*, 1973, 98.
[660] M. Das and S. E. Livingstone, *Austral. J. Chem.*, 1974, **27**, 53.

Monothiotrifluoroacetylacetonates of Ni^{II}, Pd^{II}, Pt^{II}, Zn^{II} and Co^{III} have been examined.[661] The palladium and platinum compounds are similar to the nickel derivative. The Co^{III} compound shows a pmi and base peak due to $[CoL_2]^+$; other ions observed include $[pm - L - Me]^+$, $[pm - 2L]^+$, and $[pm - 2L - CF_3]^+$.

5 Inorganic Compounds

Group I.—Some studies of high-temperature vaporization of compounds containing Li, Na, and Cs have been reported. They include lithium oxide,[662] caesium nitrate,[663] and caesium halide[664] systems and mixtures of lithium fluoride with gallium,[665] aluminium, and scandium trifluorides, sodium fluoride with beryllium, magnesium, barium, aluminium, scandium, yttrium, lanthanum, zirconium, and vanadium fluorides[666] and caesium fluoride with aluminium fluoride.[667] The spectrum of lithium perrhenate, vaporized at 480—680 °C, showed ions due to $[LiReO_n]^+$ ($n = 0$—4), $[ReO_4]^+$, and the dimeric species $[(LiReO_4)_2]^+$.[668]

Group II.— Barium oxide,[669,670] and zinc[671] and cadmium[672] sulphides and selenides have been the subjects of high-temperature studies. The identification of $[CdClBr]^+$ from mixtures of $CdCl_2$ and $CdBr_2$ vaporized at 450—500 °C has been reported.[673] Mixtures of beryllium chloride with aluminium chloride,[674] beryllium chloride with chlorides of indium, iron, zirconium, or uranium, and cadmium chloride with thallium chloride, and zinc chloride with indium chloride[675] have been studied. Barium perrhenate gave $[BaReO_4]^+$ as the ion of greatest mass as well as the ions $[BaReO_n]^+$ ($n = 1$—3) and $[BaO]^+$.[676]

---

[661] R. Belcher, W. I. Stephen, I. J. Thomson, and P. C. Uden, *J. Inorg. Nuclear Chem.*, 1972, **34**, 1017.
[662] D. L. Hildenbrand, *J. Chem. Phys.*, 1972, **57**, 4556.
[663] L. L. Ames, J. L.-F. Wang, and J. L. Margrave, *Inorg. Nuclear Chem. Letters*, 1973, **9**, 1243.
[664] J. Berkowitz, J. L. Dehmer, and T. E. H. Walker, *J. Chem. Phys.*, 1973, **59**, 3645.
[665] L. N. Siderov, N. A. Zhegul'skaya, and M. V. Korobov, *Russ. J. Phys. Chem.*, 1973, **47**, 759.
[666] L. N. Siderov and V. B. Shol'ts, *Internat. J. Mass. Spectrometry Ion Phys.*, 1972, **8**, 437.
[667] E. N. Kolosov, T. N. Tuvaeva, L. N. Siderov, and P. A. Akishin, *Russ. J. Phys. Chem.*, 1973, **47**, 603.
[668] K. Skudlarski and W. Lukas, *J. Less-Common Metals*, 1973, **33**, 171.
[669] I. G. Panchenov, A. V. Gusanov, and L. N. Goronkov, *Russ. J. Phys. Chem.*, 1973, **47**, 55.
[670] G. A. Semenov, O. S. Popkov, A. I. Solveichik, and S. N. Persiyaninova, *Russ. J. Phys. Chem.*, 1972, **46**, 898.
[671] Yu. M. Korenov, N. M. Karasev, I. A. Timoshin, T. A. Volkova, L. N. Siderov, and A. V. Novoselova, *Russ. J. Phys. Chem.*, 1972, **46**, 984.
[672] N. M. Karasev, Yu. M. Korenov, I. A. Timoshin, L. N. Siderov, and A. V. Novoselova, *Russ. J. Phys. Chem.*, 1972, **46**, 986.
[673] H. Bloom and R. G. Anthony, *Austral. J. Chem.*, 1972, **25**, 23.
[674] R. Rabeneck and H. Schafer, *Z. anorg. Chem.*, 1973, **395**, 69.
[675] M. Binneuries and H. Schafer, *Z. anorg. Chem.*, 1973, **395**, 77.
[676] G. A. Semenov, E. N. Nikolaev, and I. G. Opendak, *Russ. J. Inorg. Chem.*, 1972, **17**, 943.

Group III.—The only boron-containing negative ion observed from BF_3 was $[BF_2]^-$.[677] Reaction of boron trihalides with $[SF_6]^-$ gave SF_5 and $[BX_3F]^-$ (X = F or Cl). The characterization of B_8F_{12} and related compounds,[678] and the fragmentation of a number of metal chlorides, including aluminium and gallium trichlorides,[679] have been reported. The positive and negative ions from some indium chlorides and the ionization and appearance potentials of some of the ions have been reported[680] as have vaporization studies of mixtures of chlorides of indium with uranium, and thallium with lead, thorium, and uranium.[675] Other systems studied include alumina,[681] aluminium carbides,[682] gallium bismuthide,[683] boron phosphide,[684] thallium metaborate and mixtures of aluminium trifluoride and thallium monofluoride.[685]

Boron Hydrides and Related Compounds. From a study of the relative abundances of doubly-charged ions, $[B_nH_x]^{2+}$, derived from a number of boron hydrides (B_nH_y, where $x \leqslant y$) and some derivatives, it was concluded that, in general, there was a direct correlation between the ratio of the sum of all intensities of doubly-charged ions with n B atoms to the sum of all intensities of their singly-charged counterparts, $\sum_0^x [B_nH_x]^{2+} / \sum_0^x [B_nH_x]^+$, and the number of possible resonance structures of the parent compound.[686] An exception appeared to be $B_{10}H_{16}$ which had five times the ratio found for $B_{10}H_{14}$ but fewer resonance structures (84 compared with 111). It was suggested that $B_{10}H_{16}$, containing two boron cages, could stabilize the double charge by a charge separation mechanism. The presence of a donor ligand such as PR_3 or SR_2 also appeared to increase the ratio.

Low-temperature (-190 to $-130\,^\circ C$) CI mass spectra of B_2H_6, B_4H_{10}, and B_5H_9 have been obtained with methane as reactant gas.[687] Ions formed by addition of a proton to the parent molecule were observed and the proton affinities were calculated for B_2H_6 ($6.35 \pm 0.2\,eV$) and B_4H_{10} ($6.25 \pm 0.2\,eV$). A comparison with values for other compounds revealed that the saturated boron hydrides are weaker bases than unsaturated ones (see also Chapter 3).

The characterization of B_3H_7,[688] suggested to be a reactive intermediate in several reactions of boron hydrides, has been achieved with a molecular beam mass spectrometric technique. Triborane(7) was produced by pyrolysing $(Me_2N)F_2PB_3H_7$ at 175—200 °C. The small pmi observed for triborane(7) suggested the presence of a BH_2 group and this was later confirmed by an X-ray

---

[677] J. A. Stockdale, D. R. Nelson, F. J. Davis, and R. N. Compton, *J. Chem. Phys.*, 1972, **56**, 3336.
[678] R. W. Kirk, D. L. Smith, W. Airey, and P. L. Timms, *J.C.S. Dalton*, 1972, 1392.
[679] H. Preiss, *Z. anorg. Chem.*, 1972, **389**, 280.
[680] A. Sh. Sultanov, *Russ. J. Inorg. Chem.*, 1972, **17**, 309.
[681] M. Farber, R. D. Srivastava, and O. M. Uy, *J.C.S. Faraday I*, 1972, 249.
[682] C. A. Stearus and F. J. Kohl, *J. Phys. Chem.*, 1973, **77**, 136.
[683] V. Piacente and A. Desideri, *J. Chem. Phys.*, 1972, **57**, 2213.
[684] K. A. Gingerich, *J. Chem. Phys.*, 1972, **56**, 4239.
[685] D. H. Feather and A. Buchler, *J. Phys. Chem.*, 1973, **77**, 1600.
[686] L. C. Ardini and T. P. Fehlner, *Internat. J. Mass Spectrometry Ion Phys.*, 1972/73, **10**, 489.
[687] R. C. Pierce and R. F. Porter, *J. Amer. Chem. Soc.*, 1973, **95**, 3849.
[688] R. T. Paine, G. Sodeck, and F. E. Stafford, *Inorg. Chem.*, 1972, **11**, 2593.

structure analysis of B_3H_7CO.[689] However, the carbonyl adduct did not show a pmi, unlike either the $(Me_2N)F_2P$ or F_3P adducts. The pyrolytic formation of the unstable nonaborane B_9H_{13} has been reported,[690] and a number of derivatives $B_9H_{13}L$ have been characterized including $B_9H_{13}CO$.[691]

The dicarbonyl adduct $B_2H_4(CO)_2$ has been reported to show both a pm and $[pm - H]^+$ ion.[692] The existence of the heptaborane, B_7H_{11}, has been asserted from mass spectral evidence but others claimed to be 'B_7H_{12}' and 'B_7H_{13}' are probably mixtures.[693]

Mass spectrometry has been used to study intramolecular hydrogen exchange reactions of B_4H_{10}[694] and B_9H_{15},[695] and reactions of BH_3 with B_2H_6 and B_5H_9 to give B_3H_y and B_6H_y.[696]

Derivatives of boranes for which data have been reported include a novel series of cyclic cyanoboranes $(BH_2CN)_n$ $(n = 4$—$9)$,[697] aminoboranes $H_2B(NMe_2)BH_2(NMe_2)BH_3$ and $H_2B(NMe_2)_2Al(BH_4)_2$,[698] and the adducts F_2XPBH_3 and $F_2XPB_4H_8$ (X = Cl, Br, or I),[699] $B_{18}H_{20}(C_5H_5N)_2$ and $B_{18}H_{20}$-$CNH_2C_6H_{11}$.[700]

Boron–Nitrogen Compounds, including Borazines. The spectra of some unsymmetrically B-substituted borazines have been discussed with regard to the relative stabilities of the pmi's and the ions formed by loss of a B- or N-substituent or H.[701] The ions formed by loss of a B-substituent may be stabilized

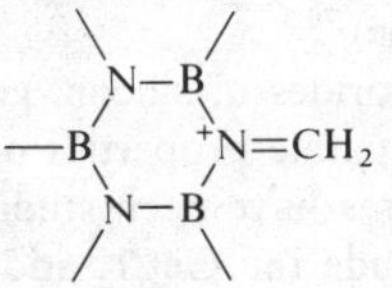

(97)

by release of electron density from adjacent nitrogen atoms. By loss of H, borazines with *N*-methyl groups may form an immonium type ion (97), which can be stabilized by electron-releasing groups on adjacent boron atoms. The calculation of the ion abundances in the above paper[701] has been criticized[702]

[689] J. D. Glore, J. W. Rathke, and R. Schaeffer, *Inorg. Chem.*, 1973, **12**, 2209.
[690] L. C. Ardini and T. P. Fehlner, *Inorg. Chem.*, 1973, **12**, 799.
[691] R. Schaeffer and E. Walter, *Inorg. Chem.*, 1973, **12**, 2209.
[692] J. W. Rathke and R. Schaeffer, *Inorg. Chem.*, 1974, **13**, 760.
[693] E. McLaughlin and R. W. Rozett, *Inorg. Chem.*, 1972, **11**, 2567.
[694] R. Schaeffer and L. G. Sneddon, *Inorg. Chem.*, 1972, **11**, 3098.
[695] R. Schaeffer and L. G. Sneddon, *Inorg. Chem.*, 1972, **11**, 3102.
[696] S. A. Fridmann and T. P. Fehlner, *Inorg. Chem.*, 1972, **11**, 396.
[697] B. F. Spielvogel, R. F. Bratton, and C. G. Moreland, *J. Amer. Chem. Soc.*, 1972, **94**, 8597.
[698] P. C. Keller, *J. Amer. Chem. Soc.*, 1972, **94**, 4020.
[699] R. T. Paine and R. W. Parry, *Inorg. Chem.*, 1972, **11**, 1237.
[700] R. L. Sneath and L. J. Todd, *Inorg. Chem.*, 1973, **12**, 44.
[701] L. A. Melcher, J. L. Adcock, G. A. Anderson, and J. J. Lagowski, *Inorg. Chem.*, 1973, **12**, 601.
[702] J. M. Miller and G. L. Wilson, *Inorg. Chem.*, 1974, **13**, 498.

because the deconvolution procedure required to obtain accurate ion abundances from the spectra was not used with enough rigour. However, although the ion abundances were changed, the conclusions were not materially affected.

Another series of borazines containing an N-cyclohexyl group have all been reported to show pm and base peak ions due to $[pm - C_3H_7]^+$.[703] Other compounds studied include the cyclic trimers $[RNC(X)HBH_2]_3$ (R = Me or Ph, X = O; R = Me, X = S),[704] B-substituted derivatives of 2,5-dimethyltetrazaboroline,[705] and H_2NBCl_2.[706]

Group IV.—A detailed study of the positive and negative ions formed from germanium tetrafluoride has been reported.[707,708] Appearance potentials of the ions were measured and modes of formation discussed. From the data, bond dissociation energies of several species were deduced and trends noted for the fluorides of carbon, silicon, and germanium were compared with those reported for the analogous hydrides. In the case of the fluorides, the trends in bond dissociation energies were correlated with suggested changes in electron configurations.

The fragmentation of SiH_4 and the appearance potentials of the ions $[SiH_x]^+$ ($x = 0$—3) have been reported.[709] Ion–molecule reactions of species derived from SiH_4[710] and GeH_4[711] produce ions containing up to four Si atoms and three Ge atoms. Silicon hydrides and derivatives which have been characterized include linear Si_5H_{12},[712] cyclic $(SiH_2)_5$,[713] $H_3SiNHPF_2$,[714] and ammonium salts $NH_4(XSiH_3)$ (X = S or Se).[715]

The spectra of the tetrachlorides of silicon, germanium, and tin have been reported.[679] Some thermodynamic properties of $SnCl_2$, $SnBr_2$, and of some species produced from mixtures have been studied.[716] Other compounds for which data are available include the $GeCl_2$ adducts of benzothiazole and N-methylbenzimidazole,[717] R_2NSiF_3,[718] $(F_3Si)_2NH$,[719] and compounds with

[703] P. Powell, *Inorg. Chem.*, 1973, **12**, 913.
[704] R. Molinelli, S. R. Smith, and J. Tanaka, *J.C.S. Dalton*, 1972, 1363.
[705] B. Hessett, J. B. Leach, J. H. Morris, and P. G. Perkins, *J.C.S. Dalton*, 1972, 131.
[706] M. L. Pinsky and A. C. Bond, *Inorg. Chem.*, 1973, **12**, 605.
[707] P. W. Harland, S. Cradock, and J. C. J. Thynne, *Internat. J. Mass Spectrometry, Ion Phys.*, 1972/73, **10**, 169.
[708] P. W. Harland, S. Cradock, and J. C. J. Thynne, *Inorg. Nuclear Chem. Letters*, 1973, **9**, 53.
[709] J. D. Morrison and J. C. Traeger, *Internat. J. Mass Spectrometry Ion Phys.*, 1973, **11**, 277.
[710] T. Y. Yu, T. M. H. Cheng, V. Kempter, and F. W. Lampe, *J. Phys. Chem.*, 1972, **76**, 3321.
[711] J. K. Northrop and F. W. Lampe, *J. Phys. Chem.*, 1973, **77**, 30.
[712] F. Hofler and R. Jannach, *Inorg. Nuclear Chem. Letters*, 1973, **9**, 723.
[713] E. Hengge and C. Bauer, *Angew. Chem. Internat. Edn.*, 1973, **12**, 316.
[714] D. E. J. Arnold, E. A. V. Ebsworth, H. F. Jessep, and D. W. H. Rankin, *J.C.S. Dalton*, 1972, 1681.
[715] S. Cradock, E. A. V. Ebsworth, and H. F. Jessep, *J.C.S. Dalton*, 1972, 359.
[716] S. Ciach, D. J. Knowles, A. J. C. Nicholson, and D. L. Swingler, *Inorg. Chem.*, 1973, **12**, 1443.
[717] P. Jutzi, H. J. Hoffmann, D. J. Brauer, and C. Kruger, *Angew. Chem. Internat. Edn.*, 1973, **12**, 1003.
[718] B. J. Aylett, I. A. Ellis, and C. J. Porritt, *J.C.S. Dalton*, 1973,83.
[719] M. Allan, B. J. Aylett, I. A. Ellis, and C. J. Porritt, *J.C.S. Dalton*, 1973, 2675.

silicon–silicon bonds $F_2Si_3H_6$, XSi_3H_7 (X = F, Cl,[720] or Br),[721] and $H_nSi(SiF_3)_{4-n}$ (n = 2 or 3).[722]

High-temperature studies of the molecules Si_2NO[723] and GeO_2[724] have been reported. The gas-phase hydration of $[Pb]^{+}_{+}$ by up to eight molecules of water has been studied.[725]

Group V.—The use of a quadrupole mass spectrometer in an electron-impact study of PH_3 and NH_3 has been reported.[726] Ionization and appearance potentials for all $[MH_n]^+$, (n = 0—3) ions were measured. Phosphylene (PH) has been identified as a pyrolysis product of H_3MPH_2 (M = Si or Ge).[727]

A comparatively large number of papers deal with Main Group V halides or their derivatives, particularly F_2PX compounds. Aspects of the spectra of penta- and tri-halides of phosphorus and antimony, the mixed halide $SbCl_4F$, compounds PCl_4PF_6, $AsCl_4,SbCl_6,AsCl_3$ and $PCl_4,(MCl_6),AsCl_3$ (M = Sb, Nb, or Ta), and adducts such as MF_5CH_3CN (M = P, As, or Sb) have been discussed.[679] Only monomeric ions were present in the spectra of phosphorus and arsenic pentafluorides, but ions corresponding to $[M_3F_{14}]^+$ were found from the bismuth and antimony compounds and, in the case of SbF_5, traces of higher polymeric species were observed.[728,729] Negative ions from reactions between $[SF_6]^-$ and AsF_3 or AsF_5 have been examined.[730]

The positive- and negative-ion spectra of PF_2NCO, PF_2NCS,[731] and PF_2CN[731,732] have been reported, together with the ionization and appearance potentials of principal ions from PF_2CN.[732] The positive-ion spectra showed pmi's, $[PF_2]^+$ as base peak, and a number of rearranged ions including $[PN]^+$ from F_2PCN, $[PF_nO]^+$ (n = 0—2) from F_2PNCO, and $[PC]^+$ from both the cyanate and thiocyanate. In each spectrum, doubly-charged ions were in relatively high abundance, the most abundant being the doubly-ionized parent molecules. A study of positive and negative ions from bis(difluorophosphine) carbodi-imide produced no secure evidence for the carbodi-imide rather than cyanimide structure[733] although several interesting fragmentations supported by metastable peaks were observed, including loss of PF_3 from the pmi.

[720] R. L. Jenkins, A. J. Vanderwielen, S. P. Ruis, S. R. Gird, and M. A. Ring, *Inorg. Chem.*, 1973, **12**, 2968.
[721] T. C. Geisler, C. G. Cooper, and A. D. Norman, *Inorg. Chem.*, 1972, **11**, 1710.
[722] D. Sloan and A. B. Burg, *Inorg. Chem.*, 1972, **11**, 1253.
[723] D. W. Muenow, *J. Phys. Chem.*, 1973, **77**, 970.
[724] E. K. Kazenas, D. M. Chizhikov, Yu. V. Tsevetkov, and Yu. V. Vasyuta, *Russ. J. Phys. Chem.*, 1973, **47**, 389.
[725] I. N. Tang and A. W. Castleman, *J. Chem. Phys.*, 1972, **57**, 3638.
[726] J. D. Morrison and J. C. Traeger, *Internat. J. Mass Spectrometry Ion Phys.*, 1973, **11**, 277.
[727] L. E. Elliot, P. Estucio, and M. A. Ring, *Inorg. Chem.*, 1973, **12**, 2193.
[728] M. J. Vasile and W. E. Falconer, *Inorg. Chem.*, 1972, **11**, 2282.
[729] M. J. Vasile, G. R. Jones, and W. E. Falconer, *Internat. J. Mass Spectrometry Ion Phys.*, 1972/73, **10**, 457.
[730] T. C. Rhyme and J. G. Dillard, *Inorg. Chem.*, 1974, **13**, 322.
[731] D. W. H. Rankin, P. W. Harland, and J. C. J. Thynne, *Inorg. Nuclear Chem. Letters*, 1972, **8**, 1101.
[732] P. W. Harland, D. W. H. Rankin, and J. C. J. Thynne, *Inorg. Chem.*, 1973, **12**, 1442.
[733] D. W. H. Rankin, *J.C.S. Dalton*, 1972, 869.

Some fluorophosphine derivatives giving pmi's include $F_2P(X)N_3$ (X = O or S),[734] F_2PN_3,[734,735] and the borane adduct,[735] $(CF_3)_2NOPF_2$ and $(CF_3)_2NOP(O)F_2$,[736] and $(MeS)_nPF_{3-n}$ (n = 1 or 2).[737] Compounds for which data have been reported but which fail to show pmi's include H_2NPF_4,[738] Bu^tOPF_2,[739] $\{(CF_3)_2NO\}_2PF_3$ and $\{(CF_3)_2NO\}_3PF_2$.[736]

Other compounds containing Main Group V atoms which have been studied include $Cl_2PN(NSOF)_2$ and $Cl_2PN(NSOCl)NSOPh$,[740] and $(Cl_2PN)_2NSOF$.[741]

A study of the esters (98)—(101) included the measurement of ionization and appearance potentials.[742] General conclusions from this study were that the

$(MeO)_3PO$	$(MeO)_2(MeS)PO$	$(MeO)(MeS)_2PO$	$(MeO)_2(MeS)PS$
(98)	(99)	(100)	(101)

presence of one or more sulphur atoms considerably altered the fragmentation patterns and lowered the ionization potentials of compounds (99)—(101) compared with (98). A number of ions were suggested as being formed by rearrangement of H or CH_3, *e.g.* formation of (102) in the spectrum of (101) (Scheme 38). Similar rearrangements had been noted previously.[743]

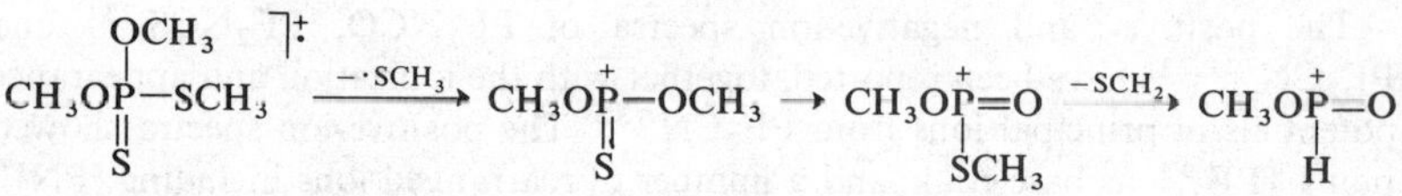

Scheme 38

Other compounds containing P—O bonds for which data have been reported include $(MeO)_2PH$,[744] and a series of phosphoramidic acid esters.[745] (Other P—O bonded compounds containing P—C bonds are discussed in Section 2).

A number of phosphorus sulphides and selenides and derivatives have been characterized. Ionization potential measurements were used to confirm the presence of pmi's from the series P_4S_x (x = 3, 5, 7, or 10).[746] No pmi's were

---

[734] S. R. O'Neill and J. M. Shreeve, *Inorg. Chem.*, 1972, **11**, 1629.
[735] E. L. Lines and L. F. Centofanti, *Inorg. Chem.*, 1972, **11**, 2269.
[736] C. S.-C. Wang and J. M. Shreeve, *Inorg. Chem.*, 1973, **12**, 81.
[737] R. Forester and K. Cohn, *Inorg. Chem.*, 1972, **11**, 2590.
[738] A. H. Cowley and J. R. Schweiger, *J.C.S. Chem. Comm.*, 1972, 560.
[739] E. L. Lines and L. F. Centofanti, *Inorg. Chem.*, 1973, **12**, 2718.
[740] H. H. Baalmann and J. C. Van de Grampel, *Rec. Trav. chim.*, 1973, **92**, 716.
[741] H. H. Baalmann and J. C. Van de Grampel, *Rec. Trav. chim.*, 1973, **92**, 1237.
[742] E. Santoro, *Org. Mass Spectrometry*, 1973, **7**, 589.
[743] See ref. 3, pp. 412, 414.
[744] L. F. Centofanti, *Inorg. Chem.*, 1973, **12**, 1131.
[745] P. Jakobsen, S. Treppendahl, and J. Wieczorkowski, *Org. Mass Spectrometry*, 1972, **6**, 1303.
[746] D. W. Muenow and J. L. Margrave, *J. Inorg. Nuclear Chem.*, 1972, **34**, 89.

observed either in a series of amine derivatives of P_4Se_7,[747] or in the compounds $P_2Se_nI_4$ ($n = 1$ or 2) and $PSeI_3$.[748]

High-temperature studies of the species in vaporized Sb_2O_3,[749] Bi_2O_3,[750] and elemental antimony[751] have been reported.

Phosphorus–Nitrogen Compounds, including Phosphonitrile Derivatives. The di-iminophosphorane (103) has a pmi and expected fragment ions such as the $[pm - Me]^+$ ion (as base peak).[752] Other compounds reported include phosphino-derivatives of hydrazine,[753] a series of arylaminobisdichlorophosphines

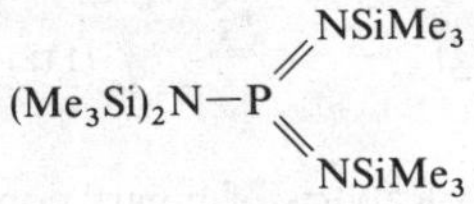

(103) 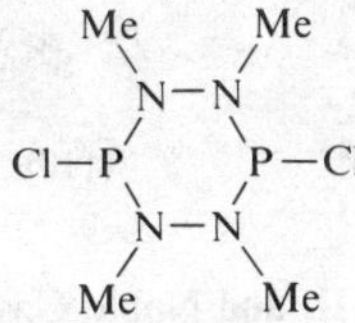 (104)

$RN(PCl_2)_2$, some compounds with four-membered ring systems, dichlorodiazadiphosphetidines $(RN-PCl)_2$,[754] a number of compounds containing six-membered ring systems (104),[753] (105) [R = $Me_3Sn(Me)N-$ or $(O)PF_2-(Me)N-$],[755] (106),[756] (107),[757] (108; R = Me, Pr, or Bu),[758] (109; X = Cl or

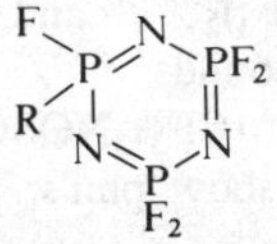

(105) 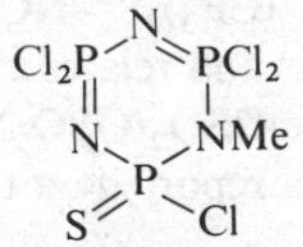 (106) 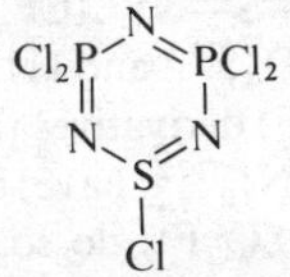 (107) 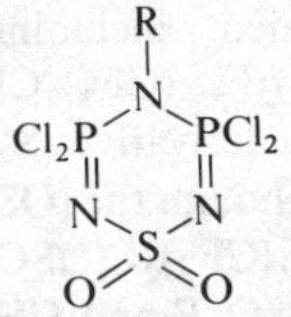 (108)

[747] C. D. Mickey and R. A. Zingaro, *Inorg. Chem.*, 1973, **12**, 2115.
[748] M. Baudler, B. Volland, and H. W. Valpertz, *Chem. Ber.*, 1973, **106**, 1049.
[749] E. K. Kazenas, D. M. Chizhikov, Yu. V. Tsvetkov, and M. V. Ol'shevsku, *Russ. J. Phys. Chem.*, 1973, **47**, 879.
[750] E. K. Kazenas, D. M. Chizhikov, Yu. V. Tsvetkov, and M. V. Ol'shevsku, *Proc. Acad. Sci. U.S.S.R.*, 1973, **207**, 867.
[751] B. Cabaud, A. Hoareau, P. Nounon, and R. Uzan, *Internat. J. Mass Spectrometry Ion Phys.*, 1973, **11**, 157.
[752] E. Niecke and W. Flick, *Angew. Chem. Internat. Edn.*, 1974, **13**, 134.
[753] H. Noth and R. Ullmann, *Chem. Ber.*, 1974, **107**, 1019.
[754] A. R. Davies, A. T. Dronsfield, R. N. Haszeldine, and D. R. Taylor, *J.C.S. Perkin I*, 1973, 379.
[755] H. W. Roesky and H. Wiezer, *Chem. Ber.*, 1973, **106**, 280.
[756] H. W. Roesky, *Chem. Ber.*, 1972, **105**, 1439.
[757] H. W. Roesky, *Angew. Chem. Internat. Edn.*, 1972, **11**, 642.
[758] U. Klingebiel and O. Glemser, *Chem. Ber.*, 1972, **105**, 1510.

NR^{1}R^2),[759] and (110),[760] and some eight-membered ring systems N$_4$P$_4$Cl$_n$X$_{8-n}$, ($n = 4$, X = thioalkyl or thioaryl;[761] $n = 3$—7, X = isothiocyanate).[762] Other compounds containing ring systems have included (111),[763] in which the pmi is the base peak and which fragments by loss of NMe, N$_2$Me$_2$, and phosphorus. Fused-ring systems such as in N$_5$S$_3$PF$_2$,[764] and nitrilohexaphosphonitrilic chloride N$_7$P$_6$Cl$_9$,[765] have been described.

(109)   (110)   (111)   (112)

Groups VI, VII, and Noble Gas Compounds.—Reports of binary compounds of Main Group VI elements include a thermodynamic study of the species SF$_n$ ($n = 6, 4, 2,$ or 1),[766] the characterization of S$_8$O,[767] and the spectrum of H$_2$Se obtained with an ICR spectrometer.[768] This last study included ionization and appearance potential measurements and calculation of proton affinities of [HSe]$^-$ (1416 ± 21 kJ mol^{-1}) and H$_2$Se (710 ± 13 kJ mol^{-1}). Negative-ion formation from MF$_6$ (M = S, Se, or Te) has been studied and trends in bond-dissociation energies in [MF$_n$]$^-$ ions ($n = 4$—6) have been compared.[769]

The spectra of a number of compounds containing (S=N) bonds have been recorded, including F$_2$S=NC(O)NCO,[770] (CF$_3$)$_2$C=NC(CF$_3$)$_2$N=S=NC-(CF$_3$)$_2$F, (CF$_3$)$_2$CFN=S=NFC(CF$_3$)$_2$, (CF$_3$)$_2$S=NCF(CF$_3$)$_2$,[771] and the cyclic derivatives S$_4$N$_4$O$_4$,[772] and (112)[773] and related compounds.

Although the (OSO$_2$F) derivatives (F$_5$Se)OSO$_2$(OSO$_2$F),[774] (CF$_3$)$_2$NOSO$_2$F, and {(CF$_3$)$_2$C(OSO$_2$F)N}$_2$[775] have been reported not to show pmi's, both MeOSO$_2$F and CH$_2$(OSO$_2$F)$_2$ do so.[775]

[759] U. Klingebiel, T.-P. Lin, B. Buss, and O. Glemser, *Chem. Ber.*, 1973, **106**, 2969.
[760] H. W. Roesky and L. F. Grimm, *Angew. Chem. Internat. Edn.*, 1972, **11**, 642.
[761] A. P. Carroll, R. A. Shaw, and M. Woods, *J.C.S. Dalton*, 1973, 2737.
[762] R. L. Dieck and T. Moeller, *Inorg. Nuclear Chem. Letters*, 1972, **8**, 763.
[763] R. Goetze, H. Noth, and D. S. Payne, *Chem. Ber.*, 1972, **105**, 2637.
[764] H. W. Roesky and O. Peterson, *Angew. Chem. Internat. Edn.*, 1973, **12**, 415.
[765] R. T. Oakley and N. L. Paddock, *Canad. J. Chem.*, 1973, **51**, 520.
[766] D. L. Hildenbrand, *J. Phys. Chem.*, 1973, **77**, 897.
[767] R. Steudel and M. Rebsck, *Angew. Chem. Internat. Edn.*, 1972, **11**, 302.
[768] D. A. Dixon, D. Holtz, and J. L. Beauchamp, *Inorg. Chem.*, 1972, **11**, 960.
[769] J. C. J. Thynne and P. W. Harland, *Internat. J. Mass. Spectrometry Ion Phys.*, 1973, **11**, 137; P. W. Harland and J. C. J. Thynne, *Inorg. Nuclear Chem. Letters*, 1973, **9**, 265.
[770] A. F. Clifford, J. S. Harman, and C. A. McAuliffe, *Inorg. Nuclear Chem. Letters*, 1972, **8**, 567.
[771] R. M. Swindell and J. M. Shreeve, *J. Amer. Chem. Soc.*, 1972, **94**, 5713.
[772] H. W. Roesky and O. Peterson, *Angew. Chem. Internat. Edn.*, 1972, **11**, 918.
[773] H. W. Roesky and B. Kuhtz, *Chem. Ber.*, 1974, **107**, 1.
[774] K. Seppelt, *Chem. Ber.*, 1972, **105**, 3131.
[775] R. L. Kirchmeier and J. M. Shreeve, *Inorg. Chem.*, 1973, **12**, 2886.

Compounds containing selenium or tellurium oxyfluoride groups have received attention. Selenium oxytetrafluoride gives a pmi and ions $[SeOF_n]^+$ and $[SeF_n]^+$ ($n = 0$—3).[776] The ion at highest mass in the spectrum of $(SeOF_4)_2$ is $[Se_2O_2F_7]^+$ whereas for the tellurium analogue it is the pmi.[777] Several F_5SeOX derivatives have been reported with X = H,[778] Cl, SeF_5, Xe,[779] $OSeF_5$,[779,780] XeF, $XeOSeF_5$,[781] and $COCF_3$,[774] and so has $(F_5SeO)_3Br$.[779] Two of the xenon derivatives (X = Xe and XeF) had pmi's. Similar tellurium compounds have been studied, F_5TeOX (X = Cl, $OTeF_5$, $HgOTeF_5$,[780] or $B(OTeF_5)_2$[782]]; all gave pmi's. Other tellurium compounds studied were the amino-derivatives, H_2NTeF_5 and $Me_3Si(H)NTeF_5$, and a polymer $\{HgN(TeF_5)\}_x$.[783]

Two groups have reported the spectrum of $ClOF_3$.[784,785] No pmi was observed and the base peak was either $[ClOF_2]^{+\cdot}$ [784] or $[ClFO]^{+\cdot}$.[785] The negative-ion spectrum of $ClOF_5$ had $[ClOF_4]^-$ as the ion at highest mass, and other ions included $[ClOF_n]^-$ ($n = 0$—3) and $[ClF_n]^-$ ($n = 0$—4).[784] The spectra of compounds IF_4OR (R = Me or Et) are reported to have pmi's and a base peak $[IF_2O]^+$.[786]

Transition Metal Groups.—A specially constructed mass spectrometer with a molecular beam source has been used to study several pentafluorides MF_5 (M = V, Nb, Ta, Cr, Mo, W, Re, Ru, Os, Rh, Ir, and Pt).[729] For the majority (M = Nb, Ta, Mo, Re, Os, Rh, Ir, and Pt) there was evidence of associated species in the gas phase. Dimeric and trimeric ions were most readily detected and tetrameric ions were found for M = Nb, Mo, Ru, Rh, and Ir. Although the extent of association in the neutral species was difficult to ascertain quantitatively because of the fragmentation of the ions, it was possible to show that the ions observed did not originate from either ion–molecule reactions or clustering during isotropic expansion. Of the other pentafluorides studied, it seems probable that those of V and Cr are not associated in the gas phase. Tungsten pentafluoride was found to disproportionate to WF_6 and WF_4 to such an extent that nothing definitive could be said about the gas-phase composition of WF_5 under the conditions used. Another report of the spectra of NbF_5 and TaF_5 confirms the formation of polymeric ions.[679] Ions with up to four metal atoms are reported for both Nb and Ta whereas in the molecular beam study tetrameric species were observed only from NbF_5. The spectra of the mixed halides, MCl_4F (M = Nb or Ta), and the

---

[776] K. Seppelt, *Angew. Chem. Internat. Edn.*, 1974, **13**, 91.
[777] K. Seppelt, *Angew. Chem. Internat. Edn.*, 1974, **13**, 92.
[778] K. Seppelt, *Angew. Chem. Internat. Edn.*, 1972, **11**, 630.
[779] K. Seppelt, *Chem. Ber.*, 1973, **106**, 157.
[780] K. Sepplet and D. Nothe, *Inorg. Chem.*, 1973, **12**, 2727.
[781] K. Seppelt, *Angew. Chem. Internat. Edn.*, 1972, **11**, 723.
[782] F. Sladky, H. Kropshofer, and O. Leitzke, *J.C.S. Chem. Comm.*, 1973, 134.
[783] K. Seppelt, *Inorg. Chem.*, 1973, **12**, 2837.
[784] K. Zuchner and O. Glemser, *Angew. Chem. Internat. Edn.*, 1972, **11**, 1094.
[785] D. Pilipovich, C. B. Lindahl, C. J. Schack, R. D. Wilson, and K. O. Christie, *Inorg. Chem.*, 1972, **11**, 2189.
[786] G. Oates and J. M. Winfield, *Inorg. Nuclear Chem. Letters*, 1972, **8**, 193.

chlorides, $ZrCl_4$, MCl_5 (M = Nb, Ta, or Mo), and WCl_6, were discussed together with some measured ionization and appearance potentials.[679]

The formation of negative ions from tungsten hexafluoride has been discussed.[769] The electron affinities of WF_4 ($\geqslant 2.3$ eV) and WF_5 (0.8 eV), and the fluoride affinity of WF_4 (2.8 eV) were estimated. Some trends in bond-dissociation energies in the ions $[MF_n]^-$ were compared for M = S, Se, Te, and W for $n = 4$—6.

Other transition-metal halides for which data have been published include CrF_5 and CrF_4,[787] rhenium halides, Re_3X_9 (X = Cl, Br,[788,789] and I[789]), OsF_4,[790] IrF_4,[791] and mixtures of CuCl with chlorides of thorium and uranium.[675] Vapour-phase studies of CuCl,[792] CuI,[793] AgCl,[794] $ScCl_3$,[795] $NdCl_3$, and $GdCl_3$[796] have been reported. The heats of sublimation and ionization potentials of the tri-iodides of a number of lanthanides have been measured.[797]

A study has been reported of the oxotetrahalides, $MOCl_4$ (M = Mo, W, Re, or Os), $WOBr_4$, the dioxodihalides MO_2Cl_2 (M = Cr or Mo), MoO_2Br_2, and $WSCl_4$ using a molecular beam mass spectrometer.[798] In the gas phase, these compounds were shown to be predominantly monomeric. For some compounds, interpretation of their fragmentation patterns was difficult because of disproportionation and surface reactions within the mass spectrometer. A number of new species were observed as a result of these reactions including ions from MoO_2BrCl, $WOSCl_2$, and WS_2Cl_2. Ionization and appearance potential measurements were reported but with large uncertainties (± 1 or ± 0.5 eV). Comparison with other data for some main-group and transition-metal halides and oxyhalides was interpreted as suggesting that the electron removed on ionization comes primarily from an oxygen atom in oxyfluorides but from a halogen atom in chloro- or bromo-compounds. Tungsten oxotribromide has been reported to undergo simultaneous evaporation and disproportionation at 140—420 °C to give a number of products, mainly WO_2Br_2, but also $WOBr_4$ and WBr_x ($x = 3, 4,$ or 5).[799] The heat of formation of gaseous WO_2Br_2 has been obtained using a mass spectrometric method.[800]

---

[787] A. J. Edwards, W. E. Falconer, and W. A. Sunder, *J.C.S. Dalton*, 1974, 541.
[788] M. A. Bush, P. M. Druce, and M. F. Lappert, *J.C.S. Dalton*, 1972, 500.
[789] H. Schafer, K. Rinke, and H. Rabeneck, *Z. anorg. Chem.*, 1974, **403**, 23.
[790] W. E. Falconer, R. D. Burbank, G. R. Jones, W. A. Sunder, and M. J. Vasile, *J.C.S. Chem. Comm.*, 1972, 1080.
[791] W. A. Sunder and W. E. Falconer, *Inorg. Nuclear Chem. Letters*, 1972, **8**, 537.
[792] M. Guido, G. Grigli, and G. Balducci, *J. Chem. Phys.*, 1972, **57**, 3731.
[793] T. E. Joyce and E. J. Rolinski, *J. Phys. Chem.*, 1972, **76**, 2310.
[794] L. C. Wagner and R. T. Grimley, *J. Phys. Chem.*, 1972, **76**, 2819.
[795] Yu. B. Patrikeev, V. A. Morozova, G. P. Dudchik, O. G. Polyachenok, and G. I. Norikov, *Russ. J. Phys. Chem.*, 1973, **47**, 152.
[796] S. Caich, A. J. C. Nicholson, D. L. Swingler, and P. J. Thistlethwaite, *Inorg. Chem.*, 1973, **12**, 2072.
[797] C. Hirayama and P. M. Castle, *J. Phys. Chem.*, 1973, **77**, 311.
[798] D. E. Singleton and F. E. Stafford, *Inorg. Chem.*, 1972, **11**, 1208.
[799] S. K. Gupta, *Inorg. Chem.*, 1973, **12**, 1622.
[800] J. H. Dettingmeijer and B. Minders, *Z. anorg. Chem.*, 1973, **400**, 10.

Other oxyhalides studied include VOF_3, which was shown to be dimeric in the gas phase,[801] CrO_2F_2 and $CrOF_4$,[787] $ReOF_4$ and ReO_2F_2,[802] ReO_3X (X = Cl, Br, or I),[803] $OsOF_4$ and $OsOF_3$,[790] and yttrium oxofluorides.[804]

Several papers have appeared on transition-metal oxides including thermodynamic studies on oxides of titanium,[805,806] vanadium,[807] oxygen-rich vanadium–tungsten species,[808] rhenium,[809] cobalt,[810] the rare earths,[811] and uranium.[812] Reported values of D_0^0(Ti—O) do not agree within experimental error.[805,806] The formation of negative ions such as $[CrO_3]^-$ and $[HCrO_3]^-$ in flames containing chromium and potassium has been studied.[813]

A quadrupole mass spectrometer was used to study $Cu(NO_3)_2$ and $Ti(NO_3)_4$.[814] At 185 °C a weak pmi was observed from the copper compound and the major fragment ions were $[CuNO_3]^+$ and $[CuO]^+$. The titanium compound did not have a pmi, only $[Ti(NO_3)_3]^+$, $[TiO(NO_3)]^+$, and $[TiO_n]^+$ (n = 1—3) being found.

High-temperature studies have been reported concerning the vaporization of cobalt,[815] ytterbium,[816] Rh_2 and TiRh,[817] ThIr and ThPt,[818] carbides of Rh, Ti,[819] Ru, Ir, and Pt,[820] VN,[821] EuCN,[822] the CuN–Ag system,[823] lanthanide oxide sulphates,[824] and thorium and uranyl nitrates.[825]

---

[801] A. J. Edwards and D. R. Lloyd, *J.C.S. Chem. Comm.*, 1972, 719.
[802] R. T. Paine, *Inorg. Chem.*, 1973, **12**, 1457.
[803] H. Rabeneck, K. Rinke, and J. Schafer, *Z. anorg. Chem.*, 1973, **397**, 112.
[804] E. I. Smagina and P. I. Ozhegov, *Russ. J. Phys. Chem.*, 1973, **47**, 605.
[805] H. Y. Wu and P. G. Wahlbeck, *J. Chem. Phys.*, 1972, **56**, 4534.
[806] G. Balducci, G. De Maria, M. Guido, and V. Piacente, *J. Chem. Phys.*, 1972, **56**, 3422.
[807] M. Farber, O. M. Uy, and R. D. Srivastava, *J. Chem. Phys.*, 1972, **56**, 5312.
[808] S. L. Bennett, S.-S. Lin, and P. W. Gilles, *J. Phys. Chem.*, 1974, **78**, 266.
[809] H. B. Skinner and A. W. Searcy, *J. Phys. Chem.*, 1973, **77**, 1578.
[810] D. M. Chizhikov, Yu. V. Tsvetzov, E. K. Kazenas, and V. K. Tagirov, *Russ. J. Inorg. Chem.*, 1972, **17**, 465.
[811] L.-D. Nguyen and M. De Saint Simon, *Internat. J. Mass Spectrometry Ion Phys.*, 1972, **9**, 299.
[812] P. E. Blackburn and P. M. Danielson, *J. Chem. Phys.*, 1972, **56**, 6156.
[813] W. J. Miller, *J. Chem. Phys.*, 1972, **57**, 2354.
[814] L. Dauerman and G. E. Salser, *J. Inorg. Nuclear Chem.*, 1973, **35**, 304.
[815] F. M. Wachi and D. E. Gilmartin, *J. Chem. Phys.*, 1972, **57**, 4713.
[816] M. Guido and G. Balducci, *J. Chem. Phys.*, 1972, **57**, 5611.
[817] K. A. Gingerich and D. L. Cocke, *J.C.S. Chem. Comm.*, 1972, 536.
[818] K. A. Gingerich, *Chem. Phys. Letters*, 1973, **23**, 270.
[819] D. L. Cocke and K. A. Gingerich, *J. Chem. Phys.*, 1972, **57**, 3654.
[820] K. A. Gingerich, *J.C.S. Chem. Comm.*, 1974, 199.
[821] M. Farber and R. D. Srivastrava, *J.C.S. Perkin I*, 1973, 390.
[822] D. L. Cocke, K. A. Gingerich, and J. Kordis, *J.C.S. Chem. Comm.*, 1973, 561.
[823] J. Kordis and K. A. Gingerich, *J. Phys. Chem.*, 1973, **77**, 700.
[824] E. I. Smagina, V. S. Kutzev, N. G. Abdullina, and A. A. Grizik, *Russ. J. Phys. Chem.*, 1973, **47**, 598.
[825] K. G. Neumann, *Internat. J. Mass Spectrometry Ion Phys.*, 1972, **9**, 315.

6
Natural Products

BY D. E. GAMES

1 Introduction

This article covers selected references from the *Mass Spectrometry Bulletin* (July 1972—May 1974) and from the major biochemical and chemical journals to June 1974. A number of areas of study not included in previous Reports are dealt with.

Extensive use continues to be made of electron impact (EI) mass spectrometry and of gas chromatography–mass spectrometry (g.c.m.s.) (see Chapter 8) in structural studies of natural products. The major advances during the past two years have been the emergence of alternative methods of ionization (chemical ionization (CI), field ionization (FI) and field desorption (FD) as techniques which may be routinely applied to studies in this area. The development of an external atmospheric pressure source using ^{63}Ni or a discharge as a source of electrons has been reported and shows considerable promise in providing improved high sensitivities for the detection of natural compounds.[1,2] Preliminary reports have appeared of the combination of high-pressure liquid chromatography (h.p.l.c.) with mass spectrometry,[1,3,4,5] and in one case underivatized peptides have been successfully studied.[5] However, considerable further development is required before these systems are routinely usable for the study of compounds which are not amenable to g.c.m.s. An alternative approach has been to monitor h.p.l.c. fractions by taking their FD spectra;[6] although not a direct coupling, this technique does enable the advantages which h.p.l.c. has in the study of thermally labile and involatile compounds to be exploited.

[1] E. C. Horning, M. G. Horning, D. I. Carroll, I. Dzidic, and R. N. Stillwell, *Analyt. Chem.*, 1973, **45**, 936; and in 'Advances in Biochemical Psychopharmacology', ed. E. Costa and B. Holmstedt, Raven Press, New York, 1973, Vol. 7, p. 15.
[2] D. I. Carroll, I. Dzidic, R. N. Stillwell, M. G. Horning, and E. C. Horning, *Analyt. Chem.*, 1974, **46**, 706.
[3] R. E. Lovins, S. R. Ellis, G. D. Tolbert, and C. R. McKinney, *Analyt. Chem.*, 1973, **45**, 1553; and in 'Advances in Mass Spectrometry', ed. A. R. West, Applied Science, London, 1974, Vol. 6, p. 457.
[4] V. Tal'rose, V. E. Skurat, I. G. Gorodetskii, and N. B. Zolotoi, *Russ. J. Phys. Chem.*, 1972, **46**, 456.
[5] M. A. Baldwin and F. W. McLafferty, *Org. Mass Spectrometry*, 1973, **7**, 1111; P. Arpino, M. A. Baldwin, and F. W. McLafferty, *Biomedical Mass Spectrometry*, 1974, **1**. 80.
[6] H.-R. Schulten and H. D. Beckey, *J. Chromatog.*, 1973, **83**, 315.

Curie point pyrolysis in direct combination with low voltage EI,[7] FI, or FD[8] provides a useful method for the direct analysis of non-volatile organic materials, e.g. proteins, polysaccharides, bacteria, and DNA, yielding characteristic fingerprints of the material under investigation.

A particularly useful volume[9] containing reviews of the application of mass spectral methods to studies of fatty acids, lipids, steroids, bile acids, carbohydrates, terpenes, amino-acids, oligopeptides, nucleic acids, antibiotics, vitamins, hormones, tetrapyrroles, alkaloids, flavour components, and semiochemicals has appeared since the last Report, and the mass spectrometry review in *Analytical Chemistry*[10] contains a number of sections on the mass spectrometry of natural products.

2 Alkaloids

Mass spectral studies of alkaloids in general[11] and indole[12] and steroidal[13] alkaloids in particular have been reviewed, and the technique continues to play a key role in structural studies of alkaloids.[14] However, the recent structural revision of vindoline from (1) to (2) provides a note of caution.[15]

Detailed mass spectral studies using ^{2}H labelling and/or high-resolution methods have been carried out on bisbenzylisoquinoline,[16] Amaryllidaceae,[17]

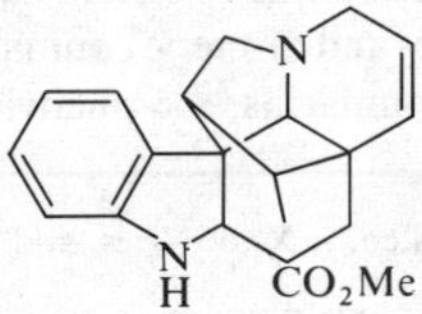

(1) 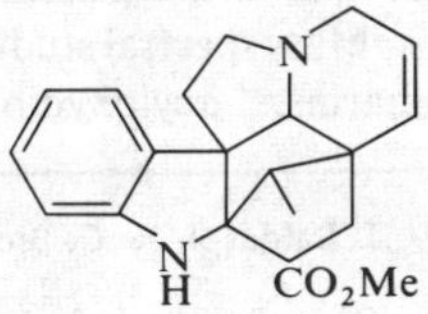   (2)

---

[7] H. L. C. Meuzelaar and P. G. Kristemaker, *Analyt. Chem.*, 1973, **45**, 587; H. L. C. Meuzelaar, M. A. Posthumus, P. G. Kristemaker, and J. Kristemaker, *Analyt. Chem.*, 1973, **45**, 1546; M. A. Posthumus, A. J. H. Boerboom, and H. L. C. Meuzelaar, in 'Advances in Mass Spectrometry', ed. A. R. West, Applied Science, London, 1974, Vol. 6, p. 397.

[8] H.-R. Schulten, H. D. Beckey, H. L. C. Meuzelaar, and A. J. H. Boerboom, *Analyt. Chem.*, 1973, **45**, 191, 2358; H.-R. Schulten and H. D. Beckey, in 'Advances in Mass Spectrometry', ed. A. R. West, Applied Science, London, 1974, Vol. 6, p. 499.

[9] 'Biochemical Applications of Mass Spectrometry', ed. G. R. Waller, Wiley, New York, 1972.

[10] A. L. Burlingame, R. E. Cox, and P. J. Derrick, *Analyt. Chem.*, 1974, **46**, 248R.

[11] S. D. Sastry, in 'Biochemical Applications of Mass Spectrometry', ed. G. R. Waller, Wiley, New York, 1972, p. 655.

[12] M. Hesse, in 'Progress in Mass Spectrometry', ed. H. Budzikiewicz, Verlag Chemie, Berlin, 1974, Vol. 1, Pts. 1 and 2.

[13] H. Budzikiewicz, in ref. 9, p. 283.

[14] e.g. R. Tschesche, E. U. Kaussmann, and G. Eckhardt, *Tetrahedron Letters*, 1973, 2577; J. Naranjo, M. Pinar, M. Hesse, and H. Schmid, *Helv. Chim. Acta*, 1972, **55**, 752.

[15] A. Ahond, M.-M. Janot, N. Langlois, G. Lukacs, P. Potier, P. Rasoanaivo, M. Sangré, M. Neuss, M. Plat, J. Le Men, E. W. Hagaman, and E. Wenkert, *J. Amer. Chem. Soc.*, 1974, **96**, 633.

[16] J. Baldas, I. R. C. Bick, T. Ibuka, R. S. Kapil, and Q. N. Porter, *J.C.S. Perkin I*,

quinolizidine,[18] and peptide[19] alkaloids and lysergic acid derivatives.[20] The nature of the $[M - 29]^+$ ion of nicotine has been investigated using ^{2}H labelling and DADI[21] and the mass spectra of the methines of benzylisoquinoline alkaloids have been studied.[22] Most of the fragment ions in the mass spectrum of tetracetyl spermine (3) have been shown to arise by neighbouring group participation.[23]

$$CH_3CONH(CH_2)_3\underset{COCH_3}{N}(CH_2)_4\underset{COCH_3}{N}(CH_2)_3NHCOCH_3$$

(3)

Low- and high-resolution FI have been used in the study of crude *Erythrina* alkaloid mixtures[24] and, when used in combination with g.c.m.s.,[25] provide a rapid and efficient method for the study of mixtures of this type.

Distinction between isomeric compounds on the basis of their EI spectra can often be difficult, particularly if only very small amounts of sample are available. CI using methane as reagent gas readily distinguishes between morphine and its isomer morphone, morphine showing a characteristic loss of OH from the molecular ion whereas morphone shows a loss of 42 mass units from the molecular ion.[26]

3 Oxygen Heterocycles and Phenols

The mass spectra of natural chromans, coumarins, xanthones, and flavanoids have been reviewed.[27] Mass spectral studies of mono- and di-meric coumarins,[28] 3,4-dialkoxyfurocoumarins,[29] acyloxydihydrofurocoumarins,[30] 4-phenylcoumarins,[31]

1972, 592, 599; J. Baldas, I. R. C. Bick, M. R. Falco, J. X. de Vries, and Q. N. Porter, *ibid.*, p. 597.

[17] P. Longevialle, D. H. Smith, A. L. Burlingame, H. M. Fales, R. J. Highet, *Org. Mass Spectrometry*, 1973, **7**, 401; P. Longevialle, H. M. Fales, R. J. Highet and A. L. Burlingame, *ibid.*, p. 417.

[18] E. Fujita and Y. Saeki, *J.C.S. Perkin I*, 1973, 301.

[19] A. O. Mascaretti, V. M. Merkuza, G. M. Ferraro, E. A. Ruveda, C.-J. Chang, and E. Wenkert, *Phytochemistry*, 1972, **11**, 1133.

[20] T. Inoue, Y. Nakahara, and T. Niwaguchi, *Chem. and Pharm. Bull (Japan)*, 1972, **20**, 409.

[21] J. G. Liehr, P. Schulze, and W. J. Richter, *Org. Mass Spectrometry*, 1973, **7**, 45.

[22] L. Dolejs and J. Slavik, *Org. Mass Spectrometry*, 1973, **7**, 775.

[23] E. Schopp and M. Hesse, *Helv. Chim. Acta*, 1973, **56**, 124.

[24] D. E. Games, A. H. Jackson, and D. S. Millington, *Tetrahedron Letters*, 1973, 3063.

[25] D. S. Millington, D. H. Steinman, and K. L. Rinehart, jun., *J. Amer. Chem. Soc.*, 1974, **96**, 1909.

[26] E. M. Chait, C. Blanchard, and V. H. Adams, in 'Advances in Mass Spectrometry', ed. A. R. West, Applied Science, London, 1974, Vol. 6, p. 471.

[27] S. E. Drewes, in 'Progress in Mass Spectrometry', ed. H. Budzikiewicz, Verlag Chemie, Berlin, 1974, Vol. 2.

[28] J. P. Kutney, G. Eigendorf, T. Inaba, and D. L. Dreyer, *Org. Mass Spectrometry*, 1971, **5**, 249.

[29] J. K. MacLeod and M. Nakayama, *Org. Mass Spectrometry*, 1972, **6**, 293; M. Nakayama, S. Eguchi, A. Matsuo, S. Havashi, S. Hishida, and Y. Kato, *Mass Spectrometry (Japan)*, 1972, **20**, 89.

[30] P. J. Zaharov, P. B. Terentjev, G. K. Nikonov, and A. J. Banjkowsky, *Khim. prirod. Soedinenii*, 1972, 275, 431.

[31] V. V. S. Murti, P. S. S. Kumar, and T. S. Seshadri, *Indian J. Chem.*, 1972, **10**, 19.

4-hydroxycoumarins,[32] and 7-methoxycoumarin derivatives,[33] having various isoprenoid side-chains at C-6, have been reported. The technique has been extensively applied in the identification of 4-alkyl- and 4-phenyl-coumarins from members of the Guttiferae family.[34] Molecular ions of coumarins with isoprenoid ether groups are often difficult to identify. FI mass spectra solve this problem and, in addition, structurally significant fragment ions are present.[35]

Isotope labelling and metastable ion analyses have been used to investigate the fragmentation of monohydroxy- and monomethoxy-xanthones.[36] Studies of the mass spectra of 2'-hydroxychalcones and the corresponding 2-phenylchrom-4-ones suggest that an intramolecular equilibrium exists between a chalcone-type and a flavanone-type molecular ion.[37] The spectra of isoprenylated flavones,[38] aurone epoxides,[39] and pterocarpans[40] have also been studied. The methane CI spectra of flavanones and 3-hydroxyflavanones provide information complementary to that available from their EI spectra. However, the technique was found to have little application for structural studies of flavones and flavonols.[41] Other CI studies of flavanoids have used isobutane with hydrogen as reagent gas, enabling D_2 to be substituted for H_2 thus providing information about some of the ionization processes.[42]

Glycosides, particularly O-glycosides, are not readily amenable to EI mass spectrometry, since the molecules usually eliminate the sugar moiety and yield only the mass spectra of the aglycone.[43] Techniques developed for the study of oligosaccharides have been successfully applied to flavanoid mono-, di-, and tri-saccharides. The glycosides are first perdeuteriomethylated[44] or permethylated;[45] the mass spectra of these derivatives allow the structural type of aglycone, the sugar sequence, the position of the interglycosidic linkage and the position of substitution on the aglycone to be established. Similar derivatives of flavanoid

---

[32] H. Nakata, A. Tatematsu, H. Yoshizumi, and S. Naga, *J.C.S. Perkin I*, 1972, 1924.
[33] J. K. MacLeod, *Org. Mass Spectrometry*, 1972, **6**, 1011.
[34] D. E. Games and A. H. Jackson, in 'Proceedings of the International Symposium on Gas Chromatography Mass Spectrometry', ed. A. Frigerio, Tamburini, Milan, 1972, p. 261; L. Crombie, D. E. Games, N. J. Haskins, and G. F. Reed, *J.C.S. Perkin I*, 1972, 2241, 2248, 2255; D. E. Games, *Tetrahedron Letters*, 1972, 3187.
[35] D. E. Games, A. H. Jackson, D. S. Millington, and M. J. Rossiter, *Biomedical Mass Spectrometry*, 1974, **1**, 5.
[36] P. Arends, P. Helboe, and J. Moller, *Org. Mass Spectrometry*, 1973, **6**, 667.
[37] C. Van de Sande, J. W. Serum, and M. Vandewalle, *Org. Mass Spectrometry*, 1972, **6**, 1333.
[38] A. V. Rama Rao, S. S. Rathi, and K. Venkataraman, *Indian J. Chem.*, 1972, **10**, 987.
[39] B. A. Brady, W. I. O'Sullivan, and A. M. Duffield, *Org. Mass Spectrometry*, 1972, **6**, 199.
[40] M. Nakayama, S. Eguchi, A. Matsuo, S. Hayashi, S. Hishida, and Y. Kato, *Mass Spectrometry (Japan)*, 1972, **20**, 239.
[41] D. G. I. Kingston and H. M. Fales, *Tetrahedron*, 1973, **29**, 4083.
[42] J. W. Clarke-Lewis, C. N. Harwood, M. J. Lacy, and J. S. Shannon, *Austral. J. Chem.*, 1973, **26**, 1577.
[43] E.g. A. Prox, *Tetrahedron*, 1968, **24**, 3697.
[44] R. D. Schmid, *Tetrahedron*, 1972, **28**, 3259; R. D. Schmid, P. Varenne, and R. Paris, *ibid.*, p. 5037: R. D. Schmid and J. B. Harborne, *Phytochemistry*, 1973, **12**, 2269.
[45] H. Wagner and O. Seligmann, *Tetrahedron*, 1973, **29**, 3029.

aglycones also provide useful EI spectra.[46] FD mass spectrometry has recently been shown to provide structurally informative spectra of flavanoid disaccharides without the necessity of derivatization.[47]

The mass spectra of depsidones have been discussed[48] and used extensively in the structural elucidation of two new depsidones from a lichen.[49] Identification of phloroglucinols in plants using mass spectral data has been reviewed,[50] and studies of the mass spectral fragmentations of this type of compound have been reported.[51] The mass spectral fragmentation of five-membered ring hop bitter acid derivatives[52] and usnic acid (4) and related compounds[53] has also been investigated.

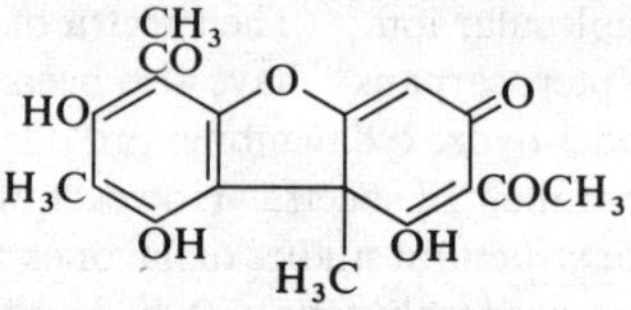

(4)

Using o- and p-1,2-tetrahydrocannibinol and their methylated homologues the origin of the ion at m/e 231 in the mass spectrum of trans-1,2-tetrahydrocannibinol has been shown to be predominantly formed by phenolic proton transfer

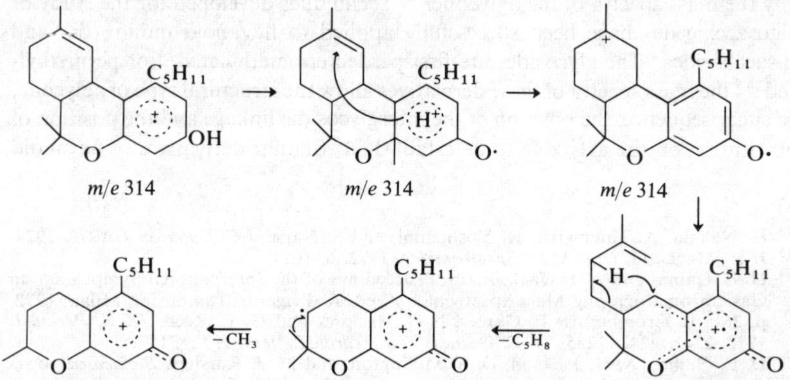

Scheme 1

[46] R. D. Schmid, R. Mues, J. H. McReynolds, G. V. Velde, N. Nakatani, E. Rodriguez, and T. J. Mabry, *Phytochemistry*, 1973, **12**, 2765.
[47] H.-R. Schulten and D. E. Games, *Biomedical Mass Spectrometry*, 1974, **1**, 120.
[48] E. Martinez and R. Mestres, *An. Quim.*, 1972, **68**, 1321.
[49] T. M. Cresp, J. A. Elix, S. Kurokawa, and M. V. Sargent, *Austral. J. Chem.*, 1972, **25**, 2167.
[50] M. Lounasmaa, *Planta Med.*, 1973, **24**, 148.
[51] M. Lounasmaa, A. Karjalainen, C.-J. Widen, and A. Huhtikangas, *Acta Chem. Scand.*, 1972, **26**, 89; M. Lounasmaa, C.-J. Widen, and T. Reichstein, *Helv. Chim. Acta*, 1973, **56**, 1133.
[52] E. Cant, M. Vandewalle, and M. Verzele, *Org. Mass Spectrometry*, 1972, **6**, 977.
[53] J. P. Kutney, I. H. Sanchez, and T. H. Yee, *Org. Mass Spectrometry*, 1974, **8**, 129.

to the 1,2-double bond (Scheme 1), with a smaller contribution from double migration to the 1,6-position (Scheme 2).[54]

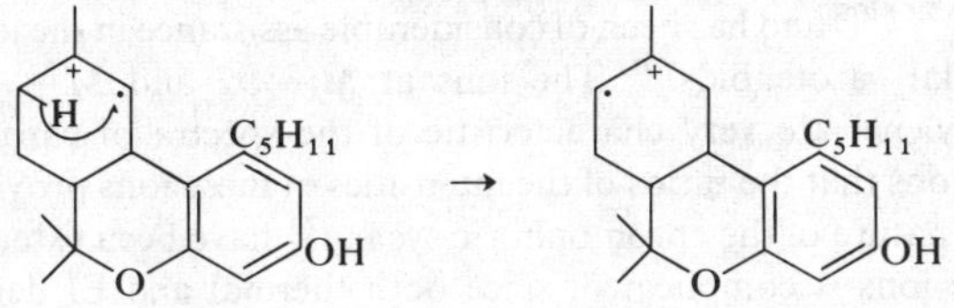

Scheme 2

4 Isoprenoids

The mass spectra of terpenes, terpenoids,[55] and triterpenoids[56] have been reviewed. Studies of the mass spectral fragmentation of natural pyrethrins,[57] diterpenoid acetals,[58] sesquiterpene lactones,[59] and some terpenoid bitter principles[60] have been reported. FI has proved useful in establishing the molecular weights of some terpenes.[26,61] EI spectra continue to play an important role in structural studies in this area, and have been used e.g. in the structural elucidations of neocembrene A (5), a termite trail pheromone,[62] and of new sesquiterpene alcohols[63,64] and lactones;[65] CI was used to obtain molecular weights in the latter case.

The mass spectra of trimethylsilylmethoximes of ecdysone and 20-hydroxyecdysone have been described[66] and high-resolution and ^{2}H-labelling methods

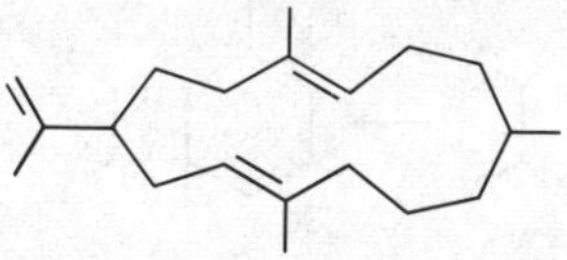

(5)

[54] T. B. Vree and N. M. M. Nibbering, *Tetrahedron*, 1973, **29**, 3849.
[55] C. R. Enzell, R. A. Appleton, and I. Wahlberg, in ref. 9, p. 351.
[56] M. J. Kulshreshtha, D. K. Kulshreshtha, and R. P. Rastogi, *Phytochemistry*, 1972, **11**, 2369.
[57] G. Pattenden, L. Crombie, and P. Hemesley, *Org. Mass Spectrometry*, 1973, **6**, 719.
[58] R. C. Cambie, A. F. Preston, and P. D. Woodgate, *Org. Mass Spectrometry*, 1974, **8**, 161.
[59] S. Matsueda, *J. Pharm. Soc. Japan*, 1972, **92**, 905.
[60] T. Murae, A. Sugie, Y. Moriyama, T. Tsuyuki, and T. Takahashi, *Org. Mass Spectrometry*, 1974, **8**, 297.
[61] D. E. Games, A. H. Jackson, D. S. Millington, and M. Rossiter, in 'Advances in Mass Spectrometry', ed. A. R. West, Applied Science, London, 1974, Vol. 6, p. 137.
[62] A. J. Birch, W. V. Brown, J. E. T. Corrie, and B. P. Moore, *J.C.S. Perkin I*, 1972, 2653.
[63] K. Ohara, Y. Ohta, and Y. Hirose, *Bull. Chem. Soc. Japan*, 1973, **46**, 641.
[64] M. A. Irwin and T. A. Geissman, *Phytochemistry*, 1973, **12**, 849.
[65] S. M. Kupchan, R. L. Baxter, C.-K. Chiang, C. J. Gilmore, and R. F. Bryan, *J.C.S. Chem. Comm.*, 1973, 842.
[66] E. D. Morgan and A. P. Woodbridge, *Org. Mass Spectrometry*, 1973, **7**, 102.

have been used to study the fragmentations of terpenoid esters of the juvenile hormone class.[67]

The role of mass spectrometry in the structural elucidation of carotenoids has been reviewed[55,68,69] and has been of considerable assistance in the identification of new bacterial carotenoids.[70] The ions at $M - 92$ and $M - 106$ (loss of toluene and xylene) are very characteristic of the spectra of carotenoids and early observations that the ratios of the intensities of these ions provide information about the nature of the chromophoric system[71] have been extended.[72] The origin of these ions is complicated, since both thermal and EI degradation is involved and a number of mechanisms have been proposed.[73] ^{2}H-Labelling studies favour the mechanisms shown in Scheme 3 for the origin of the $M - 92$ and $M - 106$ ions.[69]

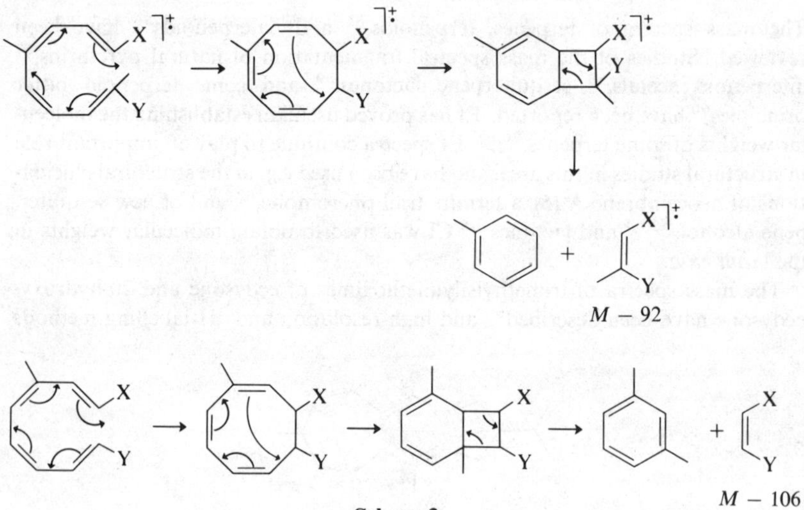

Scheme 3

Mass spectrometry has also been used to study the role of carotenoid epoxides in photosynthesis in green plants using ^{18}O.[69]

The EI fragmentations of eleven analogues of vitamin A have been studied.[74]

---

[67] R. J. Liedtke and C. Djerassi, *J. Org. Chem.*, 1972, **37**, 2111.
[68] S. Liaaen-Jensen, *Pure Appl. Chem.*, 1973, **35**, 81.
[69] H. Budzikiewicz, in 'Advances in Mass Spectrometry', ed. A. R. West, Applied Science, London, 1974, Vol. 6, p. 163.
[70] A. G. Andrews and S. Liaaen-Jensen, *Acta Chem. Scand.*, 1972, **26**, 2194; N. Arpin, S. Liaaen-Jensen and M. Trouilloud, *ibid.*, p. 2524.
[71] C. R. Enzell, G. W. Francis, and S. Liaaen-Jensen, *Acta Chem. Scand.*, 1968, **22**, 1054.
[72] G. W. Francis, *Acta Chem. Scand.*, 1972, **26**, 1443; 1974, **28B**, 244.
[73] U. Schwieter, G. Englert, N. Rigassi, and W. Vetler, *Pure Appl. Chem.*, 1969, **20**, 365.
[74] R. Reid, E. C. Nelson, E. D. Mitchell, M. L. McGregor, G. R. Waller, and K. V. John, *Lipids*, 1973, **10**, 558.

5 Steroids

Comprehensive reviews of the mass spectrometry of steroids[75,76] and bile acids[77] have appeared.

Enhanced molecular or quasi-molecular ions may be obtained for many steroids using FD, FI, or CI. FD has been used with h.p.l.c. to study steroid mixtures,[6] and CI studies of TMS ethers of bile acid esters,[78] 17-hydroxy-,[79] 17-keto-,[80] and other steroids,[81] including steroidal amino-alcohols,[82] have been reported. In the last case, CI enables structural isomers to be distinguished; specific fragmentations, e.g. loss of water from the $(M + 1)^+$ ion, are discussed in terms of conformational equilibria and the conclusions are consistent with those obtained from i.r. data.

Considerable efforts are still being expended on elucidating the specificity of fragmentations of the steroid skeleton in relation to such features as the substituents present, position of the double bond, and stereochemical features.[83] Most of these investigations have used both high-resolution and ^{2}H-labelling methods. Some of the knowledge obtained from such studies has been used in the development of computer programmes for the automatic interpretation of high-resolution mass spectra, and for the interpretation of high-resolution mass spectra of oestrogens.[84] Steroidal 6-one ethylene acetals[85] and steroidal lactones[86] have been examined.

Because of their importance in g.c.m.s. studies, silyl derivatives of steroids have been the subject of a number of detailed investigations. Halogeno-methyl-

---

[75] C. J. W. Brooks and B. S. Middleditch, in 'Modern Methods of Steroid Analysis', ed. E. Heftmann, Academic Press, New York, 1973, p. 140.
[76] H. Budzikiewicz, in ref. 9, p. 251.
[77] W. H. Elliott, in ref. 9, p. 291.
[78] B. Jelus, B. Munson, and C. Fenselau, *Analyt. Chem.*, 1974, **46**, 729.
[79] J. Michnowicz and B. Munson, *Org. Mass Spectrometry*, 1972, **6**, 765.
[80] J. Michnowicz and B. Munson, *Org. Mass Spectrometry*, 1974, **8**, 49.
[81] J. L. Smith and W. J. A. Van den Heuvel, *Analyt. Letters*, 1972, **5**, 51.
[82] P. Longevialle, G. W. A. Milne, and H. M. Fales, *J. Amer. Chem. Soc.*, 1973, **95**, 6666.
[83] M. Ende and G. Spiteller, *Tetrahedron*, 1973, **29**, 2457; F. J. Hammerschmidt and G. Spiteller, *Tetrahedron*, 1973, **29**, 2456, 3995; H. Klein and C. Djerassi, *Chem. Ber.*, 1973, **106**, 1897; C. Marazano and P. Longevialle, *Compt. rend.*, 1973, **276**, *C*, 175; M. Moet-Ner, E. Premuzic, S. R. Lipsky, and W. J. McMurray, *Steroids*, 1972, **19**, 493; R. R. Muccino and C. Djerassi, *J. Amer. Chem. Soc.*, 1973, **95**, 8726; 1974, **96**, 556; A. Rotman, A. Mandelbaum, and Y. Mazur, *Tetrahedron*, 1973, **29**, 1303; L. Tokés and B. A. Amos, *J. Org. Chem.*, 1972, **37**, 4221; H. Budzikiewicz and Linscheid, *Org. Mass Spectrometry*, 1974, **9**, 88; E. Zietz and G. Spiteller, *Tetrahedron*, 1974, **30**, 585, 597; J. Jovanovic and G. Spiteller, *Tetrahedron*, 1973, **29**, 4017; I. Midgley and C. Djerassi, *J.C.S. Perkin I*, 1972, **21**, 2771; P. Toft, B. A. Lodge, and Simard, *Canad. J. Pharm. Sci.*, 1972, **7**, 53; G. Adam, D. Voigt, K. Schreiber, M. V. Ardenne, R. Tummler, and K. Steinfelder, *J. prakt. Chem.*, 1973, **315**, 125.
[84] D. H. Smith, B. G. Buchanan, R. S. Engelmore, H. Adlercreutz, and C. Djerassi, *J. Amer. Chem. Soc.*, 1973, **95**, 6078; D. H. Smith, B. G. Buchanan, R. S. Engelmore, A. M. Duffield, A. Yeo, E. A. Feigenbaum, J. Lederberg, and C. Djerassi, *ibid.*, 1972, **94**, 5962; D. H. Smith, B. G. Buchanan, W. C. White, E. A. Feigenbaum, J. Lederberg, and C. Djerassi, *Tetrahedron*, 1973, **29**, 3117.
[85] M. S. Ahmad and G. A. S. Ansari, *Org. Mass Spectrometry*, 1972, **6**, 1095.
[86] M. S. Ahmad and M. Mushfiq, *Org. Mass Spectrometry*, 1972, **6**, 1109; M. S. Ahmad, M. Mushfiq, and Shafiullah, *Steroids*, 1973, **21**, 181.

dimethylsilyl ethers, in particular chloromethyldimethylsilyl ethers, are alternatives to TMS derivatives for hydroxy-steroids and their mass spectra enable differentiation of stereoisomers of the 3-hydroxy-group or at the A/B ring junction.[87] The mass spectra of di-TMS-enol-TMS derivatives of corticosteroids possessing dihydroxyacetone side-chains have been studied.[88] These derivatives offer possible alternatives to the methoxime-TMS derivatives normally used with this type of steroid. Methods have been developed for the preparation of mixed TMS and [^{2}H$_9$]TMS derivatives of hydroxy-steroids in which the [^{2}H$_9$]TMS group occupies a specific position. The differing silylation rates of sterically hindered and unhindered hydroxy-groups were used in these preparations and the products have proved very useful in the interpretation of the mass spectra of TMS derivatives of hydroxy-steroids.[89] An example is the discovery of reciprocal exchange of entire trimethylsilyl groups on EI (Scheme 4).[90]

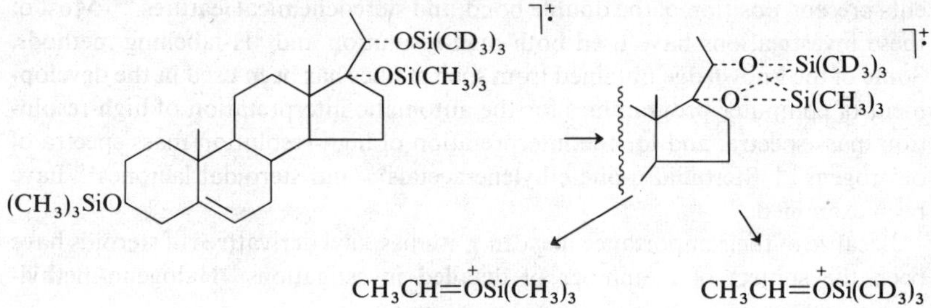

Scheme 4

Detailed fragmentation studies[91] of a wide range of TMS ethers of Δ^5-3β-hydroxy-steroids confirm the validity of many previously proposed fragmentations. A number of diagnostic ions are not as specific as had formerly been supposed, e.g. the $[M - 56]^+$ ion in the spectra of saturated 16- and 17-ketones arises by fragmentation of ring D. High-resolution and [^{18}O + ^{2}H]-labelling studies show that the same ion in Δ^5-androstanols arises by an EI-induced rearrangement followed by fragmentation of ring A (Scheme 5),[92] and TMS derivatives of 11-hydroxy-steroids have been similarly studied.[93]

---

[87] J. R. Chapman and E. Bailey, *Analyt. Chem.*, 1973, **45**, 1636; C. J. W. Brooks and B. S. Middleditch, *Analyt. Letters*, 1972, **5**, 611.
[88] E. M. Chambaz, C. Madani, and A. Ros, *J. Steroid Biochem.*, 1972, **3**, 741; E. M. Chambaz, G. Defaye, and C. Madani, *Analyt. Chem.*, 1973, **45**, 1090.
[89] P. Vouros and D. J. Harvey, *Analyt. Chem.*, 1973, **45**, 7.
[90] P. Vouros, *J. Org. Chem.*, 1973, **38**, 3555.
[91] C. J. W. Brooks, D. J. Harvey, B. S. Middleditch, and P. Vouros, *Org. Mass Spectrometry*, 1973, **7**, 925.
[92] C. J. W. Brooks, D. J. Harvey, and B. S. Middleditch, *J. Org. Chem.*, 1972, **37**, 3365; I. Björkhem, J.-Å. Gustafsson, and J. Sjovall, *Org. Mass Spectrometry*, 1973, **7**, 277.
[93] P. Vouros and D. J. Harvey, *J.C.S. Perkin I*, 1973, 727.

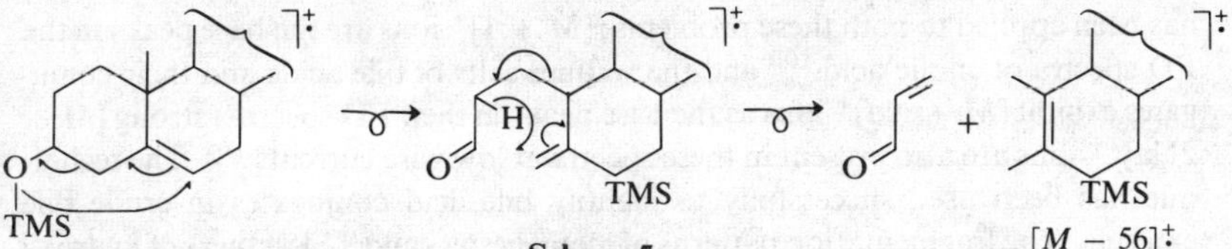

Scheme 5

The structures of a number of novel steroids isolated from marine sources have been elucidated with the aid of mass spectral data.[94] In some cases these compounds are $\Delta^{17(20)}$- or $\Delta^{20(22)}$-steroidal olefins as with (6) and (7) isolated from starfish toxins. This has resulted in a study of differentiation of model $\Delta^{17(20)}$-, $\Delta^{20(21)}$-, and $\Delta^{20(22)}$-steroidal olefins using mass spectrometric and other methods.[95]

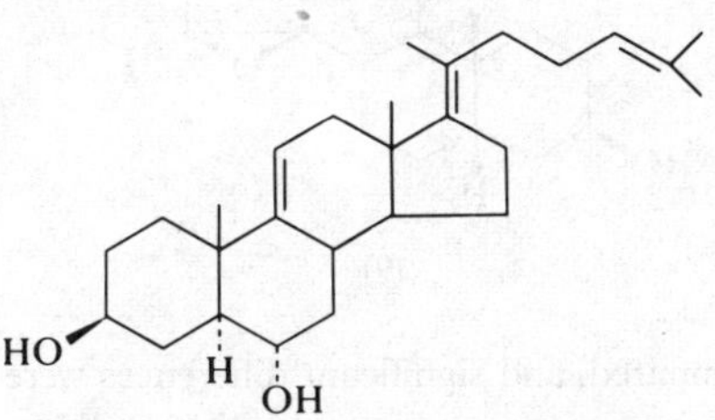

The mass spectrometry of cardenolides has been reviewed, with particular reference to the complementary information available from their EI and FI spectra.[96] EI spectra of TMS derivatives of cardiac aglycones and monoglycosides show abundant molecular ions and structurally informative fragment ions.[97] A detailed study of the fragmentation of bufadienolides shows that their high-resolution mass spectra enable distinction to be made between those with 14β-hydroxy and 14β,15β-epoxy functions, e.g. (8) and (9); other structural features are readily ascertainable.[98] The structure of the steroidal lactone (10) from *Dunalia australis* has been elucidated with considerable aid from its mass spectrum and those of some of its derivatives.[99]

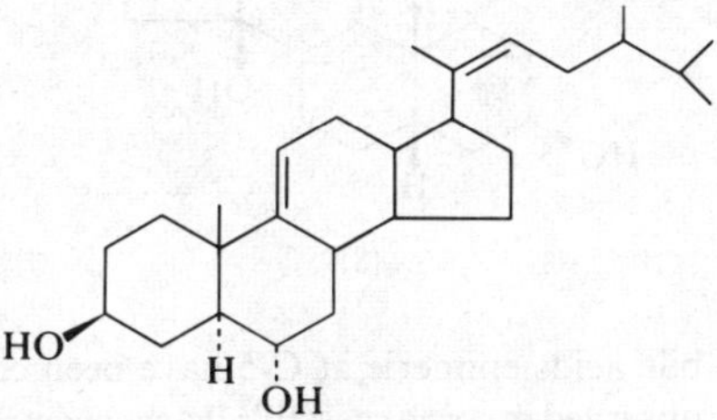

Difficulties often arise in the identification of the molecular ion in the spectra of free bile acids and in obtaining mass spectra of bile acid conjugates.[77] FD

---

[94] Y. M. Sheikh, B. M. Tursch, and C. Djerassi, *J. Amer. Chem. Soc.*, 1972, **94**, 3278; Y. M. Sheikh, B. M. Tursch, and C. Djerassi, *Tetrahedron Letters*, 1972, 3721; S. Ikegami, Y. Kamiya, znd S. Tamurz, *Tetrahedron Letters*, 1972, 1601; Y. Shimizu, *J. Amer. Chem. Soc.*, 1972, **94**, 4051; D. S. H. Smith, A. B. Turner, and A. M. Mackie, *J.C.S. Perkin I*, 1973, 1745; T. R. Erdman and R. H. Thomson, *Tetrahedron*, 1972, **28**, 5163; M. Kobayashi, R. Tsuru, K. Todo, and H. Mitsuhasi, *Tetrahedron*, 1973, **29**, 1193.
[95] Y. M. Sheikh and C. Djerassi, *J. Org. Chem.*, 1973, **38**, 3545.
[96] P. Braun, F. Bruschweiler, and G. R. Petit, *Helv. Chim. Acta*, 1972, **55**, 531.
[97] F. C. Falkner, J. Frolich, and J. Throck-Watson, *Org. Mass Spectrometry*, 1973, **7**, 141.
[98] P. Brown, V. Kamano, and G. R. Petit, *Org. Mass Spectrometry*, 1972, **6**, 47, 613.
[99] G. Adam and M. Hesse, *Tetrahedron*, 1972, **28**, 3527.

has been applied to both these problems; $[M + 1]^+$ ions are the base peaks in the FD spectra of cholic acid,[100] and the sodium salts of bile acids and their conjugates exhibit $[M + Na]^+$ ions as the base peaks in their FD spectra; strong $[M + 2Na]^{2+}$ ions are also present in these spectra at low wire currents.[101] The technique has been used successfully to identify bile acid conjugates in crude bile extracts. The fragmentation patterns of methyl esters and TMS ethers of hydroxy

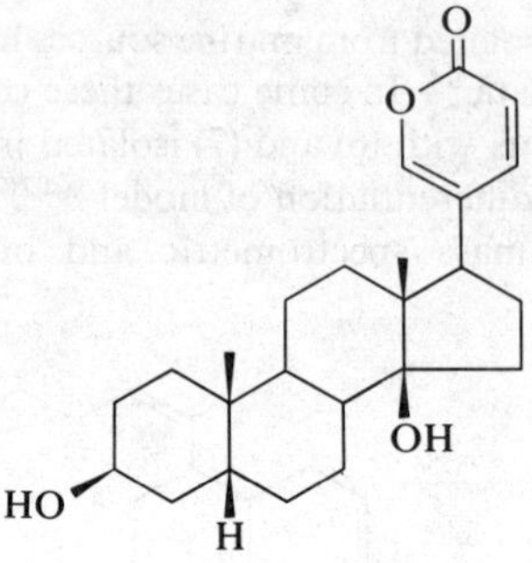

(8)

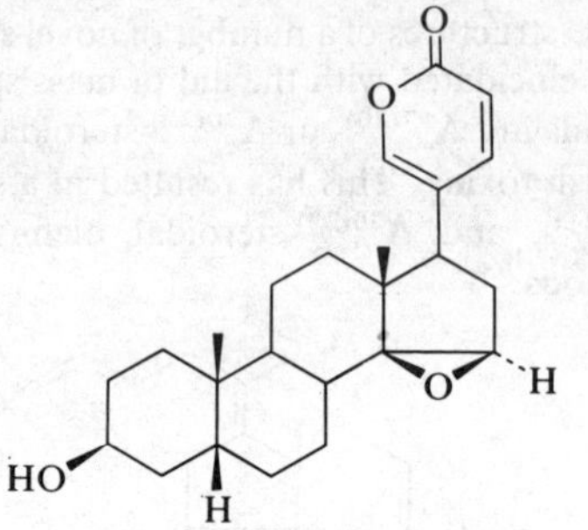

(9)

bile acids epimeric at C-5 have been compared, and significant differences were observed in some cases.[102] Mass spectrometry continues to assist in the identification of new bile acids; recent examples have been isolated from kite bile,[103] lizard bile,[104] and *Arapaima gigas* (fish).[105] The bile acids formed in cholestatic liver disease have also been studied.[106]

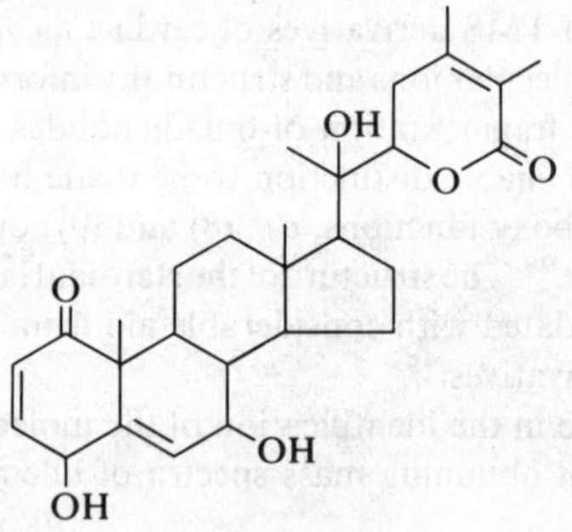

(10)

[100] K. H. Maurer and U. Rapp, Twenty-first Annual Conference on Mass Spectrometry and Allied Topics, San Francisco, California, 1973, paper N.2.
[101] D. E. Games, M. P. Games, A. H. Jackson, A. H. Olavesen, M. Rossiter, and P. J. Winterburn, *Tetrahedron Letters*, 1974, 2377.
[102] W. H. Elliott and P. M. Hyde, Twentieth Annual Conference on Mass Spectrometry and Allied Topics, Dallas, Texas, June 1972, p. 77.
[103] T. Kuramoto and T. Hoshita, *J. Biochem. (Japan)*, 1972, **72**, 199.
[104] K. Okuda, M. G. Horning, and E. C. Horning, *J. Biochem. (Japan)*, 1972, **71**, 885.
[105] G. A. D. Haslewood and L. Tokes, *Biochem. J.*, 1972, **126**, 1161.
[106] G. M. Murphy, F. H. Jansen, and B. H. Billing, *Biochem. J.*, 1972, **129**, 491.

6 Antibiotics

A comprehensive review has appeared on the use of mass spectral methods for the study of antibiotics.[107]

Many antibiotics have low volatility or thermal instability and hence EI spectra are often difficult to obtain and when obtained do not provide molecular weight information. Derivatization can be used to overcome these difficulties, but it is preferable if possible to study underivatized material. Both CI and FD methods have been tested as means of providing solutions to these problems. The EI and FD spectra of erythromycin, filipin, dermostatin, streptovaricins, neomycin, streptolydigin, and novobiocin have been compared.[108] CI spectra could only be obtained with erythromycin, novobiocin, and streptolydigin and, in these cases, the spectra were similar to the EI spectra, with enhanced ion intensities at higher mass; molecular weight information was only obtainable for erythromycin.[108] The FD spectra showed strong $[M]^{+\cdot}$ and $[M + H]^+$ ions for all the compounds examined, showing it to be the method of choice for molecular weight determination. At higher wire currents the FD spectra also provided considerable structural information, complementing that obtained from the EI and CI spectra. Three of the materials examined, streptovaricins, filipin, and dermostatin, are mixtures of compounds and in these cases the FD spectra gave a semi-quantitative estimate of the composition of the mixtures. These studies have recently been extended to the neutral macrolide antibiotics lankamycin, chalcomycin, neutramycin, and a number of related compounds of unknown structure, and similar results were obtained.[109] FD also provides molecular weight and structural information for salts of penicillins and for amino-cycloside antibiotics.[110]

A number of CI studies of antiobiotics have been reported. Celesticetin (11), with ammonia as reagent gas, has a protonated molecular ion base peak and undergoes the fragmentations indicated;[111] a markedly different spectrum, with

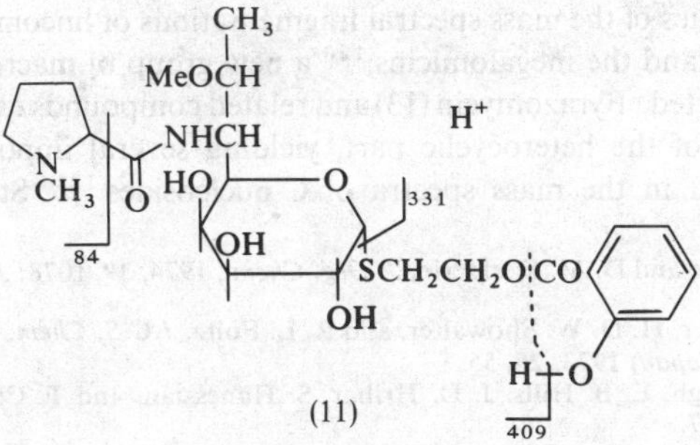

(11)

[107] K. L. Rinehart, jun. and G. E. Van Lear, in ref. 9, p. 499.
[108] K. L. Rinehart, jun., J. C. Cook, jun., K. H. Maurer, and U. Rapp, *J. Antibiotics*, 1974, 1.
[109] R. T. Hargreaves, J. C. Cook, jun, and K. L. Rinehart, jun, Twenty-second Annual Conference on Mass Spectrometry and Allied Topics, Philadelphia, May 1974, paper O2.
[110] D. E. Games, A. H. Jackson, and H. R. Schulten, unpublished work.
[111] D. Horton, J. D. Wander, and R. G. Foltz, *Analyt. Biochem.*, 1973, **55**, 123.

its major peak at m/e 139, is obtained using isobutane instead of ammonia. These differences are attributed to the differing reactions of the heteroatomic centres towards each ionizing agent and should prove useful for structural studies of related compounds. Ammonia CI has also proved useful in combination with EI in studying the composition of the antimycin A complex, enabling the molecular ion to be identified. Ammonolysis of lactone, ester, and amide bonds under the CI conditions provides valuable structural information.[112] CI studies of macrolide antibiotics have also been reported.[113]

Derivatization is an alternative for improvement of mass spectral data. A thorough study of the EI fragmentations of the N-acetyl, NO-methyl, and N-acetyl-O-trimethylsilyl derivatives of kanamycin A has been reported;[114] CI proved particularly useful for the recognition of structural features in the former derivatives, e.g. the sequence of sugar units. TMS derivatives of anthracycline antibiotics,[115] tetracycline antibiotics,[116] and the polyene macrolide antibiotics nystatin, amphotericin B, and pimaricin[117] have also been studied in detail.

The mass spectra of secalonic acids (12) show features closely related to the stereochemistry at the six chiral centres in these compounds.[118] Analysis of the

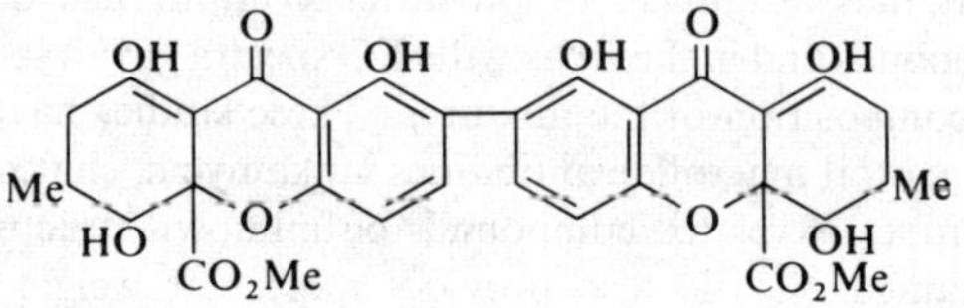

(12)

spectra suggests that the mass spectrum could be used to define much of the stereochemistry of any new secalonic acid provided source temperatures are maintained below 180 °C.

Detailed studies of the mass spectral fragmentations of lincomycin and related compounds[119] and the megalomicins,[120] a new group of macrolide antibiotics, have been reported. Pyrazomycin (13) and related compounds exhibit an unusual fragmentation of the heterocyclic part, yielding several important peaks not normally found in the mass spectra of C-nucleosides.[121] Studies on model

[112] K. D. Haegele and D. M. Desiderio, *J. Org. Chem.*, 1974, **39**, 1078; *J. Antibiotics*, 1973, **26**, 215.
[113] L. A. Mitscher, H. D. W. Showalter, and R. L. Foltz, *J.C.S. Chem. Comm.*, 1972, 796; *Antibiotics (Japan)* 1973, **26**, 55.
[114] D. C. DeJongh, E. B. Hills, J. D. Hribar, S. Hanessian, and T. Chang, *Tetrahedron*, 1973, **29**, 2707.
[115] J. Roboz, Twenty-second Annual Conference on Mass Spectrometry and Allied Topics, Philadelphia, May, 1974, paper G6.
[116] K. Tsuji and J. H. Robertson, *Analyt. Chem.*, 1973, **45**, 2136.
[117] K. D. Haegele and D. M. Desiderio, *Biomedical Mass Spectrometry*, 1974, **1**, 20.
[118] C. C. Howard and R. A. W. Johnstone, *J.C.S. Perkin I*, 1973, 2033.
[119] F. Kagan and M. F. Grostic, *Org. Mass Spectrometry*, 1972, **6**, 1217.
[120] R. S. Jaret, A. K. Mallams, and H. F. Vernay, *J.C.S. Perkin I.*, 1973, 1389.
[121] P. F. Crain, J. A. McCloskey, A. F. Lewis, K. H. Schram, and L. B. Townsend, *J. Heterocyclic Chem.*, 1973, **10**, 843.

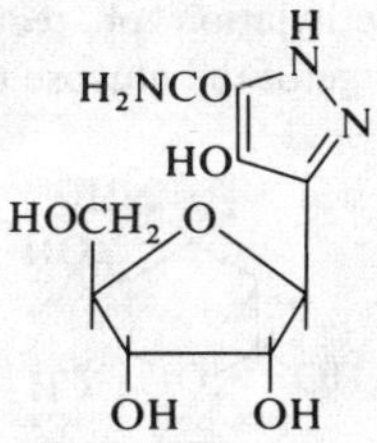

(13)

compounds show that these unusual ions result from juxtaposition of the exocyclic hydroxy- and carboxamido-groups of the aglycone, e.g. the expulsion of ammonia from the ion m/e 156 to yield m/e 139 can be rationalized as shown.

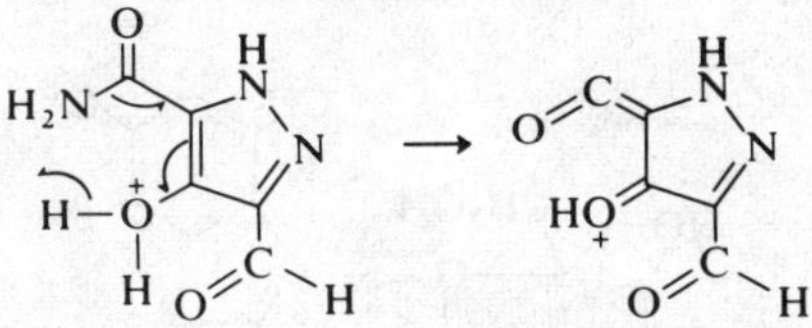

m/e 156 m/e 139

Flavofungin, an antibiotic from *Streptomyces flavofungini*, has been shown to be a mixture (14a and b);[122] chainin[123] and oleficin[124] are other polyene antibiotics for which structural assignments have been considerably assisted by mass spectra. Further features of the structure of vancomycin, a complex antibiotic, have been revealed.[125] A combination of FI, high resolution, and deuterium ^{2}H-labelling assisted in the identification of a degradation product from this antibiotic as the amino-sugar (15).[126] Permethylation and hydrolysis of vancomycin,

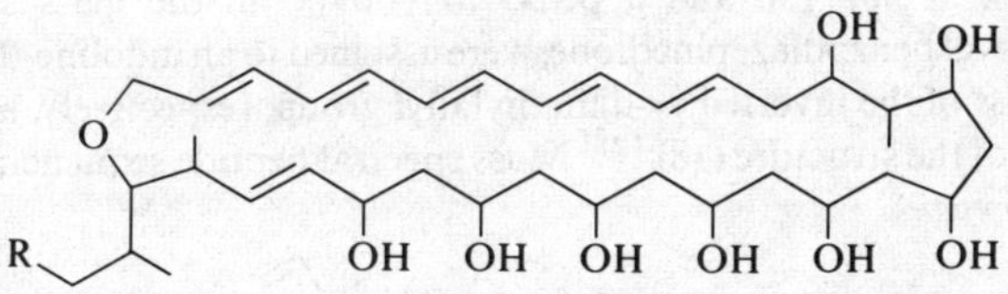

(14) a; R = H
     b; R = Me

[122] R. Bognár, S. Maklier, K. Zsupán, B. O. Brown, W. J. S. Lockley, T. P. Toube, and B. C. L. Weedon, *J.C.S. Perkin I*, 1972, 1848.
[123] R. C. Pandey, N. Narasimhachari, K. L. Rinehart, jun., and D. S. Millington, *J. Amer. Chem. Soc.*, 1972, **94**, 4306.
[124] Gy. Horvath, J. Cjyimesi, and Zs. Mehesfalvi-Vajna, *Tetrahedron Letters*, 1973, 3634.
[125] W. D. Weringa, D. H. Williams, J. Feeney, J. P. Brown, and R. W. King, *J.C.S. Perkin I*, 1972, 443; A. W. Johnson, R. M. Smith, and R. D. Guthrie, *J.C.S. Perkin I*, 1972, 2153.
[126] P. J. Roberts, O. Kennard, K. A. Smith, and D. H. Williams, *J.C.S. Chem. Comm.*, 1973, 772.

followed by either further methylation or reduction and acetylation have established the position of linkage of the glucose to the amino-sugar and to the

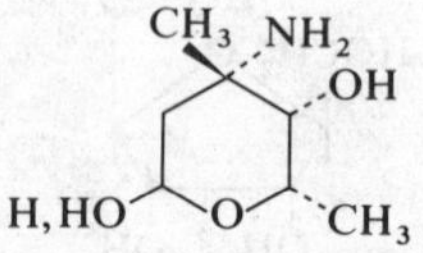

(15)

aryl skeleton resulting in the partial formulation of vancomycin as (16). Permethylation and perdeuteriomethylation of hikizimycin, a new antibiotic from *Streptomyces* A-5, followed by mass spectral examination, suggest the presence of cytosine and 3-amino-3-deoxy-D-glucose residues in its structure.[127] An

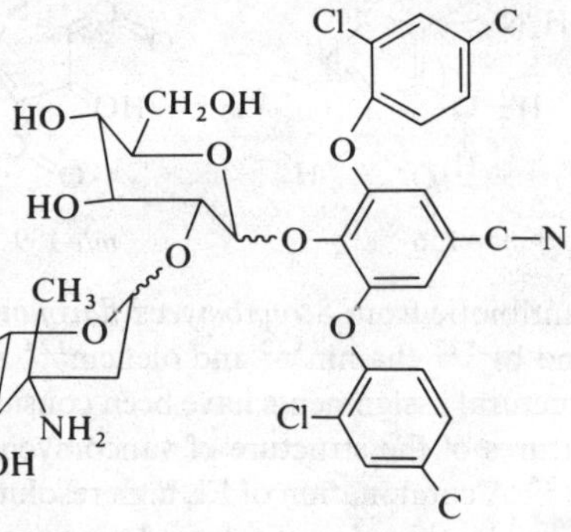

(16)

abundant peak at m/e 130 and a peak $[M - 69]^+$, in the mass spectrum of LL-S490β, a novel benzodiazepinedione, were assigned to an indoline-3-methylene ion (17) and loss of the inverted γγ-dimethylallyl group, respectively, assisting the establishment of the structure (18).[128] Mass spectral peptide sequencing methods

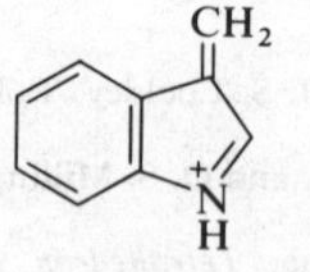

(17)

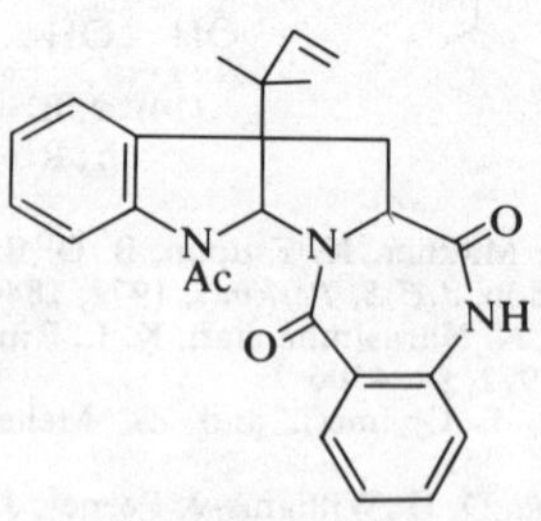

(18)

[127] B. C. Das, J. Defaye, and K. Uchida, *Carbohydrate Research*, 1972, **22**, 293.
[128] G. A. Ellestad, P. Mirando, and M. P. Kunstmann, *J. Org. Chem.*, 1973, **38**, 4204.

have been applied to the structural investigation of an antibiotic tripeptide from *Ketatmophyton terreum*,[129] and the EI mass spectra of transformation products of the antiviral antibiotics kikumycin A and B established their amino-acid sequences (19).[130] High-resolution mass spectral data were extensively used in studies of the constitution of primycin (20).[131] Mass spectra have assisted in structural studies on a number of macrolide antibiotics.[132]

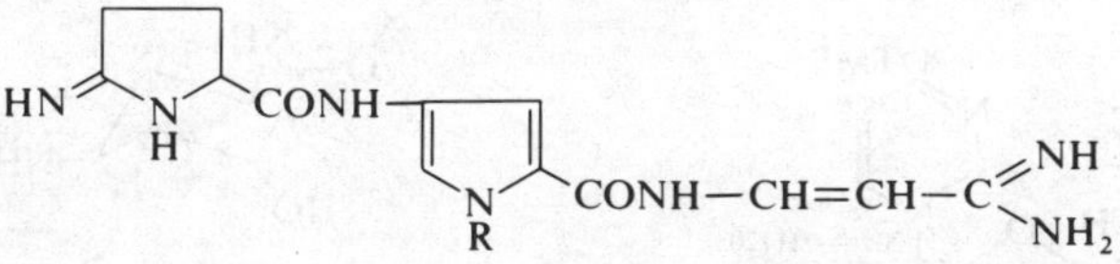

(19)

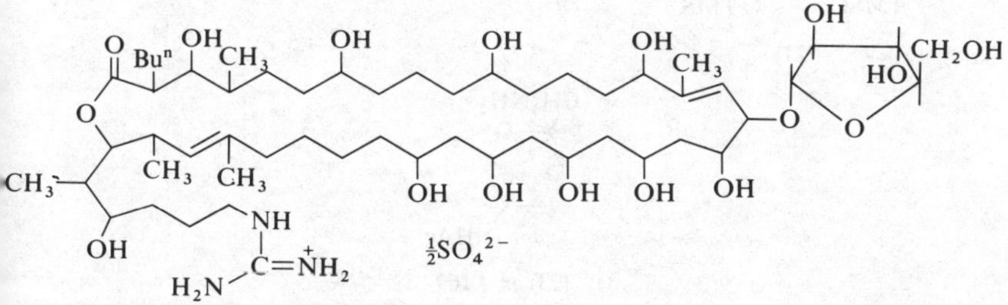

(20)

Study of the biosynthesis of aberrant nucleoside antibiotics has involved the feeding 5-fluorouracil to *Streptomyces cacaoi*. The products obtained from these experiments were converted into their TMS derivatives and subjected to g.c.m.s., and their characteristic fragmentations, *e.g.* (21), enabled structural assignments to be made.[133] In another investigation where bacterial strains resistant to aminoglycoside antibiotics were being studied, mass spectrometry was used to show that these organisms contained an enzyme which acetylates the 2-amino-group of the aminohexose ring of the antibiotic. Peaks at m/e 169 and m/e 404

---

[129] W. A. König, W. Loeffler, W. H. Meyer, and R. Uhmann, *Chem. Ber.*, 1973, **106**, 816.
[130] T. Takaishi, Y. Sugawara, and M. Suzuki, *Tetrahedron Letters*, 1972, 1873.
[131] J. Aberhart, R. C. Jain, T. Fehr, P. de Mayo, and I. Szilagyi, *J.C.S. Perkin I*, 1974, 816; D. E. F. Gracey, L. Baczynskyji, T. I. Martin, and D. B. Maclean, *J.C.S. Perkin I*, 1974, 827; R. Fehr, R. C. Jain, P. de Mayo, O. Motl, I. Szilagyi, L. Baczynskyji, D. E. F. Gracey, H. L. Holland, and D. B. MacLean, *J.C.S. Perkin I*, 1974, 836.
[132] M. Brufani, L. Cellai, C. Musu, and W. Keller-Schierlein, *Helv. Chem. Acta*, 1972, **55**, 2329; R. Muntwyler and W. Keller-Schierlein, *Helv. Chem. Acta*, 1972, **55**, 2071; H. Reimann and R. S. Jaret, *J.C.S. Chem. Comm.*, 1972, 1270; S. M. Kupchan, Y. Komoda, G. J. Thomas, and H. P. J. Hintz, *J.C.S. Chem. Comm.*, 1972, 1065; A. Kinumaki and M. Suzuki, *J.C.S. Chem. Comm.*, 1972, 744.
[133] K. Isono, P. F. Crain, T. J. Odiorne, J. A. McCloskey and R. J. Suhadolnik, *J. Amer. Chem. Soc.*, 1973, **95**, 5788.

in the mass spectrum of the product obtained from sisomicin (22) were assigned to ions (23) and (24), the former giving the location of the acetate on the unsaturated sugar ring and the latter, formed by a retrodiene cleavage, indicating the position in the molecule.[134]

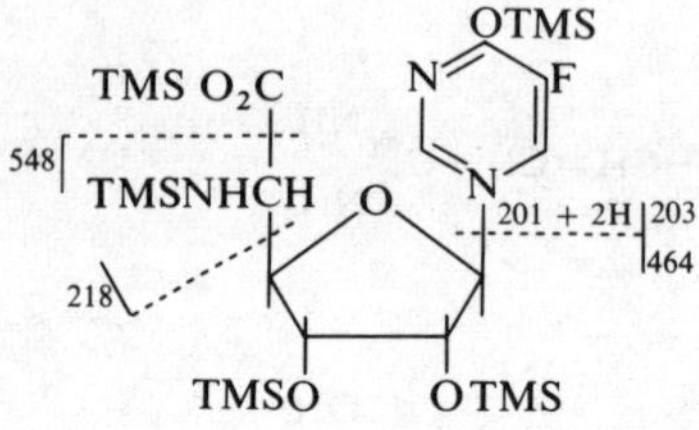

(21)

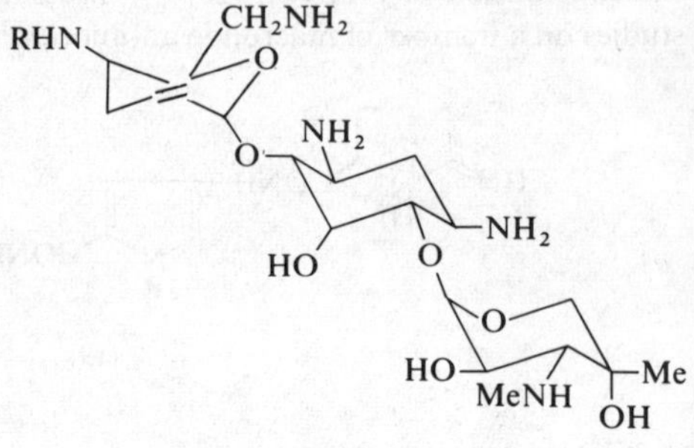

(22)

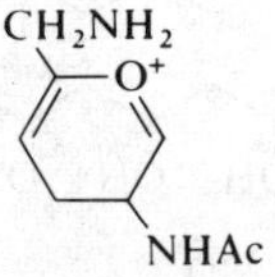

(23) m/e 169

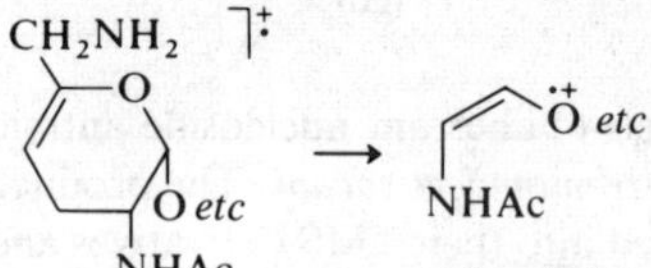

(24) m/e 404

7 Nucleic Acid Components

Mass spectral methods for nucleotide sequence analysis have been reviewed,[69] and a more comprehensive review covering all aspects of the mass spectrometry of nucleic acids and derivatives has appeared.[135]

Studies of the utility of softer ionization methods in structural investigation of these classes of compound have been reported. The FD spectra of adenine and guanine have only molecular ions in their spectra and no fragment ions.[136] The nucleosides, guanosine, adenosine, and thymidine have been studied by high-resolution FD and all show strong molecular ions and fragment ions, providing

---

[134] M. Chevereau, P. J. L. Daniels, J. Davies, and F. Le Goffic, *Biochemistry*, 1974, **13**, 598.
[135] C. Hignite, in ref. 9, p. 429.
[136] H. R. Schulten and H. D. Beckey, *Org. Mass Spectrometry*, 1972, **6**, 885.

information about the base and sugar [see (25)].[137] These studies were extended to nucleotides; adenosine-5'-monophosphate had a strong $[M + 1]^+$ ion, and ions which provided information about the sugar and base were present. Similar results were obtained for 5'-deoxycytidylic, 5'-thymidylic, and 5'-deoxyguanylic

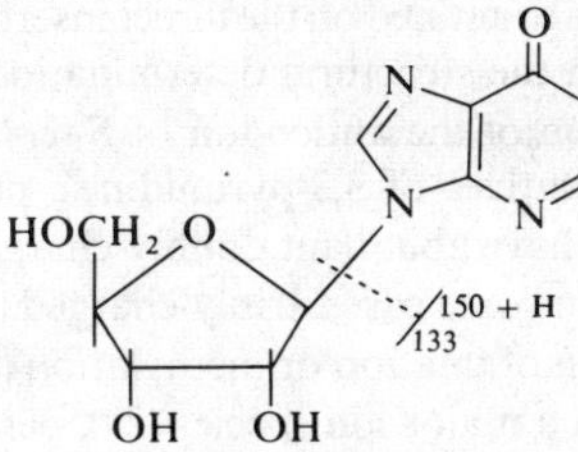

(25)

acids,[137] and recently the FD spectrum of the sodium salt of cyclic adenosine monophosphate has been reported.[101] Pyrolysis–FD has been carried out on herring DNA resulting in the identification of cytosine, methylcytosine, thymine, adenine, and guanine as well as some nucleosides, nucleotides, and dinucleotides.[8]

CI studies of nucleosides with various reagent gases gave several different principal ions.[138] The extent of fragmentation was dependent on reagent gas basicity, being considerable with methane and very small with trimethylamine. The technique has also been used to compare glycosyl bond cleavages of the isomeric 7- and 9-β-D-ribofuranosyl purines (26) and (27) with similar cleavages

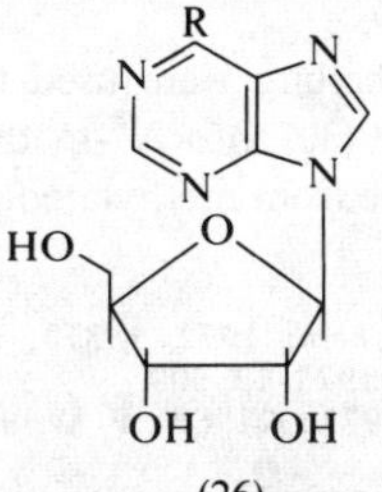

(26) 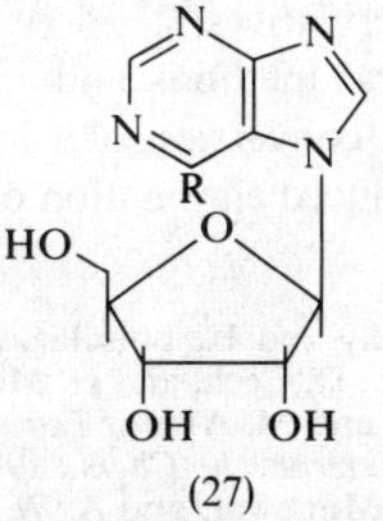   (27)

observed in solution.[139] Using ammonia as reagent gas only protonated molecular ions and protonated free base $(BH_2)^+$ were observed. The spectra enable differentiation of isomers and, as in solution, the glycosidic bonds of the 7-isomers showed greater tendency to fragment than did the 9-isomers.

---

[137] H. R. Schulten and H. D. Beckey, *Org. Mass Spectrometry*, 1973, **7**, 861.
[138] M. H. Wilson and J. A. McCloskey, Twenty-first Annual Conference on Mass Spectrometry and Allied Topics, San Francisco, May, 1973, p. 276.
[139] J. A. McCloskey, J. H. Futrell, and T. A. Elwood, K. H. Schram, B. P. Panzica, and L. B. Townsend, *J. Amer. Chem. Soc.*, 1973, **95**, 5762.

Studies of the EI fragmentations of purines,[140] pteridines,[141] and substituted phenyluracil and thiouracil[142] derivatives are reported. *NO*-Permethyl derivatives of nucleosides are amenable to g.c.m.s. investigation, giving good molecular and base-containing ions, which carry a relatively large proportion of the total ion current; their disadvantages are relatively poor g.c. characteristics and extraneous peaks with samples examined on the direct insertion probe.[143] Derivatives of this type were used in the structural determination of N^4-acetylcytidine isolated from the first position of the anticodon of *E. coli* tRNA$^{\text{Met-12}}$.[144] The fragmentations of TMS derivatives of 3,3-pyrimidines, purines, and pteridines have been studied[145,146] and have abundant doubly-charged $M - 30$ ions due to loss of two methyl radicals. In some cases, triply-charged $[M - 3\text{Me}]^{3+}$ ions are also observed and the position of thiation or methylation (C-5 *vs* C-6) in yrimidines can be established from a major ion species (29), derived from ion (28).[145]

$$\underset{(28) \ X = O, S, \text{ or NH}}{\text{TMSO}\underset{N}{\overset{+}{\underset{\|}{N}}}\overset{\overset{XSiMe_2}{|}}{\underset{5}{\overset{4}{\diagup}}}R} \longrightarrow \underset{(29)}{Me_2 \overset{+}{Si}X\underset{4}{C}\equiv\underset{5}{C}R}$$

The 3'- and 5'-*o*-t-butyldimethylsilyl derivatives of thymidine and deoxyadenosine give characteristically different mass spectra enabling the isomerically substituted nucleosides to be differentiated.[147] The fragmentations of derivatized and underivatized cyclonucleosides have also been studied and differentiation between isomers is often possible on the basis of their fragmentations or those of appropriate derivatives.[148]

Mass spectral methods and ^{15}N- and ^{2}H-labelling were used to study the base-catalysed conversion of 1-methyladenosine (30) into N^6-methyladenosine (31).[149] EI-induced elimination of methylenimine from methylated purine bases

---

[140] J. S. Connolly and H. Linschiz, *J. Heterocyclic Chem.*, 1972, **9**, 379; H. Beerbaum, K. H. Grupe, D. Cech, and H. Meinert, *Z. Chem.*, 1973, **13**, 309.
[141] Y. Iwanami and M. Akino, *Tetrahedron Letters*, 1972, 3219; V. P. Williams and J. E. Ayling, *J. Heterocyclic Chem.*, 1973, **10**, 827.
[142] J. Clark, Z. Munawar, and A. W. Tims, *J.C.S. Perkin II*, 1972, 233.
[143] D. L. von Minden and J. A. McCloskey, *J. Amer. Chem. Soc.*, 1973, **95**, 7480.
[144] Z. Ohashi, K. Murao, T. Yahagi, D. L. von Minden, J. A. McCloskey, and S. Nishimura, *Biochim. Biophys. Acta*, 1972, **262**, 209.
[145] E. White, V. P. M. Krueger, and J. A. McCloskey, *J. Org. Chem.*, 1972, **37**, 430.
[146] W. J. A. Van den Heuvel, J. L. Smith, P. Haug, and J. L. Beck, *J. Heterocyclic Chem.*, 1972, **9**, 451.
[147] K. K. Ogilvie and D. J. Iwacha, *Tetrahedron Letters*, 1973, 317; M. A. Quillam, K. K. Ogilvie, and J. B. Westmore, *Org. Mass Spectrometry*, 1974, **1**, 78.
[148] J. B. Westmore, D. C. K. Lin, K. K. Ogilvie, H. Wayborn, and J. Berestiansky, *Org. Mass Spectrometry*, 1972, **6**, 1243; J. B. Westmore, D. C. K. Lin, and K. K. Ogilvie, *Org. Mass Spectrometry*, 1973, **7**, 317; E. G. Lovett and D. Lipkin, *J. Amer. Chem. Soc.*, 1973, **95**, 2312; D. C. K. Lin, L. Slotin, K. K. Ogilvie, and J. B. Westmore, *J. Org. Chem.*, 1973, **38**, 1118.
[149] M. H. Wilson and J. A. McCloskey, *J. Org. Chem.*, 1973, **38**, 2247.

or nucleosides is useful in the structural characterization of these compounds and mechanistic studies have been reported.[150]

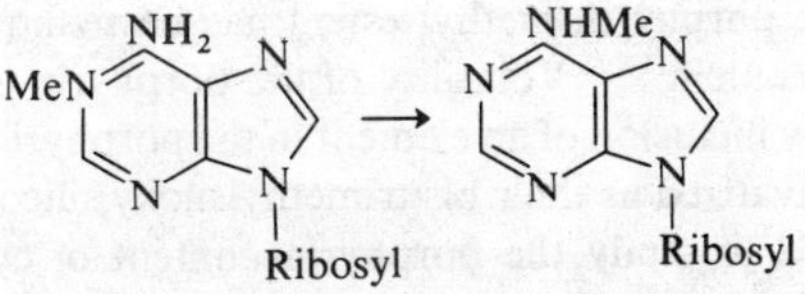

(30) → (31)

High- and low-resolution data assisted in the assignment of structure (32) to a new pyrimidine from the nucleic acid of *Bacillus subtilis*.[151] Studies designed to assist in the identification of modified adenosine derivatives in RNA's and cytokinins have been reported.[152] Examination of the mass spectrum of the TMS derivative of a modified nucleoside isolated from *E. coli* tRNA^{Thr} gave considerable assistance in its formulation as (33),[153] and similar methods have been used to identify other modified nucleosides.[154]

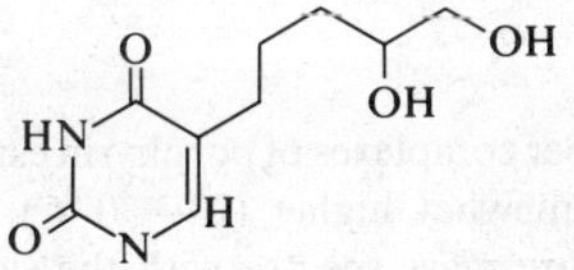

8 Pyrrole Pigments

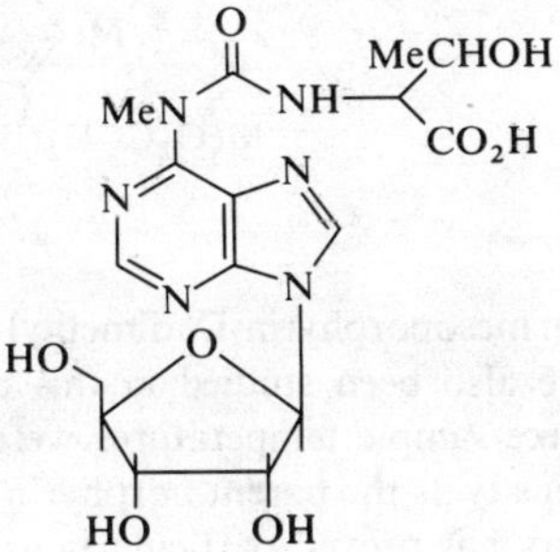

Mass spectral studies in this area have been comprehensively reviewed.[155]

The porphyrins involved in normal and abnormal heme metabolism are relatively involatile, and in most cases satisfactory mass spectra may be obtained

[150] D. L. von Minden, J. G. Liehr, M. H. Wilson, and J. A. McCloskey, *J. Org. Chem.*, 1974, **39**, 285.
[151] C. Brandon, P. M. Gallop, J. Marmur, H. Hayashi, and K. Nakanishi, *Nature*, 1972, **239**, 70.
[152] S. M. Hecht and J. J. McDonald, *Analyt. Biochem.*, 1972, **47**, 157.
[153] F. Kimura-Harada, D. L. von Minden, J. A. McCloskey, and S. Nishimura, *Biochemistry*, 1972, **11**, 3910.
[154] S. Nichimura, Z. Ohashi, M. Maeda, J. G. Liehr, P. F. Grain, and J. A. McCloskey, Twenty-second Annual Conference on Mass Spectrometry and Allied Topics, Philadelphia, May 1974, paper P-1.
[155] R. C. Dougherty, in ref. 9, p. 591.

only following conversion into the more volatile methyl esters.[156] In the case of hydroxyporphyrins further derivatization is necessary to obtain molecular weights since their methyl esters lose H_2O in the mass spectrometer.[157] Preparation of TMS ethers of the porphyrin methyl esters has been shown to be effective in overcoming this problem.[157] Volatility of the porphyrin methyl esters may be improved further by inclusion of an element in the porphyrin ring. Alkyl porphyrins have been derivatized as their bis(trimethylsiloxy)silicon complexes and the technique was used to study the porphyrin content of crude oils by g.c. and mass spectral methods.[158] These derivatives have recently been shown to be amenable to combined g.c.m.s. study,[159] but porphyrins containing carboxylic acid groups were not so amenable. A bis(diethoxy)silicon complex of mesoporphyrin-IX dimethyl ester (34) has been prepared and is considerably more volatile

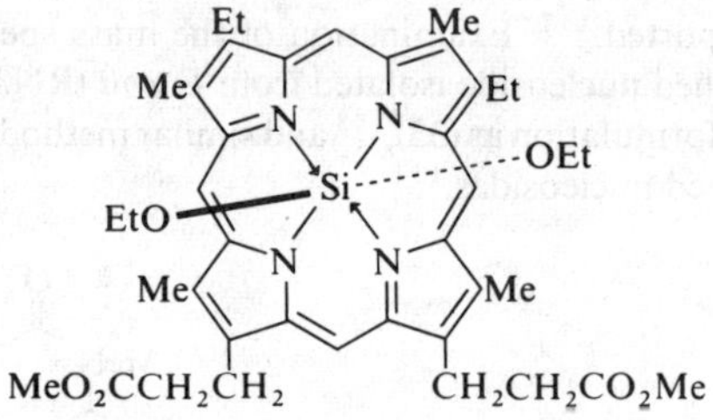

(34)

than mesoporphyrin-IX dimethyl ester.[159] Copper complexes of porphyrin esters have also been studied in this context but somewhat higher (20—30 °C) ion source/sample temperatures were required before mass spectra with the same intensity as the parent porphyrin ester could be obtained.[160] FD mass spectrometry has proved particularly useful here, since good spectra of free porphyrin acids are obtainable; in general, molecular and quasi-molecular ions were obtained with few fragment ions.[161,162] Mixtures of porphyrin methyl esters, obtained from the urine of porphyrics, and pheophytins have been studied using a combination of FD with high-pressure liquid chromatography.[162]

[156] A. H. Jackson, G. W. Kenner, K. M. Smith, R. T. Aplin, H. Budzikiewicz, and C. Djerassi, *Tetrahedron*, 1965, **21**, 2913.
[157] J. R. Chapman and G. H. Elder, *Org. Mass Spectrometry*, 1972, **6**, 991.
[158] D. B. Boylan and M. Calvin, *J. Amer. Chem. Soc.*, 1967, **89**, 5472; D. B. Boylan, Y. I. Alturki, and G. Eglington, in 'Advances in Organic Geochemistry', ed. P. A. Schenck and I. Havenaar, Pergamon, New York, 1969.
[159] D. E. Games, A. H. Jackson, and D. S. Millington, in 'Mass Spectrometry in Biochemistry and Medicine', ed. A. Frigerio and N. Castagnoli, Raven Press, New York, 1974, p. 257.
[160] J. Møller and T. K. With, *Org. Mass Spectrometry*, 1974, **9**, 443.
[161] D. E. Games, A. H. Jackson, D. S. Millington, and M. Rossiter, in 'Advances in Mass Spectrometry', ed. A. R. West, Applied Science, London, 1974, Vol. 6, p. 137.
[162] N. Evans, D. E. Games, A. H. Jackson, S. A. Matlin, M. Rossiter, and R. G. Saxton, Twenty-second Annual Conference on Mass Spectrometry and Allied Topics, Philadelphia, May 1974, paper O-6.

Examination of the reductive degradation products from isocoproporphyrin tetramethyl ester by g.c.m.s. enabled formulation of this porphyrin as (35) rather than (36).[163] Oxidative and reductive degradation in combination with mass spectrometry and g.c.m.s. for structural studies of pyrrole pigments has been

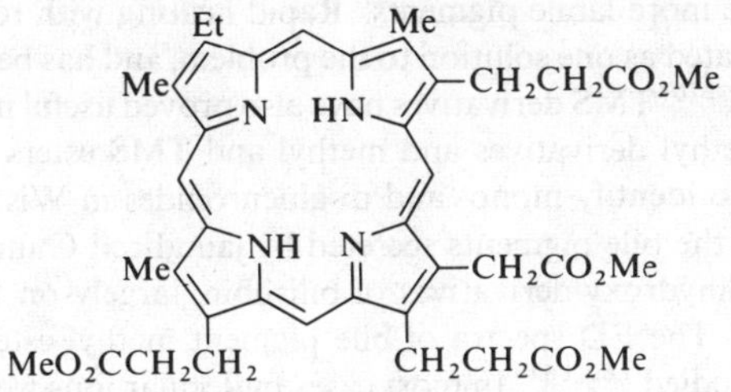

(35)

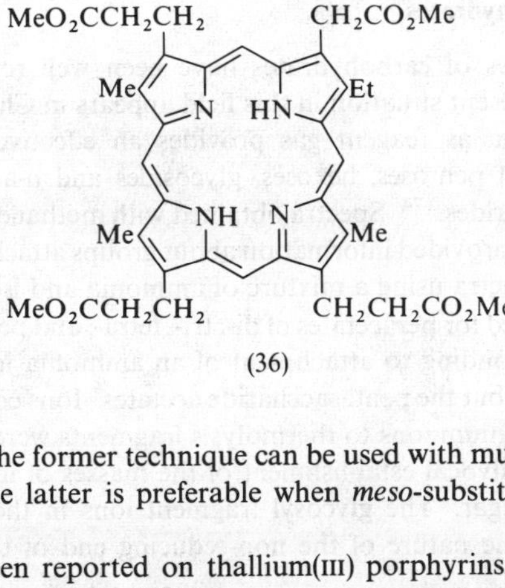

(36)

investigated.[164] The former technique can be used with much smaller quantities of material but the latter is preferable when *meso*-substituted compounds are being studied.

Studies have been reported on thallium(III) porphyrins,[165] *meso*-substituted porphyrins,[166] and chlorins.[167] Ruthenium octaethylporphyrin carbonyl has been found to react with perfluorokerosene and perfluorotributylamine with ions of the type $[M - CO]$ and $[C_nF_{2n}]$, $n = 1—4$, being formed.[168]

[163] M. S. Stoll, G. H. Elder, D. E. Games, P. J. O'Hanlon, D. S. Millington, and A. H. Jackson, *Biochem. J.*, 1973, **131**, 429.
[164] A. H. Jackson, D. S. Millington, and D. E. Games, in 'Advances in Mass Spectrometry', ed. A. R. West, Applied Science, London 1974, Vol. 6, p. 215.
[165] K. M. Smith, *Org. Mass Spectrometry*, 1972, **6**, 1401.
[166] M. Meot-Ner, A. D. Adler, and J. H. Green, *Org. Mass Spectrometry*, 1974, **9**, 72.
[167] M. Meot-Ner, A. D. Adler, and J. H. Green, *Org. Mass Spectrometry*, 1973, **7**, 1395.
[168] D. Rosenthal, F. R. Hopf, D. G. Whitten, and M. M. Bursey, *Org. Mass Spectrometry*, 1973, **7**, 497.

EI mass spectrometry continues to find applications in structural studies of bile and plant pigments.[169] Like porphyrins, bile pigments are more amenable to mass spectral study as their methyl esters. Problems often arise in assigning the correct molecular weights to these compounds, both in the free and derivatized forms, owing to disproportionation in the mass spectrometer or to decomposition in the case of the more labile pigments. Rapid heating with repetitive scanning has been investigated as one solution to the problem, and has been used in studies of plant pigments.[170] TMS derivatives have also proved useful in this context and, together with methyl derivatives and methyl and TMS esters of azodipyrroles, have been used to identify mono- and di-glucuronides in Wistar rats.[171] TMS derivatization of the bile pigments secreted by jaundiced Gunn rats led to their formulation as dihydroxy-derivatives of bilirubin, largely on the basis of mass spectral data.[172] The FD spectra of bile pigment methyl esters and free acids have also been studied.[162,164] In most cases molecular ions were the base peaks and little or no disproportionation was present. The more stable pigments showed few or no fragment ions but more labile pigments provided structurally useful fragment ions.

9 Carbohydrates

Mass spectral studies of carbohydrates have been well reviewed,[173] and an assessment of the present situation in this field appears in Chapter 10.

CI using ammonia as reagent gas provides an effective determination of molecular weights of pentoses, hexoses, glycosides and o-acetyl derivatives of mono- and di-saccharides.[174] Spectra obtained with methane gave characteristic fragment ions which provided information about groups attached to the pyranose ring.[174] CI mass spectra using a mixture of ammonia and isobutane as reagent gas have been reported for peracetates of di-, tri-, tetra-, and penta-saccharides.[175] Intense ions corresponding to attachment of an ammonia ion to the molecule were observed for all but the pentasaccharide acetates. Ions corresponding to the attachment of ammonium ions to thermolysis fragments were usually dominant and permitted unequivocal establishment of the masses of individual sub-units within the parent sugar. The glycosyl fragment ions in the spectra provided information about the nature of the non-reducing end of the oligosaccharide chain.

[169] D. J. Chapman, H. Budzikiewicz, and H. W. Siegelman, *Experientia*, 1972, **28**, 876; S. D. Killitea and P. O'Carra, *Biochem. J.*, 1972, **129**, 1179, R. Bonnett and A. F. McDonagh, *J.C.S. Perkin I*, 1973, 881.
[170] R. J. Beuhler, L. Friedman, and H. W. Siegelman, Twenty-second Annual Conference on Mass Spectrometry and Allied Topics, Philadelphia, May 1974, paper H-7.
[171] M. Salmon and C. Fenselau, Twenty-second Annual Conference on Mass Spectrometry and Allied Topics, Philadelphia, May 1974, P-2.
[172] C. S. Berry, J. E. Zarembo, and J. D. Ostrow, *Biochem. Biophys. Res. Comm.*, 1972, **49**, 1366.
[173] T. Radford and D. C. De Jongh, ref. 9, p. 313.
[174] A. M. Hogg and T. L. Nagabushan, *Tetrahedron Letters*, 1972, 4827.
[175] R. C. Dougherty, J. D. Roberts, W. W. Binkley, O. S. Chizhov, V. I. Kadentsev, and A. A. Solov'yov, *J. Org. Chem.*, 1974, **39**, 451.

The 1,6-anhydro-3,4-O-isopropylidene-β-D-talopyranose (37),[176] the isomeric 2,3-acetal (38),[177] and 2-acetamido-1,3,4,6-tetra-O-acetyl-2-deoxy-α-D-glucopyranose[178] have been the subject of detailed mass spectral study. FD spectra

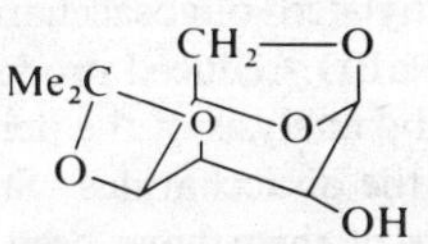

(37)

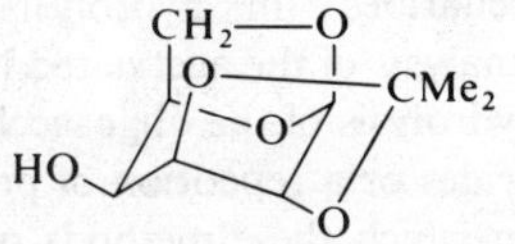

(38)

of disodium D-glucose-6-phosphate[179] and stereoisomeric glycosides[180] have been described and, in the latter case, the relative intensities of fragment ions can be used to differentiate the isomers. A combination of EI and FD assisted in the assignment of structure (39) to the anthraquinone glycoside, frangulin B.[181]

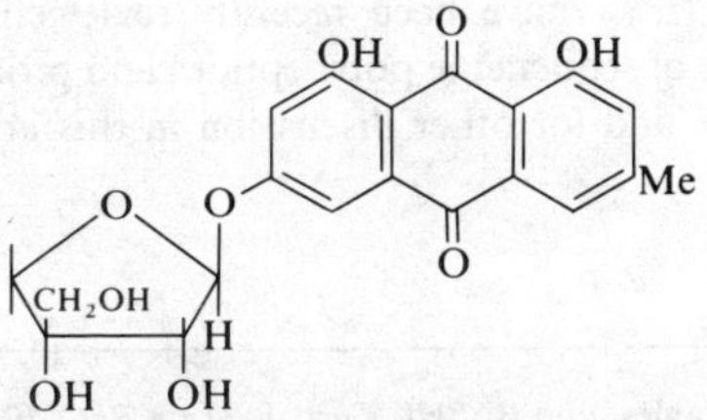

(39)

Mass spectral studies continue to be carried out on a large number of derivatized carbohydrates,[182–197] because of the importance of these compounds in structural studies using g.c.m.s. (see also Chapter 8).

---

[176] D. Horton, J. S. Jewell, E. K. Just, J. D. Wander, and R. L. Foltz, *Biomedical Mass Spectrometry*, 1974, **1**, 145.
[177] D. Horton, E. K. Just, and J. D. Wander, *Org. Mass Spectrometry*, 1972, **6**, 1121.
[178] R. C. Dougherty, D. Horton, K. D. Phillips, and J. D. Wander, *Org. Mass Spectrometry*, 1973, **7**, 805.
[179] H.-R. Schulten, H. D. Beckey, E. M. Bessell, A. B. Foster, M. Jarman, and J. H. Westwood, *J.C.S. Chem. Comm.*, 1973, 416.
[180] W. D. Lehmann, H.-R. Schulten, and H. D. Beckey, *Org. Mass Spectrometry*, 1973, **7**, 1103.
[181] H. Wagner and G. Demuth, *Tetrahedron Letters*, 1972, 5013.
[182] T. Matsubara and A. Hayashi, *Biomedical Mass Spectrometry*, 1974, **1**, 62.
[183] J. P. Kamerling, J. F. G. Vliegenthart, and J. Vink, *Carbohydrate Res.*, 1974, **33**, 297.
[184] K. Axberg, H. Bjorndal, A. Pilotti, and S. Svensson, *Acta Chem. Scand.*, 1972, **26**, 1319.
[185] V. Kovacik and P. Kovac, *Carbohydrate Res.*, 1972, **24**, 23.
[186] R. M. Caprioli and W. E. Seifert, *Biochim. Biophys. Acta*, 1973, **297**, 213.
[187] R. M. Caprioli and E. J. Heron, *Biochim. Biophys. Acta*, 1973, **296**, 321.
[188] R. A. Laine and C. C. Sweley, *Carbohydrate Res.*, 1973, **27**, 199.
[189] S. Inouye, *Chem. and Pharm. Bull. (Japan)*, 1972, **29**, 2320.

Almost complete sequence information can be provided for oligosaccharides by various combinations of the following procedures: (i) derivatization, permethylation, peracetylation, or persilylation followed by mass spectral analysis; (ii) complete hydrolysis of the oligosaccharide and g.c.m.s. analysis of the derivatized monosaccharides; (iii) hydrolysis of the permethylated oligosaccharide and g.c.m.s. analysis of the acetylated $NaBH_4$ and/or $NaBD_4$ reduced products; (iv) partial hydrolysis of the oligosaccharide followed by analysis of the derivatized disaccharides or a repetition of procedure (iii) on the disaccharides. Structural studies in which these methods or minor variants of them have been applied include pneumococcus type-II capsular polysaccharides,[198] polysaccharides from the bark of white willow,[199] disaccharides from human urine,[200] polysaccharides of *Klebsiella* strains,[201] of *Clostridium botulinum*,[202] and a tetrasaccharide from horse chestnut seed.[203]

10 Amino-acids, Biogenic Amines, and Peptides

The mass spectrometry of amino-acids[204] and the use of mass spectral methods in peptide chemistry[205,206] have been recently reviewed. For an assessment of the present methods of sequencing polypeptides and proteins by mass spectrometry, see Chapter 10, and for other discussion in this area, see Chapters 4, 7, and 8.

---

[190] D. S. Robinson, J. Eagles, and R. Self, *Carbohydrate Res.*, 1973, **26**, 204; P. J. Wood, and I. B. Siddiqui, *ibid.*, 1974, **32**, 97.
[191] I. R. McKinley, H. Weigel, C. B. Barlow, and R. D. Guthrie, *Carbohydrate Res.*, 1974, **32**, 187.
[192] V. N. Reinhold and K. Biemann, Twenty-first Annual Conference on Mass Spectrometry and Allied Topics, San Francisco, May 1973, p. 285.
[193a] A. Vigevani, B. Gioia, and G. Cassinelli, *Carbohydrate Res.*, 1974, **32**, 321.
[193b] A. Buchs and E. Charollais, *Helv. Chim. Acta*, 1973, **56**, 207.
[194] J. P. Kamerling, J. F. G. Vliegenthart, J. Vink, and J. J. de Ridder, *Tetrahedron*, 1972, **28**, 4375.
[195] K. G. Das and B. Thayumanavan, *Org. Mass Spectrometry*, 1972, **6**, 1063.
[196] N. Kashimura, K. Yoshida, and K. Onodera, *Carbohydrate Res.*, 1972, **25**, 264.
[197] O. S. Chizhov, N. N. Malysheva, and N. K. Kochetkov, *Izvest. Akad. Nauk S.S.S.R., Ser. khim.*, 1973, **5**, 1022, 1030, 1246; *Carbohydrate Res.*, 1973, **28**, 21.
[198] O. Larm, B. Lindberg, and S. Svensson, *Carbohydrate Res.*, 1972, **22**, 391.
[199] R. Toman, Š. Karácsonyi, and V. Kováčik, *Carbohydrate Res.*, 1972, **25**, 371.
[200] A. Lundblad and S. Svensson, *Biochemistry*, 1973, **12**, 306.
[201] B. Lindberg, J. Lönngren, and W. Nimmich, *Acta Chem. Scand.*, 1972, **26**, 2231; B. Lindberg, J. Lönngren, J. Thompson, and W. Nimmich, *Carbohydrate Res.*, 1972, **25**, 49; B. Lindberg, K. Samuelsson, and W. Nimmich, *Carbohydrate Res.*, 1973, **30**, 63; Y.-M. Choy and G. G. S. Dutton, *Canad. J. Chem.*, 1973, **51**, 3015, 3021.
[202] J. N. C. Whyte and G. A. Strasdine, *Carbohydrate Res.*, 1972, **25**, 435.
[203] J. P. Kamerling, J. F. G. Vliegenthart, W. Kahl, A. Roszkowski, and A. Zurowska, *Carbohydrates Res.*, 1972, **25**, 293.
[204] W. Vetter, ref. 9, p. 387.
[205] K. Biemann, in ref. 9, p. 405.
[206] M. M. Shemyakin, Yu. A. Ovchinnikov, and A. A. Kiryushkin, in 'Mass Spectrometry, Techniques and Applications', ed. G. W. A. Milne, Wiley-Interscience, New York, 1971, p. 289.

Amino-acids and Biogenic Amines.—Amino-acids, their amides, methyl esters, and N^a-acetyl derivatives have been the subjects of detailed CI studies.[207-209] Enhanced $[M + 1]^+$ ion intensities have been obtained in the FD spectra of creatine and arginine by use of an indirect heating device.[210] A mixture of twelve amino-acids has been examined using FD; the amino-acids were readily identified but quantitation will require further refinements in technique.[211]

Dimethylaminomethylene alkyl esters of amino-acids are a useful additional derivative giving simultaneous derivatization of both amino and carboxyl functions.[212] They are readily prepared from dimethylformamide dialkylacetal, generally show molecular ions, and give good g.c. peaks. These derivatives have been used in the analysis of amino-acid mixtures using a direct insertion probe and a temperature programmed source.[213] Identification of the amino-acids in a mixture of arginine, citrulline, cystine, histidine, and tryptophan was effected and a semiquantitative estimation of the amounts of each amino-acid present were obtained. The mass spectra of the thiohydantoin derivatives of amino-acids obtained from Edman degradation of peptides have been discussed.[214,215] *N*-Methyl- and *N*-phenyl-thiourea derivatives of amino-acids and peptides have been shown to rearrange to methyl- and phenyl-thiohydantoin derivatives in the mass spectrometer.[216] Difficulties are experienced in the quantitation of the phenylthiohydantoin of cysteine because of its weak molecular ion and the overlap of its predominant fragment ion with one from the phenylthiohydantoin of serine; use of *p*-bromophenylthiohydantoins overcomes the problem.[217] *N*-Trifluoroacetyl-*O*-n-butyl ester derivatives have been used for the quantitation of ten amino-acids in soil,[218] and the mass spectral fragmentation of tris-TMS amino-acids,[219] *N*-quaternary amino-acids[220] and esters[221] and sulphur-containing amino-acids[222] have been reported.

---

[207] P. A. Leclercq and D. M. Desiderio, *Org. Mass Spectrometry*, 1973, **7**, 515.
[208] G. W. A. Milne, T. Axenrod, and H. M. Fales, *J. Amer. Chem. Soc.*, 1970, **92**, 5170.
[208a] M. Moet-Ner and F. H. Field, *J. Amer. Chem. Soc.*, 1973, **95**, 7207.
[209] G. W. A. Milne, H. M. Fales, and R. W. Colburn, *Analyt. Chem.*, 1973, **45**, 1952.
[210] H. U. Winkler and H. D. Beckey, *Org. Mass Spectrometry*, 1973, **7**, 1007.
[211] H. U. Winkler, Twenty-second Conference on Mass Spectrometry and Allied Topics, Philadelphia, May 1974, p. 5.
[212] J. P. Thenot and E. C. Horning, *Analyt. Letters*, 1972, **5**, 519.
[213] I. Horman and F. J. Hesford, *Biomedical Mass Spectrometry*, 1974, **1**, 115.
[214] T. Suzuki, S. Matsui, and K. Tuzimura, *Agric. and Biol. Chem. (Japan)*, 1972, **36**, 1061.
[215] F. Richards and R. E. Lovins, in 'Methods in Enzymology', ed. C. H. W. Hirs and S. N. Timasheff, Academic Press, New York, 1972, Vol. 25(B), p. 314.
[216] T. Fairwell, S. Ellis, and R. E. Lovins, *Analyt. Biochem.*, 1973, **53**, 115.
[217] H. Tschesche, M. Schneider, and E. Wachter, *F.E.B.S. Letters*, 1972, **23**, 367.
[218] W. E. Pereira, Y. Hoyano, W. E. Reynolds, R. E. Summons, and A. M. Duffield, *Analyt. Biochem.*, 1973, **55**, 236.
[219] G. Mischer, *Z. analyt. Chem.*, 1972, **262**, 81.
[220] K. Undheim and T. Laerum, *Acta Chem. Scand.*, 1973, **27**, 589.
[221] D. F. Biggs, R. T. Coutts, and D. B. Henderson, *Canad. J. Pharm. Sci.*, 1972, **7**, 90.
[222] H. Nishimura, S. Tahara, H. Okuyama, and J. Mitzutoni, *Tetrahedron*, 1972, **28**, 4503.

The mass spectral fragmentation of the aminolactone hydrochloride (40) from thermozymocidin, an antiobiotic from *Mycolia sterilia*, assisted the assignment of its structure (41).[223]

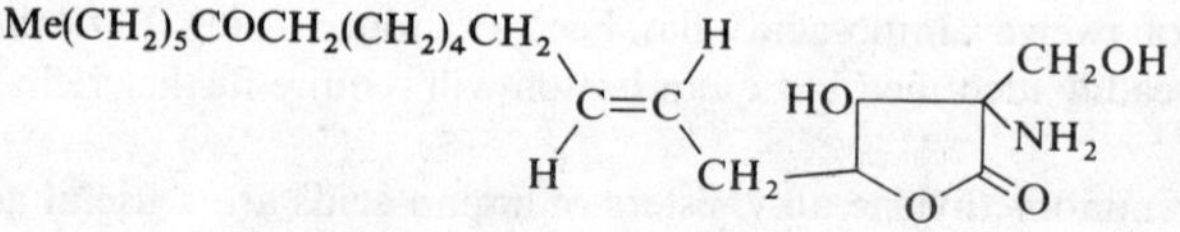

$$\text{Me(CH}_2)_5\text{COCH}_2\text{(CH}_2)_4\text{CH}_2\diagdown\text{C}=\text{C}\diagup\text{H} \quad \text{HO}\diagdown\diagup\text{CH}_2\text{OH},\text{NH}_2$$

(40)

$$\text{Me(CH}_2)_5\text{CO(CH}_2)_6\diagdown\text{C}=\text{C}\diagup\text{H},\text{CH}_2\text{CHOHCHOHCCO}_2^- \text{ with CH}_2\text{OH and }\overset{+}{\text{NH}_3}$$

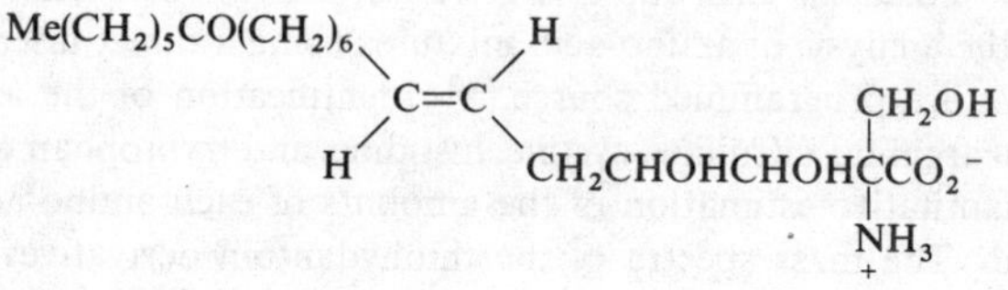

(41)

3-Amino-12-methyltetradecanoic and 3-amino-12-methyltridecanoic acids have been isolated from iturin A, a peptide antiobiotic from *Bacillus subtilis*.[224] Their structures were determined by g.c.m.s. of their N-acetylmethyl esters. The new amino-acids (42)[225] and (43)[226] have been isolated from a New Guinea fungus and *Fagopyrum esculentum*, respectively, and a homologus series of n-acyl-2-methylene-β-alanine methyl esters have been reported from the sponge *Fasciospongia cavernosa*.[226a]

$$\text{CH}_2=\text{CHCH(CH}_3)\text{CH}_2\text{CH(NH}_2)\text{CO}_2\text{H}$$

(42)

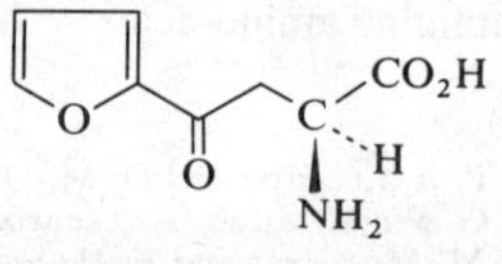

(43)

Peptides.—The CI spectra of 25 dipeptides have been studied.[227] By deposition of sample on the surface of an extended tip of a conventional sample probe and introduction into the ion plasma of a CI source, spectra of underivatized peptides

---

[223] F. Aragozzini, P. L. Manachini, R. Craveri, B. Rindone, and C. Scolastico, *Tetrahedron*, 1972, **28**, 5493.
[224] F. Peypoux, F. Guinand, G. Michel, L. Delcambe, B. C. Das, P. Varenne, and E. Lederer, *Tetrahedron*, 1973, **29**, 3455.
[225] R. Rudzats, E. Gelbert, and B. Halpern, *Biochem. Biophys. Res. Comm.*, 1972, **47**, 290.
[226] A. Ichihara, H. Hasegawa, H. Sato, M. Koyama, and S. Sakamura, *Tetrahedron Letters*, 1973, 37.
[226a] Y. Kashman, L. Fishelson, and I. Ne'eman, *Tetrahedron*, 1973, **29**, 3655.
[227] D. V. Bowen and F. H. Field, Twenty-first Annual Conference on Mass Spectrometry and Allied Topics, San Francisco, May 1973, p. 299; *Internat. J. Peptide Protein Res.*, in the press.

can be obtained at temperatures at least 150 °C lower than those usually required;[228] 21 underivatized peptides, including two pentapeptides, were examined this way. A second group[229] has confirmed the enhanced volatility obtained by this technique but found that, in the majority of cases, underivatized peptides did not provide sequence information. CI studies of proline dipeptides using ammonia as reagent gas have shown that ammonolysis of the base-sensitive amide bonds occurs, and this could prove useful in sequence studies.[112]

FD mass spectrometry remains the method of choice for obtaining molecular weight information from underivatized peptides; $[M + H]^+$ ions were almost the only peaks at optimal emitter currents in the FD spectra of a cyclic decapeptide and a linear nonapeptide, both with molecular weights in excess of 1000. The latter peptide had arginine, histidine, and methionine present in its structure. From studies reported to date,[210,230,231a,b] it appears that FD provides little sequencing information for underivatized peptides and that its main utility in this area is going to be in providing a criterion of purity, molecular weight determination, and the study of mixtures.

EI studies have been reported for a number of different types of peptides.[232–234]

The use of 3-hydroxylalkanoyl and Δ^2-acyl groups as N-protecting groups has been assessed for sequence analysis.[235] Dec(OH)- and Δ^2-dec-peptide derivatives are both amenable to permethylation, and were found to be superior to previously reported derivatives of this type. In another study, 34 N-terminal derivatives were compared in terms of relative intensities of molecular and N-terminal sequence ions.[236] 5-(NN-Dimethylamino)naphthalene sulphonyl, p-dimethylaminobenzylidene, and 4-(NN-dimethylamino)naphthylidene derivatives gave the best results. These studies have recently been extended to N-terminal prolyl peptides, where the use of Schiff base derivatives was again found to be useful.[237] N-Methyl- and N-phenyl-thiourea derivatives have been favoured by another group for the sequencing of small peptides.[238] In the mass

[228] M. A. Baldwin, and F. W. McLafferty, *Org. Mass Spectrometry*, 1973, **7**, 1353.
[229] D. V. Bowen and F. H. Field, Twenty-second Annual Conference on Mass Spectrometry and Allied Topics, Philadelphia, May 1974, P-4.
[230] U. Rapp, A. Buck, and K. H. Maurer, Twenty-second Annual Conference on Mass Spectrometry and Allied Topics, Philadelphia, May 1974, P-7.
[231a] H. U. Winkler and H. D. Beckey, *Biochem. Biophys. Res. Comm.*, 1972, **46**, 391.
[231b] N. Evans, D. E. Games, M. J. E. Hewlins, J. F. J. Hughes, A. H. Jackson, J. R. Jackson, M. Rossiter, H. A. Swaine, and K. T. Taylor, Twenty-second Annual Conference on Mass Spectrometry and Allied Topics, Philadephia, May 1974, O-5.
[232] C. Bogentoft, J. K. Chang, H. Sievertsson, B. Currie, and K. Folkers, *Org. Mass Spectrometry*, 1972, **6**, 735.
[233] C. N. C. Drey and J. Lowbridge, *Org. Mass Spectrometry*, 1973, **7**, 779.
[234] K. D. Haegele and D. M. Desiderio, Twenty-second Annual Conference on Mass Spectrometry and Allied Topics, Philadelphia, May 1974, P-6.
[235] K. Okada, S. Nagai, T. Uyehara, and M. Hiramoto, *Tetrahedron*, 1974, **30**, 1175; K. Okada, T. Uyehara, M. Hiramoto, H. Kato, and T. Suzuki, *Chem. and Pharm. Bull. (Japan)*, 1973, **21**, 2217.
[236] R. A. Day, H. Falter, J. P. Lehman, and R. E. Hamilton, *J. Org. Chem.*, 1973, **38**, 782; G. V. Patil, R. E. Hamilton, and R. A. Day, *Org. Mass Spectrometry*, 1973, **17**, 817.
[237] R. E. Hamilton, G. V. Patil, K. Jayasimhulu, and R. A. Day, *Org. Mass Spectrometry*, 1974, **9**, 211.
[238] S. Ellis, T. Fairwell, and R. E. Lovins, *Biochem. Biophys. Res. Comm.*, 1972, **49**, 1407; T. Fairwell, S. Ellis, and R. E. Lovins, *Analyt. Biochem.*, 1973, **53**, 115.

spectrometer these derivatives gave methyl- and phenyl-thiohydantoins, as shown, of the N-terminal amino-acid which can then be detected. Repetition of the cycle and derivatization of the remaining peptide, followed by mass spectral examination, provides sequence information; however, problems arise with larger peptides.

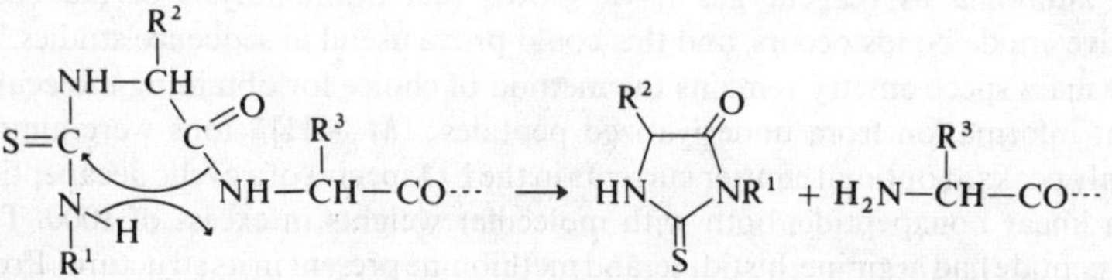

$$S=C\begin{matrix}NH-CH\\ \\C=O\\ \\N\\R^1\end{matrix}\begin{matrix}R^2\\ \\ \\H\end{matrix}NH-CH-CO\cdots \;\longrightarrow\; HN\begin{matrix}R^2\\ \\ \\S\end{matrix}NR^1 + H_2N-CH-CO\cdots$$

Improved permethylation procedures for methionine-, cysteine-, and arginine-containing peptides have been reported.[239] Controlled permethylation[240] has been used successfully in the sequencing of α-melanocyte stimulating hormone.[241] The amino-acid composition of the peptides formed by tryptic digestion and the mass spectra of the derivatized cyanogen bromide and chymotryptic cleavage products of the hormone were used. A simplified procedure for the conversion of peptides into volatile derivatives suitable for mass spectral sequencing has been described. The peptide is heated with trimethylanilinium hydroxide and either pivaldehyde, acetylacetone,[242] or methyl trifluoroacetate[243] in the solid probe tip of the mass spectrometer. Mass spectra obtained by this pyrolysis procedure were identical with reference spectra obtained by conventional procedures. In addition to the normal sequencing peaks, permethylated peptides containing Tyr-Tyr or Trp-Tyr units have additional peaks present 30 mass units higher.[244] Their formation is rationalized as shown in Scheme 6.

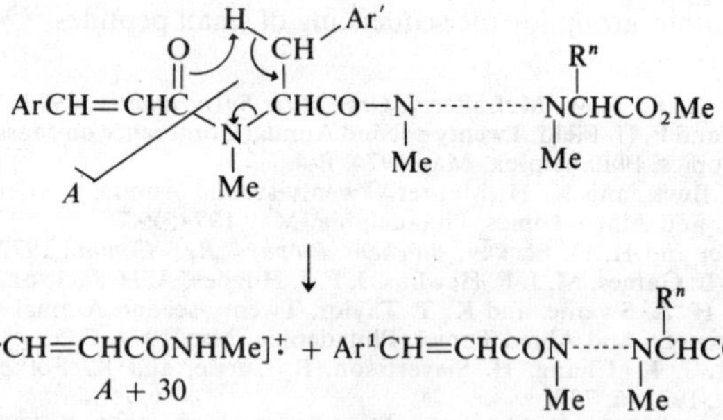

$$ArCH=CHC\begin{matrix}O\\ \\ \\N\\ \\Me\end{matrix}\begin{matrix}H\\ \\CH\\ \\ \end{matrix}\begin{matrix}Ar'\\ \\ \\ \end{matrix}CHCO-N\begin{matrix}\\ \\ \\ \\Me\end{matrix}\cdots\cdots NCHCO_2Me\begin{matrix}R''\\ \\ \\ \\Me\end{matrix}$$

A

↓

$$[ArCH=CHCONHMe]^{\ddagger} + Ar^1CH=CHCON\cdots\cdots NCHCO_2Me$$
$$A + 30 \qquad\qquad\qquad Me\quad Me\quad R''$$

Scheme 6

[239] H. R. Morris, R. J. Dickinson, and D. H. Williams, *Biochem. Biophys. Res. Comm.*, 1973, **51**, 247.
[240] M. L. Polan, W. J. McMurray, S. R. Lipsky, and S. Lande, *Biochem. Biophys. Res. Comm.*, 1970, **38**, 1127.
[241] M. L. Polan, W. J. McMurray, S. R. Lipsky, and S. Lande, *J. Amer. Chem. Soc.*, 1972, **94**, 2847.
[242] G. M. Schier, J. Korth, and B. Halpern, *Tetrahedron Letters*, 1972, 4621.
[243] G. M. Schier and B. Halpern, *Austral. J. Chem.*, 1974, **27**, 393.
[244] B. C. Das and R. D. Schmid, *F.E.B.S. Letters*, 1972, **25**, 253.

Peptide sequencing has been carried out by degradation to diketopiperazines[245] and identification of these compounds by g.c.m.s. has been reported.[246]

The characteristic peaks arising from simple cleavage of the polyamide chain have been used in the mass spectral elucidation of amino-acid sequences in oligopeptides. Metastable ion spectra have been used for this and for sequencing mixtures.[206,247,248] Collisional activation spectra have been shown to provide more sequence information than the metastable ion spectra.[249] These studies have been extended to other peptides and a combination of metastable ion and collisional activation spectra has been shown to provide unambiguous differentiation between leucine and isoleucine residues in peptides.[250]

The amino-acid sequence of a cyanogen bromide fragment derived from the C-terminus of rabbit skeletal muscle actin has been determined by computer-assisted g.c.m.s.[251] The N-terminal residue was determined by conventional procedures and the position of glutamine by cleavage of the polypeptide with trypsin. Input of the N-terminal residue with the oligopeptides determined by the g.c.m.s. experiment gave only one sequence in agreement with the known amino-acid composition and sequence.[252]

An alternative g.c.m.s. approach to polypeptide sequencing has used dipeptidylamino-peptidase I which sequentially removes dipeptides from the N-terminus of the polypeptide.[253—255] The individual dipeptides are identified by g.c.m.s. of their pentafluoropropionyl methyl esters.[253] To test the technique, the sequence of porcine insulin A chain was determined. The enzyme cannot hydrolyse peptide bonds involving proline, and dipeptides cannot be hydrolysed from a polypeptide when either lysine or arginine is in the N-terminal position. These problems can be partially overcome by subjecting the enzyme digest to a one-step Edman degradation followed by further enzyme treatment which should give rise to 'post-proline' dipeptide.[254] An alternative solution to both problems is the use of another enzyme system; elastase has been proposed, and it gives rise

---

[245] N. F. Albertson and F. C. McKay, *J. Amer. Chem. Soc.*, 1953, **75**, 5323; A. B. Mauger, *Chem. Comm.*, 1971, 39.
[246] R. A. W. Johnstone, T. J. Povall and J. D. Baty, *J.C.S. Chem. Comm.*, 1973, 392.
[247] F. W. McLafferty, R. Venkataraghavan, and P. Irving, *Biochem. Biophys. Res. Comm.*, 1970, **39**, 274.
[248] H.-K. Wipf, P. Irving, M. McCamish, R. Venkataraghavan, and F. W. McLafferty, *J. Amer. Chem. Soc.*, 1973, **95**, 3369.
[249] F. W. McLafferty, R. Kornfeld, W. F. Haddon, I. Sakai, P. F. Bente, S.-C. Tsai, and H. D. R. Schüddemage, *J. Amer. Chem. Soc.*, 1973, **95**, 3886.
[250] K. Levson, H.-K. Wipf, and F. W. McLafferty, *Org. Mass Spectrometry*, 1974, **8**, 117.
[251] H. Nau, J. A. Kelly, and K. Biemann, *J. Amer. Chem. Soc.*, 1973, **95**, 7162.
[252] M. Elzinga and J. H. Collins, in 'The Mechanism of Muscle Contraction', Cold Spring Harbor Symposium on Quantitative Biology, Cold Spring Harbor Lab., New York, 1973, Vol. 37, p. 1.
[253] R. M. Caprioli, W. E. Seifert, jun., and D. E. Sutherland, *Biochem. Biophys. Res. Comm.*, 1973, **55**, 67.
[254] H. Nau, J. A. Kelly, H. J. Forster, J. E. Biller, T. R. Smith, and K. Biemann, Twenty-first Annual Conference on Mass Spectrometry and Allied Topics, San Francisco, May 1973, p. 293.
[255] H. J. Forster, J. A. Kelly, H. Nau, and K. Biemann, in 'Chemistry and Biology of Peptides', ed. J. Meienhofer, Ann Arbor Science Publishers, Ann Arbor, 1972, p. 679.

to mixtures of peptides in the 200—1200 molecular weight range.[256] Using this enzyme the sequences of the major component peptides of an elastase digest of dihydrofolate reductase have been obtained by mass spectral examination of the derivatized mixture of peptides.

An alternative approach to the sequencing of peptides in mixtures has been reported, using low- and high-resolution, chemical ionization, fractional vaporization and deuteriated derivatives.[248] Some of the problems involved in studying peptide mixtures by mass spectrometry have been discussed, and correct sequences were obtained for a mixture of two hexapeptides by examination of the mass spectra obtained from the mixture after (i) acetylation and permethylation, (ii) deuterioacetylation and deuteriopermethylation and (iii) one cycle of Edman degradation followed by acetylation and permethylation.[257]

The natures of two histidine-containing cross-links from collagen have been established as (44) and (45).[258] The compounds were obtained from the cyanogen bromide digest of sodium borohydride-reduced collagen. High- and low-resolution mass spectra of trifluoroacetyl and pentafluoropropionyl methyl and ethyl

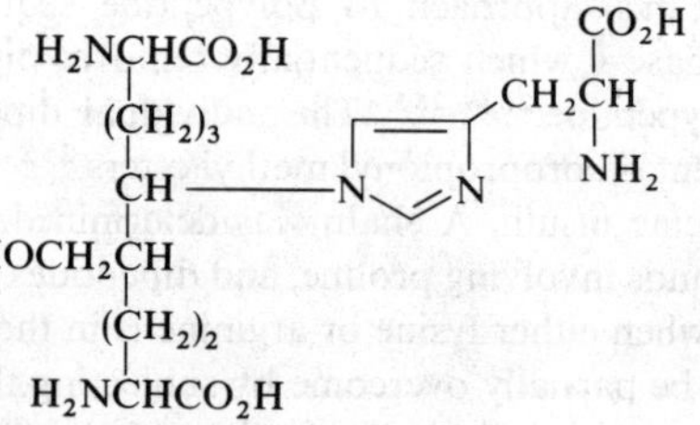

$$H_2NCHCO_2H$$
$$(CH_2)_3$$
$$CH$$
$$HOCH_2CH$$
$$(CH_2)_2$$
$$H_2NCHCO_2H$$

$$CO_2H$$
$$CH_2CH$$
$$NH_2$$

(44)

$$H_2N \quad CO_2H$$
$$CH$$
$$(CH_2)_3$$
$$CH$$
$$CO_2H$$
$$CH(CH_2)_2CHOHCH_2NHCH_2CH$$
$$NH_2$$
$$(CH_2)_2$$
$$CH$$
$$H_2N \quad CO_2H$$

$$CO_2H$$
$$CH_2CH$$
$$NH_2$$

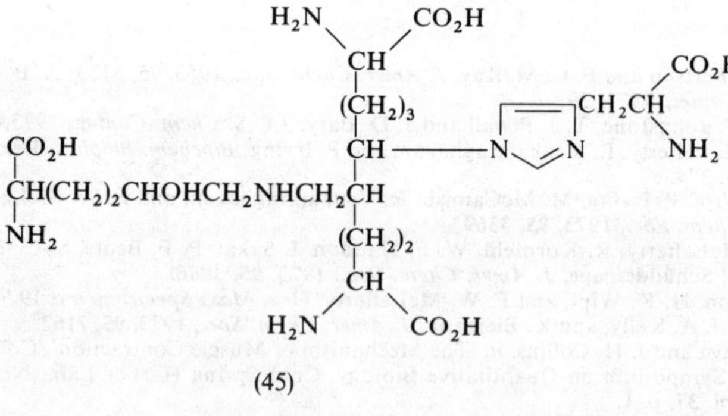

(45)

[256] H. R. Morris, K. E. Batley, N. G. L. Harding, R. A. Bjur, J. G. Dann, and R. W. King, *Biochem. J.*, 1974, **137**, 409.
[257] P. Roepstorff and K. Brunfeldt, in 'Advances in Mass Spectrometry', ed. A. R. West, Applied Science, London, 1974, Vol. 6, p. 199.
[258] M. L. Tanzer, T. Housley, L. Berube, R. Fairweather, C. Franzblau, and P. M. Gallop, *J. Biol. Chem.*, 1973, **248**, 393.

esters of (44) and (45), together with ^{1}H n.m.r. data, were the basis for the structural assignments. The structure of (45) has been confirmed using ^{1}H and ^{13}C n.m.r. and mass spectral data.[259]

High-resolution mass spectrometry and metastable defocusing were used to determine the amino-acid sequence (46) in CY 1-2, a plant growth inhibitor from *Cylindrocladium scoparium*.[260] Mass spectrometry has also assisted in assigning

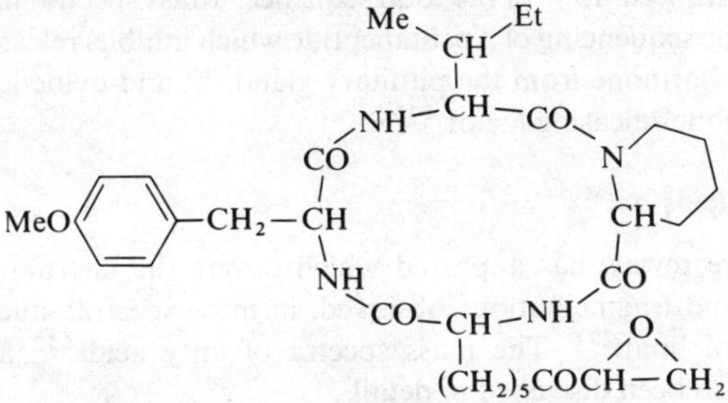

(46)

structures to cyclic peptides from the lichen *Rocella canariensis*[261] and from the posterior lobe of pig pituitaries,[262] and to tentoxin, a phytoxic cyclic tetrapeptide.[263]

Mass spectral study of the acetylated and permethylated pentapeptide obtained from a tryptic digest of somatostatin, a hypothalamic polypeptide which inhibits the secretion of growth hormone, together with information obtained by other methods enabled the sequence: H-Ala-Gly-Cys-Lys-Asn-Phe-Phe-Trp-Lys-Thr-Phe-Thr-Ser-Cys-OH to be assigned.[264]

Scotophobin is a pentadecapeptide which is claimed to induce dark avoidance in untrained animals. The whole peptide was trifluoroacetylated and then treated with diazomethane, and the derivatized material examined by high-resolution mass spectrometry. Two tryptic fragments were treated in a similar manner. Analysis of peaks in the low-mass region, together with information from other experiments, enabled a tentative sequence to be formulated for the polypeptide; however, three positions could not be fixed. Synthesis resolved[265] the sequence as Ser-Asp-Asn-Asn-Gln-Gln-Gly-Lys-Ser-Ala-Gln-Gln-Gly-Gly-Tyr-NH$_2$. A critical appraisal, particularly of the mass spectral methods used in this study, has appeared.[266]

[259] E. Hunt and H. R. Morris, *Biochem. J.*, 1973, **135**, 833.
[260] A. Hirota, A. Suzuki, K. Aizawa, and S. Tamura, *Biomedical Mass Spectrometry*, 1974, **1**, 15.
[261] G. Bohman-Lindgren, *Tetrahedron*, 1972, **28**, 4625.
[262] D. A. Holwerda, *European J. Biochemistry*, 1972, **28**, 340.
[263] M. Koncewicz, P. Mathiaparanam, T. F. Uchytil, L. Sparapano, J. Tam, D. H. Rich, and R. D. Durbin, *Biochem. Biophys. Res. Comm.*, 1973, **53**, 653.
[264] N. Ling, R. Burgus, J. Rivier, W. Vale, and P. Brazeau, *Biochem. Biophys. Res. Comm.*, 1973, **50**, 127.
[265] G. Ungar, D. M. Desiderio, and W. Parr, *Nature*, 1972, **238**, 198.
[266] W. W. Stewart, *Nature*, 1972, **238**, 202.

The way in which mass spectral techniques can complement conventional techniques is well illustrated by a study of the peptide from chicken muscle triose phosphate isomerase.[267] Methods developed by other workers were used[268] and the results obtained show the applicability of the techniques to peptides containing histidine and carboxymethylcysteine and to mixtures produced by proteolytic digestion of small quantities of peptide. Mass spectral data were obtained on 35 residues representing *ca.* 15% of the total sequence. Mass spectral methods have also assisted in the sequencing of a pentapeptide which inhibits release of melanocyte-stimulating hormone from the pituitary gland,[269] and ovine hypothalamic luteinizing hormone-releasing factor.[270]

11 Lipids

A comprehensive review has appeared which covers the instrumentation and methods used, and fragmentations observed, in mass spectral studies of many different types of lipid.[271] The mass spectra of fatty acids[272] and complex lipids[273] have also been discussed in detail.

Simple Lipids.—Alkene mixtures have been analysed by capillary g.c.m.s. using a combined CI–EI source.[273] CI was used to determine the molecular weight of the alkene, and examination of the EI spectrum of the TMS derivative of the *vic*-diol enabled the position of the double bond to be established. The reaction of acrylonitrile with unsaturated fatty acids has been suggested as a method for determination of double bond position; however, the formation of a complex mixture of bis-acryloamide derivatives mitigates against its use in this way.[274] Formation of hexafluoroacetone ketals is a more viable alternative,[275] but the ketals lack abundant molecular ions and stereochemical assignments are not possible. CI studies, using isobutane as a reagent gas, of monoenoic fatty acids, their methyl esters, and monoepoxy fatty acid methyl esters indicate that the CI spectra of the epoxides can be used to determine the position of the double bond but do not provide information about its stereochemistry.[276] In comparison with the other compounds studied, the epoxides exhibited enhanced fragmentations which were explained in terms of a proton complex (47) formed from the protonated molecular ion. An alternative method for determination of the position of cyclopropane rings in fatty acids involves methoxylation with BF_3 methanol and mass

---

[267] J. D. Priddle, *Biochem. J.*, 1974, **139**, 23; A. J. Furth, J. D. Milman, J. D. Priddle, and R. E. Offord, *Biochem. J.*, 1974, **139**, 11.
[268] H. R. Morris, D. H. Williams, and R. P. Ambler, *Biochem. J.*, 1971, **125**, 189; H. R. Morris, *F.E.B.S. Letters*, 1972, **22**, 257.
[269] R. M. G. Nair, A. J. Kastin, and A. V. Schally, *Biochem. Biophys. Res. Comm.*, 1972, **47**, 1420; 1972, **48**, 253.
[270] R. Burgus, M. Butcher, M. Amoss, N. Ling, M. Monahan, J. Rivier, R. Fellows, R. Blackwell, W. Vale, and R. Guillemin, *Proc. Nat. Acad. Sci. U.S.A.*, 1972, **69**, 278.
[271] A. Zeman and H. Scharmann, *Fette. Seifen. Anstrighm*, 1972, **74**, 509; 1973, **75**, 32, 170.
[272] G. Odham and E. Stenhagen, in ref. 9, pp. 211, 229.
[273] W. Blum and W. J. Richter, *Tetrahedron Letters*, 1973, 835.
[274] S. Blum, S. Gertler, S. Sarel, and D. Sinnreich, *J. Org. Chem.*, 1972, **37**, 3114; S. Blum and S. Sarel, *ibid.*, p. 3121.
[275] B. M. Johnson and J. W. Taylor, *Analyt. Chem.*, 1972, **44**, 1438.
[276] R. J. Weinkam, *J. Amer. Chem. Soc.*, 1974, **96**, 1032.

spectral examination of the resulting isomeric mixture.[277] Conversion of long-chain compounds with monomethyl chain-branching into alcohols, followed by g.c.m.s. examination of their methyl ethers, has been used to establish the site of branching.[278]

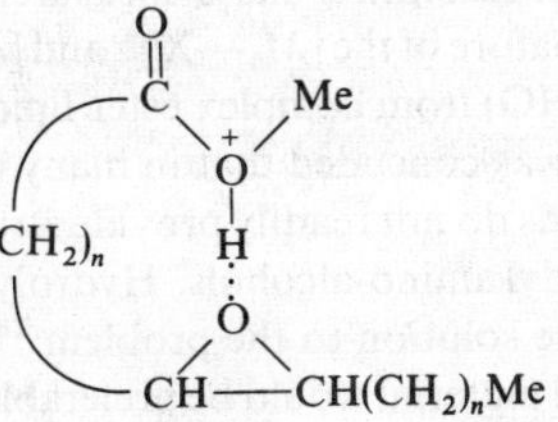

(47)

Mass spectral studies of fatty acid methyl and ethyl esters with 41, 50, 60, and 69 carbon atoms,[279] methylepoxyoctadecanoates,[280] and unsaturated and oxygenated methyl esters[281] have been reported.

Many of the structural studies in this area used g.c.m.s. methods and are dealt with in Chapter 8. A novel method has been developed for the identification of the monomer units of cutin. The technique involves reduction of the polymer with deuteriated and nondeuteriated lithium aluminium hydride and sodium borohydride followed by g.c.m.s. of the resulting polyols as TMS derivatives.[282] The method used is well illustrated by the identification of 16-oxo-9-hydroxy-hexadecanoic acid as a component of cutin in embryonic *Vicia faba* (Scheme 7).

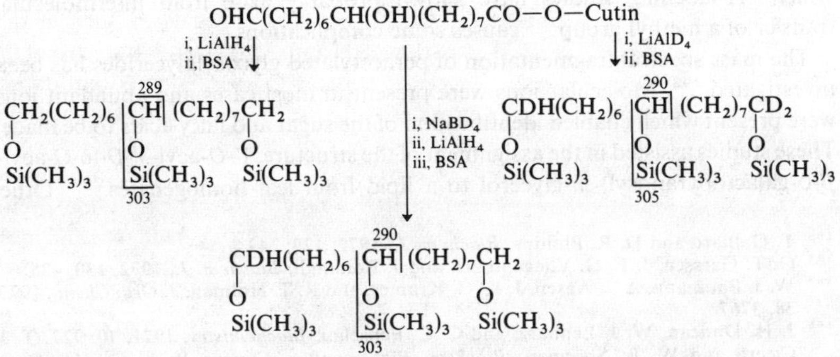

Scheme 7

[277] D. E. Minnikin, *Lipids*, 1972, **7**, 398.
[278] K.-A. Karlsson, B. E. Samuelsson, and G. O. Steen, *Chem. Phys. Lipids*, 1973, **11**, 17.
[279] A. Raal, G. Stallberg, and E. Stenhagen, *Chem. Scripta*, 1973, **3**, 125.
[280] F. D. Gunstone and F. R. Jacobsberg, *Chem. Phys. Lipids*, 1972, **9**, 26.
[281] R. Kleiman and G. F. Spencer, *J. Amer. Oil Chem. Soc.*, 1973, **50**, 31.
[282] P. E. Kolattukudy, *Biochem. Biophys. Res. Comm.*, 1972, **49**, 1040; *Biochemistry*, 1974, **13**, 1354; P. E. Kolattukudy and T. J. Walton, *Biochemistry*, 1972, **11**, 1897.

The characterizations of 9-(nona-1′,3′-dienoxy)non-8-enoic acid in potato tuber homogenates[283] and of dimeric acids[284] formed anaerobically from linoleic and L-13-hydroperoxy-octadec-*cis*-9-*trans*-11-dienoic acids have been assisted by mass spectrometry.

Complex Lipids.—High-resolution mass measurements and ^{2}H-labelling have been used to study the nature of the $[M - X]^+$ and $[M - XH]^+$ ions (X = RCO_2, RO, RCH=CHO, or HO) from complex ester lipids of glycerol and other polyhydric alcohols.[285] It was concuded that in many cases these ions are cyclic.

Mass spectral methods do not readily provide structural information about the polar part of phosphatidyl amino-alcohols. Hydrolysis followed by derivatization and g.c.m.s. study is one solution to the problem.[286] Spectral information from the intact underivatized material would be preferable. EI studies have shown that a weak molecular ion can be detected for 1,2-dioleoyl-*sn*-glycero-3-phosphorylcholine but not for other phosphatidyl amino-alcohols.[287] Extension of these studies by metastable scanning in the first field-free region has enabled molecular ions to be detected for 1,2-diacylglycerylphosphorylcholines. Use of low-voltage spectra and examination of the spectrum obtained in the m/e 25—100 region allows the amino-alcohol part to be identified.[288] This latter technique was used to show the presence of a 3-aminopropane-1,2-diol group in a sphingolipid from sheep rumen.[289] FD studies show that this technique is very useful for this class of compound. A number of phosphatidyl amino-alcohols have been studied;[231b,290—292] in all cases molecular or quasimolecular ions were the base peaks, and important structural features could be established from the fragment ions. The technique has been shown to be applicable to the study of mixtures of phosphatidylcholines.[231b,291] However, the presence in the spectra of peaks which ^{2}H-labelling studies have shown probably arise from intermolecular transfer of a methyl group[292] causes some complications.

The mass spectral fragmentation of peracetylated glycosylglycerides has been investigated.[293] Molecular ions were present in most cases and abundant ions were present which enabled identification of the sugar and fatty acids to be made. These studies assisted in the assignment of the structure, 1′-*O*-acyl-3′-*O*-(6-*O*-acyl-β-D-galactopyranosyl)-*sn*-glycerol to a lipid from leaf homogenates.[294] Other

---

[283] T. Galliard and D. R. Phillips, *Biochem. J.*, 1972, **129**, 743.
[284] G. T. Garssen, J. F. G. Vliegenthart, and J. Boldingh, *Biochem. J.*, 1972, **130**, 435.
[285] W. J. Baumann, A. J. Aasen, J. K. G. Kramer, and R. T. Holman, *J. Org. Chem.*, 1973, **38**, 3767.
[286] J. H. Duncan, W. J. Lennarz, and C. C. Fenselau, *Biochemistry*, 1971, **10**, 927; T. J. Cicero and W. R. Sherman, *Biochem. Biophys. Res. Comm.*, 1971, **43**, 451; S.-G. Karlander, K.-A. Karlsson, and I. Pascher, *Biochim. Biophys. Acta*, 1973, **326**, 174.
[287] R. A. Klein, *J. Lipid Res.*, 1971, **12**, 123, 628.
[288] R. A. Klein, *J. Lipid Res.*, 1972, **13**, 672.
[289] P. Kemp, R. M. C. Dawson, and R. A. Klein, *Biochem. J.*, 1972, **130**, 221.
[290] G. W. Wood and P.-Y. Lau, *Biomedical Mass Spectrometry*, 1974, **1**, 154.
[291] N. Evans, D. E. Games, J. L. Harwood, and A. H. Jackson, *Trans. Biochem. Soc.*, in press.
[292] G. W. Wood, P.-Y. Lau, and J. Teubner, Twenty-second Annual Conference on Mass Spectrometry and Allied Topics, Philadelphia, May 1974, O-4.
[293] H. Budzikiewicz, J. Rullkötter, and E. Heinz, *Z. Naturforsch.*, 1973, **28c**, 499.
[294] C. Critchley and E. Heinz, *Biochim. Biophys. Acta*, 1973, **326**, 184.

structural studies of glycolipids have used information obtained from mass spectral examination of acetylated[295] or TMS derivatives.[296]

Acetyl[297] and TMS[298] derivatives of synthetic monoglycosylceramides have been the subjects of mass spectral studies and the results used in the characterization of natural products.[299,300] Glycolipids with more than one carbohydrate have usually been studied as their TMS ethers.[298] Methylated and methylated-plus-reduced glycolipids (reduction of amide groups of ceramide and amino-sugars to the corresponding amines) are improved derivatives for the mass spectral study of these compounds.[301] Use of both these derivatives gives molecular weight information for up to six sugar units. Carbohydrates can be identified, their sequence established, and the composition of the long-chain base and fatty acid can be determined. The technique has been applied to the characterization of the major monosialoganglioside from brain[302] and should prove an important complement to the g.c.m.s. studies of degraded molecules, necessary to establish the stereochemistry and binding positions of individual sugar.

Mass spectral methods have been developed to enable the carbohydrate ring-size to be determined in intact cerebrosides.[299,303] The method employed is outlined in Scheme 8. The technique has been used for the identification of the major molecular species (48) of one of the cerebrosides from the salt gland of the herring gull.[299] Mass spectral methods have been used in numerous other structural studies of cerebrosides[304] and ceramides,[305] and also in investigations of the biosynthesis of cerebrosides.[306]

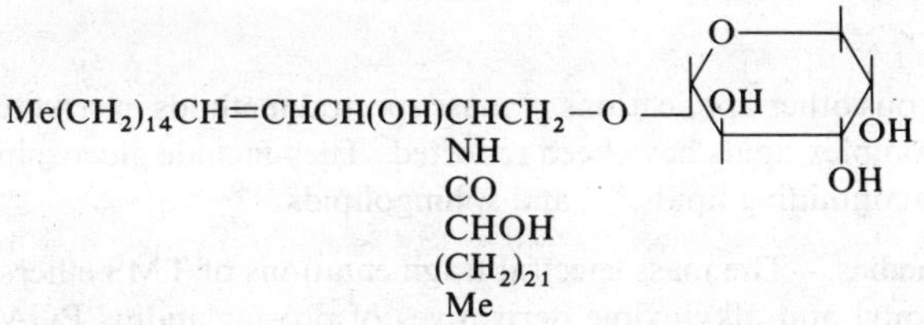

$$Me(CH_2)_{14}CH=CHCH(OH)CHCH_2-O-\text{[sugar ring with OH, OH, OH]}$$
$$\begin{array}{c} NH \\ CO \\ CHOH \\ (CH_2)_{21} \\ Me \end{array}$$

(48)

[295] M. Gastambide-Odier, *Org. Mass Spectrometry*, 1973, **7**, 845; T. W. Esders and R. J. Light, *J. Lipid Res.*, 1972, **13**, 663.
[296] R. A. Laine, P. F. S. Griffin, C. C. Sweeley, and P. J. Brennan, *Biochemistry*, 1972, **11**, 2267; F. Lambein and C. P. Wolk, *Biochemistry*, 1973, **12**, 791.
[297] B. Å. Anderson, K.-A. Karlsson, I. Pascher, B. E. Samuelsson, and G. O. Steen, *Chem. Phys. Lipids*, 1972, **9**, 89.
[298] K.-A. Karlsson, I. Pascher, B. E. Samuelsson, and G. O. Steen, *Chem. Phys. Lipids*, 1972, **9**, 230.
[299] K.-A. Karlsson, B. E. Samuelsson, and G. O. Steen, *J. Lipid Res.*, 1972, **13**, 169.
[300] K.-A. Karlsson, B. E. Samuelsson, and G. O. Steen, *Biochem. Biophys. Acta*, 1973, **306**, 317.
[301] K.-A. Karlsson, I. Pascher, W. Pimlott, and B. E. Samuelsson, *Biomedical Mass Spectrometry*, 1974, **1**, 49.
[302] K.-A. Karlsson, *F.E.B.S. Letters*, 1973, **32**, 317.
[303] L. R. Bjorkman, K.-A. Karlsson, I. Pascher, and B. E. Samuelsson, *Biochem. Biophys. Acta*, 1972, **270**, 260.
[304] F. J. Schmitz and F. J. McDonald, *J. Lipid Res.*, 1974, **15**, 158.

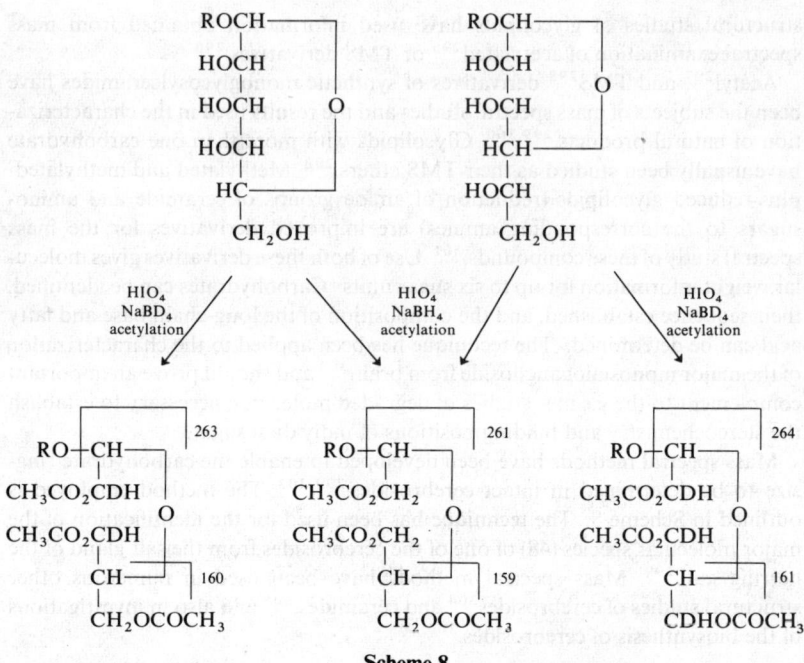

Scheme 8

Numerous other applications of mass spectral methods in structural studies of various complex lipids have been reported. They include glucosphingolipids,[307] ornithine-containing lipids,[308] and sphingolipids.[309]

Prostaglandins.—The mass spectral fragmentations of TMS ethers and esters of free carbonyl and alkyloxime derivatives of prostaglandins PGA,[310] PGB,[311]

---

[305] K.-A. Karlsson, B. E. Samuelsson, and G. O. Steen, *Biochem. Biophys. Acta*, 1973, **316**, 336; T. Matsubara and A. Hayashi, *Biochem. Biophys. Acta*, 1973, **296**, 171; *J. Biochem.*, 1973, **74**, 853; R. Laine, C. C. Sweeley, Y.-T. Li, A. Kisic, and M. M. Rapport, *J. Lipid Res.*, 1972, **13**, 519; K. Stellner, K. Watanabe, and S. Hakomori, *Biochemistry*, 1973, **12**, 656; M. Royer and J. L. Foote, *Chem. Phys. Lipids*, 1972, **7**, 266; W. Krivit and S. Hammarstrom, *J. Lipid Res.*, 1972, **13**, 525; B. Siddiqui, J. Kawanami, Y.-T. Li, and S. Hakomori, *J. Lipid Res.*, 1972, **13**, 657.
[306] S. Hammarstrom and B. Samuelsson, *J. Biol. Chem.*, 1972, **247**, 1001.
[307] P. D. Snyder, W. Krivit, and C. C. Sweeley, *J. Lipid Res.*, 1972, **13**, 128; W. J. Esselman, J. R. Ackermann, and C. C. Sweeley, *J. Biol. Chem.*, 1973, **248**, 7310; W. Stoffel and P. Hanfland, *Z. physiol. Chem.*, 1973, **354**, 21.
[308] A. Gorchein, *Biochem. Biophys. Acta*, 1973, **306**, 127.
[309] K. A. Ferguson, R. L. Conner, F. B. Mallory, and C. W. Mallory, *Biochem. Biophys. Acta*, 1972, **270**, 111; S.-G. Karlander, K.-A. Karlsson, H. Leffler, A. Lilja, B. E. Samuelsson, and G. O. Steen, *Biochem. Biophys. Acta* 1972, **270**, 117.
[310] B. S. Middleditch and D. M. Desiderio, *Prostaglandins*, 1973, **4**, 31.
[311] B. S. Middleditch and D. M. Desiderio, *Lipids*, 1973, **8**, 267.

PGE,[312] and PGF[313] have been studied by low- and high-resolution mass spectrometry, g.c.m.s., and specific [^{2}H$_9$]TMS labelling. The general spectral characteristics have been compared for representative members of the four classes.[314] It was found that certain fragmentation processes were common to all spectra and that, in addition, each type of prostaglandin had a fragmentation unique to itself. The methylester-TMS, TMS ethers and esters, methyl ester-t-butyl-dimethylsilylether, methylester-methoxime-TMS, methylester-methoxime-acetate and methylester-methoxime derivatives of prostaglandins A and B have also been the subjects of a detailed comparative fragmentation study, and it was concluded that methyl ester-silyl ethers or persilylated derivatives are preferable for identification and quantification of these prostaglandins.[315] New prostaglandins have been identified in humans mainly on the basis of mass spectral fragmentations of derivatives.[316]

---

[312] B. S. Middleditch and D. M. Desiderio, *J. Org. Chem.*, 1973, **38**, 2204.
[313] B. S. Middleditch and D. M. Desiderio, *Analyt. Biochem.*, 1973, **55**, 509.
[314] B. S. Middleditch and D. M. Desiderio, in 'Advances in Mass Spectrometry', ed. A. R. West, Applied Science, London, 1974, Vol. 6, p. 173.
[315] J. T. Watson and B. J. Sweetman, *Org. Mass. Spectrometry*, 1974, **9**, 39.
[316] E. Granström and B. Samuelsson, *J. Amer. Chem. Soc.*, 1972, **94**, 4380; B. S. Middleditch and D. M. Desiderio, Twenty-second Annual Conference on Mass Spectrometry and Allied Topics, Philadelphia, May 1974, H-8.

7
Reactions of Organic Functional Groups: Positive and Negative Ions

BY J. H. BOWIE

1 General Introduction

This Report follows the general format adopted in previous Volumes.[1] A functional group approach is used; this provides a summary of mass spectral processes for the general reader, and a review of progress for the specialist.*

The stage has now been reached where the majority of positive ion fragmentations of functional groups are either known or predictable. During the past two years a large number of papers have been published in which fragmentation processes are described without recourse to high-resolution data or labelling studies. Such reports add little to our knowledge at this time in the development of organic mass spectrometry. Over two thousand papers were scanned for this section and, in the Reporter's opinion, those chosen for inclusion do make a positive contribution to the general field of mass spectrometry.

Many books[2-14] and reviews[15-24] have been published during the review period. Of particular note are the introductory text by Johnstone,[6] the more

---

[1] J. H. Bowie, in 'Mass Spectrometry', ed. D. H. Williams (Specialist Periodical Reports), The Chemical Society, London, 1971, Vol. 1, p. 91; 1973, Vol. 2, p. 90.
[2] 'Advances in Mass Spectrometry', ed. A. Quayle, Institute of Petroleum, London, and Elsevier, Amsterdam, 1971, Vol. 5.
[3] 'Mass Spectrometry: Techniques and Applications', ed. G. W. A. Milne, Wiley Interscience, New York, 1971.
[4] Proceedings of the International School of Mass Spectrometry, Ljubljana, Yugoslavia, August 1969, ed. J. Marsal, J. Stefan Inst., Ljubljana Publ., 1971.
[5] W. Simon and T. Clerc, 'Structural Analysis of Organic Compounds by Spectroscopic Methods', MacDonald and Co., London, and American Elsevier, New York, 1971.
[6] R. A. W. Johnstone, 'Mass Spectrometry for Organic Chemists', Cambridge University Press, London, 1972.
[7] H. C. Hill, 'Introduction to Mass Spectrometry' (Second Edition), Heyden and Son Ltd., London, 1972.
[8] D. H. Williams and I. Howe, 'Principles of Organic Mass Spectrometry', McGraw Hill, London, 1972.
[9] M. C. Hamming and N. G. Foster, 'Interpretation of Mass Spectra of Organic Compounds', Academic Press, New York, 1972.
[10] S. Facchetti, 'Mass Spectrometry. EUR 4765 f.i.e. Book. Procs. 2nd Conf. of Mass Spectrometry, Ispra, Italy, Sept 1971, EURATOM, CID, Luxembourg, 1972.
[11] 'Organic Spectral Problems,' Foundations of Modern Chemistry Series, ed. K. L. Rinehart, Prentice Hall, New Jersey, 1972.

* Owing to the time lag for arrival of some journals in Australia, this review covers the period July 1972—March 1974.

advanced treatment of Williams and Howe,[8] the book 'Metastable Ions' by Cooks, Beynon, Caprioli and Lester,[13] and the M.T.P. review 'Mass Spectrometry'.[12] These are recommended as additions to the libraries of all mass spectrometrists.

2 Reactions of Positive Ions

Hydrocarbons (including Hydrocarbon Cations).—*Alkanes and Alkenes.* The mass spectrum of methane has been considered in detail.[25] INDO calculations for possible activated complexes which may explain hydrogen scrambling in the ethane molecular ion indicate that the intermediate (1) is more stable than one in which hydrogens on the same side of the molecule exchange positions.[26] The

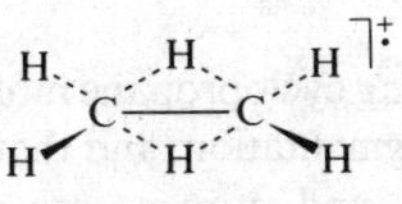

---

[12] 'Mass Spectrometry', M.T.P. International Review of Science, Physical Chemistry Series 1, ed. A. Maccoll, Butterworths, London, and Univ. Park Press, Baltimore, 1973, Vol. 5.
[13] R. G. Cooks, J. H. Beynon, R. M. Caprioli, and G. R. Lester, 'Metastable Ions,' Elsevier, Amsterdam, 1973.
[14] 'Ion Molecule Reactions', ed. J. L. Franklin, Butterworths, London, 1973, Vols. 1 and 2.
[15] D. H. Williams, 'Mass Spectrometry', in 'Determination of Organic Structures by Physical Methods', ed., F. C. Nachod and J. J. Zuckerman, Academic Press, New York, 1971, Vol. 3, 247.
[16] R. A. W. Johnstone and F. A. Mellon, *Ann. Reports (B)*, 1972, **68**, 5; 1973, **69**, 7.
[17] K. Biemann, in 'Techniques of Chemistry. Elucidation of Organic Structures by Physical and Chemical Methods' (Second Edition), ed. K. W. Bentley and G. W. Kirby, Wiley Interscience, New York, 1972, Vol. 1, Chapter 5.
[18] C. J. Creswell, O. A. Runquist, and M. M. Campbell, 'Mass Spectroscopy', in 'Spectral Analysis of Organic Compounds' Second Edition, Longmans, London, and Burgess Publ. Co., U.S.A., 1972, p. 237.
[19] M. M. Campbell and O. A. Runquist, 'Fragmentation Mechanisms in Mass Spectrometry', *J. Chem. Educ.*, 1972, **49**, 104.
[20] D. H. Smith, 'Mass Spectrometry', in 'Guide to Modern Methods of Instrumental Analysis', ed. T. H. Gouw, Wiley Interscience, New York, 1972, Chap. 10.
[21] J. Jalonen and K. Pihlaga, 'Ionization and Appearance Potentials in Structure Analysis', *Org. Mass Spectrometry*, 1973, **7**, 1203.
[22] A. L. B. Robertson, 'Mass Spectrometric Analysis, Analytical Chemistry', M.T.P. International Review of Science, Physical Chemistry Series 1, ed. T. S. West, Butterworths, London, and Univ. Park Press, Baltimore, 1973, Vol. 13, p. 127.
[23] J. M. Wilson, 'Mass Spectrometry', in 'Structure Determination in Organic Chemistry', M.T.P. International Review of Science, Organic Chemistry Series 1, Butterworths, London, and Univ. Park Press, Baltimore, 1973, Vol. 1, p. 26.
[24] G. P. Arsenault, 'Mass Spectrometry of Biomolecules', in 'Experimental Methods in Biophysical Chemistry', ed. C. Nicolan, Wiley Interscience, London and New York, 1973, Chap. 1.
[25] J. D. Morrison and J. C. Traeger, *Internat. J. Mass Spectrometry Ion Phys.*, 1973, **11**, 289.
[26] J. A. Pople, D. L. Beveridge, and P. A. Dobosh, *J. Chem. Phys.*, 1967, **47**, 2026.

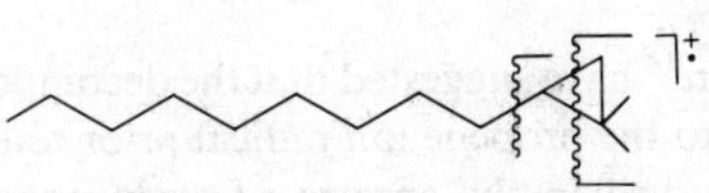

major decomposition modes in the spectrum of 1,1-dimethyl-2-n-nonylcyclopropane are shown in (2).[27] The loss of a methyl radical from the molecular ion of (3) may be explained by ring-opening followed by allylic cleavage (4)→(5), accompanied by specific hydrogen randomization between positions 1–5 and 2–6.[28] Adamantanes,[29–31] bicyclo[3,3,1]nonanes,[32] and methylbicyclo[4,4,0]decanes[33] have been studied.

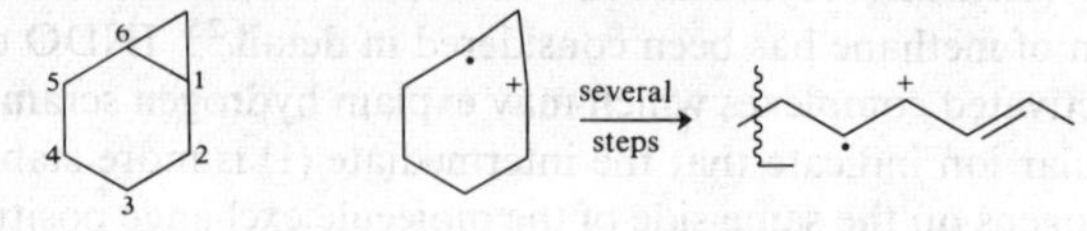

(3)   (4)   (5)

Gross and Lin[34] have suggested that the decomposing cyclopropane molecular ion isomerizes to the propene ion radical prior to fragmentation, and that small differences observed in the spectra of cyclopropane and propene are due to different internal energy distributions in the decomposing ions. This is in contrast with ICR results[35,36] which indicate that the two non-decomposing molecular ions have different structures. Labelling studies have shown[37] that the loss of a methyl radical from the isobutene molecular ion occurs with prior carbon and hydrogen scrambling. It is proposed that the carbon scrambling may be rationalized in terms of a molecular ion rearrangement to form (6), which may then isomerize by cleavage of the 2,3 and 2,4 bonds to form linear structures which may then eliminate Me·.

Carbon atoms 3 and 5 are involved in the elimination of C_2H_4 from the 2,4-dimethylpent-1-ene molecular ion, indicative of an internal rearrangement.[38]

Field ionization mass spectrometry has been used to show[39] that the H–D randomization which competes with the retro Diels–Alder cleavage of (7), starts at 10^{-11} s. and is complete at 10^{-9} s. The scrambling is rationalized in terms of a series of 1,3-allylic rearrangements, e.g. (7)→(8). But see Chapter 1 for discussion of FIMS.

[27] C. E. Parker, M. M. Bursey, and L. G. Pedersen, *Org. Mass Spectrometry*, 1973, **7**, 1077.
[28] J. R. Dias and C. Djerassi, *Org. Mass Spectrometry*, 1973, **7**, 753.
[29] O. S. Chizov, S. S. Novikov, N. F. Karpenko, and A. G. Yurchenko, *Izvest. Akad. Nauk S.S.S.R., Ser khim.*, 1972, 1510.
[30] A. A. Polyakova, E. V. Khramova, Ye. I. Bagrii, N. N. Tsetsugina, I. M. Lukashenko, T. Yu. Frid, and P. I. Sanin, *Neftekhimiya*, 1973, **13**, 9.
[31] K. K. Khullar, C. L. Bell and L. Bauer, *J. Org. Chem.*, 1973, **38**, 1042.
[32] R. Nicoletti and A. Gambacorta, *Org. Mass Spectrometry*, 1973, **7**, 699.
[33] I. M. Lukashenko, Ye. S. Brodskii, I. A. Musaev, E. Kh. Kurachova, V. G. Lebederskaya, and P. I. Sanin, *Neftekhimiya*, 1973, **13**, 163.
[34] M. L. Gross and P.-H. Lin, *Org. Mass Spectrometry*, 1973, **7**, 795.
[35] M. L. Gross and F. W. McLafferty, *J. Amer. Chem. Soc.*, 1971, **93**, 1267.
[36] M. L. Gross, *J. Amer. Chem. Soc.*, 1972, **94**, 3744.
[37] M. S.-H. Jin and A. G. Harrison, *Canad. J. Chem.*, 1974, **52**, 1813.
[38] A. Stefani, *Org. Mass Spectrometry*, 1973, **7**, 17.
[39] P. J. Derrick, A. M. Falick, and A. L. Burlingame, *J. Amer. Chem. Soc.*, 1973, **94**, 6794.

The fragmentations of nonenes[40] and decenes[41] have been discussed. The heats of formation of the molecular, $C_5H_7^+$ and $C_4H_5^+$ ions in the spectra of the C_5H_8 isomers cyclopentene, penta-1,3-diene, and 2-methylbuta-1,3-diene suggest that these ions fragment through common intermediates. Spiropentane behaves in a different manner from the above isomers. For example, [1,1,2,2-$^2H_4$]-spiropentane loses $CH_3\cdot:CH_2D\cdot:CHD_2\cdot:CD_3\cdot$ in the ratio 1.28:3.20:3.01:1.00 from the molecular ion at 70 eV, suggesting reactions of the type (9) → (10) and (9) → (12).[42] ^{13}C-Labelling shows that the molecular ion of (13) undergoes the retro Diels–Alder reaction with a minimum of carbon scrambling.[43]

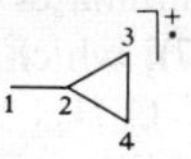

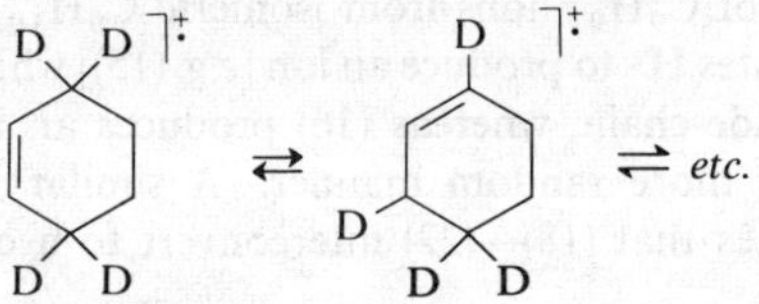

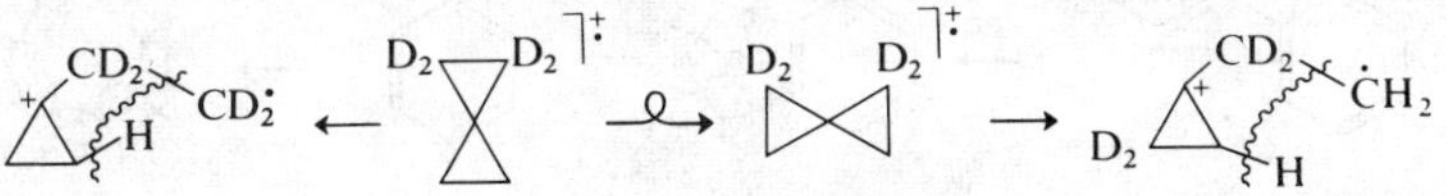

Aromatic. The full paper[44] describing the work of the Purdue group on the scrambling of benzene has appeared (see Vol. 2, p. 94). The kinetic energy released by the $C_6H_6^{\dagger}$ ions generated (i) by direct ionization of benzene, and (ii) by charge exchange of $C_6H_6^{++}$, suggests that the reactive form of the benzene molecular ion has an acyclic structure.[45]

Photodissociation of variously deuteriated toluene parent ions by loss of H· and D· occurs with complete H–D scrambling and with a strong isotope effect.[46] Ion kinetic energy spectrometry indicates that the $C_7H_8^{\dagger}$ ions formed from

---

[40] S. A. Rang, O. G. Éizen, A. A. Polyakova, I. M. Lukashenko, E. S. Brodskii, M. A. Myurisepp, R. K. Siniyaev, M. T. Vyalemits, and T. V. Tiikmaa, *Zhur. org. Khim.*, 1972, **8**, 1993.
[41] B. M. Johnson and J. W. Taylor, *Internat. J. Mass Spectrometry Ion Phys.*, 1972, **10**, 1.
[42] J. L. Holmes, *Org. Mass Spectrometry*, 1974, **8**, 247.
[43] R. A. Davidson and P. S. Skell, *J. Amer. Chem. Soc.*, 1973, **95**, 6843.
[44] J. H. Beynon, R. M. Caprioli, W. O. Perry, and W. E. Baitinger, *J. Amer. Chem. Soc.*, 1972, **94**, 6828.
[45] T. Keough, T. Ast, J. H. Beynon, and R. G. Cooks, *Org. Mass Spectrometry*, 1973, **7**, 245.
[46] R. C. Dunbar, *J. Amer. Chem. Soc.*, 1973, **95**, 472.

toluene, cycloheptatriene, n-butylbenzene, the three methylanisoles, methyl tropyl ether, and benzyl methyl ether, all eliminate a hydrogen atom from a common structure.[47] The $C_7H_7^+$ ions derived from these precursors appear to eliminate acetylene from both benzyl and tropylium structures.[47] ^{13}C-Labelling evidence demonstrates that the $C_7H_7^+$ ion produced from cycloheptatriene[43] and from tropylium iodide[48] eliminate C_2H_2 (to form $C_5H_5^+$) with complete carbon scrambling. It is not known whether this scrambling involves a tropylium–benzyl equilibrium or valence tantomeric shifts in the seven-membered ring structure.[48] Substituted benzenes have been studied,[49] and partial randomization of hydrogen precedes the loss of Me· from tri- and tetra-methylbenzenes.[50]

A study of $C_9H_9^+$ ions from isomeric C_9H_{10} compounds has shown[51] that (14) eliminates H· to produce an ion [e.g. (15)] which specifically eliminates C_2H_2 from the side-chain, whereas (16) produces an ion [perhaps (17)] which loses C_2H_2 in a more random manner. A similar investigation of $C_9H_{11}^+$ ions demonstrates that (18)—(22) interconvert to a common intermediate prior to

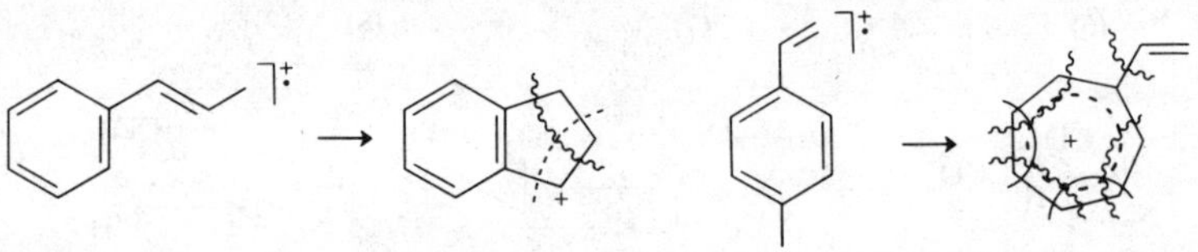

(14)   (15)   (16)   (17)

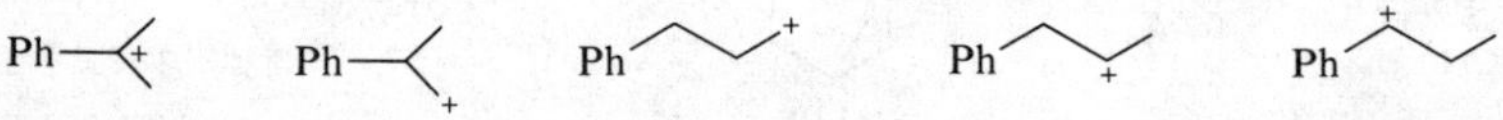

Ph—⟨   Ph—⟨   Ph∕∖+   Ph∕∖   Ph∕∖+
(18)   (19)   (20)   (21)   (22)

elimination of C_2H_4, C_3H_6, and C_6H_5·, but that isomeric ions produced from dialkylbenzene precursors do not rearrange to this intermediate.[52] ^{13}C-Labelling suggests that the $C_9H_{11}^+$ ions from 2,5-dimethylbenzylbromide[53] and 1-chloro-1-phenylpropane[54] are best represented as substituted tropylium ions.

Phenylcyclobutanes fragment by cleavage of the cyclobutane ring,[55] substituted diphenylmethanes have been studied,[56] and the differences observed between the spectra of *threo-* and *erythro-*2,3-diphenylbutane [i.e. (23) → (24)

---

[47] R. G. Cooks, J. H. Beynon, M. Bertrand, and M. K. Hoffman, *Org. Mass Spectrometry*, 1973, 7, 1303.
[48] A. Siegel, *J. Amer. Chem. Soc.*, 1974, 96, 1251.
[49] F. Benoit, *Org. Mass Spectrometry*, 1972, 6, 1377.
[50] R. M. Dawson and R. G. Gillis, *Org. Mass Spectrometry*, 1972, 6, 1003.
[51] C. Köppel, H. Schwarz, and F. Bohlmann, *Org. Mass Spectrometry*, 1974, 8, 25.
[52] N. A. Uccella and D. H. Williams, *J. Amer. Chem. Soc.*, 1972, 94, 8778.
[53] H. Schwarz, C. Köppel, and F. Bohlmann, *Org. Mass Spectrometry*, 1973, 7, 1211.
[54] C. Köppel, H. Schwarz, and F. Bohlmann, *Org. Mass Spectrometry*, 1973, 7, 869.
[55] K. A. Chochua, O. S. Chizhov, and Yu. S. Shabarov, *Zhur. org. Khim.*, 1972, 8, 1867.
[56] G. Innorta, S. Torroni, S. Pignataro, and V. Mancini, *Org. Mass Spectrometry*, 1973, 7, 1399.

but (25) → (26)] are ascribed to enthalpy differences between the two decomposing molecular ions.[57] *ortho*-Effects are noted in the spectra of monosubstituted stilbenes,[58] the $C_{11}H_9^+$ ions produced from chloromethylnaphthalenes and 1-chloro-5-phenylpent-2-en-4-yne possibly have the benztropylium structure,[59,60] and it has been proposed that the naphthyl end-group of α-naphthyl–$(CH_2)_n$–α-naphthyl [n = 2—16] stabilizes the molecular ion by the formation of

```
    Ph         Ph                                    H        ⎤⁺˙
     \       /                              ⎡            ⎤
      CH—CH          →  C₈H₉⁺              ⎢           CH₂ ⎥
     /       \                              ⎣  CH—CH      ⎦
    Me        Me                             Me       Ph
        (23)              (24)                    (25)

                                                    ↓

       ⌐(CH₂)ₙ⌐
     Naph⁺    Naph                             C₈H₈⁺˙
         (27)                                    (26)
```

an intramolecular dimer (27).[61] Strain energies in methylphenanthrenes have been used to rationalize differences in fragmentation,[62] and mass spectrometry has been used to detect aromatic hydrocarbons as air pollutants.[63]

Halides.—Metastable ions from C_2F_6 and C_3F_8 have been studied in detail,[64] appearance potentials and heats of formation of ions from chlorofluoroethylenes have been listed,[65] and the cleavages of fluorinated aliphatic compounds[66] and neopentyl halides[67] have been reported. The different abundances of $M -$ Br· ions in the *cis*- and *trans*-bromohalogeno-cyclopentanes and -cyclohexanes have been rationalized in terms of enthalpy differences between the decomposing ions.[68]

Complete and independent randomization of C, H, and F accompanies or precedes the loss of C_2H_2 from the decomposing fluorobenzene molecular ion.[69] A comparison of the relative abundances of metastable ions for the losses of HF and C_2H_2 from $C_7H_6F^+$ ions from benzyl fluoride and the isomeric fluorotoluenes, led Jennings and Whiting[70] to propose that decomposition was

[57] J. M. Péchiné, *Org. Mass Spectrometry*, 1972, **6**, 805.
[58] H. Güsten, L. Klasinc, V. Kramer, and J. Marsel, *Org. Mass Spectrometry*, 1974, **8**, 323.
[59] H. Schwarz and F. Bohlmann, *Org. Mass Spectrometry*, 1973, 7, 23.
[60] H. Schwarz and F. Bohlmann, *Org. Mass Spectrometry*, 1973, 7, 395.
[61] P. Caluwe, K. Shimada, and M. Szwarc, *J. Amer. Chem. Soc.*, 1973, **95**, 1433.
[62] J. Jalonen and K. Pihlaja, *Org. Mass Spectrometry*, 1972, **6**, 1293.
[63] R. C. Lao, R. S. Thomas, J. L. Monkman, and R. F. Pottie, 'International Symposium on Identification and Measurement of Environmental Pollutants', ed. B. Westley, N.R.C.C. Ottawa, 1971, p. 144.
[64] T. Su and L. Kevan, *Internat. J. Mass Spectrometry Ion Phys.*, 1973, **11**, 57.
[65] B. G. Syrvatka, M. M. Gil'bura and A. L. Bel'yerman, *Zhur. org. Khim.*, 1972, **8**, 1553.
[66] E. Santoro and P. Piccardi, *Org. Mass Spectrometry*, 1973, 7, 123.
[67] J. L. Holmes, D. C. M. Tong, and R. T. B. Rye, *Org. Mass Spectrometry*, 1972, **6**, 897.
[68] Y. Gounelle, J. M. Péchiné, and D. Solgadi, *Org. Mass Spectrometry*, 1973, 7, 1287.
[69] R. J. Dickinson and D. H. Williams, *J.C.S. Perkin II*, 1972, 1363.
[70] K. R. Jennings and A. Whiting, *Org. Mass Spectrometry*, 1972, **6**, 917.

proceeding through the fluorotropylium ion. Small differences in metastable ratios were explained by differing internal energy effects. These systems were also examined by Safe,[71] who suggested that metastable data pointed to incomplete F randomization in $C_7H_6F^+$, and that this observation could be rationalized by incomplete fluorobenzyl → fluorotropylium conversion. The spectrum of $C_7H_7^+$ is produced when tropylium halides are introduced into a mass spectrometer.[48,72] The spectrum of 3,4-dichlorocycloheptatriene has been interpreted in terms of an equilibrium reaction occurring between the dichloroheptatriene and dichlorotoluene ion radicals.[73]

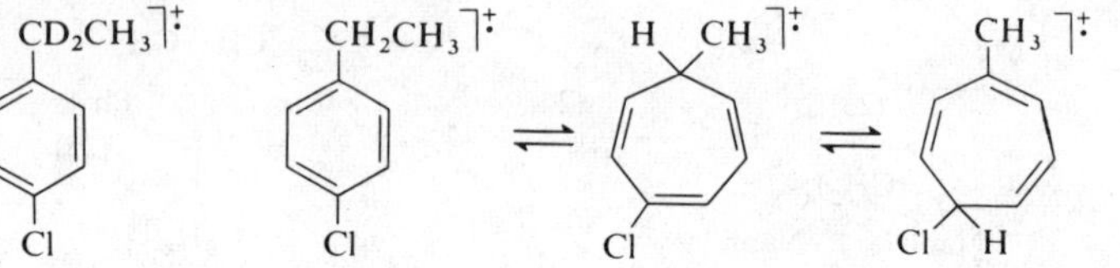

In an elegant study, Uccella and Williams[74] showed that the $M - Cl\cdot$ processes from (28) and (29) exhibit a deuterium isotope effect of 2.7 in favour of (28). This demands a rearrangement reaction prior to the elimination of $Cl\cdot$, and a ring expansion [(29) $\rightleftharpoons$ (30)] followed by H scrambling [e.g. (30) $\rightleftharpoons$ (31)] is proposed. Chlorotoluenes[75] and various perfluoroaromatic compounds[76,77] have been studied. Ion kinetic energy spectra of DDT isomers suggest that the chlorine substituents retain their positional identity.[78] Di- and tri-chlorobenzenes show incomplete chlorine randomization,[79] whereas the chloro-substituents of dichlorodiphenylacetylenes show complete equilibration.[80] Partial hydrogen randomization is observed for the dichlorodiphenylacetylene system.[80]

Alcohols and Phenols.—Labelling data support the proposal that the $C_2H_5O^+$ species formed from propan-2-ol is produced initially as $CH_3C\overset{+}{=}OH$ but then isomerizes (via consecutive 1,2-hydrogen shifts) to $CH_2=CH-\overset{+}{O}H_2$ prior to dissociation to $C_2H_3^+$ and H_3O^+.[81] The structures of the decomposing $C_3H_7O^+$ and $C_4H_9O^+$ ions depend upon the nature of the precursor alcohols.[82—84]

[71] S. Safe, W. D. Jamieson, and D. J. Embree, *Canad. J. Chem.*, 1974, **52**, 867.
[72] G. Hvistendahl, K. Undheim, and P. Györösi, *Org. Mass Spectrometry*, 1973, **7**, 903.
[73] M. K. Hoffmann and T. L. Amos, *Tetrahedron Letters*, 1972, 5235.
[74] I. Howe, N. A. Uccella, and D. H. Williams, *J.C.S. Perkin II*, 1973, 76.
[75] A. P. Krasnoshcheck, *Zhur. org. Khim.*, 1972, **8**, 2100.
[76] J. L. Cotter, *Org. Mass Spectrometry*, 1972, **6**, 905.
[77] J. L. Cotter, *Org. Mass Spectrometry*, 1973, **7**, 11.
[78] S. Safe, O. Hutzinger, W. D. Jamieson, and M. Cook, *Org. Mass Spectrometry*, 1973, **7**, 217.
[79] S. Safe, O. Hutzinger, and W. D. Jamieson, *Org. Mass Spectrometry*, 1973, **7**, 169.
[80] S. Safe, *Org. Mass Spectrometry*, 1973, **7**, 1329.
[81] B. G. Keyes and A. G. Harrison, *Org. Mass Spectrometry*, 1974, **9**, 221.
[82] C. W. Tsang and A. G. Harrison, *Org. Mass Spectrometry*, 1973, **7**, 1377.
[83] T. J. Mead and D. H. Williams, *J.C.S. Perkin II*, 1972, 876.
[84] F. W. McLafferty and I. Sakai, *Org. Mass Spectrometry*, 1973, **7**, 971.

^{13}C- and ^{2}H-labelling studies show that neopentyl alcohol fragments mainly by simple cleavage [e.g. (32)] but that the loss of Me· probably occurs after the hydrogen migration (32) → (33).[67]

Cyclobutanols have been studied by several groups.[85—87] The two compounds (34) and (38) were used in order to ascertain whether the 'double McLafferty rearrangement' (see Vol. 2, p. 103) proceeds via hydrogen rearrangement to carbon or oxygen.[86] The ion (34) fragments to (36) which may undergo the double McLafferty rearrangement to yield (37) by hydrogen rearrangement either to carbon or to oxygen [through (41)]. On the other hand, (38) fragments to (40) which can only produce (37) through (41). The ion (37) does not appear in the spectrum of (38), favouring hydrogen migration to carbon.

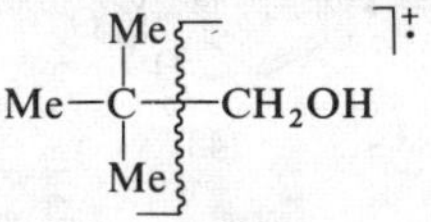

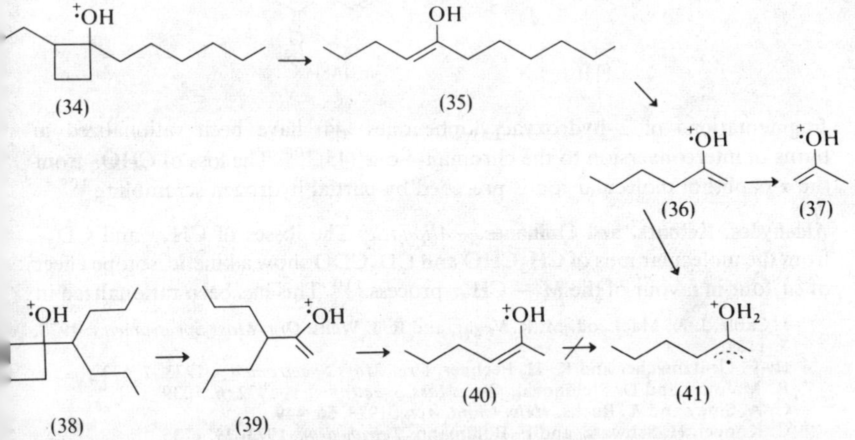

The study of $M - H_2O$ processes from cyclic alcohols has continued. Systems investigated include cyclohexanol,[88] bicyclo[2,2,1]heptan-2-ols,[89] exo- and endo-norborneol,[90,91] cyclohexyl-2-norbornanols,[92] bicyclo[3,3,1]nonan-9-ols,[93] bi-

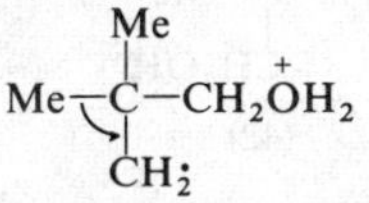

[85] J. L. Holmes and R. T. B. Rye, Canad. J. Chem., 1973, 51, 2342.
[86] G. Eadon, J. Amer. Chem. Soc., 1972, 94, 8938.
[87] G. Eadon, Org. Mass Spectrometry, 1973, 7, 1345.
[88] J. L. Holmes, D. McGillivray, and R. T. B. Rye, Org. Mass Spectrometry, 1973, 7, 347.
[89] D. R. Dimmel and J. M. Seipenbusch, J. Amer. Chem. Soc., 1973, 94, 6211.
[90] J. L. Holmes and D. McGillivray, Org. Mass Spectrometry, 1973, 7, 559.
[91] R. Robbiani and J. Seibl, Org. Mass Spectrometry, 1973, 7, 1153.
[92] P. H. Chen, W. F. Kuhn, D. C. Kleinfelter, and J. M. Miller, Org. Mass Spectrometry, 1972, 6, 785.
[93] J. K. MacLeod and R. J. Wells, J. Amer. Chem. Soc., 1973, 95, 2387.

cyclo[3,3,1]nonanols,[94] decalin diols,[95] and dimethanonaphthalene compounds[96] Cyclopentene diols[97] and ethylenic and acetylenic alcohols[98,99] have been studied. It has been proposed[98] that primary and secondary acetylenic alcohols isomerize to ketones before fragmentation.

The 2-methyl-2-phenylpropane-1,3-diol molecular ion undergoes the eliminations $M - CH_2OH\cdot - C_2H_2O$, and labelling data suggest the operation of a rearrangement (42) → (43).[100] The spectra of aminobenzyl alcohols,[101] p-methyl-1-phenylethanol,[102] cinnamyl alcohol,[103] phenyltetralols,[104] napthyl alcohols,[105] and tetrahydronaphthalene diols[106,107] have been reported. The

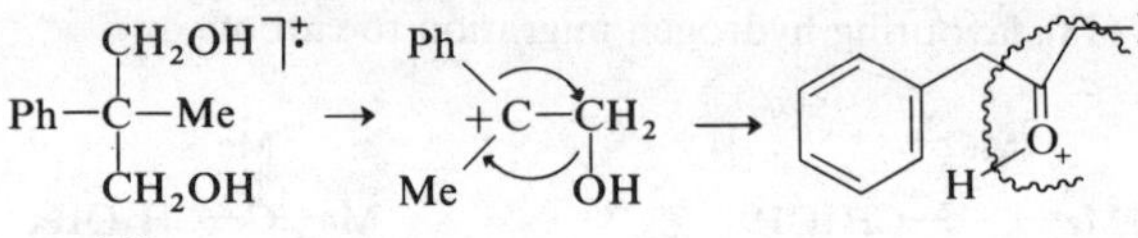

fragmentations of 2'-hydroxyacrylophenones (44) have been rationalized in terms of interconversion to the chroman-4-one (45).[108] The loss of CHO· from the α-naphthol molecular ion is preceded by partial hydrogen scrambling.[109]

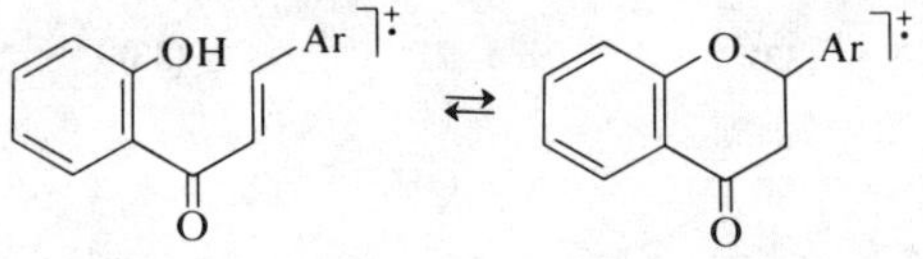

Aldehydes, Ketones, and Quinones.—*Aliphatic.* The losses of $CH_3\cdot$ and $CD_3\cdot$ from the molecular ions of CH_3CHO and CD_3CDO show a kinetic isotope effect of ca. four in favour of the $M - CH_3\cdot$ process.[110] This has been rationalized in

[94] J. Cable, J. K. MacLeod, M. R. Vegar, and R. J. Wells, *Org. Mass Spectrometry*, 1973, **7**, 1137.
[95] H.-F. Grützmacher and K.-H. Fechner, *Org. Mass Spectrometry*, 1973, **7**, 573.
[96] R. Malojčić and D. Stefanović, *Org. Mass Spectrometry*, 1972, **6**, 1039.
[97] G. A. Singy and A. Buchs, *Helv. Chim. Acta*, 1973, **56**, 449.
[98] C. Köppel, H. Schwarz, and F. Bohlmann, *Tetrahedron*, 1973, **29**, 1735.
[99] J. Kossanyi, J. Chuche, N. Manisse, and M. Vanderwalle, *Bull. Soc. chim. belges.*, 1973, **82**, 767.
[100] M. A. Th. Kerkhoff and N. M. M. Nibbering, *Org. Mass Spectrometry*, 1973, **7**, 37.
[101] G. Eckhardt and P. Welzel, *Org. Mass Spectrometry*, 1974, **9**, 125.
[102] C. Köppel, H. Schwarz, and F. Bohlmann, *Org. Mass Spectrometry*, 1973, **7**, 875.
[103] H. Schwarz, C. Köppel, and F. Bohlmann, *Org. Mass Spectrometry*, 1973, **7**, 881.
[104] S. Sasaki, H. Abe, M. Suzuki, Y. Sakurabu, and M. Ohtomo, *Mass Spectrometry (Japan)*, 1973, **21**, 195.
[105] H. Schwarz and F. Bohlmann, *Org. Mass Spectrometry*, 1973, **7**, 29.
[106] P. Perros, J.-P. Morizur, J. Kossanyi, and A. M. Duffield, *Bull. Soc. chim. France*, 1973, 2105.
[107] P. Perros, J.-P. Morizur, J. Kossanyi, and A. M. Duffield, *Org. Mass Spectrometry*, 1973, **7**, 357.
[108] C. van de Sande and M. Vanderwalle, *Bull. Soc. chim. belges*, 1973, **82**, 775.
[109] L. Klasinc and H. Güsten, *Z. Naturforsch.*, 1972, **27a**, 168.
[110] J. G. Pritchard, *Org. Mass Spectrometry*, 1974, **8**, 103.

terms of the hydrogen transfer process (46) ⇌ (47) occurring prior to or during the elimination of $CH_3\cdot$. The spectra of nonan-4-one[111] and acetylenic ketones[112] have been reported.

Bursey and colleagues[113] have used ICR spectroscopy to show that the enol ion (49) is produced initially by the McLafferty rearrangement from (48) when the ion transit time in the ICR cell is 4×10^{-4} s, but that (49) rearranges to the keto

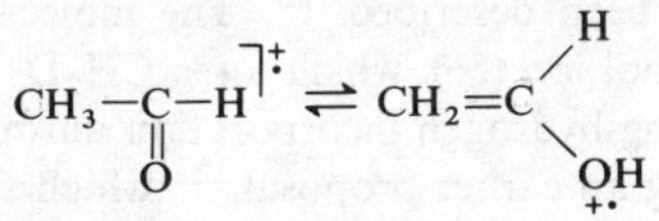

form (50) when the transit time is increased to 1×10^{-3} s. The spectra of adamantanones,[114] long-chain β-diketones[115] and 5-cyclo-octene-1,2-dione[116] have been discussed. The molecular ion of (51) undergoes the retro-Diels–Alder process only when the junction of the central ring is cis.[117] This has been explained[117] in terms of symmetry rules.[118] This stereoselectivity should be contrasted with the retro-Diels–Alder reactions of bicyclo-Δ^2-olefins, which are not dependent upon the stereochemistry of the central bond[119] (for further comment, see Chapter 2, Section 9). The cleavages of the decomposing molecular ions of cyclopentane-1,2,4-triones indicate the prevalence of the enol ion (52),

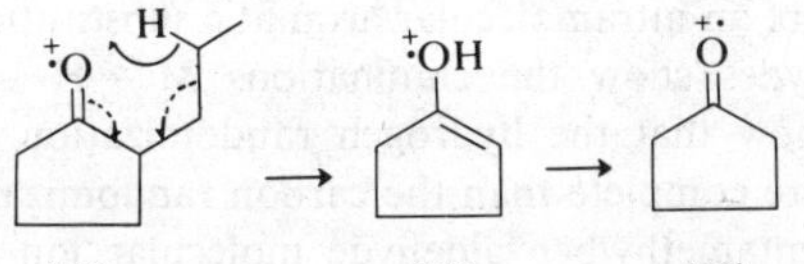

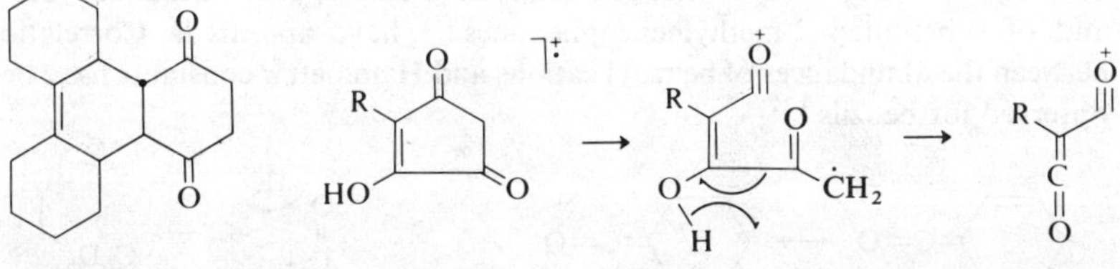

[111] J. H. Beynon, R. M. Caprioli, and R. G. Cooks, Org. Mass Spectrometry, 1974, 9, 1.
[112] É. M. Auvinen, V. V. Takhistov, A. P. Misharev, and I. E. Favorskayn, Zhur. org. Khim., 1972, 8, 2196.
[113] J. R. Hass, M. M. Bursey, D. G. I. Kingston, and H. P. Tannenbaum, J. Amer. Chem. Soc., 1972, 94, 5095.
[114] G. Eckhardt and H.-W. Fehlhaber, Org. Mass Spectrometry, 1973, 7, 485.
[115] A. Trka and M. Streibl, Coll. Czech. Chem. Comm., 1974, 29, 468.
[116] P. Yates, E. G. Lewars, and P. H. McCabe, Canad. J. Chem., 1972, 26, 1548.
[117] A. Karpati, A. Rave, J. Deutsch, and A. Mandelbaum, J. Amer. Chem. Soc., 1973, 95, 4244.
[118] R. C. Dougherty, J. Amer. Chem. Soc., 1968, 90, 5780, 5788.
[119] S. Hammerum and C. Djerassi, J. Amer. Chem. Soc., 1973, 95, 5806.

which decomposes characteristically to form (53).[120] Dienamines derived from αβ-unsaturated aldehydes and ketones have been studied.[121]

Aromatic. It is proposed that the measured heats of formation of $C_7H_5O^+$ ions from benzaldehyde, acetophenone, and benzophenone point to the formation of the ground-state structure $Ph-C\equiv O^+$ from acetophenone, but that an excited form [perhaps (54)] is formed from benzaldehyde and benzophenone.[122] Butyrophenones have been described.[123] The molecular ion (55) eliminates ethylene to form the enol ion (56), which loses $CH_2D\cdot$ and $CH_3\cdot$ in the ratio 9 : 1.[124] This shows ring-hydrogen incorporation during the loss of the methyl radical, thus disproving an earlier proposal,[125] which suggested elimination of $CH_3\cdot$ after 1–4 migration of the hydroxyl hydrogen to carbon. The elimination of $H\cdot$ from the molecular ion of 2-benzylidenecyclohexanone to form (57) may be further example of an intramolecular aromatic substitution.[126] The isomeric trimethylbenzaldehydes show the eliminations $M - H\cdot - CO - C_2H_4$, and labelling studies show that the hydrogen randomization which precedes the loss of C_2H_4 is more complete than the carbon randomization.[127] The loss of $CH_3\cdot$ from the pentamethylbenzaldehyde molecular ion originates from the 2, 4, and 6 positions.[128,129] Reports of benzyl ethers of phenylacetaldehydes[130] and of substituted 2-methylbenzophenones[131] have appeared. Correlations between the abundances of benzoyl cations and Hammett σ constants have been observed for benzils.[132]

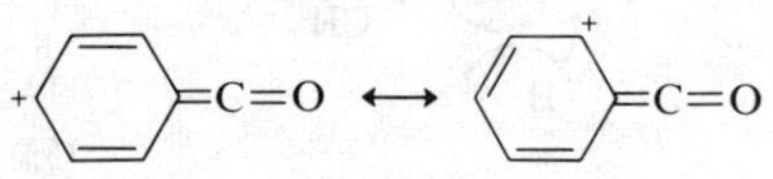

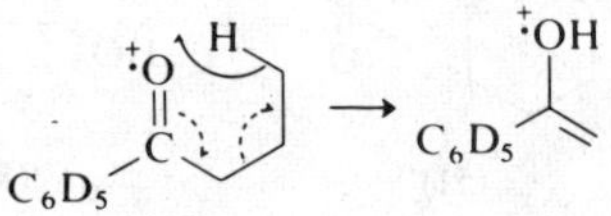

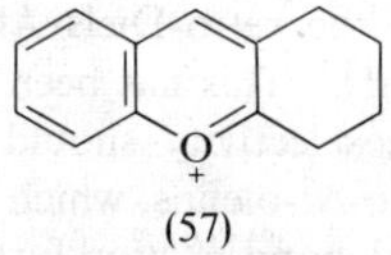

[120] E. Cant and M. Vanderwalle, *Org. Mass Spectrometry*, 1972, **6**, 681.
[121] P. W. Hickmott and C. T. Yoxall, *J.C.S. Perkin II*, 1972, 890.
[122] F. Benoit, *Org. Mass Spectrometry*, 1973, **7**, 1407.
[123] A. Frigerio and C. Rovere, *Org. Mass Spectrometry*, 1974, **8**, 103.
[124] K. B. Tomer and C. Djerassi, *Org. Mass Spectrometry*, 1972, **6**, 1285.
[125] J. H. Beynon, R. M. Caprioli, and T. W. Shannon, *Org. Mass Spectrometry*, 1971, **5**, 967.
[126] P. J. Smith, J. R. Dimmock, and W. A. Turner, *Canad. J. Chem.*, 1973, **51**, 1458.
[127] H. Schwarz and F. Bohlmann, *Org. Mass Spectrometry*, 1972, **6**, 815.
[128] H. Schwarz, F. Bohlmann, and B. Russ, *Org. Mass Spectrometry*, 1973, **7**, 1001.
[129] H. Schwarz, F. Bohlmann, and W. Vorlaender, *Org. Mass Spectrometry*, 1973, **7**, 1005.
[130] A. V. Danks and R. Hodges, *Austral. J. Chem.*, 1972, **25**, 2721.
[131] J. Grimshaw, C. S. Sell, and R. J. Haslett, *Org. Mass Spectrometry*, 1974, **8**, 381.
[132] N. Einolf and B. Munson, *Org. Mass Spectrometry*, 1973, **7**, 155.

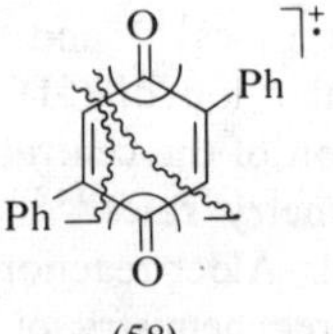

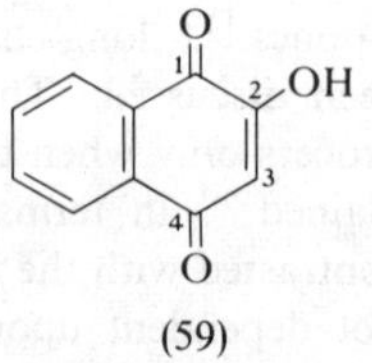

2,5-Diphenyl-*p*-benzoquinone fragments as shown in (58).[133] Juglone[134] and 2-methylnaphthoquinone[135] derivatives have been reported. ^{12}C-Labelling shows that the initial loss of CO from 2-hydroxynaphthoquinone (59) is exclusively from the 2 position,[136] and it is probable that the elimination occurs from the triketo form of the molecular ion. This disproves the original suggestion[137] that the loss originated from the 1 position. The spectrum of pentacene-5,7,12,14-diquinone shows elimination of four CO units,[138] while triapentafulvalene quinone fragments by successive losses of two formyl radicals.[139]

Acids, Esters, Anhydrides, and Carbonates.—Metastable decompositions from the formic, oxalic, acetic, and malonic acid molecular ions have been studied.[140] Acyclic $\beta\gamma$-unsaturated acids have been investigated,[141] an INDO MO approach has been used to rationalize the *ortho*-H/carboxyl-H scrambling in benzoic acid derivatives,[142] and differing substituent effects have been reported for substituted *m*- and *p*-benzoic acids.[143]

The rate of the McLafferty rearrangement from (60) decreases when a substituent X (X = phenyl or Me$_2$NH) of low ionization potential is moved further from the ester function. This is taken to indicate that the charge is localized on X.[144] The cleavages of methyl-12-dimethylsilyloxyoctadecanoate have been described,[145] $\alpha\beta$-unsaturated esters fragment mainly as shown in (61),[146] and it is proposed that partial isomerism occurs between the molecular ions of $\alpha\beta$- and

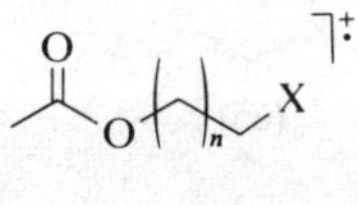

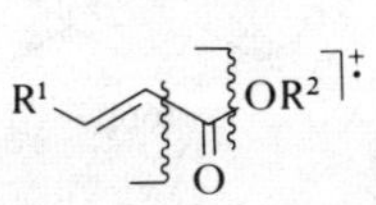

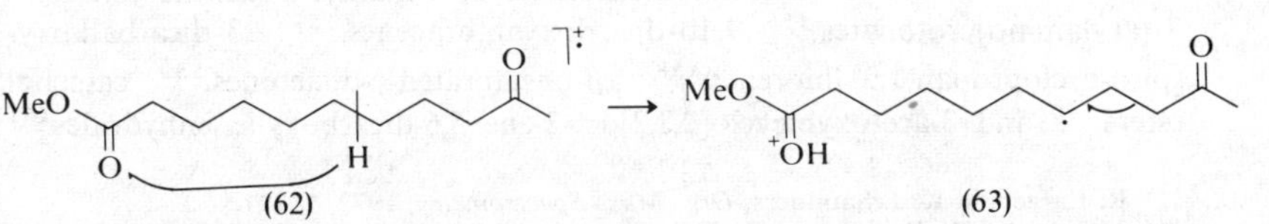

[133] M. Lounasmaa and A. Karjalainen, *Acta Chem. Scand.*, 1973, **27**, 3427.
[134] T. L. Folk, H. Singh and P. J. Scheuer, *J.C.S. Perkin II*, 1973, 1781.
[135] H. J. Kallmayer, *Arch. Pharm.*, 1973, **306**, 707.
[136] R. E. Moore, M. R. Brennan, and J. S. Todd, *Org. Mass Spectrometry*, 1972, **6**, 603.
[137] J. H. Bowie, D. W. Cameron, and D. H. Williams, *J. Amer. Chem. Soc.*, 1965, **87**, 5094.
[138] A. D. Vasil'eva, I. A. Rotérmé, N. N. Artamonova, L. A. Gaeva, A. A. Dubroven, and B. N. Kolokolov, *Zhur. org. Khim.*, 1972, **8**, 2109.
[139] I. Agranat, *Org. Mass Spectrometry*, 1973, **7**, 907.
[140] J. L. Holmes, *Org. Mass Spectrometry*, 1973, **7**, 341.
[141] R. G. Alexander, D. B. Bigley, and J. F. J. Todd, *Org. Mass Spectrometry*, 1972, **6**, 1153.
[142] C. E. Parker, M. M. Bursey, and L. G. Pedersen, *Org. Mass Spectrometry*, 1974, **9**, 204.
[143] F. Benoit, *Org. Mass Spectrometry*, 1973, **7**, 295.
[144] G. Wood, A. M. Falick, and A. L. Burlingame, *Org. Mass Spectrometry*, 1974, **8**, 279.
[145] D. H. Hunneman and W. J. Richter, *Org. Mass Spectrometry*, 1972, **6**, 909.
[146] W. Lauwers, J. W. Serum, and M. Vanderwalle, *Org. Mass Spectrometry*, 1973, **7**, 1027.

$\beta\gamma$-unsaturated esters.[146] The spectra of TMS esters of maleic and fumaric acids show differences in the abundances of ions.[147] The cleavages of allene esters have been reported.[148] The molecular ion of methyl 10-oxoundecanoate eliminates $CH_3COCH_2\cdot$ but the corresponding methyl undecanoate does not lose $C_3H_7\cdot$. It has been proposed that the driving force for this elimination is hydrogen transfer [see (62)], which is then followed by the cleavage shown in (63).[149]

Further aspects of the decompositions of β-keto-esters have been described,[150] and the fragmentations of cyclic *gem*-diesters[151] and β-phenylpropionic esters[152] have been recorded. The loss of water from the benzyl benzoate molecular ion has been rationalized by the scheme (64) → (66), with scrambling of the three *ortho* and the two benzylic hydrogens [in (65)] preceding the elimination of water.[153]

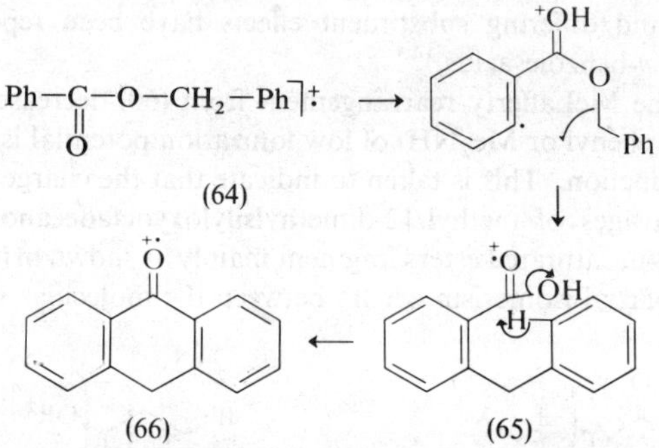

The two $C_7H_7^+$ ions from benzyl phenylacetate have been shown to have different reactivities.[154] The fragmentations of 4-n-alkyl-trimellitic esters,[155] 3-(arylamino)crotonates,[156] 9,10-diacyloxyanthracenes,[157] 2,3-dicarbalkoxy-spiro-cyclopropan-1,9′-fluorenes,[158] $\alpha\beta$-unsaturated-γ-dilactones,[159] catechol esters,[160] and 3-acetoxybicyclo[2,2,2]oct-2-ene 5,6-dicarboxylic anhydrides[161]

[147] R. Large and K. J. Saunders, *Org. Mass Spectrometry*, 1973, **7**, 291.
[148] J. G. Liehr, W. Runge, and W. J. Richter, *Org. Mass Spectrometry*, 1972, **6**, 853.
[149] W. J. Richter and J. G. Liehr, *Helv. Chim. Acta*, 1972, **55**, 2421.
[150] L. Weiler, *Canad. J. Chem.*, 1972, **26**, 2707.
[151] K. Jankowski, J. Couturier, and J.-Y. Daigle, *Canad. J. Chem.*, 1972, **26**, 1535, 1539.
[152] V. I. Kadentsev, O. S. Chizhov, L. A. Yanovskaya, and V. F. Kucherov, *Izvest. Akad. Nauk S.S.S.R., Ser. khim.*, 1972, 2440.
[153] J. H. Beynon, R. M. Caprioli, R. H. Shapiro, K. B. Tomer, and C. W. J. Chang, *Org. Mass Spectrometry*, 1972, **6**, 863.
[154] M. K. Hoffmann and J. C. Wallace, *J. Amer. Chem. Soc.*, 1973, **95**, 5064.
[155] S. Meyerson, I. Puskas, and E. K. Fields, *J. Amer. Chem. Soc.*, 1973, **95**, 6056.
[156] A. N. Kost, Z. F. Solomko, Yu. A. Ustynyuk, P. A. Sharbatyan, and V. S. Tkachenko, *Zhur. org. Khim.*, 1972, **8**, 2040.
[157] A. D. Filyngina, B. N. Kolokolov, S. I. Bragina, I. A. Rotérmél, and A. A. Dubroven, *Zhur. org. Khim.*, 1972, **8**, 2103.
[158] W. J. Richter and A. M. Braun, *Helv. Chim. Acta*, 1973, **56**, 569.
[159] P. Kolsaker, *Org. Mass Spectrometry*, 1973, **7**, 535.
[160] L.-A. Svensson, *Acta Chem. Scand.*, 1972, **26**, 2663.
[161] J. Cassan, S. Geribaldi, G. Torri, and M. Azzaro, *Org. Mass Spectrometry*, 1974, **8**, 11.

Reactions of Organic Functional Groups: Positive and Negative Ions 275

have been reported. Collision-induced dissociations (Chapter 1), which are suggested to be *independent* of the initial internal energy of the decomposing ion,[162,163] have been used to show that the $C_{12}H_{10}O]^{+\cdot}$ ions from both diphenylcarbonate and diphenyl ether have the same structure, but one which differs from that of the decomposing molecular ion of phenylphenol.[164]

Ethers.—The mass spectra of methoxy- and dimethoxy-cycloalkanes have been reported;[165,166] *e.g.* 1,1-dimethoxycyclohexane decomposes as shown in (67).[165] The fragmentations of decalin-1,5-dimethyl ethers,[167] 1,3-dioxalanes,[168] hexafluoroacetone ketals of N-alkenes,[169] and 1,2,4-trioxepans[170] have been described.

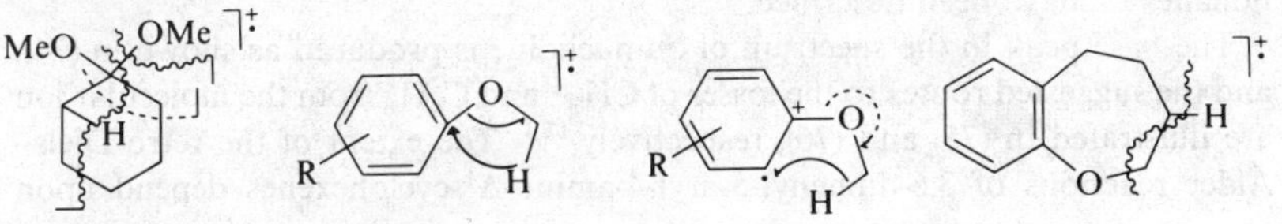

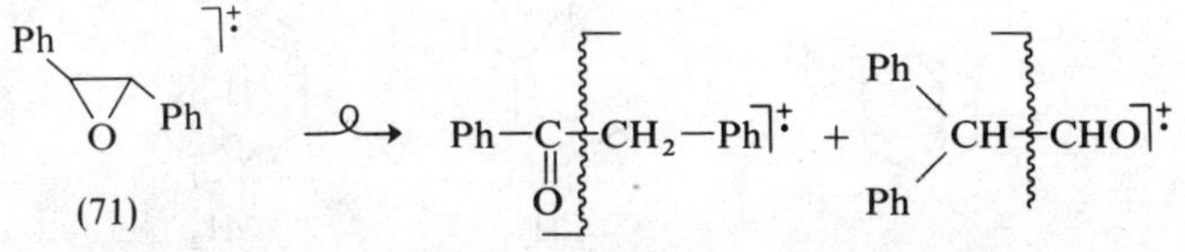

Metastable peak profiles show that there are two distinct processes by which formaldehyde is eliminated from ionized anisoles, and the two mechanisms depicted in (68) and (69) have been proposed.[171] ICR spectroscopy has been used to demonstrate that ion–molecule reactions of the non-decomposing $C_6H_6O^{+\cdot}$ ion from ionized phenetole are consistent with a phenol radical ion structure.[172] The loss of $CH_3\cdot$ from 2,3,4,5-tetrahydrobenzoxepin involves the atoms shown in (70).[173] Details of the fragmentations of nitro-substituted diphenyl ethers,[174]

[162] F. W. McLafferty, P. F. Bente, R. Kornfeld, S.-C. Tsai, and I. Howe, *J. Amer. Chem. Soc.*, 1973, **95**, 2120.
[163] F. W. McLafferty, R. Kornfeld, W. F. Haddon, K. Levsen, I. Sakai, P. F. Bente, S.-C. Tsai, and H. D. R. Schuddemage, *J. Amer. Chem. Soc.*, 1973, **95**, 3886.
[164] K. Levsen and F. W. McLafferty, *Org. Mass Spectrometry*, 1974, **8**, 353.
[165] P. E. Manni, R. D. Cooper, and C. L. Hardesty, *Org. Mass Spectrometry*, 1972, **6**, 949.
[166] R. B. Jones, E. S. Waight, and J. E. Herz, *Org. Mass Spectrometry*, 1973, 7, 781.
[167] H. F. Grützmacher and K.-H. Fechner, *Org. Mass Spectrometry*, 1974, **9**, 152.
[168] L. S. Golovkena, N. S. Wulfson, G. V. Fridlyansky, and G. N. Zakharova, *Izvest. Akad. Nauk S.S.S.R., Ser. khim.*, 1972, 1809.
[169] B. M. Johnson and J. W. Taylor, *Org. Mass Spectrometry*, 1973, 7, 259.
[170] W. Adam and N. Duran, *J.C.S. Chem. Comm.*, 1972, 799.
[171] R. G. Cooks, M. Bertrand, J. H. Beynon, M. E. Rennekamp, and D. W. Setser, *J. Amer. Chem. Soc.*, 1973, **95**, 1732.
[172] N. M. M. Nibbering, *Tetrahedron*, 1973, **29**, 385.
[173] W. J. Richter, J. G. Liehr, and A. L. Burlingame, *Org. Mass Spectrometry*, 1973, 7, 479.
[174] J. F. Jáuregui and P. A. Lehmann, *Org. Mass Spectrometry*, 1974, **9**, 58.

2,6-dimethoxynaphthalenes,[175] and various epoxides[176,177] have been published. Particular fragment ions present in the spectra of aryl-substituted epoxides point to a molecular ion rearrangement to carbonyl systems,[177] e.g. the fragment ions shown in (72) and (73) are present in the spectrum of (71).[177]

Amines, Amides, and Related Systems.—Collision-induced dissociations of the ions $C_2H_6N^+$ and $C_3H_8N^+$ show that the structures of these ions depend upon the structure of the precursor amine.[178] Stable forms of $C_2H_6N^+$ are $CH_3-CH=\overset{+}{N}H_2$ and $CH_3-\overset{+}{N}H=CH_2$.[178] The fragmentations of secondary alkylamines,[179] 1,3-diaminopropane derivatives,[180] 7-azabicyclo[2,2,1]-heptanes,[181] aza-6-bicyclo[3,2,1]octanes[182] and 3-aza-3-methylbicyclo[3,3,1]-nonanes[183] have been described.

The base peak in the spectrum of quinuclidine is produced as shown in (74), and the suggested routes to the losses of $CH_3\cdot$ and C_4H_8 from the molecular ion are illustrated in (75) and (76), respectively.[184] The extent of the retro-Diels–Alder reactions of 3,6-diphenyl-5-aryl-4-amino-Δ^1-cyclohexenes depend upon

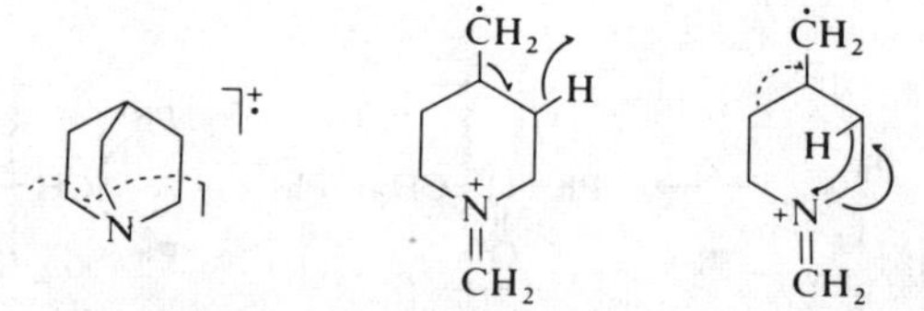

(74)   (75)   (76)

the substituent on the 5-aryl group. The amount of diene produced increases as σ^+ increases.[185] Quaternary ammonium systems undergo thermal reactions prior to ionization.[186,187] The spectrum of an amine picrate is that of the amine superimposed on the spectrum of picric acid.[188]

---

[175] J. Castonguay, A. Rossi, J. C. Richer, and Y. Rousseau, *Org. Mass Spectrometry*, 1972, **6**, 1225.
[176] O. S. Chizhov, S. A. Shamshurina, B. M. Zolotarev, L. A. Yanovskaya, and B. Umirzahov, *Izvest Akad. Nauk S.S.S.R., Ser. khim.*, 1973, 1389.
[177] M. Fetizon, Y. Henry, G. Aranda, H.-E. Audier, and H. de Luzes, *Org. Mass Spectrometry*, 1974, **8**, 201.
[178] K. Levsen and F. W. McLafferty, *J. Amer. Chem. Soc.*, 1974, **96**, 139.
[179] M. Yamamoto, I. Fujita, M. Itoh, and K. Hirota, *Bull. Chem. Soc. Japan*, 1972, **45**, 3520.
[180] E. Lerch and M. Hesse, *Helv. Chim. Acta*, 1972, **55**, 1883.
[181] K. G. Das, P. S. Kulkarni and S. K. Roy, *Org. Mass Spectrometry*, 1973, **7**, 1419.
[182] R. Furstoss, A. Heumann, B. Waegell, and J. Gore, *Org. Mass Spectrometry*, 1972, **6**, 1207.
[183] A. Z. Britten and J. O'Sullivan, *Org. Mass Spectrometry*, 1974, **8**, 109.
[184] J. Paleček and J. Mitera, *Org. Mass Spectrometry*, 1972, **6**, 1353.
[185] K. G. Das and K. P. Madhusudan, *Org. Mass Spectrometry*, 1973, **7**, 619.
[186] G. Hvistendahl and K. Undheim, *Org. Mass Spectrometry*, 1973, **7**, 627.
[187] P. J. Smith, *Canad. J. Chem.*, 1974, **42**, 365.
[188] J. Deutsch and B. Sklarz, *Israel J. Chem.*, 1972, **10**, 51.

The loss of $CH_3\cdot$ from the molecular ion of (77) comes from the $N(CHMe_2)_2$ groups.[189] The spectra of maleic and fumaric diamides may be differentiated by the occurrence of a pronounced ion corresponding to (78) in that of the maleic isomer.[190] The retro process shown in (79) produces the base peak of the spectrum when the imide substituent is *endo*, but it yields a much smaller peak for the

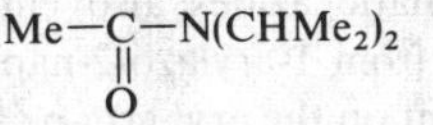

(77), 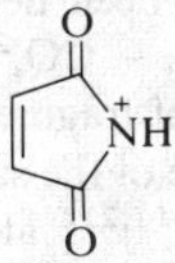 (78), 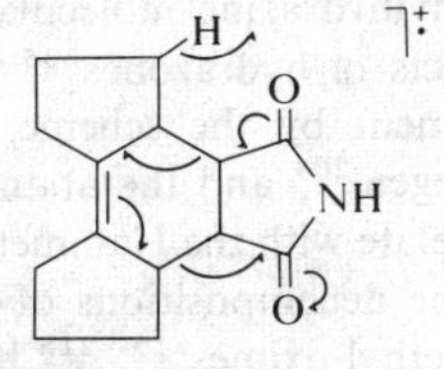 (79)

exo-isomer.[191] The losses of H· from the molecular ions of benzamide and naphthamide originate mainly from the *ortho*-hydrogens.[192] The spectra of acetanilides,[193,194] hydroxysalicylanilides,[195] and amides of 2,2-diaminodiphenylmethanes[196] have been described. The molecular ion (80) undergoes concerted elimination of the elements of phenylacetic acid.[197]

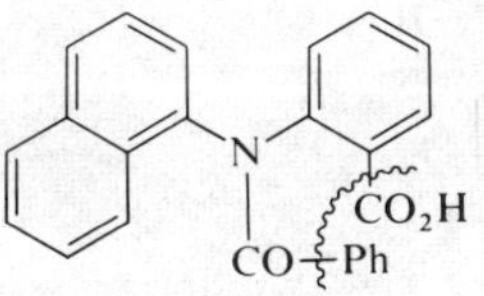

(80), 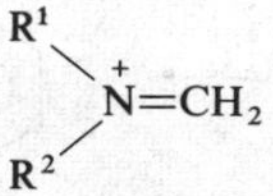 (81), 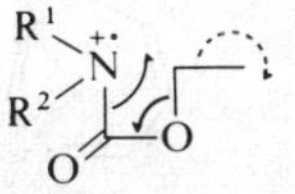 (82), 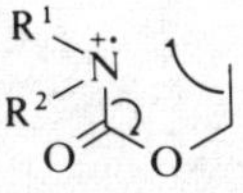 (83)

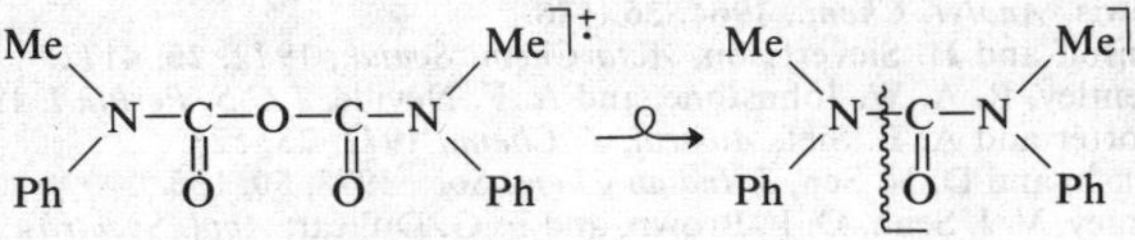

(84), (85)

[189] R. G. Kostyanovsky, V. G. Plekhanov, Kh. Khafizov, L. M. Zagurskaya, G. K. Kadorkina, and Yu. I. Elnatanov, *Org. Mass Spectrometry*, 1973, 7, 1113.
[190] J. L. Holmes, *Org. Mass Spectrometry*, 1973, 7, 335.
[191] T. Lesman and J. Deutsch, *Org. Mass Spectrometry*, 1973, 7, 1321.
[192] A. M. Duffield, G. DeMartino, and C. Djerassi, *Org. Mass Spectrometry*, 1974, 9, 137.
[193] S. Hammerum and K. B. Tomer, *Org. Mass Spectrometry*, 1972, 6, 1369.
[194] S. A. Benezra and M. M. Bursey, *J.C.S. Perkin II*, 1972, 1537.
[195] W. J. A. VandenHeuvel, *Analyt. Letters*, 1973, 6, 51.
[196] G. Eckhardt, H.-W. Fehlhaber, H. Volk, and P. Welzel, *Org. Mass Spectrometry*, 1974, 9, 68.
[197] V. D. Goldberg and M. M. Harris, *J.C.S. Perkin II*, 1973, 1303.

Carbamates[198-200] have been studied by several groups. Labelling studies show that the ion (81)[cf. 201] is produced by two processes. Suggested mechanisms are shown in (82) and (83).[200] Phenylcarbamic acid anhydride undergoes the rearrangements illustrated in (84) and (85).[202]

The —C=N—, —CN, >N—N< and —N=N— Groups.—The NN'-dibenzoylphenylhydrazine molecular ion eliminates benzoic acid,[203, cf. 204] and further aspects of hydrazones[205, 206] have been described.[207] Ionized quinone azides fragment by the scheme $M - N_2 - CO$,[208] aromatic azides also eliminate nitrogen,[209] and the abundances of fragment ions from 1-arylazo-2-naphthols correlate with the Hammett σ value of the substituent on the aryl group.[210]

The decompositions of oximes,[211,212] aldoximes,[213,214] and of a variety of O-methyl oximes[215-218] have been described. The measurement of the kinetic energy released occurring during the loss of HCN from Ar—CH=N—OMe has

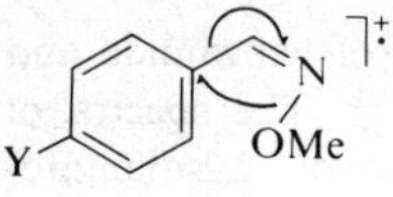

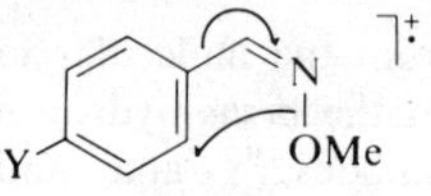

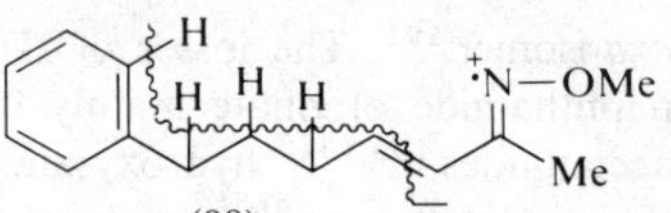

[198] P. A. Tardella and L. Pellacani, *Gazzetta*, 1973, **103**, 337.
[199] R. T. Coutts, *J. Pharm. Sci.*, 1973, **62**, 769.
[200] C. Wünsche, *Org. Mass Spectrometry*, 1973, **7**, 1253.
[201] C. P. Lewis, *Analyt. Chem.*, 1964, **36**, 176.
[202] C. Bogentoff and H. Sievertsson, *Acta Chem. Scand.*, 1972, **26**, 4172.
[203] T. W. Bentley, R. A. W. Johnstone, and A. F. Neville, *J.C.S. Perkin I*, 1973, 449.
[204] Q. N. Porter and A. E. Sief, *Austral. J. Chem.*, 1972, **25**, 523.
[205] S. K. Bhasin and D. N. Sen, *J. Indian Chem. Soc.*, 1973, **50**, 155.
[206] J. B. Stanley, V. J. Senn, D. F. Brown, and F. G. Dollear, *Appl. Spectroscopy*, 1973, **27**, 141.
[207] J. Cassan, M. Azarro, and R. I. Reed, *Org. Mass Spectrometry*, 1972, **6**, 1023.
[208] B. Adler, H. G. O. Becker, and H. Bottcher, *J. prakt. Chem.*, 1972, **314**, 37.
[209] R. T. M. Fraser, N. C. Paul, and M. J. Bagley, *Org. Mass Spectrometry*, 1973, **7**, 83.
[210] K. Kobayashi, K. Kurihara, and K. Hirose, *Bull. Chem. Soc. Japan*, 1972, **45**, 3551.
[211] P. J. Smith, J. R. Dimmock, and W. A. Turner, *Canad. J. Chem.*, 1973, **51**, 1471.
[212] R. T. Coutts and J. L. Malicky, *Org. Mass Spectrometry*, 1973, **7**, 985.
[213] E. V. Brown, L. B. Hough, and A. C. Plasz, *Org. Mass Spectrometry*, 1973, **7**, 1337.
[214] K. G. Das and P. S. Kulkarni, *Org. Mass Spectrometry*, 1973, **7**, 715.
[215] Y. M. Sheikh, R. J. Liedtke, A. M. Duffield, and C. Djerassi, *Canad. J. Chem.*, 1972, **50**, 2776.
[216] J. H. Beynon, M. Bertrand, and R. G. Cooks, *Org. Mass Spectrometry*, 1973, **7**, 785.
[217] R. J. Liedtke, Y. M. Sheikh, A. M. Duffield, and C. Djerassi, *Org. Mass Spectrometry*, 1972, **6**, 1271.
[218] W. M. Leyshon and D. A. Wilson, *Org. Mass Spectrometry*, 1973, **7**, 251.

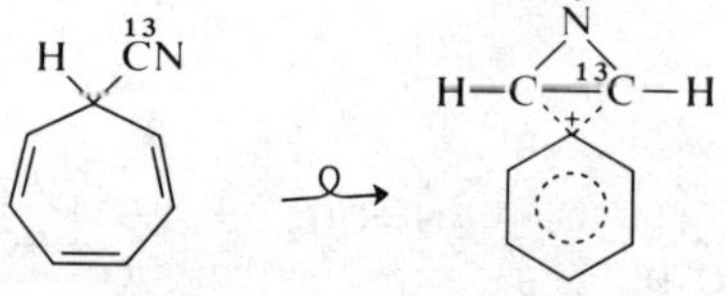

been interpreted in terms of the two mechanisms depicted in (86) and (87).[216] The first example of a quadruple hydrogen transfer process has been discovered in the spectrum of 7-phenylhept-3-en-2-one O-methyl oxime ether, and the hydrogens involved are shown in (88).[217]

The losses of H· and HCN from ionized propionitrile are preceded by 1,2-hydrogen migration to yield $CH_3-\overset{+}{C}H-CH=N·$.[219] The molecular ion of (89) loses both HCN and $H^{13}CN$ at both 9 eV and in the second field-free region of the mass spectrometer, and it has been proposed that ring-contraction to (90) may explain the elimination of HCN.[220]

The N—O Group.—The mass spectra of benzaldehyde-N-methyl nitrone,[221] nitrosamines,[222,223] and nitroalkanes[224,225] have been reported. The $(M - NO·)^+$ ion derived from (91) decomposes to both CH_3CO^+ and CD_3CO^+, and structure (92) is proposed for the decomposing $(M - NO·)^+$ species.[225] The loss of HNO_2 from methyl γ-nitrobutyrate occurs as shown in (93).[226]

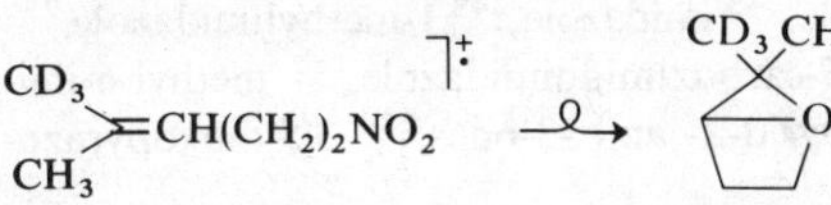

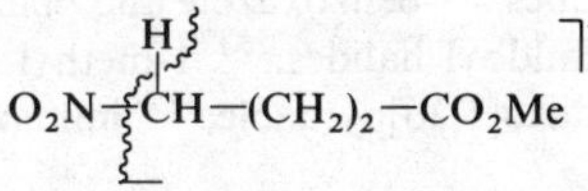

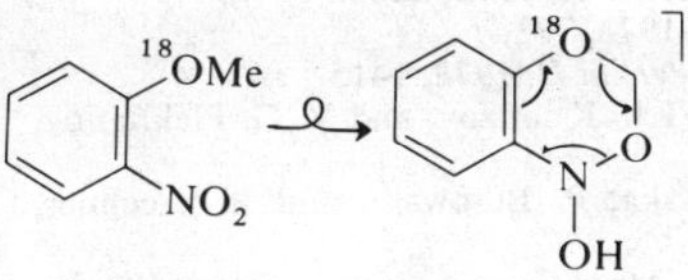

Metastable peak shapes and kinetic energy release measurements show the loss of NO· from aromatic nitro-compounds to occur by two unimolecular pathways. One elimination may proceed *via* a three-membered transition state, the other by oxygen rearrangement to an *ortho* position.[227] The molecular ion of (94) eliminates unlabelled CH_2O, and a plausible reason for this is shown in

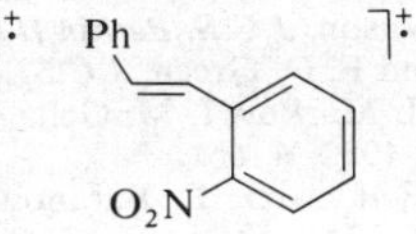

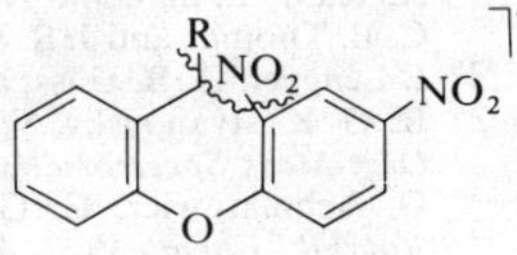

---

[219] S. Meyerson and G. J. Karabatsos, *Org. Mass Spectrometry*, 1974, **8**, 289.
[220] A. Venema, N. M. M. Nibbering, and Th. J. de Boer, *Org. Mass Spectrometry*, 1972, **6**, 675.
[221] H. Stamm, J. Hoenicke, and H. Steuder, *Arch. Pharm.*, 1972, **305**, 625.
[222] K. W. Yang and E. V. Brown, *Analyt. Letters*, 1972, **5**, 293.
[223] S. Billets, H. H. Jaffé, and F. Kaplan, *Org. Mass Spectrometry*, 1973, **7**, 431.
[224] P. Kriemler and S. E. Buttrill, *J. Amer. Chem. Soc.*, 1973, **95**, 1365.
[225] R. L. Carney, *Org. Mass Spectrometry*, 1972, **6**, 1239.
[226] T. A. Molenaar-Langeveld and N. M. M. Nibbering, *Org. Mass Spectrometry*, 1974, **9**, 257.
[227] J. H. Beynon, M. Bertrand, and R. G. Cooks, *J. Amer. Chem. Soc.*, 1973, **95**, 1739.

(95).[228] The base peak in the spectrum of (96) is produced by an $(M - \text{PhCHO})^{+\cdot}$ species.[57] The driving force for the loss of RNO_2 from (97) must be the stability of the product ion formed by the cyclization process.[174] The cleavages of aryl nitro-compounds have been reported.[229—232]

Heterocyclic Systems (excluding Sulphur Compounds).—Because of the large number of references in this section, and because most of the fragmentation patterns are predictable, references to more important mechanistic studies are listed at the beginning of each subsection. Cleavages of particular interest are mentioned subsequent to that list.

Three-, Four-, and Five-membered Rings. The spectra of the following compounds have been reported: N-t-alkylaziridine-1-carboxamides,[233] various azetidines,[234] 2,5-dimethoxy-2,5-dihydrofurans,[235] acyl 2,2′-difurylmethanes,[236] pyrrolidine nitroxides,[237,238] N-arylpyrroles,[239] dihydroindoles,[240] porphrins,[241—243] 3,5-dimethylisoxazoles,[244] 3-aryl-5-methylisoxazole-4-carboxylic acid,[245] isoxazole-5(4H)-ones,[246] 4-arylhydrazono-3-methylisoxazol-5-ones,[247] benzoxazole and benzisoxazole,[248] imidazole,[249] 1-methylimidazole,[250] imidoyl halides,[251] 1-methyl-4-nitro-5-carboximidoimidazole,[252] methyl-6-imidazo[1,2b]pyrazole,[253] imidazo[1,2a]pyrid-2- and -3-ones,[254] 3,5-dioxypyrazo-

---

[228] K. B. Tomer, T. Gebreyesus, and C. Djerassi, *Org. Mass Spectrometry*, 1973, **7**, 383.
[229] E. K. Fields and S. Meyerson, *J. Org. Chem.*, 1972, **37**, 3861.
[230] S. Meyerson and R. W. Van der Haar, *J. Org. Chem.*, 1972, **37**, 4114.
[231] A. Kalir, E. Sali, and A. Mandelbaum, *J.C.S. Perkin II*, 1972, 2262.
[232] C. B. Thomas and J. S. Wilson, *J.C.S. Perkin II*, 1972, 778.
[233] I. Lengyel, D. B. Uliss, and F. D. Green, *J.C.S. Perkin II*, 1972, 1415.
[234] R. G. Kostyanovsky, V. I. Markov, I. M. Gella, Kh. Khafizov, and V. G. Plekhanov, *Org. Mass Spectrometry*, 1972, **6**, 661.
[235] O. Achmatowicz, G. Grynkiewicz, E. Baranowska, P. Bulowski, and B. Szechner, *Roczniki Chem.*, 1972, **46**, 2241.
[236] A. Ferretti, V. P. Flanagan, and J. M. Ruth, *Org. Mass Spectrometry*, 1974, **8**, 403.
[237] A. P. Davies, A. Morrison, and M. D. Barratt, *Org. Mass Spectrometry*, 1974, **8**, 43.
[238] B. A. Andersson and G. Fölsch, *Chemica Scripta*, 1972, **2**, 21.
[239] H. El. Khadem, L. A. Kemler, Z. M. El-Shafei, M. M. A. Abel-Rahman, and S. El. Sadany, *J. Heterocyclic Chem.*, 1972, **9**, 1413.
[240] G. S. d'Alcontres, G. Cum, and N. Uccella, *Org. Mass Spectrometry*, 1973, **7**, 1173.
[241] J. R. Chapman and G. H. Elder, *Org. Mass Spectrometry*, 1972, **6**, 991.
[242] M. Meot-Ner, A. D. Alder, and J. H. Green, *Org. Mass Spectrometry*, 1973, **7**, 1395.
[243] M. Meot-Ner, A. D. Alder, and J. H. Green, *Org. Mass Spectrometry*, 1974, **9**, 72.
[244] K. K. Zhigulev, S. D. Sokolov, and R. A. Khmelnitskii, *Khim. geterotsikl. Soedinenii*, 1972, 1336.
[245] K. K. Zhigulev, R. A. Khmelnitskii, M. A. Panina, I. I. Grandberg, and B. M. Zolotarev, *Khim. geterotsikl. Soedinenii*, 1972, 889.
[246] G. Cum, P. D. Giannetto, and N. Uccella, *J.C.S. Perkin II*, 1973, 2038.
[247] R. G. Fenwick and H. G. Garg, *Org. Mass Spectrometry*, 1973, **7**, 683.
[248] A. Maquestiau, Y. van Haverbeke, C. DeMeyer, and R. Flammang, *Org. Mass. Spectrometry*, 1974, **9**, 149.
[249] K. J. Klebe, J. J. van Houte, and J. van Thuijl, *Org. Mass Spectrometry*, 1972, **6**, 1363.
[250] J. van Thuijl, K. J. Klebe, and J. J. van Houte, *Org. Mass Spectrometry*, 1973, **7**, 1165.
[251] J. Gal, B. A. Phillips, and R. Smith, *Canad. J. Chem.*, 1973, **51**, 132.
[252] G. H. Lord, B. J. Millard, and J. Memel, *J.C.S. Perkin I*, 1973, 572.
[253] J. Elguero, R. Jacquier, and S. Mignonae-Mondon, *J. Heterocyclic Chem.*, 1973, **10**, 411.
[254] W. K. Anderson and A. E. Friedman, *Org. Mass Spectrometry*, 1972, **6**, 797.

lidines,[255] 1-methylpyrazole,[256] 5-alkoxypyrazoles,[256] pyrazolin-5-ones,[256–258] trimethylsilylpyrazoles,[259] 2-isoxazolines,[260] 2,3-dihydro-1,3,4-oxadiazoles,[261] mesoionic derivatives of oxadiazole and triazoles,[262] 3,5-diphenyl-1,2,4-oxadiazole,[263] 1,2,3-triazole,[264] 1,2,4-triazoles,[264–266] 1,2,4-triazol-4-N-imines,[267] 1,2,5-triphenyl-1,2,4-triazole,[268] and benzotriazoles.[269]

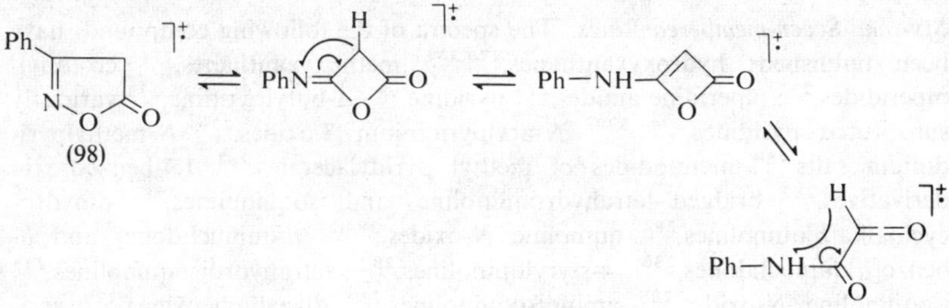

The possible formation of the aniline ion-radical by elimination of C_3O_2 from (98) has been rationalized in terms of the sequence (98) → (99).[246] The loss of H· from the molecular ion of imidazole comes from the 4(5) position but HCN loss is not specific, with the hydrogen originating from $2 > 4(5) \gg 1$.[249] The major

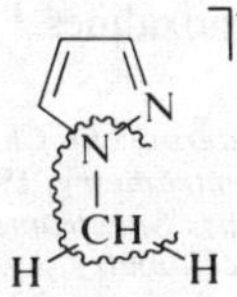

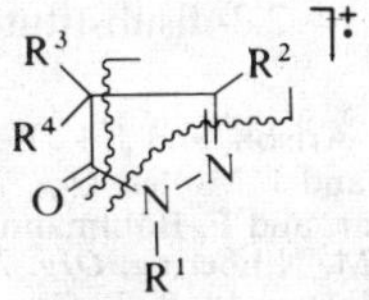

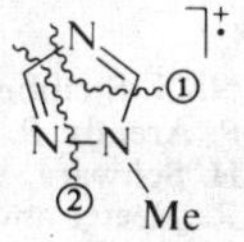

---

[255] B. Unterhalt, *Arch. Pharm.*, 1972, **305**, 334.
[256] A. Maquestiau, Y. van Haverbeke, and A. Bruyere, *Bull. Soc. chim. belges*, 1973, **82**, 747.
[257] A. Maquestiau, Y. van Haverbeke, and A. Bruyere, *Bull. Soc. chim. belges*, 1973, **82**, 757.
[258] E. Larsen, I. H. Qureshi, and J. Møller, *Org. Mass Spectrometry*, 1973, **7**, 89.
[259] L. Birkofer, M. Franz, and G. Schmidtberg, *Org. Mass Spectrometry*, 1974, **8**, 347.
[260] A. Selva and U. Vettori, *Gazzetta*, 1973, **103**, 223.
[261] S. Hammerum and P. Wolkoff, *J. Org. Chem.*, 1972, **37**, 3965.
[262] K. T. Potts, R. Armbruster, E. Houghton, and J. Kane, *Org. Mass Spectrometry*, 1973, **7**, 203.
[263] A. Selva, L. F. Zerilli, B. Cavalleri, and G. G. Gallo, *Org. Mass Spectrometry*, 1972, **6**, 1347.
[264] A. Maquestiau, Y. van Haverbeke, R. Flammang, and J. Elguero, *Org. Mass Spectrometry*, 1973, **7**, 271.
[265] A. Maquestiau, Y. van Haverbeke, and R. Flammang, *Org. Mass Spectrometry*, 1972, **6**, 1139.
[266] A. J. Blackman and J. H. Bowie, *Org. Mass Spectrometry*, 1973, **7**, 57.
[267] H.-J. Timpe, *J. Prakt. Chem.*, 1973, **315**, 775.
[268] J. L. Cotter, *Org. Mass Spectrometry*, 1972, **6**, 1071.
[269] A. Maquestiau, Y. van Haverbeke, M. Flammang, M. C. Pardo, and J. Elguero, *Org. Mass Spectrometry*, 1973, **7**, 1267.

losses of HCN from ionized 1-methylimidazole and 1-methylpyrazole are shown in (100) and (101), respectively.[250] The main fragmentations of 2-pyrazolin-5-ones are illustrated in (102).[257,258] ^{15}N- and ^{2}H-labelling demonstrates that the successive elimination of two HCN units from the 1-methyl-1,2,4-triazole molecular ion are specific, and come from the positions indicated in (103).[266] Benzotriazole fragments by the processes $M - N_2 - HCN$.[269]

Six- and Seven-membered Rings. The spectra of the following compounds have been published: hydroxyxanthanes,[270,271] methoxyxanthanes,[271] crotonylpiperidides,[272] piperidine amides,[273] pyridine,[69] 4-t-butylpyridine,[273] variously substituted pyridines,[274—278a] N-arylpyridinium 3-oxides,[279] N-methylpyridinium salts,[280] methiodides of methyl pyridylacetates,[281] 1,4-benzodioxin derivatives,[282] bridged tetrahydroquinolines and isoquinolines,[283] dihydrocycloalka[b]quinolines,[284] quinoline N-oxides,[284,285] β-quinuclidones and β-benzo[b]quinuclidones,[286] 4-styrylquinolines,[287] tetrahydroisoquinolines,[288] isoquinoline N-oxides,[285] aminoisoquinolines,[289] diazabiphenylene,[290] acridine,[291] phenoxazines and azaphenoxazines,[292] various pyrimidines,[293—295] 2,5-diketopiperazines,[296] 2,3-disubstituted quinoxalines,[297] 3-amino-1-[2-

[270] N. G. Steinberg, B. A. Arison, and J. L. Beck, *J. Heterocyclic Chem.*, 1972, **9**, 1181.
[271] P. Arends, P. Helboe, and J. Møller, *Org. Mass. Spectrometry*, 1973, **7**, 667.
[272] H. Schwarz, W. Mathar, and F. Bohlmann, *Org. Mass Spectrometry*, 1974, **9**, 84.
[273] R. Neeter and N. M. M. Nibbering, *Org. Mass Spectrometry*, 1973, **7**, 1091.
[274] R. G. Cooks, R. N. McDonald, P. T. Cranor, H. E. Petty, and N. Lee Wolfe, *J. Org. Chem.*, 1973, **38**, 1114.
[275] T. Grønneberg, *Chemica Scripta*, 1973, **3**, 139.
[276] K. Undheim and T. Hurum, *Acta Chem. Scand.*, 1972, **26**, 2075.
[277] P. B. Terent'yev, R. A. Khmelnitskii, N. A. Klyuyev, A. B. Belikov, and V. V. Dorogov, *Khim. geterotsikl. Soedinenii*, 1973, 70.
[278] T. Grønneberg and K. Undheim, *Org. Mass Spectrometry*, 1972, **6**, 823.
[278a] J. H. Bowie, *Austral. J. Chem.*, 1973, **26**, 1043.
[279] K. Undheim and P. E. Hansen, *Org. Mass Spectrometry*, 1973, **7**, 635.
[280] P. Salsmans and G. van Binst, *Org. Mass Spectrometry*, 1974, **8**, 357.
[281] H. Hvistendahl and K. Undheim, *J.C.S. Perkin II*, 1972, 2030.
[282] M. Ricard, P. Dizabo, and J. Dagaut, *Org. Mass Spectrometry*, 1974, **9**, 247.
[283] S. Shiofani and K. Mitsuhashi, *Chem. and Pharm. Bull. Japan*, 1972, **20**, 1980.
[284] C. W. Koch, R. M. Milberg, and J. H. Markgraf, *J. Heterocyclic Chem.*, 1973, **10**, 973.
[285] A. M. Duffield and O. Buchardt, *Acta Chem. Scand.*, 1972, **26**, 2423.
[286] A. I. Yermakov, Yu. N. Sheinker, Ye. Ye. Mikhlina, L. I. Mastafanova, V. Ya. Vorobyova, A. D. Yanina, L. N. Yakhontov, and R. G. Kostyanovskii, *Khim. geterotsikl. Soedinenii*, 1972, 1404, 1411.
[287] H. Güsten, L. Klasinc, and D. Stefanović, *Org. Mass Spectrometry*, 1973, **7**, 1.
[288] G. A. Charnock and A. H. Jackson, *J. Chem. Soc. Perkin II*, 1972, 856.
[289] E. V. Brown and S. R. Mitchell, *Org. Mass Spectrometry*, 1972, **6**, 943.
[290] J. Kramer and R. S. Berry, *J. Amer. Chem. Soc.*, 1972, **94**, 8336.
[291] R. M. Acheson, R. T. Aplin, and R. G. Bolton, *Org. Mass Spectrometry*, 1974, **8**, 95.
[292] J. Cassan, M. Rouillard, M. Azzaro, and R. L. Mital, *Org. Mass Spectrometry*, 1974, **9**, 19.
[293] L. Yu. Ivanovskaya, M. I. Gorfinkel, V. F. Sedova, and Z. D. Dubovenko, *Org. Mass Spectrometry*, 1973, **7**, 911.
[294] J. T. Watson and F. C. Falkner, *Org. Mass Spectrometry*, 1973, **7**, 1227.
[295] G. Bourgeois, A. Brachet-Liermain, and L. Ferrus, *Org. Mass. Spectrometry*, 1974, **9**, 53.
[296] K. Jankowski and R. Luce, *Bull. Acad. polon. Sci. Ser. Sci. chim.*, 1973, **21**, 175.
[297] V. Kováčik, M. Fedoroňko, and I. Ježo, *Org. Mass Spectrometry*, 1973, **7**, 449.

(2-ethoxy)ethoxy]benzo[f]quinoxaline.[298] 2-methoxycarbonyl-2-methoxycarbonylmethyl-2,3-dihydro-4(1H)-quinazolinones,[299] aminopteridine and pteridinones,[300] 3,4-hydrated 4-trifluoromethylpteridines,[301] chlorinated purines,[302] 1,2-dihydro-5-trifluoromethylpyrimido(5,4-E)-*asym*-triazines,[303] 2,3-dihydrobenzoxepin,[304] tetrahydro-1,5-benzodiazepines,[305] 1,5-benzodiazepin-2-ones,[306] and dibenz[bf]azepine.[307]

Complete and independent randomization of carbon and hydrogen precede the elimination of HCN from ionized pyridine.[69] Loss of Me· from the t-butylpyridine ion-radical occurs from the side-chain without scrambling, but loss of C_2H_4 from the $(M - \text{Me·})^+$ ion occurs after extensive C and H scrambling.[273] Aryl participation occurs during elimination of XO· (X = H, Ac, or tosyl) from (104).[274] The base peak in the spectrum of 2-methylaminopyridine is produced by the process $M - CH_3N$· [see (105)].[275] The loss of a phenyl radical from the

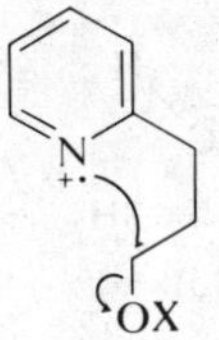

(104) 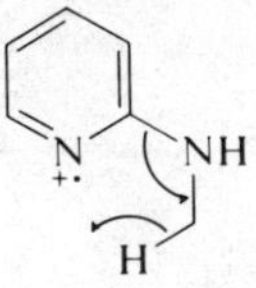 (105) 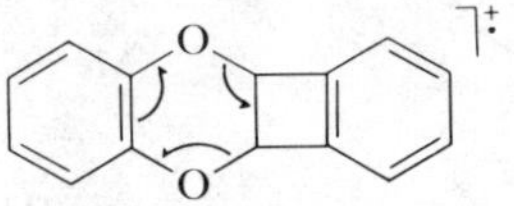 (106) 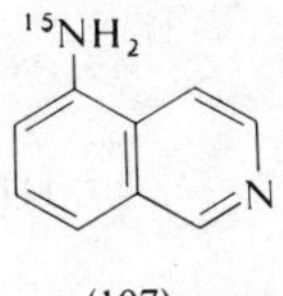 (107)

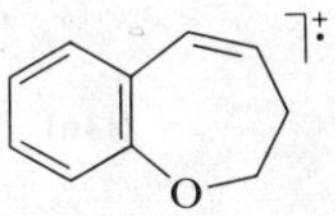

(108) 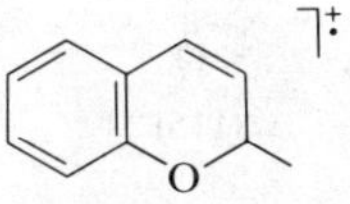 (109) 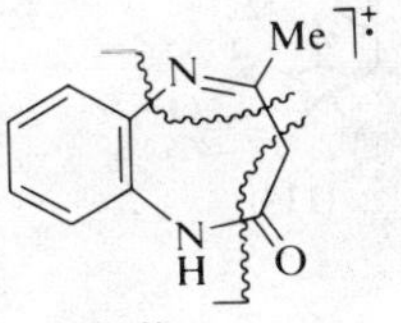 (110)

molecular ion of 2,4,6-triphenylpyridine is a slow process which occurs from the 2 position with negligible scrambling of the pyridine carbon atoms and with little or no migration of the phenyl groups. Considerable randomization of all seventeen hydrogens precedes or accompanies this fragmentation.[278a] The retro-Diels–Alder reaction of (106) gives charge retention only on the hydrocarbon fragment.[282] Ionized (107) eliminates $HC^{15}N$ and HCN in the ratio 4 : 1.[289] It is

[298] S. K. Sengupta, H. K. Protopapa, E. J. Modest, and B. C. Das, *J. Heterocyclic Chem.*, 1973, **10**, 59.
[299] T. F. Lemke, H. W. Snady, and N. D. Heindel, *J. Org. Chem.*, 1972, **37**, 2337.
[300] V. P. Williams and J. E. Ayling, *J. Heterocyclic Chem.*, 1973, **10**, 827.
[301] J. Clark and A. E. Cuncliffe, *Org. Mass Spectrometry*, 1973, **7**, 737..
[302] H. Beerham, K.-H. Grupe, D. Cech, and H. Meinert, *Z. Chem.*, 1973, **13**, 309.
[303] J. Clark, *Org. Mass Spectrometry*, 1973, **7**, 225.
[304] W. J. Richter, J. G. Liehr, and P. Schulze, *Tetrahedron Letters*, 1972, 4503.
[305] P. W. W. Hunter and G. A. Webb, *Tetrahedron*, 1972, **28**, 5573.
[306] A. N. Kost, P. A. Sharbatyan, P. B. Terent'ev, Z. F. Solomko, V. S. Tkachenko, and L. G. Gergel, *Zhur. org. Khim.*, 1972, **8**, 2113.
[307] K. M. Baker and A. Frigerio, *J.C.S. Perkin II*, 1973, 648.

suggested that (108) contracts to (109) prior to fragmentation.[304] The major fragmentations of 4-methyl-1,5-benzodiazepin-2-one are shown in (110).[306]

Sulphur Compounds.—Aromatic sulphides[308] and pentafluorophenylthio-derivatives[309,310] have been described. It has been shown that the $C_2H_4S^{+\cdot}$ ion derived from (111) is consistent with (112) when hydrogen is transferred from the 2 position, and (113) when the hydrogen comes from position 3 or 4.[311] Sulphinate esters have been examined,[312,313] alkylsulphinylamines undergo a McLafferty type rearrangement [see (114)],[314] and the spectra of a variety of sulphones and sulphoxides have been reported.[315—317] The sulphones (115; R^1 = H or Ph, R^2 = alkyl or aryl) undergo rearrangement of $R^1CH=CH$ (in preference to R) to form (116).[316,317] The base peak in the spectra of the sulphonates (117; n = 0—3) is produced by the ion (118).[318] Toluene-o-sulphonates exhibit *ortho* effects in their spectra; *e.g.* (119) eliminates methanol.[319] Benzene sulphonylhydrazides fragment α to the sulphonyl residue.[320]

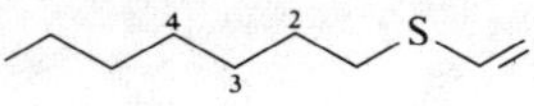

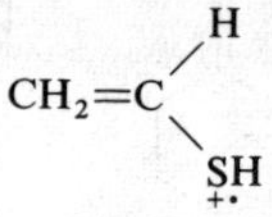

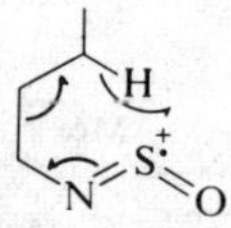

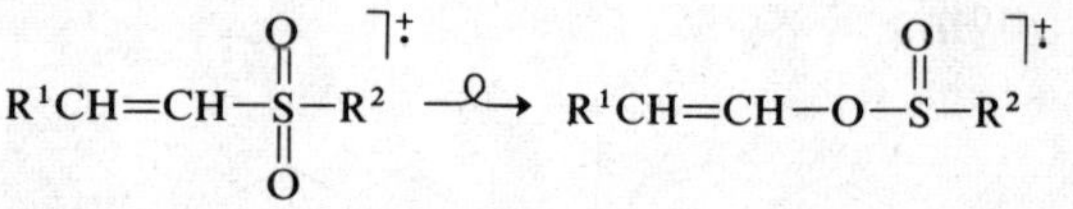

[308] I. Granoth, *J.C.S. Perkin II*, 1972, 1503.
[309] W. D. Jamieson and M. E. Peach, *J. Fluorine Chem.*, 1972, **2**, 119.
[310] W. D. Jamieson and M. E. Peach, *Org. Mass Spectrometry*, 1974, **8**, 147.
[311] K. B. Tomer and C. Djerassi, *J. Amer. Chem. Soc.*, 1973, **95**, 5335.
[312] W. G. Filby, R.-D. Penzhorn, and L. Steinglitz, *Org. Mass Spectrometry*, 1974, **8**, 409.
[313] W. H. Baarschers and B. W. Krupay, *Canad. J. Chem.*, 1973, **51**, 156.
[314] J. R. Grunwell and A. K. Kochan, *J. Org. Chem.*, 1973, **38**, 1610.
[315] I. Pratanata, L. R. Williams, and R. N. Williams, *Org. Mass Spectrometry*, 1974, **8**, 175.
[316] T. H. Kinstle and W. R. Oliver, *Org. Mass Spectrometry*, 1972, **6**, 699.
[317] R. J. Soothill and L. R. Williams, *Org. Mass Spectrometry*, 1972, **6**, 1145.
[318] J. G. Pomonis and D. R. Nelson, *Org. Mass Spectrometry*, 1973, **7**, 1039.
[319] V. Kubelka and J. Mostecký, *Org. Mass Spectrometry*, 1973, **7**, 283.
[320] F. Balza and N. Duran, *Org. Mass Spectrometry*, 1974, **8**, 413.

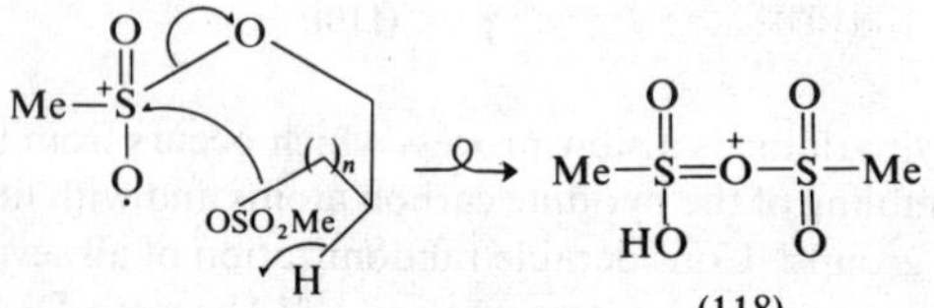

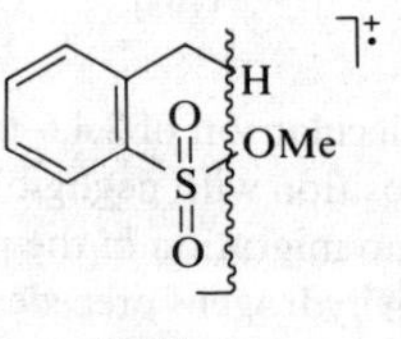

The labelled S-ethylthiobenzoate (120) fragments by the scheme $M - C_2H_4 - CHO\cdot - CS$ [see (120) → (121)].[321] The spectra of S-phenyl methylthiocarbamates[322] and thioamides[323] have been published. o-Nitrothiobenzoic acid S-p-tolyl ester undergoes the rearrangement (122) → (123).[324] The ion (124) fragments as indicated.[325]

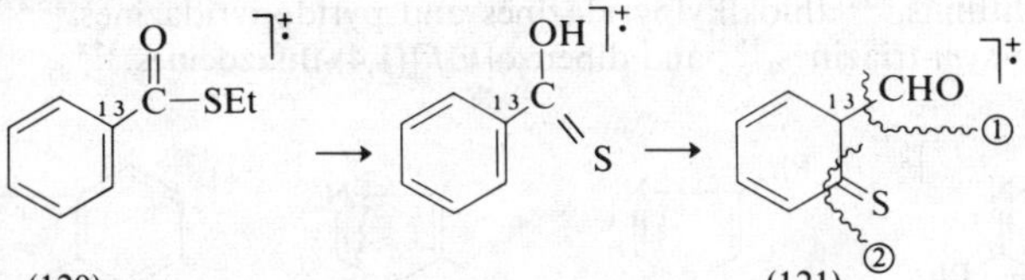

The spectra of the following sulphur heterocyclic systems have been reported: phthalide sulphur analogues,[326] thiophen and alkylthiophens,[327,329,330] alkylthiazoles,[328] benzo[b]thienothiophens,[331] dihydrothiazolo[3,2a]pyridinium derivatives,[332] thiazolo[3,2a]pyridinium 3-oxides,[333] dihydrothiazolo[3,2a]-pyridinium carboxylates,[334] thiazolo[2,3f]theophyllines,[335] 2-(2'-benzothiazoyl)-furans,[336] thiazolo[3,2d]tetrazolium salts,[337] phenylthiazoles,[338] benzothiazole, benzimidazole, and benzoxazole 2-thiones,[339] bisalkylthio-3,5-isothiazoles,[340]

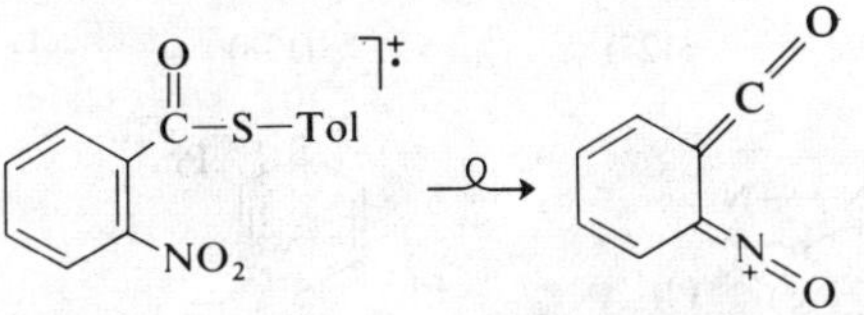

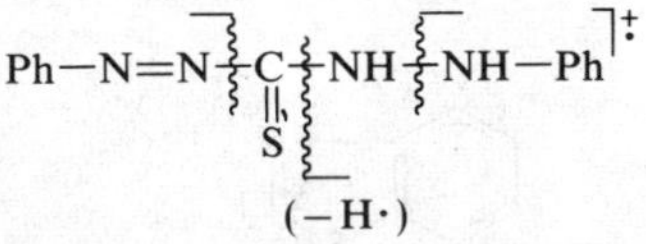

---

[321] K. B. Tomer and C. Djerassi, *Org. Mass Spectrometry*, 1973, **7**, 771.
[322] Lj. Stamboliga and D. Stefanović, *Org. Mass Spectrometry*, 1973, **7**, 1415.
[323] F. C. V. Larsson, S.-O. Lawesson, J. Møller, and G. Schroll, *Acta Chem. Scand.*, 1973, **27**, 747.
[324] J. Martens, K. Praefcke, and H. Schwarz, *Org. Mass Spectrometry*, 1974, **8**, 317.
[325] P. A. Alsop and H. M. N. H. Irving, *Analyt. Chim. Acta*, 1973, **65**, 202.
[326] W. J. McMurray, S. R. Lipsky, R. J. Cushley, and H. G. Mautner, *J. Heterocyclic Chem.*, 1972, **9**, 1093.
[327] R. M. Dawson and R. G. Gillis, *Austral. J. Chem.*, 1972, **25**, 1221.
[328] M. E. Rennekamp, W. O. Perry, and R. G. Cooks, *J. Amer. Chem. Soc.*, 1972, **94**, 4985.
[329] P. J. M. Wakkers, M. J. Janssen, and W. D. Weringa, *Org. Mass Spectrometry*, 1972, **6**, 963.
[330] R. G. Buttery, L. C. Ling and R. E. Lunken, *J. Agric. Food Chem.*, 1973, **21**, 488.
[331] P. Jacquignon, A. Croisy, M. Renson, E. Iteke, and L. Christiaens, *Org. Mass Spectrométry*, 1973, **7**, 1235.
[332] K. Undheim and T. Hurum, *Acta Chem. Scand.*, 1972, **26**, 2385.
[333] P. E. Fjeldstad and K. Undheim, *Org. Mass Spectrometry*, 1973, **7**, 639.
[334] R. Lie and K. Undheim, *Acta Chem. Scand.*, 1972, **26**, 3459.
[335] H. Uno, A. Irie, and K. Hino, *Chem. and Pharm. Bull. Japan*, 1972, **20**, 2603.
[336] I. Oprean, V. Fărcăsan, and F. Paiu, *Rev. Roumaine Chim.*, 1972, **17**, 1901.
[337] H. Alper and R. W. Stout, *J. Heterocyclic Chem.*, 1973, **10**, 5.
[338] J. P. Aune and J. Metzger, *Bull. Soc. chim. France*, 1972, 3537.
[339] D. C. DeJongh and M. L. Thomson, *J. Org. Chem.*, 1973, **38**, 1356.
[340] J. Julien, E.-J. Vincent, J.-C. Poite, and J. Roggero, *Org. Mass Spectrometry*, 1973, **7**, 463.

1,2-benzoisothiazoles,[341] benzothiazoles,[342] 2,6-bis-(1,2-dithiol-3-ylidene)cyclohexanethiones,[343] 2,1,3-benzothiadiazoline and 2,1,3-benzothiadiazine 2,2-dioxides,[344] 1,3,4-thiadiazolines,[345] 2-aryl-5-chloromethyl-1,2,3-oxathiazolidine-2-oxides,[346] 1-alkyl-2-substituted-5-nitroimidazoles,[347] 3-amino-1,2,4-thiadiazoles,[348] 1-phenyl-5-mercaptotetrazoles,[349] methyl 1,3-oxathianes,[350] 2,5-diaryl-1,4-dithiins,[351] thioalkylpyridazines and pyridopyridazines,[352] 2-amino-4-mercapto-*sym*-triazines,[353] and dibenzo[b,f](1,4)-thiazocines.[354]

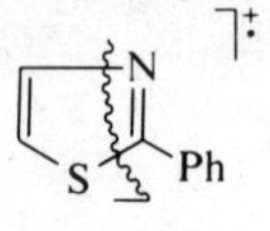

(125) 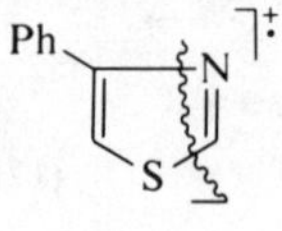   (126) 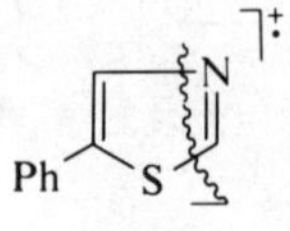   (127) 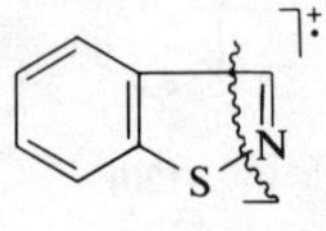   (128)

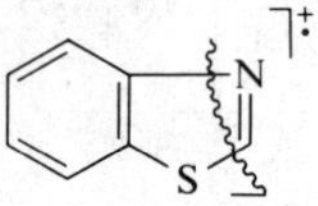

(129) 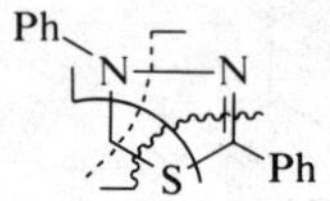   (130) 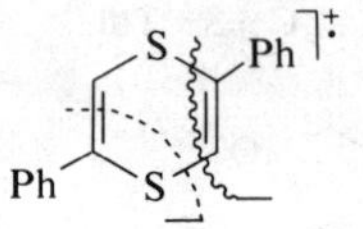   (131)

The following fragmentations of sulphur heterocycles are of particular interest. The three monophenylthiazoles undergo the primary fragmentations shown in (125), (126), and (127).[388,cf,355] The losses of HCN from the 1,2-benzoisothiazole[341] and benzothiazole[356] species occur mainly (77 and 92%, respectively, at 70 eV) as indicated in (128) and (129). The 1,3,4-thiadiazoline (130) fragments as shown; in addition, the main fragment ion is produced by the rearrangement $M - PhS\cdot$.[345] The ring cleavages of ionized 2,5-diphenyl-1,4-dithiin are depicted in (131).[351]

[341] A. Selva and E. Gaetani, *Org. Mass Spectrometry*, 1973, **7**, 327.
[342] A. Croisy, P. Jacquignon, R. Weber, and M. Renson, *Org. Mass Spectrometry*, 1972, **6**, 1321.
[343] C. Th. Pedersen, M. Stavaux, and J. Møller, *Acta Chem. Scand.*, 1972, **26**, 3875.
[344] K. C. C. Bancroft, L. H. M. Guindi, A. F. Temple, and B. J. Millard, *Org. Mass Spectrometry*, 1972, **6**, 1313.
[345] P. Wolkoff and S. Hammerum, *Org. Mass Spectrometry*, 1974, **9**, 181.
[346] T. Nishiyama and F. Yamada, *Bull. Chem. Soc. Japan*, 1973, **46**, 2166.
[347] G. E. van Lear, *Org. Mass Spectrometry*, 1972, **6**, 1117.
[348] A. Shoeb, S. P. Popli, and R. Gopalchari, *Org. Mass Spectrometry*, 1973, **7**, 555.
[349] A. Antonowa, R. Borsdorf, R. Herzschuh, G. Fischer, and H. Engelmann, *J. prakt. Chem.*, 1973, **315**, 313.
[350] J. Jalonen, P. Pasanen, and K. Pihlaja, *Org. Mass Spectrometry*, 1973, **7**, 949.
[351] C. M. Buess, V. O. Brandt, R. C. Srivastava, and W. R. Carper, *J. Heterocyclic Chem.*, 1972, **9**, 887.
[352] V. Kramer, M. Medved, B. Stanovnik, and M. Tišler, *Org. Mass Spectrometry*, 1974, **8**, 31.
[353] J. de Lannory and R. Nasielski-Hinkens, *Bull. Soc. chim. belges*, 1972, **81**, 587.
[354] B. J. Millard, A. Saunders and J. M. Sprake, *Org. Mass Spectrometry*, 1973, **7**, 65.
[355] G. M. Clark, R. Grigg, and D. H. Williams, *J. Chem. Soc. (B)*, 1966, 339.
[356] R. G. Cooks, I. Howe, S. W. Tam, and D. H. Williams, *J. Amer. Chem. Soc.*, 1968, **90**, 4064.

3 Doubly-charged Positive Ions

When doubly-charged ions pass through a collision region containing neutral molecules (usually the first field-free region of a double-focusing mass spectrometer) a charge exchange reaction of the type $R^{++} + N \rightarrow R^+ + N^+$ may occur, where N represents any neutral species, R^{++} is any doubly-charged ion, and where R^+ and N^+ are singly-charged (odd or even electron) species (see Chapters 1, 2, 3). Beynon and his colleagues have shown[357,358] that the product ions will be transmitted by the electric sector if the sector voltage is changed from its normal voltage (E) to approximately double ($2E$) the original value. This gives rise to a 'doubly-charged ion' spectrum produced by the singly-charged product ions, a spectrum which mirrors the spectrum of doubly-charged ions produced in the source region.[357,358] The singly-charged product ions may fragment further to produce metastable transitions which may be detected in the first[359,360] and second[357–361] field-free regions of the mass spectrometer. Although such a technique does not constitute a viable alternative to the conventional positive ion approach, it does yield important information concerning both the formation of doubly-charged ions, and the decompositions of singly-charged product ions.[357–365]

Aromatic compounds give suitable 'doubly-charged ion' spectra,[357–366] and it has been suggested that those containing two functional groups remote from each other give the most abundant doubly-charged species.[367] The following results have been obtained since the last Report. The decomposing $C_6H_6^{+}$ ion from $C_6H_6^{++}$ may have acyclic structure.[45] Differences have been noted in the behaviour of the M^{++} and M^{+} ions of aromatic amines, where M^{++} generally eliminates C_2H_2, whereas M^{+} loses HCN.[364] A survey of skeletal rearrangement processes in 'doubly-charged ion' spectra of aromatic compounds has been carried out.[359] Twenty systems were chosen which all undergo pronounced skeletal rearrangement from conventional singly-charged molecular ions either by the reaction $ABC^+ \rightarrow AC^+ + B$ or by complex molecular ion reorganization. Many examples were found where the same rearrangement occurred from both singly- and doubly-charged ions, but there were also examples of skeletal rearrangements which originated from only a singly-charged, or only a doubly-charged species.[359]

---

[357] J. H. Beynon, R. M. Caprioli, W. E. Baitinger, and J. W. Amy, *Org. Mass Spectrometry*, 1970, **3**, 455.
[358] J. H. Beynon, A. Mathias, and A. E. Williams, *Org. Mass Spectrometry*, 1971, **5**, 303.
[359] T. Blumenthal and J. H. Bowie, *Org. Mass Spectrometry*, 1972, **6**, 1083.
[360] J. H. Bowie, *Austral. J. Chem.*, 1973, **26**, 195.
[361] T. Blumenthal, J. H. Bowie, and S. G. Hart, *Austral. J. Chem.*, 1973, **26**, 2019.
[362] T. Ast, J. H. Beynon, and R. G. Cooks, *Org. Mass Spectrometry*, 1972, **6**, 741.
[363] T. Ast, J. H. Beynon, and R. G. Cooks, *Org. Mass Spectrometry*, 1972, **6**, 749.
[364] T. Ast and J. H. Beynon, *Org. Mass Spectrometry*, 1973, **7**, 503.
[365] H. Sakurai, A. Tatematsu, and H. Nakata, *Org. Mass Spectrometry*, 1973, **7**, 1109.
[366] A. Tatematsu, H. Sakurai, H. Nakata, and T. Goto, *Mass Spectrometry (Japan)*, 1972, **20**, 331.
[367] R. Engel, D. Halpern, and B.-A. Funk, *Org. Mass Spectrometry*, 1973, **7**, 177.

Anthranilates and salicylates undergo the same *ortho* effect from both M^{++} and M^{+} [see (132; X = O or NH)], but rearrangements compete with simple cleavage for the *m*- and *p*-analogues. For example, (133) decomposes by the

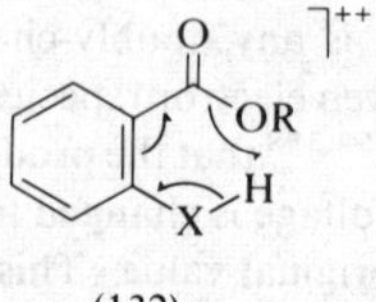

(132) 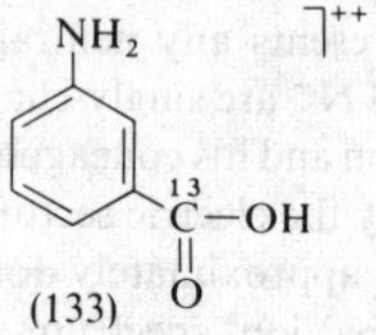   (133)

processes $M^{++} - HO\cdot - {}^{13}CO - HCN$ and $M^{++} - {}^{13}CO$ (base peak).[360] Hydrogen scrambling processes are generally more complete for doubly-charged ions than for singly-charged species produced by direct ionization.[361]

4 Reactions of Negative Ions

Introduction.—A brief introduction to negative-ion mass spectrometry is included here because only specific fragmentations have been covered in the two previous Reports.

Studies of gaseous negative ions carried out before 1960 have been discussed by Melton;[368] work since that time is described in several books,[369,370] and review articles.[1,371—377] The latest comprehensive review is that of Dillard[377] which covers the literature until the end of 1971. A review entitled 'Negative Ion Mass Spectrometry of Organic, Organometallic, and Co-ordination Compounds' should appear in early 1975.[378]

---

[368] C. E. Melton, in 'Mass Spectrometry of Organic Ions', ed. F. W. McLafferty, Academic Press, New York, 1963, Chap. 3.
[369] C. E. Melton, 'Principles of Mass Spectrometry and Negative Ions', Marcel Dekker, New York, 1970.
[370] M. von Ardenne, K. Steinfelder, and R. Tümmler, 'Electronenanlagerungs Massenspektrographic organisher Substanzen', Springer Verlag, Berlin, Heidelberg, New York, 1971.
[371] G. Briegleb, *Angew. Chem. Internat. Edn.*, 1964, **3**, 317.
[372] J. C. J. Thynne, K. A. G. MacNeil, and K. J. Caldwell, in 'Time of Flight Mass Spectrometry', ed. D. Price and J. E. Williams, Pergamon, London, 1969, 117.
[373] J. M. Wilson, in 'Mass Spectrometry', ed. D. H. Williams (Specialist Periodical Reports), The Chemical Society, London, 1970, Vol. 1, p. 12.
[374] R. W. Kiser, in 'Recent Developments in Mass Spectrometry', Proc. Internat. Conf. Kyoto, Japan, Sept. 1969, ed. K. Ogata and T. Hayakawa, Univ. Park Press, Baltimore, 1970, 844.
[375] P. W. Harland, K. A. G. MacNeil, and J. C. J. Thynne, in 'Dynamic Mass Spectrometry', ed. D. Price and J. E. Williams, Heyden and Son, London, 1970, Vol. 1, p. 105.
[376] J. C. J. Thynne, in 'Dynamic Mass Spectrometry', ed. D. Price, Heyden and Son, London, 1972, Chap. 3.
[377] J. G. Dillard, *Chem. Rev.*, 1973, **73**, 589.
[378] J. H. Bowie and B. D. Williams, in 'Mass Spectrometry', M. T. P. International Review of Science, Physical Chemistry, Series 2, ed. A. Maccoll, Butterworths, London, 1975, Vol. 5.

Reviews which cover aspects of negative-ion mass spectrometry beyond the scope of this report are listed below. Atomic negative ions,[379—382] inorganic systems,[377,378] negative ions and the magnetron,[383] radiation processes,[384] negative ion–molecule reactions with respect to flowing afterglow studies,[382] tandem mass spectrometric studies,[385] ions in flames,[386] negative ion–neutral reactions,[387] and ICR spectroscopy;[388—391] electron beam and swarm experiments,[384] the formation of negative ions on metallic surfaces,[392—397] field ionization,[398] appearance potentials,[372,375,376,399] and lifetimes of negative ions.[376]

Apart from the pioneering work of von Ardenne and co-workers on large organic molecules,[370] the majority of investigations carried out with negative ions between 1950 and 1965 were concerned with theoretical rather than practical aspects, and this feature is reflected in many review articles.[372,373,375—377] Since 1965 there has been a continuing interest in the development of negative-ion mass spectrometry as an aid to structure determination, and previous Reports[1] have reviewed progress in this area.

Modes of Formation of Negative Ions.—The mechanism of formation of negative ions depends on both the energy of the electron and the nature of the molecule. One of three processes may occur and these are summarized for a diatomic system AB as follows:

---

[379] B. L. Moiseiwitsch, *Adv. Mol. Phys.*, 1965, **1**, 61.
[380] B. M. Smirnov, *High Temp.* (*U.S.S.R.*), 1965, **3**, 716.
[381] R. S. Berry, *Chem. Rev.*, 1969, **69**, 533.
[382] E. E. Ferguson, in 'Ion Molecule Reactions', ed. J. L. Franklin, Butterworths, London, 1972, Vol. 2, p. 363.
[383] F. M. Page and G. C. Goode, 'Negative Ions and the Magnetron', Wiley Interscience, London, 1969.
[384] L. G. Christophorou, 'Atomic and Molecular Radiation Physics', Wiley Interscience, London, 1971.
[385] J. H. Futrell and T. O. Tiernan, in 'Ion Molecule Reactions', ed. J. L. Franklin, Butterworths, London, 1972, Vol. 2, 485.
[386] H. F. Calcote, in 'Ion Molecule Reactions', ed. J. L. Franklin, Butterworths, London, 1972, Vol. 2, p. 673.
[387] J. F. Paulson, in 'Ion Molecule Reactions', ed. J. L. Franklin, Butterworths, London, 1972, Vol. 1, p. 77.
[388] J. D. Balderschweiler and S. S. Woodgate, *Accounts Chem. Res.*, 1971, **4**, 114.
[389] J. H. Futrell, in 'Dynamic Mass Spectrometry', ed. D. Price, Heyden and Son, London, 1971, Vol. 2, p. 97.
[390] J. M. S. Henis, in 'Ion Molecule Reactions', ed. J. L. Franklin, Butterworths, London, 1972, Vol. 2, p. 395.
[391] C. J. Drewery, G. C. Goode, and K. R. Jennings, 'Mass Spectrometry', M.T.P. International Review of Science, Physical Chemistry Series 1, ed. A. Maccoll, Butterworths, London, 1973, Vol. 5, p. 123.
[392] I. N. Bakulina and N. I. Ionov, *Russ. J. Phys. Chem.*, 1959, **33**, 286.
[393] V. E. Krohn, *J. Appl. Phys.*, 1962, **33**, 3523.
[394] M. D. Scheer and J. Fine, *Phys. Rev. Letters*, 1966, **17**, 283.
[395] M. D. Scheer and J. Fine, *J. Chem. Phys.*, 1967, **46**, 3998.
[396] A. Persky, E. F. Greene, and A. Kuppermann, *J. Chem. Phys.*, 1968, **49**, 2347.
[397] M. D. Scheer, *J. Res. Nat. Bur. Stand.* (*A*), 1970, **74**, 37.
[398] A. J. B. Robertson and P. Williams, *Proc. Chem. Soc.* (*London*), 1964, 286.
[399] J. D. Morrison, in 'Mass Spectrometry', M.T.P. International Review of Science, Physical Chemistry Series 1, ed. A. Maccoll, Butterworths, London, 1973, Vol. 5, Chap. 2.

Resonance Capture

$$AB + e \rightarrow AB^{\bar{\cdot}}$$

Dissociative Resonance Capture

$$AB + e \begin{array}{c} \nearrow A^- + B\cdot \\ \searrow A\cdot + B^- \end{array}$$

Ion Pair Production

$$AB + e \begin{array}{c} \nearrow A^- + B^+ + e \\ \searrow A^+ + B^- + e \end{array}$$

Resonance capture forms a parent ion near 0 eV, dissociative resonance capture occurs in the range 0—15 eV, and ion pair production is noted above 10 eV. Many organic molecules produce molecular anions when the electron beam energy is in the range 20—80 eV. It has been shown that this is due to the capture of secondary electrons (electrons with near thermal energies), which may be produced either from the ionization process of molecular cations (*i.e.* $M + e \rightarrow M^{+\cdot} + 2e$) or from electrode surfaces.[1] These low-energy molecular anions may undergo unimolecular decompositions,[1,400] which include both simple cleavage processes and more complex decompositions often involving hydrogen rearrangements and/or skeletal rearrangements. Many of these have been reviewed.[1,378] Simple cleavages generally occur as indicated in (134) and (135),[400,401] either α to the charge-retaining unit, or α to some atom (or group of atoms) in conjugation with the centre of charge.

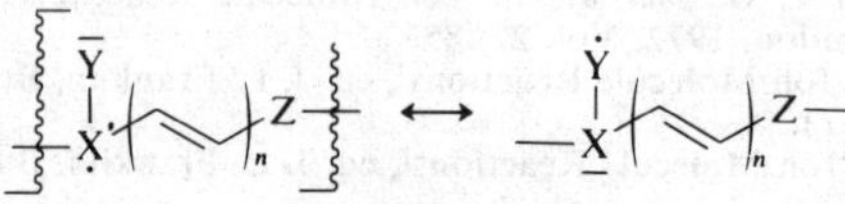

(134)

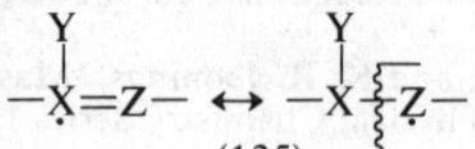

(135)

Ions produced by rearrangement reactions are common in negative-ion spectra and, when observed, are generally more abundant than those produced by

[400] J. H. Bowie and A. C. Ho, *Austral. J. Chem.*, 1973, **26**, 2009 (and references cited therein).
[401] R. G. Alexander, D. B. Bigley, and J. F. J. Todd, *Org. Mass Spectrometry*, 1973, **7**, 643.

competing cleavage reactions. This is a direct consequence of the low internal energy of the decomposing molecular anion, a situation analogous to that observed for the decompositions of low-energy positive ions.[402,403] Rearrangement processes often occur where two functional groups are adjacent, or where a hydrogen transfer can proceed through a favoured six-membered transition state. Rearrangements involving o-substituted compounds are useful for the characterization of such systems; in these cases the negative-ion spectra are often more diagnostic than the corresponding positive-ion spectra.[1] It is premature at present to write general mechanisms for the rearrangement processes; these have been considered individually in past Reports.[1,378]

Instrumentation.—Mass spectrometers used to measure negative-ion spectra fall into four main categories: (a) conventional mass spectrometers; (b) time-of-flight instruments; (c) spectrometers with high-pressure sources; and (d) ICR spectrometers.

Many conventional mass spectrometers either have a negative-ion capability or may be modified to allow negative-ion measurement. A problem associated with such instruments is that of collection efficiency. The first dynode of the electron multiplier is maintained at a high negative potential, and it repels negative ions. In order to overcome this problem the negative-ion accelerating potential should be maintained at a higher value than that of the first dynode. Advantages of using conventional double-focusing mass spectrometers are that they may be used to measure metastable decompositions, negative-ion kinetic energy spectra, and collision-induced dissociations. The modes of negative-ion formation are kept to a minimum by using the secondary electron capture method, and ion–molecule products are restricted by keeping the source pressure less than 10^{-6} Torr.

The investigation of negative ions by time-of-flight mass spectrometry has been reviewed.[404,405] Such instruments are particularly useful for the determination of appearance potential measurements,[372,375,376] and for the measurement of auto-detachment lifetimes (*i.e.* the time required for a negative ion to lose an electron).[376,406]

Von Ardenne has described a high-pressure ion source which has been used primarily for naturally occurring compounds. This source utilizes an argon gas discharge (at 10^{-2} Torr) to produce slow electrons. This method has been used successfully to produce molecular anions for complex molecules, when the corresponding molecular cation is absent.[370,407]

---

[402] D. H. Williams and R. G. Cooks, *Chem. Comm.*, 1968, 663.
[403] I. Howe, in 'Mass Spectrometry', ed. D. H. Williams (Specialist Periodical Reports), The Chemical Society, London, 1970, Vol. 1, pp. 31 and 38.
[404] D. Price, *Chem. in Britain*, 1968, **4**, 255.
[405] Various authors in 'Dynamic Mass Spectrometry', ed. D. Price and J. E. Williams, Heyden and Son, London, 1969, 1970, and 1972, Vols. 1, 2, and 3.
[406] P. W. Harland and J. C. J. Thynne, *Internat. J. Mass Spectrometry Ion. Phys.*, 1972, **10**, 11.
[407] M. von Ardenne, *Z. angew. Phys.*, 1959, **11**, 121.

ICR spectrometers[388-391] are particularly useful for negative-ion measurement, as they measure negative and positive ions with equal efficiency. ICR spectrometers are mainly used for the study of ion–molecule reactions.

Metastable Ions and Collision-induced Dissociations.—Metastable ions are abundant in the negative-ion spectra of organic compounds when low-energy molecular ions undergo decomposition.[408] Metastable ions for rearrangement processes are generally more abundant than those observed for simple cleavage processes. Most metastable peaks in negative-ion spectra, formed in the second field-free region of the mass spectrometer, are gaussian but some may be broad and flat-topped (e.g. the loss of NO· from substituted nitrobenzene molecular anions[409,410]). The metastable defocusing technique (using variable electrostatic potentials[411]) has been adapted for negative ions.[408,412] Not only can this method be used to assign unequivocally fragmentation pathways, but it can also be used to detect multi-degradation sequences of metastable ions.[408] Negative-ion kinetic spectra may be obtained by monitoring the ion beam passing through the β-slit as the sector voltage is varied continuously at constant accelerating voltage.[408]

Keough, Beynon, and Cooks[413] have described a method whereby negative ions may be formed from the collision process $M^+ + N \rightarrow M^- + N^{++}$, where N is any neutral gas. As an example, the ions C_nH^- ($n = 1$—6) are formed from benzene. Such reactions clearly have little analytical application, but they may provide important information about the structures of negative ions.

Many molecular anions produced by the capture of secondary electrons do not have enough internal energy to allow decomposition. Collisional excitation[cf. 414,415] enables initially unreactive molecular anions to gain enough internal energy for cleavage to occur,[416] and this technique may also be used to observe the fragmentations of initially unreactive fragment ions.[416] Sample pressures of ca. 1×10^{-6} Torr are used in the source (with differential pumping), and the collision gas (generally nitrogen) is introduced in the first field-free region to ca. 10^{-5} Torr. This technique has been used recently to examine the decompositions of quinones,[416] ketones,[416] amides,[417] and carboxylic acids.[418]

A qualitative estimate of relative rates of negative ion decompositions may be determined by observing the changes in abundances of collision induced peaks (in negative IKE spectra) as the pressure of collision gas is increased (i.e. with

---

[408] J. H. Bowie and S. G. Hart, *Internat. J. Mass Spectrometry Ion Phys.*, 1974, **13**, 319.
[409] J. H. Bowie, *Org. Mass Spectrometry*, 1971, **5**, 945.
[410] C. L. Brown and W. P. Weber, *J. Amer. Chem. Soc.*, 1970, **92**, 5775.
[411] A. H. Struck and H. W. Major, paper presented to the 12th Annual Conference on Mass Spectrometry, A.S.T.M.E. 14 Dallas, 1969.
[412] R. W. Kiser, R. E. Sullivan, and M. S. Lupin, *Analyt. Chem.*, 1969, **41**, 1958.
[413] T. Keough, J. H. Beynon, and R. G. Cooks, *J. Amer. Chem. Soc.*, 1973, **95**, 1695.
[414] K. R. Jennings, *Internat. J. Mass Spectrometry Ion Phys.*, 1968, **1**, 227.
[415] W. F. Haddon and F. W. McLafferty, *J. Amer. Chem. Soc.*, 1968, **90**, 4745.
[416] J. H. Bowie, *J. Amer. Chem. Soc.*, 1973, **95**, 5795.
[417] J. H. Bowie, *Austral. J. Chem.*, 1973, **26**, 2719.
[418] J. H. Bowie, *Org. Mass Spectrometry*, 1974, **9**, 304.

increasing internal energy of the decomposing ion).[416] If a molecular anion fragments by competing cleavage and rearrangement processes, the rate of the simple cleavage reaction increases more rapidly than that of the rearrangement reaction. The same effect is noted for decomposing positive ions.[102,103]

Formation and Fragmentation of Negative Ions.—Cyclo-octatetraene, acenaphthalene, and fluoranthrene produce molecular anions near 0 eV,[419] negative ions have been reported from polyene hydrocarbons,[420] and Cl_2^- ions have been observed in the spectra of dichloroalkanes.[421] The negative chemical ionization spectra of polycyclic chlorinated insecticides have been determined using methane, isobutane or methylene chloride as collision gas.[422] All compounds yield molecular anions, but many peaks [e.g. $(M + Cl\cdot)^-$] were obtained at masses greater than that of the molecular anion.

The ions F^- and $(M - F)^-$ are observed in the spectra of perfluoroalkanes,[423,424] and F^{--} ions occur in the spectrum of CF_3Cl.[425] Perfluorocyclobutane forms a molecular anion between 0 and 1 eV.[406,424] The major fragment ion in this spectrum is F^-, which shows collision cross-section maxima at 3.7, 6.6, 7.5, and 10.0 eV.[406] The major negative ion from methanol is MeO^-.[426] The negative-ion mass spectra of 20-chloro-steroids have been measured using the von Ardenne source.[427]

Nitroethylene yields a small molecular anion at 100 eV, but the main fragment ion is NO_2^-.[428] Further studies of aryl nitro-compounds show the formation of molecular anions by secondary electron capture.[429—431] The loss of HO· from (136) occurs as indicated.[431] Christophorou and colleagues[432—435] have confirmed that NO_2-containing benzene derivatives, CN-substituted organic molecules, higher aromatic hydrocarbons, strained structures and organic molecules containing the functional groups —CO—CO—, —CO(CH)OH—,

---

[419] W. F. Frey, R. N. Compton, W. T. Naff, and H. C. Schweinler, *Internat. J. Mass Spectrometry Ion Phys.*, 1973, **12**, 19.
[420] V. I. Khvostenko, V. P. Yur'ev, I. Kh. Aminev, G. A. Tolstikov, and S. R. Rafikov, *Doklady Chem.*, 1972, **202**, 128.
[421] A. Ito, K. Matsumoto, and T. Takeuchi, *Org. Mass Spectrometry*, 1972, **6**, 1045.
[422] R. C. Dougherty, J. Dalton, and F. J. Biros, *Org. Mass Spectrometry*, 1972, **6**, 1171.
[423] T. Su and L. Kevan, *J. Phys. Chem.*, 1973, **77**, 148.
[424] C. Lifshitz and R. Grajower, *Internat. J. Mass Spectrometry Ion Phys.*, 1972, **10**, 25.
[425] J. E. Ahnell and W. S. Koski, *Nature Phys. Chem.*, 1973, **245**, 30.
[426] J. C. J. Thynne, *Org. Mass Spectrometry*, 1973, **7**, 899.
[427] G. Adam, D. Voigt, K. Schreiber, M. von Ardenne, R. Tummler, and K. Steinfelder, *J. prakt. Chem.*, 1973, **315**, 125.
[428] T. Shiga, H. Yamaoka, K. Arakawa, and T. Suguira, *Bull. Chem. Soc. Japan*, 1972, **45**, 2065.
[429] J. Yinon and H. G. Boettger, *Internat. J. Mass Spectrometry Ion Phys.*, 1972, **10**, 161.
[430] J. Yinon, H. G. Boettger, and W. P. Weber, *Analyt. Chem.*, 1972, **44**, 2235.
[431] J. F. J. Todd, R. B. Turner, B. C. Webb, and C. H. J. Wells, *J.C.S. Perkin II*, 1973, 1167.
[432] A. Hadjiantoniou, L. G. Christophorou, and J. G. Carter, *J.C.S. Faraday II*, 1973, **69**, 1691.
[433] A. Hadjiantoniou, L. G. Christophorou, and J. G. Carter, *J.C.S. Faraday II*, 1973, **69**, 1704.
[434] L. G. Christophorou, A. Hadjiantoniou, and J. G. Carter, *J.C.S. Faraday II*, 1973, **69**, 1713.
[435] V. I. Khvostenko, I. Kh. Aminev, and I. I. Furlei, *Teor. i eksp. Khim.*, 1973, **9**, 99.

—CO_2H, and —$CHCHO$ capture thermal electrons and form long-lived parent negative ions *via* nuclear-excited Feshbach resonances.

The major fragment ions in the spectra of 2-aryl-1,3-oxathianes are produced by the elimination $(M - C_3H_6S)$.[400] This ion is produced at 70 eV (by fragmentation of a low-energy molecular anion) after statistical equilibration of the hydrogens at the 2 and 4 positions, *viz.* (137) → (139) [*cf.* the analogous scrambling in

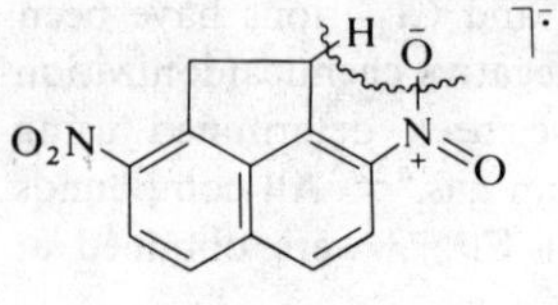

(136) 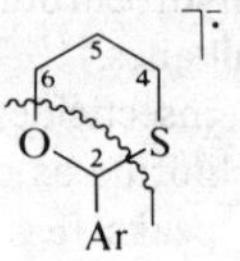   (137) 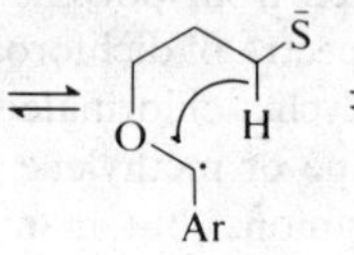   (138) 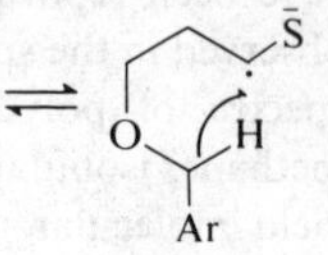   (139)

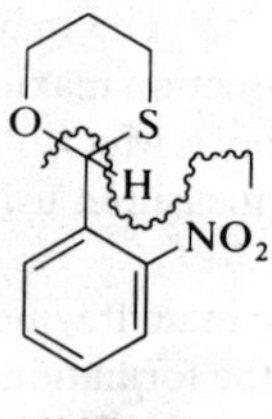

(140) 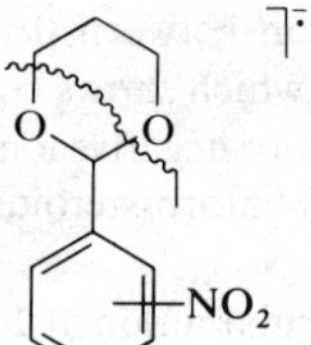   (141) 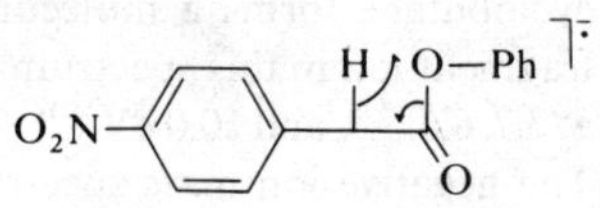   (142)

1,3-dithianes (Vol. 2, p. 141)]. The ionization efficiency plots for ions from 2-(*p*-nitrophenyl)-1,3-oxathiane show that the molecular ion is produced *only* by secondary electron capture above 15 eV, whereas the $(M - C_3H_6S)^{\bar{.}}$ is produced by a dissociative mechanism at *ca.* 5 eV, and *directly* from the molecular anion above 15 eV. The $(M - C_3H_6S)^{\bar{.}}$ ion is produced *without* scrambling at 5 eV, but with *complete* scrambling above 15 eV. This demonstrates that the decomposing molecular anion produced by secondary electron capture has a *lower* internal energy than the species that fragments at 5 eV. The base peak from *o*-nitrophenyl-1,3-oxathiane is produced (without H scrambling) as shown in (140). All three nitrophenyldioxans fragment by *specific* loss of C_3H_6O as illustrated in (141). No *ortho*-effect is observed for the *o*-nitro-derivative.

The main elimination from the phenyl(*p*-nitrophenyl)acetate molecular anion (142) is loss of PhOH.[436] Appearance potential measurements suggest that the similarities in ρ-values obtained[437] for the fragmentation of aryl *m*- and *p*-nitrobenzoates are probably due to correspondence between Hammett σ and appearance potential plots in the two series.[436] The spectra of aliphatic dicarboxylic acids contain molecular anions of small abundance, and pronounced $(M - H\cdot)^-$ and $(M - CO_2H\cdot)^-$ ions.[438] The spectra of all three benzene dicarboxylic

[436] J. H. Bowie and B. Nussey, *Org. Mass Spectrometry*, 1974, **9**, 310.
[437] J. H. Bowie and B. Nussey, *Org. Mass Spectrometry*, 1972, **6**, 429.
[438] A. Ito, K. Matsumoto, and T. Takeuchi, *Org. Mass Spectrometry*, 1973, **7**, 1279.

acids show $(M - CO_2H\cdot)^-$ ions, only that of phthalic acid exhibits an $M - H_2O)^-$ ion.[438] The loss of a hydroxyl radical from *m*-nitrobenzoic acid occurs as indicated in (143).[418]

The cleavages of some functional groups attached to aromatic systems are not observed because the molecular anions do not have enough internal energy to fragment. One such group is an *isolated* carboxy-group, and collisional activation

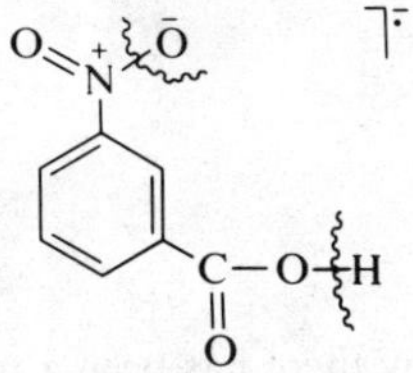

(143)

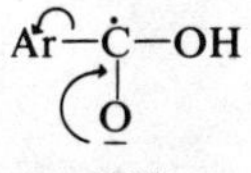

(144)

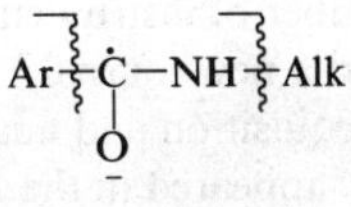

(145)

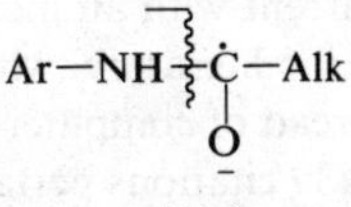

(146)

spectra (see above) of anthraquinone carboxylic acids, nitrobenzoic acids, and cyanobenzoic acids show the basic cleavage of the carboxyl group to be $M - CO_2H\cdot$ [see (144)].[418] Collision activation shows that aryl amides cleave as shown in (145) and (146).[417] The conventional spectra of quinones[416] and anhydrides[416,439] produce only molecular anions. Collisional activation of naphthoquinone produces the eliminations $M - CHO\cdot$ and $M - (CHO\cdot + CO)$; of phthalic anhydride, $M - CO - CO_2$ and $M - C_2O_3$; and of maleic anhydride, $M - CO$.[416]

---

[439] C. D. Cooper, R. N. Compton, H. C. Schweinler, and V. E. Anderson, 20th Annual Conference on Mass Spectrometry and Allied Topics, Dallas, Texas, June 1972, A.S.T.M. Comm. E 14, NB 2, pp. 35-6, 1972.

8
Gas Chromatography–Mass Spectrometry

BY C. J. W. BROOKS AND B. S. MIDDLEDITCH

1 General Considerations

Introduction.—During the period under review there has been a marked rise in the rate of appearance of publications on gas chromatography–mass spectrometry (g.c.m.s.), concurrent with an increase in the number of instruments now in use. This growth of the literature, which has spawned a new journal,[1] has also been aided by the spread of computer-assisted data acquisition and analysis systems. In 1973 alone, 439 citations pertaining to g.c.m.s. appeared in the *Mass Spectrometry Bulletin*,[2] 433 publications were abstracted in *Gas Chromatography—Mass Spectrometry Abstracts*[3] and 247 in *Gas Chromatography Literature—Abstracts and Index*.[4]

Detailed coverage of computerized data acquisition and handling is provided in Chapter 4, and drug metabolism is reviewed in Chapter 9. Routine applications of g.c.m.s. to organic chemistry are largely excluded from this chapter since, as in previous reports,[5] emphasis is placed on analytical applications in bio-organic chemistry, medicine, food chemistry, geochemistry, and environmental chemistry.

Pertinent books and reviews are cited in the Table, with brief indications of their scope.

Table Books and reviews pertaining to g.c.m.s.

Principal topic	References
Techniques of m.s. and g.c.m.s. and applications in organic analysis	a—d
Advances in m.s. (Symposia)	e, f
Applications of chromatography including g.c.m.s. (Symposium)	g
Stable isotopes in clinical and metabolic studies	h, i
Measurements of isotope ratios in organic molecules by m.s.	j

[1] *Biomedical Mass Spectrometry*, Heyden and Son Ltd., London, 1974, Vol. 1.
[2] *Mass Spectrometry Bulletin*, H.M.S.O., London, 1973, Vol. 7.
[3] *Gas Chromatography–Mass Spectrometry Abstracts*, Science and Technology Agency, London, 1973, Vol. 4.
[4] *Gas Chromatography Literature—Abstracts and Index*, Preston Technical Abstracts Co., Niles, Illinois, 1973, Vol. 6.
[5] (*a*) C. J. W. Brooks in 'Mass Spectrometry', ed. D. H. Williams (Specialist Periodical Reports), The Chemical Society, London, 1971, Vol. 1, Chapter 7; (*b*) C. J. W. Brooks and B. S. Middleditch, *ibid.*, 1973, Vol. 2, Chapter 7.

Table—*continued*

Principal topic	References
G.c.m.s. in biochemistry and medicine (Symposia)	k, l
G.c.m.s. in neurobiology	m
G.c.m.s. in forensic toxicology	n
G.c.m.s. in pharmacology	o
G.c.m.s. in environmental studies	p, q
G.c.m.s. in geochemistry	r
Mass fragmentography	s—u
M.s. and g.c.m.s. of lipids	v
G.c.m.s. of lipids	w
G.c.m.s. of polar lipids	x
G.c. of steroids	y
M.s. and g.c.m.s. of steroids	z, aa—ee
M.s. and g.c.m.s. of amino-acids and peptides	ff
General texts	gg—oo
Course notes	pp—qq
Index to collections of m.s. data	rr

(a) A. L. Burlingame, R. E. Cox and P. J. Derrick, *Analyt. Chem.*, 1974, **46**, 248R—287R (1647 references); (b) W. H. McFadden, 'Techniques of Combined Gas Chromatography/Mass Spectrometry', Wiley–Interscience, New York, 1973, (463 pages); (c) R. Ryhage, *Quart. Rev. Biophys.*, 1973, **6**, 311—335; (d) L. Weil, F. Frimmel and K. E. Quentin, *Z. analyt. Chem.*, 1974, **268**, 97—101; (e) 'Advances in Mass Spectrometry', ed. A. R. West, Applied Science Publishers, Barking, 1974, Vol. 6; (f) 'Dynamic Mass Spectrometry', ed. D. Price, Heyden and Son, Ltd., London, 1972, Vol. 3; (g) 'Advances in Chromatography 1973', ed. A. Zlatkis, Chromatography Symposium, University of Houston, Houston, 1973, (390 pages); (h) M. Strolin-Benedetti and P. Strolin, *J. Pharmacol. Clin.*, 1973, **1**, 15—23; (i) 'Proceedings of a Seminar on the Use of Stable Isotopes in Clinical Pharmacology', ed. P. D. Klein and L. J. Roth, U.S.A.E.C., Oak Ridge, Tennessee, 1972; (j) P. D. Klein, in ref. (i), pp. 101—131 (including discussion); (k) 'International Symposium on Mass Spectrometry in Biochemistry and Medicine', ed. A. Frigerio, Summaries: Tamburini Editore, Milan, 1973; Proceedings: Raven Press, New York, 1974; (l) 'Proceedings of the International Symposium on Gas Chromatography-Mass Spectrometry', ed. A. Frigerio, Tamburini Editore, Milan, 1972; (m) 'Advances in Biochemical Psychopharmacology', ed. E. Costa and B. Holmstedt, Raven Press, New York, 1973 (168 pages); (n) R. F. Skinner, E. J. Gallaher, J. B. Knight and E. J. Bonelli, *J. Forensic Sci.*, 1972, **17**, 189—198; (o) J. T. Watson, *Ann. Rev. Pharmacol.*, 1973, **13**, 391; (p) A. W. Garrison, L. H. Keith, and A. L. Alford, *Adv. Chem. Ser.*, 1972, **111**, 26; (q) J. Roboz in 'Water and Water Pollution Handbook', ed. L. L. Ciaccio, Marcel Dekker, New York, 1973, Vol. 4, p. 1617; (r) D. H. Smith, N. A. B. Gray, C. T. Pillinger, B. J. Kimble, and G. Eglinton, 'Advances in Organic Geochemistry', ed. H. R. von Gaertner and H. Wehner, Pergamon Press, Oxford, 1971, Vol. 33; (s) B. Holmstedt and L. Palmér, in ref. *m* pp. 1—14; and bibliography, pp. 161—168 (129 references); (t) A. E. Gordon and A. Frigerio, *J. Chromatog.*, 1972, **73**, 401; (u) B. Holmstedt and J. E. Lindgren, *Z. analyt. Chem.*, 1972, **261**, 291—297; (v) A. Zeman and H. Scharmann, *Fette, Seifen, Anstrichm.*, 1972, **74**, 509—519; 1973, **75**, 32—44, 170—180; (w) U. Pallotta, *Afinidad*, 1972, **29**, 1087, 1103; (x) C. V. Viswanathan, *J. Chromatog.*, 1974, **98**, 105—128; (y) M. Novotny and J. Janak, *Chem. listy*, 1973, **66**, 693; (z) 'Modern Methods of Steroid Analysis', ed. E. Heftmann, Academic Press, New York, 1973; (aa) B. A. Knights, in ref. (z) pp. 103—138 (222 references); (bb) S. J. Clark and H. H. Wotiz, in ref. (z) pp. 72—102 (98 references); (cc) C. J. W. Brooks and B. S. Middleditch, in ref. (z) pp. 140—198 (324 references); (dd) W. J. A. VandenHeuvel, J. L. Smith, G. Albers Schönberg, B. Plazonnet, and P. Bélanger, in ref. (z) pp. 199—219 (57 references); (ee) R. H. Thompson, N. D. Young, J. E. Harten, T. A. Springer, R. Vihko and C. C. Sweeley, 'Gas Chromatography and Mass Spectrometry of Selected C_{19} and C_{21} Steroids', Dept. of Biochemistry, Michigan State University, East Lansing, Michigan, 1973; (ff) 'New Techniques in Amino Acid, Peptide, and Protein Analysis', ed. A. Niederwieser and G. Pataki, Ann Arbor Sci. Publ. Inc., Ann Arbor, Michigan, 1971; (gg) G. P. Arsenault in 'Experimental Methods in Biophysical Chemistry', ed. C. Nicolan, Wiley–Interscience, New York, 1973, p. 3; (hh) K. Biemann in

Practical Aspects.—Further developments in the preparation and application of glass capillary columns have been reported. New injection systems have been devised.[6—8] Reliable methods for obtaining columns of consistent quality have emerged, in respect of relatively nonpolar phases such as SE-30 or OV-101. Coating is effected either by a static procedure[9] or by a dynamic method in which a hydrophobic silica ('Silanox', 6—10 μm diameter : Cabot Corp.) is incorporated into the film.[10] (Thermostable columns coated with polar phases are more difficult to prepare).[11] Columns of both types have been successfully applied to the analysis of complex mixtures of urinary steroids,[12,13] urinary sugars,[13] and drug metabolites;[13] their established use for g.c.m.s. of compounds of lower mass has been further extended, for example to volatile components of breath, urine,[14,15] and blood.[16]

The coupling of glass capillary columns to mass spectrometers has presented problems that initially limited the value of such columns, for example in urinary steroid determinations.[17] However, several methods[18—20] have been devised

---

'Techniques of Chemistry', ed. K. W. Bentley and G. W. Kirby, Wiley–Interscience, New York, Vol. 4, 2nd Edn., 1972, p. 377; (*ii*) B. Danieli and A. Frigerio, 'Notes on Mass Spectrometry (Italian)', Tamburini Editore, Milan, 1973; (*jj*) L. de Galan in 'Analytical Spectrometry', Adam Hilger, London, 1971; (*kk*) R. A. W. Johnstone, 'Mass Spectrometry for Organic Chemists', Cambridge University Press, London, 1972; (*ll*) 'Mass Spectrometry', ed. A. Maccoll (MTP International Review of Science), Physical Chemistry Series 1, Butterworths, London, 1972, Vol. 5; (*mm*) A. J. B. Robertson in 'Analytical Chemistry', ed. T. S. West, Butterworths, London, Part 2, 1973; (*nn*) 'Principles of Organic Mass Spectrometry', D. H. Williams and I. Howe, McGraw-Hill, London, 1972; (*oo*) J. M. Wilson in 'Structure Determination in Organic Chemistry', ed. W. D. Ollis, Butterworths, London, 1973; (*pp*) 'Proceedings of an International School on Mass Spectrometry', ed. J. Marsal, J. Stefan Institute, Ljubljana, Yugoslavia, 1971; (*qq*) 'Papers from the First Theoretical and Practical course in Mass Spectrometry', ed. A. Frigerio, Tamburini Editore, Milan, Italy, 1971; (*rr*) B. S. Middleditch and J. A. McCloskey, 'A Guide to Collections of Mass Spectral Data', American Society for Mass Spectrometry, Baylor College of Medicine, Houston, 1974.

[6] M. Verzele, M. Verstappe, P. Sandra, E. van Luchene, and A. Vuye, *J. Chromatog. Sci.*, 1972, **10**, 668.
[7] A. L. German and E. C. Horning, *Analyt. Letters*, 1972, **5**, 619.
[8] P. M. J. van den Berg and T. P. H. Cox, *Chromatographia*, 1972, **5**, 301.
[9] G. A. F. M. Rutten and J. A. Luyten, *J. Chromatog.*, 1972, **74**, 177.
[10] A. L. German and E. C. Horning, *J. Chromatog. Sci.*, 1973, **11**, 76.
[11] G. Alexander and G. A. F. M. Rutten, *Chromatographia*, 1973, **6**, 231.
[12] J. A. Luyten and G. A. F. M. Rutten, *J. Chromatog.*, 1974, **91**, 393.
[13] E. C. Horning, M. G. Horning, J. Szafranek, P. van Hout, A. L. German, J. P. Thenot, and C. D. Pfaffenberger, *J. Chromatog.*, 1974, **91**, 367.
[14] A. Zlatkis, W. Bertsch, H. A. Lichtenstein, A. Tishbee, F. Shunbo, H. M. Liebich, A. M. Coscia, and N. Fleischer, *Analyt. Chem.*, 1973, **45**, 763.
[15] K. E. Matsumoto, D. H. Partridge, A. B. Robinson, L. Pauling, R. A. Flath, T. R. Mon, and R. Teranishi, *J. Chromatog.*, 1973, **85**, 31.
[16] A. Zlatkis, W. Bertsch, D. A. Bafus, and H. M. Liebich, *J. Chromatog.*, 1974, **91**, 379.
[17] C. H. L. Shackleton, J.-Å. Gustafsson, and F. L. Mitchell, *Acta Endocrinol.*, 1973, **74**, 157.
[18] N. Neuner-Jehle, F. Etzweiler, and G. Zarske, *Chromatographia*, 1973, **6**, 211.
[19] K. Grob and H. Jaeggi, *Analyt. Chem.*, 1973, **45**, 1788.
[20] J. G. Leferink and P. A. Leclercq, *J. Chromatog.*, 1974, **91**, 385.

for direct coupling and one of these[20] is applicable to 'steroid profiles'.[12] Coupling via a Becker–Ryhage separator is also satisfactory for this purpose.[21] Other examples of capillary g.c.m.s. are mentioned in the sequel.

Flow programming has been employed with silicone[22] and porous silver[23] membrane molecular separators. Column 'bleed' may be reduced by extensive conditioning of g.c. columns;[24] 'bleed' associated with column coupling is eliminated by the use of polyimide ferrules.[25]

Molecular separators have been reviewed,[26] and the efficient coupling of a g.c. column to a molecular separator has been discussed.[27,28] A modified steel jet-type molecular separator incorporating an adjustable first jet has been described.[29] A detailed study of the porous silver membrane molecular separator has been made.[30] Further details of a palladium/silver separator have been published.[31,32] High-speed differential pumping of the ion source and analyser compartments of the mass spectrometer has permitted direct introduction of the total effluent from packed g.c. columns at flow rates of up to 20 ml min^{-1}.[33]

Chemical ionization (CI) g.c.m.s. can be used in several different modes. The reagent gas can be used as the carrier gas,[34] added to the gas stream at the exit of the g.c. column,[35,36] or introduced to the ion source via a separate port.[37] Conversion between the chemical and electron-impact ionization modes of operation can be effected by instrumental modification, which may be rapid,[37] or by employing helium as the carrier gas for the production of charge-exchange spectra which are similar to those obtained by electron-impact ionization.[36] A very sensitive ion source which operates at atmospheric pressure, and is potentially useful for g.c.m.s., has been described.[38] Capillary columns have been used with a chemical ionization mass spectrometer.[35] A brief evaluation of field ionization (FI) g.c.m.s. in the analysis of organic mixtures has been reported.[39]

The use of low ionization energy (17 eV) shows some promise for the differentiation of stereoisomeric sugars (as TMS ethers) by g.c.m.s.[40] Evidence of the identity of nuclear structure in two homologous cannabinoids was adduced by

[21] B. F. Maume and J. A. Luyten, *J. Chromatog. Sci.*, 1973, **11**, 607.
[22] T. A. Gough and K. S. Webb, *J. Chromatog.*, 1973, **79**, 57.
[23] M. A. Grayson, R. M. Levy, and C. J. Wolf, *Analyt. Chem.*, 1973, **45**, 373.
[24] E. Mussini, F. de Nadai, R. Fanelli, and A. Frigerio, *Analyt. Biochem.*, 1972, **38**, 635.
[25] S. P. Markey and S. L. Simons, *Analyt. Chem.*, 1973, **45**, 818.
[26] C. F. Simpson, *C.R.C. Crit. Rev. Analyt. Chem.*, 1973, **3**, 1.
[27] F. Bruner, A. di Corcia, G. Goretti, and S. Zelli, *J. Chromatog.*, 1973, **76**, 1.
[28] F. Bruner, P. Ciccioli, and S. Zelli, *Analyt. Chem.*, 1973, **45**, 1002.
[29] Ref. c (Table, p. 297).
[30] M. A. Grayson and J. J. Bellina, jun., *Analyt. Chem.*, 1973, **45**, 487.
[31] R. E. Snyder, *J. Chromatog. Sci.*, 1972, **10**, 487.
[32] W. D. Dencker, D. R. Rushneck, and G. R. Shoemake, *Analyt. Chem.*, 1972, **44**, 1753.
[33] W. Henderson and G. Steel, *Analyt. Chem.*, 1972, **44**, 2302.
[34] A. F. Fentiman, Jr., R. L. Foltz, and G. W. Kinzer, *Analyt. Chem.*, 1973, **45**, 580.
[35] W. Blum and W. J. Richter, *Tetrahedron Letters*, 1973, 835.
[36] G. P. Arsenault, *J. Amer. Chem. Soc.*, 1973, **93**, 8241.
[37] J. Yinon and H. G. Boettger, *Chem. Instr.*, 1972, **4**, 103.
[38] E. C. Horning, M. G. Horning, D. I. Carroll, I. Dzidic, and R. N. Stillwell, *Analyt. Chem.*, 1973, **45**, 936.
[39] D. E. Games, A. H. Jackson, and D. S. Millington, *Tetrahedron Letters*, 1973, 3063.
[40] I. M. Campbell and R. Bentley, *Advan. Chem. Ser.*, 1973, **117**, 1.

obtaining mass spectra at various electron energies during successive g.c. runs and evaluating the relationship between electron energy and ion abundance.[41] G.c.m.s. has been used to determine the specific activity of [^{14}C]-compounds.[42–44] Several systems have been described in which selective ion monitoring (s.i.m.)* during g.c. is aided by a digital computer,[45–49] and two analogue devices have been designed for this purpose.[50,51] Computer-aided assignment of retention index values is an important development.[52] Further details have appeared of a programmed repetitive-scanning system in which spectra are recorded on magnetic tape for processing on a large (IBM 1800) computer. The system is designed for the analysis of steroids in biological materials, and is particularly suitable for the evaluation of large numbers of analyses.[53] Off-line computer matching of mass spectra has been applied to the analyses of a wide range of compounds in body fluids.[54] The techniques of s.i.m. and repetitive scanning during g.c.m.s. have been compared.[55] Applications of s.i.m. to the quantitative analysis of drugs and metabolites are reviewed in Chapter 9.

Significant progress has been made in facilitating the interpretation of the voluminous amount of data generated by g.c.m.s. Although the *Archives of Mass Spectral Data*[56] has ceased publication, a comprehensive 'Registry of Mass Spectral Data' has appeared.[57] The Mass Spectrometry Data Centre collection is now accessible by international on-line computer search facilities.[58] Sub-sets of the latter collection which, in some cases, also contain g.c. data, are limited to

---

[41] T. B. Vree, D. D. Breimer, C. A. M. van Ginneken, and J. M. van Rossum, *J. Chromatog.*, 1972, **74**, 124.
[42] D. H. Bowen, J. MacMillan, and J. E. Graebe, *Phytochemistry*, 1972, **11**, 2253.
[43] W. J. A. VandenHeuvel, J. L. Smith, H. E. Mertel, and R. L. Ellsworth, presented at Twenty-first Annual Conference on Mass Spectrometry and Allied Topics, San Francisco, May 1973.
[44] K. J. Judy, D. A. Schooley, L. L. Dunham, M. S. Hall, B. J. Bergot, and J. B. Siddall, *Proc. Nat. Acad. Sci. U.S.A.*, 1973, **70**, 1509.
[45] K. Elkin, L. Pierrou, U. G. Ahlborg, B. Holmstedt, and J.-E. Lindgren, *J. Chromatog.*, 1973, **81**, 47.
[46] J. F. Holland, C. C. Sweeley, R. E. Thrush, R. E. Teets, and M. A. Bieber, *Analyt. Chem.*, 1973, **45**, 308.
[47] L. Baczynskyj, D. J. Duchamp, J. F. Zieserl, jun., and U. Axen, *Analyt. Chem.*, 1973, **45**, 479.
[48] W. F. Holmes, W. H. Holland, B. L. Shore, D. M. Bier, and W. R. Sherman, *Analyt. Chem.*, 1973, **45**, 2063.
[49] J. T. Watson, D. R. Pelster, B. J. Sweetman, J. C. Frölich, and J. A. Oates, *Analyt. Chem.*, 1973, **45**, 2071.
[50] R. W. Kelly, *J. Chromatog.*, 1972, **71**, 337.
[51] D. J. Jenden and R. W. Silverman, *J. Chromatog. Sci.*, 1973, **11**, 601.
[52] H. Nau and K. Biemann, *Analyt. Chem.*, 1974, **46**, 426.
[53] R. Reimendal and J. Sjövall, *Analyt. Chem.*, 1973, **45**, 1083.
[54] E. Jellum, O. Stokke, and L. Eldjarn, *Analyt. Chem.*, 1973, **45**, 1099.
[55] B. S. Middleditch and D. M. Desiderio, *Analyt. Chem.*, 1973, **45**, 806.
[56] *Archives of Mass Spectral Data*, Interscience, New York, 1970—1972.
[57] 'Registry of Mass Spectral Data', ed. E. Stenhagen, S. Abrahamsson, and F. W. McLafferty, John Wiley and Sons, New York, 1974.
[58] See, for example, *Mass Spectrometry Bull.*, 1974, **8**, v.

* The phrase 'selected ion monitoring' has been proposed as a general term by J. T. Watson, F. C. Falkner, and B. J. Sweetman (*Biomed. Mass Spectrometry*, 1974, **1**, 156).

drugs and metabolites,[59–61] C_{19} and C_{21} steroids,[62] and urinary metabolites.[63,64] These and other data collections have been reviewed.[65] A molecular weight index of the Merck Index[66] has been compiled to facilitate analysis of drugs by mass spectrometry.[67]

Stable Isotopes.—The use of stable isotopes in g.c.m.s. has expanded rapidly especially in pharmacological work.[62] Internal standards for quantitative analysis now frequently comprise isotope-labelled analogues of the compounds under study. It is usually assumed that the properties of such molecules (bearing only a few heavy-isotope atoms) will not distinguish them from the parents during sample extraction and working-up procedures. Moreover, the close similarity of gas chromatographic retention times permits the use of isotopic compounds in large excess as 'carriers' for small quantities of their parents (Vol. 2, p. 311). Applications are cited in this Chapter and Chapter 9.

Derivatives prepared from reagents labelled with stable isotopes are particularly convenient for the verification of fragmentation processes (and may, of course, be combined with labelling of the substrate itself). The selective introduction of TMS and [$^2H_9$]TMS groups into hydroxylic steroids provides an interesting approach.[68]

A novel method to facilitate the recognition of metabolites has been described by several groups of workers. Drugs, or other precursors, are partially labelled, typically with ^{15}N,[69] ^{13}C,[70,71] or ^2H,[71] in such proportions that the $[M]^{+\bullet}$ and $[M + 1]^+$ ions are of equal abundance: thus metabolites are formed with a similar ratio of molecular ion intensities, while fragment ions containing the labelled centre similarly appear as characteristic 'twin ions'. By choosing appropriate combinations of isotopes, other conspicuous 'ion clusters' can be created.[72]

---

[59] 'Final Report on the Rapid Identification of Drugs from Mass Spectra', R. L. Foltz, Battelle Columbus Laboratories, Columbus, Ohio, 1974.
[60] B. S. Finkle, R. L. Foltz, and D. M. Taylor, *J. Chromatog. Sci.*, 1974, **12**, 304.
[61] 'Mass Spectra of Drugs', ed. M.I.T. Mass Spectrometry Laboratory, Amer. Soc. Mass Spectrometry, Cambridge, Massachusetts, 1972.
[62] Ref. *i* (Table, p. 297).
[63] 'Identification of Endogenous Metabolites by Gas Chromatography-Mass Spectrometry: A Collection of Mass Spectral Data', S. P. Markey, H. A. Thobhani, and K. B. Hammond, University of Colorado Medical Center, Denver, 1972.
[64] 'Applications of Gas Chromatography-Mass Spectrometry to the Investigation of Human Disease. The Proceedings of a Workshop,' ed. O. A. Mamer, W. J. Mitchell, and C. R. Scriver, Montreal Children's Hospital, Montreal, 1974.
[65] Ref. *rr* (Table, p. 297).
[66] 'The Merck Index', ed. P. G. Stecher, Merck and Co., Rahway, N.J., 8th Edn., 1968.
[67] Tung Sun, D. J. Pedder, and H. M. Fales, *Analyt. Chem.*, 1973, **45**, 2297.
[68] P. Vouros and D. J. Harvey, *Analyt. Chem.*, 1973, **45**, 7.
[69] M. Vore, N. Gerber, and M. T. Bush, *Pharmacologist*, 1971, **13**, 220.
[70] M. T. Bush, H. J. Sekerke, M. Vore, B. J. Sweetman, and J. T. Watson, in ref. *i* (Table, p. 297), p. 233.
[71] W. E. Braselton, jun., J. C. Orr, and L. L. Engel, *Analyt. Biochem.*, 1973, **53**, 64.
[72] D. R. Knapp, T. E. Gaffney, and R. E. McMahon, *Biochem. Pharmacol.*, 1972, **21**, 425; D. R. Knapp, T. E. Gaffney, R. E. McMahon, and G. Kiplinger, *J. Pharmacol. Exp. Therap.*, 1972, **180**, 784; D. R. Knapp and T. E. Gaffney, in ref. *i* (Table, p. 297), p. 249.

Deuterium labelling is of special value for metabolic studies in human subjects because there is normally no problem of toxicity. The technique is now being widely used; examples include a study of the metabolism of human plasma glycosphingolipids,[73] a series of reports on amino-acid metabolism,[74] and an investigation of the incorporation of deuterioethanol into bile acids.[75] ^{18}O-labelling has been used to determine the biosynthetic origins of oxygen atoms in organic molecules. Examples include hydroperoxides of linoleic acid,[76] prostaglandins,[77] and chlorosulpholipids;[78] others are mentioned later.

The analysis of stable isotopes with a quadrupole mass spectrometer has been reviewed.[79]

Functional Derivatives.—Apart from isotope labelling, there have been few developments in the range of derivatives or in their mode of application. Silyl ethers continue to show their versatility. An elegant new method of stabilizing corticosteroids for g.c.m.s. involves base-catalysed enolization of the 20-oxo-group and formation of a 17,20,21-tri-TMS derivative (see p. 317).[80] Selective displacement of TMS groups by heptafluorobutyryl groups has afforded 'mixed' derivatives of zooecdysones, useful for analytical determinations.[81] Intramolecular rearrangements and intermolecular reactions of TMS groups have been studied in steroid derivatives under electron impact and chemical ionization conditions.[82] An important application of trimethylchlorosilane has been established in its reaction with labile epoxides to yield chlorohydrin TMS ethers directly amenable to g.c.m.s.[83]

Other silyl ethers have proved to be of value, for example in s.i.m. when it is necessary to produce an ion of an m/e value at which background ion interference is low. Bromomethyl(dimethyl)silyl ethers were applied in this manner to the determination of plasma testosterone: the abundant molecular ions (^{79}Br, 438; ^{81}Br, 440) were in a clear region of the spectrum, and the isotope pattern provided a check on the absence of extraneous contributions.[84] Chloromethyl(dimethyl)-silyl ethers were used in analysis of metabolites of 4,6-androstadien-3-one.[85] An interesting difference has been reported in the major fragmentations of di- and tri-methylsilyl ethers of primary aliphatic alcohols [(1) and (2)].[86] The former

---

[73] D. E. Vance and C. C. Sweeley, in ref. *i* (Table, p. 297), p. 141.
[74] H.-C. Curtius, J. A. Völlmin, and K. Baerlocher, *Analyt. Chem.*, 1973, **45**, 1107, and references there cited.
[75] D. M. Wilson, A. L. Burlingame, T. Cronholm, and J. Sjövall, *Biochem. Biophys. Res. Comm.*, 1974, **56**, 828.
[76] W. J. Esselman and C. O. Clagett, *J. Lipid Res.*, 1974, **15**, 173.
[77] P. Foss, C. Takeguchi, H. Tai, and C. Sih, *Ann. New York Acad. Sci.*, 1971, **180**, 126.
[78] J. Elovson, *Biochemistry*, 1974, **13**, 2105.
[79] R. M. Caprioli, W. F. Fies, and M. S. Story, *Analyt. Chem.*, 1974, **46**, 453A.
[80] E. M. Chambaz, G. Defaye, and C. Madani, *Analyt. Chem.*, 1973, **45**, 1090.
[81] H. Miyazaki, M. Ishibashi, C. Mori, and N. Ikekawa, *Analyt. Chem.*, 1973, **45**, 1164.
[82] P. Vouros, D. J. Harvey, and T. J. Odiorne, *Spectroscopy Letters*, 1973, **6**, 603.
[83] D. J. Harvey, D. B. Johnson, and M. G. Horning, *Analyt. Letters*, 1972, **5**, 745.
[84] J. R. Chapman and E. Bailey, *J. Chromatog.*, 1974, **89**, 215.
[85] B. W. L. Brooksbank, D. A. Wilson, and J.-Å. Gustafsson, *Steroids and Lipids Res.*, 1972, **3**, 263.
[86] W. J. Richter and D. H. Hunneman, *Helv. Chim. Acta*, 1974, **57**, 1131.

undergo more pronounced cleavage of the C–C bond adjoining the ether oxygen atom.

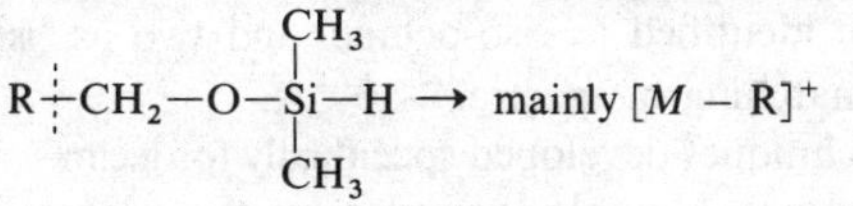

$$R\!\!\stackrel{|}{\div}\!\!CH_2\!-\!O\!-\!\underset{\underset{CH_3}{|}}{\overset{\overset{CH_3}{|}}{Si}}\!-\!H \rightarrow \text{mainly } [M-R]^+$$

(1)

$$R\!-\!CH_2\!-\!O\!-\!\underset{\underset{CH_3}{|}}{\overset{\overset{CH_3}{|}}{Si}}\!\div\!CH_3 \rightarrow \text{mainly } [M-CH_3]^+$$

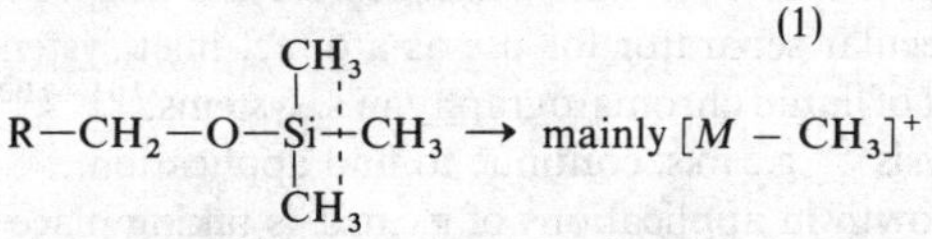

(2)

Attempts to simplify the analytical preparation of derivatives are represented by 'on-column' methylation of barbiturates,[87] fatty acids,[88] and amino-acids,[89] based on trimethylammonium or trimethylanilinium hydroxide. A simple device for preparing diazomethane without distillation has been described.[90]

Methoxymercuration–demercuration has proved to be of general value for the location of olefinic bonds.[91] In monoenoic acids, conversion into pyrrolidides is advantageous in yielding simple mass spectra indicative of the double bond position.[92] Cyclopropane rings may be located by methoxylation.[93]

Artifacts have been observed during the fluoroacylation and g.c.m.s. of biogenic amines.[94] The 19-norsteroid, norethynodrel, underwent partial aromatization during trimethylsilylation.[95]

Current Trends.—A notable trend is towards reduction of sample size. For example, comparison of the aroma concentrates from 10 g samples of 'radurized' and normal strawberries revealed the presence of eight new components,[96] whereas previous investigations of strawberry aroma by g.c.m.s. employed 2.5×10^4 g and 9×10^6 g of fruit.[97,98] Low-level analysis will also be aided by the more widespread availability (and lower cost) of isotopically labelled compounds for use as 'carriers' and/or internal standards.

[87] E. B. Solow, J. M. Metaxas, and T. R. Summers, *J. Chromatog. Sci.*, 1974, **12**, 256, and references there cited.
[88] B. S. Middleditch and D. M. Desiderio, *Analyt. Letters*, 1972, **5**, 605.
[89] K. M. Williams and B. Halpern, *Analyt. Letters*, 1973, **6**, 839.
[90] H. M. Fales, T. M. Jaouni, and J. F. Babashak, *Analyt. Chem.*, 1973, **45**, 2302.
[91] D. E. Minnikin, P. Abley, F. J. McQuillin, K. Kusamran, K. Maskens, and N. Polgar, *Lipids*, 1974, **9**, 135.
[92] B. Å. Andersson and R. T. Holman, *Lipids*, 1974, **9**, 185.
[93] D. E. Minnikin, *Lipids*, 1972, **7**, 398.
[94] G. S. King and M. Sandler, *Clinica Chim. Acta*, 1973, **49**, 295.
[95] R. M. Thompson and E. C. Horning, *Steroids and Lipids Res.*, 1973, **4**, 135.
[96] J. Schubert, E. B. Saunders, S. F. Pan, and N. Wald, *J. Agric. Food Chem.*, 1973, **21**, 684.
[97] R. Tressl, F. Drawert, and W. Heimann, *Z. Naturforsch.*, 1969, **24b**, 1201.
[98] W. H. McFadden, R. Teranishi, J. Corse, D. R. Black, and T. R. Mon, *J. Chromatog.*, 1965, **18**, 10.

Automation of data handling is leading to more widespread routine use of g.c.m.s. in many laboratories, but it does not alter the need for a critical approach to the results: in a recent paper three widely-separated discrete peaks on a gas chromatogram were identified as 'iso-octane' and two as 'iso-nonane' on the basis of computer-matching of spectra.[99]

Apparatus and techniques developed specifically for g.c.m.s. are being adapted for more widespread use. Notable examples are the adaptation of the porous membrane molecular separator for use as a direct inlet system for m.s.,[100] and the development of liquid chromatography-m.s. systems.[101—103] Pyrolysis[104,105] and laser pyrolysis[106] g.c.m.s. continue to find application.

The major growth in applications of g.c.m.s. is taking place in the areas conveniently designated by the term 'biomedical', and ranging from natural metabolism to pharmacology and pollution. The analytical power of the technique is such that certain of its important but less-developed applications, *e.g.* in studies of plant biochemistry, and in forensic toxicology, will receive increased attention. In relation to pharmacology and enzymology especially the value of g.c.m.s. to determine stereochemical selectivity is one of its qualities that merits further development.

2 Applications

Hydrocarbons.—An improved method has been devised for the collection and analysis of volatile hydrocarbons in air: the organic materials are adsorbed on Chromosorb 102, and a timed elution procedure is used for sampling into the g.c.m.s. system. An on-line computer initiates repetitive scanning, and the data may be quantitatively evaluated after calibration with reference compounds.[107] At the other extreme, a comprehensive study of the analysis of polycyclic hydrocarbons in air samples, using both packed and capillary columns (together with computer-aided data processing) has identified more than 70 such components.[108] A detailed analysis of a pyrolysis naphtha (containing C_2—C_{10} hydrocarbons), based on high-resolution gas chromatography combined with

---

[99] K. L. E. Kaiser and P. T. S. Wong, *Bull. Environ. Contam. Toxicol.*, 1974, **11**, 291.
[100] R. H. Hertel, *The U.T.I. Journal (J. Quadrupole Mass Spectrometry)*, 1974, **1**(5), 1; D. E. Green and I. S. Forrest, presented at Fifth International Congress on Pharmacology, San Francisco, July, 1972.
[101] D. I. Carroll, I. Dzidic, R. N. Stillwell, M. G. Horning, and E. C. Horning, presented at Twenty-second Annual Conference on Mass Spectrometry and Allied Topics, Philadelphia, May, 1974.
[102] R. E. Lovins, S. R. Ellis, G. D. Tolbert, and C. R. McKinney, *Analyt. Chem.*, 1973, **45**, 1553; ref. *e* (Table, p. 297), p. 457.
[103] P. Arpino, M. A. Baldwin, and F. W. McLafferty, *Biomed. Mass Spectrometry*, 1974, **1**, 80.
[104] W. G. Fischer, *Glas.-Instr.-Tech.*, 1972, **16**, 37.
[105] E. B. Higman, H. C. Higman, O. T. Chortyk, and I. Schmeltz, *J. Agric. Food Chem.*, 1973, **21**, 202.
[106] E. G. Perkins and J. C. Means, presented at Amer. Oil Chemists' Soc. Sixty-fourth Annual Spring Meeting, New Orleans, 1973.
[107] R. Perry and J. D. Twibell, *Biomed. Mass Spectrometry*, 1974, **1**, 73.
[108] R. C. Lao, R. S. Thomas, H. Oja, and L. Dubois, *Analyt. Chem.*, 1973, **45**, 908.

'mass chromatography', led to the identification of most of the 152 separated peaks.[109] Capillary g.c.m.s. has also been applied to petroleum reformates (mixtures of hydrocarbons in the C_{10}—C_{14} range).[110]

The biosynthesis of alkanes in the leaves of *Allium porrum* takes place in both parenchymal and epidermal tissue: a thorough study has been made of the incorporation of labelled fatty acids into epidermal alkanes and long-chain acids.[111] Alkanes of lichens have been examined for their value in taxonomy, which remains dubious.[112] The occurrence and distribution of alkanes in fungi has been reviewed.[113]

Alkanes (n- and branched) and alkenes of milk fat have been characterized, and phytene has been isolated from this source.[114] Volatiles emanating from marijuana,[115] and from the pine rust fungus *Cronartium fusiforme*[116] consist mainly of monoterpenes. Twelve isomeric sesquiterpenes ($C_{15}H_{24}$) of diverse types, together with cuparene ($C_{15}H_{22}$) have been characterized, by capillary g.c.m.s., in the liverwort *Scapania parvitexta*.[117] In a similar analysis of spike oil, the principal components (mainly mono- and sesqui-terpenes) were identified.[118]

The catabolism of 1-phenylnonane and 1-phenyldodecane by *Nocardia salmonicolor* has been investigated.[119]

Long-chain Compounds.—Benzyl esters, which now may be conveniently prepared using phenyldiazomethane, have been characterized by g.c.m.s.[120] They would appear to be most useful for the lower fatty acids. Improvements have been reported in the identification by g.c.m.s. of chain branching near the methyl end of long-chain compounds. Conversion of fatty acids and long-chain bases into the corresponding alcohols, and then into the methyl ethers afforded better separations on packed columns and more conclusive evidence of branching points than could be achieved with other derivatives.[121] Capillary g.c.m.s. has been applied to the separation and identification of α-branched acids (as methyl esters; C_{11}—C_{17}, and C_{19}) and β-branched primary alcohols (C_6—C_{10}).[122]

The well-established method of locating double bonds through hydroxylation and g.c.m.s. of the diol TMS ethers has been applied to a model mixture of nonenes, using capillary g.l.c. with both electron impact and chemical ionization

---

[109] E. J. Gallegos, I. M. Whittemore, and R. F. Klaver, *Analyt. Chem.*, 1974, **46**, 157.
[110] J. T. Swansiger and F. E. Dickson, *Analyt. Chem.*, 1973, **45**, 811.
[111] C. Cassagne, *Qual. Plant. Mater. Veg.*, 1972, **21**, 257.
[112] S. J. Gaskell, G. Eglinton, and T. Bruun, *Phytochemistry*, 1973, **12**, 1174.
[113] J. D. Weete, *Phytochemistry*, 1972, **11**, 1201.
[114] V. P. Flanagan and A. Ferretti, *J. Lipid Res.*, 1973, **4**, 306.
[115] L. V. S. Hood, M. E. Dames, and G. T. Barry, *Nature*, 1973, **242**, 402.
[116] J. L. Laseter, J. D. Weete, and C. H. Walkinshaw, *Phytochemistry*, 1973, **12**, 387.
[117] A. Matsuo, M. Nakayama, and S. Hayashi, *Bull. Chem. Soc. Japan*, 1973, **46**, 1010.
[118] V. Kubelka, J. Mitera, and P. Zachař, *J. Chromatog.*, 1972, **74**, 195.
[119] F. S. Sariaslani, D. B. Harper, and I. J. Higgins, *Biochem. J.*, 1974, **140**, 31.
[120] U. Hintze, H. Röper, and G. Gercken, *J. Chromatog.*, 1973, **87**, 481.
[121] K. A. Karlsson, B. E. Samuelsson, and G. O. Steen, *Chem. and Phys. Lipids*, 1973, **11**, 17.
[122] N. S. Nikitina, N. I. Vikhrestyuk, and A. E. Mysak, *J. Chromatog.*, 1974, **91**, 775.

(CI) m.s. The CI spectra gave clear ions at m/e $[M + 1]^+$ and $[M - 1]^+$, and base peaks at m/e $[M - 15]^+$.[35] The positions of oxide groups in unsaturated epoxy-acid methyl esters may be located by cleavage with BF_3–methanol and g.c.m.s. of the mixed methoxyhydrins (best as TMS ethers).[123] The identification of epoxy-acids in plant cutins is based on similar derivatives; the alkoxyhydrins may be isolated by direct treatment of cutin with alcoholic sodium alkoxide, under strictly anhydrous conditions to avoid the formation of free acids.[124] Many new cutin acids have been identified by g.c.m.s. Among these are 9,10,12,13,18-pentahydroxyoctadecanoic acid from leaves of *Rosmarinus officinalis*,[125] 9,16-dihydroxyhexadecanoic acid (notably in gymnosperms)[126] and 10-hydroxy-16-oxohexadecanoic acid from the cutin of embryonic *Vicia faba*.[127] The biosynthesis of cutin acids has been studied in the latter plant[128] and in skin slices of young apple fruits,[129] using a variety of [1-^{14}C]-labelled acids. The results are consistent with a sequence of ω-hydroxylation, epoxidation, and hydration, with oleic and linoleic acid as principal substrates. Hydroperoxy-acids formed by the action of lipoxygenases,[130,131] and hydroxy-oxo-acids resulting from subsequent enzyme-catalysed isomerization,[76] have been characterized after reduction. The diagnostic value of TMS ethers is well exemplified by the assignment of structures to further groups of mycolic acids, those of *Nocardia erythropolis*[132] and *Corynebacterium ulcerans*.[133] Methoxy-esters obtained by methoxymercuration-demercuration were used to identify *cis*-5-enoic fatty acids in *Chenopodium* seed oils.[134]

Deuteriated substrates have been used in biosynthetic studies of fatty acids. The organism *Acholeplasma laidlawii* B(PG 9), grown on media supplemented with perdeuterio-lauric acid, produced myristic and palmitic acids containing (as shown by g.c.m.s.) the intact deuteriated lauric acid unit.[135] Similar evidence of chain elongation in fatty acid biosynthesis by alkane-using micro-organisms has been obtained by using deuteriated alkanes as substrates: [^{2}H$_{22}$]-n-decane was incorporated into longer-chain acids by a *Pseudomonas*.[136]

The fatty acid compositions of tissue cultures of *Pinus elliottii* have been shown to be similar to those of intact seedlings.[137]

[123] R. Kleiman and G. F. Spencer, *J. Amer. Oil Chemists' Soc.*, 1973, **50**, 31.
[124] P. J. Holloway and A. H. B. Deas, *Phytochemistry*, 1973, **12**, 1721.
[125] C. H. Brieskorn and L. Kabelitz, *Phytochemistry*, 1971, **10**, 3195.
[126] D. H. Hunneman and G. Eglinton, *Phytochemistry*, 1972, **11**, 1989.
[127] P. E. Kolattukudy, *Biochem. Biophys. Res. Comm.*, 1972, **49**, 1040.
[128] P. E. Kolattukudy, *Biochemistry*, 1974, **13**, 1354.
[129] P. E. Kolattukudy, T. J. Walton, and R. P. S. Kushwaha, *Biochemistry*, 1973, **12**, 4488.
[130] J. P. Christopher, E. K. Pistorius, F. E. Regnier, and B. Axelrod, *Biochim. Biophys. Acta*, 1972, **289**, 82.
[131] H. W. Gardner, D. D. Christianson, and R. Kleiman, *Lipids*, 1973, **8**, 271.
[132] I. Yano, K. Saito, Y. Furukawa, and M. Kusunose, *F.E.B.S. Letters*, 1972, **21**, 215.
[133] I. Yano and K. Saito, *F.E.B.S. Letters*, 1972, **23**, 352.
[134] R. Kleiman, M. H. Rawls, and F. R. Earle, *Lipids*, 1972, **7**, 494.
[135] E. Oldfield, *J.C.S. Chem. Comm.*, 1972, 719.
[136] C. W. Bird and Y. C. Yeong, *Chem. and Ind.*, 1974, 459.
[137] J. L. Laseter, G. C. Lawler, C. H. Walkinshaw, and J. D. Weete, *Phytochemistry*, 1973, **12**, 817.

Analyses of tissue lipids *post mortem* from a case of methylmalonic aciduria indicated the presence of appreciable proportions (0.2—0.9%) of isomeric methylhexadecanoic acids, together with abnormally high concentrations of C_{15}- and C_{17}-alkanoic acids. The g.c.m.s. data provided evidence for the presence of the 2-, 6-, 10-, 12- and 14-methylhexadecanoic acids, which are among those expected from the incorporation of methyl malonyl CoA.[138] Branched acids and alcohols from uropygial gland lipids have also been identified.[139]

Natural triglyceride mixtures from various sources have been analysed directly by g.c.m.s. at column temperatures up to 330 °C.[140] Ovolecithin was studied *via* acetolysis of the phosphorylcholine group and g.c.m.s. of the acetyldiglycerides.[141] Major fragment ions from triglycerides (and esters of other polyols) correspond to the loss of an acyloxy radical from the molecular ion: convincing evidence indicates cyclic acetal structures for these ions.[142] Alkanediols obtained as minor products from natural triglycerides (by methanolysis and acid hydrolysis) have been examined,[143] and diol choline phosphatides have been found in rat liver.[144]

Base-catalysed cyclization of linolenic acid afforded complex mixtures of ω-(2-alkyl-3,5-cyclohexadienyl)alkanoic acids and the analogous aromatic acids; even with preliminary chromatography (using silver nitrate–silicic acid) the g.c.m.s. analyses reported showed poor resolution.[145]

(+)-S-Phenylpropionates have been applied to the analytical resolution of enantiomeric methyl hydroxy-octadecanoates by g.l.c. Good separations were obtained for the 3-, 15-, 16- and 17-hydroxy-ester derivatives.[146]

12- and 13-Methyltetradecan-1-ol have been found free, and in the ester form (with 2-methylbutyric acid or isovaleric acid), in the subauricular scent glands of the male pronghorn (*Antilocapra americana*); the pheromonal action appeared to reside chiefly in the isovaleric acid.[147]

Prostaglandins.—Exemplary refinement in the application of g.c.m.s. is to be found in the work of the Karolinska group led by Samuelsson and Hamberg. In two important papers[148,149] these authors describe the detection, isolation and identification of the endoperoxides (3) and (4), formed during prostaglandin

---

[138] Y. Kishimoto, M. Williams, H. W. Moser, C. Hignite, and K. Biemann, *J. Lipid Res.*, 1973, **14**, 69.
[139] A. Zeman and J. Jacob, *Fette, Seifen, Anstrichm.*, 1973, **75**, 667.
[140] T. Murata and S. Takahashi, *Analyt. Chem.*, 1973, **45**, 1816.
[141] K. Hasegawa and T. Suzuki, *Lipids*, 1973, **8**, 631.
[142] W. J. Baumann, A. J. Aasen, J. K. G. Kramer, and R. T. Holman, *J. Org. Chem.*, 1973, **38**, 3767.
[143] L. D. Bergelson, *Fette, Seifen, Anstrichm.*, 1973, **75**, 89.
[144] L. D. Bergelson, V. A. Vaver, N. V. Prokazova, A. N. Ushakov, B. V. Rozynov, K. Stefanov, L. I. Ilukhina, and T. N. Simonova, *Biochim. Biophys. Acta*, 1972, **260**, 571.
[145] E. G. Perkins and W. T. Iwaoka, *J. Amer. Oil Chemists' Soc.*, 1973, **50**, 44.
[146] S. Hammarström and M. Hamberg, *Analyt. Biochem.*, 1973, **52**, 169.
[147] D. Müller-Schwarze, C. Müller-Schwarze, A. G. Singer, and R. M. Silverstein, *Science*, 1974, **183**, 860.
[148] M. Hamberg and B. Samuelsson, *Proc. Nat. Acad. Sci. U.S.A.*, 1973, **70**, 899.
[149] M. Hamberg, J. Svensson, T. Wakabayashi, and B. Samuelsson, *Proc. Nat. Acad. Sci. U.S.A.*, 1974, **71**, 345.

biosynthesis. The intermediacy of an endoperoxide intermediate had been inferred earlier from the mode of incorporation of oxygen into prostaglandin E_1: the oxygens of the 9-oxo and 11-hydroxy groups were shown both to originate from the same oxygen molecule. In the recent work, [1-^{14}C]-arachidonic acid

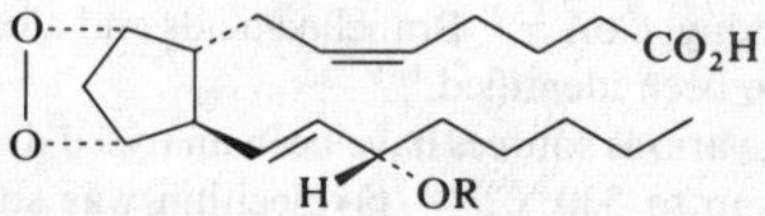

(3) Prostaglandin G_2: R = OH
(4) Prostaglandin H_2: R = H

was incubated with the microsomal fraction of sheep vesicular gland homogenates, and the presence of a labile oxygenated intermediate was shown by differential micro-analysis of the incubation mixtures (*e.g.* reduction with NaB^2H_4 to distinguish the endoperoxide from isomeric ketols). Isolation of the endoperoxide by t.l.c. led to its identification by reduction (mainly to prostaglandin $F_{2\alpha}$), and by rearrangement to a mixture of prostaglandin E_2 and 11-dehydroprostaglandin $F_{2\alpha}$. The second endoperoxide was isolated from incubations carried out in the presence of *p*-mercuribenzoate, an inhibitor of the rearrangement of the endoperoxides to ketols of the E_2 type. Both endoperoxides show high biological potency in respect of smooth muscle stimulation and platelet aggregation.

Further studies of the metabolism of prostaglandins have been made.[150—153] A new metabolite of (administered) prostaglandin $F_{2\alpha}$ in man, also detected as an endogenous metabolite, is the 15-deoxy compound, $7\alpha,9\alpha$-dihydroxy(dinor, tetranor)prost-3-ene-1,14-dioic acid (5).[150] In the metabolism of prostaglandin $F_{3\alpha}$ (6) in the rat, the Δ^{17} double bond remained intact, and in most of the urinary

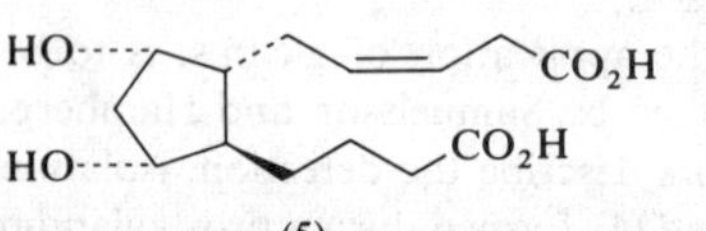

(5) 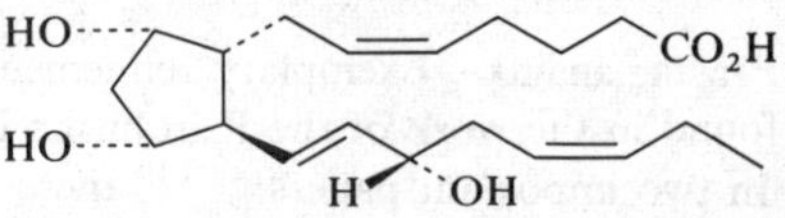   (6)

metabolites the 15-hydroxy-group remained unoxidized; the pattern of transformation was thus markedly distinct from that of prostaglandin $F_{2\alpha}$.[152] Six monocarboxylic acids were identified by g.c.m.s. in the urine of guinea-pigs treated with prostaglandin $F_{1\alpha}$.[153]

Some quantitative determinations of prostaglandins[154—157] have been based on the use of deuteriated derivatives as internal standards (Vol. 2, p. 311): the

[150] E. Granström and B. Samuelsson, *J. Amer. Chem. Soc.*, 1972, **94**, 4380.
[151] M. Hamberg, *Analyt. Biochem.*, 1973, **55**, 368.
[152] V. Dimov and K. Gréen, *Biochim. Biophys. Acta*, 1973, **306**, 257.
[153] H. Kindahl and E. Granström, *Biochim. Biophys. Acta*, 1972, **280**, 466.
[154] M. Hamberg, *Biochem. Biophys. Res. Comm.*, 1972, **49**, 720.
[155] U. Axen, L. Baczynskyj, D. J. Duchamp, and J. F. Zieserl, *J. Reprod. Med.*, 1972, **9**, 372.
[156] R. W. Kelly, *Analyt. Chem.*, 1973, **45**, 2079.
[157] K. Gréen, E. Granström, B. Samuelsson, and U. Axen, *Analyt. Biochem.*, 1973, **54**, 434.

excretion rate of the major metabolite of prostaglandins E_1 and E_2 in man has been shown to be greatly suppressed by the oral administration of salicylates.[154] Investigations of TMS and O-alkyloxime TMS derivatives of prostaglandins[158] and of their methyl esters[159] have provided useful qualitative and quantitative data. Prostaglandins of types A and B are distinguishable by the mass spectra of their derivatives e.g. by the preponderance in the B series of ions at $[M - 99]^+$ resulting from loss of $C_5H_{11}^{\cdot}$ and CO.[158,159] The TMS ether methyl esters[160] of prostaglandins A and B are detectable at picogram levels by s.i.m., and these derivatives may be applicable to the indirect estimation of prostaglandins E in human peripheral blood. Heptafluorobutyrate methyl esters are detectable at the nanogram level (as reference compounds),[161] but the use of these derivatives for the estimation of prostaglandins F in seminal plasma by electron-capture g.c.[162] seems inferior to methods based on g.c.m.s.

Sphingosine Derivatives.—The recent review by Viswanathan[163] gives a useful outline of some applications of g.c.m.s. in this field. Several papers deal with constituents of sphingolipids. Mass chromatography has been applied to identify TMS derivatives of 2-hydroxy-fatty acid methyl esters in mixtures obtained by acidic methanolysis of cat brain galactocerebrosides. Computer-generated chromatograms of m/e 73 and $[M - 59]^+$ located the hydroxy-acid derivatives, while m/e 74 and $M^{\pm}$ indicated fatty acid methyl esters.[164] Methanolysis of a yeast cerebrin phosphate afforded new natural 2,3-dihydroxy-alkanoic acids (C_{24}—C_{27}): the major acid was fully characterized as 2,3-*erythro*-dihydroxyhexacosanoic acid by g.c.m.s. of derivatives [e.g. (7) which gave the three ions shown in Scheme 1 in high relative abundance] and by synthesis.[165] Methanolysis products of cerebrosides of the marine sponge *Chondrilla nucula* have

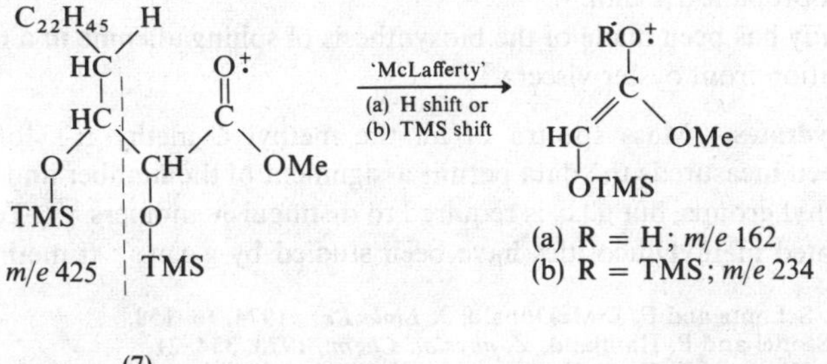

Scheme 1

[158] B. S. Middleditch and D. M. Desiderio, *Prostaglandins*, 1973, **4**, 31; *Lipids*, 1973, **8**, 267; *J. Org. Chem.*, 1973, **38**, 2204.
[159] J. T. Watson and B. J. Sweetman, *Org. Mass Spectrometry*, 1974, **9**, 39.
[160] B. J. Sweetman, J. C. Frölich, and J. T. Watson, *Prostaglandins*, 1973, **3**, 75.
[161] B. S. Middleditch and D. M. Desiderio, *Prostaglandins*, 1972, **2**, 195.
[162] M. Sugiura and K. Hirano, *J. Chromatog.*, 1974, **90**, 169.
[163] Ref. x (Table, p. 297).
[164] R. A. Laine, N. D. Young, J. N. Gerber, and C. C. Sweeley, *Biomed. Mass Spectrometry*, 1974, **1**, 10.
[165] M. Hoshi, Y. Kishimoto, and C. Hignite, *J. Lipid Res.*, 1973, **14**, 406.

been identified.[166] An improved procedure has been devised[167] for the complete methylation (using Corey's reagent) of hydroxy and acetylamino groups in glycosphingolipids that contain aminosugars. Acid hydrolysis of the methylated sphingolipids affords methylated alditol acetates and N-methylacetylamino-alditol acetates suitable for g.c.m.s.

Selective demethylation of phosphorylcholine liberated by phospholipase hydrolysis has been effected with sodium benzenethiolate in dimethylformamide:[168] the resulting O-phosphoryl-NN-dimethylethanolamine is converted to its TMS derivative for g.c.m.s. The analogous 2-(dimethylamino)ethylphosphonic acid diTMS ester is also suitable for g.c.m.s. The authors applied their procedure to choline-containing lipids of human atherosclerotic aortas; the presence of phosphonates, suggested by earlier workers, was not detected[168] (TMS derivatives of 1-aminoalkylphosphonates are not directly amenable to g.c., but conversion of the amino group to isothiocyanate renders them stable; these derivatives are also suitable for characterization of 2- and 3-aminoalkylphosphonates by g.c.m.s.[169]). Detailed analyses have been made of sphingolipids from bovine kidney, cortex, medulla and papilla.[170]

In mass spectrometric studies of acetyl derivatives of homogeneous monoglycosyl ceramides, molecular ions (m/e 713—1055) were observed, and fragment ions from the acetylated sugar were dominant.[171] Structures of free ceramides of human platelets were established by g.c.m.s. of the TMS derivatives.[172] Structures have also been determined of ceramide aminoethylphosphonates from a sea anemone[173] and from oyster adductor.[174] In the latter example, the compound was analysed directly by g.c.m.s. of its TMS derivative, and additional data were obtained for the free ceramide prepared by enzymic hydrolysis. An interesting new bacterial lipid has been shown to be a 1-(N-acylsphingophosphoryl) ester of 3-aminopropane-1,2-diol.[175]

A study has been made of the biosynthesis of sphingadienine in a microsomal preparation from oyster viscera.[176]

Carbohydrates.—Mass spectra of all the methyl O-methyl-D-xylofuranosides have been measured; the data permit assignment of the number and location of the methyl groups, but g.l.c. is required to distinguish anomers.[177] Ten partially methylated methylglucosides have been studied by g.c.m.s.: trimethylsilylation

[166] F. J. Schmitz and F. J. McDonald, *J. Lipid Res.*, 1974, **15**, 158.
[167] W. Stoffel and P. Hanfland, *Z. physiol. Chem.*, 1973, **354**, 21.
[168] S.-G. Karlander, K.-A. Karlsson, and I. Pascher, *Biochim. Biophys. Acta*, 1973, **326**, 173.
[169] D. J. Harvey and M. G. Horning, *J. Chromatog.*, 1973, **79**, 65; *Org. Mass Spectrometry*, 1974, **9**, 111.
[170] K. S. Karlsson, B. E. Samuelsson, and G. O. Steen, *Biochim. Biophys. Acta*, 1973, **316**, 317, 336.
[171] B. Å. Andersson, K. A. Karlsson, I. Pascher, B. E. Samuelsson, and G. O. Steen, *Chem. and Phys. Lipids*, 1972, **9**, 89.
[172] W. Krivit and S. Hammarström, *J. Lipid Res.*, 1972, **13**, 525.
[173] K. A. Karlsson and B. E. Samuelsson, *Biochim. Biophys. Acta*, 1974, **337**, 204.
[174] T. Matsubara and A. Hayashi, *Biochim. Biophys. Acta* 1973, **296**, 171.
[175] P. Kemp, R. M. C. Dawson, and R. A. Klein, *Biochem. J.*, 1972, **130**, 221.
[176] R. K. Hammond and C. C. Sweeley, *J. Biol. Chem.*, 1973, **248**, 632.
[177] V. Kováčik and P. Kováč, *Carbohydrate Res.*, 1972, **24**, 23.

of the free hydroxyl groups improved the distinctions between isomers.[178] The TMS ethers of 21 monoacetates and 10 diacetates of methyl D-hexopyranosides are well separated, and the positions of the acetyl groups determined, by g.c.m.s.[179] Laine and Sweeley[180] have reported further investigations of O-methyloxime TMS ethers, which are effective derivatives for structural diagnosis. A procedure for the analysis of many aldonic and some uronic acids, concurrently with common neutral sugars, has been devised by Petersson:[181] acids and lactones are converted into their sodium salts, oximation (or O-methyloximation) is effected, and the products are trimethylsilylated. The resulting acyclic derivatives are well distinguished by g.c.m.s., the occurrence of *syn* and *anti* oximes being only a minor complication. The ease of interpretation is exemplified for the isomeric derivatives (8) and (9) of glucuronic and *lyxo*-5-hexulosonic acid (M, 641) (Scheme 2).

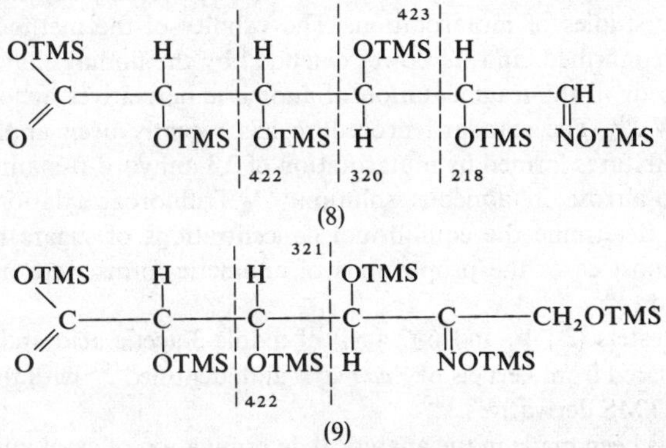

```
              |       |   423 |
   OTMS    H  | H     | OTMS  | H
     \     |  |  |    |  |    |  |
      C————C——|—C——————|—C————|—C————CH
     //    |  |  |    |  |    |  ||
    O     OTMS|OTMS| H    |OTMS  NOTMS
              |422 |320      |218
                      (8)

                   | 321|
   OTMS    H  | H  | OTMS
     \     |  |  | |  |
      C————C——|—C—|—C————C————CH₂OTMS
     //    |  |  | |  ||
    O     OTMS|OTMS|H  NOTMS
              |422
                     (9)
```

Scheme 2

Other derivatives investigated by g.c.m.s. have included trifluoroacetates,[182] aldonitrile acetates (of partially methylated xylo- and arabino-pyranoses),[183] two permethylated pseudoaldobiouronic acids,[184] and permethylated aldosylaldonic acids.[185] The latter acids were obtained by oxidation of disaccharides with buffered bromine water, and afforded improved discrimination between (1 → 3), (1 → 4), and (1 → 6) linked sugars. Sugar phosphates have been studied by g.c.m.s. of TMS and O-methyloxime TMS derivatives.[186]

Products from the formose reaction have been analysed by capillary g.c.m.s. principally as alditol derivatives.[187] Structural elucidations have been reported

[178] T. Matsubara and A. Hayashi, *Biomed. Mass Spectrometry*, 1974, **1**, 62.
[179] H. B. Borén, P. J. Garegg, L. Kenne, Å. Pilotti, S. Svensson, and C.-G. Swahn, *Acta Chem. Scand.*, 1973, **27**, 3557.
[180] R. A. Laine and C. C. Sweeley, *Carbohydrate Res.*, 1973, **27**, 199.
[181] G. Petersson, *Carbohydrate Res.*, 1974, **33**, 47.
[182] W. A. Koenig, H. Bauer, W. Voelter, and E. Bayer, *Chem. Ber.*, 1973, **106**, 1905.
[183] U. N. El'Kin, B. V. Rozynov, and A. K. Dzizenko, *Khim. Prirod. Soedinenii*, 1972, 642.
[184] C. C. Kuenzle, *Carbohydrate Res.*, 1972, **24**, 169.
[185] J. N. C. Whyte, *Canad. J. Chem.*, 1973, **51**, 3197.
[186] D. J. Harvey and M. G. Horning, *J. Chromatog.*, 1973, **76**, 51.
[187] R. D. Partridge, A. M. Weiss, and D. Todd, *Carbohydrate Res.*, 1972, **24**, 29.

from a thermophilic bacterium.[189] Alditol acetates have been applied, in conjunction with methylation, to characterize *Klebsiella* lipopolysaccharides[190] and lemon gum:[191] in the latter study the authors point out that butanediol succinate is as satisfactory a phase as ECNSS-M for packed column separations.

Urinary sugar alcohols have been analysed (after the removal of aldoses and ketoses as hydrazones) as their trifluoroacetates: fucitol was among nine alditols identified.[192] Glucuronides isolated from urine have been studied as TMS derivatives,[193] and oligosaccharides from the urine of patients with mannosidosis have been characterized.[194] Solutions of fructose for intravenous injection, when improperly controlled, may contain considerable amounts of furan derivatives resulting from decomposition: these are detectable in the urine of patients receiving the injections.[195]

A critical examination has been made of the trimethylsilylation–g.c.m.s. procedure for studies of mutarotation. The validity of the method has been convincingly confirmed, and its power extended by the finding that significant differences occur in the fragmentation of anomeric derivatives at low electron voltage (17 eV).[40] The standard procedure has recently been applied to the equilibrium mixtures formed by mutarotation of 2,3-anhydro-D-mannose and of 3,4-anhydro-D-altrose in aqueous solution.[196] Trifluoroacetylation has also been used to determine the equilibrium concentrations of sugars in aqueous solution; in most cases the proportions of anomeric forms were not affected by the acylation.[182]

Three new esters (2-, 4-, and 6-O-acyl) of indole-3-acetic acid and D-glucose have been isolated from kernels of *Zea mays*, and identified[197] with the aid of O-methyloxime TMS derivatives.[180]

Progress has been made in the analytical determination of cytokinins, as their TMS derivatives.[198] *trans*-Zeatin riboside has been identified in sycamore sap,[199] and a new cytokinin from *Populus robusta* was initially characterized by g.c.m.s. as a ribosylpurine (M^{+} 661), then fully identified by high-resolution m.s. and by synthesis.[200]

[188] B. Siddiqui, J. Kawanami, Y. Li, and S. Hakomori, *J. Lipid Res.*, 1972, **13**, 657.
[189] M. Oshima and T. Yamakawa, *Biochemistry*, 1974, **13**, 1140.
[190] B. Lindberg, J. Lonngren, and W. Nimmich, *Acta Chem. Scand.*, 1972, **26**, 2231.
[191] Y. M. Choy, G. G. S. Dutton, K. B. Gibney, S. Kabir, and J. N. C. Whyte, *J. Chromatog.*, 1972, **72**, 13.
[192] H. Haga, T. Imanari, Z. Tamura, and A. Momose, *Chem. and Pharm. Bull. (Japan)*, 1972, **20**, 1805.
[193] J. E. Mrochek and W. T. Rainey, *Analyt. Biochem.*, 1974, **57**, 173.
[194] N. E. Nordén, A. Lundblad, S. Svensson, and S. Autio, *Biochemistry*, 1974, **13**, 871.
[195] E. Jellum, H. C. Børresen, and L. Eldjarn, *Clinica Chim. Acta*, 1973, **47**, 191.
[196] J. G. Buchanan and D. M. Clode, *J.C.S. Perkin I*, 1974, 388.
[197] A. Ehmann, *Carbohydrate Res.*, 1974, **34**, 99.
[198] C. D. Upper, J. P. Helgeson, and C. J. Schmidt in 'Plant Growth Substances, Proceedings of the Seventh International Conference, 1970', ed. D. J. Carr, Springer, New York, 1972.
[199] R. Horgan, E. W. Hewett, J. G. Purse, J. M. Horgan, and P. F. Wareing, *Plant Science Letters*, 1973, **1**, 321.
[200] R. Horgan, E. W. Hewett, J. G. Purse, and P. F. Wareing, *Tetrahedron Letters*, 1973, 2827.

Evidence presented for the characterization of a new class of plant hormones, 'brassins' (considered to be glucose esters of fatty acids),[201] has been cogently criticized.[202]

Oxygenated Terpenoids.—A new natural insect juvenile hormone, isolated from organ cultures of corpora allata of the tobacco hornworm moth (*Manduca sexta*) has the normal C_{15} (farnesane) skeleton. The presence of this hormone [the (10R)-10,11-epoxide of methyl farnesoate] *in vivo* has been confirmed, in extracts of larval haemolymph. (S)-[*Me*-^{14}C]-methionine of high isotopic purity (48.4 Ci mol^{-1} ≡ 77.5 atom % ^{14}C) was used in the culture medium, and g.c.m.s. of the isolated hormones indicated incorporation only into the ester methyl group (*m/e* 114 → 116; *cf.* Vol. 2, p. 316).[44]

The previously described juvenile hormones are based on homo- and bishomo-farnesane skeletons. In an important study of condensations catalysed by farnesyl pyrophosphate synthetase from pig liver, Koyama *et al.* achieved the enzymatic synthesis of bishomofarnesyl pyrophosphate (12) by two routes; (i), from bishomo-geranyl pyrophosphate (10) and isopentenyl pyrophosphate (11); and (ii) directly from 3-ethylbut-3-enyl pyrophosphate (13), *cis*-3-methylpent-2-enyl pyrophosphate (14), and (11) (Scheme 3). In the latter instance, (12) and the trishomofarnesyl pyrophosphate were formed in about equal amounts.[203]

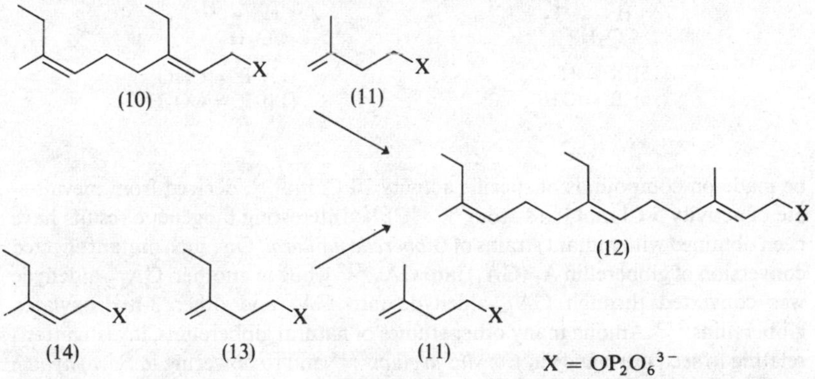

$X = OP_2O_6^{3-}$

Scheme 3

G.c.m.s. was of considerable value in verifying the structures of degradation products of petasin during an investigation of the biosynthesis of this sesquiterpenoid in *Petasites hybridus*.[204] The presence of fukinone as a minor component of leaf extracts of this plant was shown with the aid of deuterium labelling *in transitu*.[205] Three acyclic monoterpenoids present in the oil of *Ledum palustre*

---

[201] N. Mandava and J. W. Mitchell, *Chem. and Ind.*, 1972, 930.
[202] B. V. Milborrow and R. J. Pryce, *Nature*, 1973, **243**, 46.
[203] T. Koyama, K. Ogura, and S. Seto, *Chemistry Letters*, 1973, 401.
[204] C. J. W. Brooks and R. A. B. Keates, *Phytochemistry*, 1972, **11**, 3235.
[205] R. A. B. Keates, G. M. Anthony, and C. J. W. Brooks, *Phytochemistry*, 1973, **12**, 879.

were identified with the aid of 'carbon skeleton gas chromatography' (hydrogenation–hydrogenolysis) together with g.c.m.s.[206] Among other analytical studies of essential oils, a notable example is that of leaf extracts of *Carphephorus* species: more than 130 compounds, including many terpenoids, were identified by capillary g.c.m.s. using three different stationary phases.[207] Abscisic acid has been characterized in chloroplasts of *Pisum sativum*.[208] The terpenoid reagents drimanoyl chloride and chrysanthemoyl chloride have been applied to the gas-phase resolution of enantiomeric alcohols and amines.[209] G.c. correlations, derived from data for menthyl esters of branched-chain acids of known configuration, provide a basis for determining the stereochemistry of acyclic isoprenoid acids.[210]

The determination of the specific activity of [^{14}C]-labelled compounds by g.c.m.s. has been demonstrated and discussed by MacMillan and his colleagues,[42] with reference to the biosynthesis of the gibberellin precursors (15)—(18) in a cell-free enzyme system from *Cucurbita pepo*.[211] Satisfactory measurements could

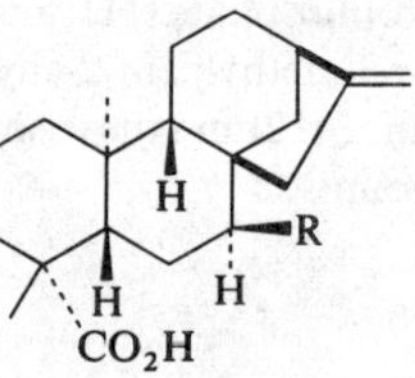

(15) R = H
(16) R = OH

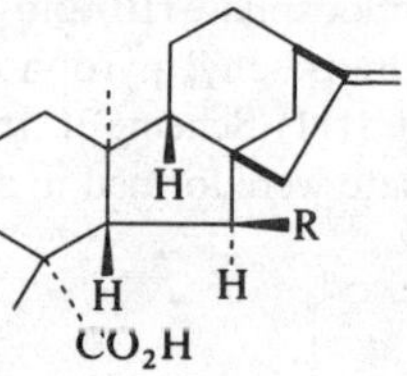

(17) R = CHO
(18) R = CO_2H

be made on compounds of specific activity 20 Ci mol^{-1}, derived from mevalonate of activity 5 Ci mol^{-1} (8 atom % ^{14}C).[42] Interesting biogenetic results have been obtained with mutant strains of *Gibberella fujikuroi*. One such mutant effected conversion of gibberellin A_1 (GA_1) into GA_3,[212] while in another, GA_{12}-aldehyde was converted through GA_{14}-aldehyde into GA_3 and other 3-hydroxylated gibberellins.[213] Among many other studies of natural gibberellins, investigations relating to seed dormancy in *Corylus avellans*,[214] and to flowering in *Bryophyllum daigremontianum*,[215] are representative. The aqueous decomposition of gibberellic acid proceeds *via* gibberellenic acid; on boiling with ^2H_2O gibberellenic acid

[206] M. von Schantz, K.-G. Widén, and R. Hiltunen, *Acta Chem. Scand.*, 1973, **27**, 551.
[207] K. Karlsson, I. Wahlberg, and C. R. Enzell, *Acta Chem. Scand.*, 1972, **26**, 3839
[208] I. D. Railton, D. M. Reid, P. Gaskin, and J. MacMillan, *Planta*, 1974, **117**, 179.
[209] C. J. W. Brooks, M. T. Gilbert, and J. D. Gilbert, *Analyt. Chem.*, 1973, **45**, 896.
[210] R. G. Ackman, R. E. Cox, G. Eglinton, S. N. Hooper, and J. R. Maxwell, *J. Chromatog. Sci.*, 1972, **10**, 392.
[211] J. E. Graebe, D. H. Bowen, and J. MacMillan, *Planta*, 1972, **102**, 261.
[212] J. R. Bearder, J. MacMillan, and B. O. Phinney, *Phytochemistry*, 1973, **12**, 2655.
[213] P. Hedden, J. MacMillan, and B. O. Phinney, *J.C.S. Perkin I*, 1974, 587.
[214] P. M. Williams, J. W. Bradbeer, P. Gaskin, and J. MacMillan, *Planta*, 1974, **117**, 101.
[215] P. Gaskin, J. MacMillan, and J. A. D. Zeevaart, *Planta*, 1973, **111**, 347.

afforded [9-^{2}H]-allogibberic acid, its 9-epimer, and 9,11-didehydroallogibberic acid, all of which were identified by g.c.m.s.[216]

Changes in the concentrations of the major diterpenoids, and of enmein and oridonin, have been explored during the growth of *Isodon trichocarpus*.[217] Two epoxylabdenones have been identified in Turkish tobacco.[218] The esterifying alcohols of pumpkin seed chlorophylls have been shown to comprise all the possible C_{20}-diterpenoid alcohols between geranyl-geraniol and phytol.[219]

G.l.c. and m.s. were separately used in the elucidation of the structure of the interesting C_{25} compounds diumycinol (19) and isodiumycinol (20): it remains to be determined whether the unusual side-chains are of isoprenoid origin.[220]

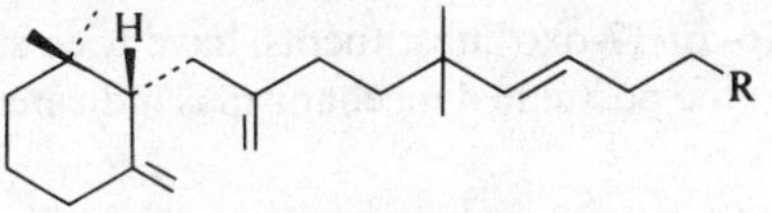

(19) R = CMe=CHCH$_2$OH
(20) R = C(OH)Me·CH=CH$_2$

Triterpenoids of *Buxus sempervirens* have been studied.[221] A molluscicidal saponin of *Phytolacca docecandra* has been identified as a diglucosyl glucoside of oleanolic acid.[222] C_{35}-compounds isolated from *Acetobacter xylinum* are hopane derivatives, substituted in the side-chain by a 2,3,4,5-tetrahydroxypentyl residue.[223]

Steroids: (A) Reference Compounds.—*Alcohols.* G.c.m.s. data for TMS ethers of a number of sterols and steroid diols have been compiled,[224] and a brief report has been made on dimethylsilyl ethers of sterols.[225]

E- and *Z*-isomers of 24-ethylidene sterols are distinguishable by g.l.c. and g.c.m.s. of free sterols, acetates and TMS ethers.[226]

Chloromethyl(dimethyl)silyl ethers are useful alternative derivatives[84,227,228] by virtue of their distinctive retention increments and of the ease of recognition of fragment ions containing the chlorinated group.

---

[216] R. J. Pryce, *J.C.S. Perkin I*, 1974, 1179.
[217] E. Fujita, Y. Nagao, S. Nakano, Y. Masada, K. Hashimoto, and T. Inouye, *Yakugaku Zasshi*, 1972, **92**, 1400.
[218] A. J. Aasen, B. Kimland, S. Almquist, and C. R. Enzell, *Acta Chem. Scand.*, 1972, **26**, 832.
[219] R. K. Ellsworth and C. A. Nowak, *Analyt. Biochem.*, 1974, **57**, 534.
[220] W. A. Slusarchyk, J. A. Osband, and F. L. Weisenborn, *Tetrahedron*, 1973, **29**, 1465.
[221] D. Abramson, L. J. Goad, and T. W. Goodwin, *Phytochemistry*, 1973, **12**, 2211.
[222] R. M. Parkhurst, D. W. Thomas, W. A. Skinner, and L. W. Cary, *Phytochemistry*, 1973, **12**, 1437.
[223] H. J. Förster, K. Biemann, W. G. Haigh, N. H. Tattrie, and J. R. Colvin, *Biochem. J.*, 1973, **135**, 133.
[224] C. J. W. Brooks, W. Henderson, and G. Steel, *Biochim. Biophys. Acta*, 1973, **296**, 431.
[225] D. H. Hunneman, in ref. *k* (Table, p. 297), Summaries, p. 31.
[226] C. J. W. Brooks, B. A. Knights, W. Sucrow, and B. Radüchel, *Steroids*, 1972, **20**, 487.
[227] C. J. W. Brooks and B. S. Middleditch, *Analyt. Letters*, 1972, **5**, 611.
[228] J. R. Chapman and E. Bailey, *Analyt. Chem.*, 1973, **45**, 1636.

G.c.m.s. has been applied to show that the reduction of Δ^5- and Δ^4-3β-hydroxy-sterols by *Eubacterium* ATCC 21,408 proceeded stereospecifically to yield the 5β-stanols.[229]

A detailed analytical study of the monoacetates of 5α-androstane-$3\alpha,17\beta$-diol and 5α-androstane-$3\beta,17\beta$-diol, in connection with their isolation from tissue extracts, led to satisfactory separation (notably as TMS ethers on SILAR-5-CP stationary phase) and characterization of the four isomers.[230]

Fragmentation modes of TMS ethers of Δ^5-3β-hydroxy C_{19}-steroids (with and without various other nuclear substituents) have been investigated with the aid of ^{2}H- and ^{18}O-labelling and high-resolution mass measurements.[231] Ions at m/e $[M - 56]^{+}$, observed during g.c.m.s. of many Δ^5-3β-trimethylsilyloxy steroids possessing 16- or 17-oxo substituents, have been shown[232,233] to result from A-ring cleavage: the postulated mechanism is indicated in Scheme 4.

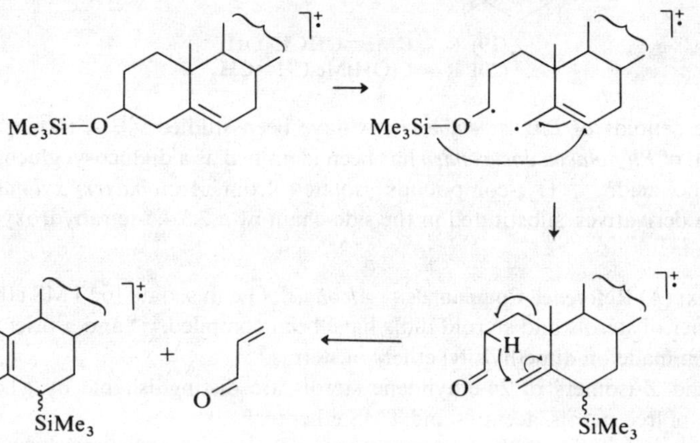

Scheme 4

The selective introduction of trimethylsilyl and perdeuteriotrimethylsilyl groups in steroid diols affords 'mixed' derivatives that are of value in interpretation of mass spectral fragmentations.[68]

Ketones. A further study of the incidence of Beckmann fission of O-methyloximes during g.l.c. has confirmed that this side-reaction can be avoided.[234] O-Butyl-oximes and O-pentyloximes are of value for g.c.m.s. of ketonic steroids, because

---

[229] H. J. Eyssen, G. G. Parmentier, F. C. Compernolle, G. De Pauw, and M. Piessens-Denef, *European J. Biochem.*, 1973, **36**, 411.
[230] F. L. Berthou, R. F. Morfin, D. Picart, and L. G. Bardou, *J. Chromatog.*, 1974, **88**, 271.
[231] C. J. W. Brooks, D. J. Harvey, B. S. Middleditch, and P. Vouros, *Org. Mass Spectrometry*, 1973, **7**, 925.
[232] I. Björkhem, J.-Å. Gustafsson, and J. Sjövall, *Org. Mass Spectrometry*, 1973, **7**, 277.
[233] C. J. W. Brooks, D. J. Harvey, and B. S. Middleditch, *J. Org. Chem.*, 1972, **37**, 3365.
[234] J.-P. Thenot and E. C. Horning, *Analyt. Letters*, 1972, **5**, 801.

their retention and mass increments (intermediate between those of O-methyloximes and O-benzyloximes) afford sequential elution of non-ketonic, monoketonic and diketonic derivatives.[235]

Halogenated 3(enol),17-diesters of testosterone have been characterized by g.c.m.s. in conjunction with electron-capture detection.[236]

A valuable survey has been made of the g.c.m.s. properties of catechol oestrogens and of their methyl and TMS ethers.[237] Characteristic properties (including g.c.m.s. data) of 18-hydroxyoestrone, 18-noroestrone, and 18-hydroxyoestradiols (17α and 17β), together with their TMS ethers, have been recorded. The 17β,18-diol gave a fragment ion at m/e $[M - 48]^{+\cdot}$, ascribed to a cyclic elimination of H_2O and CH_2O, and was also distinguished from the 17α-epimer by a shorter retention time (H-bonding) and by the formation of a cyclic 'dimethylsiliconide'.[238]

Corticosteroids. A new method of stabilization of the dihydroxyacetone sidechain for g.l.c. has been introduced by Chambaz, Madani, and Ros:[239] base-catalysed trimethylsilylation effects complete conversion into a tris-TMS derivative that appears to be the 17α,21-di-TMS 20-enol TMS ether.[80] Products containing this grouping are very satisfactory for g.c.m.s., as confirmed by further model studies using capillary columns,[240] but application of the method to urinary steroid extracts is subject to some difficulties, *e.g.* the variable formation of enol TMS ethers from 17-oxo-steroids. The 16α-methyl substituent in dexamethasone hinders methoximation of the 20-ketone, but the O-methyloxime tri-TMS ether may be prepared in virtually quantitative yield under conditions devised on the basis of a study by g.c.m.s.[241]

Other Steroids. Mass spectra of TMS derivatives of individual cardiac aglycones and monoglycosides have been recorded by g.c.m.s. using a short (0.25 m) column.[242] The enol TMS ethers formed from the butenolide group give rise to ions at m/e 170 and 183. Methods for the gas-phase separation and characterization of ecdysones have been further studied. The TMS ethers are satisfactory for g.c.m.s., and the groups at the 2- and 26-positions can be selectively replaced by heptafluorobutyryl groups,[81] yielding mixed derivatives suitable for electron-capture detection. Estimation of subnanogram amounts of α-ecdysone and ecdysterone is possible by s.i.m. of the TMS ethers at m/e 564 and 561 respectively. O-Methyloxime TMS ethers of ecdysone and ecdysterone have been studied by g.c.m.s.: *syn-* and *anti-*oximes are formed, giving rise to two g.l.c. peaks from each compound.[243]

---

[235] T. A. Baillie, C. J. W. Brooks, and E. C. Horning, *Analyt. Letters*, 1972, **5**, 351.
[236] L. Dehennin, A. Reiffsteck, and R. Scholler, *J. Chromatog. Sci.*, 1972, **10**, 224.
[237] H. O. Hoppen and L. Siekmann, *Steroids*, 1974, **23**, 17.
[238] J. K. Findlay, L. Siekmann, and H. Breuer, *Biochem. J.*, 1974, **137**, 263.
[239] E. M. Chambaz, C. Madani, and A. Ros, *J. Steroid Biochem.*, 1972, **3**, 741.
[240] S. Z. Nicosia, G. Galli, A. Fiecchi, and A. Ros, *J. Steroid Biochem.*, 1973, **4**, 417.
[241] J.-P. Thenot and E. C. Horning, *Analyt. Letters*, 1972, **5**, 905.
[242] F. C. Falkner, J. Frölich, and J. T. Watson, *Org. Mass Spectrometry*, 1973, **7**, 141.
[243] E. D. Morgan and A. P. Woodbridge, *Org. Mass Spectrometry*, 1974, **9**, 102.

Steroids: (B) in Biological Material.—*Sterols*. Investigation of natural sterols, especially from marine organisms, has been very active. In many instances, sterols have been isolated in pure form, and g.c.m.s. has played a subsidiary role: the technique has, however, been increasingly used in biosynthetic studies. Two sterols, aplysterol (21) and its 24(28)-dehydro analogue, possessing the novel

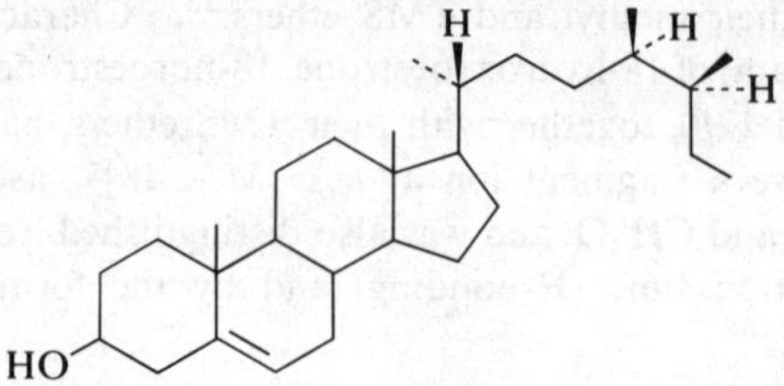

(21) aplysterol

24,26-dimethylcholestane skeleton, have been isolated from *Verongia aerophoba*:[244] in a survey of 25 species of Porifera, these sterols were found only in the family Verongidae.[245] Other new sponge sterols (from *Hymeniacidon perleve*)[246] include 24-methylenecholestanol and 5α-cholest-22-en-3β-ol. The occurrence of C_{26}-sterols is now seen to be widespread, as indicated by studies of Annelida,[247] gorgonians,[248] echinoderms,[249,250] jelly fish,[251] sponges,[246] and algae.[252]

New marine sterols with oxygenated side-chains include a 25-hydroxy-24-methylcholesterol isolated from a soft coral,[253] and a series of $\Delta^{9(11)}$-steroids isolated as aglycones of toxic starfish saponins; e.g. 3β,6α-dihydroxy-5α-cholesta-9(11),20(22)-dien-23-one from *Acanthaster planci* ('crown of thorns' starfish)[254] and the related $\Delta^{9(11)}$- and $\Delta^{9(11),24}$-sterols from *Marthasterias glacialis*.[255,256] The analogous C_{21}-steroid, 3β,6α-dihydroxy-5α-pregn-9(11)en-20-one, was also isolated from *A. planci*.[254,257] The brown alga, *Fucus evanescens*, afforded 5,(E)-24(28)-stigmastadien-3β,7α-diol, together with 3,5,(E)-24(28)-stigmastatrien-7-one (possibly an artifact).[258]

---

[244] P. De Luca, M. De Rosa, L. Minale, and G. Sodano, *J.C.S. Perkin I*, 1972, 2132.
[245] M. De Rosa, L. Minale, and G. Sodano, *Comp. Biochem. Physiol*, 1973, **B46**, 823.
[246] T. R. Erdman and R. H. Thomson, *Tetrahedron*, 1972, **28**, 5163.
[247] M. Kobayashi, M. Nishizawa, K. Todo, and H. Mitsuhashi, *Chem. and Pharm. Bull. (Japan)*, 1973, **21**, 323.
[248] J. H. Block, *Steroids*, 1974, **23**, 421.
[249] A. G. Smith, I. Rubinstein, and L. J. Goad, *Biochem. J.*, 1973, **135**, 443.
[250] J. S. Grossert, P. Mathiaparanam, G. D. Hebb, P. Price, and I. M. Campbell, *Experientia*, 1973, **29**, 258.
[251] S. Yasuda, *Comp. Biochem. Physiol.*, 1974, **B48**, 225.
[252] J. P. Ferezou, M. Devys, J. P. Allais, and M. Barbier, *Phytochemistry*, 1974, **13**, 593.
[253] J. P. Engelbrecht, B. M. Tursch, and C. Djerassi, *Steroids*, 1972, **20**, 121.
[254] Y. M. Sheikh, B. M. Tursch, and C. Djerassi, *J. Amer. Chem. Soc.*, 1972, **94**, 3278.
[255] D. S. H. Smith, A. B. Turner, and A. M. Mackie, *J.C.S. Perkin I*, 1973, 1745.
[256] S. Ikegami, Y. Kamiya, and S. Tamura, *Agric. and Biol. Chem. (Japan)*, 1973, **37**, 367.
[257] Y. Shimizu, *J. Amer. Chem. Soc.*, 1972, **94**, 4051.
[258] N. Ikekawa, M. Morisaki, and K. Hirayama, *Phytochemistry*, 1972, **11**, 2317.

The alga *Scenedesmus obliquus* has been grown in both H_2O and ^2H_2O: two of the six sterols normally produced were shown by g.c.m.s. to be absent from the alga cultured in ^2H_2O. Retention times of the deuteriated sterols were about 10% lower than for the undeuteriated compounds.[259] A difference in the pattern of sterol production was also noted for the alga (lichen symbiont) *Trebouxia*: in cultures grown in the presence of [Me-$^2H_3$]-methionine, the proportion of C_{28}-sterol was markedly increased in comparison with that found for normal methionine.[260] In these experiments, three deuterium atoms were found to be incorporated into the side-chain of ergost-5-en-3β-ol. In a parallel study of the fungal symbiont, *Xanthoria parietina*, g.c.m.s. showed that only two deuterium atoms were incorporated into the C_{28}-sterols, indicating the intermediacy of 24-methylenesterols, as observed for sterol biosynthesis in other fungi.[261a] A new sterol, ergosta-5,8,22-trien-3β-ol ('lichesterol') was identified in *X. parietina*.[261b] The sterol composition of the lichen *Pseudevernia furfuracea* has been determined, largely by t.l.c., g.l.c., and g.c.m.s. of sterol acetates.[262] The effect of inhibitors on sterol biosynthesis in *Chlorella* has been studied[263] and a new 14α-methylsterol, 24-methylenepollinastanol, isolated from cultures treated with triparanol.[264] This interesting sterol [14α-methyl-9β,19-cyclo-5α-ergost-24(28)-en-3β-ol] has also been found in banana peel[265] and in the alga *Astasia longa*.[266]

There is growing interest in the possible role of sterols in the physiology and development of plants.[267] Applications of g.c.m.s. in this area are exemplified by analysis of sterols in the fungus *Rhizopus arrhizus*[268] at various stages of its growth period, and in *Pinus elliotti* callus tissue cultures, seeds, and seedlings.[269]

Sterol esters of *Digitalis purpurea* are of interest in that the major acid moiety appears to be a C_{10}-enoic acid, possibly αβ-unsaturated.[270]

Sterol glycosides of *Cheiranthus cheiri* seed have been studied as tetra-TMS ethers and as trifluoroacetyl and heptafluorobutyryl esters; the trifluoroacetates are the most convenient derivatives for g.c.m.s.[271]

---

[259] P. Bélanger, J. A. Zintel, W. J. A. VandenHeuvel, and J. L. Smith, *Canad. J. Chem.*, 1973, **51**, 3294.
[260] L. J. Goad, F. F. Knapp, J. R. Lenton, and T. W. Goodwin, *Biochem. J.*, 1972, **129**, 219.
[261] (a) J. R. Lenton, L. J. Goad, and T. W. Goodwin, *Phytochemistry*, 1973, **12**, 2249; (b) *ibid.*, p. 1135.
[262] Z. A. Wojciechowski, L. J. Goad, and T. W. Goodwin, *Phytochemistry*, 1973, **12**, 1433.
[263] L. G. Dickson and G. W. Patterson, *Lipids*, 1972, **7**, 635.
[264] P. J. Doyle, G. W. Patterson, S. R. Dutky, and M. J. Thompson, *Phytochemistry*, 1972, **11**, 1951.
[265] F. F. Knapp, D. O. Phillips, L. J. Goad, and T. W. Goodwin, *Phytochemistry*, 1972, **11**, 3497.
[266] M. Rohmer and R. D. Brandt, *European J. Biochem.*, 1973, **36**, 446.
[267] B. A. Knights, *Chem. in Britain*, 1973, 106.
[268] J. D. Weete, G. C. Lawler, and J. L. Laseter, *Arch. Biochem. Biophys.*, 1973, **155**, 411.
[269] J. L. Laseter, R. Evans, C. H. Walkinshaw, and J. D. Weete, *Phytochemistry*, 1973, **12**, 2255.
[270] F. J. Evans, *J. Pharm. Pharmacol.*, 1973, **25**, 156.
[271] B. A. Knights, *Analyt. Letters*, 1973, **6**, 495.

Cholest-5-ene-3β,26-diol has been isolated from human brain,[272] and occurs in human atherosclerotic plaques partly in diesterified form.[273] Cholesterol linoleate hydroperoxides of atherosclerotic plaques have been characterized.[274]

Steroidal Acids. 3β-Hydroxy-5-cholenoic acid is an important constituent of the bile acids of human meconium, occurring mainly as the sulphate esters of its glycine and taurine conjugates.[275] The occurrence in petroleum of 5α- and 5β-cholanoic acid, and of 5α-pregnane-20-carboxylic acids,[276] has been considered as evidence of animal origin, but 5β-cholanoic acid has recently been isolated from a plant source—the seeds of *Abrus precatorius*.[277]

Hormonal Steroids and Metabolites in the Human. The volume of work in this field is such that only a selection of references can be included. The structural elucidation of new metabolites continues, but the major applications of g.c.m.s. are now to quantitative determinations: the analysis of steroids labelled with stable isotopes is increasingly important in studies of metabolic pathways.

An interesting new steroid found in the urine of hypertensive patients has been partially characterized by g.c.m.s. of its periodate oxidation product.[278] A thorough study, based on g.c.m.s. and double isotope dilution techniques, has been made of 5α-androstane-3β,17β-diol, a testosterone metabolite present in human urine; useful chromatographic and mass spectrometric data are also reported for 27 C_{19}- and C_{21}-steroid diols.[279] 3α,16α-Dihydroxyandrost-5-en-17-one has been isolated from the urine of a patient with adrenal carcinoma: the free steroid was not separable from its 3β-epimer on either of the phases (SE-30, QF-1) used for g.l.c., but the acetates were well resolved.[280] Another 3α-hydroxy-steroid, pregn-5-ene-3α,16α,20α-triol, which is excreted in large amounts in many cases of adrenal carcinoma, occurs at lower levels in normal urine; no direct comparison with the 3β-epimer was made in this instance.[281]

Improvements have been made in the determination of urinary steroid 'profiles' by g.c.m.s.[10,12,13,17,282] The inadequacy, for this purpose, of the separations achieved on packed columns has stimulated further work on glass capillary

---

[272] L. L. Smith, J. D. Wells, and N. I. Pandya, *Texas Repts. Biol. Med.*, 1973, **31**, 37; A. G. Smith, J. D. Gilbert, W. A. Harland, and C. J. W. Brooks, *Biochem. J.*, 1974, **139**, 793.
[273] J. D. Gilbert, C. J. W. Brooks, and W. A. Harland, *Biochim. Biophys. Acta*, 1972, **270**, 149.
[274] W. A. Harland, J. D. Gilbert, and C. J. W. Brooks, *Biochim. Biophys. Acta*, 1973, **316**, 378.
[275] P. Back and K. Ross, *Z. physiol. Chem.*, 1973, **354**, 83.
[276] W. K. Seifert, E. J. Gallegos, and R. M. Teeter, *J. Amer. Chem. Soc.*, 1972, **94**, 5880.
[277] N. Mandava, J. D. Anderson, S. R. Dutky, and M. J. Thompson, *Steroids*, 1974, **23**, 357.
[278] H. Adlercreutz, R. Hekali, and O. Wahlroos, *Brit. Med. J.*, 1973, **3**, 499.
[279] F. Berthou, L. Bardou, and H. H. Floch, *J. Steroid Biochem.*, 1972, **3**, 819.
[280] V. Fantl, *J. Endocrinol.*, 1973, **56**, 615.
[281] V. Fantl, M. Booth, and C. H. Gray, *J. Endocrinol.*, 1973, **57**, 135.
[282] J. Desgres, R. J. Bégué, and P. Padieu, *Clinica Chim. Acta*, 1973, **46**, 277.

columns,[9—13,283] and these are coming into general use.[13,21,284] Preliminary fractionation of urinary steroids by gel chromatography is of value.[17] Procedures mentioned earlier (p. 300) for computer analysis of g.c.m.s. data of steroids are applicable to urinary 'profiles' (*cf.* also ref. 285).

Extensive studies have been reported on neutral steroids of foetal meconium,[286] of normal female urine,[287] and of the urine of children with steroid 21-hydroxylase deficiency.[288] Many quantitative analyses have been based on s.i.m. (*cf.* Chapter 9). Notable examples include the quantitation of major and minor oestrogens in pregnancy urine $(50 \, \mu l)$[289] and the determination of plasma aldosterone as its acetal heptafluorobutyrate $(M^{+\cdot}, 538)$.[290]

Hormonal Steroids of Animals or Plants. Steroidal excretion patterns have been determined for ovariectomized and adrenalectomized rats.[291] Certain synthetic steroids (*e.g.* 16β-bromo-3β,17α-dihydroxy-5α-pregnane-11,20-dione) are powerful inhibitors of testosterone biogenesis in the rat.[292] Corticosterone and nine of its reduced metabolites in rat liver have been determined quantitatively by g.c.m.s. of the derived *O*-methyloxime pertrimethylsilyl derivatives: the amounts of metabolites differed markedly between male and female adult rat livers.[293] The main steroids present in boar testis tissue were found to be 3α- and 3β-hydroxy-5α-androst-16-enes; 19-nortestosterone was also detected (for the first time in biological material).[294]

The micro-estimation of ecdysone and ecdysterone based on 'mass fragmentography'[81] has been applied to demonstrate the biosynthesis of ecdysterone by silkworm prothoracic glands in organ culture.[295]

Metabolism of Exogenous Steroids. G.c.m.s. is widely used for studies of steroid biotransformations *in vivo* and *in vitro*. The metabolism of 4-androstene-3,17-dione by a rat liver nuclear 5α-reductase preparation yielded 5α-androstane-3,17-dione under standard conditions; with longer reaction times 7α-hydroxylated products were formed.[296] The major metabolite of 4,16-androstadien-3-one in man was shown to be 5α-androstenol,[85] while incubation of the same steroid

[283] M. Novotny, R. Segura, and A. Zlatkis, *Analyt. Chem.*, 1972, **44**, 9.
[284] B. F. Maume, P. Bournot, J. C. Lhuguenot, C. Baron, F. Barbier, G. Maume, M. Prost, and P. Padieu, *Analyt. Chem.*, 1973, **45**, 1073.
[285] J. D. Baty and A. P. Wade, *Analyt. Biochem.*, 1974, **57**, 27.
[286] I. Huhtaniemi and R. Vihko, *J. Endocrinol.*, 1973, **57**, 143.
[287] L. Viinikka and O. Jänne, *Clinica Chim. Acta*, 1973, **49**, 277.
[288] L. Viinikka, O. Jänne, J. Perheentupa, and R. Vihko, *Clinica Chim. Acta*, 1973, **48**, 359.
[289] H. Adlercreutz and D. H. Hunneman, *J. Steroid Biochem.*, 1973, **4**, 233.
[290] H. Breuer, H. Kaulhausen, W. R. Külpmann, L. Nocke-Fink, and L. Siekmann, *Z. klin. Chem. klin. Biochem.*, 1973, **11**, 99; L. Siekmann, B. Spiegelhalder, and H. Breuer, *Z. analyt. Chem.*, 1973, **261**, 377.
[291] J.-Å. Gustafsson and Å. Pousette, *Biochim. Biophys. Acta*, 1972, **280**, 182.
[292] A. S. Goldman, J.-Å. Gustafsson, and S. A. Gustafsson, *Acta Endocrinol.*, 1973, **73**, 146.
[293] P. Bournot, B. F. Maume, and P. Padieu, *Biomed. Mass Spectrometry*, 1974, **1**, 29.
[294] A. Ruokonen and R. Vihko, *J. Steroid Biochem.*, 1974, **5**, 33.
[295] H. Chino, S. Sakurai, T. Ohtaki, N. Ikekawa, H. Miyazaki, M. Ishibashi, and H. Abuki, *Science*, 1974, **183**, 529.
[296] J.-Å. Gustafsson and Å. Pousette, *Biochemistry*, 1974, **13**, 875.

with human foetal liver microsomes afforded 16β,17α-dihydroxy-4-androsten-3-one, possibly *via* epoxidation.[297] A comparison of the 7α-hydroxylation of cholesterol and sitosterol in cell-free enzyme preparations of rat liver indicated that the latter sterol was a very poor substrate.[298] An elegant analysis has been carried out of intermediates in the aromatization of androstenedione and testosterone by human placental microsomes: doubly-labelled substrates were used in conjunction with the 'twin-ion' technique (p. 301).[71] 16α-Hydroxylation occurred during incubation of 18-hydroxy-deoxycorticosterone with human adrenal glands,[299] and of the 'retro'-steroid trengesterone (6-chloro-9β,10α-1,4,6-pregnatriene-3,20-dione) with liver slices of rats.[300] On the other hand, a major urinary metabolite of the synthetic steroid, (±)-norgestrel, in women was the 16β-hydroxy derivative.[301] Urinary metabolites of the contraceptive steroids dimethisterone [secrosterone: 17β-hydroxy-6α-methyl-17α-(1-propynyl)-4-androsten-3-one] and norethisterone (17α-ethynyl-17β-hydroxy-4-oestren-3-one) were mainly ring A tetrahydro derivatives.[302] A 20(22)-dihydrodigoxin has been identified as a metabolite of digoxin in human plasma and urine: the tridigitoxoside was hydrolysed by treatment with heptafluorobutyric anhydride, which afforded dihydrodigoxigenin heptafluorobutyrate (apparently a monoacyl derivative) suitable for g.c.m.s.[303]

Amines: Reference Compounds and Reaction Products.*—Work in this area is dominated by investigations relating to 'biogenic' amines. Certain amines may be directly examined by g.c.m.s. (*e.g.* 2,5-dimethoxy-4-methylamphetamine)[304] but more commonly the preparation of derivatives is necessary. The choice of reagent depends on the effects required: among these are chromatographic stability and separation, sensitivity of detection, particular modes of fragmentation, and the resolution of enantiomers. The selective conversion of primary amines to isothiocyanates permits a simple distinction from analogous secondary amines.[305] For the isomeric 3- and 4-methyl ethers of 'dopamine', the isothiocyanates afford chromatographic separation, while the isothiocyanate O-TMS derivatives show distinguishable mass spectra.[306] Fluoroacylation is a convenient process especially for hydroxylic amines, and the products can be detected at low concentrations, both by mass spectrometry and by electron capture.

[297] A. Rane and J.-Å. Gustafsson, *Clin. Pharmacol. Therap.*, 1973, **14**, 833.
[298] L. Aringer and P. Eneroth, *J. Lipid Res.*, 1973, **14**, 563.
[299] S. L. Dale and J. C. Melby, *Steroids*, 1973, **21**, 617.
[300] H. Breuer, D. E. Kime, and R. Knuppen, *Acta Endocrinol.*, 1973, **74**, 127.
[301] S. F. Sisenwine, H. B. Kimmel, A. L. Liu, and H. W. Ruelius, *Acta Endocrinol.*, 1973, **73**, 91.
[302] W. G. Stillwell, E. C. Horning, M. G. Horning, R. N. Stillwell, and A. Zlatkis, *J. Steroid Biochem.*, 1972, **3**, 699.
[303] E. Watson, D. R. Clark, and S. M. Kalman, *J. Pharmacol. Exp. Therap.*, 1973, **184**, 424.
[304] A. Frigerio, R. Fanelli, and B. Danieli, *Chem. and Ind.*, 1972, 769.
[305] N. Narasimhachari, *J. Chromatog.*, 1974, **90**, 163.
[306] N. Narasimhachari and P. Vouros, *J. Chromatog.*, 1972, **70**, 135; R. L. Lin and N. Narasimhachari, *Analyt. Biochem.*, 1974, **57**, 46.

* Amines in biological material are included under Natural Metabolites and Drugs: see also Chapter 9 for quantitative studies.

Pentafluoropropionyl[307] and trifluoroacetyl[308,309] derivatives are among those used, and the fragmentations of N-trifluoroacetamides of 1-phenylisopropylamines have been elucidated with deuterium-labelled compounds.[309] Data for N-(trifluoroacetylamino)tetrahydronaphthalenes and related compounds have been reported.[310] The use of stable isotopes for securing additional structural characterization, *via* peak matching during multiple ion monitoring, has been evaluated with trifluoroacetyl derivatives of nortriptylene and its $^2H_2$ and ^{15}N analogues.[311]

TMS ethers of adrenolutin (22) and 5,6-dihydroxy-1-methylindole (23) have been examined by g.c.m.s. The mass spectrum of (22) consists largely of the

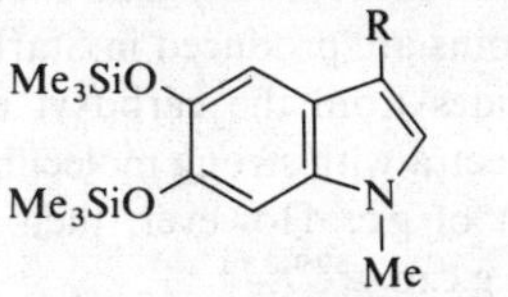

Me$_3$SiO

Me$_3$SiO

(22) R = OTMS; $M = 395$
(23) R = H; $M = 307$

Me

molecular ion and an interesting ion at m/e 291 ($[M - 104]^{+}_{.}$) on which no comment is made by the authors.[312] Losses of 103 units are well known in similar examples.[313] Trimethylsilylation of 7-methylguanosine was shown to be accompanied by formation of the amide (7-methyl-8-oxoguanosine): the oxygen atom is derived from dissolved oxygen.[314] Tertiary β-aminopropiophenones related to natural metabolites have been subjected (before and after borohydride reduction and trimethylsilylation) to g.c.m.s. with chemical ionization.[315] Separation of TMS derivatives of tetrahydropterins has been briefly reported.[316]

Methods for the gas-phase resolution of enantiomers have been further explored. Among chiral reagents applied to amines are the imidazolides of (−)-(S)-N-pentafluorobenzoylproline[317,318] and (+)-α-pentafluorophenyl-α-methoxypropionic acid;[319] N-trifluoroacetyl-S-prolyl chloride;[320,321] and the

---

[307] F. Karoum, F. Cattabeni, E. Costa, C. R. J. Ruthven, and M. Sandler, *Analyt. Biochem.*, 1972, **47**, 550.
[308] D. A. Garteiz and T. Walle, *J. Pharm. Sci.*, 1972, **61**, 1728.
[309] B. Lindeke and A. K. Cho, *Acta Pharm. Suecica*, 1973, **10**, 171.
[310] J. P. Chaytor, B. Crathorne, and M. J. Saxby, *J. Chromatog.*, 1972, **70**, 141.
[311] D. R. Knapp, T. E. Gaffney, and K. R. Compson, *Adv. Biochem. Psychopharmacol.*, 1973, **7**, 83.
[312] R. A. Heacock and J. E. Forrest, *J. Chromatog.*, 1973, **81**, 57.
[313] J. L. Smith, J. L. Beck, and W. J. A. VandenHeuvel, *Org. Mass Spectrometry*, 1971, **5**, 473.
[314] D. L. von Minden, R. N. Stillwell, W. A. Koenig, K. J. Lyman, and J. A. McCloskey, *Analyt. Biochem.*, 1972, **50**, 110.
[315] E. O. Oswald, L. Fishbein, B. J. Corbett, and M. P. Walker, *J. Chromatog.*, 1972, **73**, 43, 59.
[316] R. Weber, W. Frick, and M. Viscontini, *Helv. Chim. Acta*, 1973, **56**, 2919.
[317] S. B. Matin, M. Rowland, and N. Castagnoli, *J. Pharm. Sci.*, 1973, **62**, 821.
[318] K. S. Marshall and N. Castagnoli, *J. Med. Chem.*, 1973, **16**, 266.
[319] L. R. Pohl and W. F. Trager, *J. Med. Chem.*, 1973, **16**, 475.
[320] B. Halpern and J. W. Westley, *Chem. Comm.*, 1966, 34.
[321] A. H. Beckett and B. Testa, *J. Pharm. Pharmacol.*, 1973, **25**, 382.

acid chlorides of (+)-*trans*-chrysanthemic and drimanoic acids.[209] The chiral amine (+)-(*R*)-α-phenylethylamine is suitable for the microanalysis of the enantiomeric composition of α-phenylbutyric anhydride, in conjunction with Horeau's method of determining configurations of secondary alcohols.[322]

Amino-acids and Peptides.—A critical appraisal of the status of and current trends in the sequencing of oligopeptides by mass spectrometry is presented in Chapter 10. Discussion in this section is limited to aspects of the application of g.c.m.s. to sequencing. Useful reviews have appeared of the g.c. of amino-acids[323] and of the g.c.[324] and m.s.[325] of peptides. Studies of various derivatives of reference compounds have continued. In a delayed paper,[326] good separations of trimethylsilylated methylthiohydantoins on short (4.5 m) glass capillary columns are reported. Unsubstituted thiohydantoins are produced in Stark's method[327] for the sequential degradation of peptides from the carboxyl end. The free thiohydantoins give informative mass spectra with strong molecular ions,[328,329] but are inconveniently polar in respect of g.c. However, their *NN*-bis-TMS derivatives are shown to be amenable to g.c.m.s.[329,330]

Dimethylformamide dialkylacetals have been shown to react with both functional groups of amino-acids, forming *N*-dimethylaminomethylene alkyl esters suitable for g.c.m.s.[331] Isopropylation[332] and trimethylsilylation[333] have been further studied, as one-step derivatization methods.

N-Neopentylidene ethyl esters have been applied to the study of amino-acids in serum and urine; the use of a Schiff base rather than an *N*-acyl derivative is advocated to avoid the contamination of samples that arises from the acylation of other metabolites. The method is stated to allow detection of fifteen of the known inborn errors of amino-acid metabolism.[334] Tiglylglycine was identified by g.c.m.s. of its methyl ester (and that of 2-methylbutyrylglycine obtained on hydrogenation) in the urine of a child with β-methylcrotonylglycinuria.[335] Urinary dipeptides characterized from a case of dermatological purpura were consistent with a collagen abnormality.[336] Other urinary analyses have concerned

---

[322] C. J. W. Brooks and J. D. Gilbert, *J.C.S. Chem. Comm.*, 1973, 194; J. D. Gilbert and C. J. W. Brooks, *Analyt. Letters*, 1973, **6**, 639.
[323] J. R. Coulter and C. S. Hann, in ref. *ff* (Table, p. 297), p. 75.
[324] B. Kolb, in ref. *f* (Table, p. 297), p. 129.
[325] B. C. Das and E. Lederer, in ref. *ff* (Table, p. 297), p. 175.
[326] J. Eyem and J. Sjöquist, *Analyt. Biochem.*, 1973, **52**, 255.
[327] G. R. Stark, *Biochemistry*, 1968, **7**, 1796.
[328] T. Sun and R. E. Lovins, *Org. Mass Spectrometry*, 1972, **6**, 39; T. Suzuki, S. Matsui, and K. Tuzimura, *Agric. and Biol. Chem. (Japan)*, 1972, **36**, 1061.
[329] R. Burgus, N. Ling, M. Butcher, and R. Guillemin, *Proc. Nat. Acad. Sci., U.S.A.*, 1973, **70**, 684.
[330] M. Rangarajan, R. E. Ardrey, and A. Darbre, *J. Chromatog.*, 1973, **87**, 499.
[331] J.-P. Thenot and E. C. Horning, *Analyt. Letters*, 1972, **5**, 519.
[332] B. Blessington and N. I. Y. Fiagbe, *J. Chromatog.*, 1972, **68**, 259; 1973, **78**, 343.
[333] G. Mischer, *Z. analyt. Chem.*, 1972, **262**, 81.
[334] K. M. Williams and B. Halpern, *Austral. J. Biol. Sci.*, 1973, **26**, 831.
[335] D. Gompertz and G. H. Draffan, *Clinica Chim. Acta*, 1972, **37**, 405.
[336] R. A. W. Johnstone, T. J. Povall, J. D. Baty, J.-L. Pousset, C. Charpentier, and A. Lemonnier, *Clinica Chim. Acta*, 1974, **52**, 137.

anthranilic acid (converted into methyl salicylate for g.c.m.s.),[337] hippuric acid[338] and β-aminoisobutyric acid.[339] Deuteriated amino-acids (e.g. phenylalanine, tyrosine and leucine) are suitable for elucidating metabolic transformations in normal subjects and patients; g.c.m.s. is used for analysis of the isotopic content of metabolites.[74]

Acetylcholine esters may be studied indirectly by g.c.m.s. after demethylation by sodium benzenethiolate in butanone at 80 °C. Quantitative analyses based on deuteriated internal standards have been applied to various tissues.[340,341] 5-Hydroxyindole-3-acetic acid has been estimated in cerebrospinal fluid,[342] and pyroglutamic acid in normal and psoriatic epidermis.[343] Hydrolysates from actinomycin Z_5 (from *Streptomyces fradiae*) contain *cis*-5-methylproline.[344] Amino-acids in soil extracts have been determined by g.c.m.s. of N-trifluoroacetyl n-butyl esters.[345]

The scope of g.c.m.s. in determining amino-acid sequences in individual peptides has been explored with reference compounds. Selective cleavage of dipeptide units from the N-terminus may be effected by the action of dipeptidyl aminopeptidase I ('DAP I'). Application of the enzyme to one sample directly, and to another after removal of the N-terminal amino-acid (by Edman degradation) yields two sets of dipeptides corresponding to even and odd cleavage points as exemplified in Scheme 5.[346] In simple examples the overlapping data obtained

Val ┆ Gly ┆ Gly ┆ Val ┆ Glu ┆ Ser ┆ Leu ┆ Gly ┆ Gly ┆ Thr ┆ Gly ┆ Ala ┆ Leu ┆ Arg

-------- cleavage points by direct treatment with DAP I
~~~~~~~ cleavage points after initial Edman degradation

Scheme 5

by this 'domino'[347] procedure may determine the sequence: otherwise, it may be combined with kinetic studies of analysis of partially-degraded peptides.[348] The dipeptides can be characterized by g.c.m.s. as N-trifluoroacetyl[346] or N-pentafluoropropionyl methyl esters,[348] or as N-acetyl permethyl derivatives.[349] Some difficulties remain, notably the fact that DAP I does not cleave prolyl peptides. In

---

[337] K. Hirano, M. Naruse, S. Kawai, and T. Ohno, *J. Chromatog.*, 1972, **70**, 53.
[338] U. Langenbeck and J. E. Seegmiller, *J. Chromatog.*, 1973, **80**, 81.
[339] W. E. Pereira, R. E. Summons, W. E. Reynolds, T. C. Rindfleisch, and A. M. Duffield, *Clinica Chim. Acta*, 1973, **49**, 401.
[340] D. J. Jenden, M. Roch, and R. A. Booth, *Analyt. Biochem.*, 1973, **55**, 438.
[341] D. J. Jenden, *Adv. Biochem. Psychopharmacol.*, 1973, **7**, 69.
[342] L. Bertilsson, A. J. Atkinson, J. R. Althaus, Å. Härfäst, J. E. Lindgren, and B. Holmstedt, *Analyt. Chem.*, 1972, **44**, 1434.
[343] S. Marstein, E. Jellum, and L. Eldjarn, *Clinica Chim. Acta*, 1973, **49**, 389.
[344] E. Katz, K. T. Mason, and A. B. Mauger, *Biochem. Biophys. Res. Comm.*, 1973, **52**, 819.
[345] W. E. Pereira, Y. Hoyano, W. E. Reynolds, R. E. Summons, and A. M. Duffield, *Analyt. Biochem.*, 1973, **55**, 236.
[346] Yu. A. Ovchinnikov and A. A. Kiryushkin, *F.E.B.S. Letters*, 1972, **21**, 300.
[347] R. J. Rowlands and H. Lindley, *Biochem. J.*, 1972, **126**, 683.
[348] R. M. Caprioli, W. E. Seifert, and D. E. Sutherland, *Biochem. Biophys. Res. Comm.*, 1973, **55**, 67.
[349] D. H. Calam, *J. Chromatog.*, 1972, **70**, 146.

a related non-enzymic method, briefly outlined for a model pentapeptide (and for the tetrapeptide derived from Edman degradation), thermal treatment affords diketopiperazines suitable for direct analysis by g.c.m.s.[350]

A comprehensive approach to sequence determination, involving g.c.m.s. of derivatives of the oligopeptides generated by various hydrolytic procedures, has been described by Biemann's group,[351] who have applied it to a dodecapeptide isolated as the C-terminal cyanogen bromide fragment of rabbit skeletal muscle actin.[352] The sequence deduced was the same as that found by conventional techniques. In the method, oligopeptides are converted into the corresponding polyaminoalcohols, and these are O-trimethylsilylated for g.c.m.s. The resulting derivatives afford abundant ions, arising from simple C–C cleavages along the ethylenediamine chain, and indicating the sequence of side-chains. The use of a computer to correlate mass spectra, 'mass chromatograms', and retention indices greatly aids the interpretation of the data. Peptides containing arginine, tryptophan, and cysteine are amenable to the general procedure, but certain histidine-containing moieties are unsuitable for g.c.m.s. and require separate mass spectrometry.

Natural Metabolites and Drugs: (A) Reference Compounds.*—Mixtures of methyl isovalerate and methyl α-methylbutyrate may be analysed by monitoring characteristic ions at m/e 74 and 88.[353] Trimethylsilylation of dicarboxylic acids (malonic to suberic) and their esters has been studied: malonic acid yielded a tri-TMS derivative.[354] Quinoxalinol TMS ethers derived from pyruvic acid have been characterized.[355] Data for TMS derivatives of the six dihydroxybenzoic acids and related acids have been recorded.[356] The separation of seven tetracyclines as TMS derivatives has been studied on six stationary phases, though mass spectra were obtained by direct probe sampling.[357] Synthetic cannabinoids[358] and isomers of tetrahydrocannabinols[359] have been characterized.

In the period under review, the significance of N-oxygenation in the metabolism of aralkyl amines has become apparent. The separation and identification of hydroxylamines, oximes and nitrones so produced is greatly aided by g.c.m.s.[360] Benzodiazepine drugs have been studied with special reference to transformations

---

[350] R. A. W. Johnstone, T. J. Povall, and J. D. Baty, *J.C.S. Chem. Comm.*, 1973, 392.
[351] H.-J. Förster, J. A. Kelley, H. Nau, and K. Biemann, in 'Chemistry and Biology of Peptides', ed. J. Meienhofer, Ann Arbor Sci. Publ., Ann Arbor, Michigan, 1972, p. 679.
[352] H. Nau, J. A. Kelley, and K. Biemann, *J. Amer. Chem. Soc.*, 1973, **95**, 7162.
[353] K. Tanaka and G. M. Yu, *Clinica Chim. Acta*, 1973, **43**, 151.
[354] O. A. Mamer and S. S. Tjoa, *Clinical Chem.*, 1973, **19**, 58.
[355] A. Frigerio, P. Martelli, K. M. Baker, and P. A. Biondi, *J. Chromatog.*, 1973, **81**, 139.
[356] H. Morita, *J. Chromatog.*, 1972, **71**, 149.
[357] K. Tsuji and J. H. Robertson, *Analyt. Chem.*, 1973, **45**, 2136.
[358] T. B. Vree, D. D. Breimer, C. A. M. van Ginneken, and J. M. van Rossum, *J. Chromatog.*, 1972, **74**, 209.
[359] T. B. Vree, D. D. Breimer, C. A. M. van Ginneken, J. M. van Rossum, and N. M. M. Nibbering, *J. Chromatog.*, 1973, **79**, 81.
[360] A. H. Beckett, R. T. Coutts, and F. A. Ogunbona, *J. Pharm. Pharmacol.*, 1973, **25**, 708.

* The material in this section has been restricted, to avoid undue overlap with Chapter 9.

occurring during g.l.c.:[361] for example, carbamazepine epoxide is degraded to 9-acridine-carboxaldehyde.[362] Useful data have been recorded for 21 barbiturates as methyl derivatives (prepared by 'on-column' methylation with trimethylanilinium hydroxide).[363] Products of permethylation with $CH_3SOCH_2^- - CH_3I$ (including seco-amides) have been characterized and their fragmentations elucidated with the aid of perdeuteriomethylation.[364] Good separations of natural methylated xanthines have been recorded, with mass spectra of free compounds and derivatives.[365] Data for tetramethyluric acids were reported without identification of the isomers.[366]

Natural Metabolites and Drugs: (B) in Biological Material.—*Non-nitrogenous compounds*. A supposed connection between *trans*-3-methyl-2-hexenoic acid and schizophrenia has been disproved.[367] 2-Hydroxybutyric acid accompanies lactic acid in the urine of patients with lactic acidosis.[368] Other urinary hydroxy-acids identified by g.c.m.s. are β-(*m*-hydroxyphenyl)hydracrylic acid,[369] and β-(2-methoxyphenoxy)lactic acid, a metabolite of glyceryl guaiacolate.[370] Urinary glutarate may be determined with the aid of an isotopic internal standard.[371] α-Keto-isovaleric acid and other branched-chain α-keto-acids have been characterized from abnormal urine by g.l.c. of the free compounds.[372] A general procedure for determining metabolic profiles involving keto-acids in urine or serum is based on conversion into TMS-oxime TMS esters.[373] *O*-Methoxime TMS esters were applied to the analysis of bacterial and yeast keto-acids.[374] For metabolic profiles of urinary acids,[375] methyl esters are more satisfactory than TMS esters because the latter are too liable to hydrolysis.[376]

The biogenesis of homovanillic acid in brain was studied in rats exposed to air enriched in $^{18}O_2$; the incorporation of only one oxygen atom, shown by g.c.m.s. (of the methyl ester heptafluorobutyrate), indicated that tyrosine, and not phenylalanine, was the major precursor.[377] Procedures for determining homovanillic acid from various sources employ 'mass fragmentography' of fluoroacyl

[361] A. Frigerio, K. M. Baker, and G. Belvedere, *Analyt. Chem.*, 1973, **45**, 1846.
[362] K. M. Baker, A. Frigerio, P. L. Morselli, and G. Pifferi, *J. Pharm. Sci.*, 1973, **62**, 475.
[363] R. F. Skinner, E. G. Gallaher, and D. B. Predmore, *Analyt. Chem.*, 1973, **45**, 574.
[364] R. M. Thompson and D. M. Desiderio, *Org. Mass Spectrometry*, 1973, **7**, 989.
[365] K. Kamei and A. Momose, *Chem. and Pharm. Bull. (Japan)*, 1973, **21**, 1228.
[366] U. Langenbeck and J. E. Seegmiller, *Analyt. Biochem.*, 1973, **56**, 34.
[367] S. G. Gordon, K. Smith, J. L. Rabinowitz, and P. R. Vagelos, *J. Lipid Res.*, 1973, **14**, 495.
[368] J. E. Petterson, S. Landaas, and L. Eldjarn, *Clinica Chim. Acta*, 1973, **48**, 213.
[369] J. H. Duncan, M. W. Couch, G. Gotthelf, and K. N. Scott, *Biomed. Mass Spectrometry*, 1974, **1**, 40.
[370] W. J. A. VandenHeuvel, J. L. Smith, and R. H. Silber, *J. Pharm. Sci.*, 1972, **61**, 1997.
[371] C. R. Lee and R. J. Pollitt, *Biochem. Med.*, 1972, **6**, 536.
[372] D. Gompertz and G. H. Draffan, *Clinica Chim. Acta*, 1972, **40**, 5.
[373] H. J. Sternowsky, J. Roboz, F. Hutterer, and G. Gaull, *Clinica Chim. Acta*, 1973, **47**, 371.
[374] I. Andersson, B. Norkrans, and G. Odham, *Analyt. Biochem.*, 1973, **53**, 629.
[375] T. A. Witten, S. P. Levine, J. O. King, and S. P. Markey, *Clinical Chem.*, 1973, **19**, 586; T. A. Witten, S. P. Levine, M. T. Killian, P. J. R. Boyle, and S. P. Markey, *ibid.*, p. 963.
[376] I. Gan, J. Korth, and B. Halpern, *J. Chromatog.*, 1974, **92**, 435.
[377] G. Sedvall, A. Mayevsky, C.-G. Fri, B. Sjöquist, and D. Samuel, *Adv. Biochem. Psychopharmacol.*, 1973, **7**, 57.

derivatives, with deuteriated[378] or isomeric[379] compounds as internal standards. Similar methods are used for the analogous 4-hydroxy-3-methoxyphenylglycol[380] and 4-hydroxy-3-methoxyphenylethanol.[381]

G.c.m.s. has continued to reveal new natural cannabinoids;[358] minor components detected in hashish include cannabicyclol-C_3,[41] and other lower homologues of the common cannabinoids.[382] Metabolites found in human plasma, after oral administration of Δ^9-tetrahydrocannabinol (Δ^9-THC), were examined as their TMS ethers; 11-hydroxy- and 8α,11-dihydroxy-Δ^9-THC were positively identified, and two other components appeared to be 8α- and 8β-hydroxy-Δ^9-THC.[383] Similar allylic hydroxylations had been observed in animal metabolism.

6-Ethylsalicylic acid was observed as a new metabolite in cultures of *Mycobacterium phlei* grown on media supplemented with propionate; the incorporation of β,β,β-trideuteriopropionate was demonstrated by g.c.m.s.[384]

Nitrogenous compounds. Biogenic amines occurring at low concentrations have been analysed by s.i.m. of various derivatives, *e.g.* isothiocyanates,[385] TMS ethers[386] and pentafluoropropionates.[387] N-Acetylserotonin and melatonin afford cyclized products during fluoroacylation.

The metabolism of phenylalkylamines, long known to involve C-hydroxylation[388] in the aromatic ring or side-chain, also occurs *via* N-hydroxylation: this may be a major pathway in some species.[389] The hydroxylamines formed from primary amines are readily characterized by g.c.m.s. as the corresponding oximes (formed during g.l.c.)[360] or as their more stable TMS ethers, *e.g.* (24).[390]

$$Ar-CH_2-CMe_2-NHOTMS$$
(24)

[378] B. Sjöquist and E. Änggård, *Analyt. Chem.*, 1972, **44**, 2297; B. Sjöquist, B. Lindström, and E. Änggård, *Life Sci.*, 1973, **13**, 1655.
[379] N. Narasimhachari, *Biochem. Biophys. Res. Comm.*, 1974, **52**, 36.
[380] L. Bertilsson, *J. Chromatog.*, 1973, **87**, 147; C. Braestrup, *Analyt. Biochem.*, 1973, **55**, 420.
[381] F. Karoum, H. Lefevre, L. B. Bigelow, and E. Costa, *Clinica Chim. Acta*, 1973, **43**, 127.
[382] L. Strömberg, *J. Chromatog.*, 1972, **68**, 248.
[383] M. E. Wall, D. R. Brine, C. G. Pitt, and M. Perez-Reyes, *J. Amer. Chem. Soc.*, 1972, **94**, 8579.
[384] J. G. Dain, L. A. Ernst, I. M. Campbell and R. Bentley, *Biomed. Mass Spectrometry*, 1974, **1**, 57.
[385] H. Brandenberger and D. Schnyder, *Z. analyt. Chem.*, 1972, **261**, 297.
[386] F. P. Abramson, M. W. McCaman, and R. E. McCaman, *Analyt. Biochem.*, 1974, **57**, 482.
[387] F. Cattabeni, S. H. Koslow, and E. Costa, *Science*, 1972, **178**, 166; S. H. Koslow and A. R. Green, *Adv. Biochem. Psychopharmacol.*, 1973, **7**, 33.
[388] A. K. Cho, *Res. Comm. Chem. Pathol. Pharmacol.*, 1974, **7**, 67, and references there cited.
[389] A. H. Beckett and S. Al-Sarraj, *J. Pharm. Pharmacol.*, 1972, **24**, 174.
[390] A. K. Cho, B. Lindeke, and B. J. Hodshon, *Res. Comm. Chem. Pathol. Pharmacol.*, 1972, **4**, 519; B. Lindeke, A. K. Cho, T. L. Thomas, and L. Michelson, *Acta Pharm. Suecica*, 1973, **10**, 493.

Further metabolic oxidation to *C*-nitroso- and nitro-compounds has been observed.[391] Pethidine *N*-oxide has been identified in human urine.[392]

Confirmation of the importance of epoxides in the biotransformations of aromatic and olefinic substrates has stimulated the study of these intermediates by g.c.m.s. As many epoxides are too reactive to be analysed directly, it is convenient to effect quantitative cleavage, for example to the chlorohydrin TMS ethers;[83] urinary epoxides of allyl-substituted barbiturates have been characterized in this way.[393]

Applications of g.c.m.s. to the detection of simple metabolic changes such as *N*-demethylation are numerous, and noteworthy examples are the identification of normorphine as a urinary metabolite of codeine,[394] and the identification of methadone metabolites that contribute to the analgesic activity of the drug.[395] Comprehensive analyses of the structure and distribution of urinary drug metabolites are exemplified by studies of glutethimide (25),[396] methsuximide (26),[397]

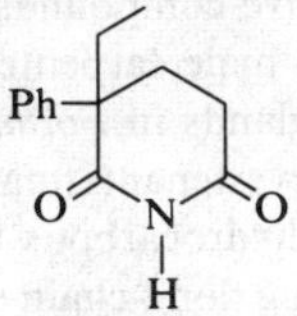

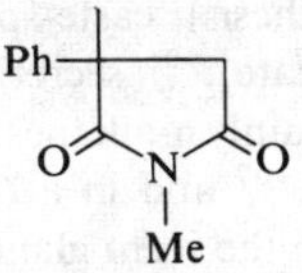

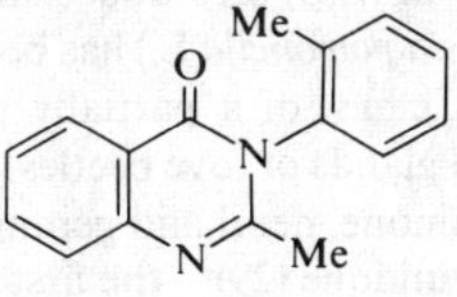

methaqualone (27),[398] and propranolol (28),[399] The identification of five monohydroxylated urinary metabolites of methaqualone was aided by reference data for ten synthetic isomers.[398] The propranolol metabolites found in urine

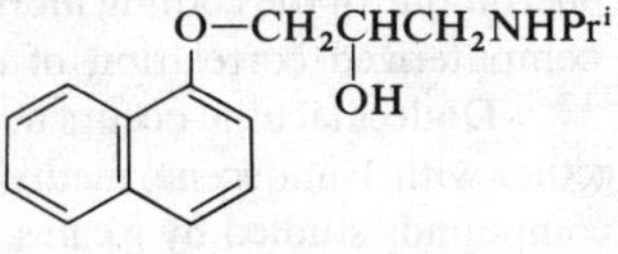

---

[391] A. H. Beckett and P. M. Bélanger, *J. Pharm. Pharmacol.*, 1974, **26**, 205.
[392] M. Mitchard, M. J. Kendall, and K. Chan, *J. Pharm. Pharmacol.*, 1972, **24**, 915.
[393] D. J. Harvey, L. Glazener, C. Stratton, D. B. Johnson, R. M. Hill, E. C. Horning, and M. G. Horning, *Res. Comm. Chem. Path. Pharmacol.*, 1972, **4**, 247.
[394] W. O. R. Ebbighausen, J. Mowat, and P. Vestergaard, *J. Pharm. Sci.*, 1973, **62**, 146.
[395] H. R. Sullivan, S. E. Smits, S. L. Due, R. E. Booher, and R. E. McMahon, *Life Sci. Part 1*, 1972, **11**, 1093.
[396] W. G. Stillwell, M. Stafford, and M. G. Horning, *Res. Comm. Chem. Pathol. Pharmacol.*, 1973, **6**, 579.
[397] M. G. Horning, C. Butler, D. J. Harvey, R. M. Hill, and T. E. Zion, *Res. Comm. Chem. Pathol. Pharmacol.*, 1973, **6**, 565.
[398] R. Bonnichsen, C.-G. Fri, C. Negoita, and R. Ryhage, *Clinica Chim. Acta*, 1972, **40**, 309.
[399] T. Walle and T. E. Gaffney, *J. Pharmacol. Exp. Therap.*, 1972, **182**, 83.

after oral administration were similar to those produced in rat liver microsomal preparations.[400]

An important application of g.c.m.s. has been made to the characterization of haem moieties in cytochromes P450 and P448. Reductive degradation with HI afforded substituted pyrroles which were readily identified as opso-, crypto-, haemo- and phyllo-pyrrole, suggesting that the haem is protoporphyrin IX as in haemoglobin.[401] G.c.m.s. was applied to the characterization of new *Erythrina* alkaloids,[402] and of a family of antibiotics (celestosaminides) produced by *Streptomyces caelestis*.[403]

Insect Pheromones and Other Secretions.—Much of the work in this area has been based on the examination of samples isolated by chromatography. G.c.m.s. has played a part in the instances cited. 3-Octanol and 3-octanone have been identified directly in the volatile secretions of *Crematogaster* species.[404] Other insect samples analysed, generally without preliminary purification of extracts, are exemplified by the following (only representative compounds identified are cited in parenthesis): caste-specific components in male carpenter ants (methyl 6-methylsalicylate);[405] secretions from Dufour's glands in *Formica* species (46 compounds, mainly n-alkanes and olefins),[406] in harvester ants (mainly branched-chain alkanes),[407] and in a *Camponotus* species (hydrocarbons C_{10}—C_{17});[408] secretions from the scent glands of the milkweed bug (long-chain esters);[409] and volatile cephalic substances from stingless bees (*Trigona* spp.) (aliphatic alcohols and ketones).[410] *trans*-β-Farnesene has been shown to be an alarm pheromone of several species of aphids.[411,412] The presence of (*E*,*E*)-8,10-dodecadien-1-ol as a sex pheromone of the codling moth (*Laspeyresia pomonella* L.) has been verified by computerized correlation of data from g.c.m.s. of a partially purified extract.[413] γ-Dodecalactone occurs in the pygidial glands of rove beetles (*Bledius* spp.) together with 1-undecene, methyl-*p*-benzoquinone, neral and geranial.[414]

New compounds studied by g.c.m.s. include gyrinidone (29)—the first natural cyclopentanoid norsesquiterpenoid to be identified—from the gyrinid beetle

---

[400] G. L. Tindell, T. Walle, and T. E. Gaffney, *Life Sci. Part II*, 1972, **11**, 1029.
[401] M. D. Maines and M. W. Anders, *Arch. Biochem. Biophys.*, 1973, **159**, 201.
[402] D. S. Millington, D. H. Steinman, and K. L. Rinehart, *J. Amer. Chem. Soc.*, 1974, **96**, 1909.
[403] T. F. Brodasky and A. D. Argoudelis, *J. Antibiotics*, 1973, **26**, 131.
[404] U. P. Schlunegger and R. H. Leuthdold, *Insect Biochem.*, 1972, **2**, 150.
[405] J. M. Brand, R. M. Duffield, J. G. MacConnel, M. S. Blum, and H. M. Fales, *Science*, 1972, **179**, 388.
[406] G. Bergström and J. Löfqvist, *J. Insect. Physiol.*, 1973, **19**, 877.
[407] F. E. Regnier, M. Nieh, and B. Hölldobler, *J. Insect Physiol.*, 1973, **19**, 981.
[408] J. J. Brophy, G. W. K. Cavill, and J. S. Shannon, *J. Insect. Physiol.*, 1973, **19**, 791.
[409] D. E. Games and B. W. Staddon, *Experientia*, 1973, **29**, 532.
[410] J. A. Luby, F. E. Regnier, E. T. Clarke, E. C. Weaver, and N. Weaver, *J. Insect. Physiol.*, 1973, **19**, 1111.
[411] W. S. Bowers, L. R. Nault, R. E. Webb, and S. R. Dutky, *Science*, 1972, **177**, 1121.
[412] W. H. J. M. Wietjens, A. C. Lakwijk, and T. van der Marel, *Experientia*, 1973, **29**, 658.
[413] M. Beroza, B. A. Bierl, and H. R. Moffitt, *Science*, 1974, **183**, 89.
[414] J. W. Wheeler, G. M. Happ, J. Araujo, and J. M. Pasteels, *Tetrahedron Letters*, 1972, 4635.

Dineutes discolor,[415] and three pheromones produced by Pharaoh's ants (*Monomorium pharaonis* L.). The gross structure (30) of one of the latter compounds was confirmed by synthesis.[416]

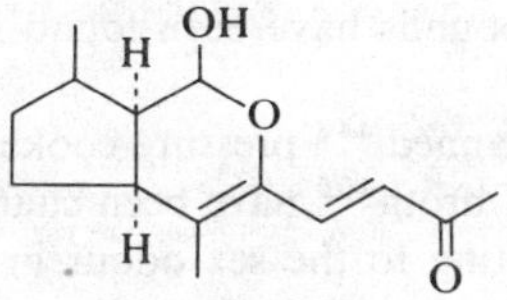

(29) gyrinidone 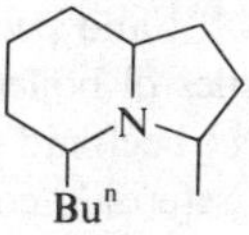   (30)

The defensive oral secretion of larvae of the sawfly *Neodiprion sertifer* is chemically similar to the resin of the host plant (*Pinus sylvestris*): the presence of α- and β-pinene and seven diterpenoid acids has been demonstrated.[417]

Pheromones of *Camponotus ligniperda* worker ants are discussed in a short review of g.c.m.s. techniques used in this field.[418]

Food Flavours and Aromas.—Refinements in technique and apparatus have facilitated analyses of small quantities of flavour volatiles.[419] Efficient fractionation methods have also been developed[420,421] and comprehensive comparative studies are now possible. These include analysis of acids from ten fruit juices and seven wines,[422] and determinations of diphenyl and *p*-phenylphenol in 18 citrus fruits.[423] Other fruit aromas which have been studied are apple,[424] grapefruit,[425] loganberry,[426] Meyer lemon,[427] muskmelon,[428] orange,[429] papaya,[430] passion fruit,[431] strawberry,[432] and tangerine.[433]

---

[415] J. W. Wheeler, S. K. Oh, E. F. Benfield, and S. E. Neff, *J. Amer. Chem. Soc.*, 1972, **94**, 7589.
[416] E. Talman and F. J. Ritter, in ref. *k* (Table, p. 297), Summaries, p. 50.
[417] T. Eisner, J. S. Johnessee, J. Carrel, L. B. Hendry, and J. Meinwald, *Science*, 1974, **184**, 997.
[418] S. Ställberg-Stenhagen, E. Stenhagen, and G. Bergström, *Zoon, Suppl.*, 1973 (*Suppl.* 1), 77; (*Chem. Abs.*, 1974, **80**, 92 786p).
[419] J. A. Yeransian, K. G. Sloman, and A. F. Foltz, *Analyt. Chem.*, 1973, **45**, 77R.
[420] J. K. Palmer, *J. Agric. Food Chem.*, 1973, **21**, 923.
[421] K. E. Murray, J. Shipton, and F. B. Whitfield, *Austral. J. Chem.*, 1972, **25**, 1921.
[422] J. J. Ryan and J. A. Dupont, *J. Agric. Food Chem.*, 1973, **21**, 45.
[423] H. Beernaert, *J. Chromatog.*, 1973, **77**, 331.
[424] H. E. Nursten and M. L. Woolfe, *J. Sci. Food Agric.*, 1972, **23**, 803.
[425] R. L. Coleman, E. D. Lund, and P. E. Shaw, *J. Agric. Food Chem.*, 1972, **20**, 100.
[426] P. H. Miller, L. M. Libbey, and H. Y. Yang, *J. Agric. Food Chem.*, 1973, **21**, 508.
[427] M. G. Moshonas, P. E. Shaw, and M. K. Veldhuis, *J. Agric. Food Chem.*, 1972, **20**, 751.
[428] T. R. Kemp, D. E. Knavel, and L. P. Stoltz, *Phytochemistry*, 1972, **11**, 3321.
[429] M. G. Moshonas, E. D. Lund, R. E. Berry, and M. K. Veldhuis, *J. Agric. Food Chem.*, 1972, **20**, 688.
[430] H. T. Chan, jun., R. A. Flath, R. R. Forrey, C. G. Cavaletto, T. O. M. Nakayama, and J. E. Brekke, *J. Agric. Food Chem.*, 1973, **21**, 566.
[431] T. H. Parliment, *J. Agric. Food Chem.*, 1972, **20**, 1043.
[432] J. Schubert, E. B. Saunders, S. F. Pan, and N. Wald, *J. Agric. Food Chem.*, 1973, **21**, 684.
[433] C. J. Muller, R. E. Kepner, and A. D. Webb, *J. Agric. Food Chem.*, 1972, **20**, 193.

Volatile components of raw[434] and baked[435] potatoes, potato flakes,[436] and potato crisps[437] have been characterized. Flavours of ginger,[438] mung beans,[439] mushrooms,[440] parsley,[441] pumpkin,[219] sweet potatoes,[442] and tomatoes[443] have also been investigated.

Several aldehydes and other compounds have been found in the aromas of filberts,[444] peanuts,[445] and pecans.[446]

Numerous volatiles of boiled,[447] canned,[448] pressure-cooked,[449] roasted,[450] and shallow-fried[451] beef and of beef broth[452] have been characterized. Some additional Δ^{16}-C_{19}-steroids contributing to the sex odour in pork have been found.[453] Further work has been carried out on the flavours of sterile,[454] stale,[455] and feed-flavoured[456] milk. It has been suggested that some of these components arise from enzymatic hydrolysis of newly-identified conjugates.[457] G.c.m.s. was employed to determine the absorption of ketones by freeze-dried cream during a flavour-enhancement process.[458] Volatiles of cheese[459] have been identified. Extensive studies have been made of the volatiles of stored fish[460] and of fish upon which micro-organisms have acted.[461]

---

[434] R. G. Buttery and L. C. Ling, *J. Agric. Food Chem.*, 1973, **21**, 746.
[435] S. R. Pareles and S. R. Chang, *J. Agric. Food Chem.*, 1974, **22**, 339.
[436] G. M. Sapers, O. Panasuik, F. B. Talley, and S. F. Osman, *J. Food Sci.*, 1972, **37**, 579.
[437] R. G. Buttery, *J. Agric. Food Chem.*, 1973, **21**, 31.
[438] Y. Masada, T. Inoue, K. Hashimoto, M. Fujioka, and K. Shiraki, *Yakugaku Zasshi*, 1973, **93**, 318.
[439] J. M. C. Geuns, *Phytochemistry*, 1973, **12**, 103.
[440] A. F. Thomas, *J. Agric. Food Chem.*, 1973, **21**, 955; S. M. Picardi and P. Issenberg, *ibid.*, p. 959.
[441] R. Kasting, J. Andersson, and E. von Sydow, *Phytochemistry*, 1972, **11**, 2277.
[442] M. R. Boyd and B. J. Wilson, *J. Agric. Food Chem.*, 1972, **20**, 428.
[443] H. Sato and S. Sakamura, *Agric. and Biol. Chem. (Japan)*, 1973, **37**, 225.
[444] R. M. Sheldon, R. C. Lindsay, and L. M. Libbey, *J. Food Sci.*, 1972, **37**, 313.
[445] H. M. B. Ballschmieter, *Fette, Seifen, Anstrichm.*, 1972, **74**, 112.
[446] Pao-Shui Wang and G. V. Odell, *J. Agric. Food Chem.*, 1972, **20**, 206.
[447] C. Hirai, K. O. Herz, J. Pokorny, and S. S. Chang, *J. Food Sci.*, 1973, **38**, 393.
[448] T. Persson and E. von Sydow, *J. Food Sci.*, 1973, **38**, 377.
[449] C. J. Mussinan, R. A. Wilson, and I. Katz, *J. Agric. Food Chem.*, 1973, **21**, 871.
[450] H. M. Liebich, D. R. Douglas, A. Zlatkis, F. Müggler-Chavan, and A. Donzel, *J. Agric. Food Chem.*, 1972, **20**, 96.
[451] K. Watanabe and Y. Sato, *J. Agric. Food Chem.*, 1972, **20**, 174.
[452] H. W. Brinkman, H. Copier, J. J. M. de Leeuw, and Sing Boen Tjan, *J. Agric. Food Chem.*, 1972, **20**, 177.
[453] R. H. Thompson, jun., A. M. Pearson, and K. A. Banks, *J. Agric. Food Chem.*, 1972, **20**, 185.
[454] O. W. Parks and C. Allen, jun., *J. Dairy Sci.*, 1973, **56**, 328.
[455] A. Ferretti and V. P. Flanagan, *J. Agric. Food Chem.*, 1972, **20**, 695.
[456] D. T. Gordon and M. E. Morgan, *J. Dairy Sci.*, 1972, **55**, 905.
[457] C. R. Brewington, O. W. Parks, and D. P. Schwartz, *J. Agric. Food Chem.*, 1973, **21**, 38; 1974, **22**, 293.
[458] Z. J. Hawrysh and C. M. Stine, *J. Food Sci.*, 1973, **38**, 7.
[459] D. J. Manning and H. M. Robinson, *J. Dairy Res.*, 1973, **40**, 63.
[460] A. Miller, R. A. Scanlan, J. S. Lee, and L. M. Libbey, *J. Fisheries Res. Board Canada*, 1972, **29**, 1125.
[461] A. Miller, R. A. Scanlan, J. S. Lee, and L. M. Libbey, *Appl. Microbiol.*, 1973, **25**, 952; **26**, 18.

Fifty-seven compounds were identified in the aroma of Ceylon tea,[462] and the 'flavoury' and 'non-flavoury' components were compared.[463] A characteristic aroma component of a red table wine was found to be 4-ethoxy-4-hydroxybutyric acid γ-lactone (31).[464] Sixty-one compounds were found in vinegars.[465] The

(31)

odour of white bread has been investigated.[466] 6-Chloro-o-cresol was found to be responsible for a 'disinfectant' taint in biscuits.[467]

G.c.m.s. has been particularly useful in the identification and measurement of potential carcinogens such as diethylstilbestrol[468] and patulin.[469,470] The nitrosamine controversy has continued.[471] G.c. alone is insufficient for unequivocal identification of nitrosamines, g.c.m.s. affords more specificity; a timely compilation of g.c. and m.s. data for 25 nitrosamines has appeared.[472] G.c.m.s. using high-resolution m.s. affords even greater specificity,[22,473] but there have been no reports of applications of this technique. N-Nitrosodimethylamine[474] and N-nitrosopyrrolidine[475] have been detected in preparations produced under laboratory conditions which were designed to mimic the preparation of frankfurters and bacon, respectively. Nitrosamines were not found in a culture medium and did not inhibit the growth of *Clostridia*.[476] Concentrations of N-nitrosodimethylamine in the range 0.3—0.45 p.p.m. in fish meal intended for animal consumption were measured by g.c.m.s.[477] The detection limit was stated to be 0.1 p.p.m. The same group also claimed to have identified this compound

---

[462] T. Yamanishi, Y. Kita, K. Watanabe, and Y. Nakatani, *Agric. and Biol. Chem. (Japan)*, 1972, **36**, 1153.
[463] R. L. Wickremasinghe, E. L. Wick, and T. Yamanishi, *J. Chromatog.*, 1973, **79**, 75.
[464] C. J. Muller, R. E. Kepner, and A. D. Webb, *J. Agric. Food Chem.*, 1972, **20**, 193.
[465] J. H. Kahn, G. B. Nickol, and H. A. Conner, *J. Agric. Food Chem.*, 1972, **20**, 214.
[466] E. J. Mulders, M. C. ten Noever de Brauw, and S. van Straten, *Z. Lebensm. Untersuch.-Forsch.*, 1973, **150**, 305.
[467] N. M. Griffiths and D. G. Land, *Chem. and Ind.*, 1973, 904.
[468] C. J. Mirocha, C. M. Christensen, G. Davis, and G. H. Nelson, *J. Agric. Food Chem.*, 1973, **21**, 135.
[469] P. M. Scott and B. P. C. Kennedy, *J. Assoc. Offic. Analyt. Chemists*, 1973, **56**, 813.
[470] J. Harwig, Y.-K. Chen, B. P. C. Kennedy, and P. M. Scott, *Canad. Inst. Food Sci. Technol. J.*, 1973, **6**, 22.
[471] L. S. du Plessis and J. R. Nunn, *N-Nitroso Compounds Newsletter*, 1973, **4**, 2.
[472] J. W. Pensabene, W. Fiddler, C. J. Dooley, R. C. Doerr, and A. E. Wasserman, *J. Agric. Food Chem.*, 1972, **20**, 274.
[473] T. A. Bryce and G. M. Telling, *J. Agric. Food Chem.*, 1972, **20**, 910.
[474] W. Fiddler, E. G. Piotrowski, J. W. Pensabene, R. C. Doerr, and A. E. Wasserman, *J. Food Sci.*, 1973, **37**, 668.
[475] D. D. Bills, K. I. Hildrum, R. A. Scanlan, and L. M. Libbey, *J. Agric. Food Chem.*, 1973, **21**, 876.
[476] A. E. Wasserman and C. N. Huhtanen, *J. Food Sci.*, 1972, **37**, 785.
[477] N. P. Sen, L. A. Schwinghamer, B. A. Donaldson, and W. F. Miles, *J. Agric. Food Chem.*, 1972, **20**, 1281.

(at a concentration of 0.03 p.p.m.) in uncooked bacon, and N-nitrosopyrrolidine (0.025 p.p.m.) in fried bacon.[478]

The principal compounds which interfere with nitrosamine analysis by g.c. and polarography are pyrazines. G.c.m.s. is able to distinguish between these classes of compounds, and the latter are found in a number of roasted foods. They have now been identified in cocoa beans,[479] fish muscle incubated with *Pseudomonas perolens*,[461] ponerine ants (as alarm pheromones),[480] white bread,[466] and roasted beef,[450] filberts,[444] and pecans.[446] They are also produced by heating casein,[481] some aminohydroxy-compounds,[482] and mixtures of amino-acids with sugars.[455,483] Some additional products of the last reaction have been characterized.[484,485]

G.c.m.s. has also been used in the characterization of volatile sulphur compounds in foods,[486] volatile components of thermally degraded thiamine,[487] a vitamin K_3 supplement,[488,489] sugars in cocoa beans,[490] the essential oil of mace,[491] volatiles of olive oil,[492,493] and sterols[494,495] in vegetable oils.

Pesticides and Pollutants.—Organic compounds have been identified in various samples of water,[496,497] including bay water,[498] river water,[499,500] drinking water,[501—503] and sewage wastewater (before and after laboratory chlorina-

---

[478] N. P. Sen, B. Donaldson, J. R. Iyengar, and T. Panalaks, *Nature*, 1973, **241**, 473.
[479] G. A. Reineccius, P. G. Kenney, and W. Weissberger, *J. Agric. Food Chem.*, 1972, **20**, 202.
[480] J. W. Wheeler and M. S. Blum, *Science*, 1973, **182**, 501.
[481] H. Kato, F. Hayase, and M. Fujimaki, *Agric. and Biol. Chem. (Japan)*, 1972, **36**, 951.
[482] Pao-Shui Wang and G. V. Odell, *J. Agric. Food Chem.*, 1973, **21**, 868.
[483] M. Fujimaki, M. Tajima, and H. Kato, *Agric. and Biol. Chem. (Japan)*, 1972, **36**, 663.
[484] A. Ferretti and V. P. Flanagan, *J. Agric. Food Chem.*, 1973, **21**, 35; R. A. Scanlan, S. G. Kayser, L. M. Libbey, and M. E. Morgan, *ibid.*, p. 673.
[485] S. Kato, T. Kurata, and M. Fujimaki, *Agric. and Biol. Chem. (Japan)*, 1973, **37**, 539.
[486] H. Nishimura, S. Koike, and J. Mizutani, *Agric. and Biol. Chem. (Japan)*, 1973, **37**, 1219.
[487] B. K. Dwivede and R. G. Arnold, *J. Food Sci.*, 1973, **38**, 450.
[488] V. W. Winkler and J. M. Yoder, *J. Assoc. Offic. Analyt. Chemists*, 1972, **55**, 1219.
[489] V. W. Winkler, F. E. Regnier, J. M. Yoder, and L. R. Macy, *J. Pharm. Sci.*, 1972, **61**, 1462.
[490] G. A. Reineccius, D. A. Andersen, T. E. Kavanagh, and P. G. Keeney, *J. Agric. Food Chem.*, 1972, **20**, 199.
[491] J. E. Forrest, R. A. Heacock, and T. P. Forrest, *J. Chromatog.*, 1972, **69**, 115.
[492] E. Fedeli, D. Baroni, and G. Jacini, *Riv. Ital. Sostanze Grasse*, 1973, **50**, 38.
[493] R. A. Flath, R. R. Forrey, and D. G. Guadagni, *J. Agric. Food Chem.*, 1973, **21**, 948.
[494] T. Itoh, T. Tamura, and T. Matsumoto, *J. Amer. Oil Chemists' Soc.*, 1973, **50**, 122, 427; *Lipids*, 1974, **9**, 173.
[495] E. Fedeli, A. Daghetta, D. Baroni, and N. Cortesi, *Riv. Ital. Sostanze Grasse*, 1972, **49**, 159.
[496] L. E. Harris, *Analyt. Instr.*, 1973, **11**, 183.
[497] E. J. Bonelli and R. D. Smith, *Effluent Water Treatment J.*, 1972, **12**(2), 87.
[498] B. R. Simoneit, D. H. Smith, G. Eglinton, and A. L. Burlingame, *Arch. Environ. Contam. Toxicol.*, 1973, **1**, 193.
[499] J. P. Mieure and M. W. Dietrich, *J. Chromatog. Sci.*, 1973, **11**, 559.
[500] R. A. Hites and K. Biemann, *Science*, 1972, **178**, 158.
[501] R. D. Kleopfer and B. J. Fairless, *Environ. Sci. Technol.*, 1972, **6**, 1036.
[502] J. Novák, J. Żlutický, V. Kubelka, and J. Mostecký, *J. Chromatog.*, 1973, **76**, 45.
[503] K. Grob, *J. Chromatog.*, 1973, **85**, 255.

tion).[504] n-Hexanal was found to be responsible for the odour of lake water containing a diatom.[505] G.c.m.s. has confirmed the presence of geosmin [racemic (32)] as the source of musty odour in a public water supply. Geosmin is a metabolite of *Actinomyces*.[506,507]

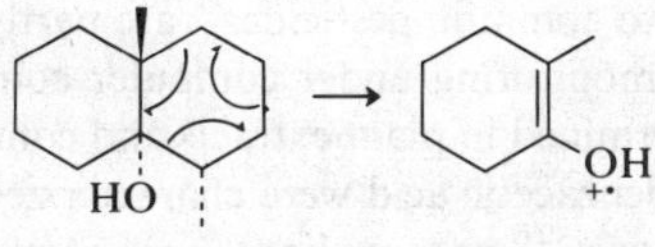

(32) $M^{\ddagger}$ 182 m/e 112 (100%)

More than 70 polynuclear aromatic hydrocarbons have been identified in polluted air.[108] Volatile compounds in air have been trapped on activated charcoal for analysis by g.c.m.s.[508] Bis-(chloromethyl)ether has been determined in air *via* concentration of organic components on Chromosorb 101.[509] Insecticides in air have been determined.[510] Detection of squalene by g.c.m.s. has aided the identification of skin in airborne particulate matter.[511] Smoke components have also been identified by g.c.m.s.[512,513]

The mass spectra of heptachlor and 1-hydroxychlordene obtained by g.c.m.s. and by using a direct insertion probe have been compared.[514] Chlordane residues in milk and fat of cows,[515] oxychlordane in human tissue,[516] and chlorinated hydrocarbons in human plasma[517] have been determined. Photolysis products of methoxychlor have been identified.[518] Chloroazobenzenes and chlorohydrazobenzenes are formed from certain phenylamide herbicides.[519] Metabolites of the fungicide BAS 3191 have been identified.[520] Problems associated with the generation and degradation of dibenzodioxins and dibenzofurans have

---

[504] W. H. Glaze, J. E. Henderson, quat., J. E. Bell, and V. A. Wheeler, *J. Chromatog. Sci.*, 1973, **11**, 580.
[505] T. Kikuchi, T. Mimura, Y. Moriwaki, M. Ando, and K.-I. Negoro, *Yakugaku Zasshi*, 1972, **92**, 1567.
[506] T. Kikuchi, T. Mimura, Y. Masada, and T. Inoue, *Chem. and Pharm. Bull. (Japan)*, 1973, **21**, 1847.
[507] T. Kikuchi, T. Mimura, Y. Moriwaki, K.-I. Negoro, S. Nakazawa, and H. Ono, *Yakugaku Zasshi*, 1972, **92**, 652.
[508] M. Rollet and M. Moisson, *Rev. Inst. Pasteur*, 1972, **5**, 439.
[509] L. A. Shadoff, G. J. Kallos, and J. S. Woods, *Analyt. Chem.*, 1973, **45**, 2341.
[510] K. Beyermann and W. Eckrich, *Z. analyt. Chem.*, 1973, **265**, 4.
[511] R. P. Clark and S. G. Shirley, *Nature*, 1973, **246**, 39.
[512] M. R. Kornreich and P. Issenberg, *J. Agric. Food Chem.*, 1972, **20**, 1109.
[513] C. MacFarlane, J. B. Lee, and M. B. Evans, *J. Inst. Brewing*, 1973, **79**, 202.
[514] A. Demayo and M. Comba, *Bull. Environ. Contam. Toxicol.*, 1972, **8**, 212.
[515] H. W. Dorough and R. W. Hemken, *Bull. Environ. Contam. Toxicol.*, 1973, **10**, 208.
[516] F. J. Biros and H. F. Enos, *Bull. Environ. Contam. Toxicol.*, 1973, **10**, 257.
[517] L. Palmér and B. Kolmodin-Hedman, *J. Chromatog.*, 1972, **74**, 21.
[518] J. D. MacNeil, R. W. Frei, S. Safe, and O. Hutzinger, *J. Assoc. Offic. Analyt. Chemists*, 1972, **55**, 1270.
[519] L. M. Bordeleau, J. D. Rosen, and R. Bartha, *J. Agric. Food Chem.*, 1972, **20**, 573.
[520] P. R. Wallnöfer, M. Königer, S. Safe, and O. Hutzinger, *J. Agric. Food Chem.*, 1972, **20**, 20.

been fully investigated.[521,522] Acid hydrolysis products of DDD and DDT precursors were identified,[523] and the degradation of DDT was studied.[524] G.c.m.s. has been used extensively in the analysis of polychlorinated biphenyls,[525—534] polychlorinated terphenyls,[535] and pentachlorophenol.[536] Problems posed by polychlorinated biphenyls—their complex isomeric composition and similarity to common pesticides—are partly solved by a technique based on selective ion monitoring under computer control.[537] Rotenone and deguelin have been determined in plant extracts and commercial insecticides.[538] Metabolites of naphthaleneacetic acid were characterized.[539] Studies have also been made of metaldehyde,[540] organophosphorus compounds,[541] and organomercury compounds.[542—544] The use of glass capillary columns in the analysis of tobacco smoke by g.c.m.s. has been discussed,[545] and alkaloids[546] have been found in tobacco smoke.

Volatile organic micropollutants in air and water have been studied by low temperature capillary g.c.m.s.[547] and an improved method of concentrating such components has been reported.[548]

Organic Geochemistry.—The loss of momentum in this subject concomitant with the hiatus in the Apollo project is reflected in a decrease in the number of

[521] W. B. Crummett and R. H. Stehl, *Environ. Health Perspect.*, 1973, **5**, 15; G. W. Bowes, B. R. Simoneit, A. L. Burlingame, B. W. Lappe, and R. W. Risebrough, *ibid.*, p. 191; D. G. Crosby, K. W. Moilanen, and A. S. Wong, *ibid.*, p. 259; O. Hutzinger, S. Safe, B. R. Wentzell, and V. Zitko, *ibid.*, p. 267.
[522] T. J. N. Webber and D. G. Box, *Analyst*, 1973, **98**, 181.
[523] B. L. Jensen and R. E. Counsell, *J. Org. Chem.*, 1973, **38**, 835.
[524] F. K. Pfaender and M. Alexander, *J. Agric. Food Chem.*, 1972, **20**, 842.
[525] L. O. Ruzo, M. J. Zabik, and R. D. Schuetz, *Bull. Environ. Contam. Toxicol.*, 1972, **8**, 217.
[526] M. Moron, G. Sundström, and C. A. Wachtmeister, *Acta Chem. Scand.*, 1972, **26**, 830.
[527] J. W. Rote and W. J. Morris, *J. Assoc. Offic. Analyt. Chemists*, 1973, **56**, 188.
[528] P. Moza, I. Weisgerber, W. Klein, and F. Korte, *Chemosphere*, 1973, 217.
[529] C. E. Bagley and E. Cromartie, *J. Chromatog.*, 1973, **75**, 219.
[530] D. G. Shaw, *Bull. Environ. Contam. Toxicol.*, 1972, **8**, 208.
[531] R. G. Webb and A. C. McCall, *J. Chromatog. Sci.*, 1973, **11**, 366.
[532] L. Fishbein, *J. Chromatog.*, 1972, **68**, 345.
[533] R. A. Carnes, J. V. Doerger, and H. L. Sparks, *Arch. Environ. Contam. Toxicol.*, 1973, **1**, 27.
[534] M. Ahnoff and B. Josefsson, *Analyt. Letters*, 1973, **6**, 1083.
[535] J. Freudenthal and P. A. Greve, *Bull. Environ. Contam. Toxicol.*, 1973, **10**, 108.
[536] J. R. Plimmer, *Environ. Health Perspect.*, 1973, **5**, 41.
[537] J. W. Eichelberger, L. E. Harris, and W. L. Budde, *Analyt. Chem.*, 1974, **46**, 227.
[538] N. E. Delfel, *J. Assoc. Offic. Analyt. Chemists*, 1973, **56**, 1343.
[539] W. W. Shindy, L. S. Jordan, V. A. Jolliffe, C. W. Coggins, jun., and J. Kamamoto, *J. Agric. Food Chem.*, 1973, **21**, 629.
[540] S. Selim and J. N. Seiber, *J. Agric. Food Chem.*, 1973, **21**, 430.
[541] G. Pellegrini and R. Santi, *J. Agric. Food Chem.*, 1972, **20**, 944.
[542] G. L. Baughman, M. H. Carter, N. L. Wolf, and R. G. Zepp, *J. Chromatog.*, 1973, **76**, 471.
[543] S. Ohkoshi, T. Takahashi, and T. Sato, *Bunseki Kagaku*, 1973, **22**, 593.
[544] P. Mushak, *Environ. Health Perspect.*, 1973, **4**, 55.
[545] K. Grob, *Chem. and Ind.*, 1973, 248.
[546] E. V. Brown and I. Ahmad, *Phytochemistry*, 1972, **11**, 3485.
[547] K. H. Bergert, V. Betz, and D. Pruggmayer, *Chromatographia*, 1974, **7**, 115.
[548] B. J. Tyson and G. C. Carle, *Analyt. Chem.*, 1974, **46**, 568.

publications. Some additional reports on analysis of lunar data have appeared,[549] and further details of the g.c.m.s. instrument which is to be included in the first proposed Viking mission have been given.[550] The identification of organic compounds in meteorites[551] and the detection of extraterrestrial biopolymers[552] have been reviewed, and further analyses of the Murchison meteorite[553] and carbonaceous chondrites[554] have been performed.

Aromatic hydrocarbons[555,556] and C_{28}- and C_{29}-stanols[557] have been found in Green River shale, and alicyclic hydrocarbons in Israeli[558] and African[559] shales. Methods were described for the characterization of amino-acids,[345] humic acids,[560] and polyphenols[561] in soil, and carboxylic acids in tasmanite.[562] Lignin oxidation products[563] and lignite components[564] have been identified. Numerous organic compounds have been found in various sediments[565—568] and bitumens,[569,570] and aliphatic hydrocarbons in weathered limestone have been characterized.[571]

G.c.m.s. has been of considerable use in the petroleum industry, with applications in exploration[572] and analysis of petroleum[573,574] and petroleum

---

[549] B. Nagy, M. A. J. Mohammed, and V. E. Modzeleski, *Space Life Sci.*, 1972, **3**, 323; C. W. Gehrke, *ibid.*, p. 342; E. K. Gibson and C. B. Moore, *ibid.*, p. 404; R. C. Murphy, *ibid.*, p. 450; D. A. Flory and B. R. Simoneit, *ibid.*, p. 457.
[550] B. Miller, *Aviation Week Space Technol.*, 1972, **97**, (13) 67.
[551] E. Anders, R. Hayatsu, and M. H. Studier, *Science*, 1973, **182**, 781.
[552] W. Henderson, W. C. Kray, and M. Calvin, *Space Life Sci.*, 1973, **4**, 45.
[553] R. L. Levy, M. A. Grayson, and C. L. Wolf, *Geochim. Cosmochim. Acta*, 1973, **37**, 467; G. U. Yuen and K. A. Kvenvolden, *Nature*, 1973, **246**, 302.
[554] C. F. Folsome, J. G. Lawless, M. Romiez, and C. Ponnamperuma, *Geochim. Cosmochim. Acta*, 1973, **37**, 455.
[555] D. E. Anders, F. G. Doolittle, and W. D. Robinson, *Geochim. Cosmochim. Acta*, 1973, **37**, 1213.
[556] E. J. Gallegos, *Analyt. Chem.*, 1973, **45**, 1399.
[557] G. Steel and W. Henderson, *Nature*, 1972, **238**, 148.
[558] R. Ikan and A. Bortinger, *Israel J. Chem.*, 1971, **9**, 679.
[559] C. Spyckerelle, P. Arpino, and G. Ourisson, *Tetrahedron*, 1972, **28**, 5703.
[560] S. U. Khan and M. Schnitzer, *Israel J. Chem.*, 1971, **9**, 667.
[561] H. Morita, *Geochim. Cosmochim. Acta*, 1973, **37**, 1587.
[562] B. R. Simoneit and A. L. Burlingame, *Geochim. Cosmochim. Acta*, 1973, **37**, 595.
[563] M. Erickson, S. Larsson, and G. E. Miksche, *Acta Chem. Scand.*, 1973, **27**, 127.
[564] J. Kubat and V. Vcelak, *Fette, Seifen, Anstrichm.*, 1973, **75**, 28.
[565] R. Ikan, M. J. Baedecker, and I. R. Kaplan, *Nature*, 1973, **244**, 154.
[566] G. Kovacev, D. Delova, and K. Stransky, *Coll. Czech. Chem. Comm.*, 1972, **37**, 4106.
[567] Z. Aizenshtat, M. J. Baedecker, and I. R. Kaplan, *Geochim. Cosmochim. Acta*, 1973, **37**, 1881.
[568] J. R. Maxwell, R. E. Cox, G. Eglinton, C. T. Pillinger, R. G. Ackman, and S. N. Hooper, *Geochim. Cosmochim. Acta*, 1973, **37**, 297.
[569] H. S. Hertz, B. D. Andersen, M. V. Djuričic, and K. Biemann, *Phytochemistry*, 1973, **37**, 1687.
[570] D. W. Nooner, W. S. Updegrove, D. A. Flory, J. Oró, and G. Müller, *Chem. Geol.*, 1973, **11**, 189.
[571] D. W. Nooner, J. Oró, J. M. Gibert, V. L. Ray, and J. E. Mann, *Geochim. Cosmochim. Acta*, 1972, **36**, 953.
[572] D. H. Welte, *J. Geochem. Exploration*, 1972, **1**, 117.
[573] B. Wallaert, *Chimie et Industrie*, 1971, **104**, 1531.
[574] H. J. Coleman, J. E. Dooley, D. E. Hirsch, and C. J. Thompson, *Analyt. Chem.*, 1973, **45**, 1724.

products.[575—577] Degradation of petroleum by micro-organisms has been studied.[578,579]

Miscellaneous.—Short-chain diols (C_2—C_5) have been studied as cyclic acetals of hexadecanol[580] and as TMS ethers.[581] Eicosatrienoic acid ω-9 was found in serum lipids of one in every three patients with hepatocellular carcinoma, but was absent from other cases and from normal serum.[582] Lipids of plant tissue cultures grown with lunar and terrestrial soil have been compared.[583] The anal sac secretion of the red fox has been analysed.[584] New natural coumarins have been identified,[585] and the characterization of perdeuteriomethylated flavonoid aglycones by g.c.m.s. has been studied.[586] Characteristic fragmentations of methylsterols have been collated.[587]

G.c.m.s. has been applied to determine traces of permanent gases.[588] Data have been reported for methyl cyclosiloxanes[589] and for S-alkyl derivatives of NN-dialkyldithiocarbamates.[590] The contents of aerosol projectors have been analysed.[591]

Biomedical applications of g.c.m.s. were presented at a recent symposium.[592]

---

[575] J. T. Swansiger and F. E. Dickson, *Analyt. Chem.*, 1973, **45**, 1724.
[576] V. Kubelka, J. Mitera, and P. Zachař, *J. Chromatog.*, 1972, **74**, 195.
[577] I. Ötvös, S. Iglewski, D. H. Hunneman, B. Bartha, Z. Baltházar, and G. Pályi, *J. Chromatog.*, 1973, **78**, 309.
[578] N. J. L. Bailey, A. M. Dobson, and M. A. Rogers, *Chem. Geol.*, 1973, **11**, 203.
[579] D. Volfova and K. Pecka, *Folia Microbiol. (Prague)*, 1973, **18**, 286.
[580] E. Schupp and W. J. Baumann, *J. Lipid Res.*, 1973, **14**, 121.
[581] K. Hasegawa, M. Murata, and T. Suzuki, *Yukagaku*, 1972, **21**, 383.
[582] N. Okazaki and E. Araki, *Clinica Chim. Acta*, 1974, **53**, 11.
[583] J. L. Laseter, J. D. Weete, P. S. Baur, and C. H. Walkinshaw, *Space Life Sci.*, 1973, **4**, 352.
[584] E. S. Albone, G. Eglinton, J. M. Walker, and G. C. Ware, *Life Sci.*, 1974, **14**, 387.
[585] D. E. Games, *Tetrahedron Letters*, 1972, 3187.
[586] R. D. Schmid, R. Mues, J. H. McReynolds, G. Vander Velde, N. Nakatani, E. Rodríguez, and T. J. Mabry, *Phytochemistry*, 1973, **12**, 2765.
[587] T. Iida, T. Tamura, K. Satomi, C. Hirai, Y. Sasaki, and T. Matsumoto, *Yukagaku*, 1974, **23**, 233.
[588] L. Giry, M. Chaigneau, and L. P. Ricard, *Analusis*, 1973, **2**, 163, 475.
[589] W. J. A. VandenHeuvel, J. L. Smith, R. A. Firestone and J. L. Beck, *Analyt. Letters*, 1972, **5**, 285.
[590] F. I. Onuska and W. R. Boos, *Analyt. Chem.*, 1973, **45**, 967.
[591] A. A. Casselman, R. A. B. Bannard, and R. F. Pottie, *J. Chromatog.*, 1973, **80**, 155.
[592] Summaries of the Second International Symposium on Mass Spectrometry in Biochemistry and Medicine, Milan, June 1974, ed. A. Frigerio and A. Leonardi.

9
Drug Metabolism

BY B. J. MILLARD

1 Introduction

Although some scattered references to drug metabolism appeared in Volume 1, and applications of gas chromatography–mass spectrometry (g.c.m.s.) in this field were covered in Volume 2, a comprehensive review of this subject is new to this series. Because of this, it has been felt advisable to cover rather more than the usual two year period. References in the literature which follow those mentioned in an earlier comprehensive review[1] have been covered to May 1974. However, in the case of more obscure journals, only references appearing in the *Mass Spectrometry Bulletin* up to the July 1974 issue have been included. In order to avoid excessive overlap with Chapter 8, references which utilize g.c.m.s. in a qualitative sense appear in that Chapter, while quantitative aspects are covered here.

The major emphasis in the use of mass spectrometry in the field of drug metabolism has moved away from the identification of new metabolites to their quantitative estimation. Where this is carried out by multiple ion monitoring in a g.c.m.s. system, more attention is being paid to the proper choice of internal standard. As well as ^{2}H-labelled standards, several papers report the use of the more metabolically secure ^{13}C label, foreshadowing its use as a tracer. The use of chemical ionization or field desorption as alternative ionization methods is also becoming more widespread as more laboratories update their instruments.

Recent reviews and textbooks include applications of mass spectrometry in drug research,[1,2] drug metabolism[3] and problems in medicine and biochemistry,[4] an account by Hammar of his work with Holmstedt on mass fragmentography,[5] mass fragmentography,[6] the use of stable isotopes in pharmacology and clinical

---

[1] B. J. Millard, in 'Advances in Drug Research', ed. N. J. Harper and A. B. Simmonds, Academic Press, New York, 1971, Vol. 6, pp. 157–231.
[2] 'Gas Chromatography–Mass Spectrometry in the Study of Drugs', ed. A. Frigerio, Tamburini Editore, Milan, 1973.
[3] M F. Grostic, in 'Biochemical Applications of Mass Spectrometry', ed. G. R. Waller, Wiley-Interscience, New York, 1971, pp. 573–590.
[4] 'Mass Spectrometry, Techniques and Applications', ed. G. W. A. Milne, Wiley-Interscience, New York, 1971, pp. 327–372.
[5] C.-G. Hammar, *Acta Pharm. Suecica*, 1971, **8**, 129.
[6] A. E. Gordon and A. Frigerio, *J. Chromatog.*, 1972, **73**, 401.

pharmacology,[7] pharmacological applications of g.c.m.s.,[8,9] a book with chapters on mass fragmentography, chemical ionization, and mass spectrometry applied to a study of brain alkylamines,[10] and the published proceedings of the first International Symposium on Mass Spectrometry in Medicine and Biochemistry, held in Milan.[11] A new journal introduced in 1974, *Biomedical Mass Spectrometry*, contains a substantial proportion of papers devoted to drug metabolism.

2 Low-resolution Mass Spectrometry

Reference Compounds.—There are still many adherents to the view that the mass spectra of a series of drugs and synthetic possible metabolites should be investigated before studying biological material. Certainly a knowledge of the mass spectrum of a possible metabolite enables the correct ions for maximum sensitivity and selectivity to be chosen in multiple ion monitoring studies. Although the results of such synthetic studies are usually not published, a few papers have appeared recently.

The characteristic mass spectrum of methaqualone (1) has been used to identify this substance in organ material obtained after autopsy.[12] In a study of the fragmentation of the monohydroxy-derivatives,[13] it was shown that much information as to the position of hydroxylation could be obtained from the mass spectra. The mass spectra of two hydroxymethyl analogues of (1) have also been described.[14]

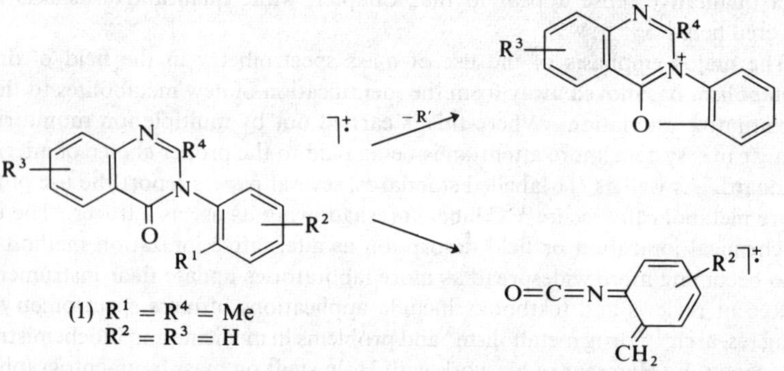

(1) $R^1 = R^4 = Me$
    $R^2 = R^3 = H$

---

[7] D. R. Knapp and T. E. Gaffney, *Clin. Pharm. Exp. Ther.*, 1972, **13**, 307.
[8] D. J. Jenden and A. K. Cho, *Ann. Rev. Pharmacol.*, 1973, **13**, 371.
[9] 'Techniques of Combined Gas Chromatography–Mass Spectrometry. Applications in Organic Analysis', ed. W. H. McFadden, Wiley–Interscience, New York, 1973.
[10] 'Advances in Biochemical Psychopharmacology', ed. E. Costa and B. Holmstedt, Raven Press, New York, 1973, Vol. 7.
[11] 'Proceedings of the International Symposium on Mass Spectrometry in Biochemistry and Medicine,' Milan, 1973, ed. A. Frigerio, Raven Press, New York, 1974.
[12] G. Bohn and G. Rucker, *Z. klin. Chem. klin. Biochem.*, 1972, **10**, 346.
[13] C. Bogentoft, O. Ericsson, B. Danielsson, J.-E. Lindgren, and B. Holmstedt, *Acta Pharm. Suecica*, 1972, **9**, 151.
[14] C. Bogentoft, O. Ericsson, and B. Danielsson, *Acta Pharm. Suecica*, 1974, **11**, 59.

Several papers report the mass spectra of hallucinogens such as 2,5-dimethoxy-amphetamine[15,16] and 2,5-dimethoxy-4-methylamphetamine.[17] In the latter case as little as 100 pg could be detected by the use of single ion monitoring. The mass spectra of the N-trifluoroacetyl derivatives of amphetamine, phentermine, and p-methoxyamphetamine have also been reported.[18] The mass spectrum of 1-(2-thioxopyrrolidino)-4-pyrrolidinobut-2-yne, a thiolactam analogue of tremorine, has been described.[19] Most of the vast amount of mass spectrometric research on cannabis has entailed the use of g.c.m.s., and is reported in Chapter 8, but the direct-inlet mass spectra of various metabolites of Δ^8- and Δ^9-tetrahydrocannabinol, e.g. 11-hydroxy-Δ^9-THC, have been reported.[20]

The mass spectra of a group of tranquillizers, including haloperidol and related compounds, have been discussed,[21] as have the mass spectra of five aryloxy β-blocking drugs, including propranolol (2) and its hydroxylated (3) and de-isopropylated (4) metabolites.[22]

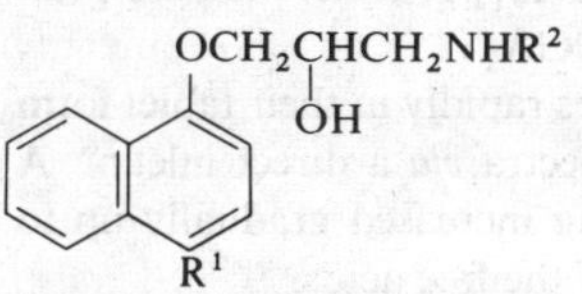

(2) $R^1 = H, R^2 = Pr^i$
(3) $R^1 = OH, R^2 = Pr^i$
(4) $R^1 = R^2 = H$

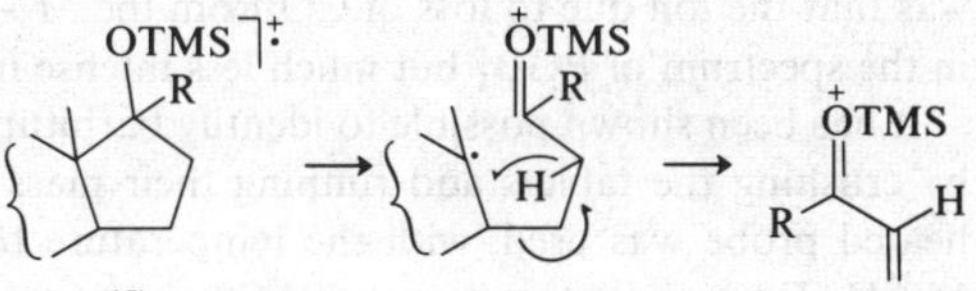

m/e (128 + R)

The mass spectrum of a hydroxy metabolite of levallorphan agreed with the structure (from X-ray analysis) being $(-)$-N-allyl-3,6β-dihydroxymorphinan.[23] The ring D fragmentation of a number of steroid drugs and metabolites which contain a 17-hydroxy-group and a second substituent such as CH_3, C_2H_5, and C_2H have been investigated as trimethylsilyl ethers.[24] The fragmentation of (5) for example, the mechanism of which was elucidated by deuterium labelling, gives a characteristic 128 + R ion, and therefore the nature of R can be determined.

[15] K. Bailey, Analyt. Chim. Acta, 1972, 60, 287.
[16] R. C. Shaler and J. J. Padden, J. Pharm. Sci., 1972, 61, 1851.
[17] A. Frigerio, R. Fanelli, and B. Danielli, Chem. and Ind., 1972, 769.
[18] B. Lindeke and A. K. Cho, Acta Pharm. Suecica, 1973, 10, 171.
[19] B. Lindeke, B. Karlen, R. Dahlbom, R. Green, and J. D. Jenden, J. Pharm. Pharmacol., 1972, 24, 25.
[20] C. G. Pitt, F. Hauser, R. L. Hawks, S. Sathe, and M. E. Wall, J. Amer. Chem. Soc., 1972, 94, 8578.
[21] B. Blessington, Org. Mass Spectrometry, 1971, 5, 1113.
[22] D. A. Garteiz and T. Walle, J. Pharm. Sci., 1972, 61, 1728.
[23] J. F. Blount, E. Mohacsi, F. M. Vane, and G. J. Mannering, J. Medicin. Chem., 1973, 16, 352.
[24] B. S. Middleditch, P. Vouros, and C. J. W. Brooks, J. Pharm. Pharmacol., 1973, 25, 143.

Although the mass spectra of prostaglandins have been obtained almost invariably *via* a g.c.m.s. system, the direct inlet spectra of the TMS ester TMS ether derivatives of prostaglandins B_1 (6) and B_2 (7) and of the corresponding *O*-methyl and *O*-ethyl oximes have been described.[25] Although the major ions were formed by the same processes as in the PGA series, the major difference

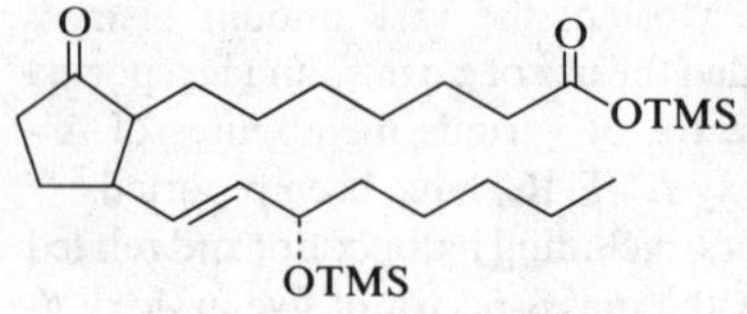

(6) 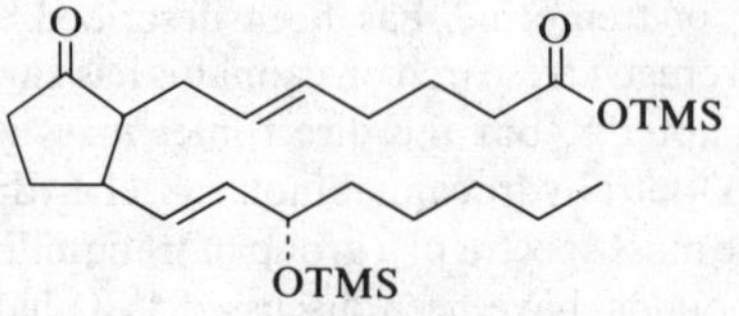   (7)

was that the ion due to loss of CO from the $M - C_5H_{11}$ ion was the base peak in the spectrum of PGB_1 but much less intense in PGA_1.

It has been shown possible to identify barbiturates rapidly in their tablet form by crushing the tablets and running their mass spectra *via* a direct inlet.[26] A heated probe was used, with the temperature being increased gradually up to 150°C. Even sodium salts gave the mass spectra of the free acids.

Chemical ionization mass spectrometry has been used to study 17-hydroxy-steroids[27] and some macrolide antibiotics such as erythromycin B.[28] In the latter case, besides the prominent MH^+ ions formed, some fragment ions were present which had been produced by cleavage at the glycosidic bonds.

Metabolites Separated by Thin-layer Chromatography.—Although the application of g.c.m.s. to drug metabolic studies steals more than its fair share of the limelight, many drugs that possess good g.c. properties give rise to metabolites that are not amenable to this technique. Tertiary amines can give rise to *N*-oxides, secondary amines may give hydroxylamines,[29] primary amines may form oximes *via* hydroxylamines, and oximes may also be formed from imines, as demonstrated for 2,4,6-trimethylacetophenone imine.[30]

After separation by t.l.c., the major *in vitro* metabolite of fenfluramine (8) in microsomal fractions of guinea-pig liver was shown by mass spectrometry to be the nitrone (9)[31] having a weak but detectable molecular ion at m/e 230 (3%). Other important ions were (10) (84%), (11) (100%), and (12) (63%). The hydroxylamine (13) was also found to be a metabolite. Another recent study of phenylalkylamines using g.c.m.s. of the TMS derivatives has also shown hydroxylamines

---

[25] B. S. Middleditch and D. M. Desiderio, *Lipids*, 1973, **8**, 271.
[26] J. D. McChesney, D. K. Beal, and R. M. Fox, *J. Pharm. Sci.*, 1972, **61**, 310.
[27] J. Michnowicz and B. Munson, *Org. Mass Spectrometry*, 1972, **6**, 765.
[28] L. A. Mitscher and H. D. H. Showalter, *J.C.S. Chem. Comm.*, 1972, 796.
[29] A. H. Beckett, J. M. Van Dyk, H. H. Chissick, and J. W. Gorrod, *J. Pharm. Pharmacol.*, 1971, **23**, 809.
[30] C. J. Pavli, N. Wang, and R. E. McMahon, *J. Biol. Chem.*, 1971, **246**, 6953.
[31] A. H. Beckett, R. T. Coutts, and F. A. Ogunbona, *J. Pharm. Pharmacol.*, 1973, **25**, 190.

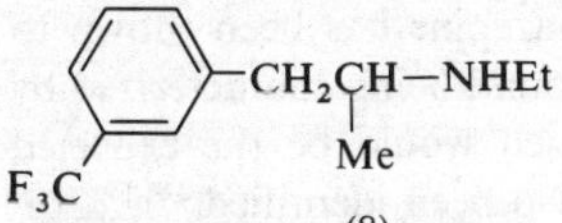

[Structures (8), (9), (11), (10), (12), (13) shown]

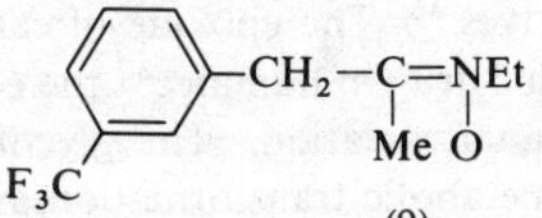

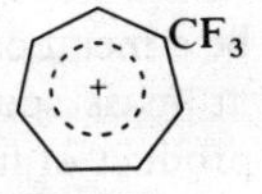

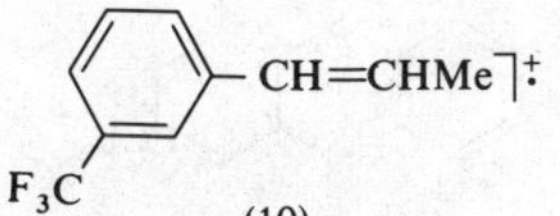

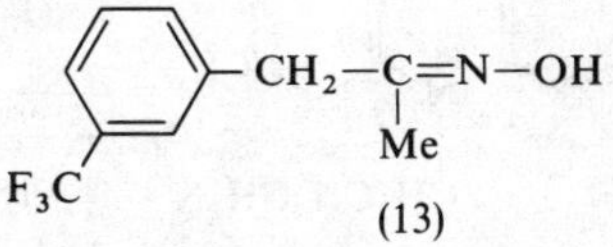

to be the major products of microsomal oxidation.[32] Hydroxylamines are now recognized as important metabolites, and the hydroxylamines (14) and (15), derived from N-demethylchlorpromazine and its sulphoxide, were found to be present in the red blood cells of patients taking chlorpromazine.[33] The structures were deduced from the presence of an ion corresponding to (16) in each spectrum.

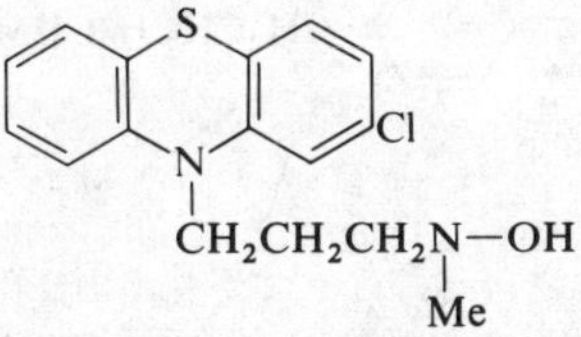

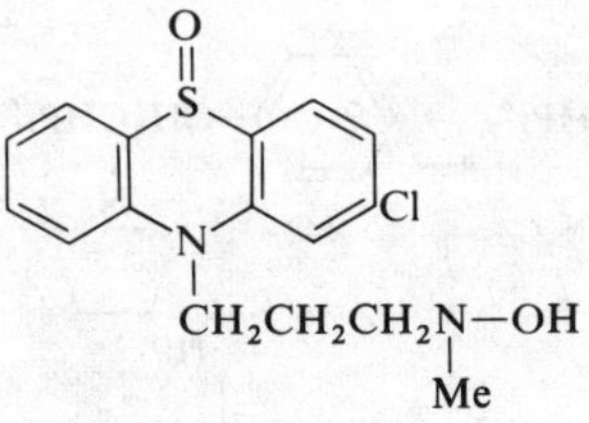

[Structures (14), (15), (16) shown]

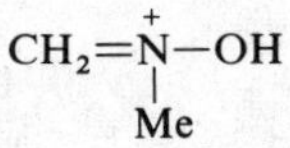

In the case of mescaline, (3,4,5-trimethoxyphenylethylamine), another phenylalkylamine, the N-acetyl derivative, and 3,4,5-trimethoxybenzoic acid were found to be the major *in vivo* metabolites.[34]

In a number of phenothiazine derivatives in which the side-chain nitrogen atom was present either as a dimethylamino-group or as part of a piperazine ring, such as fluphenazine (17), the primary amine sulphoxide such as (18) was produced *in vivo*.[35]

Recently much work has centred on the detection of epoxides as intermediates in the formation of glycols. In the case of barbiturates such as alphenal, secobarbital, and allobarbital, the epoxides found in rat urine were identified as

---

[32] B. Lindeke, A. K. Cho, J. L. Thomas, and L. Michelson, *Acta Pharm. Suecica*, 1973, **10**, 493.
[33] A. H. Beckett and E. E. Essien, *J. Pharm. Pharmacol.*, 1973, **25**, 188.
[34] N. Suler and L. Demisch, *Biochem. Pharmacol.*, 1974, **23**, 259.
[35] U. Breyer, H. J. Gaertner, and A. Prox, *Biochem. Pharmacol.*, 1974, **23**, 313.

chloro-TMS derivatives.[36] The epoxide of carbamazepine has been shown to be a metabolite of this drug in humans,[37] the compound being characterized by its mass spectrum after isolation. The glycol, which would be the expected product of further metabolic transformation, has also been identified.[38,39]

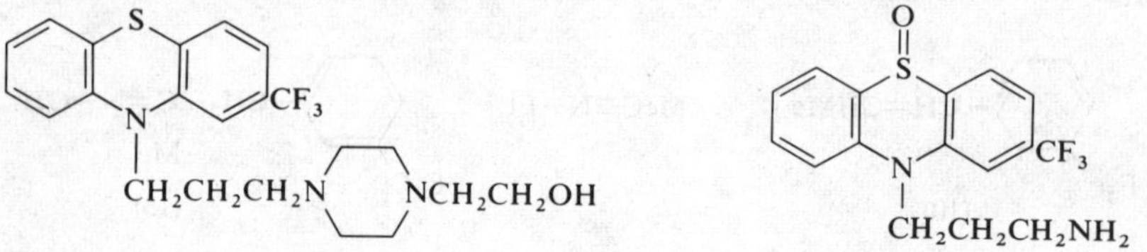

(17)

(18)

The biotransformation of pyrrolidine compounds such as nicotine and tremorine to pyrrolidone derivatives has been known for several years.[40] Similar derivatives have been found recently for prolintane (19). The simple metabolite (20) has been isolated by t.l.c. after oxidation of ^{14}C-labelled (19) by rabbit liver microsomes,[41] while *in vivo* studies in the rabbit have resulted in the isolation of compounds (20)—(24), with (22)—(24) as conjugates.[42] The tetrahydrofurfuryl

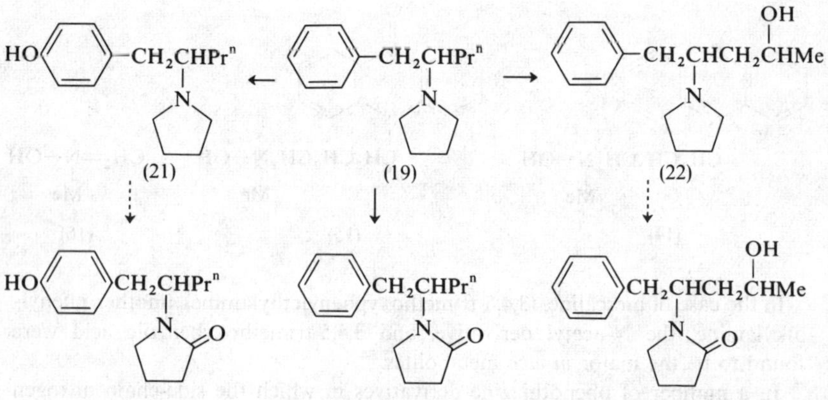

(21)   (19)   (22)

(24)   (20)   (23)

[36] M. G. Horning, D. J. Harvey, E. C. Horning, and R. M. Hill, Proceedings of the International Symposium of Gas Chromatography–Mass Spectrometry, Elba, 1972, ed. A. Frigerio, Tamburini Editore, Milan, 1972, pp. 405—429.
[37] A. Frigerio, R. Fanelli, P. Biandrate, G. Passerini, P. C. Morselli, and S. Garattini, *J. Pharm. Sci.*, 1972, **61**, 1144.
[38] K. M. Baker, J. Csetenyi, A. Frigerio, P. C. Morselli, F. Parravicini, and G. Pifferi, *J. Medicin. Chem.*, 1973, **16**, 703.
[39] S. Goenechea and E. Hecke-Seibicke, *Z. klin. Chem. klin. Biochem.*, 1972, **10**, 112.
[40] C.-G. Hammar, W. Hammer, B. Holmstedt, B. Karlen, F. Sjoqvist, and J. Vessman, *Biochem. Pharmacol.*, 1969, **18**, 1549.
[41] H. B. Hucker, S. C. Stauffer, and R. E. Rhodes, *Experientia*, 1972, **28**, 430.
[42] S. Yoshihara and H. Yoshimura, *Chem. and Pharm. Bull. (Japan)*, 1972, **20**, 1906.

Drug Metabolism

derivative (25) underwent the equivalent oxidation to the lactone (26) before ring opening to (27).[43] Mass spectrometry has also been used as additional confirmation of the structures of metabolites of DL-methadone in man and rat.[44]

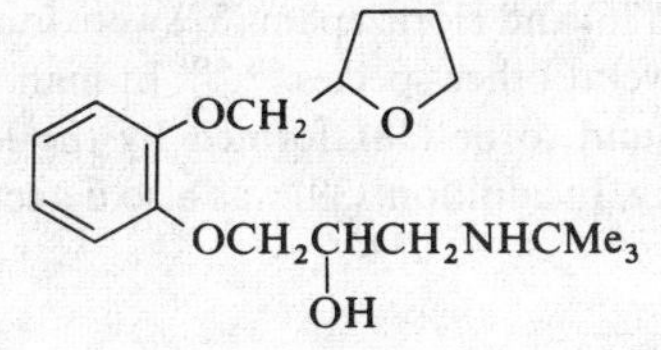

(25)

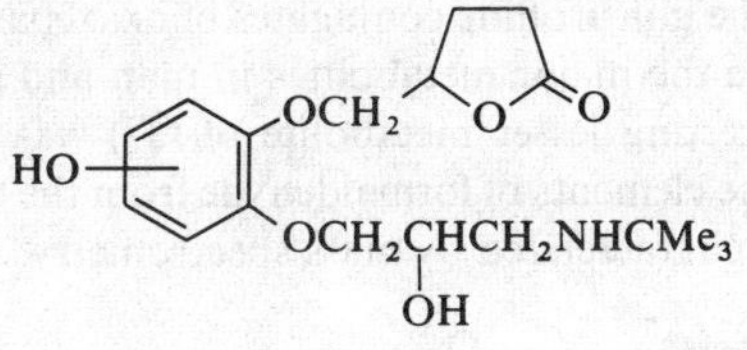

(26)

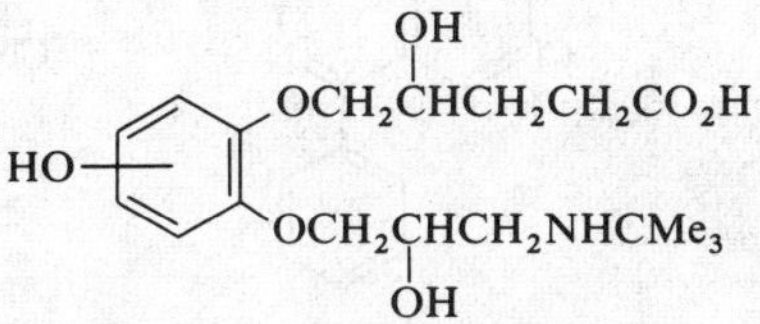

(27)

The metabolism of simple acetamidothiazoles such as (28)—(30) has been investigated.[45] The unexpected ring-opened metabolites (31)—(33) were identified by their mass spectra. Compounds (28) and (29) also gave rise to (34). Another simple heterocyclic compound, 4-methylpyrazole, behaved more normally and did not give ring-opened metabolites. Compounds such as 4-carboxypyrazole, 4-hydroxymethylpyrazole, and 4-methylpyrazole-*N*-glucosiduronic acid were formed.[46] Another substituted pyrazole, benzydamine (35),[47] has

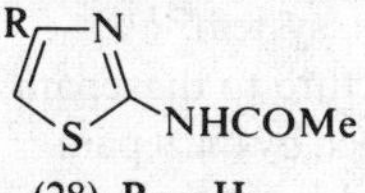

MeCONHCSNHCHCO$_2$H
    |
    R

MeCONHCSNH$_2$
(34)

(28) R = H
(29) R = Me
(30) R = Ph

(31) R = H
(32) R = Me
(33) R = Ph

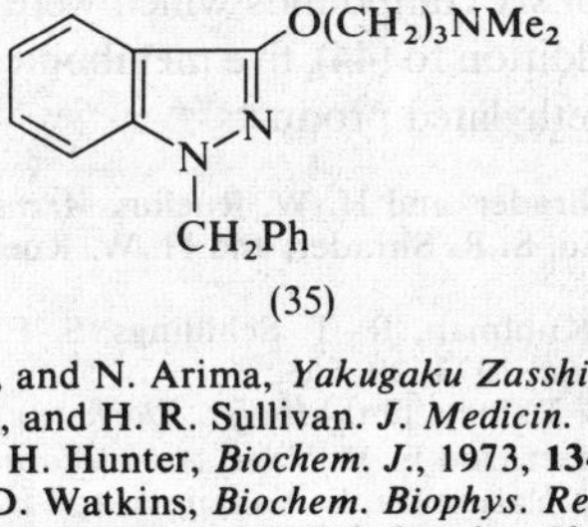

(35)

[43] M. Nakanishi, Y. Kato, and N. Arima, *Yakugaku Zasshi*, 1972, **92**, 299.
[44] A. Pohland, H. E. Boaz, and H. R. Sullivan. *J. Medicin. Chem.*, 1971, **14**, 194.
[45] D. H. Chatfield and W. H. Hunter, *Biochem. J.*, 1973, **134**, 869.
[46] R. C. Murphy and W. D. Watkins, *Biochem. Biophys. Res. Comm.*, 1972, **49**, 283.
[47] S. Kataoka, K. Taira, T. Ariyoshi, and E. Takabataka, *Chem. and Pharm. Bull. (Japan)*, 1973, **21**, 358.

also been investigated using t.l.c. and mass spectrometry. In the rabbit, the major metabolites were found to be the N-oxide and the p-hydroxybenzyl compound. Other minor metabolites were formed by N-demethylation and N-debenzylation.

The glucuronide conjugates of oxazepam (36) and clorazepam (37) were found to be the major metabolites in man and several other species.[48,49] In man, an interesting lesser metabolite of (37) was found to be (38), formed by the loss of the elements of formaldehyde from the ring. In addition, (39) was also detected and characterized by mass spectrometry.

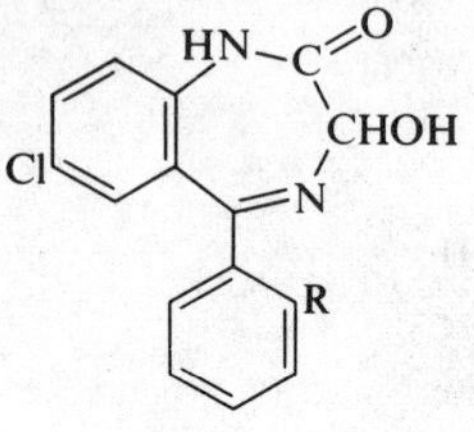

(36) R = H
(37) R = Cl
(38) 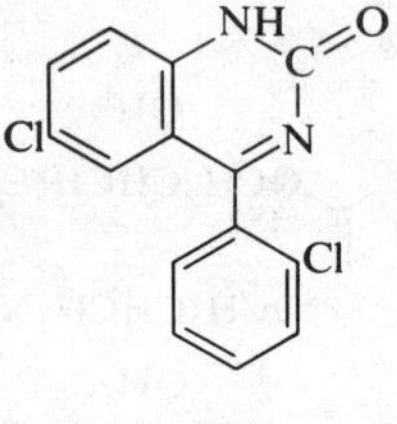
(39) 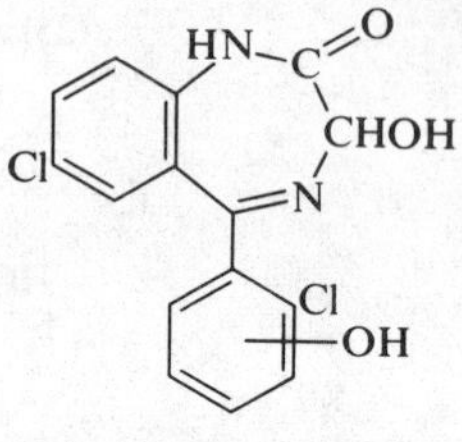

Drugs (36) and (37) were characterized after initial enzymatic cleavage of the glucuronide linkage. In a recent paper,[50] an MS 902 fitted with a chemical ionization source was used to obtain spectra of the intact glucuronides. The latter had been obtained from urine by the use of an XAD-2 column. After formation of the pentatrimethylsilyl derivatives, the appropriate MH$^+$ ions were readily obtained at m/e 823 and 857, respectively. It is to be expected that much research will be carried out on the analysis of intact conjugates by mass spectrometry, especially as a recent paper has shown that the TMS derivatives of some simple glucuronides can be passed through a g.c.m.s. system.[51]

In a study of cloxazolam (40), a compound similar in structure to diazepam, in addition to ring-hydroxylated products similar to those formed by diazepam,[52] the ring-opened metabolites (41), (42), and (43) were identified from their mass spectra.[53] In the case of dibenzepine (44), a preliminary t.l.c. separation gave four spots which were eluted and acetylated before being subjected to g.c. This resulted in the trapping of six compounds which were examined by direct-inlet mass spectrometry. In addition to (44), five metabolites were identified as being ring- and side-chain-demethylated products.[54]

[48] R. T. Schillings, S. R. Shrader, and H. W. Ruelius, *Arzneim.-Forsch.*, 1971, **21**, 1059.
[49] S. F. Sisenwine, C. O. Tio, S. R. Shrader, and H. W. Ruelius, *Arzneim.-Forsch.*, 1972, **22**, 682.
[50] T. T. L. Chang, C. F. Kuhlman, R. T. Schillings, S. F. Sisenwine, C. O. Tio, and H. W. Ruelius, *Experientia*, 1973, **29**, 653.
[51] J. E. Mrochek and W. T. Rainey, jun., *Analyt. Biochem.*, 1974, **57**, 173.
[52] M. A. Schwartz, P. Bommer, and F. M. Vane, *Arch. Biochem. Biophys.*, 1967, **121**, 508.
[53] H. Murata, K. Kongo, A. Yasumura, E. Nakajima, and H. Shundo, *Chem. and Pharm. Bull. (Japan)*, 1973, **21**, 404.
[54] A. DeLeenheer and A. Heyndrickx, *J. Pharm. Sci.*, 1973, **62**, 31.

A combination of t.l.c. followed by g.c.m.s. was utilized in the elucidation of the metabolism of (45) in rats and humans.[55] Surprisingly, demethylation was

(40) (41) (43)

(44) (42)

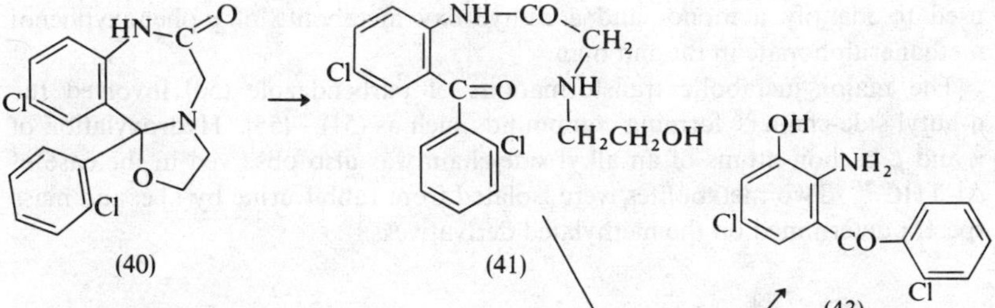

not observed, but the reduced acetylated compound (46) was identified. With another pyrimidine compound, piribedil (47), hydroxylation of the 5-position of the pyrimidine ring was the major metabolic pathway in humans.[56] In the rat the *N*-oxide (48) was formed. Synthetic samples of the two *N*-oxides (48)

(45) (46)

(47)

(48)

(49)

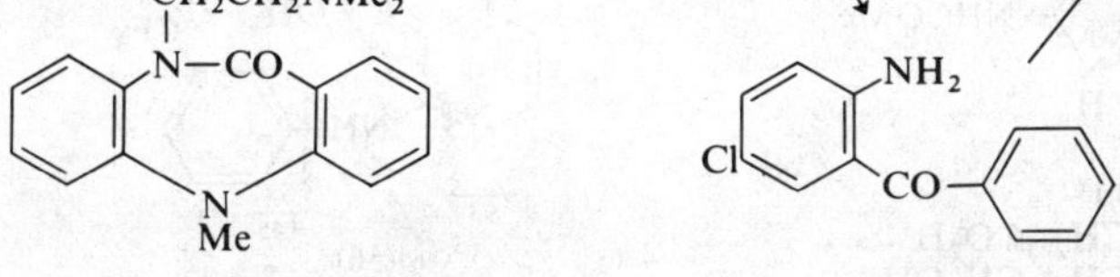

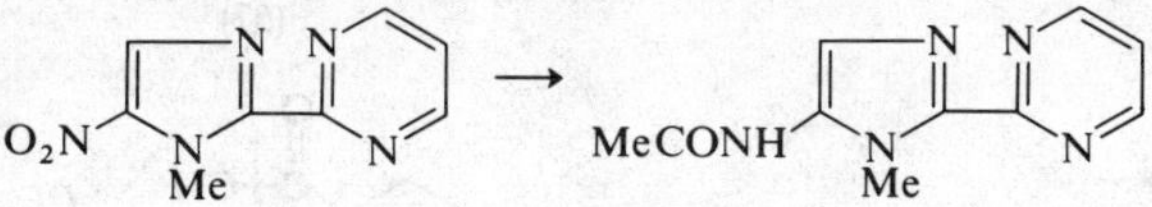

[55] K. M. Baker, M. Coerezza, L. Del Corona, A. Frigerio, G. G. Massaroli, and G. Sekules, *J. Pharm. Sci.*, 1974, **63**, 293.

[56] B. J. Millard, D. B. Campbell, P. Jenner, and A. R. Taylor, Proceedings of the International Symposium on Mass Spectrometry in Biochemistry and Medicine, Milan, 1973, Raven Press, New York, 1974, p. 1.

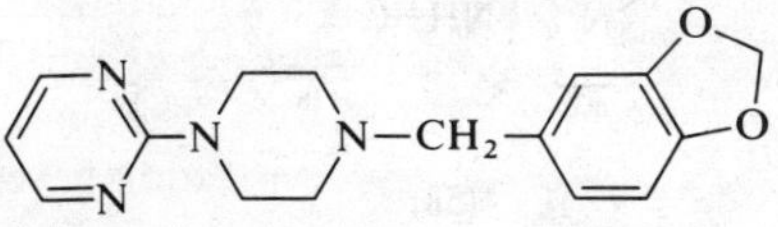

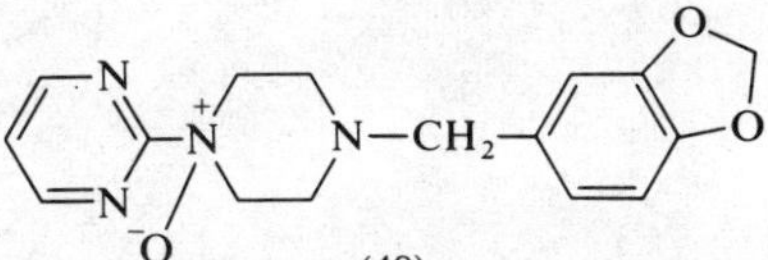

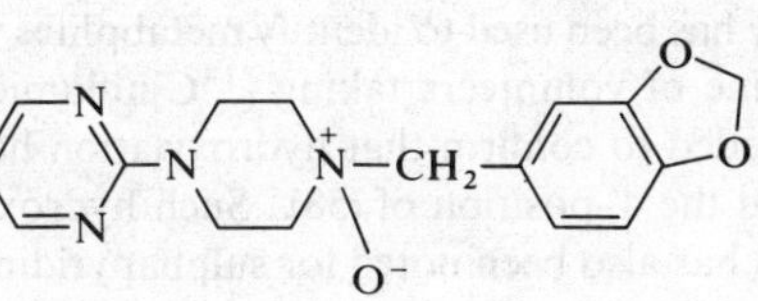

and (49) gave surprisingly different mass spectra, (49) being so stable that loss of oxygen, hydroxyl, or water did not occur.

A preliminary separation by t.l.c. followed by g.c.m.s. was also the procedure used to identify a mono- and a di-hydroxy metabolite of *p*-phenoxyphenol methanesulphonate in rat and man.[57]

The major metabolic transformations of parbendazole (50) involved the n-butyl side-chain,[58] forming compounds such as (51)—(55). Hydroxylation of α and β carbon atoms of an alkyl side-chain was also observed in the case of Δ^1-THC.[59] Two metabolites were isolated from rabbit urine by t.l.c. and mass spectra determined on the methylated derivatives.

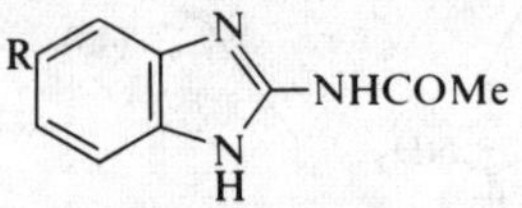

(50) R = Bu^n
(51) R = (CH$_2$)$_3$CO$_2$H
(52) R = (CH$_2$)$_3$CH$_2$OH
(53) R = CH(OH)Et
(54) R = CH(OH)Pr^n
(55) R = (CH$_2$)$_3$COMe

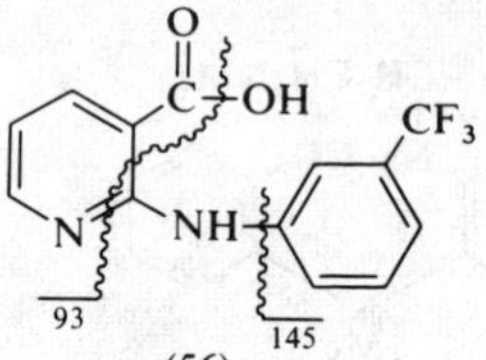

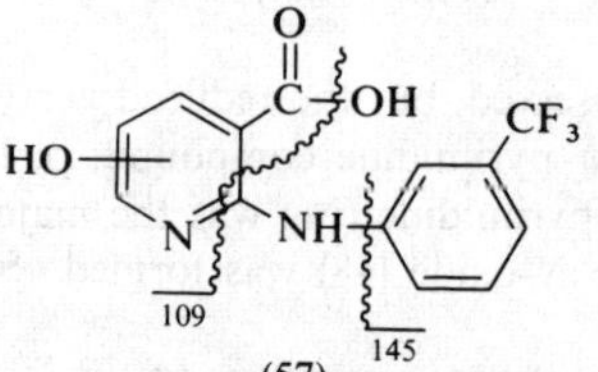

Mass spectrometry has been used to identify metabolites which were separated by t.l.c. from the urine of volunteers taking [^{14}C]niflumic acid (56).[60] N.m.r. spectroscopy was needed to confirm that hydroxylation had taken place at the 5-position of (57) and the 4'-position of (58). Such hydroxylation of a pyridine ring in the 5-position has also been noted for sulphapyridine.[61] The metabolite,

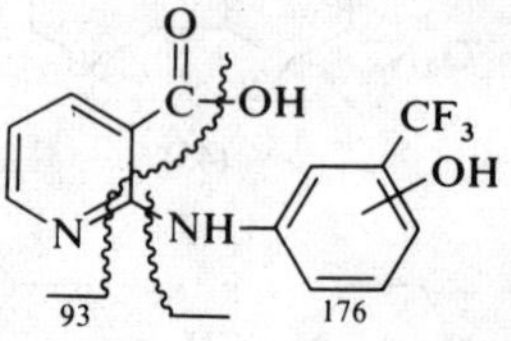

[57] F. F. Sun, *J. Pharm. Sci.*, 1973, **62**, 1780.
[58] G. L. Duma, G. Gallagher, L. D. Davis, and J. R. E. Hoover, *J. Medicin. Chem.*, 1973, **16**, 996.
[59] S. Burnstein, J. Rosenfeld, and T. Wittstruck, *Science*, 1972, **176**, 422.
[60] A. I. Cohen, I. Weliky, S. J. Lan, and S. D. Levine, *Biomed. Mass Spectrometry*, 1974, **1**, 1.
[61] H. Schröder and B. M. Schröder, *Acta Pharm. Suecica*, 1973, **10**, 263.

excreted mainly as the glucuronide in humans, was identified by mass spectrometry after hydrolysis.

In addition to the metabolites theophylline, theobromine, and paraxanthine isolated earlier from the urine of rats after dosage with caffeine, two new metabolites (59) and (60) have since been isolated by t.l.c. and identified by mass spectrometry.[62]

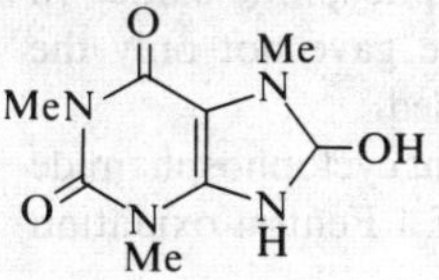

(59) 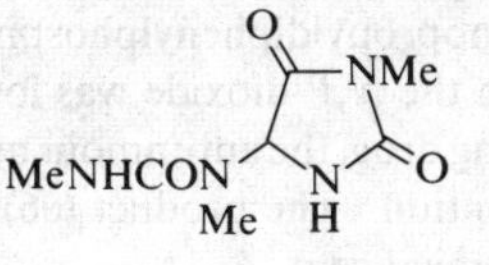   (60) 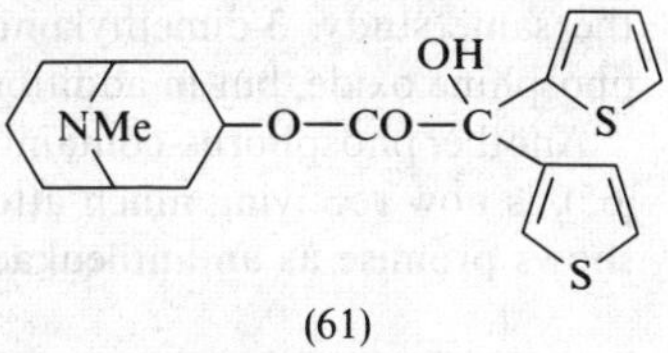   (61)

Although some ten metabolites were isolated after biotransformation of (61), nine of these were derived from the two products of ester hydrolysis.[63] The only metabolite to possess the intact ester linkage was the N-demethyl compound. In the case of another drug, nicergoline, which also has an ester group, hydrolysis also occurred.[64] N-Demethylation of the resulting alcohol was observed.

Seven metabolites of the coronary vasodilator (62) were isolated from the urine and faeces of dogs by solvent extraction and t.l.c. Besides demethylation

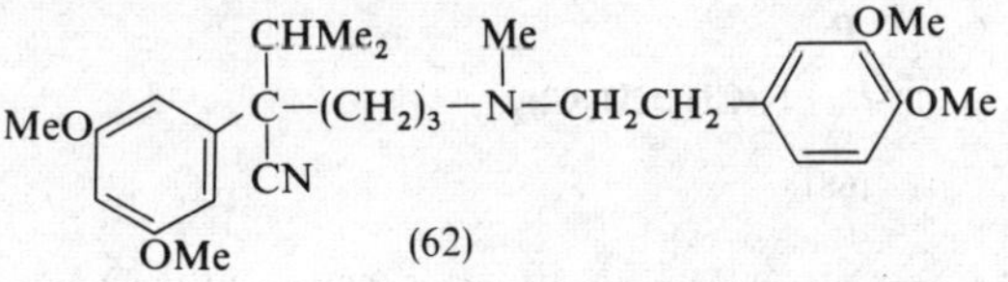

(62) 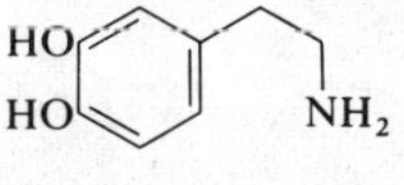   (63)

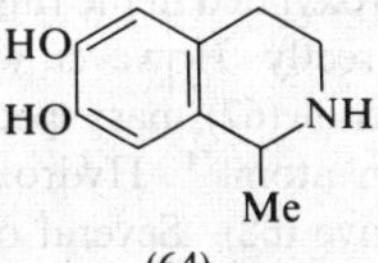

(64)

of a methoxy-group from each ring, three metabolites had also lost the dimethoxyphenylethyl moiety.[65]

In an interesting study of L-dopa metabolism in humans,[66] patients were given L-dopa followed by ethanol, the latter to act as a source of acetaldehyde. The compound salsolinol (64), a condensation product of the primary metabolite

[62] K. L. Khanna, G. S. Rao, and H. H. Cornish, *Toxicol. Appl. Pharmacol.*, 1972, **23**, 720; G. S. Rao, K. L. Khanna, and H. H. Cornish, *Experientia*, 1973, **29**, 953.
[63] T. Meshi, S. Nakamura, and Y. Sato, *Chem. and Pharm. Bull (Japan)*, 1972, **20**, 1687.
[64] F. Arcamone, A. G. Glasser, J. Grafnetterova, A. Minghetti, and V. Nicolella, *Biochem. Pharmacol.*, 1972, **21**, 2205.
[65] H. M. McIlhenny, *J. Medicin. Chem.*, 1971, **14**, 1178.
[66] M. Sandler, S. Bonham Carter, K. R. Hunter, and G. M. Stern, *Nature*, 1973, **241**, 439.

dopamine (63) with acetaldehyde, was identified by its mass spectrum. It was noted, however, that (64) was formed even without the addition of ethanol, suggesting that a high concentration of acetaldehyde is present in the body.

o,p'-Dichlorodiphenylacetic acid was identified as a metabolite of mitotane from the mass spectrum of both the free acid and the anilide.[67]

In a study of the oxidation of diphenylmethylphosphine by rat liver microsomes,[68] mass spectrometry showed the product to be the phosphine oxide. In the same study, 3-dimethylaminopropyldiphenylphosphine gave not only the phosphine oxide, but in addition the *N,P*-dioxide was formed.

Another phosphorus-containing drug, the antitumour agent cyclophosphamide (65), is now receiving much attention. The product (66) of a Fenton oxidation shows promise as an antileukaemic agent.[69]

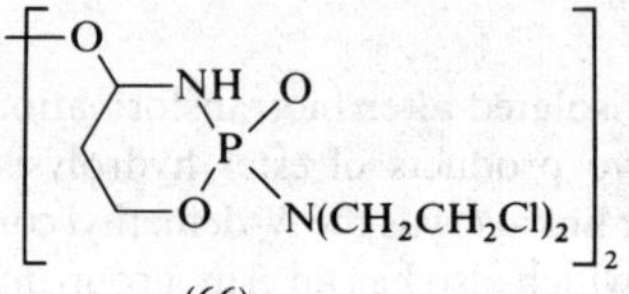

$$\left[\begin{array}{c} \text{O} \\ \text{NH} \\ \text{P} \\ \text{O} \quad \text{N(CH}_2\text{CH}_2\text{Cl)}_2 \end{array} \right]_2$$

(66)

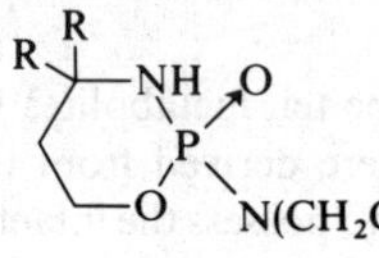

R\quad R
\\—NH\quad O
\quad P
O\quad N(CH$_2$CH$_2$Cl)$_2$

(65) R = H
(67) R = D

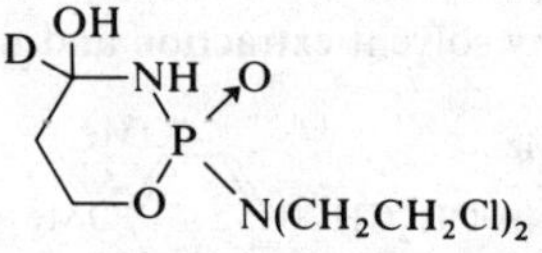

D\quad OH
\\—NH\quad O
\quad P
O\quad N(CH$_2$CH$_2$Cl)$_2$

(68)

Although the rat liver microsomic oxidation product of (65) could be shown by mass spectrometry to be hydroxylated in the ring,[70] the position of hydroxylation could not be determined directly. However, when the same experiment was carried out on the labelled substrate (67), mass spectrometry showed the product to contain only one deuterium atom.[71] Hydroxylation must therefore have occurred at the 4-position to give (68). Several other microsomal metabolites have been reported[70,72] and the distribution of cyclophosphamide in humans has also been studied.[73]

[67] J. E. Sinsheimer, J. Guilford, L. J. Bobrin, and D. E. Schteingart, *J. Pharm. Sci.*, 1972, **61**, 314.
[68] R. A. Wiley, L. A. Sternson, H. A. Sasame, and J. R. Gillette, *Biochem. Pharmacol.*, 1972, **21**, 3235.
[69] R. F. Struck, M. C. Thorpe, W. C. Coburn, and W. R. Laster, *J. Amer. Chem. Soc.*, 1974, **96**, 313.
[70] T. A. Connors, P. J. Cox, P. B. Farmer, A. B. Foster, and M. Jarman, *Biochem. Pharmacol.*, 1974, **23**, 115.
[71] T. A. Connors, P. J. Cox, P. B. Farmer, A. B. Foster, M. Jarman, and J. K. MacLeod, *Biomed. Mass Spectrometry*, 1974, **1**, 130.
[72] M. Colvin, C. A. Padgett, and C. Fenselau, *Cancer Res.*, 173, **33**, 915.
[73] J. H. Duncan, M. Colvin, and C. Fenselau, *Toxicol. Appl. Pharmacol.*, 1973, **24**, 317.

Drug Metabolism

The biotransformation of the oxazoline compound (69) was shown to lead to the unusual metabolite (70) in the dog.[74] However, in the rate the more normal product (71) was produced.

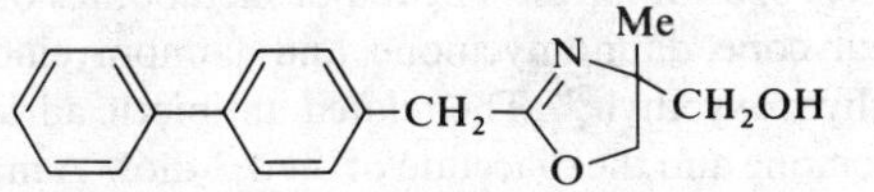

(69) Ph-C6H4-CH2-C(=N)-O-C(Me)(CH2OH)- (4,4-disubstituted oxazoline)

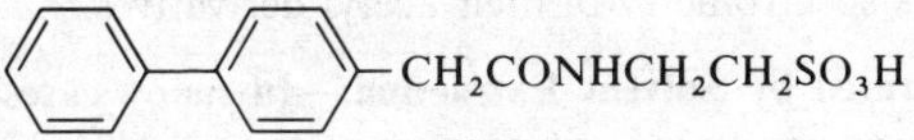

(70) Ph-C6H4-CH2CONHCH2CH2SO3H

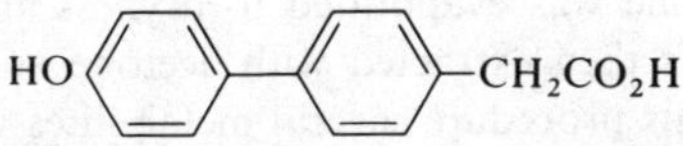

(71) HO-C6H4-C6H4-CH2CO2H

The production of artificial isotope clusters to aid the identification and interpretation of mass spectra was pioneered in the peptide field several years ago.[75] Such an approach has been used in studying the metabolism of (72).[76] After extraction of the metabolites from a t.l.c. plate, successive spectra were taken *via* a direct-inlet system. Since an equimolar mixture of (72) and the same compound labelled with ^{13}C at the 2-position of the pyrimidine ring was used, the spectra were inspected for the presence of $M, M + 1$ doublets. By this means (73) was readily identified. Simultaneous labelling of (72) with ^{13}C and ^{14}C at the 2-position has also been utilized.[77]

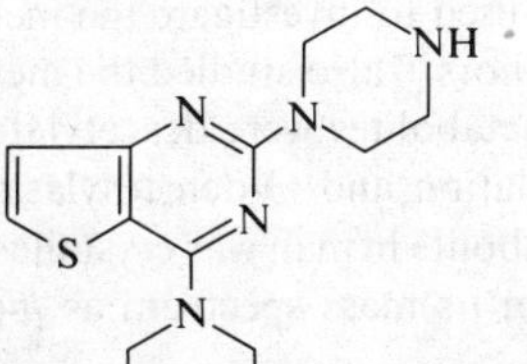

(72) thieno-pyrimidine with piperazine and morpholine substituents

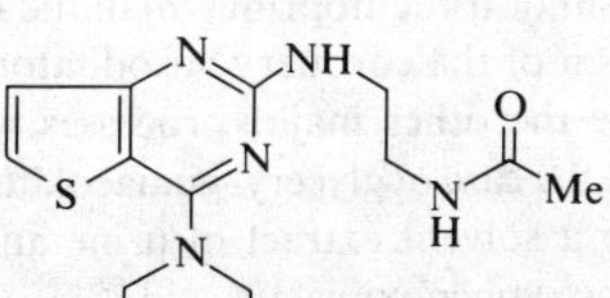

(73) thieno-pyrimidine with NHCH2CH2NHC(O)Me and morpholine substituents

---

[74] L. B. Turnbull, C. P. Johnson tert., Y. H. Chen, L. F. Sancilio, and R. B. Bruce, *J. Medicin. Chem.*, 1974, **17**, 45.
[75] J. Van Heijenoort, E. Bricas, B. C. Das, E. Lederer, and W. A. Wolstenholme, *Tetrahedron*, 1967, **23**, 3403.
[76] A. Prox, A. Zimmer, and H. Machleidt, *Xenobiotica*, 1973, **3**, 103.
[77] A. Zimmer, A. Prox, H. Pelzer, and R. Henkwitz, *Biochem. Pharmacol.*, 1973, **22**, 2213.

The product of metabolism of tritiated digitoxigenin by rabbit liver homogenate was shown by mass spectrometry to be 6β-hydroxy-3-epidigitoxigenin.[78]

Only a few papers report the use of mass spectrometry in the structural identification of metabolites of antibiotics. The major metabolites of daunomycin were identified as the aglycone, daunomycinone, and daunomycinone with the acetyl group reduced to hydroxyethyl.[79] The related antibiotic adriamycin yielded the aglycone adriamycinone and the diacetate of the 1,2-diol. A macrolide antibiotic, SF-837, was shown to undergo depropionylation and hydroxylation in the rat.[80] Five metabolites of lankacidin C 14-propionate in rats and mice have been identified by mass spectrometry of their acetyl derivatives.[81]

Metabolites Separated by Solvent Extraction.—In many cases metabolites are present in such large amounts in urine, or may have such superior solubility in certain solvents that separation by t.l.c. is not necessary. Thus in the case of sulfisomezole (74) the urine was evaporated to dryness and a mass spectrum taken.[82] The residue was then extracted with acetone and a spectrum taken from this solution. By this procedure several metabolites were identified. The same workers used a similar procedure with various solvents to identify metabolites of (75) in the rat.[83]

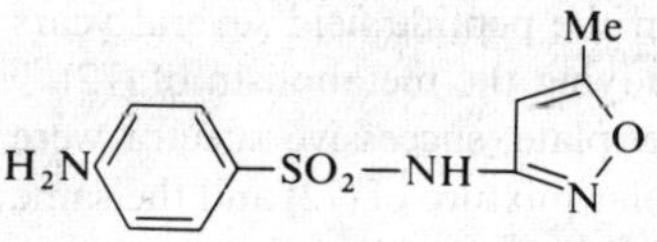

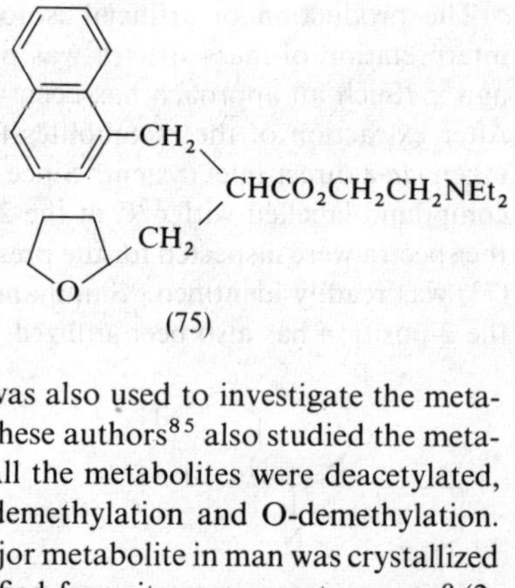

Solvent extraction of evaporated urine was also used to investigate the metabolism of trimethoprim (76) in the rat.[84] These authors[85] also studied the metabolism of the coronary vasodilator (77). All the metabolites were deacetylated, while the other major processes were N-demethylation and O-demethylation.

In the case of glyceryl guaiacolate, the major metabolite in man was crystallized from a solvent extract of urine and identified from its mass spectrum as β-(2-methoxyphenoxy)lactic acid.[86]

---

[78] W. H. Bulger, S. J. Stoks, and D. M. S. Wheeler, *Biochem. Pharmacol.*, 1974, **23**, 921.
[79] F. J. Bullock, R. J. Bruni, and M. A. Asbell, *J. Pharmacol. Exp. Therap.*, 1972, **182**, 70.
[80] S. Inouye, T. Shomura, T. Tsuruoka, S. Omoto, T. Niida, and K. Umemura, *Chem. and Pharm. Bull. (Japan)*, 1972, **20**, 2366.
[81] S. Harada, S. Tanayama, and T. Kishi, *J. Antibiotics*, 1973, **26**, 658.
[82] A. Tatematsu, T. Nadai, and H. Yoshizumi, *Yakugaku Zasshi*, 1972, **92**, 655.
[83] A. Tatematsu, T. Nadai, and H. Yoshizumi, *Biomed. Mass Spectrometry*, 1974, **1**, 66.
[84] T. Meshi and Y. Sato, *Chem. and Pharm. Bull. (Japan)*, 1972, **20**, 2079.
[85] T. Meshi, J. Sugihara, and Y. Sato, *Chem. and Pharm. Bull. (Japan)*, 1971, **19**, 1546.
[86] W. J. A. Vandenheuvel, J. L. Smith, and R. H. Silber, *J. Pharm. Sci.*, 1972, **61**, 1997.

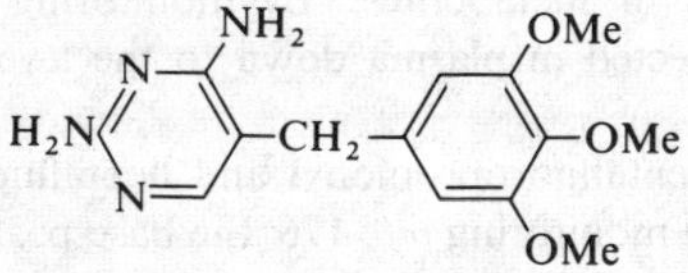

(76) [structure: 2,4-diamino-5-(3,4,5-trimethoxybenzyl)pyrimidine]

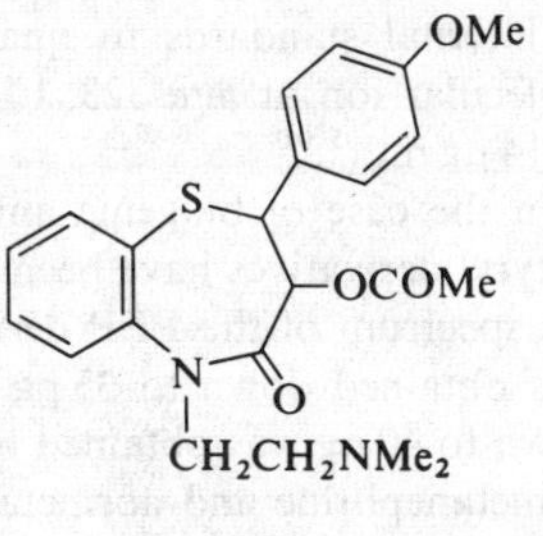

(77) [structure: benzothiazepine with OMe, OCOMe, CH₂CH₂NMe₂]

Metabolites Trapped from a Gas Chromatograph.—The use of combined g.c.m.s. is covered in another chapter, but there are a few reports of the use of preparative gas chromatography followed by the use of direct-inlet mass spectrometry. The case of dibenzepine (44) has already been mentioned.[54] The same authors trapped three metabolites of methotrimeprazine by g.c. of an extract of the urine of psychiatric patients.[87] The metabolites were identified as the sulphoxide, the N-demethyl compound, and the sulphoxide of this.

The unusual metabolite (79) of lidocaine (78) was identified by mass spectrometry after trapping from a gas chromatograph.[88]

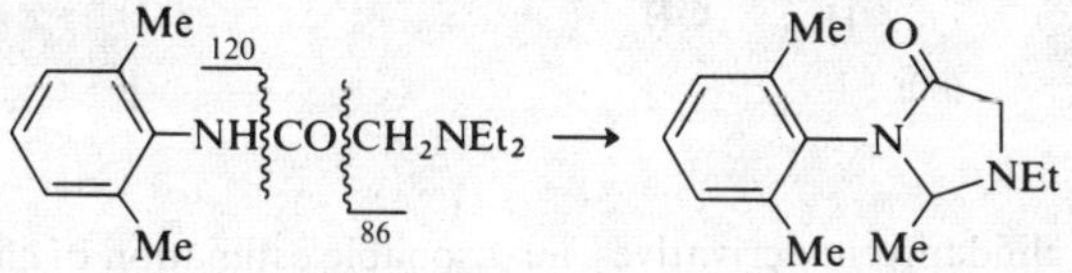

(78) → (79)

3 Quantitative Gas Chromatography–Mass Spectrometry

More and more effort is now being put into the use of mass spectrometry as a quantitative tool. Most of this effort is being applied to the use of multiple-ion monitoring in g.c.m.s. systems. It is still not clear, for example, whether it is better to use non-labelled internal standards giving the same ion as the compound to be measured, or to use a labelled version of the compound being quantified.

Single Ion Monitoring.—Although the greatest sensitivity is obviously obtained by monitoring the base peak in a mass spectrum, sensitivity often has to be sacrificed to selectivity. Thus in the case of the hallucinogen STP (2,5-dimethoxy-4-methylamphetamine),[89] instead of the base peak at m/e 44, an ion at m/e 166 was monitored. Less than 100 pg of STP could be detected in this way. Surprisingly, most papers report the use of single ion monitoring without the addition

---

[87] A. De Leenheer and A. Heyndrickx, *J. Pharm. Sci.*, 1972, **61**, 914.
[88] G. D. Breck and W. T. Trager, *Science*, 1972, **173**, 544.
[89] A. Frigerio, R. Fanelli, and B. Danieli, *Chem. and Ind.*, 1972, 769.

of internal standards to quantify drugs or metabolites. By monitoring the molecular ion at m/e 323, LSD was detected in plasma down to the level of 0.014 µg ml^{-1}.[90]

In the case of biogenic amines, the pentafluoropropionyl and heptafluorobutyryl derivatives have been used.[91] By monitoring m/e 176, the base peak in the spectrum of the PFP derivative of noradrenaline, a linear calibration line was obtained down to 33 pg. For the PFP derivative of dopamine, linearity down to 40 pg was obtained by monitoring m/e 428. The partial derivatization of metanephrine and normetanephrine with EtOH–HCl yielded O-ethyl ethers which could be determined quantitatively.[92]

In a determination of clindamycin (80) as the 2-palmitate,[93] the ion (81) was monitored in a trimethylsilylated human serum extract. Since m/e 126 is the

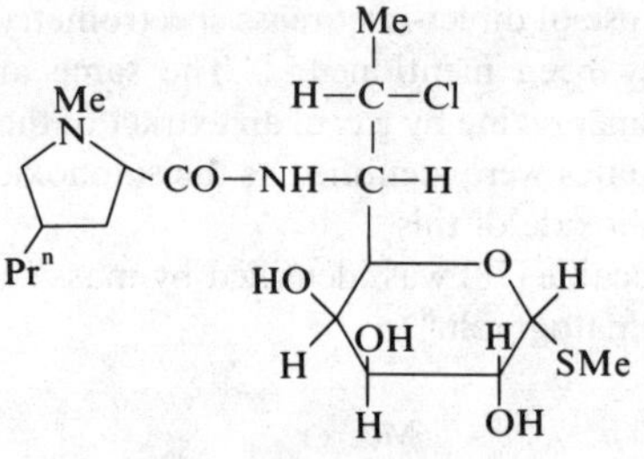

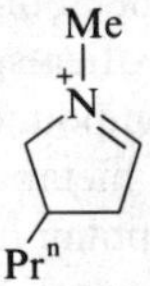

base peak in all clindamycin derivatives, a reasonable estimation of clindamycin plus its metabolites was obtained; down to 1 µg (ml of plasma)$^{-1}$ could be detected. It was pointed out that the lower limit of sensitivity could not be determined since the mass spectrometer source sensitivity diminished rapidly due to fouling. This is an interesting observation in view of controversy at recent meetings as to whether silylating agents contaminate sources to any great extent.

By the use of butobarbitone (82) as internal standard, it was found possible to determine amylobarbitone (83) in plasma down to 10 ng ml^{-1} by monitoring m/e 169.[94] In order to measure the hydroxy-metabolite (84), it was necessary to use hydroxypentobarbitone, which had a more nearly similar retention time.

Multiple Ion Monitoring.—*Non-labelled Standards.* Although multiple ion monitoring presents no problems of instrumental stability and limited mass range when quadrupole mass spectrometers are used, the use of accelerating voltage alternators has had its difficulties. The mass range over which they can

[90] K. Watanabe and Y. Makino, *Yakugaku Zasshi*, 1972, **92**, 517.
[91] F. Karoum, F. Cattabeni, E. Costa, C. R. J. Ruthven, and M. Sandler, *Analyt. Biochem.*, 1972, **47**, 550.
[92] N. Narasimhachari, *J. Chromatog.*, 1974, **90**, 163.
[93] T. F. Brodasky and F. F. Sun, *J. Pharm. Sci.*, 1974, **63**, 360.
[94] G. H. Draffen, R. A. Clare, and F. M. Williams, *J. Chromatog.*, 1973, **75**, 45.

be used is now less limited,[95] while an approach to the stability problem has been to use a computer to add or subtract a small offset voltage continually to

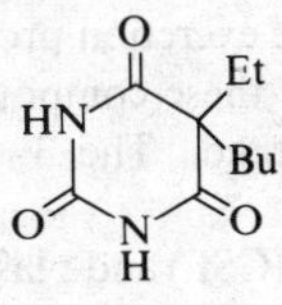

(82), 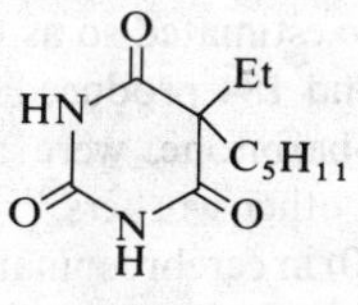 (83), 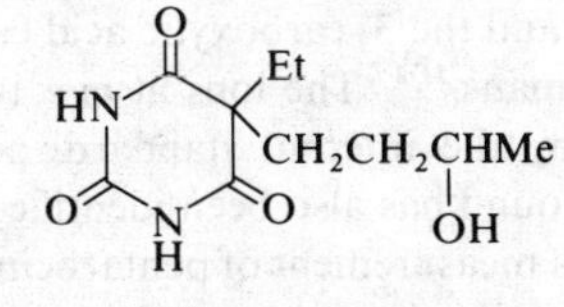 (84)

the coarse voltage chosen by the alternator.[96] This improved the precision in measuring stable isotope abundances to better than 1% on 100 ng samples of prostaglandin methyl esters.

In a quantitative determination of nortriptyline (85) and demethylnortriptyline in humans,[97] the internal standard (87) was used. The lower limit of a few nanograms per ml of plasma could be quantified with a standard deviation of 5.8% by monitoring the ion at m/e 232 produced from the heptafluorobutyryl derivatives of (85) and (86) and the ion at m/e 218 from (87). Another tricyclic antidepressant drug, imipramine, has been quantified by using promazine as the

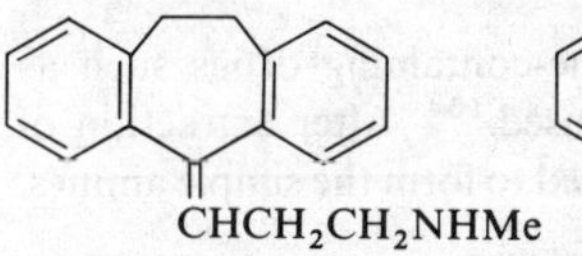

(85) CHCH$_2$CH$_2$NHMe

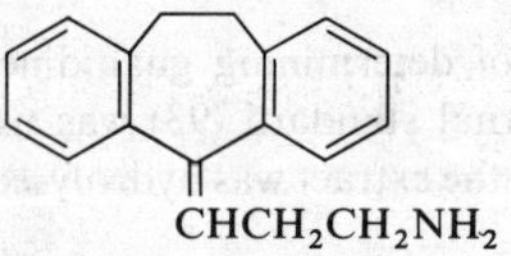

(86) CHCH$_2$CH$_2$NH$_2$

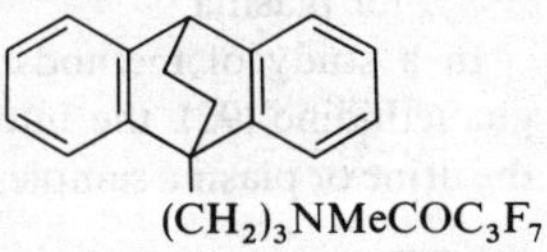

(87) (CH$_2$)$_3$NMeCOC$_3$F$_7$

internal marker.[98] Concentrations in plasma of 10 ng ml^{-1} could be detected. The dihydro-analogue of carbamazepine has also been used as an internal standard.[99]

The wider mass range of the quadrupole was utilized to measure simultaneously the plasma concentrations of lidocaine (78) and its desethylated metabolite (88) by using trimecaine (89) as internal standard. A precision of 3.1% for (78) in

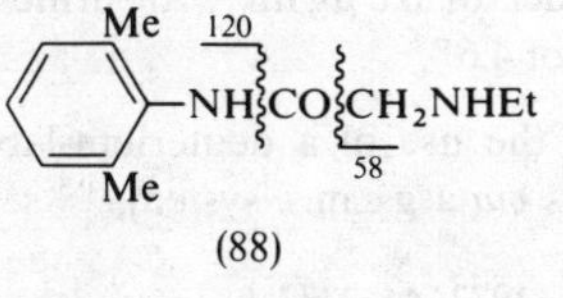

(88), 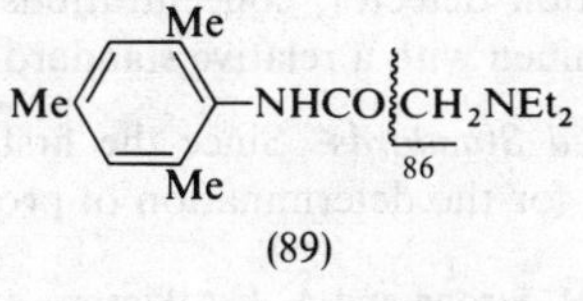 (89)

[95] P. D. Klein, J. R. Haufmann, and W. J. Eisler, *Analyt. Chem.*, 1972, **44**, 490.
[96] J. F. Holland, C. C. Sweeley, R. E. Thrush, R. E. Teets, and M. A. Beiber, *Analyt. Chem.*, 1973, **45**, 308.
[97] O. Borga, L. Palmer, A. Linnarsson, and B. Holmstedt, *Analyt. Letters*, 1971, **4**, 837.
[98] A. Frigerio, G. Belvedere, F. De Nadai, C. Fanelli, C. Pantarotto, F. Riva, and C. Morselli, *J. Chromatog.*, 1972, **74**, 201.
[99] L. Palmer, L. Bertilsson, P. Collote, and M. Rawlins, *Clin. Pharmacol. Therap.*, 1973, **14**, 827.

the range 0.5—10 µg ml⁻¹ of plasma and 7.4% for (88) in the range 0.3—5 µg ml⁻¹ of plasma was obtained.[100]

The metabolites of butobarbitone (82) such as 3'-hydroxy-, 3'-oxo-butobarbitone, and the 3'-carboxylic acid can be estimated so as to give excretion profiles in humans.[101] The ions at m/e 169 and 184 produced by all these compounds and by the internal standard, pentobarbitone, were monitored. The 3'-oxo-compound has also been identified by other workers.[102]

In a measurement of pentazocine (90) in cerebrospinal fluid (CSF) and plasma, cyclazocine (91) was used as an internal standard.[103] The standard curves

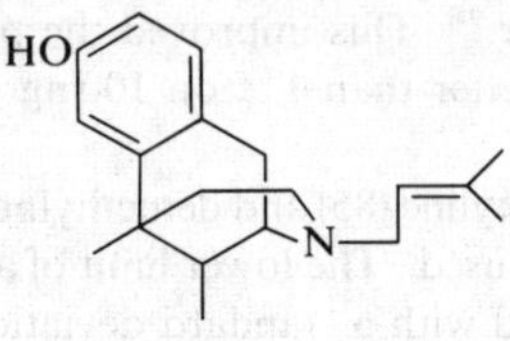

(90) 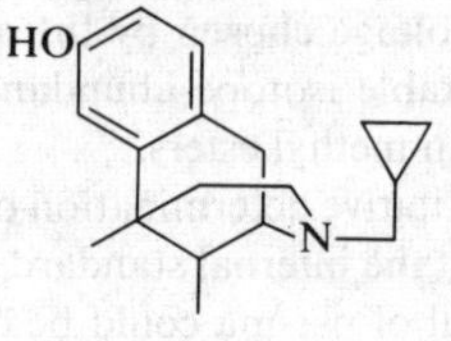   (91)

derived from peak height were linear down to 2 ng ml⁻¹ for CSF and 20 ng ml⁻¹ for plasma. At these levels the standard deviations were ±12% for CSF and ±7% for plasma.

In a study of methods of determining guanidine-containing drugs such as guanethidine (92), the internal standard (93) was used.[104] After extraction of the urine or plasma samples the extract was hydrolysed to form the simple amines.

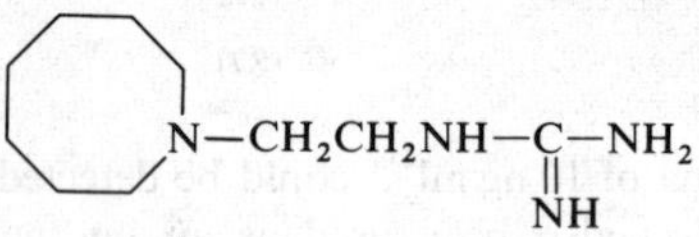

(92) 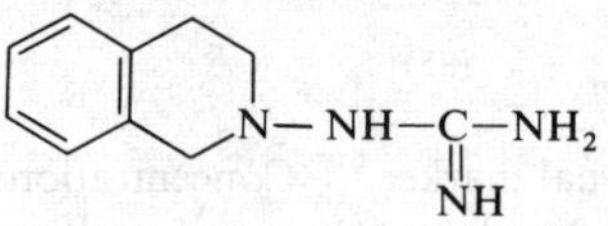   (93)

After derivatization by formation of the trifluoroacetates it was possible to measure the equivalent of 20 ng ml⁻¹ of plasma of (92) with a relative standard deviation of 8.6% when multiple ion monitoring was used. Using a flame ionization detector, concentrations of the order of 0.6 µg ml⁻¹ of urine were determined with a relative standard deviation of 4.6%.

Labelled Standards. Since the first report of the use of a deuterium-labelled carrier for the determination of prostaglandins *via* a g.c.m.s. system,[105] several

---

[100] J. M. Strong and A. J. Atkinson, *Analyt. Chem.*, 1972, **44**, 2287.
[101] J. N. T. Gilbert and J. W. Powell, *Biomed. Mass Spectrometry*, 1974, **1**, 142.
[102] J. Grove, P. A. Toseland, G. H. Draffan, R. A. Clare, and F. M. Williams, *J. Pharm. Pharmacol.*, 1974, **26**, 175.
[103] S. Agurell, L. O. Boreus, E. Gordon, J.-E. Lindgren, M. Ehrnebo, and U. Lönroth, *J. Pharm. Pharmacol.*, 1974, **26**, 1.
[104] J. H. Hengstmann, F. C. Falkner, J. T. Watson, and J. Oates, *Analyt. Chem.*, 1974, **46**, 34.
[105] B. Samuelsson, M. Hamberg, and C. C. Sweeley, *Analyt. Biochem.*, 1970, **38**, 301.

groups of workers have used ^{2}H-, ^{13}C-, or ^{15}N-labelled drugs or metabolites as internal standards for multiple ion monitoring. Typical of prostaglandin work are the determination of picomole quantities of PGE$_2$ and PGF$_{2\alpha}$[106] and the determination of PGF$_{2\alpha}$ with a precision of $\pm 3.7\%$ on 400 pg using the [3,3,4,4-^{2}H$_4$] compound as carrier.[107] A method has been developed for the quantitative determination of the major urinary metabolite 5,7-dihydroxy-11-ketotetranorprostane-1,16-dioic acid from PGF$_{1\alpha}$ and PGF$_{2\alpha}$ by use of the O-methyloxime of the [10,10,12,12-^{2}H$_4$] derivative.[108] Down to 0.51 μg could be determined with a standard deviation of $\pm 4\%$.

In a study of Δ^1-THC in the plasma of cannabis smokers, levels of 5 ng ml^{-1} of plasma have been determined using a dideuteriated derivative as internal standard.[109] In a study of nortriptyline (85), both ^{2}H and ^{15}N labelling were used.[110] The trifluoroacetyl derivatives of nortriptyline (94) and the two labelled compounds (95) and (96) were used in the determination, where 50 pg of (94) were readily detectable.

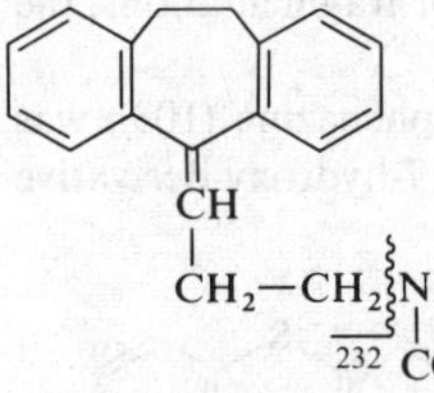

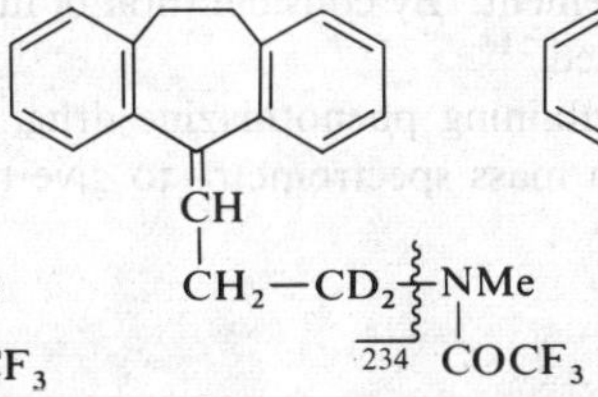

Amphetamine and phentermine were also readily estimated as their trifluoroacetyl derivatives (97) and (98) by using the labelled standards (99) and (100).[111] With 500 pmol of (99) and (100), at the 95% confidence limit, 2—5 pmol of amphetamine and 3—8 pmol of phentermine per ml of plasma could be determined.

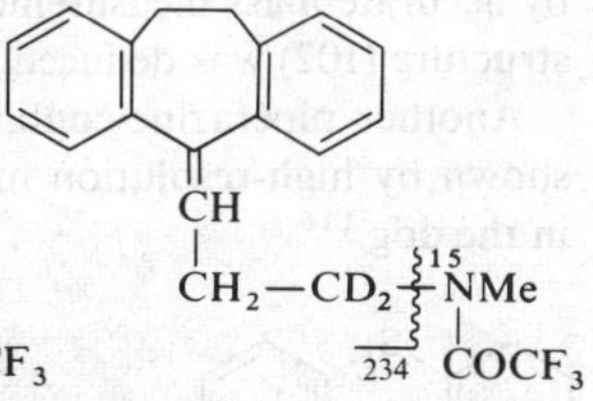

With a quadrupole mass spectrometer linked to a data system it was found possible to monitor 25 ions simultaneously.[112] This facility was applied to a

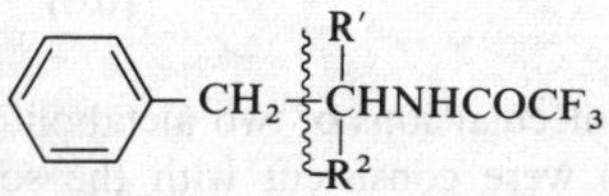

(97) R^1 = H, R^2 = Me
(98) R^1 = CH$_3$, R^2 = Me
(99) R^1 = H, R^2 = CD$_3$
(100) R^1 = Me, R^2 = CD$_3$

[106] U. Axen, K. Green, D. Horlin, and B. Samuelsson, *Biochem. Biophys. Res. Comm.*, 1971, **45**, 519.
[107] K. Green, E. Granstrom, B. Samuelsson, and U. Axen, *Analyt. Biochem.*, 1973, **54**, 434.
[108] M. Hamberg, *Analyt. Biochem.*, 1973, **55**, 368.
[109] S. Agurell, B. Gustafsson, B. Holmstedt, K. Leander, J.-E. Lindgren, I. Nilsson, F. Sandberg, and M. Asberg, *J. Pharm. Pharmacol.*, 1973, **25**, 554.
[110] T. E. Gaffney, C.-G. Hammar, B. Holmstedt, and R. E. McMahon, *Analyt. Chem.*, 1971, **43**, 307.
[111] A. K. Cho, B. J. Hodshon, B. Lindeke, and G. T. Miwa, *J. Pharm. Sci.*, 1973, **62**, 1491.
[112] R. E. Sammons, W. E. Pereira, W. E. Reynolds, T. C. Rindfleisch, and A. M. Duffield, *Analyt. Chem.*, 1974, **46**, 582.

mixture of amino-acids, as their n-butyl ester *N*-trifluoroacetyl derivatives, extracted from plasma. Deuteriated acids were used as internal standards, and down to 1 ng of various amino-acids were determined with a standard deviation of less than 10%.

^{13}C has also been used as a label for internal standards. Thus [2,4,5-^{13}C$_3$]-diphenylhydantoin and [2,4(6),5-^{13}C$_3$]pentobarbital have been used in the quantitative determination of diphenylhydantoin and various barbiturates.[113] Diphenylhydantoin at the level of 11.6 μg (ml of plasma)$^{-1}$ was determined with a reproducibility of ±5%. [2,4,5-^{13}C$_3$]phenobarbital and [1-C^{2}H$_3$]valium have also been used to determine the unlabelled drugs.[114] Quantities of the order of 80 pg (ml of plasma)$^{-1}$ were determined with a precision of ±5%.

4 High-resolution Mass Spectrometry

Qualitative Applications.—A major metabolite of perazine (101) in the rat gave a molecular ion at *m/e* 299 whose composition was shown to be C$_{17}$H$_{21}$N$_3$S by accurate mass measurement. By consideration of major fragmentations, the structure (102) was deduced.[115]

Another piperazine-containing phenothiazine drug, fluphenazine (103), was shown by high-resolution mass spectrometry to give the 7-hydroxy-derivative in the dog.[116]

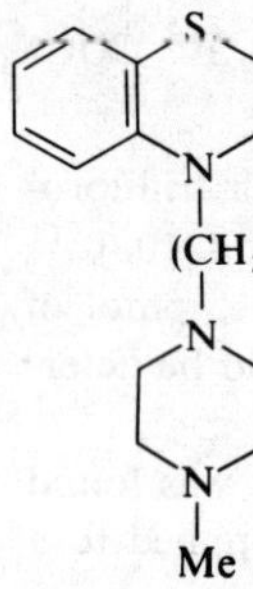

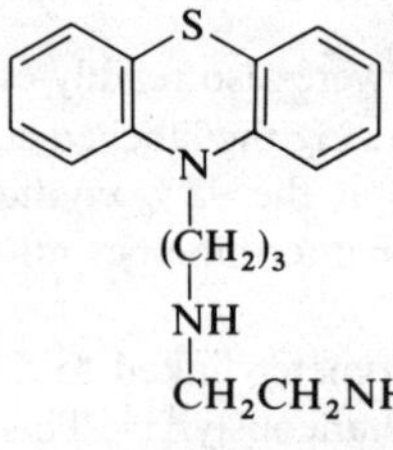

Accurate mass measurements of the molecular ions of two metabolites of the peripheral vasodilator hydralazine (104) were consistent with the somewhat unexpected structures (105) and (106).[117]

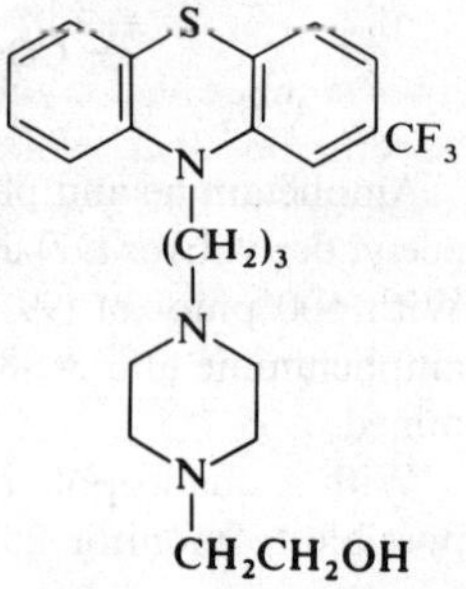

High-resolution measurement on metabolites of clindamycin (80) extracted by t.l.c. from rat and dog urine showed that, as well as unchanged (80), and in the

---

[113] M. G. Horning, J. Nowlin, K. Lertratanangkoon, R. N. Stillwell, W. G. Stillwell, and R. M. Hill, *Clinical Chem.*, 1973, **19**, 845.
[114] M. G. Horning, W. G. Stillwell, J. Nowlin, K. Lertratanangkoon, D. Carroll, I. Dzidic, R. N. Stillwell, and E. C. Horning, *J. Chromatog.*, 1974, **91**, 413.
[115] U. Breyer, D. Krauss, and J. C. Jochims, *Experientia*, 1972, **28**, 312.
[116] J. Dreyfuss and A. I. Cohen, *J. Pharm. Sci.*, 1971, **60**, 826.
[117] S. B. Zak, T. G. Gilleran, J. Karliner, and G. Lukas, *J. Medicin. Chem.*, 1974, **17**, 381.

dog its glucuronide conjugate, the formation of the sulphoxide and the N-demethyl analogue were the major metabolic pathways.[118]

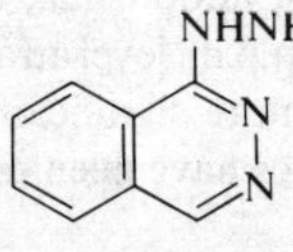

(104) 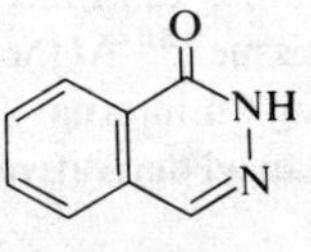 (105) 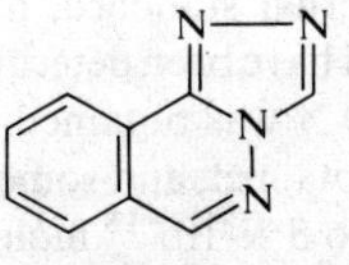 (106)

2-Ethyl-2-methyl-3-hydroxysuccinimide was identified with the aid of accurate mass measurement as a major metabolite in humans of ethosuximide (2-ethyl-2-methylsuccinimide).[119] In developing a g.c. method for the analysis of diphenylhydantoin, a methylation procedure was used.[120] High-resolution accurate mass measurements showed that a mixture of 3-methyl and 1,3-dimethyl-diphenylhydantoin was formed.

The major metabolite of probenecid was the glucuronide conjugate, which was identified by g.c.m.s. at high resolution following enzymatic hydrolysis.[121] High-resolution CI mass spectrometry was used to confirm the elemental composition of the methylated derivative of α-methyldopamine, a metabolite of α-methyldopa in man.[122]

In a study of the decomposition of injection solutions of phenylephrine (107) two compounds were isolated by t.l.c. After accurate mass measurement of several ions, these were identified as the products (108) and (109) of an alternative cyclization.[123]

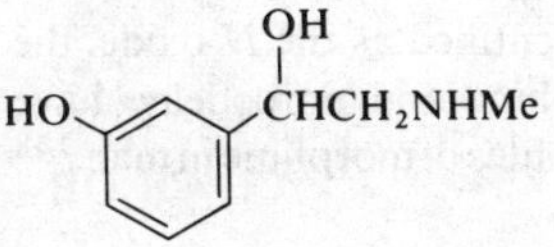

(107) 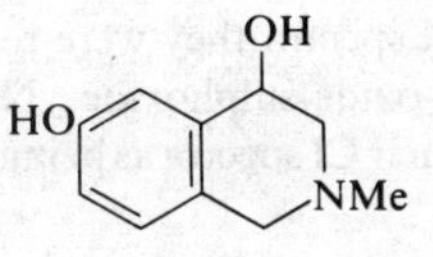 (108) 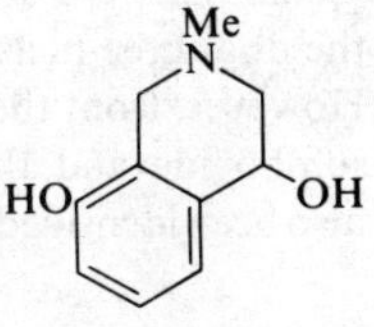 (109)

The identification of drugs by use of high-resolution mass spectrometry has also been outlined.[124]

[118] F. F. Sun, *J. Pharm. Sci.*, 1973, **62**, 1657.
[119] P. G. Preste, C. E. Westerman, N. P. Das, B. J. Wilder, and J. H. Duncan, *J. Pharm. Sci.*, 1974, **63**, 467.
[120] K. Sabih and K. Sabih, *J. Pharm. Sci.*, 1971, **60**, 1216.
[121] K. Sabih, C. D. Klaasen, and K. Sabih, *J. Pharm. Sci.*, 1971, **60**, 745.
[122] K. S. Marshall and N. Castagnioli, *J. Medicin. Chem.*, 1973, **16**, 266.
[123] B. J. Millard, E. Shotton, and D. J. Priaulx, *J. Pharm. Pharmacol.*, 1973, **25**, Suppl., p. 24.
[124] D. A. Brent and D. A. Yeowell, *Chem. and Ind.*, 1973, 190.

Quantitative Applications.—Although the 'integrated ion current' technique[125—127] has been used frequently for the quantitative estimation of biogenic amines,[128,129] there are few reports of its use in drug metabolism. Without using an internal standard, hypoxanthine, xanthine, uric acid, allopurinal, and oxipurinol have been determined in tissue.[130] At the 10—100 p.p.m. level an accuracy of $\pm 20\%$ was obtained. By using an internal standard more significant results were obtained, and some monodansyl derivatives of amines have been estimated down to 5×10^{-15} mole.[131]

By using 2-chlorophenothiazine as internal standard and monitoring the intense $M - C_2H_5$ ions, heptabarbitone (110) and its two major metabolites (111) and (112) could be determined down to 2 ng with a coefficient of variation of 5%.[132] Piribedil (47) and its 5-hydroxy metabolite have also been determined down to this level.[56]

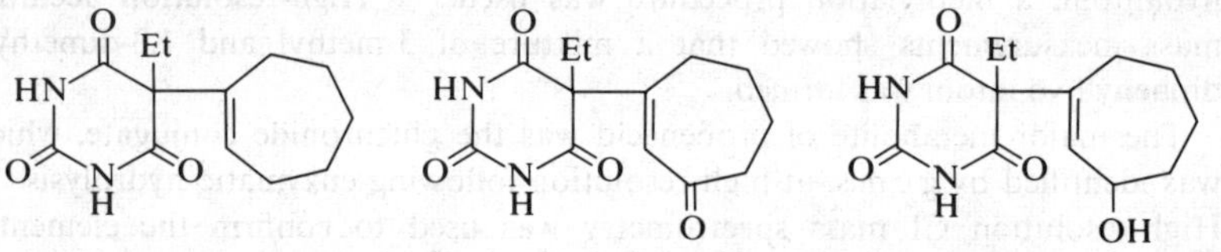

(110)   (111)   (112)

5 Chemical Ionization

Chemical ionization mass spectrometry has been used frequently for the identification of drugs of abuse[133—136] but in only a few cases for the identification of drug metabolites. In a study of the metabolism of the antiarrhythmic agent (113) in the dog three metabolites gave very weak molecular ions in their EI spectra.[137] However, from their CI spectra they were readily identified as the N-oxide, the sulphoxide and the N-oxide-sulphoxide. Normorphine and norcodeine have also been identified by their CI spectra as being metabolites of morphine in man.[138]

---

[125] A. A. Boulton and J. R. Majer, *J. Chromatog.*, 1970, **48**, 322.
[126] A. A. Boulton and J. R. Majer, *Nature*, 1970, **225**, 658.
[127] A. A. Boulton and J. R. Majer, *Canad. J. Biochem.*, 1971, **49**, 993.
[128] D. A. Durden, S. R. Philips, and A. A. Boulton, *Canad. J. Biochem.*, 1973, **51**, 995.
[129] A. A. Boulton, S. R. Philips, and D. A. Durden, *J. Chromatog.*, 1973, **82**, 137.
[130] W. Snedden and R. B. Parker, *Analyt. Chem.*, 1971, **43**, 1651.
[131] D. A. Durden, B. A. Davis, and A. A. Boulton, *Biomed. Mass Spectrometry*, 1974, **1**, 83.
[132] B. J. Millard, in 'Gascromatografia Spettrometria di Mass nello Studio di Farmaci', ed. A. Frigerio, Tamburini Editore, Milan, 1973, p. 101.
[133] D. F. Hunt and J. F. Ryan, *Analyt. Chem.*, 1972, **44**, 1306.
[134] R. L. Foltz, M. W. Crouch, M. Geer, K. N. Scott, and C. M. Williams, *Biochem. Med.*, 1972, **6**, 294.
[135] R. Saferstein and J. Chao, *J. Assoc. Offic. Analyt. Chemists*, 1973, **56**, 1234.
[136] J. Chao, R. Saferstein, and J. Manura, *Analyt. Chem.*, 1974, **46**, 296.
[137] A. I. Cohen, J. Dreyfuss, and H. M. Fales, *J. Medicin. Chem.*, 1972, **15**, 542.
[138] U. Boerner, R. L. Roe, and C. E. Becker, *J. Pharm. Pharmacol.*, 1974, **26**, 393.

Chemical ionization has recently been used to estimate the amounts of drugs and metabolites by use of a direct inlet system.[139] Lidocaine (78) and its de-ethylated metabolite MEGX (88) and a dide-ethylated metabolite were quantified by using deuterium-labelled standards. The range of detection was from 5 ng to 4 µg of (78) and 0.1 µg to 1 µg of MEGX.

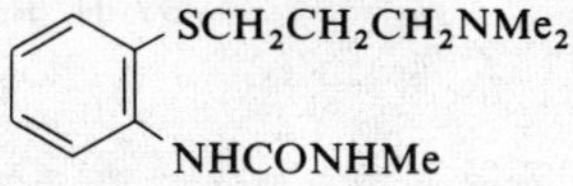

SCH$_2$CH$_2$CH$_2$NMe$_2$

NHCONHMe

(113)

---

[139] W. A. Garland, W. F. Trager, and S. D. Nelson, *Biomed. Mass Spectrometry*, 1974, 1, 124.

10
Protein and Carbohydrate Sequence Analysis

BY H. R. MORRIS AND A. DELL

1 Introduction

In this chapter, the subject has been dealt with briefly, in an informal but critical manner, selecting only those papers which appear to have a relevance to the application of mass spectrometry to the sequence analysis of proteins or carbohydrates. Based upon a survey of titles, this may appear to exclude a number of valuable papers, but as these will have been dealt with in Chapters 6 and 8 we have felt obliged to omit those studies whose titles are misleading, or whose applicability to genuine problems appears to be tenuous.

In order to provide a perspective for this Report the reader is referred to the chapter on Natural Products in Volume 2[1] and also to two review articles which have appeared within the past two years.[2,3]

2 Protein Structure

Derivatization.—Apart from field desorption studies (see below) it is still generally necessary to prepare volatile derivatives prior to mass spectrometric examination. The choice of derivative and method of preparation remain important questions, since the viability of the method rests upon the rapid and total sequence assignment of small quantities of protein-derived peptides. Derivatives possessing poor fragmentation characteristics or low preparative yields are unlikely to be of more than academic interest.

The acetyl permethyl derivative remains the most widely used, and two further studies on the previously reported modifications of the permethylation reaction, namely 'equimolar'[4] and 'short'[5] permethylation have appeared. The 'equimolar' reaction involving the use of molar equivalents of base and methyl iodide

---

[1] T. J. Mead, H. R. Morris, J. H. Bowie, and I. Howe, 'Natural Products', in 'Mass Spectrometry', ed. D. H. Williams, (Specialist Periodical Reports), The Chemical Society, London, 1973, Vol. 2, p. 143.
[2] N. N. Kochetkov, *Pure Appl. Chem.*, 1973, **33**, 53.
[3] H. R. Morris, 'Mass Spectrometry in Protein Sequence Analysis', in 'New Techniques in Biophysics and Cell Biology', ed. R. H. Pain and B. J. Smith, Wiley, London, 1973, Vol. 1, p. 149.
[4] P. A. Leclercq and D. M. Desiderio, *Biochem. Biophys. Res. Comm.*, 1971, **45**, 308.
[5] H. R. Morris, *F.E.B.S. Letters*, 1972, **22**, 257.

in order to prevent salt formation[6] has been applied in a study of a histidine-containing peptide of known sequence, α-melanocyte-stimulating hormone.[7] The sequence of this peptide, including the position of the histidine residue, was confirmed by enzymic digestion and mass spectrometric analysis of the isolated peptides. The 'short' permethylation reaction, involving the use of excess reagents and a reaction time of 1—3 min, has been extended to include methionine- and cysteine-containing peptides.[8] In the same study, modified procedures for the conversion of arginine into pyrimidyl ornithine or ornithine are described, and examples given of the application of all these procedures to small quantities of protein-derived peptides of unknown sequence.

Clearly, although the above procedures are now finding application in a number of laboratories, no derivative-forming reaction can be thought of as ideal at present, and the search for alternative modification schemes continues. Two such studies are now assessed.

Firstly, a novel and interesting approach to acyl peptide ester formation has been described[9,10] involving the formation of the acyl peptide trimethylanilinium salt in a one-step procedure in the probe tip of the mass spectrometer. This is effected by treatment of the peptide with either acetylacetone or methyltrifluoroacetate and trimethylanilinium hydroxide. The probe tip is then heated in the conventional manner in the ion source of the mass spectrometer, where pyrolysis methylation takes place, monitored simply by the liberation of dimethylaniline. The method has the attractive feature of cutting out sample-transfer and workup steps, factors leading to lower yields in other processes. It is unfortunate that the method was demonstrated using 0.4 µmol of peptide since this is far larger than would be needed for other procedures (*e.g.* acetyl permethyl), and fails to capitalize on the potential strength of the method. It is appreciated that the fragmentation characteristics of acyl peptide esters are more complex than permethyl derivatives, and also that the method is limited by volatility considerations to quite small peptides, thus restricting applicability at present. However, the approach is attractive and possibilities of permethylating peptides in an analogous manner are certainly worth some consideration.

A series of papers has appeared purporting to show the advantages of Schiff-base peptide ester derivatives in peptide sequence analysis.[11—13] Data on twenty N-terminal blocking groups ranging from acetylacetonyl to β-indolylmethylidene, including N- and C-terminal and internal fragment ion abundances, are presented

---

[6] D. W. Thomas, B. C. Das, S. D. Géro, and E. Lederer, *Biochem. Biophys. Res. Comm.*, 1968, **32**, 199.
[7] M. L. Polan, W. J. McMurray, S. R. Lipsky, and S. Landé, *J. Amer. Chem. Soc.*, 1972, **94**, 2847.
[8] H. R. Morris, R. J. Dickinson, and D. H. Williams, *Biochem. Biophys. Res. Comm.*, 1973, **51**, 247.
[9] G. M. Schier, J. Korth, and B. Halpern, *Tetrahedron Letters*, 1972, 4621.
[10] G. M. Schier and B. Halpern, *Austral. J. Chem.*, 1974, **27**, 393.
[11] G. V. Patil, R. E. Hamilton, and R. A. Day, *Org. Mass Spectrometry*, 1973, **7**, 817.
[12] R. A. Day, H. Falter, J. P. Lehman, and R. E. Hamilton, *J. Org. Chem.*, 1973, **38**, 782.
[13] R. E. Hamilton, G. V. Patil, K. Jayasimhulu, and R. A. Day, *Org. Mass Spectrometry*, 1974, **9**, 211.

in the three different papers. Whilst the object of the study is admirable, *i.e.* to ascertain possible benefits of such derivatives in increasing molecular ion and sequence ion abundances, the presentation and assessment of the data leave much to be desired. A worthwhile analysis of such derivatives should include comparative data on other derivatives previously proposed, *e.g.* acetyl permethyl for *each* peptide studied, in order to make clear the advantages of the new procedure. Unqualified statements on the usefulness of such derivatives are made when data on volatilization temperature[11,13] and derivative yields[11—13] are lacking. From the published spectra of simple synthetic peptides, these derivatives do not appear to be as useful as the simple acetyl permethyl derivatives which fragment in a much more predictable manner. The minimum criterion for a good derivative in the present state of the method must be based on the question: can 0.1 μmol (or less) of a small peptide (say 5—10 residues) of *unknown* structure be sequenced by this procedure? If the answer is no, there can be no general advantage in such a procedure where quantities of peptide available are limited, as is frequently so in practice.

Some other aspects of derivative formation will be mentioned in the context of their specific applications below.

Sequence Analysis.—*Small Peptides.* Several approaches to peptide or protein sequence analysis which can be characterized as small molecule studies are now considered.

The attractive procedure of combined gas chromatography–mass spectrometry of small peptide derivatives has been well demonstrated on a 20-residue peptide isolated from rabbit actin.[14] The peptide was degraded into small fragments by partial acid hydrolysis, followed by acetylation, reduction to polyaminoalcohols, and trimethylsilylation. One problem reported was that a histidine-containing peptide was not amenable to the g.c. analysis. A separate direct probe experiment was used to obtain the sequence of the tetrapeptide Ser-Ile-Val-His. The main peptide was examined knowing its amino-acid composition, but without a prior knowledge of its sequence, and two possible sequences were obtained by computerized overlapping of the small fragments produced. These could be readily resolved by manual methods. Clearly, this method has potential as a a rapid procedure for sequencing isolated pure peptides, but two problems remain to be solved before it can be applied in general. Firstly, we know from our own and other studies that many sequences are not amenable to partial acid degradation without loss of information, *e.g.* quantitative release of free amino-acids or rearrangement. Secondly, valuable information concerning the identity of amides or acids as Asx or Glx is forfeited. With small peptides this problem may be resolved by electrophoretic mobility, but in a general method of structure determination this remains a problem.

Apart from the cyanogen bromide cleavage at methionine, the Edman degradation is the only specific chemical hydrolysis procedure used in protein chemistry today. An analogous and novel chemical method has been suggested for the

[14] H. Nau, J. A. Kelley, and K. Biemann, *J. Amer. Chem. Soc.*, 1973, **95**, 7162.

sequencing of small peptides,[15] involving the promotion of diketopiperazine formation with concomitant chain cleavage at the second amide bond from the *N*-terminus. The diketopiperazine(s) formed is identified by g.c.m.s., and following one step of Edman degradation, the procedure is repeated. The total sequence may thus be built up as demonstrated on the simple peptide shown in (Scheme 1).

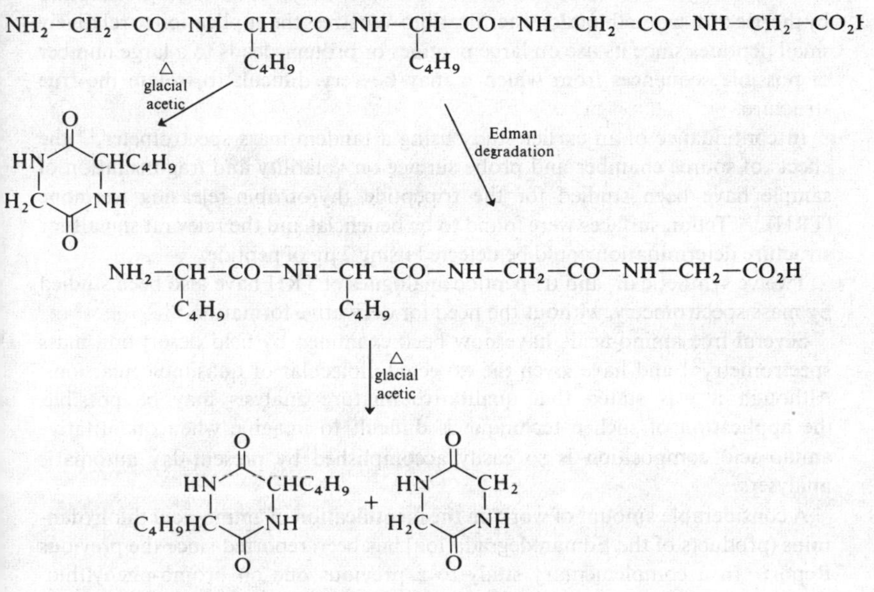

$$NH_2-CH_2-CO-NH-CH(C_4H_9)-CO-NH-CH(C_4H_9)-CO-NH-CH_2-CO-NH-CH_2-CO_2H$$

$\xrightarrow{\Delta, \text{glacial acetic}}$

Diketopiperazine: HN-CHC$_4$H$_9$ / H$_2$C-NH (with two C=O)

$\xrightarrow{\text{Edman degradation}}$

$$NH_2-CH(C_4H_9)-CO-NH-CH(C_4H_9)-CO-NH-CH_2-CO-NH-CH_2-CO_2H$$

$\xrightarrow{\Delta, \text{glacial acetic}}$

Diketopiperazines: HN-CHC$_4$H$_9$ / C$_4$H$_9$HC-NH + HN-CH$_2$ / H$_2$C-NH

Scheme 1

Although sequence information is forfeited by the formation of the diketopiperazine, overlap data are created by examination of the Edman-degraded peptide. This was a preliminary communication, and we must now await further work to see whether any problems or ambiguities will arise in looking at larger or more polar peptides.

Further development of the use of an aminodipeptidase in sequence analysis has been described.[16] These authors chose the pentafluoropropionyl methyl ester derivatives for g.c.m.s. analysis of the dipeptides produced by the enzyme. Following one step of Edman degradation on another portion of the peptide, the product is again digested with the aminodipeptidase and the products examined by g.c.m.s. A *silanized* glass frit separator was found to be important for producing good spectra with minimal thermal decomposition, and the procedure was evaluated using the insulin A chain (21 residues) as a standard. The method

---

[15] R. A. W. Johnstone, T. J. Povall, and J. D. Baty, *J.C.S. Chem. Comm.*, 1973, 392.
[16] R. M. Caprioli, W. E. Seifert, and D. E. Sutherland, *Biochem. Biophys. Res. Comm.*, 1973, **55**, 67.

suffers from the problems of enzyme specificity reported previously (its action is often blocked by Lys, Arg, or Pro), in addition to the destruction of the amides of Asn and Gln upon esterification. The aminodipeptidase enzyme provides a very attractive alternative route to the sequence analysis of small peptides, and it is interesting to note that the above-mentioned problems have been partially alleviated[17] by the simultaneous addition of other enzymes during digestion, and by the use of diazomethane for esterification. We stress the application to relatively small peptides since its use on large peptides or proteins leads to a large number of possible sequences from which it may be very difficult to isolate the true structure.

In continuance of an earlier study using a tandem mass spectrometer,[18] the effects of source chamber and probe surface on volatility and fragmentation of sample have been studied for the tripeptide thyrotropin-releasing hormone (TRH).[19] Teflon surfaces were found to be beneficial and the relevant signals for structure determination could be detected using 2 μg of peptide.

Twelve synthetic di- and tri-peptide analogues of TRH have also been studied by mass spectrometry, without the need for derivative formation.[20]

Several free amino-acids have now been examined by field desorption mass spectrometry[21] and have given the expected molecular or quasimolecular ions. Although it was stated that qualitative mixture analysis may be possible, the application of such a technique is difficult to imagine when quantitative amino-acid composition is so easily accomplished by present-day automatic analysers.

A considerable amount of work on the identification of amino-acid thiohydantoins (products of the Edman degradation) has been reported since the previous Report. In a complementary study to a previous one on bromo-phenylthiohydantoin derivatives (Br-PTH),[22] six cysteine derivatives have been examined.[23] This amino-acid normally gives rise to dehydroalanine, which is also observed in the spectrum of Br-PTH serine, thus making distinction between the two difficult. In contrast it was found that the S-methyl derivative gave an abundant molecular ion thus enabling facile identification. An alternative method of quantitation was also suggested in this paper, using a tetradeuterio-PTH derivative as standard. This would appear to be an improvement upon the use of ^{15}N reference compounds for quantitation since, with a 4 a.m.u. shift, there need be no correction for isotope contributions, and the acids (Asp and Glu) and amides (Asn and Gln) can be differentiated easily.

[17] R. M. Caprioli and W. E. Seifert, Paper presented to the Second International Symposium on Mass Spectrometry in Biochemistry and Medicine, Milan, June 1974, in press.
[18] R. J. Beuhler, L. J. Greene, and L. Friedman, *J. Amer. Chem. Soc.*, 1971, **93**, 4307.
[19] R. J. Beuhler, E. Flanigan, L. J. Greene, and L. Friedman, *Biochem. Biophys. Res. Comm.*, 1972, **46**, 1082.
[20] C. Bogentoft, J.-K. Chang, H. Sievertsson, B. Currie, and K. Folkers, *Org. Mass Spectrometry*, 1972, **6**, 735.
[21] H. U. Winkler and H. D. Beckey, *Org. Mass Spectrometry*, 1972, **6**, 655.
[22] F. Weygand and R. Obermeier, *European J. Biochem.*, 1971, **20**, 72.
[23] H. Tschesche, M. Schneider, and E. Wachter, *F.E.B.S. Letters*, 1972, **23**, 367.

The g.c.m.s. analysis of PTH derivatives has also been used to confirm some of the amino-acid residues in the sequence of ovine hypothalamic luteinizing hormone-releasing factor.[24]

Further work on the study of methyl thiohydantoins has now been reported. The subject of a paper of somewhat misleading title (Quantitative Protein Sequencing) was the thermally induced fragmentation of thiohydantoins from the thiourea amino-acid or peptide in the mass spectrometer ion source.[25] Slow heating of the probe was used to effect the transformation shown in Scheme 2.

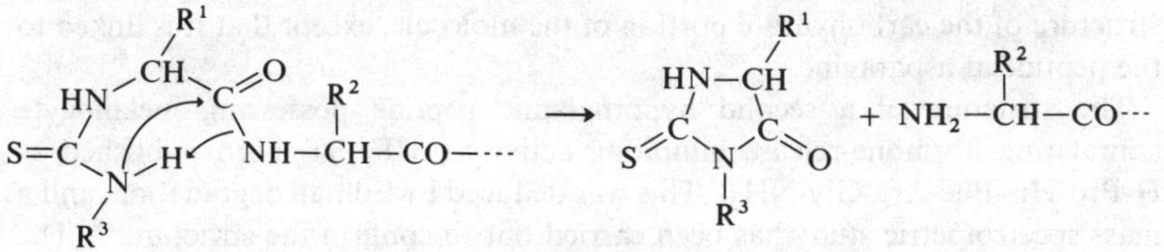

Scheme 2

The method was recommended for N-terminal amino-acid analysis, although for the quantity quoted (few µg) it offers no advantage over the dansyl technique.

A more interesting paper describes a sequencing strategy based upon the analysis of methylthiohydantoins.[26] The peptide is treated with methylisothiocyanate and an aliquot of the resulting thiourea examined by mass spectrometry (as the thiohydantoin[25]). The remainder of the thiourea peptide is treated with trifluoroacetic acid (Edman degradation) and the cycle repeated. Three tripeptides were examined, using between 19 and 50 µg at each step for mass spectrometric examination. The method is analogous to the dansyl-Edman procedure, and although more rapid, would appear to be less sensitive from the figures given; the reason for this is not clear.

An interface for the coupling of a mass spectrometer to an automatic spinning-cup sequenator has now been described in some detail.[27] This has been designed for the transfer of methylaminothiazolinones to the mass spectrometer without their prior isolation. In fact the system is not completely automatic, and requires some manual intervention, although this is not detrimental to the procedure as a whole. Solvent is removed by a flash evaporation technique, leaving the thiazolinones trapped in a gold gauze, ready for insertion into the ion source. This technique avoids the problems of preparation of volatile derivatives for the g.c. analysis. For those groups which have the facilities it could represent a very efficient means of identifying the products of Edman degradation from the sequenator, since the conventional g.c. and t.l.c. identification of these products often gives rise to problems.

[24] R. Burgus, M. Butcher, M. Amoss, N. Ling, M. Monahan, J. Rivier, R. Fellows, R. Blackwell, W. Vale, and R. Guillemin, *Proc. Nat. Acad. Sci. U.S.A.*, 1972, **69**, 278.
[25] T. Fairwell, S. Ellis, and R. E. Lovins, *Analyt. Biochem.*, 1973, **53**, 115.
[26] S. Ellis, T. Fairwell, and R. E. Lovins, *Biochem. Biophys. Res. Comm.*, 1972, **49**, 1407.
[27] R. E. Lovins, J. Craig, F. Thomas, and C. McKinney, *Analyt. Biochem.*, 1972, **47**, 539.

Polypeptide and Protein Structures. Several groups of workers have used mass spectrometry to confirm the structures of a number of small peptide hormones. Since many of these compounds are blocked at the *N*- and *C*-terminus a mass spectrometric study should be relatively simple and rapid compared to classical approaches to the structure determination of such molecules.

The peptide portion of a glycopeptide (17 residues) from the posterior lobe of pig pituitaries has been studied by mass spectrometry, and some of the sequence deduced by Edman degradation has been confirmed.[28] The ethoxycarbonyl permethylated peptide derivatives were studied. No comment was given on the structure of the carbohydrate portion of the molecule, except that it is linked to the peptide at asparagine.

The structure of a second hypothalamic peptide possessing melanocyte stimulating hormone-release inhibiting activity (MIF) has been published as H-Pro-His-Phe-Arg-Gly-NH$_2$. This was deduced by Edman degradation, and a mass spectrometric study has been carried out to confirm the structure.[29] The peptide was treated with hydrazine, acetylated and permethylated, but the spectra as published are somewhat mystifying both in their lack of normally expected cleavage fragments from the aromatic and heterocyclic amino-acids, and in the absence of the blocking group on the proline residue. Examination of the published spectra leaves some doubt as to whether the peptide has in fact got the structure proposed, or whether the spectra could be interpreted in some other manner, *e.g.* a cyclic precursor, although this would not correlate with the Edman data! Since no comment was made on the unusual nature of the spectra, we conclude that either the authors are unaware of the expected fragmentation of such molecules, or alternatively that these unusual fragmentations are anomalies to be dealt with in a future publication.

Mass spectrometry has also been used to deduce the sequence of somatostatin, a hypothalamic 14-residue peptide which inhibits secretion of growth hormone.[30] In a combined classical–mass spectrometric approach to this problem, a peptide was isolated having *N*-terminal asparagine, *C*-terminal lysine and an amino-acid composition of Asn, Lys, Phe$_2$, Trp. The mass spectrum showed the structure Asn-Phe-Phe-Trp-Lys, and together with the classical data, a total structure for somatostatin has been proposed.

Based upon previous findings that single protein-derived peptides or peptide mixtures of *unknown* amino-acid composition or structure can be readily sequenced as acetyl permethyl derivatives using low-resolution mass spectrometry,[31] further developments in strategy have now been reported. Firstly, in a study based on the proposed method of specific enzymic digestion followed by gel filtration and *direct* analysis of low molecular weight mixtures,[31] a considerable

---

[28] D. A. Holwerda, *European J. Biochem.*, 1972, **28**, 340.
[29] R. M. G. Nair, A. J. Kastin, and A. V. Schally, *Biochem. Biophys. Res. Comm.*, 1972, **47**, 1420.
[30] N. Ling, R. Burgus, J. Rivier, W. Vale, and P. Brazeau, *Biochem. Biophys. Res. Comm.*, 1973, **50**, 127.
[31] H. R. Morris, D. H. Williams, and R. P. Ambler, *Biochem. J.*, 1971, **125**, 189.

proportion (25%) of the milk protein α-lactalbumin has been sequenced.[32] This was achieved by the examination of only three consecutive mixture fractions from a Sephadex column effluent.

In two recent studies on the enzyme triose phosphate isomerase (TIM), low-resolution mass spectrometry of acetyl permethyl peptides has been used to give independent confirmation of the structures of classically determined peptides.[33,34] In one case where the Edman procedure had failed it proved possible to extend the peptide sequence of a fragment isolated from the active site of the enzyme.[34]

Further details of a mass spectrometric study on the sequence of the enzyme ribitol dehydrogenase (RDH) have now been reported.[35,36] This work, undertaken independently of the classical approach to the sequence, has resulted in the determination of >80% of the structure of the molecule (>200 residues sequenced), some of it predating the classically determined sequence. The study involved the use of both specific and non-specific enzymic digestion and low-resolution mixture analysis, and a number of useful fragmentation phenomena have been reported.[36] The sequence obtained by mass spectrometry is summarized in Scheme 3.

Met-Lys-His-Ser-Val-Ser-Ser-Met-Asn-Thr-Ser-Leu-(3 residues)-Val-Ala-Ala-Leu-Thr-Gly-Ala-Ala-Ser-Gly-Leu-Gly-Leu-Glu-(1 residue)-Ala-Arg-Thr-Leu-Leu-Gly-Ala-Gly-Ala-Lys-Val-Val-Leu-(2 residues)-Arg-Glu-Gly-Glu-Lys-(1 residue)-Asn-Lys-Leu-Val-Ala-Glu-Leu-Gly-Gln-Asn-Ala-Phe-Ala-Leu-Gln-Val-Asp-Leu-Met-(1 residue)-Ala-Asp-Gln-Val-Asp-Asn-Leu-Leu-Gln-Gly-Leu-Leu-Gln-Leu-Thr-Gly-Arg-Leu-Asp-Leu-Phe-His-Ala-Asn-Ala-Gly-Ala-Tyr-Leu-Gly-Gly-Pro-Val-Ala-(3 residues)-Pro-Asp-Val-(1 residue)-Asp-Arg-Val-Leu-His-Leu-Asn-Leu-Asn-Ala-Ala-Phe-Arg-Cmc-Val-Arg-Ser-Val-Leu-Pro-His-Leu-Leu-Ala-Gln-Lys-Ser-Gly-Asp-Leu-Leu-Phe-Thr-(10 residues)-Trp-Glu-Pro-Val-Tyr-Thr-Ala-Ser-Lys-Phe-Ala-Val-Gln-Ala-Phe-Val-His-Thr-Thr-Arg-(2 residues)-Val-Ala-Gln-Tyr-(3 residues)-Val-Gly-Ala-Val-Leu-Pro-Gly-Pro-Val-(2 residues)-Ala-Leu-Leu-(2 residues)-Trp-Pro-Lys-(2 residues)-Met-Asp-Glu-Ala-Leu-Ala-(4 residues)-Met-Gln-Pro-Leu-Glu-Val-Ala-(5 residues)-Met-Val-Thr-Arg-(2 residues)-Asn-Val-Thr-Val-(2 residues)-Leu-Val-Leu-Leu-Pro-Asn-Ser-Val-Asp-Leu.

Scheme 3 *Residues in brackets indicate those not found by mass spectrometry in this study*

A preliminary report on a new strategy applied to the structure determination of dihydrofolate reductase (DHFR), an enzyme of unknown sequence, has been given.[37] This involves the use of the non-specific enzyme, elastase, in a low-resolution mixture analysis approach. There is now strong evidence from both this and another mass spectrometric study on chloramphenicol transacetylase,

---

[32] J. R. Bacon and G. N. Graham, *Biochem. J.*, 1972, **127**, P76.
[33] J. D. Priddle, *Biochem. J.*, 1974, **139**, 23.
[34] J. D. Priddle and R. E. Offord, *F.E.B.S. Letters*, 1974, **39**, 349.
[35] B. S. Hartley, B. D. Burleigh, G. G. Midwinter, C. H. Moore, H. R. Morris, P. W. J. Rigby, M. J. Smith, and S. S. Taylor, *Proc. Eighth F.E.B.S. Meeting*, 1972, **29**, 151.
[36] H. R. Morris, D. H. Williams, G. G. Midwinter, and B. S. Hartley, *Biochem. J.*, 1974, **141**, 701.
[37] H. R. Morris, K. E. Batley, N. G. L. Harding, R. A. Bjur, J. G. Dann, and R. W. King, *Biochem. J.*, 1974, **137**, 409.

where >70% (170 residues) of the enzyme has been placed in sequence,[38] that elastase has some size-specificity, and breaks proteins down into fragments of ideal length for mass spectrometry i.e. 2—12 residues.

Work involving extensive mass spectrometric confirmation and assistance in the classical sequence analysis of a virus coat protein has been progressing,[39] using predominately low-resolution analysis of acetyl permethyl peptides.

From the above studies it can be seen that low-resolution mass spectrometry is now playing a valuable role in either confirmatory studies or direct analysis of unknown proteins. A just criticism of results to date is that the final 10% of a protein sequence is often the most difficult to obtain. It remains to be seen whether or not the random approach of a non-specific enzyme will alleviate this problem, but it is clear meanwhile that the combined use of classical and mass spectrometric techniques can certainly speed the acquisition of a total protein structure.

Miscellaneous. Some simple peptide mixtures of known sequence have been studied using the combined results of low- and high-resolution mass spectrometry, metastable ion detection, chemical ionization, and computer interpretation of data to obtain sequence information.[40] Several acetylated peptide esters were examined in addition to some permethylated acetylated peptide esters; the latter procedure is not good practice since β-keto-sulphoxides may be formed upon permethylation. It was claimed that unambiguous sequence assignment could not be made using low-resolution data alone, although judging from the volume of work on more complex unknown mixtures where unambiguous assignments have been made (see preceding section) this seems to be highly contentious. Of course ambiguities can arise, but these are normally resolved simply by deuterioacetylation or permethylation. In one of the mixtures studied (Ala-Phe-Ile-Gly-Leu-Met, Pro-Phe-His-Leu-Leu, and Pro-Phe-Asp), the complete sequences of the penta- and hexa-peptides were not obtained. We would attribute this to poor C-terminal regions in the spectra, due to the by-product formation mentioned earlier and also to the use of a conventional permethylation reaction which will have formed the sulphonium salt of the methionine residue. Alternative derivatization techniques[4,8] must be used if better results are to be obtained.

Since the introduction of the permethylation reaction into protein sequencing, a long-standing problem has been the differentiation of leucine from isoleucine. This has now been achieved on deuteriopermethylated peptides using metastable ion and neutral collisional activation techniques.[41] The distinction depends upon the decomposition of the immonium ion $CD_3NH=CH-C_4H_9$ (m/e 103) which is found in the spectra of peptides containing Leu or Ile. The species derived from leucine fragments to give an abundant signal at m/e 61 (m^*, 36.1),

---

[38] A. Dell and H. R. Morris, Paper presented to the Second International Symposium on Mass Spectrometry in Biochemistry and Medicine, Milan, June 1974, in press.
[39] R. Self, personal communication.
[40] H.-K. Wipf, P. Irving, M. McCamish, R. Venkataraghavan, and F. W. McLafferty, *J. Amer. Chem. Soc.*, 1973, **95**, 3369.
[41] K. Levsen, H.-K. Wipf, and F. W. McLafferty, *Org. Mass Spectrometry*, 1974, **8**, 117.

whereas that derived from isoleucine gives rise to a signal at m/e 69 (m^*, 46.3), as shown in Scheme 4.

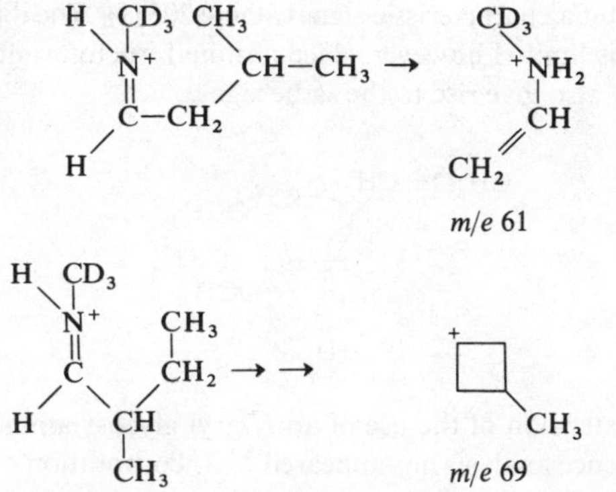

Scheme 4

The initial interest shown in the field desorption analysis of peptides continues. The method is very sensitive, and molecules of considerable polarity may be ionized and volatilized without derivative formation. However, recent studies on a variety of peptides ranging in size from two to nine residues, including free peptides,[42] acetyl,[42,43] and acetyl permethyl derivatives,[43] have shown that the optimism for sequence analysis generated by earlier reports[44,45] is partly unfounded. Whilst excellent molecular or quasimolecular ions may be obtained, it is difficult and in some cases impossible to obtain full sequence information using the thermal degradation technique. Further study directed to imparting more energy to the field-desorbed ions is now essential in order to exploit fully the advantages of this technique.

3 Carbohydrate Sequence Analysis

Mass spectrometry has been used to study six disaccharides, two trisaccharides, and a tetrasaccharide containing the biologically important fructofuranose ring system.[46] Methyl ether derivatives were studied, and it was found that the relative

---

[42] N. Evans, D. E. Games, M. J. E. Hewlins, J. F. J. Hughes, A. H. Jackson, J. R. Jackson, M. N. Khan, S. A. Matlin, M. Rossiter, R. G. Saxton, H. A. Swaine, and K. T. Taylor, Paper presented to the Second International Symposium on Mass Spectrometry in Biochemistry and Medicine, Milan, June 1974, in press.
[43] H. R. Morris and M. R. Thompson, Paper presented to the Second International Symposium on Mass Spectrometry in Biochemistry and Medicine, Milan, June 1974, in press.
[44] H. U. Winkler and H. D. Beckey, *Biochem. Biophys. Res. Comm.*, 1972, **46**, 391.
[45] H. U. Winkler and H. D. Beckey, Proceedings of the Twentieth Annual Conference on Mass Spectrometry and Allied Topics, Dallas, June 1972, p. 164.
[46] K. G. Das and B. Thayumanavan, *Org. Mass Spectrometry*, 1972, **6**, 1063.

abundances of signals at m/e 108 and 88 (analogues of m/e 217 and 204 in TMS derivatives) *cannot* be used to distinguish between furanose and pyranose ring forms. It was also shown that fructofuranose units which are linked *via* C-2 to another sugar exhibit a characteristic signal at m/e 205 (1). The diagnostic usefulness of this signal is limited however, since terminal fructofuranose units linked *via* C-1 or C-6 may also give rise to the same signal.

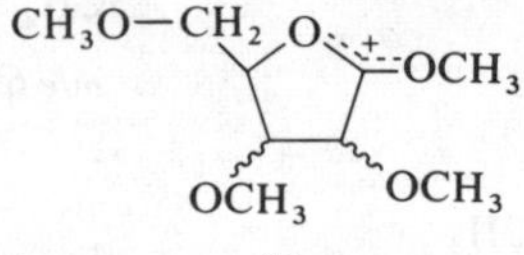

(1)

An interesting extension of the use of an *N*-aryl glycosylamine substituent in carbohydrate sequence analysis has appeared.[47] Incorporation of such aromatic units into the terminal sugars has previously been shown to lead to increased molecular ion abundances.[48,49] *N*-Aryl glycosylamine derivatives of the *O*-acetates of some tri-, tetra-, and penta-saccharides were studied, and in most cases allowed determination of molecular weight, and the nature and sequence of the monosaccharide units. Common losses from the molecular ions included keten, acetic acid, and acetic anhydride, and in some cases partial data on interglycosidic linkage could be deduced.

Structural studies using mass spectrometry have been made on an O Group lipopolysaccharide from *Klebsiella*.[50] Analysis of methylated material, or products of partial acid hydrolysis of the methylated material followed by reduction and deuteriopermethylation, was carried out. The g.c.m.s. identification of the permethylated disaccharide alditol and methylated methyl glycosides was facilitated by earlier studies on these compounds.[51,52] *O*-Acetyl groups were located by treating the lipopolysaccharide with methyl vinyl ether, thus protecting free hydroxy-groups as acetals. The product was then methylated, replacing OAc by OMe groups. Acid hydrolysis gave a mixture of sugars and methylated sugars from which the position of the initial *O*-acetyl groups could be located. It was deduced that the side-chains of the lipopolysaccharide are composed of pentasaccharide repeating units, the proposed structure of which is shown in Scheme 5.

The mass spectrum of six TMS disaccharides of the type aldohexosyl-(1→x)-fructose, in which x varies from 1 to 6, have been described.[53] It proved possible

[47] O. S. Chizhov, N. N. Malysheva, and N. K. Kochetkov, *Carbohydrate Res.*, 1973, **28**, 21.
[48] G. S. Johnson, W. S. Ruliffson, and R. G. Cooks, *Carbohydrate Res.*, 1971, **18**, 233.
[49] N. K. Kochetkov, O. S. Chizhov, N. N. Malysheva, and A. I. Shiyonok, *Org. Mass Spectrometry*, 1971, **5**, 481.
[50] B. Lindberg and J. Lönngren, *Carbohydrate Res.*, 1972, **23**, 47.
[51] J. Kärkkäinen, *Carbohydrate Res.*, 1970, **14**, 27.
[52] N. K. Kochetkov and O. S. Chizhov, *Adv. Carbohydrate Chem.*, 1966, **21**, 39.
[53] J. P. Kamerling, J. F. G. Vliegenthart, J. Vink, and J. J. de Ridder, *Tetrahedron*, 1972, **28**, 4375.

on the basis of fragmentation to subdivide these into two groups, *i.e.* (1→1)-, (1→2)-linked and (1→3)-, (1→4)-, (1→5)-, (1→6)-linked disaccharides. The mass spectra of two di-, six tri-, and three tetra-saccharides containing one or more

D-Gal f-(1→3)-D-Gal p-(1→3)-D-Gal f-(1→2)
D-Gal f-(3←1)-D-Gal p-(3←1)-D-Gal f-(3←1)   ⎤
| | D-Gal p-(3←1)-D-Gal p
(1→2)-D-Gal p-(1→ |
| ⎦$_{n \approx 7}$
D-Gal p-(1→3)

Some of Gal p are substituted with OAc at C-6
Some of Gal f are substituted with OAc at C-2 and C-6

Scheme 5

$(x \to 2)$-β-D-fructofuranose units ($x = 1$ *or* 6) were also studied,[53] and it was shown that the presence of such a unit gives rise to very abundant signals at *m/e* 437 and/or 815 in the spectra. The formation of these species for fructofuranosyl-(2→1)-fructofuranosyl-(2→6)-glucose is depicted in Scheme 6. It was concluded

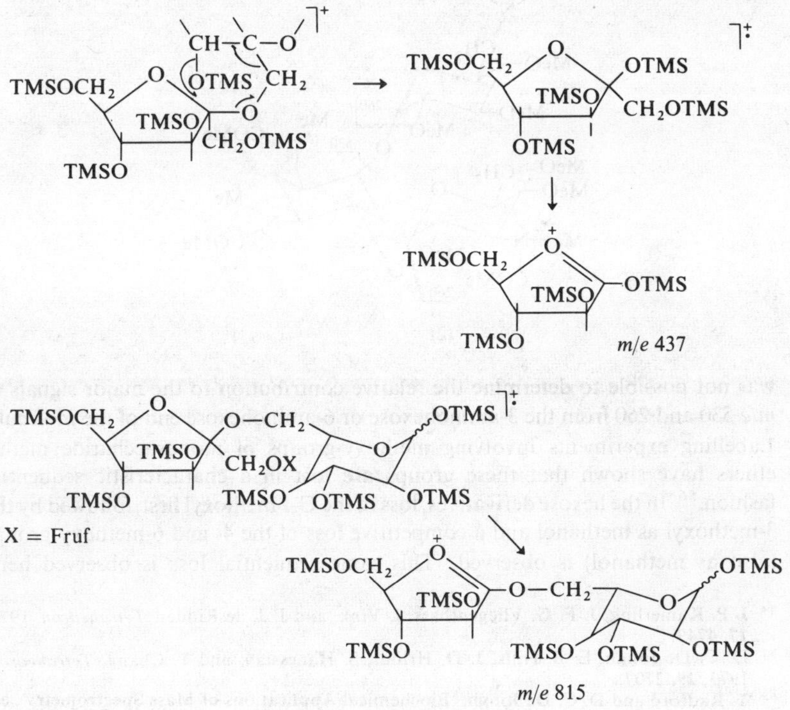

m/e 437

X = Fruf

m/e 815

Scheme 6

that the relative abundance ratio of m/e 217 to 204 is an unreliable guide to pyranose or furanose ring systems, since this ratio is governed to some extent by the type of glycosidic linkage.[54] In fact, the ratio has been used as a diagnosis for the linkage in studies of 2-acetamido-2-deoxyaldohexosyl-aldohexoses,[54] *i.e.* with the amino-sugar at the non-reducing end. Care must be taken in drawing conclusions about such ratios, as is shown in a recent study of (1→3)- and (1→4)-linked aldohexosyl-2-acetamido-2-deoxyaldohexoses[43] in which the 217:204 ratio is the reverse of that for the analogous compounds above. In this study,[43] the structure of a small glycopeptide (mol. wt. 2500) from blood serum was determined by analysis of the permethyl and TMS derivatives of the intact molecule and its hydrolysis products.

The high- and low-resolution electron impact spectra of the *N*-acetyl-*NO*-methyl and *N*-acetyl-*O*-trimethylsilyl derivatives of an aminocyclitol antibiotic Kanamycin A have been determined.[55] Interpretation was aided by the analysis of deuteriated analogues. The spectrum of the *N*-acetyl-*NO*-methyl derivative showed a weak M^+ at 806 (0.4%) which was greatly improved in the isobutane chemical ionization spectrum, $M + H^+$ (13%). The sequential arrangement of the saccharide units was readily apparent from the cleavages indicated in (2). It

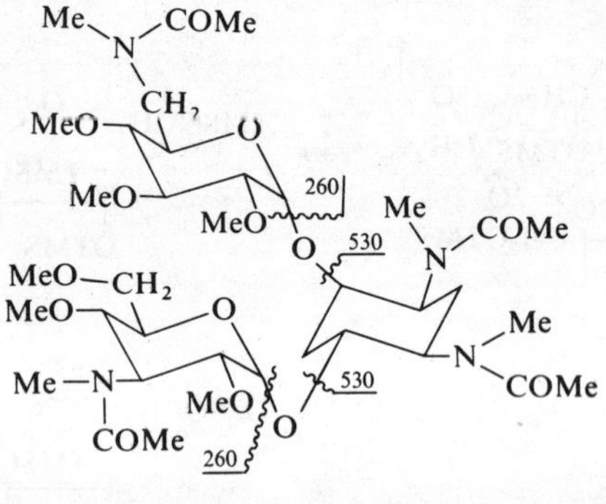

(2)

was not possible to determine the relative contribution to the major signals at m/e 530 and 260 from the 3-aminohexose or 6-aminohexose end of the molecule. Labelling experiments involving methoxy-groups of monosaccharide methyl ethers have shown that these groups are lost in a characteristic sequential fashion.[56] In the hexose derivatives, loss of the C-1 methoxyl first, followed by the 3-methoxyl as methanol and a competitive loss of the 4- and 6-methoxy-groups (also as methanol) is observed. This same sequential loss is observed here,

[54] J. P. Kamerling, J. F. G. Vliegenthart, J. Vink, and J. J. de Ridder, *Tetrahedron*, 1971, **27**, 4749.
[55] D. C. De Jongh, E. B. Hills, J. D. Hribar, S. Hanessian, and T. Chang, *Tetrahedron*, 1973, **29**, 2707.
[56] T. Radford and D. C. De Jongh, 'Biochemical Applications of Mass Spectrometry', ed. G. R. Waller, Wiley, New York, 1972, Chap. 12.

commencing with m/e 260. The loss of 73 a.m.u. corresponding to N-methylacetamide is consistent with the presence of an N-methylacetamido-group in position 3 at the 3-aminohexose end of Kanamycin, while the loss of 32 a.m.u. as methanol results from the presence of a 3-methoxy-substituent at the 6-aminohexose end. Competitive loss of 32 a.m.u. and 73 a.m.u. from m/e 228 is characteristic of the presence of a 6-N-methylacetamido-group, confirming that m/e 260 is derived from two different ends of the molecule. The N-acetyl-O-TMS derivative of Kanamycin A also showed prominent signals useful for sequencing (3). In the

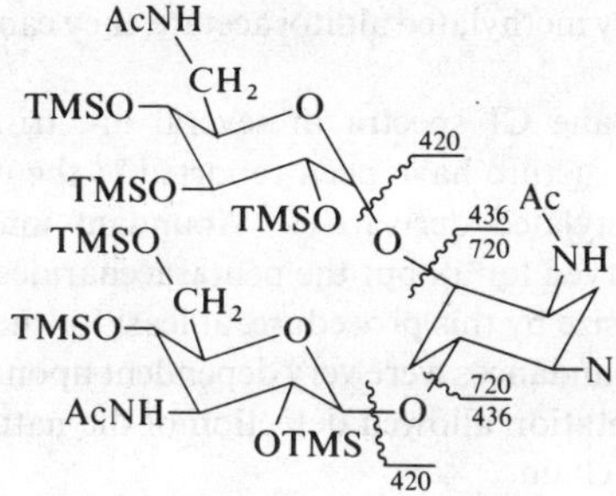

(3)

isobutane CI mass spectrum of this compound, the quasimolecular ion was relatively intense, and abundant sequence ions were noted.

The mass spectra of 13 perdeuteriomethylated flavonoid disaccharides comprising flavones, flavanones, flavonols, and an isoflavone have been reported.[57] Fragmentation allowed unequivocal differentiation according to structural type of aglycone and sugar sequence. The principal fragmentations are shown in (4).

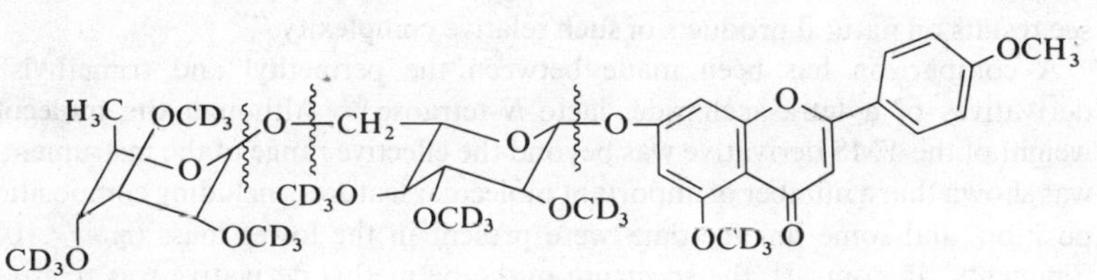

(4)

Sequence information can clearly be obtained provided that a molecular ion is observed. This study was limited, however, since only compounds containing glucose as the second residue were available. It was claimed that the position of the glycosidic linkage could be determined in these compounds since, in (1→6)-linked compounds, the sequence signal S is formed with the transfer of one H, or without H transfer; in (1→2)-linked compounds a 2H transfer is observed. However, the data were very limited, and this conclusion is tenuous.

In a similar study, the major part of the structures of three flavanoid triosides, xanthorhamnin, alaternin, and catharticin have been determined by mass

[57] R. D. Schmid, *Tetrahedron*, 1972, **28**, 3259.

spectrometry.[58] Examination before and after hydrolysis is put forward as a general technique for the elucidation of such structures. Each flavonol trisaccharide was previously known to contain one residue of galactose and two residues of rhamnose, but only partial information on the position of sugar attachment and interglycosidic linkage was available. From the spectra, the sugar sequence could clearly be established as rha-rha-gal-aglycone, but the criteria for linkage assignment used previously[57] were not applicable. Hydrolysis followed by conversion into alditol acetates[59] and examination by g.c.m.s. gave the required data in all but one case; since a 4-linked hexopyranose and a 5-linked hexofuranose give the same partially methylated alditol acetate, they cannot be differentiated by this method.

The ammonia–isobutane CI spectra of several di-, tri-, tetra-, and pentasaccharides of known structure have been reported,[60] the carbohydrates being examined as the peracetylated derivatives. Abundant ions corresponding to $(M + NH_4)^+$ were observed for all but the pentasaccharides; this would appear to be the upper limit on size by this procedure, at least for the acetate derivatives. Relative fragment ion abundances were very dependent upon source temperature, but thermolysis fragmentation allowed detection of the nature of the individual monosaccharides in the chain.

The mass spectrum of a methylated, reduced, and trimethylsilylated derivative of a pentaglycosylceramide from bovine brain has been described.[61] Signals indicative of the presence of three hexoses (one an N-acetylated hexosamine) and one N-acetylated neuraminic acid were assigned and the sequence deduced. The ceramide portion of the molecule, which had been characterized previously, also gave rise to easily identifiable fragments in the mass spectrum. Although no details of the data used for interpretation were given in this paper (high-resolution mass measurements, studies on other derivatives *etc.*), it is most encouraging to see results on natural products of such relative complexity.

A comparison has been made between the permethyl and trimethylsilyl derivatives of a tetrasaccharide, lacto-*N*-tetraose.[62] Although the molecular weight of the TMS derivative was beyond the effective range of the instrument, it was shown that a number of important molecular features, including composition, position, and some linkage data were present in the lower mass ($m/e < 1000$) fragments. In contrast, the spectrum of the permethyl derivative was relatively uninformative, but it does of course offer several advantages including lower molecular weight, stability, and a suitability for further degradative studies.

The overall indications are that, despite the complexity of potential structures involved, and whilst in this respect the majority of studies are still only 'scratching the surface', mass spectrometry will play a vital role in future years in structure elucidation of small quantities of these biologically important materials.

[58] R. D. Schmid, P. Varenne, and R. Paris, *Tetrahedron*, 1972, **28**, 5037.
[59] H. Bjoerndal, C. G. Hellerqvist, B. Lindberg, and S. Svensson, *Angew. Chem. Internat. Edn.*, 1970, **9**, 610.
[60] R. C. Dougherty, J. D. Roberts, W. W. Binkley, O. S. Chizhov, V. I. Kadentsev, and A. A. Solov'yov, *J. Org. Chem.*, 1974, **39**, 451.
[61] K.-A. Karlsson, *F.E.B.S. Letters*, 1973, **32**, 317.
[62] H. Egge, H. V. Nicolai, and F. Zilliken, *F.E.B.S. Letters*, 1974, **39**, 341.

Author Index

Aaronson, M. J., 161, 162
Aasen, A. J., 258, 307, 315
Abdullina, N. G., 223
Abe, H., 270
Abel, E. W., 198, 212
Abel-Rahman, M. M. A., 280
Aberhart, J., 239
Abley, P., 303
Ablov, A. V., 207
Abraham. K. M., 201
Abramson, D., 315
Abramson, F P., 138, 328
Abuki, H., 321
Acheson, R. M., 282
Achmatowicz, O., 280
Ackermann, J. R., 260
Ackman, R. G., 314, 337
Acton, N., 188, 189
Adam, G., 90, 231, 233, 293
Adam, W., 275
Adams, V. H., 226
Adcock, J. L., 191, 215
Adler, A. D., 245
Adler, B., 180, 278
Adlercreutz, H., 231, 320, 321
Advena, J., 172
Ånggard, E., 328
Aerni, R. J., 25, 83, 102
Agafonov, I. L., 95, 96
Agam, J., 48
Agranat, I., 273
Agurell, S., 356, 357
Ahlborg, U. G., 121, 300
Ahlrichs, R., 2, 80
Ahmad, I., 336
Ahmad, M. S., 231
Ahnell, J. E., 293
Ahnoff, M., 336
Ahond, A., 225
Ainsworth, C., 157
Airey, W., 214
Aizawa, K., 67, 255
Aizenshtat, Z., 337
Akhtar, Z. M., 87
Akino, M., 242
Akishin, P. A., 213
Aksenov, V. I., 168
Albano, V. G., 178
Albers-Schönberg, G., 297
Albert, H.-J., 160
Albertson, N. F., 253
Albone, E. S., 338
Albrecht, B. H., 141
Alder, A. D., 280

Aleksander, L., 132
Alester, G., 163
Alexander, G., 298
Alexander, M., 336
Alexander, R. G., 88, 166, 273, 290
Alexandru, Gr., 16
Alford, A. L., 139, 297
Alford, K. J., 150
Allais, J. P., 318
Allan, M., 216
Allen, C. jun., 332
Allen, C. W., 163
Allkofe, J., 181
Allmann, R., 146
Almquist, S., 315
Alper, H., 194, 198, 285
Al-Sarraj, S., 328
Alsop, J. E., 185
Alsop, P. A., 285
Althaus, J. R., 325
Alturki, Y. I., 244
Alyea, E. C., 193
Ambe, F., 206
Ambe, S., 206
Ambler, R. P., 256, 368
Ames, L. L., 213
Aminev, I. Kh., 89, 293
Amos, B. A., 231
Amos, T. L., 268
Amoss, M., 256, 367
Amy, J. W., 42, 287
Anders, D. E., 337
Anders, E., 337
Anders, M. W., 330
Andersen, D. A., 334
Anderson, B. D., 337
Anderson, D. N., 128
Anderson, G. A., 215
Anderson, J. B., 18
Anderson, J. D., 320
Anderson, J. W., 157
Anderson, R. L., 193
Anderson, V. E., 295
Anderson, W. K., 280
Anderson, W. P., 195
Andersson, B. Å., 259, 280, 303, 310
Andersson, I., 327
Andersson, J., 332
Andłauer, B., 42
Ando, M., 335
Andrew, M. N., 152
Andrews, A. G., 230
Ane, D. H., 53
Ang, H. G., 167
Angelici, R. J., 204

Angus, P. C., 159
Anicich, V. G., 102, 112
Annarelli, D., 153
Ansari, G. A. S., 231
Anthony, G. M., 313
Anthony, R. G., 213
Antonowa, A., 286
Aplin, R. T., 60, 244, 282
Appel, R., 156, 171
Appelman, E. H., 51
Appleton, R. A., 229
Aragozzini, F., 250
Arakawa, K., 293
Araki, E., 338
Aranda, G., 276
Araujo, J., 330
Arcamone, F., 349
Ardenne, M. V., 231
Ardini, L. C., 214, 215
Ardrey, R. E., 168, 324
Arends, P., 227, 282
Arens, J. F., 168
Argoudelis, A. D., 330
Arhart, R. W., 189
Arima, N., 345
Aringer, L., 322
Arison, B. A., 282
Ariyoshi, T., 345
Armbruster, R., 281
Arneson, P., 123
Arnett, E. M., 106
Arnold, D. E. J., 216
Arnold, R. G., 334
Arpin, N., 230
Arpino, P., 224, 304, 337
Arsenault, G. P., 98, 127, 263, 297, 299
Artamonova, N. N., 273
Asbell, M. A., 352
Asberg, M., 357
Asbrink, L., 30
Asgarouladi, B., 149
Aspinal, M. L., 124
Ast, T., 44, 57, 85, 265, 287
Atabek, O., 15
Atkinson, A. J., 325, 356
Atkinson, R., 88
Audier, H.-E., 26, 276
Aue, D. H., 106, 111
Aumann, R., 186
Aune, J. P., 285
Ausloos, P., 59, 80, 109, 111
Autio, S., 312
Autzen, H., 208
Auvinen, E. M., 271
Averill, F. W., 5

Awerbouch, O., 168
Axberg, K., 247
Axelrod, B., 306
Axen, U., 123, 300, 308, 357
Axenrod, T., 100, 249
Aylett, B. J., 216
Ayling, J. E., 242, 283
Azzaro, M., 274, 278, 282

Baalmann, H. H., 218
Baarschers, W. H., 284
Babashak, J. F., 303
Bach, M., 165
Back, P., 320
Back, R. D., 145
Bacon, J. R., 369
Baczynskyj, L., 123, 239, 300, 308
Baedecker, M. J., 337
Baer, T., 47
Baerlocher, K., 302
Bafus, D. A., 298
Bagley, C. E., 336
Bagley, M. J., 278
Bagrii, Ye. I., 264
Baier, H., 168
Bailar, J. C., 210
Bailey, E., 232, 302, 315
Bailey, J., 194
Bailey, K., 341
Bailey, N. J. L., 338
Bailey, T. L., 13
Baillie, T. A., 317
Baitinger, W. E., 42, 265, 287
Baker, K. M., 61, 283, 326, 327, 344, 347
Bakulina, I. N., 289
Balasubramanian, H., 153
Baldas, J., 225, 226
Balderschweiler, J. D., 289
Balducci, G., 222, 223
Baldwin, M. A., 62, 96, 224, 251, 304
Ballschmieter, H. M. B., 332
Bálthazar, Z., 338
Balza, F., 284
Bancroft, K. C. C., 286
Band, S. T., 195
Banditelli, G., 208
Banjkowsky, A. J., 226
Banks, K. A., 332
Bannard, R. A. B., 338
Banzica, R. P., 99
Baranowska, E., 280
Barber, M., 61
Barber, R. C., 120
Barbier, F., 321
Barbier, M., 318
Bardou, L. G., 316, 320
Barfield, A. L., 49
Bargery, J., 120
Barker, B. A., 185
Barker, G. K., 157
Barker, J. R., 15
Barlin, J.-J., 150
Barlow, C. B., 248
Barnett, G. H., 209

Baron, C., 321
Baroni, D., 334
Barraclough, C. G., 179
Barratt, M. D., 280
Barrau, J., 159
Barry, G. T., 305
Bartha, B., 338
Bartha, R., 335
Barton, T. J., 153
Bashkirova, S. A., 154
Batley, K. E., 254, 369
Baty, J. D., 139, 253, 321, 324, 326, 365
Batzel, V., 201
Baudler, M., 155, 219
Bauer, C., 216
Bauer, E., 165
Bauer, H., 311
Bauer, L., 264
Baughman, G. L., 336
Bauman, L. E., 43, 65
Baumann, W. J., 258, 307, 338
Baumgartel, H., 51
Baur, P. S., 338
Baxter, R. L., 229
Bayer, E., 311
Bayes, K. D., 88
Beal, D. K., 342
Bearden, J. R., 314
Beauchamp, J. L., 53, 63, 83, 102, 103, 106, 107, 108, 109, 110, 220
Beck, J. L., 156, 242, 282, 323, 338
Beck, W., 198, 200
Becker, C. E., 360
Becker, H. G. O., 278
Becker, H.-J., 191
Beckett, A. H., 323, 326, 328, 329, 342, 343
Beckey, H. D., 33, 34, 90, 91, 92, 93, 94, 95, 96, 144, 224, 225, 240, 241, 247, 249, 251, 366, 371
Beer, D. C., 150, 203
Beerbaum, H., 242
Beerham, H., 283
Beernaert, H., 331
Beggs, D. P., 109
Begin, P. A., 149
Bégué, R. J., 320
Begun, G. M., 57
Behrens, H., 204
Beiber, M. A., 355
Bélanger, P., 297, 319
Bélanger, P. M., 329
Belcher, R., 211, 213
Belikov, A. B., 282
Bell, C. L., 264
Bell, J. E., 335
Bell, N. W., 125, 139
Bellina, J. J., jun., 299
Belvedere, G., 327, 355
Bel'yerman, A. L., 267
Bendat, J. S., 125
Bender, C. F., 128, 130, 131
Bender, C. O., 209
Benezra, S. A., 26, 277

Benfield, E. F., 331
Benfield, F. W. S., 189
Benn, H., 190
Bennett, C. R., 181
Bennett, M. A., 197, 198
Bennett, M. J., 203
Bennett, R. L., 182, 194
Bennett, S. L., 109, 223
Benoit, F., 54, 72, 81, 266, 272, 273
Ben-Shaul, A., 20
Ben-Shoshan, R., 185
Benson, S. W., 83
Bente, P. F., 62, 63, 253, 275
Bentley, R., 299, 328
Bentley, T. W., 53, 59, 60, 68, 69, 70, 73, 77, 278
Berestiansky, J., 242
Bergelson, L. D., 307
Bergert, K. H., 336
Bergot, B. J., 300
Bergström, G., 330, 331
Berkowitz, J., 8, 9, 51, 53, 213
Berlingin, A., 7, 78
Berniaz, A. F., 151
Bernstein, R. B., 18, 20, 22
Beroza, M., 330
Berry, A. D., 202
Berry, C. S., 246
Berry, R. E., 331
Berry, R. S., 282, 289
Berthou, F. L., 316, 320
Bertilsson, L., 325, 328, 355
Bertrand, M., 43, 44, 45, 56, 66, 67, 266, 275, 278, 279
Bertsch, W., 298
Berube, L., 254
Bessell, E. M., 94, 144, 247
Beswick, J. A., 14
Betts, T. C., 5
Betz, V., 336
Beuhler, R. J., 246, 366
Beveridge, D. L., 263
Beyermann, K., 335
Beynon, J. H., 42, 43, 44, 45, 57, 65, 66, 67, 80, 85, 88, 263, 265, 266, 271, 272, 274, 275, 278, 279, 287, 292
Bhasin, S. K., 278
Bhuiyan, A. L., 150
Biandrate, P., 344
Bick, I. R. C., 225, 226
Bickelhaupt, F., 149, 167
Bieber, M. A., 121, 300
Biemann, K., 78, 118, 127, 136, 139, 140, 248, 253, 263, 297, 300, 307, 315, 326, 334, 337, 364
Bier, D. M., 121, 300
Bierl, B. A., 330
Bigelow, L. B., 328
Biggs, D. F., 249
Bigley, D. B., 88, 166, 273, 290
Biller, J. E., 140, 253
Billets, S., 107, 111, 279

Author Index

Billing, B. H., 234
Bills, D. D., 333
Billups, W. E., 71, 185, 186
Binkley, W. W., 246, 376
Binneuries, M., 213
Biondi, P. A., 326
Birch, A. J., 229
Birchall, T., 194
Bird, C. W., 306
Birkenhager, G., 180
Birkinshaw, K., 21
Birkofer, L., 157, 281
Biros, F. J., 90, 101, 293, 335
Björkhem, I., 232, 316
Björndal, J., 247, 376
Bjorkman, L. R., 259
Bjur, R. A., 254, 369
Black, D. R., 303
Black, D. St. C., 208
Black, P. J., 119
Black, W. W., 126
Blackborow, J. R., 149
Blackburn, P. E., 223
Blackman, A. J., 89, 281
Blackmore, T., 181
Blackwell, R., 256, 367
Blair, L. K., 110
Blanchard, C., 226
Blenderman, W. G., 195
Blessel, K., 211
Blessington, B., 324, 341
Blint, R. J., 53, 83, 108, 109
Block, J. H., 90, 91, 318
Bloom, H., 213
Blount, J. F., 341
Blum, M. S., 330, 334
Blum, S., 256
Blum, W., 101, 256, 299
Blumenthal, T., 153, 287
Boaz, H. E., 345
Bobrin, L. J., 350
Bochkarev, V. N., 150, 152, 154, 157, 168
Bodor, N., 6
Boerboom, A. J. H., 95, 225
Boerner, U., 360
Børresen, H. C., 312
Boese, R., 202
Boettger, H. G., 89, 188, 293, 299
Bogentoft, C., 251, 278, 340, 366
Boggs, R. A., 193
Bognár, R., 237
Bohlmann, F., 56, 71, 73, 266, 267, 270, 272, 282
Bohman-Lindgren, G., 255
Bohn, G., 340
Boldingh, J., 258
Bolduc, E., 10
Bolton, R. G., 282
Bommer, P., 346
Bonati, F., 57, 204, 208
Bond, A. C., 216
Bonelli, E. J., 297, 334
Bonham Carter, S., 349
Bonner, R. F., 96

Bonnet, J. J., 209
Bonnett, R., 209, 246
Bonnichsen, R., 329
Booher, R. E., 329
Boos, W. R., 338
Booth, M., 320
Booth, R. A., 325
Bor, G., 179, 200
Bordeleau, L. M., 335
Borén, H. B., 311
Boreus, L. O., 356
Borga, O., 355
Borgen, O., 133
Borleshe, S. G., 167
Borsdorf, R., 286
Bortinger, A., 337
Bos, H. J. T., 168
Bos, K. D., 160
Bose, N., 51
Boston, D. R., 208
Bottcher, H., 278
Botter, R., 55
Bouchoux, G., 26
Boulton, A. A., 360
Bourchal, K., 205
Bourgeois, G., 282
Bourgeois, P., 152
Bournot, P., 321
Bowden, J. A., 179, 195
Bowen, D. H., 300, 314
Bowen, D. V., 250, 251
Bower, B. K., 180
Bowers, M. T., 25, 53, 102, 106, 111, 112
Bowers, W. S., 330
Bowes, G. W., 336
Bowie, J. H., 88, 89, 153, 262, 273, 281, 282, 287, 288, 290, 292, 294, 362
Box, D. G., 336
Boyd, M. R., 332
Boylan, D. B., 244
Boyle, P. J. R., 327
Brachet-Liermain, A., 282
Bradbeer, J. W., 314
Bradford, C. W., 200
Bradley, C., 144, 181, 193, 212
Bradley, G. F., 202
Brady, B. A., 227
Braestrup, C., 328
Bragina, S. I., 274
Brand, J. M., 330
Brandenberger, H., 328
Brandl, A., 194
Brandon, C., 243
Brandt, D., 14
Brandt, R. D., 319
Brandt, V. O., 286
Branton, G. R., 54
Braselton, W. E., jun., 301
Bratton, R. F., 215
Brauer, D. J., 216
Brauman, J. I., 108
Braun, A. M., 274
Braun, P., 233
Brazeau, P., 255, 368
Breck, G. D., 353
Breda, A. C., 110
Brehm, B., 26, 27, 87

Breimer, D. D., 300, 326
Brekke, J. E., 331
Brennan, M. R., 273
Brennan, P. J., 259
Brent, D. A., 359
Bretschneider, E. S., 163
Breuer, H., 317, 321, 322
Brewer, H. B., 100
Brewer, P., 209
Brewington, C. R., 332
Breyer, U., 343, 358
Bricas, E., 351
Brickman, J., 16
Briegleb, G., 288
Brieskorn, C. H., 306
Briggs, P., 119
Brine, D. R., 328
Brink, R. W., 204
Brinkman, H. W., 332
Brion, C. E., 7, 58, 87
Britten, A. Z., 276
Brodasky, T. F., 330, 354
Brodskii, E. S., 264, 265
Bronzert, T. J., 100
Brook, A. G., 153
Brookes, A., 202
Brookhart, M., 185
Brooks, C. J. W., 231, 232, 296, 297, 313, 314, 315, 316, 317, 320, 324, 341
Brooksbank, B. W. L., 302
Brophy, J. J., 330
Brown, B. O., 237
Brown, C., 149
Brown, C. L., 88, 149, 292
Brown, D. F., 278
Brown, E. V., 278, 279, 282, 336
Brown, H. C., 76
Brown, J. P., 237
Brown, P., 49, 93, 233
Brown, R. A., 196
Brown, W. V., 229
Brownlee, R. T. C., 53
Bruce, M. I., 143, 181, 182, 194, 199, 203
Bruce, R. B., 351
Brufani, M., 239
Bruggeman, L. E., 119
Bruins Slot, J. H. W., 139
Brune, H. A., 184, 187
Bruner, B. L., 133
Bruner, F., 299
Brunfeldt, K., 254
Bruni, R. J., 352
Brunner, H., 179, 182, 194, 202, 211
Bruschweiler, F., 233
Brtíun, T., 305
Bruyere, A., 281
Bryan, R. F., 149, 229
Bryant, P. J., 91
Bryce, T. A., 333
Bryce, W. A., 48
Brzozowski, N., 14
Buchanan, B. G., 133, 134, 231
Buchanan, J. G., 312
Buchardt, O., 282
Buchler, A., 214

Buchler, J. W., 209
Buchner, B., 168
Buchner, W., 171, 181
Buchs, A., 118, 133, 248, 270
Buck, A., 251
Buckmann, K., 151
Budde, W. L., 121, 336
Budzikiewicz, H., 60, 74, 225, 230, 231, 244, 246, 258
Buenker, R. J., 3, 17
Buess, C. M., 286
Bugerenko, E. F., 168
Bulger, W. H., 352
Bullock, F. J., 352
Bulowski, P., 280
Bulten, E. J., 160
Bunker, D. L., 17, 19
Burbank, R. D., 222
Burg, A. B., 149, 150, 180, 217
Burger, J. K., 170
Burgus, R., 255, 256, 324, 367, 368
Burke, M. T., 153
Burleigh, B. D., 369
Burlingame, A. L., 1, 33, 34, 39, 40, 61, 68, 69, 73, 78, 92, 93, 119, 124, 142, 144, 225, 226, 264, 273, 275, 297, 302, 334, 336, 337
Burlitch, J. M., 201
Burnham, R. A., 202
Burnstein, S., 348
Bursey, J. T., 67, 140, 144, 152, 209, 211, 245, 264, 271, 273, 277
Bursey, M. M., 7, 26, 67, 72, 80, 81, 112, 113
Busch, D. H., 208
Busch, G. E., 21
Busetto, L., 204
Bush, M. A., 222
Bush, M. T., 301
Buss, B., 220
Butcher, M., 256, 324, 367
Butler, C., 329
Butterfield, R. O., 119
Buttery, R. G., 285, 332
Buttrill, S. E., 102, 279
Byers, W., 199

Cabaud, B., 10, 219
Cable, J., 70, 270
Caich, S., 222
Cairns, M. A., 196, 199
Calam, D. H., 325
Calcote, H. F., 289
Caldwell, K. J., 288
Callahan, K. P., 203
Caluwe, P., 85, 267
Calvin, M., 244, 337
Cambie, R. C., 229
Cambisi, F., 185
Cambon, A., 78
Cameron, D. W., 273

Camp, M., 132
Campbell, D. B., 347
Campbell, I. M., 299, 318, 328
Campbell, M. M., 77, 263
Cann, P. F., 168
Cannon, J. B., 182
Cant, E., 228, 272
Canty, A. J., 200
Caplan, C. E., 16
Caprioli, R. M., 42, 44, 66, 247, 253, 263, 265, 271, 272, 274, 287, 302, 325, 365, 366
Carbonaro, A., 185
Carbonneau, R., 10
Cardwell, T., 211
Cariati, F., 178
Carle, G. C., 336
Carles, J., 56
Carnes, R. A., 336
Carney, R. L., 279
Carpenter, B. K., 6
Carper, W. R., 286
Carrel, J., 331
Carrick, A., 142
Carrington, T., 17
Carroll, A. P., 220
Carroll, D. I., 114, 115, 224, 299, 304, 358
Carroll, S. R., 189
Carter, D. E., 46
Carter, J. G., 57, 293
Carter, M. H., 139, 336
Carty, A. J., 183, 185, 194
Cary, L. W., 315
Caserio, M. C., 73, 104
Casey, A. I., 212
Casey, C. P., 193
Cash, G. C., 187
Casper, K., 118, 140
Cassagne, C., 305
Cassan, J., 274, 278, 282
Casselman, A. A., 338
Cassinelli, G., 248
Castagnoli, N., 323, 359
Castle, P. M., 222
Castleman, A. W., 217
Castonguay, J., 276
Cattabeni, F., 323, 328, 354
Cavagnaro, J., 135
Cavaletto, C. G., 331
Cavalleri, B., 281
Cavell, R. G., 199
Cavill, G. W. K., 330
Cech, D., 242, 283
Cederbaum, L. S., 4
Cellai, L., 239
Centofanti, L. F., 171, 218
Cerimele, B. J., 49
Cermák, V., 14
Cetini, G., 183, 185, 199, 200
Chadwick, M., 7, 78
Chaigneau, M., 338
Chait, E. M., 226
Chambaz, E. M., 232, 302, 317
Chan, A. S. K., 198

Chan, H. T., jun., 331
Chan, K., 329
Chang, C., 45
Chang, C.-J., 226
Chang, C. W. J., 66, 274
Chang, H. W., 22
Chang, J.-K., 251, 366
Chang, S. R., 332
Chang, S. S., 332
Chang, T., 236, 374
Chang, T. T. L., 346
Chantry, P. J., 58
Chao, J., 360
Chao, K.-J., 15
Chapman, D. J., 246
Chapman, J. R., 124, 232, 244, 280, 302, 315
Chapman, S., 17
Charlson, R. J., 120
Charnock, G. A., 282
Charollais, E., 248
Charpentier, C., 324
Chatfield, D. A., 211
Chatfield, D. H., 345
Chatt, J., 194
Chaytor, J. P., 323
Chen, C. T., 11
Chen, F., 157
Chen, K. N., 187
Chen, K. S., 187
Chen, P. H., 269
Chen, T.-T., 4
Chen, Y. H., 351
Chen, Y.-K., 333
Cheng, P. L., 181, 184
Cheng, T. M. H., 216
Chernyshev, E. A., 154, 157, 168
Chesnavich, W. J., 25
Chevereau, M., 240
Chia, L. S., 198
Chiang, C.-K., 229
Child, M. S., 16
Chini, P., 178
Chino, H., 321
Chissick, H. H., 342
Chizhikov, D. M., 217, 219, 223
Chizhov, O. S., 246, 248, 264, 266, 274, 276, 372, 376
Cho, A. K., 323, 328, 340, 341, 343, 357
Chochua, K. A., 266
Chong, D. P., 4
Chong, S.-L., 53, 107
Chortyk, O. T., 304
Chow, K. K., 164
Chow, W. Y., 7
Chowdhry, V., 149
Choy, Y.-M., 248, 312
Christensen, B. G., 74
Christensen, C. M., 333
Christiaens, L., 175, 285
Christianson, D. D., 306
Christie, J. R., 15
Christie, K. O., 221
Christie, W. H., 119
Christopher, J. P., 306
Christopher, R. E., 187

Author Index

Christophorou, L. G., 7, 57, 289, 293
Chu, S. Y., 5
Chua, P. T., 24
Chuche, J., 270
Chupka, W. A., 7, 9, 27, 51, 53
Ciach, S., 216
Ciccioli, P., 299
Cicero, T. J., 258
Cjyimesi, J., 237
Clagett, C. O., 302
Clare, R. A., 354, 356
Clark, D. R., 322
Clark, G. M., 286
Clark, J., 242, 283
Clark, R. P., 335
Clark, S. J., 297
Clarke, E. T., 330
Clarke-Lewis, J. W., 227
Clemens, D. F., 170
Clemens, J., 186
Clementi, E., 2, 4
Clerc, J. T., 119, 139
Clerc, T., 262
Clerk, T., 118
Clifford, A. F., 220
Clode, D. M., 312
Clow, R. P., 98
Coburn, W. C., 350
Cocke, D. L., 223
Coerezza, M., 347
Coggins, C. W., jun., 336
Cohen, A. I., 119, 348, 358, 360
Cohen, M. J., 113, 114, 115
Cohen, S. C., 182
Cohn, K., 218
Colburn, R. W., 249
Colburn, T. J., 211
Coleman, H. J., 337
Coleman, R. L., 331
Collin, J. E., 48, 56
Collins, J. H., 47, 253
Collote, P., 355
Colton, R., 179, 195
Colvin, J. R., 315
Colvin, M., 350
Comba, M., 335
Commons, C. J., 179
Compernolle, F. C., 316
Compson, K. R., 124, 323
Compton, R. N., 52, 57, 58, 214, 293, 295
Conde-Caprace, G., 56
Condon, E. U., 61
Cone, C., 127
Conner, H. A., 333
Conner, R. L., 260
Connolly, J. S., 242
Connolly, J. W., 152
Connor, J. A., 188, 193, 195, 203
Connors, T. A., 350
Conrad, B. R., 142
Contreras, J. G., 150
Cook, J. C., jun., 94, 235
Cook, M., 268
Cook, W. L., 149, 174

Cooks, R. G., 42, 43, 44, 45, 57, 65, 66, 67, 73, 80, 85, 88, 147, 263, 265, 266, 271, 275, 278, 279, 282, 285, 286, 287, 291, 292, 372
Cooper, C. D., 58, 295
Cooper, C. G., 217
Cooper, R. D., 275
Copier, H., 332
Coppens, P., 51
Copperthwaite, R. G., 193
Corbett, B. J., 323
Corderman, R. R., 83
Corey, E. R., 159
Corey, J. Y., 159
Cornish, H. H., 349
Corrie, J. E. T., 229
Corse, J., 303
Cortesi, N., 334
Corval, M., 26
Coscia, A. M., 298
Costa, E., 323, 328, 354
Costello, C. E., 139
Cotter, J. L., 268, 281
Cotton, R. J., 8
Couch, M. W., 327
Coulter, J. R., 324
Counsell, R. E., 336
Courchene, W. L., 47
Courrier, W. D., 206
Coutts, R. T., 249, 278, 326, 342
Couturier, J., 274
Cowles, R. J. H., 185
Cowley, A. H., 197, 218
Cox, P. J., 350
Cox, R. E., 1, 225, 297, 314, 337
Cox, T. P. H., 298
Coyle, T. D., 157
Cradock, S., 58, 216
Cragg, R. H., 146, 147, 148, 149
Craig, D. P., 15
Craig, J., 367
Crain, P. F., 236, 239
Cranor, P. T., 73, 282
Crathorne, B., 323
Craveri, R., 250
Crawford, L. R., 133, 137
Cresp, T. M., 228
Creswell, C. J., 263
Crim, M., 141
Critchley, C., 258
Crittenden, A. L., 120
Croisy, A., 175, 285, 286
Cromartie, E., 336
Crombie, L., 227, 229
Cronholm, T., 302
Crosby, D. G., 336
Cross, K. P., 149
Crossland, R. K., 189
Crouch, M. W., 360
Crow, W. D., 71
Crowe, A., 11
Crummett, W. B., 336
Csetenyi, J., 344
Csizmadia, I. G., 3
Cullen, W. R., 181, 198

Cum, G., 71, 280
Cummins, S. C., 209
Cuncliffe, A. E., 283
Cundy, C. S., 181, 202
Cunningham, A. J., 109
Currie, B., 251, 366
Curtis, D. M., 158
Curtis, M. D., 202
Curtiss, L. A., 5
Curtius, H.-C., 302
Cushley, R. J., 285
Cyr, C. P., 193
Czira, G., 158

Dabrowiak, J. C., 208
Dagaut, J., 282
Daghetta, A., 334
Dagragnano, V. L., 120
Dahl, L. F., 200
Dahlbom, R., 341
Daigle, J.-Y., 274
Dain, J. G., 328
d'Alcontres, G. S., 71, 280
Dale, A. J., 165
Dale, S. L., 322
Dalton, J., 90, 101, 293
Daly, N. R., 138
Dames, M. E., 305
Damrauer, R., 153
Danby, C. J., 27, 29, 86
Danieli, B., 298, 322, 341, 353
Danieli, R., 151
Daniels, P. J. L., 240
Danielson, P. M., 223
Danielsson, B., 340
Danks, A. V., 272
Dann, J. G., 254, 369
Dannappel, H. J., 154
Dannhardt, G., 167
Danzer, W., 198, 200
Darbre, A., 324
Das, B. C., 238, 250, 252, 283, 324, 351, 363
Das, G., 2
Das, K. G., 78, 248, 276, 278, 371
Das, M., 212
Das, N. P., 359
Dash, K. C., 181
Dauerman, L., 223
Davidson, A., 203
Davidson, J. L., 198
Davidson, P. J., 160, 180
Davidson, R. A., 70, 265
Davies, A. P., 280
Davies, A. R., 219
Davies, C. H., 193
Davies, J., 240
Davies, M., 168
Davis, B. A., 360
Davis, C. S., 211
Davis, F. J., 214
Davis, J. E., 140
Davis, G., 333
Davis, L. D., 348
Davis, N., 203
Davis, R., 179, 185
Davis, R. A., 153

Dawson, J. H. J., 53, 109
Dawson, R. M., 266, 285
Dawson, R. M. C., 258, 310
Day, E. D., 199
Day, J. P., 195
Day, R. A., 251, 363
Deacon, G. B., 174
Dean, F. M., 60
Dean, W. K., 199, 200
Deas, A. H. B., 306
Deberitz, J., 190, 195
Debies, T. P., 8
de Boer, Th. J., 71, 279
De Clercq, M., 160
Decora, A. W., 142
De Corpo, J. J., 195, 202
Deeming, A. J., 200
Defaye, G., 232, 302
Defaye, J., 238
de Galan, L., 298
Deganello, G., 185
de Graaf, H. G., 167
Dehennin, L., 317
Dehmer, J. L., 213
Dehmer, P. M., 9, 51
De Jongh, D. C., 84, 236, 246, 285, 314
de Lannory, J., 286
Delcambe, L., 250
Del Corona, L., 347
De Leenheer, A., 346, 353
de Leeuw, J. J. M., 332
Delfel, N. E., 336
Delfino, A. B., 118, 133
De Liefde Meyer, H. J., 192
Dell, A., 370
Dell Glover, D., 119
Delova, D., 337
De Luca, P., 318
de Luzes, H., 276
De Maria, G., 223
De Martino, G., 277
Demayo, A., 335
de Mayo, P., 239
De Meyer, C., 280
Demisch, L., 343
Demuth, G., 247
de Nadai, F., 299, 355
Dencker, W. D., 299
Denney, D. W., 115
Denniston, M. L., 149
De Pauw, G., 316
de Poorter, B., 160
de Ridder, J. J., 139, 248, 372, 374
De Rosa, M., 318
Derrick, P. J., 1, 30, 33, 34, 39, 68, 69, 73, 92, 93, 225, 264, 297
De Saint Simon, M., 223
Desgres, J., 320
Desideri, A., 214
Desiderio, D. M., 236, 249, 251, 255, 260, 300, 303, 309, 327, 342, 362
Desjardins, C. D., 174
Dessy, R. E., 119
Dettingmeijer, J. H., 222

Deutsch, J., 73, 169, 271, 276, 277
Devlin, C. J., 165
de Vries, J. X., 226
Devyatykh, G. G., 96, 189
Devys, M., 318
Dewar, M. J. S., 6, 74, 80
Dias, J. R., 264
Dibeler, V. H., 51
Dickinson, R. J., 252, 267, 363
Dickson, F. E., 305, 338
Dickson, L. G., 319
Dickson, R. S., 183
Dickstein, J. L., 168
di Corcia, A., 299
Dieck, R. L., 220
Dierdorf, D. S., 208
Dietl, M., 156
Dietrich, M. W., 334
Dijkstra, G., 120, 134, 162
Dillard, J. G., 144, 210, 217, 288
Dillon, J. P., 162
Dimmel, D. R., 69, 151, 153, 269
Dimmock, J. R., 272, 278
Dimov, V., 308
Dirscherl, K., 190
Distefano, G., 54, 57, 151, 195, 204
Dittmer, D. C., 181, 184
Dix, D., 119
Dixon, D. A., 220
Dizabo, P., 282
Djerassi, C., 34, 60, 67, 68, 73, 83, 113, 133, 134, 230, 231, 233, 244, 264, 271, 272, 277, 278, 280, 284, 285, 318
Djuričic, M. V., 337
Dobbie, R. C., 197
Dobosh, P. A., 263
Dobson, A. M., 338
Dobson, G. R., 196
Doerger, J. V., 336
Doerr, R. C., 333
Dolby, R., 200
Dolcetti, G., 204
Dole, M., 116
Dolejs, L., 226
Dollear, F. G., 278
Dolzine, T. W., 154
Domcke, W., 4
Donaldson, B. A., 333, 334
Donzel, A., 332
Dooley, C. J., 333
Dooley, J. E., 337
Doolittle, F. G., 337
Dorner, H., 199, 200
Dorogov, V. V., 282
Dorough, H. W., 335
Dougherty, R. C., 90, 101, 243, 246, 247, 271, 293, 376
Douglas, D. R., 332
Douglas, W. M., 187, 197, 200
Doyle, P. J., 319

Draffan, G. H., 324, 327, 354, 356
Drake, J. E., 157, 159
Drawert, F., 303
Dreeskamp, H., 184
Dreher, H., 197
Drew, M. G. B., 185
Drewery, C. J., 102, 289
Drewes, S. E., 226
Drews, K. A., 195
Drey, C. N. C., 251
Dreyer, D. L., 226
Dreyfuss, J., 358, 360
Driessler, F., 80
Dromey, R. G., 47, 126
Dronsfield, A. T., 219
Drowart, J., 51
Druce, P. M., 222
Dubbeldam, J., 167
Dubois, L., 304
Dubovenko, Z. D., 282
Dubovitsky, V. A., 187
Dubrin, J., 21
Dubroven, A. A., 273, 274
Duchamp, D. J,, 123, 300, 308
Duckworth, H. E., 120
Dudchik, G. P., 222
Due, S. L., 329
Dueber, J. S., 159
Duff, J. M., 153
Duffaut, N., 152
Duffield, A. M., 133, 134, 227, 231, 249, 270, 277, 278, 282, 325, 357
Duffield, R. M., 330
Duffy, N. V., 195
Duma, G. L., 348
Dunbar, R. C., 52, 63, 104, 105, 163, 265
Duncan, J. H., 258, 327, 350, 359
Dunham, L. L., 300
Dunning, T. H., 2
Dunogues, J., 152
Dunster, M. O., 212
Du Plessis, L. S., 333
Dupont, J. A., 331
Duran, N., 275, 284
Durbin, R. D., 255
Durden, D. A., 360
Dustin, D. F., 203
Dutky, S. R., 319, 320, 330
Dutton, G. G. S., 248, 312
Dutton, H. J., 119
Duyckaerts, G., 192
Dwivede, B. K., 334
Dyczmons, V., 3, 80
Dzidic, I., 97, 106, 114, 115, 224, 299, 304, 358
Dzizenko, A. K., 311

Eaborn, C., 152
Eadon, G., 69, 269
Eady, C. R., 179, 200
Eagles, J., 248
Earle, F. R., 306
Earnshaw, D. G., 142

Author Index

Eastmond, R., 152
Eaton, G. R., 209
Eaton, S. S., 209
Ebbighausen, W. O. R., 329
Ebinger, H.-M., 199
Ebsworth, E. A. V., 216
Eckhardt, G., 225, 270, 271, 277
Eckrich, W., 335
Edqvist, O., 30
Edwards, A. J., 222, 223
Edwards, D. A., 206
Efraty, A., 180, 186, 187, 197
Egge, H., 376
Egli, H., 121
Eglinton, G., 137, 244, 297, 305, 306, 314, 334, 337, 338
Eguchi, S., 226, 227
Ehmann, A., 312
Ehntholt, D. J., 184, 185
Ehrhadt, H., 11
Ehrl, W., 196, 198
Ehrnebo, M., 356
Eichelberger, J. W., 121, 336
Eichhoff, H. J., 95
Eigendorf, G., 226
Einolf, N., 97, 100, 272
Eisenstadt, A., 186
Eisler, W. J., 141, 355
Eisner, E., 202
Eisner, T., 331
Eizen, O. G., 265
Eland, J. H. D., 27, 28, 29, 86, 87
Elder, F. A., 56
Elder, G. H., 244, 245, 280
Eldjarn, L., 139, 300, 312, 325, 327
Eley, D. D., 57, 209
Elguero, J., 280, 281
Elix, J. A., 228
El Khadem, H., 280
Elkin, K., 121, 300
El'kin, U. N., 311
Elleman, D. D., 111
Ellermann, J., 170, 171, 204
Ellestad, G. A., 238
Elliot, L. E., 217
Elliott, R. M., 61
Elliott, W. H., 231, 234
Ellis, I. A., 216
Ellis, S., 249, 251, 367
Ellis, S. R., 224, 304
Ellsworth, R. K., 315
Ellsworth, R. L., 300
El-Monafi, H. M. R., 56
Elnatanov, Yu. I., 164, 165, 277
Elovson, J., 302
El-Sadany, S., 280
El-Shafei, Z. M., 280
Elson, C. M., 209
Elwood, T. A., 99, 241
Elzinga, M., 253

Embree, D. J., 66, 268
Emerson, M. T., 146
Emmel, R. H., 8, 11, 12, 80, 87
Emsley, H. E., 146
Emsley, J., 168
Endall, R., 194
Ende, M., 231
Eneroth, P., 322
Engel, J. E., 167
Engel, L. L., 301
Engel, R., 57, 287
Engelbrecht, J. P., 318
Engelman, H. D., 119
Engelmann, H., 286
Engelmann, T. R., 184
Engelmore, R. S., 134, 231
Englert, G., 230
Ennever, J. F., 163
Enos, H. F., 335
Enzell, C. R., 229, 230, 314, 315
Erdman, T. R., 233, 318
Erickson, M., 337
Ericsson, O., 340
Erman, P., 14
Erni, F., 118, 139
Ernst, L. A., 328
Esders, T. W., 259
Esselman, W. J., 260, 302
Essien, E. E., 343
Estucio, P., 217
Etzweiler, F., 298
Euwema, R. N., 4
Evans, F. J., 319
Evans, K., 16
Evans, M. B., 335
Evans, N., 244, 251, 258, 371
Evans, R., 319
Evans, W. J., 203
Eyem, J., 324
Eyler, J. R., 102, 107
Eyring, H., 18
Eyssen, H. J., 316

Facchetti, S., 262
Fackler, J. P., 163
Faerman, V. I., 95, 96
Fairless, B. J., 334
Fairweather, R., 254
Fairwell, T., 249, 251, 367
Falco, M. R., 226
Falconer, W. E., 217, 222
Fales, H. M., 100, 138, 226, 227, 231, 249, 301, 303, 330, 360
Falick, A. M., 33, 34, 39, 40, 68, 69, 73, 78, 92, 93, 264, 273
Falkner, F. C., 121, 233, 282, 300, 317, 356
Falter, H., 251, 363
Fanelli, C., 355
Fanelli, R., 299, 322, 341, 344, 353
Fantl, V., 320
Farber, M., 214, 223
Farbman, S., 135

Fărcăsan, V., 285
Farmer, J. B., 155
Farmer, P. B., 350
Farren, J., 120
Fast, P. M., 142
Faucher, A., 160
Favorskayn, I. E., 271
Feather, D. H., 214
Featherstone, J. L., 199
Fechner, K.-H., 270, 275
Fedeli, E., 334
Federoňko, M., 282
Feeney, J., 237
Feher, F., 155
Fehlhaber, H.-W., 271, 277
Fehlner, T. P., 214, 215
Fehn, J., 170
Fehr, T., 239
Feigenbaum, E. A., 133, 134, 231
Feldman, R. J., 138
Felix, R. A., 153
Fellows, R., 256, 367
Felty, W. L., 129
Fenselau, C. C., 93, 98, 231, 246, 258, 350
Fentiman, A. F., jun., 97, 99, 299
Fenwick, R. G., 280
Ferezou, J. P., 318
Ferguson, E. E., 289
Ferguson, K. A., 260
Fernando, O., 206
Ferrari, R. P., 183
Ferraro, G. M., 226
Ferrer Correia, A. J., 112
Ferretti, A., 280, 305, 332, 334
Ferrus, L., 282
Feser, M. F., 155
Fessenden, R. T., 154
Fétizon, M., 26, 276
Fiagbe, N. I. Y., 324
Fiddler, W., 333
Fiecchi, A., 317
Field, F. H., 78, 96, 97, 99, 100, 109, 249, 250, 251
Fields, E. K., 61, 68, 274, 280
Fies, W. F., 121, 302
Filby, W. G., 284
Filyngina, Λ. D., 274
Findlay, J. K., 317
Findlay, M. C., 73, 104
Fine, J., 289
Fink, W., 178
Finkle, B. S., 301
Finlayson, B. J., 88
Finney, C. D., 49
Finney, K. A., 49
Firestein, G., 182
Firestone, R. A., 74, 338
Fischel, D. L., 188
Fischer, D. E., 167
Fischer, E. O., 187, 192, 193, 199
Fischer, G., 286
Fischer, H., 193
Fischer, S. F., 14, 15
Fischer, W. G., 304

Fish, R. H., 146
Fishbein, L., 323, 336
Fishel, M. G., 51
Fishelson, L., 250
Fisher, K. J., 193
Fitzer, L., 175
Fitzgerald, M. R., 119
Fitzsimmons, B. W., 163
Fjeldstad, P. E., 56, 285
Flammang, R., 280, 281
Flanagan, V. P., 280, 305, 332, 334
Flanigan, E., 366
Flatau, K., 146
Flath, R. A., 298, 331, 334
Fleischer, N., 298
Fleming, G. R., 14
Flesch, G. D., 188
Fletcher, S. R., 179
Flick, W., 155, 219
Flippen, J. L., 187
Fliszar, S., 56
Floch, H. H., 320
Flory, D. A., 337
Fölsch, G., 280
Förster, H.-J., 140, 253, 315, 326
Foffani, A., 57, 151, 195
Folk, T. L., 273
Folkers, K., 251, 366
Folsome, C. F., 337
Foltz, A. F., 331
Foltz, R. L., 99, 235, 236, 247, 299, 301, 360
Foner, S. N., 57
Fontaine, A. E., 42
Foote, J. L., 260
Foreman, M. I., 190
Forest, M., 78
Forester, R., 218
Formosinho, S. J., 16
Fornari, L., 47
Forrest, I. S., 304
Forrest, J. E., 323, 334
Forrest, T. P., 334
Forrey, R. R., 331, 334
Forst, W., 24
Fortin, C. J., 78
Foss, P., 302
Foster, A. B., 144, 247, 350
Foster, H. B., 94
Foster, N. G., 262
Foster, M. S., 83
Foust, A. S., 179
Fox, R. M., 342
Francis, G. W., 230
Francis, J. N., 203
Frange, B., 149
Franke, R., 181, 197
Franklin, J. L., 46, 53, 57, 58, 107
Franz, D. A., 149
Franz, M., 281
Franzblau, C., 254
Franzen, J., 132
Fraser, R. T. M., 278
Freed, K. F., 4, 6, 14, 15, 16
Frei, R. W., 335
Frembs, D. W. R., 163

Fresco, J. M., 206
Freudenthal, J., 336
Freund, R., 155
Freund, R. S., 10
Frey, R., 27, 87
Frey, W. F., 52, 293
Fri, C.-G., 327, 329
Frick, W., 323
Frid, T. Yu., 264
Fridh, C., 30
Fridlyansky, G. V., 275
Fridmann, S. A., 215
Friedman, A. E., 280
Friedman, L., 246, 366
Frigerio, A., 61, 272, 283, 297, 298, 299, 322, 326, 327, 339, 341, 344, 347, 353, 355
Frimmel, F., 297
Fringuelli, F., 54
Frintrop, P. C. M., 162
Frischler, I., 184
Fritz, G., 154
Frölich, J. C., 121, 141, 233, 300, 309, 317
Froyen, P., 165, 172
Frye, C. L., 154
Fu, E. C., 52
Fu, E. W., 104
Fu, K. S., 127
Fuchs, R., 46
Fuchs, V., 27
Fuger, J., 192
Fujimaki, M., 334
Fujioka, M., 332
Fujita, E., 226, 315
Fujita, I., 7, 10, 81, 276
Fukui, K., 10
Full, R., 149
Funk, B.-A., 57, 287
Furlei, I. I., 293
Furstoss, R., 276
Furth, A. J., 256
Furukawa, Y., 306
Futrell, J. H., 41, 98, 99, 110, 241, 289

Gaertner, H. J., 343
Gaetani, E., 286
Gaeva, L. A., 273
Gaffney, T. E., 301, 323, 329, 330, 340, 357
Gage, J., 123
Gaines, D. F., 150
Gaivoronskii, P. E., 189
Gal, J., 280
Gallagher, G., 348
Gallaher, E. G., 327
Gallaher, E. J., 297
Gallegos, E. J., 48, 305, 320, 337
Galli, G., 317
Galliard, T., 258
Gallo, G. G., 281
Gallop, P. M., 243, 254
Gambacorta, A., 264
Gambino, O., 183, 185, 199, 200
Game, C. H., 193

Games, D. E., 95, 226, 227, 228, 229, 234, 235, 244, 245, 251, 258, 299, 330, 338, 371
Games, M. P., 234
Gan, I., 327
Ganglhofer, J., 145
Gansow, O. A., 186
Garattini, S., 344
Gardner, H. W., 306
Gardner, R. C. F., 182
Gardner, S. A., 181, 190
Gardner, S. R., 187
Garegg, P. J., 311
Garg, H. G., 280
Garland, W. A., 361
Garrison, A. W., 297
Garssen, G. T., 258
Garteiz, D. A., 323, 341
Gaskell, S. J., 305
Gaskin, P., 314
Gastambide-Odier, M., 259
Gaull, G., 327
Geanangel, R. A., 80
Gebreyesus, T., 280
Geer, M., 360
Gehrke, C. W., 337
Geiger, D. L., 139
Geisler, T. C., 217
Geissman, T. A., 229
Gelbart, W. M., 15, 16
Gelbert, E., 250
Gella, I. M., 280
George, M. V., 153
George, R. D., 180, 201
George, T. F., 16, 18
Gerber, J. N., 309
Gerber, N., 301
Gercken, G., 305
Gergel, L. G., 283
Geribaldi, S., 274
German, A. L., 298
Géro, S. D., 363
Gertler, S., 256
Geuns, J. M. C., 332
Ghotra, J. S., 193
Giannetto, P. D., 280
Gibert, J. M., 337
Gibney, K. B., 312
Gibson, E. K., 337
Gibson, H. W., 72
Gibson, J. F., 208
Gielen, M., 160
Gierer, P. L., 153
Giering, W. P., 186
Giessner, B., 123
Giessner, B. G., 11, 49
Gijzeman, O. L. J., 14, 15
Gilbert, B., 192
Gilbert, J. D., 314, 320, 324
Gilbert, J. R., 177, 190
Gilbert, J. N. T., 356
Gilbert, M. T., 314
Gilbert, W. C., 210
Gil'bert, M. M., 267
Gilleran, T. G., 358
Gilles, P. W., 142, 223
Gillette, J. R., 350

Author Index

Gillis, R. G., 266, 285
Gilman, H., 153
Gilmartin, D. E., 223
Gilmore, C. J., 229
Gingerich, K. A., 214, 223
Gioia, B., 248
Gird, S. R., 217
Giry, L., 338
Glasser, A. G., 349
Glaze, W. H., 335
Glazener, L., 329
Glemser, O., 155, 199, 219, 220, 221
Glick, M. D., 159
Glockling, F., 144, 145, 146, 159, 202
Glogouski, M. C., 149
Glore, J. D., 215
Glover, D., 119
Goad, L. J., 315, 318, 319
Goby, G., 119
Goddard, W. A., 2, 17
Goedken, V. L., 209
Goenechea, S., 344
Goetze, R., 220
Goffart, J., 192
Gogan, N. J., 211
Gohausen, H. J., 155, 159
Gohlke, R. S., 124
Gokel, G. W., 188
Golay, M. J. E., 122
Goldberg, V. D., 277
Golden, D. E., 47
Golden, G. M., 83
Goldman, A. S., 321
Goll, W., 177
Golovkena, L. S., 275
Gompertz, D., 324, 327
Goodall, B. L., 182
Goode, G. C., 102, 289
Goodman, G. T., 153
Goodman, L., 5
Goodwin, T. W., 315, 319
Gopalchari, R., 286
Gorchein, A., 260
Gorden, R., jun., 111
Gordon, A. E., 297, 339
Gordon, B. J., 119
Gordon, D. T., 332
Gordon, E., 356
Gordon, H. B., 181
Gordon, S. G., 327
Gordon, S. M., 48
Gore, J., 276
Goretti, G., 299
Gorfinkel, M. I., 71, 282
Gorodetskii, I. G., 224
Goronkov, L. N., 213
Gorrod, J. W., 342
Gosling, K., 150, 157
Gosney, I., 172
Goto, T., 78, 287
Gotoh, N., 209
Gotthelf, G., 327
Gotze, H.-J., 160, 163
Goubeau, J., 149
Gough, T. A., 299
Gounelle, Y., 25, 55, 267
Gracey, D. E. F., 239

Graebe, J. E., 300, 314
Grafnetterova, J., 349
Graham, G. N., 369
Graham, R. A., 135
Graham, W. A. G., 179, 202, 203
Grain, P. F., 243
Grajower, R., 57, 293
Grandberg, I. I., 280
Granoth, I., 166, 167, 169, 284
Granström, E., 261, 308, 357
Gravel, D., 78
Gray, C. H., 320
Gray, N. A. B., 137, 297
Grayson, M. A., 299, 337
Green, A. R., 328
Green, D. E., 304
Green, E. E., 168
Green, F. D., 280
Green, J. H., 245, 280
Gréen, K., 308, 357
Green, M., 186, 193, 199
Green, M. L. H., 189, 190, 193, 199
Green, R., 341
Greene, E. F., 289
Greene, L. J., 366
Greene, P. T., 149
Greiss, G., 191
Greve, P. A., 336
Grevels, F. W., 184
Griffin, G. W., 114
Griffin, P. F. S., 259
Griffiths, N. M., 333
Grigg, R., 205, 286
Grigli, G., 222
Grimes, R. N., 149, 150, 203
Grimley, R. T., 222
Grimm, L. F., 220
Grimshaw, J., 272
Grimsrud, E., 110
Grisdale, P. J., 149
Grizik, A. A., 223
Grob, K., 298, 334, 336
Grobe, J., 155
Grønneberg, T., 56, 282
Gross, K. P., 88
Gross, M. L., 25, 54, 62, 83, 102, 111, 264
Grossert, J. S., 318
Grostic, M. F., 236, 339
Grotch, S. L., 128, 134, 136
Grove, J., 356
Gruber, W. H., 170
Grützmacher, H.-F., 73, 270, 275
Grunwell, J. R., 284
Grupe, K.-H., 242, 283
Grynkiewicz, G., 280
Guadagni, D. G., 334
Guarino, A. M., 131
Güsten, H., 267, 270, 282
Guichon, G., 120
Guido, M., 222, 223
Guilbot, J. F., 120
Guilford, J., 350

Guillemin, R., 256, 324, 367
Guinand, M., 250
Guindi, L. H. M., 286
Gunstone, F. D., 257
Gunther, H., 169
Gupta, S. K., 222
Gusanov, A. V., 213
Gustafsson, B., 357
Gustafsson, J.-A., 232, 298, 302, 316, 321, 322
Gustafsson, S. A., 321
Guthrie, R. D., 237, 248
Gutshall, P. L., 91
Guyon, P. M., 8, 29, 51
Gynane, M. J. S., 150
Györösi, P., 56, 268

Haarhoff, P. C., 48
Haas, C. K., 153
Habeeb, J. J., 151
Haber, J. N., 140
Habfast, K., 125
Haddon, R. C., 6, 80
Haddon, W. F., 63, 141, 146, 157, 253, 275, 292
Hadjiantoniou, A., 57, 293
Haegele, K. D., 236, 251
Härfast, Å., 325
Hafner, K., 185
Haga, H., 312
Hagaman, E. W., 225
Hageman, H. J., 83
Hagen, A. P., 202
Hagens, W., 168
Hagerman, D., 141
Hagg, A., 171
Hagihara, N., 193
Hahnke, M., 154
Haiduc, I., 185
Haigh, W. G., 315
Hakomori, S., 260, 312
Hall, L. D., 87, 198
Hall, M. S., 300
Hallab, M., 155
Hallett, R., 120
Halpern, B., 250, 252, 303, 323, 324, 327, 363
Halpern, D., 57, 287
Hamberg, M., 307, 308, 356, 357
Hamilton, A., 209
Hamilton, R. E., 251, 363
Hammar, C.-G., 339, 344, 357
Hammarström S., 260, 307, 310
Hammer, W., 344
Hammerschmidt, F. J., 231
Hammerum, S., 67, 73, 271, 277, 281, 286
Hamming, M. C., 262
Hammond, R. K., 310
Hanessian, S., 236, 374
Haney, M. A., 46
Hanfland, P., 260, 310
Hann, C. S., 324
Hansen, P. E., 56, 282
Happ, G. M., 330

Happ, G. P., 124
Harada, S., 352
Harborne, J. B., 227
Harder, V., 197
Hardesty, C. L., 275
Harding, N. G. L., 254, 369
Hardisson, A., 15
Hardwidge, E. A., 24
Hardy, J., 139
Hargreaves, R. T., 235
Hariharan, P. C., 2, 5, 80
Harkness, A. L., 42
Harland, P. W., 46, 57, 58, 216, 217, 220, 288, 291
Harland, W. A., 320
Harlow, R. L., 181
Harman, J. S., 220
Harmon, C. A., 192
Harper, D. B., 305
Harris, D. H., 202
Harris, K. O., 198
Harris, L. E., 121, 181, 334, 336
Harris, M. M., 277
Harrison, A. G., 49, 62, 63, 71, 264, 268
Harrison, P. G., 160, 163
Hart, F. A., 193
Hart, S. G., 287, 292
Harten, J. E., 297
Hartmann, A., 169
Hartley, B. S., 369
Harvey, D. J., 97, 151, 232, 301, 302, 310, 311, 316, 329, 344
Harwig, J., 333
Harwood, C. N., 227
Harwood, J. L., 258
Hase, W. L., 17, 19
Hasegawa, H., 250
Hasegawa, K., 307, 338
Hashimoto, K., 315, 332
Haskins, N. J., 227
Haslett, R. J., 272
Haslewood, G. A. D., 234
Hass, J. R., 7, 81, 112, 271
Hasty, E. F., 211
Haszeldine, R. N., 219
Hatfield, W. E., 211
Haufmann, J. R., 355
Haug, P., 156, 242
Haumann, J. R., 141
Haupt, H.-J., 201
Hauptmann, H., 172
Hauser, F., 341
Havashi, S., 226
Hawke, J., 120
Hawks, R. L., 341
Hawrysh, Z. J., 332
Hawthorne, M. F., 203
Hay, P. J., 2, 3
Hayase, F., 334
Hayashi, A., 247, 260, 310, 311
Hayashi, H., 243
Hayashi, J., 186, 187
Hayashi, S., 227, 305
Hayatsu, R., 337
Hayes, J. M., 123

Hayes, S. E., 201
Haynes, R. M., 119
Hazelby, D., 124
Hazeldine, D. J., 57, 209
Heacock, R. A., 323, 334
Healy, M. A., 160
Heaney, H., 84
Heath, G. A., 194
Hebb, G. D., 318
Heberhold, M., 179
Heberich, G. E., 191
Hecht, S. M., 243
Hecke-Seibicke, E., 344
Hedberg, F. L., 189
Hedden, P., 314
Heerma, W., 162
Hehre, W. J., 5, 6, 80
Heier, K.-H., 167
Heil, V., 186
Heimann, W., 303
Heindel, N. D., 283
Heinen, H. J., 90
Heinesch, F., 48
Heinz, E., 258
Hekali, R., 320
Helboe, P., 227, 282
Helgeson, J. P., 312
Heller, D. F., 15, 16
Heller, S. R., 138
Hellerqvist, C. G., 376
Helling, J. F., 187
Hellwinkel, D., 165, 168
Hemesley, P., 229
Hemken, R. W., 335
Hemmings, R. T., 159
Henchman, M. J., 21
Henderson, D. B., 249
Henderson, J. E., quat., 335
Henderson, W., 299, 315, 337
Henderson, W. G., 53, 73, 104, 106, 109
Hendrickson, D. N., 211
Hendry, L. B., 331
Hengge, E., 216
Hengstmann, J. H., 121, 356
Henion, J. D., 69, 71
Henis, J. M. S., 289
Henkwitz, R., 351
Henneberg, D., 118, 140, 190
Henry, Y., 276
Herberich, G. E., 178
Herman, Z., 21
Herndon, W. C., 81
Heron, E. J., 247
Herring, F. G., 4
Herrmann, W. A., 194, 211
Herschbach, D. R., 21
Hertel, R. H., 304
Hertz, H. S., 136, 337
Herz, J. E., 275
Herz, K. O., 332
Herzschuh, R., 286
Hesford, F. J., 249
Hesse, M., 225, 226, 233, 276
Hessett, B., 216

Hester, R. E., 196
Heumann, A., 276
Heumann, K. G., 151
Hewett, E. W., 312
Hewlins, M. J. E., 251, 371
Heyer, G., 155
Heyndrickx, A., 346, 353
Hiberty, P. C., 5, 6
Hickmott, P. W., 272
Hidetsugu, A., 132
Higgins, I. J., 305
Highet, R. J., 226
Highsmith, R. E., 201
Higman, E. B., 304
Higman, H. C., 304
Hignite, C., 240, 307, 309
Higson, B. M., 208
Hildenbrand, D. L., 213, 220
Hildrum, K. I., 333
Hill, A. E., 186
Hill, A. W., 60
Hill, H. C., 262
Hill, K. E., 197
Hill, R. M., 329, 344, 358
Hillig, H., 132
Hills, E. B., 236, 374
Hilmer, R. M., 124
Hiltunen, R., 314
Hino, K., 285
Hintz, H. P., 239
Hintze, U., 305
Hipple, J. A., 61
Hirai, C., 332, 338
Hiramoto, M., 251
Hirano, K., 309, 325
Hirayama, C., 222
Hirayama, K., 318
Hirose, K., 278
Hirose, Y., 229
Hirota, A., 255
Hirota, K., 7, 81, 276
Hirsch, D. E., 337
Hisatome, M., 189
Hishida, S., 226, 227
Hites, R. A., 136, 334
Ho, A. C., 290
Hoare, M. R., 24
Hoareau, A., 10, 219
Hobson, R. F., 185
Hodges, R., 272
Hodshon, B. J., 328, 357
Hölldobler, B., 330
Hoenicke, J., 279
Hötzel, Ch., 90
Hofer, R., 155, 198
Hofetee, H. K., 192
Hoffman, G., 125
Hoffman, M. K., 43, 65, 66, 113, 152, 266, 268, 274
Hoffman, R., 6
Hoffmann, H. J., 216
Hoffmann, H. M. R., 186
Hoffmann, R., 17
Hofler, F., 216
Hofler, M., 180, 197, 198, 203
Hogan, R., 146
Hogg, A. M., 97, 246

Author Index

Hohlneicher, G., 4
Holand, A. K., 154
Hollaeder, J., 163
Holland, H. L., 239
Holland, J. F., 121, 300, 355
Holland, W. H., 121, 300
Holloway, J. H., 51
Holloway, P. J., 306
Holm, R. H., 208, 209
Holman, R. T., 258, 303, 307
Holmes, J. L., 51, 67, 71, 265, 267, 269, 273, 277
Holmes, W. F., 121, 300
Holmstedt, B., 121, 297, 300, 325, 340, 344, 355, 357
Holste, G., 207
Holtz, D., 53, 106, 108, 110, 220
Holwerda, D. A., 255, 368
Holwich, J. L., 188
Holz, J. B., 135
Homer, G. D., 154
Hong, P., 193
Honig, R. E., 49
Hood, L. V. S., 305
Hooper, S. N., 314, 337
Hoover, J. R. E., 348
Hopf, F. R., 209, 245
Hopkinson, A. C., 3
Hoppen, H. O., 317
Hopper, S. P., 146
Horgan, J. M., 312
Horgan, R., 312
Horlbeck, G., 184, 187
Horlin, D., 357
Horman, I., 249
Horning, E. C., 114, 115, 224, 234, 249, 298, 299, 303, 304, 316, 317, 322, 324, 329, 344, 358
Horning, M. G., 115, 121, 224, 234, 298, 299, 302, 304, 310, 311, 322, 329, 344, 358
Horton, D., 235, 247
Horvath, Gy., 237
Hoshi, M., 309
Hoshino, H., 26
Hoshita, T., 234
Hota, N. K., 183, 198
Hotz, H. P., 120
Hough, L. B., 278
Houghton, E., 281
Housley, T., 254
Houston, R. E., 159
Howard, C. C., 236
Howard, J., 88, 149, 202
Howe, I., 26, 59, 62, 262, 268, 275, 286, 291, 362
Howell, J. A. S., 190
Howells, D., 167, 168
Hoyano, J. K., 202
Hoyano, Y., 249, 325
Hrabak, F., 205
Hribar, J. D., 236, 374
Hrung, C. P., 209
Hsieh, A. T. T., 146, 201, 207

Huber, W. K., 121
Hubin-Franskin, M.-J., 48
Hucker, H. B., 344
Hudson, M. F., 209
Hudson, R. L., 57
Hughes, A. N., 168
Hughes, J. F. J., 251, 371
Hughes, R. P., 186
Huhtanen, C. N., 333
Huhtaniemi, I., 321
Huhtikangas, A., 228
Hull, J. R., 180, 190
Humphries, A. P., 184
Huneck, S., 90
Hunneman, D. H., 273, 302, 306, 315, 321, 338
Hunt, D. F., 177, 185, 360
Hunt, E., 255
Hunt, W. J., 2, 97, 98, 101
Hunter, K. R., 349
Hunter, P. W. W., 283
Hunter, W. H., 345
Huntress, W. T., 25
Hursthouse, M. B., 193
Hurum, T., 282, 285
Husband, J. P. N., 147, 148
Hutchins, R. O., 170
Hutterer, F., 327
Huttner, G., 192, 199
Hutzinger, O., 66, 268, 335, 336
Hvistendahl, G., 56, 268, 276, 282
Hyatt, D., 22
Hyde, P. M., 234

Ibarbia, P. A., 160
Ibers, J. A., 199, 209
Ibuka, T., 225
Ichihara, A., 250
Ichikawa, H., 7
Iglewski, S., 338
Ihrig, A. M., 70
Iida, T., 338
Ikan, R., 337
Ikegami, S., 233, 318
Ikekawa, N., 302, 318, 321
Ikuta, S., 33, 81
Ilukhina, L. I., 307
Imai, H., 188
Imanari, T., 312
Inaba, T., 226
Indriksons, A., 155
Inghram, M. G., 11
Ingle, W. M., 203
Innorta, G., 54, 55, 57, 151, 190, 195, 200, 266
Inoue, T., 226, 332, 335
Inouye, S., 247, 352
Inouye, T., 315
Ionov, N. I., 289
Iqbal, M. Z., 182
Ireton, R. C., 24
Irie, T., 285
Irving, H. M. N. H., 285
Irving, P., 134, 253, 370
Irwin, J. G., 146
Irwin, M. A., 229
Isaacs, N. S., 67

Isaev, I. S., 71
Isenhour, T. L., 118, 127, 128, 129, 142
Ishaq, M., 197
Ishibashi, M., 302, 321
Ishibashi, N., 10
Ishida, Y., 118, 132
Ishikawa, M., 155
Isolani, P. C., 107, 110
Isono, K., 239
Issenberg, P., 332, 335
Issleib, K., 196
Iteke, E., 175, 285
Ito, A., 293, 294
Itoh, M., 7, 81, 276
Itoh, T., 334
Ivanovskaya, L. Yu., 282
Iwacha, D. J., 242
Iwanami, M., 181
Iwanami, Y., 242
Iwaoka, W. T., 307
Iyengar, J. R., 334

Jacini, G., 334
Jackels, S. C., 208
Jackson, A. H., 60, 95, 226, 227, 229, 234, 235, 244, 245, 251, 258, 282, 299, 371
Jackson, J. L., 205
Jackson, J. R., 251, 371
Jacob, J., 307
Jacob, L., 120
Jacob, R. A., 160
Jacobsberg, F. R., 257
Jacquier, R., 280
Jacquignon, P., 175, 285, 286
Jaeggi, H., 298
Jänne, O., 321
Jaffé, H. H., 107, 111, 279
Jafry, S. W. S., 168
Jain, R. C., 239
Jain, S. R., 174
Jakobsen, P., 218
Jakubetz, W., 3
Jalonen, J., 55, 59, 263, 267, 286
Jamerson, J. D., 192
James, B. D., 203
Jamieson, W. D., 66, 268, 284
Janak, J., 297
Jankowski, K., 274, 282
Jannach, R., 216
Janot, M.-M., 225
Jansen, F. H., 234
Janssen, M. J., 285
Jaouni, T. M., 303
Jardine, I., 139, 193
Jaret, R. S., 236, 239
Jarman, M., 94, 144, 247, 350
Jarvie, A. W. P., 155
Jáuregui, J. F., 275
Jayasimhulu, K., 251, 363
Jefferson, R., 155
Jellum, E., 139, 300, 312, 325

Jelus, B., 98, 231
Jenden, D. J., 140, 300, 325, 340, 341
Jenkins, R. L., 217
Jenner, P., 347
Jennings, K. R., 53, 60, 61, 62, 102, 109, 112, 267, 289, 292
Jensen, B. L., 336
Jerina, D. M., 65
Jerumanis, S., 149
Jessep, H. F., 216
Jewell, J. S., 247
Ježo, I., 282
Jin, M. S.-H., 264
Job, B. E., 96
Job, R. C., 202
Jochims, J. C., 358
Joh, T., 200
Johannesen, R. B., 197
Johanson, G. A., 144
John, K. V., 230
Johnessee, J. S., 331
Johnson, A. W., 209, 237
Johnson, B. F. G., 179, 185, 186, 190, 200
Johnson, B. M., 12, 87, 256, 265, 275
Johnson, B. R., 22
Johnson, C. P., tert., 351
Johnson, D. B., 302, 329
Johnson, G. S., 372
Johnson, I. K., 174
Johnson, J. S., 152
Johnson, M. W., 119
Johnson, T. R., 152
Johnstone, R. A. W., 47, 50, 53, 54, 59, 60, 68, 69, 70, 73, 77, 82, 141, 236, 253, 262, 263, 278, 298, 324, 326, 365
Jolliffe, V. A., 336
Jones, C. J., 198, 203
Jones, E. G., 43, 65, 66
Jones, E. M., 193, 195, 203
Jones, F. M., tert., 106
Jones, G. R., 217, 222
Jones, I. T. N., 88
Jones, R. B., 275
Jonsson, B.-O., 30
Jordan, L. S., 336
Jortner, J., 15
Josefsson, B., 336
Jost, C., 119, 139
Jovanovic, J., 231
Joyce, T. E., 222
Judy, K. J., 300
Jug, K., 15
Julien, J., 285
Jull, J. T., 181
Junk, G. A., 188
Jurs, P. C., 101, 129, 131
Just, E. K., 247
Justice, J. B., 118, 127, 128
Jutzi, P., 216

Kabelitz, L., 306
Kabir, S., 312
Kablitz, H.-J., 192

Kadentsev, V. I., 246, 274, 376
Kadorkina, G. K., 165, 277
Kaesz, H. D., 182
Kagan, F., 236
Kahl, W., 248
Kahn, J. H., 333
Kaiser, K. L. E., 304
Kalbfus, W., 192
Kalir, A., 70, 280
Kallmayer, H. J., 273
Kallos, G. J., 335
Kallweit, R., 192
Kalman, S. M., 322
Kamamoto, J., 336
Kamano, V., 233
Kamei, K., 327
Kamerling, J. P., 139, 247, 248, 372, 374
Kamiya, Y., 233, 318
Kanatomi, H., 210
Kane, D. M., 115
Kane, J., 281
Kanellakopulos, B., 192
Kanigo, T., 200
Kantor, D., 48
Kapil, R. S., 225
Kaplan, F., 107, 279
Kaplan, I. R., 337
Kapoor, P. N., 165
Kärkkäinen, J., 372
Karabatsos, G. J., 70, 279
Karácsonyi, Š., 248
Karasek, F. W., 113, 114, 115, 118
Karasev, N. M., 213
Karjalainen, A., 228, 273
Karlander, S.-G., 258, 260, 310
Karlen, B., 341, 344
Karliner, J., 358
Karlsson, K.-A., 257, 258, 259, 260, 305, 310, 314, 376
Karn, R. A., 153
Karoum, F., 323, 328, 354
Karpati, A., 73, 271
Karpenko, N. F., 264
Karsch, H. H., 181, 197
Kashimura, N., 248
Kashman, Y., 168, 250
Kastin, A. J., 256, 368
Kasting, R., 332
Kataoka, S., 345
Kato, H., 251, 334
Kato, S., 334
Kato, Y., 226, 227, 345
Katovic, V., 206
Katrib, A., 8
Katz, E., 325
Katz, I., 332
Katz, T. J., 167, 188, 189
Katzhendler, J., 169
Kaulhausen, H., 321
Kaussmann, E. U., 225
Kavanagh, T. E., 334
Kawai, S., 325
Kawanami, J., 260, 312
Kay, K. G., 14, 16
Kayser, S. G., 334

Kazenas, E. K., 217, 219, 223
Keable, H. R., 194
Keat, R., 155
Keates, R. A. B., 313
Kebarle, P., 27, 53, 105, 107, 109, 110
Keeney, P. G., 334
Keith, L. H., 297
Keller, P. C., 215
Keller-Schierlein, W., 239
Kelley, J. A., 140, 326, 364
Kelly, J. A., 253
Kelly, R. W., 300, 308
Kemler, L. A., 280
Kemp, P., 258, 310
Kemp, T. R., 331
Kempter, V., 216
Kendall, M. J., 329
Kennard, O., 237
Kenne, L., 311
Kennedy, B. P. C., 333
Kennedy, J. D., 160
Kenner, G. W., 60, 244
Kenyon, G. L., 168
Keough, T., 45, 57, 66, 80, 85, 88, 265, 292
Kepner, R. E., 331, 333
Keppie, S. A., 202
Kerber, R. C., 185
Kerkhoff, M. A. Th., 270
Kerwin, C. M., 209, 211
Kevan, L., 145, 267, 293
Key, H., 33
Keyes, B. G., 63, 268
Keyser, A. J., 139
Khafizov, Kh., 165, 277, 280
Khan, M. N., 371
Khan, S. U., 337
Khanna, K. L., 349
Khariton, Kh. Sh., 207
Khmelnitskii, R. A., 280, 282
Khots, M. S., 132
Khramova, E. V., 264
Khullar, K. K., 264
Khvostenko, V. I., 89, 293
Kidani, Y., 211
Kieffer, R., 174
Kiennemann, A., 174
Kikuchi, T., 335
Killgoar, P. C., 51
Killian, M. T., 327
Killitea, S. D., 246
Kilner, M., 194
Kilpatrick, W. D., 115
Kim, J. K., 73, 104
Kim, K. C., 15, 22, 44
Kimber, R. E., 200
Kimble, B. J., 137, 297
Kime, D. E., 322
Kimland, B., 315
Kimling, H., 187
Kimmel, H. B., 322
Kimura-Harada, F., 243
Kindahl, H., 308
King, G. S., 303
King, J. O., 327

Author Index

King, R. B., 165, 180, 182, 185, 186, 187, 197
King, R. W., 237, 254, 369
Kingston, D. G. I., 67, 69, 71, 112, 144, 227, 271
Kinneberg, K. F., 12
Kinsey, J. L., 21
Kinsinger, J. A., 10
Kinstle, T. H., 284
Kinumaki, A., 239
Kinzer, G. W., 97, 299
Kiplinger, G., 301
Kira, M., 155
Kirchmeier, R. L., 220
Kirchner, R. M., 199
Kirk, R. W., 214
Kirkien, A. M., 175
Kirsch, H. P., 183
Kiryushkin, A. A., 248, 325
Kiser, E. W., 48
Kiser, R. W., 133, 288, 292
Kishi, T., 352
Kishimoto, Y., 307, 309
Kisic, A., 260
Kispert, L. D., 6
Kita, Y., 333
Kitzmura, T., 200
Kizaki, T., 67
Klaasen, C. D., 359
Klasinc, L., 267, 270, 282
Klassen, K. L., 195
Klaver, R. F., 305
Klebe, K. J., 280
Kleiman, R., 257, 306
Klein, H., 231
Klein, H.-F., 197
Klein, P. D., 141, 297, 355
Klein, R. A., 258, 310
Klein, W., 336
Kleinberg, J., 187
Kleineberg, G. A., 139
Kleinfelter, D. C., 269
Kleinman, R., 119
Kleopfer, R. D., 334
Klimowski, R. J., 123
Kline, E., 153
Klingebiel, U., 219, 220
Klosowski, J. M., 154
Klotz, C. E., 22
Klusmann, P., 171
Klynyer, N. A., 282
Knapp, D. R., 301, 323, 340
Knapp, F. F., 319
Knavel, D. E., 331
Kneidl, F., 167
Knewstubb, P. F., 22
Knight, J., 179
Knight, J. B., 297
Knights, B. A., 297, 315, 319
Knoll, L., 184, 196
Knowles, D. J., 51, 216
Knox, S. A. R., 180, 184, 190, 201, 202
Knuppen, R., 322
Kobayashi, K., 278
Kobayashi, M., 233, 318
Kobayashi, Y., 154

Kobelt, R., 197
Kober, F., 173
Koch, C. W., 282
Kochan, A., 284
Kochetkov, N. K., 248, 362, 372
König, W. A., 239, 311, 323
Königer, M., 335
Koepnick, N. G., 47
Köppel, C., 56, 71, 266, 270
Kogure, J., 155
Kohl, F. J., 214
Kohlberger, H., 204
Kohler, E., 180
Kohler, F. H., 200
Koike, H., 211
Koike, S., 334
Kolattukudy, P. E., 257, 306
Kolb, B., 324
Kollman, P. A., 80
Kollmar, H., 6, 80
Kolmodin-Hedman, B., 335
Kolokolov, B. N., 273, 274
Kolosov, E. N., 213
Kolsaker, P., 274
Komalenkova, N. G., 154
Komoda, Y., 239
Koncewicz, M., 255
Kondo, T., 189
Kongo, K., 346
Koo, D., 132
Koos, E. W., 168
Kordis, J., 223
Korenov, Yu. M., 213
Kornfeld, R., 62, 63, 253, 275
Kornreich, M. R., 335
Korobov, M. V., 213
Korte, F., 336
Korth, J., 252, 327, 363
Kortt, P. W., 208
Koski, W. S., 293
Koslow, S. H., 328
Kossanyi, J., 270
Kost, A. N., 274, 283
Kostyanovsky, R. G., 163, 164, 165, 277, 280, 282
Kotz, J. C., 147
Kováč, P., 247, 310
Kovacev, G., 337
Kováčik, V., 247, 248, 282, 310
Kovar, R. F., 187
Kowalski, B. R., 128, 129, 130, 131
Koyama, M., 250
Koyama, T., 313
Kozarich, J., 185
Kraessig, R., 51
Kramer, J., 282
Kramer, J. K. G., 258, 307
Kramer, J. M., 52, 105
Kramer, V., 267, 286
Krannich, L. K., 174
Krasnoshcheck, A. P., 268
Krauss, D., 358

Kray, W. C., 337
Kray, W. D., 154
Krebs, A., 187
Krech, F., 196
Kreis, G., 192, 193, 199
Kreissl, F. R., 192, 193
Kreiter, C. G., 187, 192, 193
Krenmayr, P., 132
Krenos, J. R., 17
Kriemler, P., 279
Krier, C., 7, 78
Krige, G. J., 48
Kristemaker, J., 225
Kristemaker, P. G., 225
Krivit, W., 260, 310
Krohn, V. E., 289
Kropshofer, H., 221
Kruck, T., 180, 184, 196, 197
Krueger, V. P. M., 156, 242
Kruger, C., 216
Krull, I. S., 186
Krupay, B. W., 284
Kruse, W., 180
Kubassek, E., 151
Kubat, J., 337
Kubelka, V., 284, 305, 334, 338
Kucherov, V. F., 274
Kuckertz, H., 208
Külpmann, W. R., 321
Kuenzle, C. C., 311
Küstler, A., 27
Kuhlman, C. F., 346
Kuhn, M., 201
Kuhn, N., 171
Kuhn, W. F., 269
Kuhtz, B., 220
Kuivila, H. G., 160
Kulkarni, P. S., 276, 278
Kulshreshtha, D. K., 229
Kulshreshtha, M. J., 229
Kumada, M., 155, 189
Kumar, P. S. S., 227
Kunau, I.-P., 180, 196
Kunstmann, M. P., 238
Kuo, Y.-N., 157
Kupchan, S. M., 229, 239
Kuppermann, A., 289
Kuramoto, T., 234
Kurashova, E. Kh., 264
Kurata, T., 334
Kurihara, K., 278
Kurokawa, S., 228
Kurras, E., 181
Kursanov, D. N., 187
Kusamran, K., 303
Kuschel, H., 73
Kushwaha, R. P. S., 306
Kustler, A., 87
Kusunose, M., 306
Kutney, J. P., 226, 228
Kutsev, V. S., 223
Kutzelnigg, W., 2, 3, 80
Kuwata, K., 10
Kwart, H., 152
Kwok, K.-S., 136, 137
Kwong, P. T. Y., 68

Lacey, M. J., 143, 205, 227
Laerum, T., 249
Lageot, C., 154, 155, 156, 158, 159
Lagow, R. J., 150, 160
Lagowski, J. J., 191, 215
Laine, R. A., 247, 259, 260, 309, 311
Lakwijk, A. C., 330
Lalezari, I., 175
Lambein, F., 259
Lambert, R. L., 153
Lampe, F. W., 101, 216
Lan, S. J., 348
Land, D. G., 333
Landaas, S., 327
Landé, S., 252, 363
Landgrebe, J. A., 187
Lange, G., 145, 146
Langenbeck, V., 325, 327
Langer, M., 179
Langhout, J. P., 194
Langlois, N., 225
Lanthier, G. F., 151
Lao, R. C., 139, 141, 267, 304
Lappe, B. W., 336
Lappert, M. F., 155, 160, 180, 181, 202, 222
Laptev, V. T., 150
Large, R., 274
Larin, N. V., 189
Larkins, J. T., 25
Larm, O., 248
Larsen, B., 133
Larsen, E., 281
Larson, C. W., 24
Larsson, F. C. V., 285
Larsson, S., 337
Laseter, J. L., 305, 306, 319, 338
Laster, W. R., 350
Lathan, W. A., 2, 5, 80
Latscha, H. P., 171
Lau, P.-Y., 258
Laufenberg, J., 184
Laurie, W. A., 99, 100
Lauwers, W., 273
Lawesson, S.-O., 285
Lawler, G. C., 306, 319
Lawless, J. G., 337
Lawson, G., 96, 148
Leach, J. B., 216
Leach, W. P., 177, 190
Leander, K., 357
Lebederskaya, V. G., 264
Leclercq, P. A., 119, 249, 298, 362
Lederberg, J., 133, 134, 231
Lederer, E., 250, 324, 351, 363
Lee, A., 21
Lee, A. G., 207
Lee, C. C., 189
Lee, C. K., 68
Lee, C. R., 327
Lee, J. B., 335
Lee, J. D., 133
Lee, J. S., 332
Lee, K. M., 196

Lee, R. C. T., 131
Lee, T. H., 8
Lee, Y. T., 21
Lee Wolfe, N., 282
Lefebvre, R., 14, 15
Leferink, J. G., 119, 298
Lefevre, H., 328
Leffler, H., 260
Le Goffic, F., 240
Lehman, J. P., 251, 363
Lehmann, P. A., 275
Lehmann, W. D., 95, 247
Lehrakuhl, H., 150
Leigh, G. J., 194
Leitsch, J., 184
Leitzke, O., 221
Le Men, J., 225
Lemke, T. F., 283
Lemonnier, A., 324
Lengyel, I., 78, 161, 162, 280
Lennarz, W. J., 258
Lenton, J. R., 319
Lerch, E., 276
Leroi, G. E., 51
Leroy, R. L., 21
Lertratanangkoon, K., 358
Lesman, T., 277
Lester, G. R., 42, 263
Leupold, M., 193
Leuthdold, R. H., 330
Levin, G., 46
Levine, R. D., 16, 18, 20, 22
Levine, S. D., 348
Levine, S. P., 327
Levitt, L. S., 73
Levsen, K., 33, 34, 63, 64, 65, 275, 276, 370
Levson, K., 253
Levy, J. B., 166, 169
Levy, R. L., 337
Levy, R. M., 299
Lewars, E. G., 271
Lewis, A. F., 236
Lewis, C. P., 278
Lewis, J., 179, 185, 186, 190, 200
Lewis, S., 34, 93
Leyshon, W. M., 278
Lhuguenot, J. C., 321
Li, E., 102
Li, Y., 312
Li, Y.-T., 260
Liaaen-Jensen, S., 230
Lias, S. G., 59, 109
Libbey, L. M., 331, 332, 333, 334
Liberton, R., 160
Lichtenburg, D. W., 181
Lichtenstein, H. A., 298
Lidy, W., 157
Lie, G. C., 2, 4
Lie, R., 285
Liebich, H. M., 298, 332
Lieder, C. A., 102, 108
Liedtke, R. J., 230, 278
Liehr, J. G., 68, 157, 226, 243, 274, 275, 283
Lieser, K. H., 151

Lifshitz, C., 11, 13, 33, 48, 54, 57, 58, 293
Light, J. C., 16
Light, R. J., 259
Lilja, A., 260
Lin, D. C. K., 206, 242
Lin, L.-P., 71, 185, 186
Lin, M. S.-H., 71
Lin, P.-H., 62, 264
Lin, R. L., 322
Lin, S. H., 14, 15, 24
Lin, S.-S., 223
Lin, T.-P., 220
Lind, W., 150
Lindahl, C. B., 221
Lindberg, B., 248, 312, 372, 376
Lindeke, B., 323, 328, 341, 343, 357
Lindgren, J.-E., 121, 297, 300, 325, 340, 356, 357
Lindholm, E., 12, 30
Lindley, H., 325
Lindner, E., 160, 163, 197, 199
Lindsay, R. C., 332
Lindström, B., 328
Lines, E. L., 171, 218
Ling, L. C., 285, 332
Ling, N., 255, 256, 324, 367, 368
Linke, K.-H., 155, 159
Linnarsson, A., 355
Linscheid, M., 74, 231
Linschiz, H., 242
Lipkin, D., 242
Lipowitz, J., 158
Lipsky, S. R., 123, 231, 252, 285, 363
Lisle, J. B., 5
Litzow, M. R., 143
Liu, A. L., 322
Liu, C. S., 157
Livingstone, S. E., 212
Lloyd, D., 172
Lloyd, D. R., 223
Lloyd, J. C., 165
Lloyd, J. P., 188, 193
Lloyd, M. K., 203
Lo, D. H., 6
Lock, G. J. L., 206
Lockhart, J. C., 149
Lockley, W. J. S., 237
Lodge, B. A., 231
Loeffler, W., 239
Löfqvist, J., 330
Lönngren, J., 248, 312, 372
Lönroth, U., 356
Loew, G., 7, 78
Lohmann, B., 186
Long, J., 107
Longevialle, P., 226, 231
Lopez, M. I., 156
Lorberth, J., 145, 146, 181
Lord, G. H., 280
Lorenz, H., 192, 199
Lorquet, J. C., 7, 78
Loskot, S., 202
Lossing, F. P., 48, 51
Loudon, A. G., 55

Lounasmaa, M., 228, 273
Lovett, E. G., 242
Lovins, R. E., 224, 249, 251, 304, 324, 367
Lowbridge, J., 251
Luby, J. A., 330
Lucas, C. R., 193
Luce, R., 282
Lukacs, G., 225
Lukas, G., 358
Lukas, W., 213
Lukashenko, I. M., 264, 265
Lukens, H. C., 141
Lund, E. D., 331
Lundblad, A., 248, 312
Lundeen, C. V., 97
Lundin, R. E., 146, 157
Lunken, R. E., 285
Lupin, M. S., 292
Luthke, W., 175
Luyten, J. A., 298, 299
Lyman, K. J., 323
Lynch, A. H., 199
Lytle, F. E., 129

Mabry, T. J., 228, 338
McAdoo, D. J., 62, 188
McAllister, T., 102
McAmis, L., 202
McArdle, P., 185
McAuliffe, C. A., 164, 220
McCabe, P. H., 271
McCall, A. C., 336
McCaman, M. W., 328
McCaman, R. E., 328
McCamish, M., 134, 253, 370
McCaskie, J. E., 184
McChesney, J. D., 342
McClaskie, J. E., 181
McCleverty, J. A., 198, 203
McCloskey, J. A., 68, 97, 99, 156, 236, 239, 241, 242, 243, 298, 323
McCombie, S. W., 209
McConkey, J. W., 11
MacConnel, J. G., 330
McCormick, A., 138
McCormick, B. J., 199
McCulloh, K. E., 51
MacDiarmid, A. G., 201, 202, 203
McDonagh, A. F., 246
Macdonald, C. G., 205
McDonald, F. J., 259, 310
McDonald, J. D., 15
McDonald, J. J., 243
McDonald, R. N., 73, 282
McDowell, M. V., 195, 201, 202
McEwen, C. N., 97, 98
McEwen, G. K., 195, 203
McFadden, W. H., 297, 303
MacFarlane, C., 335
McGillivray, D., 67, 269
McGinnis, J., 188
McGregor, A., 202

McGregor, M. L., 230
McGuire, J. M., 139
Machleidt, H., 351
McHugh, T. M., 126
McIlhenny, H. M., 349
McIntosh, C. L., 153
McIver, R. T., jun., 53, 102, 106, 107, 110
McKay, F. C., 253
Mackay, K. M., 201
McKenzie, E. D., 208
Mackenzie Peers, A., 54
Mackey, D. J., 212
Mackie, A. M., 233, 318
McKinley, I. R., 148, 248
McKinney, C. R., 224, 304, 367
McKinney, R., 182
McKown, H. S., 119
McKoy, V., 4, 5
McLafferty, F. W., 59, 62, 63, 64, 65, 96, 120, 123, 127, 134, 136, 137, 224, 251, 253, 264, 268, 275, 276, 292, 304, 370
McLaughlin, D. R., 17
McLaughlin, E., 215
Maclean, D. B., 239
MacLeod, J. K., 70, 226, 227, 269, 270, 350
McMahon, R. E., 301, 329, 342, 357
McMahon, T. B., 53, 83, 102, 108, 109
McMaster, B. N., 47, 50, 53, 82, 141
MacMillan, J., 300, 314
McMurray, W. J., 123, 231, 252, 285, 363
MacNeil, J. D., 335
MacNeil, K. A. G., 58, 110, 288
McQuillin, F. J., 193, 303
McPherron, R. V., 119
McReynolds, J. H., 228, 338
McWilliams, D., 4
Macy, L. R., 334
Madani, C., 232, 302, 317
Madden, D. P., 194
Madhusudan, K. P., 276
Maeda, M., 243
Maedi, K., 52
Mahan, B. H., 18
Maier, D. P., 124
Maines, M. D., 330
Maire, J. C., 154, 155, 156, 158, 159
Majer, J. R., 360
Major, R. W., 292
Makino, Y., 354
Maklier, S., 237
Malek, A., 206
Malicky, J. L., 278
Malisch, W., 201
Mallams, A. K., 236
Mallory, C. W., 260
Mallory, F. B., 260
Malojčić, R., 270
Maltsev, A. K., 158

Maltz, H., 185
Malysheva, N. N., 248, 372
Mamer, O. A., 326
Manachini, P. L., 250
Mancini, V., 54, 55, 266
Mandava, N., 313, 320
Mandelbaum, A., 70, 73, 78, 231, 271, 280
Manisse, N., 270
Mann, F. G., 167
Mann, J. E., 337
Mannering, G. J., 341
Manni, P. E., 275
Manning, D. J., 332
Manoussakis, G. E., 212
Manuel, G., 154, 158
Manura, J., 360
Maquestiau, A., 280, 281
Marazano, C., 231
Marcotte, R. E., 29
Marcus, R. A., 18
Maresca, L., 179
Margrave, J. L., 58, 150, 157, 213, 218
Mark, R. V., 78
Markey, S. P., 139, 141, 299, 327
Markgraf, J. H., 282
Markl, G., 167, 168, 170, 172
Markov, V. I., 280
Marmet, P., 10
Marmur, J., 243
Marquardt, G., 154
Marr, G., 188
Marsel, J., 267
Marshall, C. J., 156
Marshall, J. C., 128
Marshall, J. L., 70
Marshall, K. S., 323, 359
Marstein, S., 325
Martelli, P., 326
Martens, J., 285
Martin, A., 90
Martin, C., 170
Martin, G., 189
Martin, J. G., 209
Martin, R. L., 212
Martin, T. I., 239
Martinengo, S., 178
Martinez, E., 228
Maryanoff, E., 170
Masada, Y., 315, 332, 335
Mascaretti, A. O., 226
Masclet, P., 26
Masino, A. P., 192
Maskens, K., 303
Maslen, V. W., 42
Mason, K. T., 325
Mason, P. R., 197
Massaroli, G. G., 347
Massmol, M., 159
Mastafanova, L. I., 282
Matern, E., 154
Mathar, W., 282
Matheson, T., 200
Mathew, M., 187, 194, 203
Mathews, R. J., 132
Mathiaparanam, P., 255, 318

Mathias, A., 287
Matin, S. B., 323
Matlin, S. A., 244, 371
Matsubara,T., 247, 260, 310, 311
Matsueda, S., 229
Matsui, S., 249, 324
Matsumato, T., 334, 338
Matsumoto, K., 293, 294
Matsumoto, K. E., 298
Matsuo, A., 226, 227, 305
Matthes, D., 167
Mauger, A. B., 253, 325
Maume, B. F., 299, 321
Maume, G., 321
Maurer, K. H., 94, 125, 234, 235, 251
Mautner, H. G., 285
Maxwell, D. C., 138
Maxwell, J. R., 314, 337
Mayence, G., 160
Mayevsky, A., 327
Mays, M. J., 179, 201
Mazengo, R. Z., 55
Mazerolles, P., 154, 158, 160
Mazur, Y., 231
Mead, T. J., 62, 268, 362
Meager, J. F., 15
Means, J. C., 304
Medved, M., 286
Meek, D. W., 199
Mehesfalvi-Vajna, Zs., 337
Meier, H., 176
Meier, T., 119
Meier, W.-P., 197
Meili, T., 139
Meinert, H., 242, 283
Meinma, H. A., 162
Meinwald, J., 331
Meisel, W. S., 127
Meisels, G. G., 8, 11, 12, 49, 80, 87
Melby, E., 104
Melby, J. C., 322
Melcher, L. A., 215
Mellon, F. A., 47, 54, 73, 77, 141, 263
Melson, G. A., 209, 211
Melton, C. E., 288
Memel, J., 280
Menes, F., 55
Mennenga, H., 181
Meot-Ner, M., 78, 99, 231, 245, 249, 280
Merault, G., 152
Mercer, A. J. H., 167
Mercer, G. D., 150
Meredith, J. O., 120
Merkuza, V. M., 226
Merlet, P., 3
Merrell, P. H., 208
Merriam, J. S., 148
Merritt, C., jun., 135, 136
Mertel, H. E., 300
Mertis, K., 208
Mertschenk, B., 190, 192
Meshi, T., 349, 352
Mestres, R., 228
Metaxas, J. M., 303

Metzger, J., 285
Meuzelaar, H. L. C., 95, 225
Mews, R., 194
Meyer, P., 187
Meyer, W., 3
Meyer, W. H., 239
Meyerson, S., 61, 68, 70, 274, 279, 280
Michel, G., 250
Michelson, L., 328, 343
Michnowicz, J. A. G., 123, 231, 342
Mickey, C. D., 219
Micromastoras, E. D., 212
Middleditch, B. S., 231, 232, 260, 261, 296, 297, 298, 300, 303, 309, 315, 316, 341, 342
Middleton, R., 180, 190
Middleton, T. B., 168
Midgley, I., 231
Midwinter, G. G., 369
Mieure, J. P., 334
Migahed, M. D., 92, 96
Mignonae-Mondon, S., 280
Miiller, D. E., 70
Mikhlina, Ye. Ye., 282
Miksche, G. E., 337
Milberg, R. M., 282
Milborrow, B. V., 313
Miles, M. F., 333
Millard, B. J., 60, 280, 286, 339, 347, 359, 360
Millard, M. M., 157
Miller, A., 332
Miller, B., 337
Miller, J. L., 157
Miller, J. M., 144, 151, 215, 269
Miller, J. R., 177, 190
Miller, P. H., 331
Miller, V. R., 149, 203
Miller, W. J., 223
Millington, D. S., 95, 226, 227, 229, 237, 244, 245, 299, 330
Milman, J. D., 256
Milne, G. W. A., 100, 131, 138, 231, 249
Milone, L., 199, 200
Mimura, T., 335
Minagawa, S., 189
Minale, L., 318
Minders, B., 222
Minghetti, A., 349
Minghetti, G., 208
Minnikin, D. E., 257, 303
Mirando, P., 238
Mirocha, C. J., 333
Mironov, V. F., 152
Mischer, G., 95, 249, 324
Misharev, A. P., 271
Mishii, N., 193
Mital, R. L., 282
Mitchard, L. C., 190
Mitchard, M., 329
Mitchell, E. D., 230

Mitchell, F. L., 298
Mitchell, J. W., 313
Mitchell, S. R., 282
Mitchum, R. K., 12
Mitera, J., 276, 305, 338
Mitscher, L. A., 99, 236, 342
Mitschke, K.-H., 171
Mitsuhashi, H., 233, 318
Mitsuhashi, K., 282
Miwa, G. T., 357
Miyazaki, H., 302, 321
Mizutani, J., 249, 334
Modest, E. J., 283
Modinos, A. G. J., 198
Modzeleski, V. E., 337
Moehlmann, J. G., 15
Møller, J., 244, 281, 282, 285, 286
Moeller, T., 220
Moers, F. G., 194
Moffitt, H. R., 330
Mohacsi, E., 341
Mohammed, M. A. J., 337
Moilanen, K. W., 336
Moiseiwitsch, B. L., 289
Moisson, M., 335
Molenaar-Langeveld, T. A., 70, 279
Molinelli, R., 216
Moller, J., 172, 227
Momose, A., 312, 327
Mon, T. R., 298, 303
Monahan, M., 256, 367
Monkman, J. L., 139, 141, 267
Moore, B. P., 229
Moore, C. B., 16, 337
Moore, C. H., 369
Moore, G. J., 153
Moore, R. E., 273
Moreland, C. G., 215
Morfin, R. F., 316
Morgan, E. D., 229, 317
Morgan, M. E., 332, 334
Mori, C., 302
Moriaty, R. M., 187
Morisaki, M., 318
Morita, H., 326, 337
Moriwaki, Y., 335
Moriyama, Y., 229
Morizur, J.-P., 270
Moron, M., 336
Morozova, V. A., 222
Morris, H. R., 252, 254, 255, 256, 362, 363, 368, 369, 370, 371
Morris, J. H., 216
Morris, W. J., 336
Morrison, A., 280
Morrison, J. D., 10, 11, 47, 49, 52, 126, 137, 216, 217, 263, 289
Morrison, R. J., 144
Morse, J. G., 171
Morse, K., 171
Morselli, C., 355
Morselli, P. L., 327, 344
Morton, T. H., 107
Moser, H. W., 307

Author Index

Moshonas, M. G., 331
Moss, J. R., 193
Mostecký, J., 284, 334
Motl, O., 239
Mowat, J., 329
Moza, P., 336
Mrochek, J. E., 312, 346
Muccino, R. R., 231
Müggler-Chavan, F., 332
Müller, B., 71
Müller, G., 337
Müller, P., 84
Müller-Schwarze, C., 307
Muenow, D. W., 217, 218
Mues, R., 228, 338
Mulders, E. J., 333
Muller, C. J., 331, 333
Muller, H., 178
Muller, J., 143, 177, 178, 187, 190, 192, 193, 199, 200
Mumma, R. O., 116
Munawar, Z., 242
Munson, B., 97, 98, 100, 107, 231, 272, 342
Muntwyler, R., 239
Murae, T., 229
Murao, K., 242
Murase, I., 210
Murata, H., 346
Murata, M., 338
Murata, T., 307
Murphy, G. J., 146
Murphy, G. M., 234
Murphy, R. C., 141, 337, 345
Murray, K. E., 331
Murray, K. S., 206, 210
Murrell, J. N., 77
Murti, V. V. S., 227
Musaev, I. A., 264
Mushak, P., 336
Mushfiq, M., 231
Mussinan, C. J., 332
Mussini, E., 299
Musso, H., 146
Musu, C., 239
Myers, J. F., 208
Mysak, A. E., 305
Myurisepp, M. A., 265

Nadai, T., 352
Naff, W. T., 52, 293
Naga, S., 78, 211, 227
Nagabhushan, T. L., 97, 246
Nagai, S., 251
Nagai, Y., 100, 155
Nagao, Y., 315
Nagatani, T., 33
Nageli, D., 139
Nagy, B., 337
Nair, R. M. G., 256, 368
Nakadaira, Y., 154
Nakahara, Y., 226
Nakajima, E., 346
Nakamura, S., 349
Nakanishi, K., 243
Nakanishi, M., 345
Nakano, S., 315
Nakata, H., 78, 227, 287
Nakatani, N., 228, 338
Nakatani, Y., 333
Nakayama, M., 226, 227, 305
Nakayama, T. O. M., 331
Nakazawa, S., 335
Nappier, T. E., 199
Naranjo, J., 225
Narasimhachari, N., 237, 322, 328, 354
Naruse, M., 325
Nasielski-Hinkens, R., 286
Nasta, M. A., 202
Natile, G., 190, 200
Nau, H., 140, 253, 300, 326, 364
Nault, L. R., 330
Nazery, M., 147, 149
Ne'eman, I., 250
Neeter, R., 62, 282
Nefedov, O. M., 158
Neff, S. E., 331
Negoita, C., 329
Negoro, K.-I., 335
Nehl, H., 150
Neike, E., 155
Nekrasov, Ya. S., 187
Nelson, D. R., 214, 284
Nelson, E. C., 230
Nelson, G. H., 333
Nelson, S. D., 361
Nelson, S. M., 185
Nelson, T. R., 184
Nesmeyanov, A. N., 187
Netzel, T. L., 14
Neumann, F. W., 201
Neumann, K. G., 223
Neumann, W. P., 162, 163
Neuner-Jehle, N., 298
Neuse, E. W., 189
Neuss, N., 225
Neville, A. F., 73, 278
Newlands, M. J., 202
Nguyen, L.-D., 119, 223
Nibbering, N. M. M., 62, 70, 71, 83, 112, 229, 270, 275, 279, 282, 326
Nichols, T. L., 135
Nicholson, A. J. C., 51, 216, 222
Nicholson, B. J., 6
Nicholson, R. S., 100
Nickol, G. B., 333
Nicolai, H. V., 376
Nicolella, V., 349
Nicoletti, R., 264
Nicosia, S. Z., 317
Niecke, E., 219
Niedenzu, K., 148, 149
Nieh, M., 330
Niehaus, A., 14
Nieuwenhuis, T., 84
Niida, T., 352
Nikitina, N. S., 305
Nikolaev, E. N., 213
Nikonov, G. K., 226
Nilsson, I., 357
Nilsson, N. J., 127
Nimmich, W., 248, 312
Nishimura, H., 249, 334
Nishimura, S., 242, 243
Nishiwaki, T., 169
Nishiyama, T., 286
Nishizawa, M., 318
Nitzan, A., 15
Niwa, Y., 81
Niwaguchi, T., 226
Nixon, J. F., 155, 196, 197
Nocke-Fink, L., 321
Nogina, O. V., 187
Noltes, J. G., 160, 162
Nooner, D. W., 337
Nordén, N. E., 312
Nordlund, D. L., 142
Norikov, G. I., 222
Norkrans, B., 327
Norman, A. D., 217
Noro, K., 209
Noro, T., 209
North, B., 186
Northrop, J. K., 216
Norton, J. R., 200
Noth, H., 149, 190, 195, 219, 220
Nothe, D., 221
Notke, D., 193
Nounou, P., 10, 219
Novák, J., 334
Novikov, S. S., 264
Novoselova, A. V., 213
Novotny, M., 297, 321
Nowak, A. V., 118
Nowak, C. A., 315
Nowell, I. W., 198
Nowlin, J., 358
Nozaki, H., 189
Nunn, J. R., 333
Nursten, H. E., 331
Nussey, B., 294
Nyholm, R. S., 200

Oakley, R. T., 220
Oates, D., 221
Oates, J., 356
Oates, J. A., 121, 141, 300
Obermeier, R., 366
O'Carra, P., 246
Occolowitz, J. L., 49, 78
Odaira, A., 189
Odell, G. V., 332, 334
Odenbrett, C., 180
Odham, G., 256, 327
Odiorne, T. J., 97, 151, 239, 302
Oehme, G., 181
Otvos, I., 338
Ofele, K., 193
Offhaus, R., 193
Offord, R. E., 256, 369
Ogata, M., 7
Ogawa, S., 209
Ogawa, T., 10
Ogden, J. S., 189
Ogilvie, K. K., 242

Ogunbona, F. A., 326, 342
Ogura, K., 313
Ogura, T., 206
Oh, S. K., 331
O'Hanlon, P. J., 245
Ohara, K., 229
Ohashi, M., 67
Ohashi, Z., 242, 243
Ohkoshi, S., 336
Ohno, T., 325
Ohta, Y., 229
Ohtaki, T., 321
Ohtomo, M., 270
Oja, H., 139, 141, 304
Okabe, H., 51
Okada, K., 251
Okamoto, N., 179
Okazaki, N., 338
Okuda, K., 234
Okuyama, H., 249
Olah, G. A., 104
Olavesen, A. H., 234
Olbrysch, O., 150
Oldfield, E., 306
Oliver, A. J., 151
Oliver, J. P., 146, 154
Oliver, W. R., 284
Olsen, L. A. R., 58
Olsen, R. W., 119, 124
Ol'shevsku, M. V., 219
O'Malley, R. M., 53, 109, 112
Omoto, S., 352
Onak, T, 88, 149
O'Neill, S. R., 218
Ono, H., 335
Onodera, K., 248
Onuska, F. I., 338
Opendak, I. G., 213
Oprean, I., 285
Orchard, D. G., 198
Orlandi, G., 15
Orlov, V. Yu., 144
Oró, J., 337
Orr, J. C., 301
Osband, J. A., 315
Oshima, M., 312
Osman, S. F., 332
Ostazewski, A. P. P., 203
Ostrow, J. D., 246
O'Sullivan, J., 276
O'Sullivan, W. I., 227
Oswald, E. O., 323
Otsuka, S., 184
Ottinger, C., 14, 42
Ourisson, G., 337
Ovchinnikov, Yu. A., 248, 325
Owen, P. W., 152, 155
Ozenne, J. B., 14
Ozhegov, P. I., 223
Ozkan, I., 5

Padden, J. J., 341
Paddock, N. L., 220
Paddon-Row, M. N., 71
Padgett, C. A., 350
Padieu, P., 320, 321
Paelinck, M. T., 160

Page, F. M., 289
Page, R. P., 118
Pahaut, P., 120
Paiaro, C., 185
Paige, H. L., 175
Paine, R. T., 214, 215, 223
Painter, T. M., 155
Paisley, H. M., 142
Paiu, F., 285
Pal, P., 24
Palazzi, A., 204
Paleček, J., 276
Palenik, G. J., 187, 194, 203
Pallotta, U., 297
Palmer, D. E., 155
Palmer, J. K., 331
Palmér, L., 297, 335, 355
Palmer, T. F., 57, 209
Pályi, G., 338
Pan, S. F., 303, 331
Panalaks, T., 334
Panasuik, O., 332
Panchenov, I. G., 213
Pandey, R. C., 237
Pandya, N. I., 320
Panina, M. A., 280
Pantarotto, C., 355
Panzica, B. P., 241
Pao-Shui-Wang, 332, 334
Pardo, M. C., 281
Pareles, S. R., 332
Parfenov, B. P., 150
Paris, R., 227, 376
Parker, C. E., 7, 26, 80, 81, 264, 273
Parker, G. J., 186
Parker, R. B., 360
Parkhurst, R. M., 315
Parkins, A. W., 185
Parks, O. W., 332
Parliment, T. H., 331
Parmenter, C. S., 14, 15
Parmentier, G. G., 316
Parr, A. C., 56
Parr, W., 255
Parravicini, F., 344
Parry, R. W., 215
Parson, J. M., 21
Parson, J. R., 196
Partridge, D. H., 298
Partridge, R. D., 311
Pasanen, P., 55, 286
Pascher, I., 258, 259, 310
Pascual, C., 47
Passerini, G., 344
Passmore, J., 174, 175
Pasteels, J. M., 330
Patel, H. A., 183, 185, 194
Patel, K. S., 210
Patil, G. V., 251, 363
Patrikeev, Yu. B., 222
Patt, S. L., 108
Pattenden, G., 229
Pattengill, M. D., 19
Patterson, G. W., 319
Patterson, T. B., 6
Patterson, W. R., 96
Paul, N. C., 278
Pauling, L., 298

Paulson, J. F., 289
Pauson, P. L., 190
Pavli, C. J., 342
Payne, D. S., 220
Payzant, J. D., 109
Pazdernik, L. J., 157
Peach, M. E., 284
Pearce, R., 180
Pearson, A. M., 332
Péchiné, J.-M., 25, 26, 55, 267
Pecka, K., 338
Pedder, D. J., 301
Pedersen, C. Th., 286
Pedersen, L. G., 7, 80, 81, 264, 273
Pellacani, L., 278
Pellegrini, G., 336
Pelster, D. R., 121, 141, 300
Pelzer, H., 351
Peng, S.-M., 209
Pensabene, J. W., 333
Penzhorn, R.-D., 284
Pereira, W. E., 249, 325, 357
Peres, M., 48
Perez-Reyes, M., 328
Perheentupa, J., 321
Perkins, E. G., 304, 307
Perkins, P. G., 216
Perkinson, W. E., 170
Perrin, C. L., 131
Perros, P., 270
Perry, R., 304
Perry, W. O., 265, 285
Persiyaninova, S. N., 213
Persky, A., 289
Persson, T., 332
Pesyna, G., 136, 137
Peterson, L. K., 157, 159
Peterson, O., 220
Petersson, G., 311
Petit, G. R., 233
Petrova, T. D., 71
Petrunin, A. B., 150
Petterson, J. E., 327
Petty, H. E., 73, 282
Peyerimhoff, S. D., 3, 17
Peypoux, F., 250
Pfaender, F. K., 336
Pfaffenberger, C. D., 298
Pfeiffer, J.-P., 33, 92, 142
Pflug, J. L., 189
Pfluger, C. E., 181
Philips, S. R., 360
Phillips, B. A., 280
Phillips, D. O., 319
Phillips, D. R., 258
Phillips, K. D., 247
Phillips, R. P., 202
Phinney, B. O., 314
Piacente, V., 214, 223
Picard, J. P., 152
Picardi, S. M., 332
Picart, D., 316
Piccardi, P., 267
Pierce, R. C., 214
Pierrou, L., 121, 300
Piersol, A. G., 125

Piessens-Denef, M., 316
Pietropaolo, D., 204
Pifferi, G., 327, 344
Pignataro, S., 54, 55, 57, 190, 200, 204, 266
Pihlaja, K., 55, 59, 263, 267, 286
Pike, W. T., 132
Pilipovich, D., 221
Pillinger, C. T., 137, 297, 337
Pilotti, Å., 247, 311
Pimlott, W., 259
Pinar, M., 225
Pinsky, M. L., 216
Piotrowski, E. G., 333
Pisano, J. J., 100
Pistorius, E. K., 306
Pitman, G. B., 97
Pitt, C. G., 328, 341
Pittman, C. U., 6
Pitts, J. N., jun., 88
Pitzer, R. M., 4
Plasz, A. C., 278
Plat, M., 225
Platonov, V. E., 71
Plattner, J. R., 141
Platzner, I., 46
Plazonnet, B., 297
Plekhanov, V. G., 163, 164, 165, 277, 280
Plimmer, J. R., 175, 336
Pohl, L. R., 323
Pohland, A., 345
Poite, J.-C., 285
Pokorny, J., 332
Polan, M. L., 252, 363
Poland, J. S., 150
Polanyi, J. C., 17
Polgar, N., 303
Polivanov, A. N., 150, 152, 154, 157, 168
Pollitt, R. J., 327
Polyachenok, O. G., 222
Polyakova, A. A., 264, 265
Pomonis, J. G., 284
Ponnamperuma, C., 337
Popkie, H., 4
Popkov, O. S., 213
Pople, J. A., 2, 5, 80, 263
Popli, S. P., 286
Popovich, G. A., 207
Porritt, C. J., 216
Porter, Q. N., 225, 226, 278
Porter, R. F., 214
Posthumus, M. A., 225
Potier, P., 225
Pottie, R. F., 139, 267, 338
Potts, A. W., 7
Potts, K. T., 281
Pousette, Å., 321
Pousset, J.-L., 324
Poutsma, M. L., 160
Povall, T. J., 253, 324, 326, 365
Powell, J., 186
Powell, J. W., 356
Powell, P., 216
Praefcke, K., 285
Pratanata, I., 284

Pravica, M., 211
Predmore, D. B., 327
Preiss, H., 174, 214
Premuzic, E., 231
Preste, P. G., 359
Preston, A. F., 229
Preston, J. A., 11
Preston, R. K., 17
Preti, G., 203
Pretzer, W. R., 149
Priaulx, D. J., 359
Price, A. P., 84
Price, D., 291
Price, P., 318
Price, W. C., 7
Priddle, J. D., 256, 369
Pritchard, J. G., 270
Prokazova, N. V., 307
Prost, M., 321
Protopapa, H. K., 283
Prox, A., 227, 343, 351
Pruggmayer, D., 336
Pryce, R. J., 313, 315
Pua, C. N. K., 54
Pulfer, J. D., 6
Puppe, L., 209
Purse, J. G., 312
Puskas, I., 68, 274

Quack, M., 19
Quéméner, J. J., 10
Quentin, K. E., 297
Quillam, M. A., 242
Quin, L. D., 167
Qureshi, I. H., 281

Raal, A., 257
Rabalais, J. W., 8
Rabeneck, H., 213, 222, 223
Rabinovitch, B. S., 15, 23, 24
Rabinowitz, J. L., 327
Rademaker, W. J., 149
Radford, T., 246, 374
Radom, L., 2, 5, 6, 80
Radüchel, B., 315
Raff, L. M., 17
Raffenetti, R. C., 4
Rafikov, S. R., 89, 293
Rai, D. N., 149
Railton, I. D., 314
Rainey, W. T., jun., 312, 346
Rakita, P. E., 152
Rambold, W., 182, 194
Ramon, F., 151, 153
Randall, G. L. P., 185
Rane, A., 322
Rang, S. A., 265
Rangarajan, M., 324
Rankin, D. W. H., 216, 217
Rao, A. V. R., 227
Rao, G. S., 349
Rao, V. N. M., 2, 146

Rapp, U., 94, 234, 235, 251
Rapport, M. M., 260
Rasoanaivo, P., 225
Rastogi, R. P., 229
Rathi, S. S., 227
Rathke, J. W., 215
Rausch, M. D., 181, 187
Rave, A., 73, 271
Rawlins, M., 355
Rawls, M. H., 306
Ray, V. L., 337
Rayborn, G. H., 13
Razavi, A., 179
Reardon, E. J., 185
Rebane, E., 174, 175
Rebbert, R. E., 80
Rebsck, M., 220
Redmore, D., 168
Redolph, R. W., 149
Redwood, M. E., 167, 201
Reed, G. F., 227
Reed, R. I., 135, 278
Refaey, K. M. A., 27, 53
Regan, T. H., 149
Regitz, M., 169
Regnier, F. E., 306, 330, 334
Reichert, B. E., 201, 210
Reichstein, T., 228
Reid, D. M., 314
Reid, N. W., 46
Reid, R., 230
Reiffsteck, A., 317
Reilley, C. N., 129
Reilly, T. J., 149, 150
Reimann, H., 239
Reimendal, R., 139, 300
Reineccius, G. A., 334
Reinhardt, P. W., 58
Reinhold, V. N., 248
Reinke, D., 51
Reiss, J. G., 180
Rendall, I. F., 212
Rennekamp, M. E., 45, 67, 275, 285
Renson, M., 175, 285, 286
Rentzepis, P. M., 14
Reynolds, W. E., 249, 325, 357
Rhodes, R. E., 344
Rhyme, T. C., 217
Ricard, L. P., 338
Ricard, M., 282
Ricci, A., 151
Rice, O. K., 16, 18
Rice, S. A., 14, 16, 21
Rich, D. H., 255
Richards, F., 249
Richards, R., 206
Richardson, W. C., 14
Richer, J. C., 276
Richter, W. J., 101, 157, 226, 256, 273, 274, 275, 283, 299, 302
Riddoch, A., 124
Ridge, D. P., 53, 83, 103, 108
Ridley, T. Y., 42
Rietz, R. R., 149

Rigassi, N., 230
Rigby, P. W. J., 369
Riley, S. J., 21
Rinck, R., 198
Rindfleisch, T. C., 325, 357
Rindone, B., 250
Rinehart, K. L., jun., 94, 226, 235, 237, 330
Ring, M. A., 217
Rinke, K., 222, 223
Risby, T. H., 101
Risebrough, R. W. 336
Ritter, F. J., 331
Ritter, H.-P., 162
Riva, F., 355
Riveros, J. M., 103, 107, 108, 110
Rivier, J., 255, 256, 367, 368
Riviere, P., 159
Robbiani, R., 70, 269
Roberts, J. D., 246, 376
Roberts, P. J., 237
Robertson, A. J. B., 34, 168, 263, 289, 298
Robertson, D. H., 135, 136
Robertson, J. H., 236, 326
Robinson, A. B., 298
Robinson, B. H., 200
Robinson, D. S., 248
Robinson, H. M., 332
Robinson, W. D., 337
Roboz, J., 236, 297, 327
Roch, M., 325
Rockstroh, C., 151
Rodger, M., 157
Rodriguez, E., 228, 338
Roe, R. L., 360
Röper, H., 305
Roepstorff, P., 254
Roesky, H. W., 156, 163, 219, 220
Rogers, L. B., 140
Rogers, M. A., 338
Rogerson, P., 123
Roggero, J., 285
Rohbock, K., 209
Rohmer, M., 319
Rohwedder, W. K., 119
Rolinski, E. J., 222
Rollet, M., 335
Rollgen, F. W., 91, 92
Romeo, G., 71
Romiez, M., 337
Ronayne, J.. 188
Ros, A., 232, 317
Rosan, A., 184
Rose, J. B., 4
Rose, N. J., 208
Rose, P. D., 193
Rosen, J. D., 335
Rosenberg, H., 189
Rosenblum, M., 184, 186
Rosenbuch, H., 198
Rosenfeld, J., 348
Rosenstock, H. M., 18, 25
Rosenthal, D., 140, 209, 245
Rosenthal, U., 181

Ross, J., 16, 18
Ross, K., 320
Rossi, A., 276
Rossiter, M., 229, 234, 244, 251, 371
Rossiter, M. J., 227
Roszkowski, A., 248
Rote, J. W., 336
Rotérmél, I. A., 273, 274
Rothenberg, S., 80
Rotman, A., 231
Rottele, H., 187
Rouillard, M., 282
Rousseau, Y., 56, 78, 276
Rovere, C., 272
Rouvier, E., 78
Rowland, M., 323
Rowlands, R. J., 325
Rowley, R. J., 155
Roy, S. K., 276
Royer, M., 260
Rozett, R. W., 215
Rozynov, B. V., 307, 311
Rubinstein, I., 318
Rubottom, G. M., 156
Rucker, G., 340
Rudzats, R., 250
Ruedenburg, K., 4
Ruelius, H. W., 322, 346
Ruff, J. K., 197
Ruis, S. P., 217
Ruliffson, W. S., 372
Rullkötter, J., 258
Runge, W., 274
Runquist, O. A., 77, 263
Ruokonen, A., 321
Ruppert, I., 156
Rushneck, D. R., 299
Russ, B., 272
Russell, J. W., 101, 177, 185
Ruth, J. M., 280
Ruthven, C. R. J., 323, 354
Rutten, G. A. F. M., 298
Ruveda, E. A., 226
Ruzo, L. O., 336
Ryan, J. F., 97, 98, 360
Ryan, J. J., 331
Ryan, P. W., 41
Rye, R. T. B., 267, 269
Ryhage, R., 120, 139, 297, 329
Ryska, M., 205

Saalfeld, F. E., 195, 201, 202
Sabherwal, I. H., 180
Sabih, K., 359
Saeki, Y., 226
Safe, S., 66, 268, 335, 336
Saferstein, R., 360
Safron, S. A., 21
Saillant, R. B., 194
Saito, K., 306
Sakai, I., 63, 64, 253, 268, 275
Sakai, T., 139
Sakamoto, N., 200
Sakamura, S., 250, 332

Sakurabu, Y., 270
Sakurai, H., 78, 154, 155, 186, 187, 287
Sakurai, S., 321
Saleh, G., 171
Salentine, C. G., 203
Sales, K. D., 212
Sali, E., 70, 280
Salisova, M., 189
Salmon, M., 246
Salser, G. E., 223
Salsmans, P., 282
Salvatori, T., 178
Sammon, J. W., 131
Sammons, R. E., 357
Sams, J. R., 198
Samuel, D., 327
Samuelsson, B., 307, 308, 356, 357
Samuelsson, B. E., 257, 259, 260, 261, 305, 310
Samuelsson, K., 248
Sanchez, I. H., 228
Sancilio, L. F., 351
Sandberg, F., 357
Sandler, M., 303, 323, 349, 354
Sandra, P., 298
Sanger, A. R., 156, 199
Sangré, M., 225
Sanin, P. I., 264
Santi, R., 336
Santini-Scampucci, C., 180
Santoro, F., 57, 218, 267
Sapers, G. M., 332
Sappa, E., 200
Sara, A. N., 160, 207
Saran, H., 155
Sarel, S., 185, 256
Sargent, M. V., 228
Sariaslani, F. S., 305
Sasaki, S., 118, 132, 270
Sasaki, Y., 338
Sasame, H. A., 350
Sastry, S. D., 225
Sathe, S., 341
Sato, H., 250, 332
Sato, T., 155, 336
Sato, Y., 332, 349, 352
Satomi, K., 338
Saudi, S. K., 78
Saunders, A., 286
Saunders, E. B., 303, 331
Saunders, K. J., 274
Savitzky, A., 122
Savolainen, J., 14
Savory, C. G., 150
Sawbridge, A. C., 163
Saxby, M. J., 323
Saxton, R. G., 244, 371
Sbrignadello, G., 179
Scanlan, R. A., 332, 333, 334
Schack, C. J., 221
Schaeffer, R., 149, 215
Schafer, H., 213, 222
Schafer, J., 223
Schally, A. V., 256, 368
Schaper, K.-J., 149
Scharmann, H., 256, 297

Author Index

Schechter, J., 131
Scheer, H., 154
Scheer, M. D., 289
Scheppele, S. E., 12
Scheps, R., 16
Scherer, H., 169
Scherer, O. J., 171, 174
Scherr, P. A., 146
Scheuer, P. J., 273
Scheuren, J., 203
Schiavone, J. A., 10
Schier, G. M., 252, 363
Schildcrout, S. M., 176
Schillings, R. T., 346
Schlag, E. W., 14
Schleyer, P. von R., 5, 6, 80
Schlunegger, U. P., 330
Schmeltz, I., 304
Schmid, G., 201, 202
Schmid, H., 225
Schmid, R. D., 227, 228, 252, 338, 375, 376
Schmidbaur, H., 168, 171, 181, 197, 201
Schmidpeter, A., 171
Schmidt, C. J., 312
Schmidt, W., 77
Schmidtberg, G., 157, 281
Schmitz, F. J., 259, 310
Schmitzler, M., 197, 198
Schmock, F., 145
Schneehage, H. H., 209
Schneider, M., 249, 366
Schneider, S., 14
Schnetzle, D., 120
Schnitzer, M., 337
Schnyder, D., 328
Schodel, H., 171
Schoeller, D. A., 123
Scholer, F. R., 150
Scholler, R., 317
Schooley, D. A., 300
Schopp, E., 226
Schossner, H., 171
Schowengerdt, F. D., 47
Schramm, K. H., 99, 236, 241
Schrardt, K., 160
Schreiber, K., 90, 231, 293
Schreike, R. R., 201
Schröder, B. M., 348
Schröder, H., 348
Schroer, U., 160
Schroll, G., 133, 285
Schteingart, D. E., 350
Schubert, J., 303, 331
Schüddemage, H. D. R., 63, 253, 275
Schuetz, R. D., 336
Schulten, H.-R., 93, 94, 95, 144, 224, 225, 228, 235, 240, 241, 247
Schulz, M., 11
Schulze, G. J., 57
Schulze, P., 61, 226, 283
Schupp, E., 338
Schuster, P., 3
Schwab, W., 184
Schwartz, D. P., 332

Schwartz, M. A., 346
Schwarz, H., 56, 71, 73, 266, 267, 270, 272, 282, 285
Schwarz, J., 182
Schwarzenbach, R., 119, 139
Schwarzhaus, K. E., 201
Schweiger, J. T., 218
Schweinler, H. C., 52, 293, 295
Schwieter, U., 230
Schwinghamer, L. A., 333
Scolastico, C., 250
Scott, J. A., 110
Scott, K. N., 327, 360
Scott, P. M., 333
Scribelli, J. V., 158
Searcy, A. W., 223
Searles, S. K., 109
Sears, P. L., 207
Seddon, D., 198
Sedgewick, R., 132
Sedova, V. F., 282
Sedvall, G., 327
Seegmiller, J. E., 325, 327
Segura, R., 321
Seiber, J. N., 336
Seibl, J., 60, 70, 269
Seick, L. W., 109
Seifert, W. E., jun., 247, 253, 325, 365, 366
Seifert, W. K., 320
Seipenbusch, J. M., 69, 269
Sekerke, H. J., 301
Sekules, G., 347
Self, R., 119, 248, 370
Selhofer, H., 121
Seligmann, O., 227
Selim, S., 336
Sell, C. S., 272
Sellier, N., 120
Sellmann, D., 194
Selva, A., 281, 286
Semeluk, G. P., 51
Semenov, G. A., 213
Sen, B., 150
Sen, D. N., 278
Sen, N. P., 333, 334
Sengupta, S. K., 283
Senn, M., 127
Senn, V. J., 278
Sen Sharma, D. K., 46, 57
Seppelt, K., 220, 221
Serum, J. W., 227, 273
Seshadri, T. S., 227
Setkina, V. N., 187
Seto, S., 313
Setser, D. W., 14, 15, 22, 23, 45, 67, 275
Severin, D., 119
Severin, D. W. K., 142
Seyferth, D., 146, 153
Seyler, R. J., 190
Shabarov, Yu. S., 266
Shackleton, C. H. L., 298
Shadoff, L. A., 335
Shafiee, A., 175
Shafiullah, 231
Shainok, U., 48

Shaler, R. C., 341
Shamshurina, S. A., 276
Shannon, J. S., 143, 205, 227, 330
Shannon, T. W., 272
Shapiro, M., 16, 131
Shapiro, R. H., 45, 66, 274
Sharbatyan, P. A., 274, 283
Sharp, D. W. A., 198
Sharp, K. G., 157
Sharp, T. E., 102
Shavitt, I., 3
Shaw, D. G., 336
Shaw, M. A., 132
Shaw, P. E., 331
Shaw, R. A., 220
Shcherbinin, V. V., 152
Sheikh, Y. M., 233, 278, 318
Sheinker, Yu. N., 282
Sheldon, R. M., 332
Sheley, C. F., 188
Sheludyakov, V. D., 152
Shemyakin, M. M., 248
Shen, J., 104
Shepherd, H. D., 130
Shepherd, J. P., 188
Sherman, W. R., 121, 258, 300
Shibuya, T.-I., 4
Shida, Y., 67
Shiga, T., 293
Shih, S., 3
Shimada, K., 85, 267
Shimizu, Y., 233, 318
Shindy, W. W., 336
Shine, R. J., 175
Shiofani, S., 282
Shiokawa, T., 33, 81
Shipton, J., 331
Shirafryi, T., 189
Shiraki, K., 332
Shirley, S. G., 335
Shiyonok, A. I., 372
Shobatake, K., 21
Shoeb, A., 286
Shoemake, G. R., 299
Shol'ts, V. B., 213
Shomura, T., 352
Shore, B. L., 121, 300
Shortland, A. J., 179, 180
Shotton, E., 359
Showalter, H. D. H., 99, 236, 342
Shrader, S. R., 346
Shreeve, J. M., 163, 218, 220
Shunbo, F., 298
Shundo, H., 346
Siddall, J. B., 300
Siddiqui, B., 260, 312
Siddiqui, I. B., 248
Siderov, L. N., 213
Siebert, W., 149
Siebrand, W., 15
Sieck, L. W., 80, 111
Sief, A. E., 278
Siegel, A., 70, 266
Siegelman, H. W., 246

Siekmann, L., 317, 321
Sievers, R. E., 209
Sievertsson, H., 251, 278, 366
Sigfridsson, B., 14
Sih, C., 302
Silber, R. H., 327, 352
Silva, M. E. F., 135
Silverman, R. W., 300
Silvers, J. H., 53, 107
Silverstein, R. M., 307
Silverthorn, W. E. 190
Silvon, M. P., 184
Simard, M. B., 231
Simm, I. G., 29, 86
Simon, G. L., 200
Simon, W., 262
Simoneit, B. R., 334, 336, 337
Simonis, J., 14
Simonova, T. N., 307
Simons, J., 4
Simons, S. L., 299
Simpson, C. F., 299
Simpson, S. R., 180, 190
Sinanoglu, O., 17
Sing Boen Tjan, 332
Singer, A. G., 307
Singh, H., 273
Singh, N., 171
Singleton, D. E., 222
Singy, G. A., 270
Siniyaev, R. K., 265
Sinnreich, D., 256
Sinsheimer, J. E., 350
Sirotkina, Ye. I., 187
Sisenwine, S. F., 322, 346
Siska, P. E., 47
Sisler, H. H., 174
Sizoi, V. F., 187
Sjöquist, B., 327, 328
Sjöquist, J., 324, 327, 328, 344
Sjövall, J., 232, 300, 302, 316
Sjoqvist, F., 344
Sjovall, J. B., 139
Skapski, A. C., 179
Skell, P. S., 70, 145, 152, 155, 184, 190, 265
Skinner, H. B., 223
Skinner, R. F., 297, 327
Skinner, W. A., 315
Sklarz, B., 276
Skudlarski, K., 213
Skurat, V. E., 224
Sladky, F., 221
Slavik, J., 226
Sloan, D., 217
Sloan, M., 185
Slocum, D. W., 184
Slocum, S. W., 153
Sloman, K. G., 331
Slotin, L., 242
Slusarchyk, W. A., 315
Slutsky, J., 152
Smagina, E. I., 223
Smallwood, R. J., 193
Smart, J. B., 146
Smets, J., 51

Smirnov, B. M., 289
Smith, A. G., 318, 320
Smith, B. E., 203
Smith, D., 7, 78
Smith, D. A., 120
Smith, D. H., 119, 124, 134, 137, 226, 231, 263, 297, 334
Smith, D. L., 214
Smith, D. S. H., 233, 318
Smith, F., 153
Smith, H. E., 146
Smith, H. O., 6, 80
Smith, J. D., 150
Smith, J. L., 156, 231, 242, 297, 300, 319, 323, 327, 338, 352
Smith, K., 327
Smith, K. A., 237
Smith, K. M., 60, 209, 224, 245
Smith, L. L., 320
Smith, M. J., 369
Smith, M. R., 153
Smith, P. J., 272, 276, 278
Smith, R., 280
Smith, R. D., 334
Smith, R. G., 146, 209
Smith, R. M., 237
Smith, S. R., 216
Smith, T. R., 140, 253
Smith, W. D., 4
Smits, S. E., 329
Smolinsky, G., 151
Smyth, K. C., 10
Snady, H. W., 283
Snaith, R., 155, 203
Sneath, R. L., 215
Snedden, W., 360
Sneddon, L. G., 203, 215
Snieckus, V., 185
Snyder, P. D., 260
Snyder, R. E., 299
Sobolevskii, M. V., 157
Sodano, G., 318
Sodeck, G., 214
Sokolov, S. D., 280
Solgadi, D., 25, 55, 267
Solomko, Z. F., 274, 283
Solov'yov A. A., 246, 376
Solow, E. B., 303
Solveichik, A. I., 213
Sommer, L. H., 154
Sonogashira, K., 193
Soothill, R. J., 284
Southon, F. C. G., 120
Sovocool, G. W., 209
Spalding, T. R., 143, 160
Sparapano, L., 255
Sparks, H. L., 336
Spence, D., 57
Spencer, G. F., 257, 306
Spiegelhalder, B., 321
Spielvogel, B. F., 215
Spiteller, G., 231
Spohr, R., 51
Spokes, G. N., 83
Sprake, J. M., 286
Springer, T. A., 297

Spyckerelle, C., 337
Srivastava, R. C., 286
Srivastava, R. D., 214, 223
Sroka, W., 51
Staddon, B. W., 330
Ställberg-Stenhagen, S., 331
Staemmler, V., 80
Stafford, F. E., 214, 222
Stafford, M., 329
Staley, R. H., 83
Stallberg, G., 257
Stambolija, Lj., 68, 206, 285
Stamm, H., 279
Stanley, J. B., 278
Stanovnik, B., 286
Stanton, L., 22
Stark, G. R., 324
Stauffer, S. C., 344
Stavaux, M., 286
Stearus, C. A., 214
Stebbings, W. L., 12, 87
Steel, G., 299, 315, 337
Steen, G. O., 257, 259, 260, 305, 310
Stefani, A., 264
Stefanov, K., 307
Stefanović, D., 68, 206, 270, 282, 285
Stehl, R. H., 336
Steiger, H., 201
Stein, S. E., 23
Steinberg, N. G., 282
Steinfelder, K., 88, 90, 231, 288, 293
Steinglitz, L., 284
Steinman, D. H., 226, 330
Stellner, K., 260
Stenhagen, E., 256, 257, 331
Stephen, W. I., 211, 213
Stern, G. M., 349
Sternowsky, H. J., 327
Sternson, L. A., 350
Steudel, R., 220
Steuder, H., 279
Steward, O. W., 152
Stewart, D. W., 124
Stewart, M. A., 202
Stewart, R. P., 179 203
Stewart, W. W., 255
Stille, J. K., 168
Stillwell, R. N., 114, 115, 121, 224, 299, 304, 323
Stillwell, W. G., 121, 322, 329, 358
Stine, C. M., 332
Stobart, S. R., 145, 159, 163, 201, 202
Stobbs, L., 155
Stockbauer, R., 11, 29
Stockdale, J. A., 214
Stoffel, W., 260, 310
Stohrer, W. D., 6
Stokke, O., 139, 300
Stoklosa, H. J., 199
Stoks, S. J., 352
Stoll, M. S., 245
Stolz, L. P., 331

Author Index

Stone, F. G. A., 180, 181, 182, 186, 193, 194, 201, 202
Stonham, T. J., 132
Storch, W., 149
Story, M., 121
Story, M. S., 302
Stout, R. W., 285
Stransky, K., 337
Strasdine, G. A., 248
Stratton, C., 329
Streets, D. G., 7
Streibl, M., 271
Streitwieser, A., jun., 73, 192
Strömberg, L., 328
Strolin, P., 297
Strolin-Benedetti, M., 297
Strong, J. M., 356
Struck, A. H., 292
Struck, R. F., 350
Struve, W. S., 14
Studier, M. H., 337
Stuhler, H., 168, 171
Su, T., 102, 267, 293
Sucrow, W., 315
Sugawara, Y., 239
Sugie, A., 229
Sugihara, J., 352
Sugiura, M., 309
Suguira, T., 293
Suhadolnik, R. J., 239
Suler, N., 343
Sullivan, H. R., 329, 345
Sullivan, R. E., 292
Sultanov, A. Sh., 214
Summers, T. R., 303
Summons, R. E., 249, 325
Sun, F. F., 348, 354, 359
Sun, T., 324
Sunder, W. A., 222
Sundermeyer, W., 157
Sundström, G., 336
Sung, J. P., 49
Suplinskas, R. J., 17
Sutherland, D. E., 253, 325, 365
Sutherland, G. L., 133
Sutherland, R. G., 189
Suzuki, A., 255
Suzuki, I. H., 52
Suzuki, M., 239, 270
Suzuki, T., 249, 251, 307, 324, 338
Svec, H. J., 188
Svensson, J., 307
Svensson, L.-A., 274
Svensson, S., 247, 248, 311, 312, 376
Swahn, C.-G., 311
Swain, J. R., 197
Swaine, H. A., 251, 371
Swansiger, J. T., 305, 338
Swanwick, M. G., 199
Sweeley, C. C., 121, 247, 259, 260, 297, 300, 302, 309, 310, 311, 355, 356
Sweeney, J. J., 145
Sweetman, B. J., 121, 141, 261, 300, 309

Swindell, R. M., 220
Swingler, D. L., 216, 222
Sykes, R. J., 123
Sylvester, G., 180, 196
Symmes, C., 166
Syrvatka, B. G., 267
Szafranek, J., 298
Szechner, B., 280
Szilagyi, I., 239
Szwarc, M., 85, 267

Taagepera, M., 53, 106
Tabereaux, A., 150
Taft, R. W., 53, 106
Tagirov, V. K., 223
Tahara, S., 249
Tai, H., 302
Taira, K., 345
Tajima, M., 334
Tajima, S., 26, 54, 81
Takabataka, E., 345
Takahashi, K., 181, 229
Takahashi, N., 67
Takahashi, S., 307
Takahashi, T., 336
Takaishi, T., 239
Takats, J., 192
Takeguchi, C., 302
Taketomi, T., 184
Takeuchi, T., 293, 294
Takhistov, V. V., 271
Talley, F. B., 332
Talman, E., 331
Tal'rose, V., 224
Tam, J., 255
Tam, S. W., 286
Tamas, J., 158
Tamborski, C., 153
Tamura, S., 233, 255, 318
Tamura, T., 334, 338
Tamura, Z., 312
Tanaka, I., 10
Tanaka, J., 216
Tanaka, K., 10, 326
Tanayama, S., 352
Tang, I. N., 217
Tanger, U., 187
Tannenbaum, H. P., 112, 271
Tanner, G. M., 206
Tanzer, M. L., 254
Tardella, P. A., 278
Tashma, Z., 142, 169
Tatarczyk, H., 41
Tatematsu, A., 78, 227, 287, 352
Tatone, O. S., 115
Tattrie, N. H., 315
Taugbol, K., 207
Taylor, A. R., 347
Taylor, D. M., 301
Taylor, D. R., 219
Taylor, J. W., 10, 12, 87, 124, 256, 265, 275
Taylor, K. T., 251, 371
Taylor, L. J., 210
Taylor, S. S., 369
Teer, D., 116
Teeter, R. M., 320

Teets, R. E., 121, 300, 355
Tekaat, T., 11
Telling, G. M., 333
Temple, A. F., 286
Tennent, H. G., 180
ten Noever de Brauw, M. C., 333
Tenschert, G., 33
Teranishi, R., 298, 303
Terent'ev, P. B., 226, 282, 283
Terlouw, J. K., 162
Terroni, S., 204
Terwilliger, D. T., 42
Testa, B., 323
Teubner, J., 258
Thayumanavan, B., 248, 371
The, K. I., 157, 159
Thenn, W., 170
Thenot, J.-P., 249, 298, 316, 317, 324
Thewalt, U., 174
Thiel, G., 200
Thiele, K.-H., 180
Thistlewaite, P. J., 222
Thomas, A. F., 332
Thomas, C. B., 280
Thomas, D. W., 315, 363
Thomas, F., 367
Thomas, G. J., 239
Thomas, J. L., 180, 343
Thomas, R. S., 139, 141, 267, 304
Thomas, T. L., 328
Thompson, B. J., 189
Thompson, C. J., 337
Thompson, D. J., 186
Thompson, D. L., 17
Thompson, J., 248
Thompson, J. C., 157
Thompson, L. K., 202
Thompson, M. J., 319, 320
Thompson, M. L., 150
Thompson, M. R., 371
Thompson, R. H., jun., 297, 332
Thompson, R. M., 303, 327
Thomson, I. J., 123
Thomson, M. L., 84, 285
Thomson, R. H., 233, 318
Thorpe, M. C., 350
Throck-Watson, J., 233
Thrush, R. E., 121, 300, 355
Thynne, J. C. J., 57, 58, 216, 217, 220, 288, 291, 293
Tickner, A. W., 48
Tiedemann, P. W., 103, 107, 108, 110
Tiernan, T. O., 13, 29, 33, 80, 289
Tiikmaa, T. V., 265
Timms, P. L., 180, 190, 214
Timoshin, I. A., 213
Timpe, H.-J., 281
Tims, A. W., 242
Tindell, G. L., 330
Ting, K.-L. H., 131
Tio, C. O., 346
Tishbee, A., 298

Tišler, M., 286
Titus, J., 119
Tjoa, S. S., 326
Tkachenko, V. S., 274, 283
Todd, D., 311
Todd, J. F. J., 88, 96, 146, 147, 148, 166, 273, 290, 293
Todd, J. S., 273
Todd, K. H., 190
Todd, L. J., 150, 215
Todo, K., 233, 318
Toft, P., 231
Tokés, L., 231, 234
Tolbert, G. D., 224, 304
Tolls, E., 155
Tolstikov, G. A., 89, 293
Toma, S., 189
Toman, R., 248
Tomer, K. B., 66, 67, 68, 83, 113, 272, 274, 277, 280, 284, 285
Tomkins, I. B., 198
Tomlinson, C. H., 180, 190
Tong, D. C. M., 267
Torian, R. L., 101, 177
Torri, G., 274
Torroni, S., 55, 57, 151, 195, 266
Toseland, P. A., 356
Tou, J. C., 84
Toube, T. P., 237
Towers, C., 198
Townsend, L. B., 99, 236, 241
Toyoda, M., 10
Traegar, J. C., 10, 47, 52, 216, 217, 263
Traficante, D. D., 203
Trager, W. F., 80, 323, 361
Trager, W. T., 353
Treichel, P. M., 199, 200
Treppendahl, S., 218
Tressl, R., 303
Trippett, S., 168
Trka, A., 271
Troe, J., 19
Trouilloud, M., 230
Truex, T. J., 208
Tsai, A., 181
Tsai, B. P., 47
Tsai, S.-C., 62, 63, 253, 275
Tsai, T., 156
Tsang, C. W., 62, 268
Tschesche, H., 249, 366
Tschesche, R., 225
Tsetsugina, N. N., 264
Tsipis, C. A., 212
Tsuchiya, T., 26, 54, 81
Tsuda, S., 58
Tsuji, K., 236, 326
Tsuji, M., 10
Tsujimoto, K., 67
Tsuru, R., 233
Tsuruoka, T., 352
Tsutsui, M., 209
Tsuyuki, T., 229
Tsvetkov, Yu. V., 217, 219, 223
Tuck, D. G., 206

Tuck, D. J., 150, 151
Tümmler, R., 88, 90, 231, 288, 293
Tully, J. C., 17, 21
Tunggal, B. D., 169
Tung Sun, 301
Turnblom, E. W., 167
Turnbull, L. B., 351
Turner, A. B., 233, 318
Turner, G., 206
Turner, R. B., 88, 293
Turner, R. M., 193
Turner, W. A., 272, 278
Turney, T. W., 197
Tursch, B. M., 233, 318
Tuvaeva, T. N., 213
Tuzimura, K., 249, 324
Twibell, J. D., 304
Twine, C. E., jun., 7, 81
Tyson, B. J., 336

Uccella, N., 26, 71, 266, 268, 280
Uchida, K., 238
Uchytil, T. F., 255
Uden, P. D., 211, 213
Uhmann, R., 239
Uliss, D. B., 78, 280
Ullmann, R., 219
Umaba, T., 58
Umemura, K., 352
Umirzahov, B., 276
Underhill, M., 200
Undheim, K., 56, 249, 268, 276, 282, 285
Ungar, G., 255
Uno, H., 285
Unterhalt, B., 281
Updegrove, W. S., 337
Upham, R. A., 97
Upper, C. D., 312
Urban, W. G., 139
Urry, G., 201
Ushakov, A. N., 307
Ustynyuk, Yu. A., 274
Uy, O. M., 214, 223
Uyehara, T., 251
Uzan, R., 10, 219

Vaarkamp-Lijnse, J. S., 120, 134
Vagelos, P. R., 327
Vaglio, G. A., 183, 185, 199
Vahrenkamp, H., 196, 198
Vale, W., 255, 256, 367, 368
Valenty, S. J., 145
Valle, M., 183, 199
Valpertz, H. W., 219
van Binst, G., 282
Vance, D. E., 302
Van Dam, E. M., 184
Van de Grampel, J. C., 218
van den Berg, P. M. J., 298
Van den Bergen, A., 206
Vanden Heuvel, W. J. A., 156, 231, 242, 277, 297, 300, 319, 323, 327, 338, 352

Van der Haar, R. W., 61, 280
Van der Kooi, J. P., 168
Van der Marel, T., 330
Van der Velde, G., 338
Vanderwalle, M., 227, 228, 270, 272, 273
Vanderwielen, A. J., 217
Vanderzanden, R. J., 147
van de Sande, C., 227, 270
Van Dyk, J. M., 342
Vane, F. M., 341, 346
van Ginneken, C. A. M., 300, 326
van Halasz, S. P., 155
van Haverbeke, Y., 280, 281
Van Heijenoort, J., 351
van Hout, P., 298
van Houte, J. J., 280
van Lear, G. E., 235, 286
van Luchene, E., 298
Van Oven, H. O., 191, 192
van Rossum, J. M., 300, 326
van Santen, R. A., 16
van Straten, S., 333
van Thuijl, J., 280
Van't Klooster, H. A., 120, 134
Van Veen, R., 149
Varenne, P., 227, 250, 376
Varmuza, K., 132
Vasile, M. J., 151, 217, 222
Vasil'eva, A. D., 273
Vastola, F. J., 116
Vasyuta, Yu. V., 217
Vaver, V. A., 307
Vaziri, C., 78
Vcelak, V., 337
Veening, H., 190
Vegar, M. R., 70, 270
Velde, G. V., 228
Veldhuis, M. K., 331
Venanzi, L. M., 187
Venema, A., 71, 279
Venkataraghavan, R., 120, 123, 134, 136, 137, 253, 370
Venkataraman, K., 227
Verkouw, H. T., 191
Vermeer, H., 167
Vernay, H. F., 236
Verstappe, M., 298
Verzele, M., 228, 298
Vessman, J., 344
Vestal, M. L., 41
Vestergaard, P., 329
Vetter, W., 230, 248
Vettori, U., 281
Victor, R., 185
Vigevani, A., 248
Vihko, R., 297, 321
Viinikka, L., 321
Vikhrestyuk, N. I., 305
Villa, J. F., 2. 1
Vincent, E.-J., 285
Viney, B. W., 33, 92
Vink, J., 139, 247, 248, 372, 374

Author Index

Viscomi, A. S., 97
Viscontini, M., 323
Viswanathan, C. V., 297
Vitagliano, A., 185
Vite, J. P., 97
Vliegenthart, J. F. G., 139, 247, 248, 258, 372, 374
Völlmin, J. A., 302
Voelter, W., 311
Vogelburg, R., 121
Vogt, J., 47
Voigt, D., 90, 231, 293
Voight, E., 176
Volfova, D., 338
Volk, H., 277
Volkova, T. A., 213
Volland, B., 219
von Ardenne, M., 88, 90, 288, 291, 293
Von Dahlen, K.-H., 181
von Gustorf, E. K., 184
von Minden, D. L., 68, 242, 243, 323
von Niessen, W., 4
von Puttkamer, E., 26, 27
von Schantz, M., 314
von Sydow, E., 332
von Weyssenhoff, H., 15
von Zahn, U., 41
Vorachek, J. H., 80
Vore, M., 301
Vorlaender, W., 272
Vorobyova, V. Ya., 282
Vouros, P., 97, 151, 232, 301, 302, 316, 322, 341
Voznesensky, V. N., 164
Vree, T. B., 229, 300, 326
Vuye, A., 298
Vyalemits, M. T., 265

Waber, J. T., 5
Wachi, F. M., 223
Wachs, T., 63
Wachter, E., 249, 366
Wachtmeister, C. A., 336
Wade, A. P., 139, 321
Wade, K., 155
Wade, L. E., 74
Waegell, B., 276
Wagner, H., 227, 247
Wagner, L. C., 222
Wahl, A. C., 2
Wahlbeck, P. G., 223
Wahlberg, I., 229, 314
Wahlroos, O., 320
Wahraftig, A. L., 1, 18, 49
Waight, E. S., 275
Wakabayashi, T., 307
Wakatsuki, Y., 184
Wakkers, P. J. M., 285
Wald, N., 303, 331
Waldmann, J., 180
Walker, B. J., 165
Walker, J. A., 25, 51
Walker, J. M., 338
Walker, M. P., 323
Walker, T. E. H., 5, 213
Walkinshaw, C. H., 305, 306, 319, 338

Wall, M. E., 328, 341
Wallace, J. C., 65, 274
Wallaert, B., 337
Walle, T., 323, 329, 330, 341
Wallenstein, M. B., 18
Wallbridge, M. G. H., 150, 203
Waller, G. R., 230
Wallnöfer, P. R., 335
Walls, E. C., 124
Walter, E., 149, 215
Walther, B., 151
Walton, D. R. M., 152
Walton, T. J., 257, 306
Wander, J. D., 235, 247
Wang, C.-S., 84
Wang, C. S.-C., 163, 218
Wang, J. L.-F., 58, 213
Wang, N., 342
Wanless, G. G., 90
Wannagat, U., 208
Ward, J. E. H., 198
Ward, S. D., 117
Ware, G. C., 338
Wareing, P. F., 312
Warneck, P., 51
Warren, J. W., 48
Warren, S., 167, 168
Wasada, N., 54, 81
Wasserman, A. E., 333
Wasserman, H. H., 123
Wasson, J. R., 199
Watanabe, H., 155
Watanabe, K., 260, 332, 333, 354
Waterworth, L. G., 150
Watkins, P. M., 199
Watkins, W. D., 345
Watson, E., 322
Watson, J. T., 121, 141, 261, 282, 297, 300, 309, 317, 356
Watts, W. E., 190
Wayborn, H., 242
Weaver, E. C., 330
Weaver, N., 330
Webb, A. D., 331, 333
Webb, B. C., 86, 293
Webb, G. A., 283
Webb, H. M., 53, 106
Webb, K. S., 299
Webb, R. E., 330
Webb, R. G., 336
Webber, T. J. N., 336
Weber, J. H., 142
Weber, R., 175, 286, 323
Weber, W. P., 89, 153, 188, 292, 293
Webster, B. R., 127
Weedon, B. C. L., 237
Weeks, D. P., 100
Weete, J. D., 305, 306, 319, 338
Wegener, J., 155
Weidemüller, W., 185
Weidlein, J., 171
Weigel, H., 148, 248
Weil, L., 297
Weiler, L., 274

Weimann, B., 118
Weinberg, A., 48
Weinkam, R. J., 256
Weinstein, N. D., 21, 46
Weisenborn, F. L., 315
Weisgerber, I., 336
Weiss, A. M., 311
Weiss, K., 193
Weiss, M., 54, 58
Weiss, M. J., 54
Weissburger, W., 334
Welch, A. J., 193
Welcman, N., 198
Weliky, I., 348
Wells, C. H. J., 88, 293
Wells, D., 186
Wells, E. J., 206
Wells, J. D., 320
Wells, R. J., 70, 269, 270
Welte, D. H., 337
Welzel, P., 270, 277
Wenkert, E., 225, 226
Wentrup, C., 84
Wentzell, B. R., 336
Weringa, W. D., 167, 237, 285
Werner, A. S., 47
Werner, H., 197
West, B. O., 167, 201, 206, 210
West, R., 146, 155
Westerman, C. E., 359
Westley, J. W., 323
Westmore, J. B., 206, 242
Weston, A. F., 146, 147, 148
Westwood, J. H., 94, 144, 247
Wetzel, R. B., 168
Weyerstahl, P., 71
Weygand, F., 366
Wheeler, D. M. S., 352
Wheeler, J. W., 330, 331, 334
Wheeler, V. A., 335
White, A. H., 212
White, E., 156, 242
White, G. L., 150
White, M. J., 108
White, W. C., 134, 231
Whitehead, G. W., 169
Whitehead, M. A., 6
Whitesides, T. H., 189
Whitfield, F. B., 331
Whiting, A., 62, 267
Whittemore, I. M., 305
Whitten, D. G., 209, 245
Whitten, J. L., 2
Whyte, J. N. C., 248, 311, 312
Wick, E. L., 333
Wickremasinghe, R. L., 333
Widen, C.-J., 228
Widén, K.-G., 314
Widing, H. F., 73
Wieczorkowski, J., 218
Wien, R. W., 102
Wiersum, U. E., 83, 84
Wies, R., 174

Wieser, H., 163
Wietjens, W. H. J. M., 330
Wiezer, H., 219
Wilder, B. J., 359
Wiley, R. A., 350
Wilfinger, H.-J., 168
Wilhite, D. L., 4
Wilke, G., 190, 192
Wilkie, C. A., 151, 153
Wilkinson, G., 179, 180, 208
Wille, G., 196
Wille, H., 149
Willeford, B. R., 190
Williams, A. E., 287
Williams, B. D., 288
Williams, C. M., 360
Williams, D. H., 26, 62, 237, 252, 256, 262, 263, 266, 267, 268, 273, 286, 291, 363, 368, 369
Williams, F. M., 354, 356
Williams, J. E., 80
Williams, J. K., 168
Williams, J. R. L., 149
Williams, K. M., 303, 334
Williams, L. R., 284
Williams, M., 307
Williams, P., 120, 289
Williams, P. M., 314
Williams, R. N., 284
Williams, V. P., 242, 283
Williams-Smith, D. L., 190
Willman, K., 11
Wilmenius, P., 30
Wilson, B. J., 332
Wilson, D. A., 278
Wilson, D. A. A., 302
Wilson, D. M., 302
Wilson, G. L., 144, 215
Wilson, J., 288
Wilson, J. M., 1, 76, 263, 298
Wilson, J. S., 280
Wilson, J. W., 144
Wilson, K. R., 21
Wilson, M. H., 68, 241, 242, 243
Wilson, R. A., 332
Wilson, R. D., 221
Winfield, J. M., 221
Winkler, E., 192, 193
Winkler, H. U., 95, 249, 251, 366, 371
Winkler, J., 65
Winkler, V. W., 334
Winnik, M. A., 68
Winstead, J. A., 189
Winterburn, P. J., 234
Winters, R. E., 47
Wipf, H.-K., 134, 253, 370
Wirsam, B., 3
Wismar, K.-J., 208
With, T. K., 244
Witiak, J. L., 153
Witten, T. A., 227
Wittstruck, T., 348
Wojcicki, A., 187
Wojciechowski, Z. A., 319
Wolf, C. J., 299

Wolf, C. L., 337
Wolf, D. J., 119
Wolf, L. R., 190
Wolf, N. L., 336
Wolfe, N. E., 73
Wolfgang, R., 17, 21
Wolk, C. P., 259
Wolkoff, P., 281, 286
Wolstenholme, W. A., 351
Wong, A. S., 336
Wong, P. T. S., 304
Wong, W. H., 19, 21
Wood, G., 40, 78, 273
Wood, G. W., 258
Woodbridge, A. P., 229, 317
Woodgate, P. D., 229
Woodgate, S. D., 108
Woodgate, S. S., 289
Woods, J. S., 335
Woods, M., 220
Woodward, P., 202
Woodward, P. J., 248
Woodward, R. B., 17
Woolfe, M. L., 331
Worrall, I. J., 150
Wotiz, H. H., 297
Wreford, S. S., 203
Wrobel, G., 159
Wu, H. Y., 223
Wünsche, C., 165, 278
Wulfsberg, G., 146
Wulfson, N. S., 275

Yahagi, T., 242
Yakhontov, L. N., 282
Yalpani, M., 175
Yamada, F., 286
Yamada, Y., 67
Yamaguchi, T., 209
Yamakawa, K., 189
Yamakawa, T., 312
Yamamoto, M., 7, 81, 276
Yamamoto, Y., 181, 189
Yamanishi, T., 333
Yamaoka, H., 293
Yamazaki, H., 181, 184, 194
Yamdagni, R., 53, 105, 107, 110
Yang, H. Y., 331
Yang, K. W., 279
Yanina, A. D., 282
Yano, I., 306
Yanomoto, K., 189
Yanovskaya, L. A., 274, 276
Yarrow, D. J., 185
Yasuda, S., 318
Yasufuku, K., 181, 194
Yasumura, A., 346
Yates, K., 3
Yates, P., 271
Yeager, D. L., 4
Yee, T. H., 228
Yeo, A., 134, 231
Yeong, Y. C., 306
Yeowell, D. A., 359
Yeransian, J. A., 331

Yergey, A. L., 101
Yermakov, A. I., 282
Yeung, E. S., 16
Yinon, J., 89, 293, 299
Yoder, J. M., 334
Yokohata, A., 58
Yoshida, K., 248
Yoshida, S., 67
Yoshihara, K., 33, 81
Yoshihara, S., 344
Yoshimura, H., 344
Yoshizumi, H., 227, 352
Young, D., 189
Young, N. D., 121, 297, 309
Youren, J. W., 120
Yoxall, C. T., 272
Yu, G. M., 326
Yu, T. Y., 216
Yu, Cong, D., 149
Yurchenko, A. G., 264
Yur'ev, V. P., 89, 293

Zabielski, M. F., 126
Zabik, M. J., 336
Zach, P., 305, 338
Zagt, _. M., 164, 16_
Zaharu, P. J., 226
Zak, S. B., 358
Zakharova, G. N., 275
Zarembo, J. E., 246
Zarkadas, A., 155
Zarske, G., 298
Zbiral, E., 165
Zeevaart, J. A. D., 314
Zei, M. S., 91
Zeiss, W., 171
Zektzer, J., 208
Zelli, S., 299
Zeman, A., 256, 297, 307
Zepp, R. G., 336
Zerilli, L. F., 281
Zhakaeva, A. Zh., 187
Zhegul'skaya, N. A., 213
Zhigach, A. F., 150
Zhigalin, G. Ya., 157
Zhigulev, K. K., 280
Ziegler, E., 118, 140
Zieserl, J. F., jun., 123, 300, 308
Zietz, E., 231
Zietz, R., 51
Zilliken, F., 376
Zimmer, A., 351
Zingaro, R. A., 219
Zintel, J. A., 319
Zion, T. E., 329
Zipperer, W. C., 182, 197
Zitko, V., 336
Zlatkis, A., 298, 321, 322, 332
Zlutický, J., 334
Zolanovitch, V. I., 187
Zolotarev, B. M., 276, 280
Zolotoi, N. B., 224
Zsupán, K., 237
Zuchner, K., 221
Zurawski, B., 2
Zurowska, A., 248